Advanced Calculus with Applications in Statistics

Second Edition

Advanced Calculus with Applications in Statistics

Second Edition
Revised and Expanded

André I. Khuri

University of Florida
Gainesville, Florida

WILEY-INTERSCIENCE

A JOHN WILEY & SONS, INC., PUBLICATION

Published by John Wiley & Sons, Inc., Hoboken, New Jersey.
Published simultaneously in Canada.

For general information on our other products and services please contact our Customer Care Department within the U.S. at 877-762-2974, outside the U.S. at 317-572-3993 or fax 317-572-4002.

Wiley also publishes its books in a variety of electronic formats. Some content that appears in print, however, may not be available in electronic format.

Library of Congress Cataloging-in-Publication Data

Khuri, André I., 1940-
    Advanced calculus with applications in statistics / André I. Khuri. -- 2nd ed. rev. and expended.
        p. cm. -- (Wiley series in probability and statistics)
    Includes bibliographical references and index.
    ISBN 0-471-39104-2 (cloth : alk. paper)
    1. Calculus. 2. Mathematical statistics. I. Title. II. Series.

    QA303.2.K48 2003
    515--dc21   2002068986

Printed in the United States of America

10 9 8 7 6 5 4 3 2 1

To Ronnie, Marcus, and Roxanne
and
In memory of my sister Ninette

Contents

Preface

This edition provides a rather substantial addition to the material covered in the first edition. The principal difference is the inclusion of three new chapters, Chapters 10, 11, and 12, in addition to an appendix of solutions to exercises.

Chapter 10 covers orthogonal polynomials, such as Legendre, Chebyshev, Jacobi, Laguerre, and Hermite polynomials, and discusses their applications in statistics. Chapter 11 provides a thorough coverage of Fourier series. The presentation is done in such a way that a reader with no prior knowledge of Fourier series can have a clear understanding of the theory underlying the subject. Several applications of Fouries series in statistics are presented. Chapter 12 deals with approximation of Riemann integrals. It gives an exposition of methods for approximating integrals, including those that are multidimensional. Applications of some of these methods in statistics are discussed. This subject area has recently gained prominence in several fields of science and engineering, and, in particular, Bayesian statistics. The material should be helpful to readers who may be interested in pursuing further studies in this area.

A significant addition is the inclusion of a major appendix that gives detailed solutions to the vast majority of the exercises in Chapters 1–12. This supplement was prepared in response to numerous suggestions by users of the first edition. The solutions should also be helpful in getting a better understanding of the various topics covered in the book.

In addition to the aforementioned material, several new exercises were added to some of the chapters in the first edition. Chapter 1 was expanded by the inclusion of some basic topological concepts. Chapter 9 was modified to accommodate Chapter 10. The changes in the remaining chapters, 2 through 8, are very minor. The general bibliography was updated.

The choice of the new chapters was motivated by the evolution of the field of statistics and the growing needs of statisticians for mathematical tools beyond the realm of advanced calculus. This is certainly true in topics concerning approximation of integrals and distribution functions, stochastic processes, time series analysis, and the modeling of periodic response functions, to name just a few.

The book is self-contained. It can be used as a text for a two-semester course in advanced calculus and introductory mathematical analysis. Chapters 1–7 may be covered in one semester, and Chapters 8–12 in the other semester. With its coverage of a wide variety of topics, the book can also serve as a reference for statisticians, and others, who need an adequate knowledge of mathematics, but do not have the time to wade through the myriad mathematics books. It is hoped that the inclusion of a separate section on applications in statistics in every chapter will provide a good motivation for learning the material in the book. This represents a continuation of the practice followed in the first edition.

As with the first edition, the book is intended as much for mathematicians as for statisticians. It can easily be turned into a pure mathematics book by simply omitting the section on applications in statistics in a given chapter. Mathematicians, however, may find the sections on applications in statistics to be quite useful, particularly to mathematics students seeking an interdisciplinary major. Such a major is becoming increasingly popular in many circles. In addition, several topics are included here that are not usually found in a typical advanced calculus book, such as approximation of functions and integrals, Fourier series, and orthogonal polynomials. The fields of mathematics and statistics are becoming increasingly intertwined, making any separation of the two unpropitious. The book represents a manifestation of the interdependence of the two fields.

The mathematics background needed for this edition is the same as for the first edition. For readers interested in statistical applications, a background in introductory mathematical statistics will be helpful, but not absolutely essential. The annotated bibliography in each chapter can be consulted for additional readings.

I am grateful to all those who provided comments and helpful suggestions concerning the first edition, and to my wife Ronnie for her help and support.

ANDRÉ I. KHURI

Gainesville, Florida

Preface to the First Edition

The most remarkable mathematical achievement of the seventeenth century was the invention of calculus by Isaac Newton (1642–1727) and Gottfried Wilhelm Leibniz (1646–1716). It has since played a significant role in all fields of science, serving as its principal quantitative language. There is hardly any scientific discipline that does not require a good knowledge of calculus. The field of statistics is no exception.

Advanced calculus has had a fundamental and seminal role in the development of the basic theory underlying statistical methodology. With the rapid growth of statistics as a discipline, particularly in the last three decades, knowledge of advanced calculus has become imperative for understanding the recent advances in this field. Students as well as research workers in statistics are expected to have a certain level of mathematical sophistication in order to cope with the intricacies necessitated by the emerging of new statistical methodologies.

This book has two purposes. The first is to provide beginning graduate students in statistics with the basic concepts of advanced calculus. A high percentage of these students have undergraduate training in disciplines other than mathematics with only two or three introductory calculus courses. They are, in general, not adequately prepared to pursue an advanced graduate degree in statistics. This book is designed to fill the gaps in their mathematical training and equip them with the advanced calculus tools needed in their graduate work. It can also provide the basic prerequisites for more advanced courses in mathematics.

One salient feature of this book is the inclusion of a complete section in each chapter describing applications in statistics of the material given in the chapter. Furthermore, a large segment of Chapter 8 is devoted to the important problem of optimization in statistics. The purpose of these applications is to help motivate the learning of advanced calculus by showing its relevance in the field of statistics. There are many advanced calculus books designed for engineers or business majors, but there are none for statistics majors. This is the first advanced calculus book to emphasize applications in statistics.

The scope of this book is not limited to serving the needs of statistics graduate students. Practicing statisticians can use it to sharpen their mathematical skills, or they may want to keep it as a handy reference for their research work. These individuals may be interested in the last three chapters, particularly Chapters 8 and 9, which include a large number of citations of statistical papers.

The second purpose of the book concerns mathematics majors. The book's thorough and rigorous coverage of advanced calculus makes it quite suitable as a text for juniors or seniors. Chapters 1 through 7 can be used for this purpose. The instructor may choose to omit the last section in each chapter, which pertains to statistical applications. Students may benefit, however, from the exposure to these additional applications. This is particularly true given that the trend today is to allow the undergraduate student to have a major in mathematics with a minor in some other discipline. In this respect, the book can be particularly useful to those mathematics students who may be interested in a minor in statistics.

Other features of this book include a detailed coverage of optimization techniques and their applications in statistics (Chapter 8), and an introduction to approximation theory (Chapter 9). In addition, an annotated bibliography is given at the end of each chapter. This bibliography can help direct the interested reader to other sources in mathematics and statistics that are relevant to the material in a given chapter. A general bibliography is provided at the end of the book. There are also many examples and exercises in mathematics and statistics in every chapter. The exercises are classified by discipline (mathematics and statistics) for the benefit of the student and the instructor.

The reader is assumed to have a mathematical background that is usually obtained in the freshman–sophomore calculus sequence. A prerequisite for understanding the statistical applications in the book is an introductory statistics course. Obviously, those not interested in such applications need not worry about this prerequisite. Readers who do not have any background in statistics, but are nevertheless interested in the application sections, can make use of the annotated bibliography in each chapter for additional reading.

The book contains nine chapters. Chapters 1–7 cover the main topics in advanced calculus, while chapters 8 and 9 include more specialized subject areas. More specifically, Chapter 1 introduces the basic elements of set theory. Chapter 2 presents some fundamental concepts concerning vector spaces and matrix algebra. The purpose of this chapter is to facilitate the understanding of the material in the remaining chapters, particularly, in Chapters 7 and 8. Chapter 3 discusses the concepts of limits and continuity of functions. The notion of differentiation is studied in Chapter 4. Chapter 5 covers the theory of infinite sequences and series. Integration of functions is the theme of Chapter 6. Multidimensional calculus is introduced in Chapter 7. This chapter provides an extension of the concepts of limits, continuity, differentiation, and integration to functions of several variables (multivariable functions). Chapter 8 consists of two parts. The first part presents an overview of the various methods of optimization of multivariable functions whose optima cannot be obtained explicitly by standard advanced calculus techniques. The second part discusses a variety of topics of interest to statisticians. The common theme among these topics is optimization. Finally, Chapter 9 deals with the problem of approximation of continuous functions with polynomial and spline functions. This chapter is of interest to both mathematicians and statisticians and contains a wide variety of applications in statistics.

I am grateful to the University of Florida for granting me a sabbatical leave that made it possible for me to embark on the project of writing this book. I would also like to thank Professor Rocco Ballerini at the University of Florida for providing me with some of the exercises used in Chapters, 3, 4, 5, and 6.

ANDRÉ I. KHURI

Gainesville, Florida

CHAPTER 1

An Introduction to Set Theory

The origin of the modern theory of sets can be traced back to the Russian-born German mathematician Georg Cantor (1845–1918). This chapter introduces the basic elements of this theory.

1.1. THE CONCEPT OF A SET

A set is any collection of well-defined and distinguishable objects. These objects are called the elements, or members, of the set and are denoted by lowercase letters. Thus a set can be perceived as a collection of elements united into a single entity. Georg Cantor stressed this in the following words: "A set is a multitude conceived of by us as a one."

If x is an element of a set A, then this fact is denoted by writing $x \in A$. If, however, x is not an element of A, then we write $x \notin A$. Curly brackets are usually used to describe the contents of a set. For example, if a set A consists of the elements $x_1, x_2, \ldots, x_n$, then it can be represented as $A = \{x_1, x_2, \ldots, x_n\}$. In the event membership in a set is determined by the satisfaction of a certain property or a relationship, then the description of the same can be given within the curly brackets. For example, if A consists of all real numbers x such that $x^2 > 1$, then it can be expressed as $A = \{x | x^2 > 1\}$, where the bar | is used simply to mean "such that." The definition of sets in this manner is based on the axiom of abstraction, which states that given any property, there exists a set whose elements are just those entities having that property.

Definition 1.1.1. The set that contains no elements is called the empty set and is denoted by $\varnothing$. □

Definition 1.1.2. A set A is a subset of another set B, written symbolically as $A \subset B$, if every element of A is an element of B. If B contains at least one element that is not in A, then A is said to be a proper subset of B.

□

1

Definition 1.1.3. A set A and a set B are equal if $A \subset B$ and $B \subset A$. Thus, every element of A is an element of B and vice versa. □

Definition 1.1.4. The set that contains all sets under consideration in a certain study is called the universal set and is denoted by Ω. □

1.2. SET OPERATIONS

There are two basic operations for sets that produce new sets from existing ones. They are the operations of union and intersection.

Definition 1.2.1. The union of two sets A and B, denoted by $A \cup B$, is the set of elements that belong to either A or B, that is,

$$A \cup B = \{x \mid x \in A \text{ or } x \in B\}. \qquad \square$$

This definition can be extended to more than two sets. For example, if $A_1, A_2, \ldots, A_n$ are n given sets, then their union, denoted by $\bigcup_{i=1}^{n} A_i$, is a set such that x is an element of it if and only if x belongs to at least one of the A_i $(i = 1, 2, \ldots, n)$.

Definition 1.2.2. The intersection of two sets A and B, denoted by $A \cap B$, is the set of elements that belong to both A and B. Thus

$$A \cap B = \{x \mid x \in A \text{ and } x \in B\}. \qquad \square$$

This definition can also be extended to more than two sets. As before, if $A_1, A_2, \ldots, A_n$ are n given sets, then their intersection, denoted by $\bigcap_{i=1}^{n} A_i$, is the set consisting of all elements that belong to all the A_i $(i = 1, 2, \ldots, n)$.

Definition 1.2.3. Two sets A and B are disjoint if their intersection is the empty set, that is, $A \cap B = \varnothing$. □

Definition 1.2.4. The complement of a set A, denoted by $\bar{A}$, is the set consisting of all elements in the universal set that do not belong to A. In other words, $x \in \bar{A}$ if and only if $x \notin A$.

The complement of A with respect to a set B is the set $B - A$ which consists of the elements of B that do not belong to A. This complement is called the relative complement of A with respect to B. □

From Definitions 1.1.1–1.1.4 and 1.2.1–1.2.4, the following results can be concluded:

RESULT 1.2.1. The empty set $\varnothing$ is a subset of every set. To show this, suppose that A is any set. If it is false that $\varnothing \subset A$, then there must be an element in $\varnothing$ which is not in A. But this is not possible, since $\varnothing$ is empty. It is therefore true that $\varnothing \subset A$.

RESULT 1.2.2.   The empty set $\varnothing$ is unique. To prove this, suppose that $\varnothing_1$ and $\varnothing_2$ are two empty sets. Then, by the previous result, $\varnothing_1 \subset \varnothing_2$ and $\varnothing_2 \supset \varnothing_1$. Hence, $\varnothing_1 = \varnothing_2$.

RESULT 1.2.3.   The complement of $\varnothing$ is Ω. Vice versa, the complement of Ω is $\varnothing$.

RESULT 1.2.4.   The complement of $\bar{A}$ is A.

RESULT 1.2.5.   For any set A, $A \cup \bar{A} = \Omega$ and $A \cap \bar{A} = \varnothing$.

RESULT 1.2.6. $A - B = A - A \cap B$.

RESULT 1.2.7. $A \cup (B \cup C) = (A \cup B) \cup C$.

RESULT 1.2.8. $A \cap (B \cap C) = (A \cap B) \cap C$.

RESULT 1.2.9. $A \cup (B \cap C) = (A \cup B) \cap (A \cup C)$.

RESULT 1.2.10. $A \cap (B \cup C) = (A \cap B) \cup (A \cap C)$.

RESULT 1.2.11. $\overline{(A \cup B)} = \bar{A} \cap \bar{B}$. More generally, $\overline{\bigcup_{i=1}^{n} A_i} = \bigcap_{i=1}^{n} \bar{A}_i$.

RESULT 1.2.12. $\overline{(A \cap B)} = \bar{A} \cup \bar{B}$. More generally, $\overline{\bigcap_{i=1}^{n} A_i} = \bigcup_{i=1}^{n} \bar{A}_i$.

Definition 1.2.5.   Let A and B be two sets. Their Cartesian product, denoted by $A \times B$, is the set of all ordered pairs (a, b) such that $a \in A$ and $b \in B$, that is,

$$A \times B = \{(a, b) \mid a \in A \text{ and } b \in B\}.$$

The word "ordered" means that if a and c are elements in A and b and d are elements in B, then $(a, b) = (c, d)$ if and only if $a = c$ and $b = d$. $\square$

The preceding definition can be extended to more than two sets. For example, if $A_1, A_2, \ldots, A_n$ are n given sets, then their Cartesian product is denoted by $\times_{i=1}^{n} A_i$ and defined by

$$\underset{i=1}{\overset{n}{\times}} A_i = \{(a_1, a_2, \ldots, a_n) \mid a_i \in A_i, i = 1, 2, \ldots, n\}.$$

Here, $(a_1, a_2, \ldots, a_n)$, called an ordered n-tuple, represents a generalization of the ordered pair. In particular, if the A_i are equal to A for $i = 1, 2, \ldots, n$, then one writes A^n for $\times_{i=1}^{n} A$.

The following results can be easily verified:

RESULT 1.2.13. $A \times B = \emptyset$ if and only if $A = \emptyset$ or $B = \emptyset$.

RESULT 1.2.14. $(A \cup B) \times C = (A \times C) \cup (B \times C)$.

RESULT 1.2.15. $(A \cap B) \times C = (A \times C) \cap (B \times C)$.

RESULT 1.2.16. $(A \times B) \cap (C \times D) = (A \cap C) \times (B \cap D)$.

1.3. RELATIONS AND FUNCTIONS

Let $A \times B$ be the Cartesian product of two sets, A and B.

Definition 1.3.1.   A relations ρ from A to B is a subset of $A \times B$, that is, ρ consists of ordered pairs (a, b) such that $a \in A$ and $b \in B$. In particular, if $A = B$, then ρ is said to be a relation in A.

For example, if $A = \{7, 8, 9\}$ and $B = \{7, 8, 9, 10\}$, then $\rho = \{(a, b) | a < b, a \in A, b \in B\}$ is a relation from A to B that consists of the six ordered pairs $(7, 8)$, $(7, 9)$, $(7, 10)$, $(8, 9)$, $(8, 10)$, and $(9, 10)$.

Whenever ρ is a relation and $(x, y) \in \rho$, then x and y are said to be ρ-related. This is denoted by writing $x \rho y$.   □

Definition 1.3.2.   A relation ρ in a set A is an equivalence relation if the following properties are satisfied:

1. ρ is reflexive, that is, $a \rho a$ for any a in A.
2. ρ is symmetric, that is, if $a \rho b$, then $b \rho a$ for any a, b in A.
3. ρ is transitive, that is, if $a \rho b$ and $b \rho c$, then $a \rho c$ for any a, b, c in A.

If ρ is an equivalence relation in a set A, then for a given a_0 in A, the set

$$C(a_0) = \{a \in A | a_0 \, \rho \, a\},$$

which consists of all elements of A that are ρ-related to a_0, is called an equivalence class of a_0.   □

RESULT 1.3.1. $a \in C(a)$ for any a in A. Thus each element of A is an element of an equivalence class.

RESULT 1.3.2.   If $C(a_1)$ and $C(a_2)$ are two equivalence classes, then either $C(a_1) = C(a_2)$, or $C(a_1)$ and $C(a_2)$ are disjoint subsets.

It follows from Results 1.3.1 and 1.3.2 that if A is a nonempty set, the collection of distinct ρ-equivalence classes of A forms a partition of A.

As an example of an equivalence relation, consider that $a \rho b$ if and only if a and b are integers such that $a - b$ is divisible by a nonzero integer n. This is the relation of congruence modulo n in the set of integers and is written symbolically as $a \equiv b \pmod{n}$. Clearly, $a \equiv a \pmod{n}$, since $a - a = 0$ is divisible by n. Also, if $a \equiv b \pmod{n}$, then $b \equiv a \pmod{n}$, since if $a - b$ is divisible by n, then so is $b - a$. Furthermore, if $a \equiv b \pmod{n}$ and $b \equiv c \pmod{n}$, then $a \equiv c \pmod{n}$. This is true because if $a - b$ and $b - c$ are both divisible by n, then so is $(a - b) + (b - c) = a - c$. Now, if a_0 is a given integer, then a ρ-equivalence class of a_0 consists of all integers that can be written as $a = a_0 + kn$, where k is an integer. This in this example $C(a_0)$ is the set $\{a_0 + kn \mid k \in J\}$, where J denotes the set of all integers.

Definition 1.3.3.   Let ρ be a relation from A to B. Suppose that ρ has the property that for all x in A, if $x \rho y$ and $x \rho z$, where y and z are elements in B, then $y = z$. Such a relation is called a function.   □

Thus a function is a relation ρ such that any two elements in B that are ρ-related to the same x in A must be identical. In other words, to each element x in A, there corresponds only one element y in B. We call y the value of the function at x and denote it by writing $y = f(x)$. The set A is called the domain of the function f, and the set of all values of $f(x)$ for x in A is called the range of f, or the image of A under f, and is denoted by $f(A)$. In this case, we say that f is a function, or a mapping, from A into B. We express this fact by writing $f: A \to B$. Note that $f(A)$ is a subset of B. In particular, if $B = f(A)$, then f is said to be a function from A onto B. In this case, every element b in B has a corresponding element a in A such that $b = f(a)$.

Definition 1.3.4.   A function f defined on a set A is said to be a one-to-one function if whenever $f(x_1) = f(x_2)$ for x_1, x_2 in A, one has $x_1 = x_2$. Equivalently, f is a one-to-one function if whenever $x_1 \neq x_2$, one has $f(x_1) \neq f(x_2)$.   □

Thus a function $f: A \to B$ is one-to-one if to each y in $f(A)$, there corresponds only one element x in A such that $y = f(x)$. In particular, if f is a one-to-one and onto function, then it is said to provide a one-to-one correspondence between A and B. In this case, the sets A and B are said to be equivalent. This fact is denoted by writing $A \sim B$.

Note that whenever $A \sim B$, there is a function $g: B \to A$ such that if $y = f(x)$, then $x = g(y)$. The function g is called the inverse function of f and is denoted by f^{-1}. It is easy to see that $A \sim B$ defines an equivalence relation. Properties 1 and 2 in Definition 1.3.2 are obviously true here. As for property 3, if A, B, and C are sets such that $A \sim B$ and $B \sim C$, then $A \sim C$. To show this, let $f: A \to B$ and $h: B \to C$ be one-to-one and onto functions. Then, the composite function $h \circ f$, where $h \circ f(x) = h[f(x)]$, defines a one-to-one correspondence between A and C.

EXAMPLE 1.3.1. The relation $a\rho b$, where a and b are real numbers such that $a = b^2$, is not a function. This is true because both pairs (a, b) and $(a, -b)$ belong to ρ.

EXAMPLE 1.3.2. The relation $a\rho b$, where a and b are real numbers such that $b = 2a^2 + 1$, is a function, since for each a, there is only one b that is ρ-related to a.

EXAMPLE 1.3.3. Let $A = \{x| -1 \le x \le 1\}$, $B = \{x| 0 \le x \le 2\}$. Define $f: A \to B$ such that $f(x) = x^2$. Here, f is a function, but is not one-to-one because $f(1) = f(-1) = 1$. Also, f does not map A onto B, since $y = 2$ has no corresponding x in A such that $x^2 = 2$.

EXAMPLE 1.3.4. Consider the relation $x\rho y$, where $y = \arcsin x$, $-1 \le x \le 1$. Here, y is an angle measured in radians whose sine is x. Since there are infinitely many angles with the same sine, ρ is not a function. However, if we restrict the range of y to the set $B = \{y| -\pi/2 \le y \le \pi/2\}$, then ρ becomes a function, which is also one-to-one and onto. This function is the inverse of the sine function $x = \sin y$. We refer to the values of y that belong to the set B as the principal values of arcsin x, which we denote by writing $y = \text{Arcsin } x$. Note that other functions could have also been defined from the arcsine relation. For example, if $\pi/2 \le y \le 3\pi/2$, then $x = \sin y = -\sin z$, where $z = y - \pi$. Since $-\pi/2 \le z \le \pi/2$, then $z = -\text{Arcsin } x$. Thus $y = \pi - \text{Arcsin } x$ maps the set $A = \{x| -1 \le x \le 1\}$ in a one-to-one manner onto the set $C = \{y| \pi/2 \le y \le 3\pi/2\}$.

1.4. FINITE, COUNTABLE, AND UNCOUNTABLE SETS

Let $J_n = \{1, 2, \ldots, n\}$ be a set consisting of the first n positive integers, and let J^+ denote the set of all positive integers.

Definition 1.4.1. A set A is said to be:

1. Finite if $A \sim J_n$ for some positive integer n.
2. Countable if $A \sim J^+$. In this case, the set J^+, or any other set equivalent to it, can be used as an index set for A, that is, the elements of A are assigned distinct indices (subscripts) that belong to J^+. Hence, A can be represented as $A = \{a_1, a_2, \ldots, a_n, \ldots\}$.

3. Uncountable if A is neither finite nor countable. In this case, the elements of A cannot be indexed by J_n for any n, or by J^+.   □

EXAMPLE 1.4.1.   Let $A = \{1, 4, 9, \ldots, n^2, \ldots\}$. This set is countable, since the function $f: J^+ \to A$ defined by $f(n) = n^2$ is one-to-one and onto. Hence, $A \sim J^+$.

EXAMPLE 1.4.2.   Let $A = J$ be the set of all integers. Then A is countable. To show this, consider the function $f: J^+ \to A$ defined by

$$f(n) = \begin{cases} (n+1)/2, & n \text{ odd}, \\ (2-n)/2, & n \text{ even}. \end{cases}$$

It can be verified that f is one-to-one and onto. Hence, $A \sim J^+$.

EXAMPLE 1.4.3.   Let $A = \{x \mid 0 \leq x \leq 1\}$. This set is uncountable. To show this, suppose that there exists a one-to-one correspondence between J^+ and A. We can then write $A = \{a_1, a_2, \ldots, a_n, \ldots\}$. Let the digit in the nth decimal place of a_n be denoted by b_n $(n = 1, 2, \ldots)$. Define a number c as $c = 0 \cdot c_1 c_2 \cdots c_n \cdots$ such that for each n, $c_n = 1$ if $b_n \neq 1$ and $c_n = 2$ if $b_n = 1$. Now, c belongs to A, since $0 \leq c \leq 1$. However, by construction, c is different from every a_i in at least one decimal digit $(i = 1, 2, \ldots)$ and hence $c \notin A$, which is a contradiction. Therefore, A is not countable. Since A is not finite either, then it must be uncountable.

This result implies that any subset of R, the set of real numbers, that contains A, or is equivalent to it, must be uncountable. In particular, R is uncountable.

Theorem 1.4.1.   Every infinite subset of a countable set is countable.

Proof. Let A be a countable set, and B be an infinite subset of A. Then $A = \{a_1, a_2, \ldots, a_n, \ldots\}$, where the a_i's are distinct elements. Let n_1 be the smallest positive integer such that $a_{n_1} \in B$. Let $n_2 > n_1$ be the next smallest integer such that $a_{n_2} \in B$. In general, if $n_1 < n_2 < \cdots < n_{k-1}$ have been chosen, let n_k be the smallest integer greater than n_{k-1} such that $a_{n_k} \in B$. Define the function $f: J^+ \to B$ such that $f(k) = a_{n_k}$, $k = 1, 2, \ldots$. This function is one-to-one and onto. Hence, B is countable.   □

Theorem 1.4.2.   The union of two countable sets is countable.

Proof. Let A and B be countable sets. Then they can be represented as $A = \{a_1, a_2, \ldots, a_n, \ldots\}$, $B = \{b_1, b_2, \ldots, b_n, \ldots\}$. Define $C = A \cup B$. Consider the following two cases:

i. A and B are disjoint.
ii. A and B are not disjoint.

In case i, let us write C as $C = \{a_1, b_1, a_2, b_2, \ldots, a_n, b_n, \ldots\}$. Consider the function $f: J^+ \to C$ such that

$$f(n) = \begin{cases} a_{(n+1)/2}, & n \text{ odd,} \\ b_{n/2}, & n \text{ even.} \end{cases}$$

It can be verified that f is one-to-one and onto. Hence, C is countable.

Let us now consider case ii. If $A \cap B \neq \varnothing$, then some elements of C, namely those in $A \cap B$, will appear twice. Hence, there exists a set $E \subset J^+$ such that $E \sim C$. Thus C is either finite or countable. Since $C \supset A$ and A is infinite, C must be countable.   □

Corollary 1.4.1. If $A_1, A_2, \ldots, A_n, \ldots$, are countable sets, then $\bigcup_{i=1}^{\infty} A_i$ is countable.

Proof. The proof is left as an exercise.   □

Theorem 1.4.3. Let A and B be two countable sets. Then their Cartesian product $A \times B$ is countable.

Proof. Let us write A as $A = (a_1, a_2, \ldots, a_n, \ldots)$. For a given $a \in A$, define (a, B) as the set

$$(a, B) = \{(a, b) | b \in B\}.$$

Then $(a, B) \sim B$ and hence (a, B) is countable.

However,

$$A \times B = \bigcup_{i=1}^{\infty} (a_i, B).$$

Thus by Corollary 1.4.1, $A \times B$ is countable.   □

Corollary 1.4.2. If $A_1, A_2, \ldots, A_n$ are countable sets, then their Cartesian product $\times_{i=1}^{n} A_i$ is countable.

Proof. The proof is left as an exercise.   □

Corollary 1.4.3. The set Q of all rational numbers is countable.

Proof. By definition, a rational number is a number of the form m/n, where m and n are integers with $n \neq 0$. Thus $Q \sim \tilde{Q}$, where

$$\tilde{Q} = \{(m, n) | m, n \text{ are integers and } n \neq 0\}.$$

Since $\tilde{Q}$ is an infinite subset of $J \times J$, where J is the set of all integers, which is countable as was seen in Example 1.4.2, then by Theorems 1.4.1 and 1.4.3, $\tilde{Q}$ is countable and so is Q. $\qquad\square$

REMARK 1.4.1. Any real number that cannot be expressed as a rational number is called an irrational number. For example, $\sqrt{2}$ is an irrational number. To show this, suppose that there exist integers, m and n, such that $\sqrt{2} = m/n$. We may consider that m/n is written in its lowest terms, that is, m and n have no common factors other than unity. In particular, m and n, cannot both be even. Now, $m^2 = 2n^2$. This implies that m^2 is even. Hence, m is even and can therefore be written as $m = 2m'$. It follows that $n^2 = m^2/2 = 2m'^2$. Consequently, n^2, and hence n, is even. This contradicts the fact that m and n are not both even. Thus $\sqrt{2}$ must be an irrational number.

1.5. BOUNDED SETS

Let us consider the set R of real numbers.

Definition 1.5.1. A set $A \subset R$ is said to be:

1. Bounded from above if there exists a number q such that $x \leq q$ for all x in A. This number is called an upper bound of A.
2. Bounded from below if there exists a number p such that $x \geq p$ for all x in A. The number p is called a lower bound of A.
3. Bounded if A has an upper bound q and a lower bound p. In this case, there exists a nonnegative number r such that $-r \leq x \leq r$ for all x in A. This number is equal to $\max(|p|,|q|)$. $\qquad\square$

Definition 1.5.2. Let $A \subset R$ be a set bounded from above. If there exists a number l that is an upper bound of A and is less than or equal to any other upper bound of A, then l is called the least upper bound of A and is denoted by $\mathrm{lub}(A)$. Another name for $\mathrm{lub}(A)$ is the supremum of A and is denoted by $\sup(A)$. $\qquad\square$

Definition 1.5.3. Let $A \subset R$ be a set bounded from below. If there exists a number g that is a lower bound of A and is greater than or equal to any other lower bound of A, then g is called the greatest lower bound and is denoted by $\mathrm{glb}(A)$. The infimum of A, denoted by $\inf(A)$, is another name for $\mathrm{glb}(A)$. $\qquad\square$

The least upper bound of A, if it exists, is unique, but it may or may not belong to A. The same is true for $\mathrm{glb}(A)$. The proof of the following theorem is omitted and can be found in Rudin (1964, Theorem 1.36).

Theorem 1.5.1.   Let $A \subset R$ be a nonempty set.

1. If A is bounded from above, then $\text{lub}(A)$ exists.
2. If A is bounded from below, then $\text{glb}(A)$ exists.

EXAMPLE 1.5.1.   Let $A = \{x | x < 0\}$. Then $\text{lub}(A) = 0$, which does not belong to A.

EXAMPLE 1.5.2.   Let $A = \{1/n | n = 1, 2, \ldots\}$. Then $\text{lub}(A) = 1$ and $\text{glb}(A) = 0$. In this case, $\text{lub}(A)$ belongs to A, but $\text{glb}(A)$ does not.

1.6. SOME BASIC TOPOLOGICAL CONCEPTS

The field of topology is an abstract study that evolved as an independent discipline in response to certain problems in classical analysis and geometry. It provides a unifying theory that can be used in many diverse branches of mathematics. In this section, we present a brief account of some basic definitions and results in the so-called *point-set topology*.

Definition 1.6.1.   Let A be a set, and let $\mathscr{F} = \{B_\alpha\}$ be a family of subsets of A. Then $\mathscr{F}$ is a *topology* in A if it satisfies the following properties:

1. The union of any number of members of $\mathscr{F}$ is also a member of $\mathscr{F}$.
2. The intersection of a finite number of members of $\mathscr{F}$ is also a member of $\mathscr{F}$.
3. Both A and the empty set $\varnothing$ are members of $\mathscr{F}$.   □

Definition 1.6.2.   Let $\mathscr{F}$ be a topology in a set A. Then the pair $(A, \mathscr{F})$ is called a *topological space*.   □

Definition 1.6.3.   Let $(A, \mathscr{F})$ be a topological space. Then the members of $\mathscr{F}$ are called the *open sets* of the topology $\mathscr{F}$.   □

Definition 1.6.4.   Let $(A, \mathscr{F})$ be a topological space. A neighborhood of a point $p \in A$ is any open set (that is, a member of $\mathscr{F}$) that contains p. In particular, if $A = R$, the set of real numbers, then a neighborhood of $p \in R$ is an open set of the form $N_r(p) = \{q | |q - p| < r\}$ for some $r > 0$.   □

Definition 1.6.5.   Let $(A, \mathscr{F})$ be a topological space. A family $G = \{B_\alpha\} \subset \mathscr{F}$ is called a *basis* for $\mathscr{F}$ if each open set (that is, member of $\mathscr{F}$) is the union of members of G.   □

On the basis of this definition, it is easy to prove the following theorem.

Theorem 1.6.1. Let $(A, \mathscr{F})$ be a topological space, and let G be a basis for $\mathscr{F}$. Then a set $B \subset A$ is open (that is, a member of $\mathscr{F}$) if and only if for each $p \in B$, there is a $U \in G$ such that $p \in U \subset B$.

For example, if $A = R$, then $G = \{N_r(p) | p \in R, r > 0\}$ is a basis for the topology in R. It follows that a set $B \subset R$ is open if for every point p in B, there exists a neighborhood $N_r(p)$ such that $N_r(p) \subset B$.

Definition 1.6.6. Let $(A, \mathscr{F})$ be a topological space. A set $B \subset A$ is closed if $\bar{B}$, the complement of B with respect to A, is an open set. □

It is easy to show that closed sets of a topological space $(A, \mathscr{F})$ satisfy the following properties:

1. The intersection of any number of closed sets is closed.
2. The union of a finite number of closed sets is closed.
3. Both A and the empty set $\varnothing$ are closed.

Definition 1.6.7. Let $(A, \mathscr{F})$ be a topological space. A point $p \in A$ is said to be a limit point of a set $B \subset A$ if every neighborhood of p contains at least one element of B distinct from p. Thus, if $U(p)$ is any neighborhood of p, then $U(p) \cap B$ is a nonempty set that contains at least one element besides p. In particular, if $A = R$, the set of real numbers, then p is a limit point of a set $B \subset R$ if for any $r > 0$, $N_r(p) \cap [B - \{p\}] \neq \varnothing$, where $\{p\}$ denotes a set consisting of just p. □

Theorem 1.6.2. Let p be a limit point of a set $B \subset R$. Then every neighborhood of p contains infinitely many points of B.

Proof. The proof is left to the reader. □

The next theorem is a fundamental theorem in set theory. It is originally due to Bernhard Bolzano (1781–1848), though its importance was first recognized by Karl Weierstrass (1815–1897). The proof is omitted and can be found, for example, in Zaring (1967, Theorem 4.62).

Theorem 1.6.3 (Bolzano–Weierstrass). Every bounded infinite subset of R, the set of real numbers, has at least one limit point.

Note that a limit point of a set B may not belong to B. For example, the set $B = \{1/n | n = 1, 2, \ldots\}$ has a limit point equal to zero, which does not belong to B. It can be seen here that any neighborhood of 0 contains infinitely many points of B. In particular, if r is a given positive number, then all elements of B of the form $1/n$, where $n > 1/r$, belong to $N_r(0)$. From Theorem 1.6.2 it can also be concluded that a finite set cannot have limit points.

Limit points can be used to describe closed sets, as can be seen from the following theorem.

Theorem 1.6.4. A set B is closed if and only if every limit point of B belongs to B.

Proof. Suppose that B is closed. Let p be a limit point of B. If $p \notin B$, then $p \in \overline{B}$, which is open. Hence, there exists a neighborhood $U(p)$ of p contained inside $\overline{B}$ by Theorem 1.6.1. This means that $U(p) \cap B = \varnothing$, a contradiction, since p is a limit point of B (see Definition 1.6.7). Therefore, p must belong to B. Vice versa, if every limit point of B is in B, then B must be closed. To show this, let p be any point in $\overline{B}$. Then, p is not a limit point of B. Therefore, there exists a neighborhood $U(p)$ such that $U(p) \subset \overline{B}$. This means that $\overline{B}$ is open and hence B is closed.   □

It should be noted that a set does not have to be either open or closed; if it is closed, it does not have to be open, and vice versa. Also, a set may be both open and closed.

EXAMPLE 1.6.1. $B = \{x \mid 0 < x < 1\}$ is an open subset of R, but is not closed, since both 0 and 1 are limit points of B, but do not belong to it.

EXAMPLE 1.6.2. $B = \{x \mid 0 \leq x \leq 1\}$ is closed, but is not open, since any neighborhood of 0 or 1 is not contained in B.

EXAMPLE 1.6.3. $B = \{x \mid 0 < x \leq 1\}$ is not open, because any neighborhood of 1 is not contained in B. It is also not closed, because 0 is a limit point that does not belong to B.

EXAMPLE 1.6.4.   The set R is both open and closed.

EXAMPLE 1.6.5.   A finite set is closed because it has no limit points, but is obviously not open.

Definition 1.6.8.   A subset B of a topological space $(A, \mathscr{F})$ is *disconnected* if there exist open subsets C and D of A such that $B \cap C$ and $B \cap D$ are disjoint nonempty sets whose union is B. A set is *connected* if it is not disconnected.   □

The set of all rationals Q is disconnected, since $\{x \mid x > \sqrt{2}\} \cap Q$ and $\{x \mid x < \sqrt{2}\} \cap Q$ are disjoint nonempty sets whose union is Q. On the other hand, all intervals in R (open, closed, or half-open) are connected.

Definition 1.6.9.   A collection of sets $\{B_\alpha\}$ is said to be a *covering* of a set A if the union $\bigcup_\alpha B_\alpha$ contains A. If each B_α is an open set, then $\{B_\alpha\}$ is called an *open covering*.

Definition 1.6.10. A set A in a topological space is *compact* if each open covering $\{B_\alpha\}$ of A has a finite subcovering, that is, there is a finite subcollection $B_{\alpha_1}, B_{\alpha_2}, \ldots, B_{\alpha_n}$ of $\{B_\alpha\}$ such that $A \subset \bigcup_{i=1}^{n} B_{\alpha_i}$. □

The concept of compactness is motivated by the classical *Heine–Borel theorem*, which characterizes compact sets in R, the set of real numbers, as closed and bounded sets.

Theorem 1.6.5 (Heine–Borel). A set $B \subset R$ is compact if and only if it is closed and bounded.

Proof. See, for example, Zaring (1967, Theorem 4.78). □

Thus, according to the Heine–Borel theorem, every closed and bounded interval $[a, b]$ is compact.

1.7. EXAMPLES IN PROBABILITY AND STATISTICS

EXAMPLE 1.7.1. In probability theory, events are considered as subsets in a sample space Ω, which consists of all the possible outcomes of an experiment. A Borel field of events (also called a σ-field) in Ω is a collection $\mathscr{B}$ of events with the following properties:

i. $\Omega \in \mathscr{B}$.
ii. If $E \in \mathscr{B}$, then $\bar{E} \in \mathscr{B}$, where $\bar{E}$ is the complement of E.
iii. If $E_1, E_2, \ldots, E_n, \ldots$ is a countable collection of events in $\mathscr{B}$, then $\bigcup_{i=1}^{\infty} E_i$ belongs to $\mathscr{B}$.

The probability of an event E is a number denoted by $P(E)$ that has the following properties:

i. $0 \leq P(E) \leq 1$.
ii. $P(\Omega) = 1$.
iii. If $E_1, E_2, \ldots, E_n, \ldots$ is a countable collection of disjoint events in $\mathscr{B}$, then

$$P\left(\bigcup_{i=1}^{\infty} E_i\right) = \sum_{i=1}^{\infty} P(E_i).$$

By definition, the triple $(\Omega, \mathscr{B}, P)$ is called a probability space.

EXAMPLE 1.7.2 . A random variable X defined on a probability space $(\Omega, \mathscr{B}, P)$ is a function $X: \Omega \to A$, where A is a nonempty set of real numbers. For any real number x, the set $E = \{\omega \in \Omega | X(\omega) \leq x\}$ is an element of $\mathscr{B}$. The probability of the event E is called the cumulative distribution function of X and is denoted by $F(x)$. In statistics, it is customary to write just X instead of $X(\omega)$. We thus have

$$F(x) = P(X \leq x).$$

This concept can be extended to several random variables: Let $X_1, X_2, \ldots, X_n$ be n random variables. Define the event $A_i = \{\omega \in \Omega | X_i(\omega) \leq x_i\}$, $i = 1, 2, \ldots, n$. Then, $P(\cap_{i=1}^n A_i)$, which can be expressed as

$$F(x_1, x_2, \ldots, x_n) = P(X_1 \leq x_1, X_2 \leq x_2, \ldots, X_n \leq x_n),$$

is called the joint cumulative distribution function of $X_1, X_2, \ldots, X_n$. In this case, the n-tuple $(X_1, X_2, \ldots, X_n)$ is said to have a multivariate distribution.

A random variable X is said to be discrete, or to have a discrete distribution, if its range is finite or countable. For example, the binomial random variable is discrete. It represents the number of successes in a sequence of n independent trials, in each of which there are two possible outcomes: success or failure. The probability of success, denoted by p_n, is the same in all the trials. Such a sequence of trials is called a Bernoulli sequence. Thus the possible values of this random variable are $0, 1, \ldots, n$.

Another example of a discrete random variable is the Poisson, whose possible values are $0, 1, 2, \ldots$. It is considered to be the limit of a binomial random variable as $n \to \infty$ in such a way that $np_n \to \lambda > 0$. Other examples of discrete random variables include the discrete uniform, geometric, hypergeometric, and negative binomial (see, for example, Fisz, 1963; Johnson and Kotz, 1969; Lindgren 1976; Lloyd, 1980).

A random variable X is said to be continuous, or to have a continuous distribution, if its range is an uncountable set, for example, an interval. In this case, the cumulative distribution function $F(x)$ of X is a continuous function of x on the set R of all real numbers. If, in addition, $F(x)$ is differentiable, then its derivative is called the density function of X. One of the best-known continuous distributions is the normal. A number of continuous distributions are derived in connection with it, for example, the chi-squared, F, Rayleigh, and t distributions. Other well-known continuous distributions include the beta, continuous uniform, exponential, and gamma distributions (see, for example, Fisz, 1963; Johnson and Kotz, 1970a, b).

EXAMPLE 1.7.3. Let $f(x, \theta)$ denote the density function of a continuous random variable X, where θ represents a set of unknown parameters that identify the distribution of X. The range of X, which consists of all possible values of X, is referred to as a population and denoted by P_X. Any subset of n elements from P_X forms a sample of size n. This sample is actually an element in the Cartesian product P_X^n. Any real-valued function defined on P_X^n is called a statistic. We denote such a function by $g(X_1, X_2, \ldots, X_n)$, where each X_i has the same distribution as X. Note that this function is a random variable whose values do not depend on θ. For example, the sample mean $\overline{X} = \Sigma_{i=1}^n X_i / n$ and the sample variance $S^2 = \Sigma_{i=1}^n (X_i - \overline{X})^2 / (n-1)$

are statistics. We adopt the convention that whenever a particular sample of size n is chosen (or observed) from P_X, the elements in that sample are written using lowercase letters, for example, $x_1, x_2, \ldots, x_n$. The corresponding value of a statistic is written as $g(x_1, x_2, \ldots, x_n)$.

EXAMPLE 1.7.4. Two random variables, X and Y, are said to be equal in distribution if they have the same cumulative distribution function. This fact is denoted by writing $X \overset{d}{=} Y$. The same definition applies to random variables with multivariate distributions. We note that $\overset{d}{=}$ is an equivalence relation, since it satisfies properties 1, 2, and 3 in Definition 1.3.2. The first two properties are obviously true. As for property 3, if $X \overset{d}{=} Y$ and $Y \overset{d}{=} Z$, then $X \overset{d}{=} Z$, which implies that all three random variables have the same cumulative distribution function. This equivalence relation is useful in nonparametric statistics (see Randles and Wolfe, 1979). For example, it can be shown that if X has a distribution that is symmetric about some number μ, then $X - \mu \overset{d}{=} \mu - X$. Also, if $X_1, X_2, \ldots, X_n$ are independent and identically distributed random variables, and if $(m_1, m_2, \ldots, m_n)$ is any permutation of the n-tuple $(1, 2, \ldots, n)$, then $(X_1, X_2, \ldots, X_n) \overset{d}{=} (X_{m_1}, X_{m_2}, \ldots, X_{m_n})$. In this case, we say that the collection of random variables $X_1, X_2, \ldots, X_n$ is exchangeable.

EXAMPLE 1.7.5. Consider the problem of testing the null hypothesis H_0: $\theta \leq \theta_0$ versus the alternative hypothesis H_a: $\theta > \theta_0$, where θ is some unknown parameter that belongs to a set A. Let T be a statistic used in making a decision as to whether H_0 should be rejected or not. This statistic is appropriately called a test statistic.

Suppose that H_0 is rejected if $T > t$, where t is some real number. Since the distribution of T depends on θ, then the probability $P(T > t)$ is a function of θ, which we denote by $\pi(\theta)$. Thus $\pi: A \to [0, 1]$. Let B_0 be a subset of A defined as $B_0 = \{\theta \in A \mid \theta \leq \theta_0\}$. By definition, the size of the test is the least upper bound of the set $\pi(B_0)$. This probability is denoted by α and is also called the level of significance of the test. We thus have

$$\alpha = \sup_{\theta \leq \theta_0} \pi(\theta).$$

To learn more about the above examples and others, the interested reader may consider consulting some of the references listed in the annotated bibliography.

FURTHER READING AND ANNOTATED BIBLIOGRAPHY

Bronshtein, I. N., and K. A. Semendyayev (1985). *Handbook of Mathematics* (English translation edited by K. A. Hirsch). Van Nostrand Reinhold, New York. (Section 4.1 in this book gives basic concepts of set theory; Chap. 5 provides a brief introduction to probability and mathematical statistics.)

Dugundji, J. (1966). *Topology*. Allyn and Bacon, Boston. (Chap. 1 deals with elementary set theory; Chap. 3 presents some basic topological concepts that complements the material given in Section 1.6.)

Fisz, M. (1963). *Probability Theory and Mathematical Statistics*, 3rd ed. Wiley, New York. (Chap. 1 discusses random events and axioms of the theory of probability; Chap. 2 introduces the concept of a random variable; Chap. 5 investigates some probability distributions.)

Hardy, G. H. (1955). *A Course of Pure Mathematics*, 10th ed. The University Press, Cambridge, England. (Chap. 1 in this classic book is recommended reading for understanding the real number system.)

Harris, B. (1966). *Theory of Probability*. Addison-Wesley, Reading, Massachusetts. (Chaps. 2 and 3 discuss some elementary concepts in probability theory as well as in distribution theory. Many exercises are provided.)

Hogg, R. V., and A. T. Craig (1965). *Introduction to Mathematical Statistics*, 2nd ed. Macmillan, New York. (Chap. 1 is an introduction to distribution theory; examples of some special distributions are given in Chap. 3; Chap. 10 considers some aspects of hypothesis testing that pertain to Example 1.7.5.)

Johnson, N. L., and S. Kotz (1969). *Discrete Distributions*. Houghton Mifflin, Boston. (This is the first volume in a series of books on statistical distributions. It is an excellent source for getting detailed accounts of the properties and uses of these distributions. This volume deals with discrete distributions, including the binomial in Chap. 3, the Poisson in Chap. 4, the negative binomial in Chap. 5, and the hypergeometric in Chap. 6.)

Johnson, N. L., and S. Kotz (1970a). *Continuous Univariate Distributions—1*. Houghton Mifflin, Boston. (This volume covers continuous distributions, including the normal in Chap. 13, lognormal in Chap. 14, Cauchy in Chap. 16, gamma in Chap. 17, and the exponential in Chap. 18.)

Johnson, N. L., and S. Kotz (1970b). *Continuous Univariate Distributions—2*. Houghton Mifflin, Boston. (This is a continuation of Vol. 2 on continuous distributions. Chaps. 24, 25, 26, and 27 discuss the beta, continuous uniforms, F, and t distributions, respectively.)

Johnson, P. E. (1972). *A History of Set Theory*. Prindle, Weber, and Schmidt, Boston. (This book presents a historical account of set theory as was developed by Georg Cantor.)

Lindgren, B. W. (1976). *Statistical Theory*, 3rd ed. Macmillan, New York. (Sections 1.1, 1.2, 2.1, 3.1, 3.2, and 3.3 present introductory material on probability models and distributions; Chap. 6 discusses test of hypothesis and statistical inference.)

Lloyd, E. (1980). *Handbook of Applicable Mathematics*, Vol. II. Wiley, New York. (This is the second volume in a series of six volumes designed as texts of mathematics for professionals. Chaps. 1, 2, and 3 present expository material on probability; Chaps. 4 and 5 discuss random variables and their distributions.)

Randles, R. H., and D. A. Wolfe (1979). *Introduction to the Theory of Nonparametric Statistics*. Wiley, New York. (Section 1.3 in this book discusses the "equal in distribution" property mentioned in Example 1.7.4.)

Rudin, W. (1964). *Principles of Mathematical Analysis*, 2nd ed. McGraw-Hill, New York. (Chap. 1 discusses the real number system; Chap. 2 deals with countable, uncountable, and bounded sets and pertains to Sections 1.4, 1.5, and 1.6.)

Stoll, R. R. (1963). *Set Theory and Logic*. W. H. Freeman, San Francisco. (Chap. 1 is an introduction to set theory; Chap. 2 discusses countable sets; Chap. 3 is useful in understanding the real number system.)

Tucker, H. G. (1962). *Probability and Mathematical Statistics*. Academic Press, New York. (Chaps. 1, 3, 4, and 6 discuss basic concepts in elementary probability and distribution theory.)

Vilenkin, N. Y. (1968). *Stories about Sets*. Academic Press, New York. (This is an interesting book that presents various notions of set theory in an informal and delightful way. It contains many unusual stories and examples that make the learning of set theory rather enjoyable.)

Zaring, W. M. (1967). *An Introduction to Analysis*. Macmillan, New York. (Chap. 2 gives an introduction to set theory; Chap. 3 discusses functions and relations.)

EXERCISES

In Mathematics

1.1. Verify Results 1.2.3–1.2.12.

1.2. Verify Results 1.2.13–1.2.16.

1.3. Let A, B, and C be sets such that $A \cap B \subset \bar{C}$ and $A \cup C \subset B$. Show that A and C are disjoint.

1.4. Let A, B, and C be sets such that $C = (A - B) \cup (B - A)$. The set C is called the symmetric difference of A and B and is denoted by $A \vartriangle B$. Show that

(a) $A \vartriangle B = A \cup B - A \cap B$

(b) $A \vartriangle (B \vartriangle D) = (A \vartriangle B) \vartriangle D$, where D is any set.

(c) $A \cap (B \vartriangle D) = (A \cap B) \vartriangle (A \cap D)$, where D is any set.

1.5. Let $A = J^+ \times J^+$, where J^+ is the set of positive integers. Define a relation ρ in A as follows: If (m_1, n_1) and (m_2, n_2) are elements in A, then $(m_1, n_1) \rho (m_2, n_2)$ if $m_1 n_2 = n_1 m_2$. Show that ρ is an equivalence relation and describe its equivalence classes.

1.6. Let A be the same set as in Exercise 1.5. Show that the following relation is an equivalence relation: $(m_1, n_1) \rho (m_2, n_2)$ if $m_1 + n_2 = n_1 + m_2$. Draw the equivalence class of $(1, 2)$.

1.7. Consider the set $A = \{(-2, -5), (-1, -3), (1, 2), (3, 10)\}$. Show that A defines a function.

1.8. Let A and B be two sets and f be a function defined on A such that $f(A) \subset B$. If $A_1, A_2, \ldots, A_n$ are subsets of A, then show that:

(a) $f(\bigcup_{i=1}^{n} A_i) = \bigcup_{i=1}^{n} f(A_i)$.

(b) $f(\bigcap_{i=1}^{n} A_i) \subset \bigcap_{i=1}^{n} f(A_i)$.

Under what conditions are the two sides in (b) equal?

1.9. Prove Corollary 1.4.1.

1.10. Prove Corollary 1.4.2.

1.11. Show that the set $A = \{3, 9, 19, 33, 51, 73, \ldots\}$ is countable.

1.12. Show that $\sqrt{3}$ is an irrational number.

1.13. Let a, b, c, and d be rational numbers such that $a + \sqrt{b} = c + \sqrt{d}$. Then, either

(a) $a = c, b = d$, or

(b) b and d are both squares of rational numbers.

1.14. Let $A \subset R$ be a nonempty set bounded from below. Define $-A$ to be the set $\{-x | x \in A\}$. Show that $\inf(A) = -\sup(-A)$.

1.15. Let $A \subset R$ be a closed and bounded set, and let $\sup(A) = b$. Show that $b \in A$.

1.16. Prove Theorem 1.6.2.

1.17. Let $(A, \mathscr{F})$ be a topological space. Show that $G \subset \mathscr{F}$ is a basis for $\mathscr{F}$ in and only if for each $B \in \mathscr{F}$ and each $p \in B$, there is a $U \in G$ such that $p \in U \subset B$.

1.18. Show that if A and B are closed sets, then $A \cup B$ is a closed set.

1.19. Let $B \subset A$ be a closed subset of a compact set A. Show that B is compact.

1.20. Is a compact subset of a compact set necessarily closed?

In Statistics

1.21. Let X be a random variable. Consider the following events:

$$A_n = \{\omega \in \Omega | X(\omega) < x + 3^{-n}\}, \qquad n = 1, 2, \ldots,$$
$$B_n = \{\omega \in \Omega | X(\omega) \le x - 3^{-n}\}, \qquad n = 1, 2, \ldots,$$
$$A = \{\omega \in \Omega | X(\omega) \le x\},$$
$$B = \{\omega \in \Omega | X(\omega) < x\},$$

where x is a real number. Show that for any x,

(a) $\bigcap_{n=1}^{\infty} A_n = A$;

(b) $\bigcup_{n=1}^{\infty} B_n = B$.

1.22. Let X be a nonnegative random variable such that $E(X) = \mu$ is finite, where $E(X)$ denotes the expected value of X. The following inequality, known as *Markov's inequality*, is true:

$$P(X \geq h) \leq \frac{\mu}{h},$$

where h is any positive number. Consider now a Poisson random variable with parameter λ.

(a) Find an upper bound on the probability $P(X \geq 2)$ using Markov's inequality.

(b) Obtain the exact probability value in (a), and demonstrate that it is smaller than the corresponding upper bound in Markov's inequality.

1.23. Let X be a random variable whose expected value μ and variance σ^2 exist. Show that for any positive constants c and k,

(a) $P(|X - \mu| \geq c) \leq \sigma^2/c^2$,

(b) $P(|X - \mu| \geq k\sigma) \leq 1/k^2$,

(c) $P(|X - \mu| < k\sigma) \geq 1 - 1/k^2$.

The preceding three inequalities are equivalent versions of the so-called *Chebyshev's inequality*.

1.24. Let X be a continuous random variable with the density function

$$f(x) = \begin{cases} 1 - |x|, & -1 < x < 1, \\ 0 & \text{elsewhere}. \end{cases}$$

By definition, the density function of X is a nonnegative function such that $F(x) = \int_{-\infty}^{x} f(t)\,dt$, where $F(x)$ is the cumulative distribution function of X.

(a) Apply Markov's inequality to finding upper bounds on the following probabilities: (i) $P(|X| \geq \frac{1}{2})$; (ii) $P(|X| > \frac{1}{3})$.

(b) Compute the exact value of $P(|X| \geq \frac{1}{2})$, and compare it against the upper bound in (a)(i).

1.25. Let $X_1, X_2, \ldots, X_n$ be n continuous random variables. Define the random variables $X_{(1)}$ and $X_{(n)}$ as

$$X_{(1)} = \min_{1 \leq i \leq n} \{X_1, X_2, \ldots, X_n\},$$

$$X_{(n)} = \max_{1 \leq i \leq n} \{X_1, X_2, \ldots, X_n\}.$$

Show that for any x,

(a) $P(X_{(1)} \geq x) = P(X_1 \geq x, X_2 \geq x, \ldots, X_n \geq x)$,

(b) $P(X_{(n)} \leq x) = P(X_1 \leq x, X_2 \leq x, \ldots, X_n \leq x)$.

In particular, if $X_1, X_2, \ldots, X_n$ form a sample of size n from a population with a cumulative distribution function $F(x)$, show that

(c) $P(X_{(1)} \leq x) = 1 - [1 - F(x)]^n$,

(d) $P(X_{(n)} \leq x) = [F(x)]^n$.

The statistics $X_{(1)}$ and $X_{(n)}$ are called the first-order and nth-order statistics, respectively.

1.26. Suppose that we have a sample of size $n = 5$ from a population with an exponential distribution whose density function is

$$f(x) = \begin{cases} 2e^{-2x}, & x > 0, \\ 0 & \text{elsewhere}. \end{cases}$$

Find the value of $P(2 \leq X_{(1)} \leq 3)$.

CHAPTER 2

Basic Concepts in Linear Algebra

In this chapter we present some fundamental concepts concerning vector spaces and matrix algebra. The purpose of the chapter is to familiarize the reader with these concepts, since they are essential to the understanding of some of the remaining chapters. For this reason, most of the theorems in this chapter will be stated without proofs. There are several excellent books on linear algebra that can be used for a more detailed study of this subject (see the bibliography at the end of this chapter).

In statistics, matrix algebra is used quite extensively, especially in linear models and multivariate analysis. The books by Basilevsky (1983), Graybill (1983), Magnus and Neudecker (1988), and Searle (1982) include many applications of matrices in these areas.

In this chapter, as well as in the remainder of the book, elements of the set of real numbers, R, are sometimes referred to as scalars. The Cartesian product $\times_{i=1}^{n} R$ is denoted by R^n, which is also known as the n-dimensional Euclidean space. Unless otherwise stated, all matrix elements are considered to be real numbers.

2.1. VECTOR SPACES AND SUBSPACES

A vector space over R is a set V of elements called vectors together with two operations, addition and scalar multiplication, that satisfy the following conditions:

1. $\mathbf{u} + \mathbf{v}$ is an element of V for all $\mathbf{u}, \mathbf{v}$ in V.
2. If α is a scalar and $\mathbf{u} \in V$, then $\alpha \mathbf{u} \in V$.
3. $\mathbf{u} + \mathbf{v} = \mathbf{v} + \mathbf{u}$ for all $\mathbf{u}, \mathbf{v}$ in V.
4. $\mathbf{u} + (\mathbf{v} + \mathbf{w}) = (\mathbf{u} + \mathbf{v}) + \mathbf{w}$ for all $\mathbf{u}, \mathbf{v}, \mathbf{w}$ in V.
5. There exists an element $\mathbf{0} \in V$ such that $\mathbf{0} + \mathbf{u} = \mathbf{u}$ for all $\mathbf{u}$ in V. This element is called the zero vector.

6. For each $\mathbf{u} \in V$ there exists a $\mathbf{v} \in V$ such that $\mathbf{u} + \mathbf{v} = \mathbf{0}$.

7. $\alpha(\mathbf{u} + \mathbf{v}) = \alpha\mathbf{u} + \alpha\mathbf{v}$ for any scalar α and any $\mathbf{u}$ and $\mathbf{v}$ in V.

8. $(\alpha + \beta)\mathbf{u} = \alpha\mathbf{u} + \beta\mathbf{u}$ for any scalars α and β and any $\mathbf{u}$ in V.

9. $\alpha(\beta\mathbf{u}) = (\alpha\beta)\mathbf{u}$ for any scalars α and β and any $\mathbf{u}$ in V.

10. $1\mathbf{u} = \mathbf{u}$ for any $\mathbf{u} \in V$.

EXAMPLE 2.1.1. A familiar example of a vector space is the n-dimensional Euclidean space R^n. Here, addition and multiplication are defined as follows: If $(u_1, u_2, \ldots, u_n)$ and $(v_1, v_2, \ldots, v_n)$ are two elements in R^n, then their sum is defined as $(u_1 + v_1, u_2 + v_2, \ldots, u_n + v_n)$. If α is a scalar, then $\alpha(u_1, u_2, \ldots, u_n) = (\alpha u_1, \alpha u_2, \ldots, \alpha u_n)$.

EXAMPLE 2.1.2. Let V be the set of all polynomials in x of degree less than or equal to k. Then V is a vector space. Any element in V can be expressed as $\sum_{i=0}^{k} a_i x^i$, where the a_i's are scalars.

EXAMPLE 2.1.3. Let V be the set of all functions defined on the closed interval $[-1, 1]$. Then V is a vector space. It can be seen that $f(x) + g(x)$ and $\alpha f(x)$ belong to V, where $f(x)$ and $g(x)$ are elements in V and α is any scalar.

EXAMPLE 2.1.4. The set V of all nonnegative functions defined on $[-1, 1]$ is not a vector space, since if $f(x) \in V$ and α is a negative scalar, then $\alpha f(x) \notin V$.

EXAMPLE 2.1.5. Let V be the set of all points (x, y) on a straight line given by the equation $2x - y + 1 = 0$. Then V is not a vector space. This is because if (x_1, y_1) and (x_2, y_2) belong to V, then $(x_1 + x_2, y_1 + y_2) \notin V$, since $2(x_1 + x_2) - (y_1 + y_2) + 1 = -1 \neq 0$. Alternatively, we can state that V is not a vector space because the zero element $(0, 0)$ does not belong to V. This violates condition 5 for a vector space.

A subset W of a vector space V is said to form a vector subspace if W itself is a vector space. Equivalently, W is a subspace if whenever $\mathbf{u}, \mathbf{v} \in W$ and α is a scalar, then $\mathbf{u} + \mathbf{v} \in W$ and $\alpha\mathbf{u} \in W$. For example, the set W of all continuous functions defined on $[-1, 1]$ is a vector subspace of V in Example 2.1.3. Also, the set of all points on the straight line $y - 2x = 0$ is a vector subspace of R^2. However, the points on any straight line in R^2 not going through the origin $(0, 0)$ do not form a vector subspace, as was seen in Example 2.1.5.

Definition 2.1.1. Let V be a vector space, and $\mathbf{u}_1, \mathbf{u}_2, \ldots, \mathbf{u}_n$ be a collection of n elements in V. These elements are said to be linearly dependent if there exist n scalars $\alpha_1, \alpha_2, \ldots, \alpha_n$, not all equal to zero, such that $\sum_{i=1}^{n} \alpha_i \mathbf{u}_i = \mathbf{0}$. If, however, $\sum_{i=1}^{n} \alpha_i \mathbf{u}_i = \mathbf{0}$ is true only when all the α_i's are zero, then

$\mathbf{u}_1, \mathbf{u}_2, \ldots, \mathbf{u}_n$ are linearly independent. It should be noted that if $\mathbf{u}_1, \mathbf{u}_2, \ldots, \mathbf{u}_n$ are linearly independent, then none of them can be zero. If, for example, $\mathbf{u}_1 = \mathbf{0}$, then $\alpha \mathbf{u}_1 + 0\mathbf{u}_2 + \cdots + 0\mathbf{u}_n = \mathbf{0}$ for any $\alpha \neq 0$, which implies that the $\mathbf{u}_i$'s are linearly dependent, a contradiction. □

From the preceding definition we can say that a collection of n elements in a vector space are linearly dependent if at least one element in this collection can be expressed as a linear combination of the remaining $n - 1$ elements. If no element, however, can be expressed in this fashion, then the n elements are linearly independent. For example, in R^3, $(1, 2, -2)$, $(-1, 0, 3)$, and $(1, 4, -1)$ are linearly dependent, since $2(1, 2, -2) + (-1, 0, 3) - (1, 4, -1) = \mathbf{0}$. On the other hand, it can be verified that $(1, 1, 0)$, $(1, 0, 2)$, and $(0, 1, 3)$ are linearly independent.

Definition 2.1.2. Let $\mathbf{u}_1, \mathbf{u}_2, \ldots, \mathbf{u}_n$ be n elements in a vector space V. The collection of all linear combinations of the form $\sum_{i=1}^{n} \alpha_i \mathbf{u}_i$, where the α_i's are scalars, is called a linear span of $\mathbf{u}_1, \mathbf{u}_2, \ldots, \mathbf{u}_n$ and is denoted by $L(\mathbf{u}_1, \mathbf{u}_2, \ldots, \mathbf{u}_n)$. □

It is easy to see from the preceding definition that $L(\mathbf{u}_1, \mathbf{u}_2, \ldots, \mathbf{u}_n)$ is a vector subspace of V. This vector subspace is said to be spanned by $\mathbf{u}_1, \mathbf{u}_2, \ldots, \mathbf{u}_n$.

Definition 2.1.3. Let V be a vector space. If there exist linearly independent elements $\mathbf{u}_1, \mathbf{u}_2, \ldots, \mathbf{u}_n$ in V such that $V = L(\mathbf{u}_1, \mathbf{u}_2, \ldots, \mathbf{u}_n)$, then $\mathbf{u}_1, \mathbf{u}_2, \ldots, \mathbf{u}_n$ are said to form a basis for V. The number n of elements in this basis is called the dimension of the vector space and is denoted by $\dim V$. □

Note that a basis for a vector space is not unique. However, its dimension is unique. For example, the three vectors $(1, 0, 0)$, $(0, 1, 0)$, and $(0, 0, 1)$ form a basis for R^3. Another basis for R^3 consists of $(1, 1, 0)$, $(1, 0, 1)$, and $(0, 1, 1)$.

If $\mathbf{u}_1, \mathbf{u}_2, \ldots, \mathbf{u}_n$ form a basis for V and if $\mathbf{u}$ is a given element in V, then there exists a unique set of scalars, $\alpha_1, \alpha_2, \ldots, \alpha_n$, such that $\mathbf{u} = \sum_{i=1}^{n} \alpha_i \mathbf{u}_i$. To show this, suppose that there exists another set of scalars, $\beta_1, \beta_2, \ldots, \beta_n$, such that $\mathbf{u} = \sum_{i=1}^{n} \beta_i \mathbf{u}$. Then $\sum_{i=1}^{n} (\alpha_i - \beta_i) \mathbf{u}_i = \mathbf{0}$, which implies that $\alpha_i = \beta_i$ for all i, since the $\mathbf{u}_i$'s are linearly independent.

Let us now check the dimensions of the vector spaces for some of the examples described earlier. For Example 2.1.1, $\dim V = n$. In Example 2.1.2, $\{1, x, x^2, \ldots, x^k\}$ is a basis for V; hence $\dim V = k + 1$. As for Example 2.1.3, $\dim V$ is infinite, since there is no finite set of functions that can span V.

Definition 2.1.4. Let $\mathbf{u}$ and $\mathbf{v}$ be two vectors in R^n. The dot product (also called scalar product or inner product) of $\mathbf{u}$ and $\mathbf{v}$ is a scalar denoted by $\mathbf{u} \cdot \mathbf{v}$ and is given by

$$\mathbf{u} \cdot \mathbf{v} = \sum_{i=1}^{n} u_i v_i,$$

where u_i and v_i are the ith components of $\mathbf{u}$ and $\mathbf{v}$, respectively ($i = 1, 2, \ldots, n$). In particular, if $\mathbf{u} = \mathbf{v}$, then $(\mathbf{u} \cdot \mathbf{u})^{1/2} = (\sum_{i=1}^{n} u_i^2)^{1/2}$ is called the Euclidean norm (or length) of $\mathbf{u}$ and is denoted by $\|\mathbf{u}\|_2$. The dot product of $\mathbf{u}$ and $\mathbf{v}$ is also equal to $\|\mathbf{u}\|_2 \|\mathbf{v}\|_2 \cos\theta$, where θ is the angle between $\mathbf{u}$ and $\mathbf{v}$. $\qquad\square$

Definition 2.1.5. Two vectors $\mathbf{u}$ and $\mathbf{v}$ in R^n are said to be orthogonal if their dot product is zero. $\qquad\square$

Definition 2.1.6. Let U be a vector subspace of R^n. The vectors $\mathbf{e}_1, \mathbf{e}_2, \ldots, \mathbf{e}_m$ form an orthonormal basis for U if they satisfy the following properties:

1. $\mathbf{e}_1, \mathbf{e}_2, \ldots, \mathbf{e}_m$ form a basis for U.
2. $\mathbf{e}_i \cdot \mathbf{e}_j = 0$ for all $i \neq j$ ($i, j = 1, 2, \ldots, m$).
3. $\|\mathbf{e}_i\|_2 = 1$ for $i = 1, 2, \ldots, m$.

Any collection of vectors satisfying just properties 2 and 3 are said to be orthonormal. $\qquad\square$

Theorem 2.1.1. Let $\mathbf{u}_1, \mathbf{u}_2, \ldots, \mathbf{u}_m$ be a basis for a vector subspace U of R^n. Then there exists an orthonormal basis, $\mathbf{e}_1, \mathbf{e}_2, \ldots, \mathbf{e}_m$, for U, given by

$$\mathbf{e}_1 = \frac{\mathbf{v}_1}{\|\mathbf{v}_1\|_2}, \qquad \text{where } \mathbf{v}_1 = \mathbf{u}_1,$$

$$\mathbf{e}_2 = \frac{\mathbf{v}_2}{\|\mathbf{v}_2\|_2}, \qquad \text{where } \mathbf{v}_2 = \mathbf{u}_2 - \frac{\mathbf{v}_1 \cdot \mathbf{u}_2}{\|\mathbf{v}_1\|_2^2} \mathbf{v}_1,$$

$$\vdots$$

$$\mathbf{e}_m = \frac{\mathbf{v}_m}{\|\mathbf{v}_m\|_2}, \qquad \text{where } \mathbf{v}_m = \mathbf{u}_m - \sum_{i=1}^{m-1} \frac{\mathbf{v}_i \cdot \mathbf{u}_m}{\|\mathbf{v}_i\|_2^2} \mathbf{v}_i.$$

Proof. See Graybill (1983, Theorem 2.6.5). $\qquad\square$

The procedure of constructing an orthonormal basis from any given basis as described in Theorem 2.1.1 is known as the Gram-Schmidt orthonormalization procedure.

Theorem 2.1.2. Let $\mathbf{u}$ and $\mathbf{v}$ be two vectors in R^n. Then:

1. $|\mathbf{u} \cdot \mathbf{v}| \leq \|\mathbf{u}\|_2 \|\mathbf{v}\|_2$.
2. $\|\mathbf{u} + \mathbf{v}\|_2 \leq \|\mathbf{u}\|_2 + \|\mathbf{v}\|_2$.

Proof. See Marcus and Minc (1988, Theorem 3.4). $\qquad\square$

The inequality in part 1 of Theorem 2.1.2 is known as the *Cauchy–Schwarz inequality*. The one in part 2 is called the *triangle inequality*.

Definition 2.1.7. Let U be a vector subspace of R^n. The orthogonal complement of U, denoted by $U^\perp$, is the vector subspace of R^n which consists of all vectors $\mathbf{v}$ such that $\mathbf{u} \cdot \mathbf{v} = 0$ for all $\mathbf{u}$ in U. □

Definition 2.1.8. Let $U_1, U_2, \ldots, U_n$ be vector subspaces of the vector space U. The direct sum of these vector subspaces, denoted by $\oplus_{i=1}^n U_i$, consists of all vectors $\mathbf{u}$ that can be uniquely expressed as $\mathbf{u} = \sum_{i=1}^n \mathbf{u}_i$, where $\mathbf{u}_i \in U_i$, $i = 1, 2, \ldots, n$. □

Theorem 2.1.3. Let $U_1, U_2, \ldots, U_n$ be vector subspaces of the vector space U. Then:

1. $\oplus_{i=1}^n U_i$ is a vector subspace of U.
2. If $U = \oplus_{i=1}^n U_i$, then $\cap_{i=1}^n U_i$ consists of just the zero element $\mathbf{0}$ of U.
3. $\dim \oplus_{i=1}^n U_i = \sum_{i=1}^n \dim U_i$.

Proof. The proof is left as an exercise. □

Theorem 2.1.4. Let U be a vector subspace of R^n. Then $R^n = U \oplus U^\perp$.

Proof. See Marcus and Minc (1988, Theorem 3.3). □

From Theorem 2.1.4 we conclude that any $\mathbf{v} \in R^n$ can be uniquely written as $\mathbf{v} = \mathbf{v}_1 + \mathbf{v}_2$, where $\mathbf{v}_1 \in U$ and $\mathbf{v}_2 \in U^\perp$. In this case, $\mathbf{v}_1$ and $\mathbf{v}_2$ are called the projections of $\mathbf{v}$ on U and $U^\perp$, respectively.

2.2. LINEAR TRANSFORMATIONS

Let U and V be two vector spaces. A function $T: U \to V$ is called a linear transformation if $T(\alpha_1 \mathbf{u}_1 + \alpha_2 \mathbf{u}_2) = \alpha_1 T(\mathbf{u}_1) + \alpha_2 T(\mathbf{u}_2)$ for all $\mathbf{u}_1, \mathbf{u}_2$ in U and any scalars α_1 and α_2. For example, let $T: R^3 \to R^3$ be defined as

$$T(x_1, x_2, x_3) = (x_1 - x_2, x_1 + x_3, x_3).$$

Then T is a linear transformation, since

$$
\begin{aligned}
T[\alpha(x_1, x_2, x_3) &+ \beta(y_1, y_2, y_3)] \\
&= T(\alpha x_1 + \beta y_1, \alpha x_2 + \beta y_2, \alpha x_3 + \beta y_3) \\
&= (\alpha x_1 + \beta y_1 - \alpha x_2 - \beta y_2, \alpha x_1 + \beta y_1 + \alpha x_3 + \beta y_3, \alpha x_3 + \beta y_3) \\
&= \alpha(x_1 - x_2, x_1 + x_3, x_3) + \beta(y_1 - y_2, y_1 + y_3, y_3) \\
&= \alpha T(x_1, x_2, x_3) + \beta T(y_1, y_2, y_3).
\end{aligned}
$$

We note that the image of U under T, or the range of T, namely $T(U)$, is a vector subspace of V. This is true because if $\mathbf{v}_1, \mathbf{v}_2$ are in $T(U)$, then there exist $\mathbf{u}_1$ and $\mathbf{u}_2$ in U such that $\mathbf{v}_1 = T(\mathbf{u}_1)$ and $\mathbf{v}_2 = T(\mathbf{u}_2)$. Hence, $\mathbf{v}_1 + \mathbf{v}_2 = T(\mathbf{u}_1) + T(\mathbf{u}_2) = T(\mathbf{u}_1 + \mathbf{u}_2)$, which belongs to $T(U)$. Also, if α is a scalar, then $\alpha T(\mathbf{u}) = T(\alpha \mathbf{u}) \in T(U)$ for any $\mathbf{u} \in U$.

Definition 2.2.1. Let $T: U \to V$ be a linear transformation. The kernel of T, denoted by ker T, is the collection of all vectors $\mathbf{u}$ in U such that $T(\mathbf{u}) = \mathbf{0}$, where $\mathbf{0}$ is the zero vector in V. The kernel of T is also called the null space of T.

As an example of a kernel, let $T: R^3 \to R^3$ be defined as $T(x_1, x_2, x_3) = (x_1 - x_2, x_1 - x_3)$. Then

$$\ker T = \{(x_1, x_2, x_3) | x_1 = x_2, x_1 = x_3\}$$

In this case, ker T consists of all points (x_1, x_2, x_3) in R^3 that lie on a straight line through the origin given by the equations $x_1 = x_2 = x_3$.   □

Theorem 2.2.1. Let $T: U \to V$ be a linear transformation. Then we have the following:

1. ker T is a vector subspace of U.
2. $\dim U = \dim(\ker T) + \dim[T(U)]$.

Proof. Part 1 is left as an exercise. To prove part 2 we consider the following. Let $\dim U = n$, $\dim(\ker T) = p$, and $\dim[T(U)] = q$. Let $\mathbf{u}_1, \mathbf{u}_2, \ldots, \mathbf{u}_p$ be a basis for ker T, and $\mathbf{v}_1, \mathbf{v}_2, \ldots, \mathbf{v}_q$ be a basis for $T(U)$. Then, there exist vectors $\mathbf{w}_1, \mathbf{w}_2, \ldots, \mathbf{w}_q$ in U such that $T(\mathbf{w}_i) = \mathbf{v}_i$ $(i = 1, 2, \ldots, q)$. We need to show that $\mathbf{u}_1, \mathbf{u}_2, \ldots, \mathbf{u}_p; \mathbf{w}_1, \mathbf{w}_2, \ldots, \mathbf{w}_q$ form a basis for U, that is, they are linearly independent and span U.

Suppose that there exist scalars $\alpha_1, \alpha_2, \ldots, \alpha_p; \beta_1, \beta_2, \ldots, \beta_q$ such that

$$\sum_{i=1}^{p} \alpha_i \mathbf{u}_i + \sum_{i=1}^{q} \beta_i \mathbf{w}_i = \mathbf{0}. \tag{2.1}$$

Then

$$\mathbf{0} = T\left(\sum_{i=1}^{p} \alpha_i \mathbf{u}_i + \sum_{i=1}^{q} \beta_i \mathbf{w}_i \right),$$

where $\mathbf{0}$ represents the zero vector in V

$$= \sum_{i=1}^{p} \alpha_i T(\mathbf{u}_i) + \sum_{i=1}^{q} \beta_i T(\mathbf{w}_i)$$

$$= \sum_{i=1}^{q} \beta_i T(\mathbf{w}_i), \quad \text{since} \quad \mathbf{u}_i \in \ker T, i = 1, 2, \ldots, p$$

$$= \sum_{i=1}^{q} \beta_i \mathbf{v}_i.$$

Since the v_i's are linearly independent, then $\beta_i = 0$ for $i = 1, 2, \ldots, q$. From (2.1) it follows that $\alpha_i = 0$ for $i = 1, 2, \ldots, p$, since the u_i's are also linearly independent. Thus the vectors $u_1, u_2, \ldots, u_p$; $w_1, w_2, \ldots, w_q$ are linearly independent.

Let us now suppose that u is any vector in U. To show that it belongs to $L(u_1, u_2, \ldots, u_p$; $w_1, w_2, \ldots, w_q)$. Let $v = T(u)$. Then there exist scalars $a_1, a_2, \ldots, a_q$ such that $v = \sum_{i=1}^{q} a_i v_i$. It follows that

$$T(u) = \sum_{i=1}^{q} a_i T(w_i)$$

$$= T\left(\sum_{i=1}^{q} a_i w_i\right).$$

Thus,

$$T\left(u - \sum_{i=1}^{q} a_i w_i\right) = 0,$$

and $u - \sum_{i=1}^{q} a_i w_i$ must then belong to ker T. Hence,

$$u - \sum_{i=1}^{q} a_i w_i = \sum_{i=1}^{p} b_i u_i \tag{2.2}$$

for some scalars, $b_1, b_2, \ldots, b_p$. From (2.2) we then have

$$u = \sum_{i=1}^{p} b_i u_i + \sum_{i=1}^{q} a_i w_i,$$

which shows that u belongs to the linear span of $u_1, u_2, \ldots, u_p$; $w_1, w_2, \ldots, w_q$. We conclude that these vectors form a basis for U. Hence, $n = p + q$. $\square$

Corollary 2.2.1. $T: U \to V$ is a one-to-one linear transformation if and only if dim(ker T) = 0.

Proof. If T is a one-to-one linear transformation, then ker T consists of just one vector, namely, the zero vector. Hence, dim(ker T) = 0. Vice versa, if dim(ker T) = 0, or equivalently, if ker T consists of just the zero vector, then T must be a one-to-one transformation. This is true because if u_1 and u_2 are in U and such that $T(u_1) = T(u_2)$, then $T(u_1 - u_2) = 0$, which implies that $u_1 - u_2 \in$ ker T and thus $u_1 - u_2 = 0$. $\square$

2.3. MATRICES AND DETERMINANTS

Matrix algebra was devised by the English mathematician Arthur Cayley (1821–1895). The use of matrices originated with Cayley in connection with linear transformations of the form

$$ax_1 + bx_2 = y_1,$$
$$cx_1 + dx_2 = y_2,$$

where a, b, c, and d are scalars. This transformation is completely determined by the square array

$$\begin{bmatrix} a & b \\ c & d \end{bmatrix},$$

which is called a matrix of order 2×2. In general, let $T: U \to V$ be a linear transformation, where U and V are vector spaces of dimensions m and n, respectively. Let $\mathbf{u}_1, \mathbf{u}_2, \ldots, \mathbf{u}_m$ be a basis for U and $\mathbf{v}_1, \mathbf{v}_2, \ldots, \mathbf{v}_n$ be a basis for V. For $i = 1, 2, \ldots, m$, consider $T(\mathbf{u}_i)$, which can be uniquely represented as

$$T(\mathbf{u}_i) = \sum_{j=1}^{n} a_{ij}\mathbf{v}_j, \qquad i = 1, 2, \ldots, m,$$

where the a_{ij}'s are scalars. These scalars completely determine all possible values of T: If $\mathbf{u} \in U$, then $\mathbf{u} = \sum_{i=1}^{m} c_i \mathbf{u}_i$ for some scalars $c_1, c_2, \ldots, c_m$. Then $T(\mathbf{u}) = \sum_{i=1}^{m} c_i T(\mathbf{u}_i) = \sum_{i=1}^{m} c_i (\sum_{j=1}^{n} a_{ij}\mathbf{v}_j)$. By definition, the rectangular array

$$\mathbf{A} = \begin{bmatrix} a_{11} & a_{12} & \cdots & a_{1n} \\ a_{21} & a_{22} & \cdots & a_{2n} \\ \vdots & \vdots & \vdots & \\ a_{m1} & a_{m2} & \cdots & a_{mn} \end{bmatrix}$$

is called a matrix of order $m \times n$, which indicates that $\mathbf{A}$ has m rows and n columns. The a_{ij}'s are called the elements of $\mathbf{A}$. In some cases it is more convenient to represent $\mathbf{A}$ using the notation $\mathbf{A} = (a_{ij})$. In particular, if $m = n$, then $\mathbf{A}$ is called a square matrix. Furthermore, if the off-diagonal elements of a square matrix $\mathbf{A}$ are zero, then $\mathbf{A}$ is called a diagonal matrix and is written as $\mathbf{A} = \text{Diag}(a_{11}, a_{22}, \ldots, a_{nn})$. In this special case, if the diagonal elements are equal to 1, then $\mathbf{A}$ is called the identity matrix and is denoted by $\mathbf{I}_n$ to indicate that it is of order $n \times n$. A matrix of order $m \times 1$ is called a column vector. Likewise, a matrix of order $1 \times n$ is called a row vector.

2.3.1. Basic Operations on Matrices

1. *Equality of Matrices.* Let $\mathbf{A} = (a_{ij})$ and $\mathbf{B} = (b_{ij})$ be two matrices of the same order. Then $\mathbf{A} = \mathbf{B}$ if and only if $a_{ij} = b_{ij}$ for all $i = 1, 2, \ldots, m$; $j = 1, 2, \ldots, n$.

2. *Addition of Matrices.* Let $A = (a_{ij})$ and $B = (b_{ij})$ be two matrices of order $m \times n$. Then $A + B$ is a matrix $C = (c_{ij})$ of order $m \times n$ such that $c_{ij} = a_{ij} + b_{ij}$ $(i = 1, 2, \ldots, m; \ j = 1, 2, \ldots, n)$.

3. *Scalar Multiplication.* Let α be a scalar, and $A = (a_{ij})$ be a matrix of order $m \times n$. Then $\alpha A = (\alpha a_{ij})$.

4. *The Transpose of a Matrix.* Let $A = (a_{ij})$ be a matrix of order $m \times n$. The transpose of A, denoted by A', is a matrix of order $n \times m$ whose rows are the columns of A. For example,

$$\text{if} \quad A = \begin{bmatrix} 2 & 3 & 1 \\ -1 & 0 & 7 \end{bmatrix}, \quad \text{then} \quad A' = \begin{bmatrix} 2 & -1 \\ 3 & 0 \\ 1 & 7 \end{bmatrix}.$$

A matrix A is symmetric if $A = A'$. It is skew-symmetric if $A' = -A$. A skew-symmetric matrix must necessarily have zero elements along its diagonal.

5. *Product of Matrices.* Let $A = (a_{ij})$ and $B = (b_{ij})$ be matrices of orders $m \times n$ and $n \times p$, respectively. The product AB is a matrix $C = (c_{ij})$ of order $m \times p$ such that $c_{ij} = \sum_{k=1}^{n} a_{ik} b_{kj}$ $(i = 1, 2, \ldots, m; \ j = 1, 2, \ldots, p)$. It is to be noted that this product is defined only when the number of columns of A is equal to the number of rows of B.

In particular, if a and b are column vectors of order $n \times 1$, then their dot product $a \cdot b$ can be expressed as a matrix product of the form $a'b$ or $b'a$.

6. *The Trace of a Matrix.* Let $A = (a_{ij})$ be a square matrix of order $n \times n$. The trace of A, denoted by $\text{tr}(A)$, is the sum of its diagonal elements, that is,

$$\text{tr}(A) = \sum_{i=1}^{n} a_{ii}.$$

On the basis of this definition, it is easy to show that if A and B are matrices of order $n \times n$, then the following hold: (i) $\text{tr}(AB) = \text{tr}(BA)$; (ii) $\text{tr}(A + B) = \text{tr}(A) + \text{tr}(B)$.

Definition 2.3.1. Let $A = (a_{ij})$ be an $m \times n$ matrix. A submatrix B of A is a matrix which can be obtained from A by deleting a certain number of rows and columns.

In particular, if the ith row and jth column of A that contain the element a_{ij} are deleted, then the resulting matrix is denoted by M_{ij} $(i = 1, 2, \ldots, m; \ j = 1, 2, \ldots, n)$.

Let us now suppose that A is a square matrix of order $n \times n$. If rows $i_1, i_2, \ldots, i_p$ and columns $i_1, i_2, \ldots, i_p$ are deleted from A, where $p < n$, then the resulting submatrix is called a principal submatrix of A. In particular, if the deleted rows and columns are the last p rows and the last p columns, respectively, then such a submatrix is called a leading principal submatrix.

Definition 2.3.2. A partitioned matrix is a matrix that consists of several submatrices obtained by drawing horizontal and vertical lines that separate it into groups of rows and columns.

For example, the matrix

$$
\mathbf{A} = \left[\begin{array}{cc:cc:c} 1 & 0 & 3 & 4 & -5 \\ 6 & 2 & 10 & 5 & 0 \\ \hdashline 3 & 2 & 1 & 0 & 2 \end{array}\right]
$$

is partitioned into six submatrices by drawing one horizontal line and two vertical lines as shown above.

Definition 2.3.3. Let $\mathbf{A} = (a_{ij})$ be an $m_1 \times n_1$ matrix and $\mathbf{B}$ be an $m_2 \times n_2$ matrix. The direct (or Kronecker) product of $\mathbf{A}$ and $\mathbf{B}$, denoted by $\mathbf{A} \otimes \mathbf{B}$, is a matrix of order $m_1 m_2 \times n_1 n_2$ defined as a partitioned matrix of the form

$$
\mathbf{A} \otimes \mathbf{B} = \begin{bmatrix} a_{11}\mathbf{B} & a_{12}\mathbf{B} & \cdots & a_{1n_1}\mathbf{B} \\ a_{21}\mathbf{B} & a_{22}\mathbf{B} & \cdots & a_{2n_1}\mathbf{B} \\ \vdots & \vdots & & \vdots \\ a_{m_1 1}\mathbf{B} & a_{m_2 2}\mathbf{B} & \cdots & a_{m_1 n_1}\mathbf{B} \end{bmatrix}.
$$

This matrix can be simplified by writing $\mathbf{A} \otimes \mathbf{B} = [a_{ij}\mathbf{B}]$.   □

Properties of the direct product can be found in several matrix algebra books and papers. See, for example, Graybill (1983, Section 8.8), Henderson and Searle (1981), Magnus and Neudecker (1988, Chapter 2), and Searle (1982, Section 10.7). Some of these properties are listed below:

1. $(\mathbf{A} \otimes \mathbf{B})' = \mathbf{A}' \otimes \mathbf{B}'$.
2. $\mathbf{A} \otimes (\mathbf{B} \otimes \mathbf{C}) = (\mathbf{A} \otimes \mathbf{B}) \otimes \mathbf{C}$.
3. $(\mathbf{A} \otimes \mathbf{B})(\mathbf{C} \otimes \mathbf{D}) = \mathbf{AC} \otimes \mathbf{BD}$, if $\mathbf{AC}$ and $\mathbf{BD}$ are defined.
4. $\text{tr}(\mathbf{A} \otimes \mathbf{B}) = \text{tr}(\mathbf{A})\text{tr}(\mathbf{B})$, if $\mathbf{A}$ and $\mathbf{B}$ are square matrices.

The paper by Henderson, Pukelsheim, and Searle (1983) gives a detailed account of the history associated with direct products.

Definition 2.3.4. Let $\mathbf{A}_1, \mathbf{A}_2, \ldots, \mathbf{A}_k$ be matrices of orders $m_i \times n_i$ ($i = 1, 2, \ldots, k$). The direct sum of these matrices, denoted by $\oplus_{i=1}^{k} \mathbf{A}_i$, is a partitioned matrix of order $(\Sigma_{i=1}^{k} m_i) \times (\Sigma_{i=1}^{k} n_i)$ that has the block-diagonal form

$$
\bigoplus_{i=1}^{k} \mathbf{A}_i = \text{Diag}(\mathbf{A}_1, \mathbf{A}_2, \ldots, \mathbf{A}_k).
$$

The following properties can be easily shown on the basis of the preceding definition:

1. $\oplus_{i=1}^{k} \mathbf{A}_i + \oplus_{i=1}^{k} \mathbf{B}_i = \oplus_{i=1}^{k} (\mathbf{A}_i + \mathbf{B}_i)$, if $\mathbf{A}_i$ and $\mathbf{B}_i$ are of the same order for $i = 1, 2, \ldots, k$.

2. $[\oplus_{i=1}^{k} \mathbf{A}_i][\oplus_{i=1}^{k} \mathbf{B}_i] = \oplus_{i=1}^{k} \mathbf{A}_i \mathbf{B}_i$, if $\mathbf{A}_i \mathbf{B}_i$ is defined for $i = 1, 2, \ldots, k$.

3. $[\oplus_{i=1}^{k} \mathbf{A}_i]' = \oplus_{i=1}^{k} \mathbf{A}_i'$.

4. $\mathrm{tr}(\oplus_{i=1}^{k} \mathbf{A}_i) = \sum_{i=1}^{k} \mathrm{tr}(\mathbf{A}_i)$.   □

Definition 2.3.5. Let $\mathbf{A} = (a_{ij})$ be a square matrix of order $n \times n$. The determinant of $\mathbf{A}$, denoted by $\det(\mathbf{A})$, is a scalar quantity that can be computed iteratively as

$$\det(\mathbf{A}) = \sum_{j=1}^{n} (-1)^{j+1} a_{1j} \det(\mathbf{M}_{1j}), \qquad (2.3)$$

where $\mathbf{M}_{1j}$ is a submatrix of $\mathbf{A}$ obtained by deleting row 1 and column j ($j = 1, 2, \ldots, n$). For each j, the determinant of $\mathbf{M}_{1j}$ is obtained in terms of determinants of matrices of order $(n-2) \times (n-2)$ using a formula similar to (2.3). This process is repeated several times until the matrices on the right-hand side of (2.3) become of order 2×2. The determinant of a 2×2 matrix such as $\mathbf{b} = (b_{ij})$ is given by $\det(\mathbf{B}) = b_{11}b_{22} - b_{12}b_{21}$. Thus by an iterative application of formula (2.3), the value of $\det(\mathbf{A})$ can be fully determined. For example, let $\mathbf{A}$ be the matrix

$$\mathbf{A} = \begin{bmatrix} 1 & 2 & -1 \\ 5 & 0 & 3 \\ 1 & 2 & 1 \end{bmatrix}.$$

Then $\det(\mathbf{A}) = \det(\mathbf{A}_1) - 2\det(\mathbf{A}_2) - \det(\mathbf{A}_3)$, where $\mathbf{A}_1, \mathbf{A}_2, \mathbf{A}_3$ are 2×2 submatrices, namely

$$\mathbf{A}_1 = \begin{bmatrix} 0 & 3 \\ 2 & 1 \end{bmatrix}, \qquad \mathbf{A}_2 = \begin{bmatrix} 5 & 3 \\ 1 & 1 \end{bmatrix}, \qquad \mathbf{A}_3 = \begin{bmatrix} 5 & 0 \\ 1 & 2 \end{bmatrix}.$$

It follows that $\det(\mathbf{A}) = -6 - 2(2) - 10 = -20$.   □

Definition 2.3.6. Let $\mathbf{A} = (a_{ij})$ be a square matrix order of $n \times n$. The determinant of $\mathbf{M}_{ij}$, the submatrix obtained by deleting row i and column j, is called a minor of $\mathbf{A}$ of order $n - 1$. The quantity $(-1)^{i+j} \det(\mathbf{M}_{ij})$ is called a cofactor of the corresponding (i, j)th element of $\mathbf{A}$. More generally, if $\mathbf{A}$ is an $m \times n$ matrix and if we strike out all but p rows and the same number of columns from $\mathbf{A}$, where $p \leq \min(m, n)$, then the determinant of the resulting submatrix is called a minor of $\mathbf{A}$ of order p.

The determinant of a principal submatrix of a square matrix **A** is called a principal minor. If, however, we have a leading principal submatrix, then its determinant is called a leading principal minor.   □

NOTE 2.3.1.   The determinant of a matrix **A** is defined only when **A** is a square matrix.

NOTE 2.3.2.   The expansion of det(**A**) in (2.3) was carried out by multiplying the elements of the first row of **A** by their corresponding cofactors and then summing over j ($= 1, 2, \ldots, n$). The same value of det(**A**) could have also been obtained by similar expansions according to the elements of any row of **A** (instead of the first row), or any column of **A**. Thus if $\mathbf{M}_{ij}$ is a submatrix of **A** obtained by deleting row i and column j, then det(**A**) can be obtained by using any of the following expansions:

By row i: $\qquad \det(\mathbf{A}) = \sum_{j=1}^{n} (-1)^{i+j} a_{ij} \det(\mathbf{M}_{ij}), \qquad i = 1, 2, \ldots, n.$

By column j: $\qquad \det(\mathbf{A}) = \sum_{i=1}^{n} (-1)^{i+j} a_{ij} \det(\mathbf{M}_{ij}), \qquad j = 1, 2, \ldots, n.$

NOTE 2.3.3.   Some of the properties of determinants are the following:

  i. det(**AB**) = det(**A**)det(**B**), if **A** and **B** are $n \times n$ matrices.
  ii. If **A**′ is the transpose of **A**, then det(**A**′) = det(**A**).
  iii. If **A** is an $n \times n$ matrix and α is a scalar, then det(α**A**) = α^n det(**A**).
  iv. If any two rows (or columns) of **A** are identical, then det(**A**) = 0.
  v. If any two rows (or columns) of **A** are interchanged, then det(**A**) is multiplied by -1.
  vi. If det(**A**) = 0, then **A** is called a singular matrix. Otherwise, **A** is a nonsingular matrix.
  vii. If **A** and **B** are matrices of orders $m \times m$ and $n \times n$, respectively, then the following hold: (a) det(**A** ⊗ **B**) = [det(**A**)]^n[det(**B**)]^m; (b) det(**A** ⊕ **B**) = [det(**A**)][det(**B**)].

NOTE 2.3.4.   The history of determinants dates back to the fourteenth century. According to Smith (1958, page 273), the Chinese had some knowledge of determinants as early as about 1300 A.D. Smith (1958, page 440) also reported that the Japanese mathematician Seki Kōwa (1642–1708) had discovered the expansion of a determinant in solving simultaneous equations. In the West, the theory of determinants is believed to have originated with the German mathematician Gottfried Leibniz (1646–1716) in 1693, ten years after the work of Seki Kōwa. However, the actual development of the theory of determinants did not begin until the publication of a book by Gabriel Cramer (1704–1752) (see Price, 1947, page 85) in 1750. Other mathematicians who contributed to this theory include Alexandre Vandermonde (1735–1796), Pierre-Simon Laplace (1749–1827), Carl Gauss (1777–1855), and Augustin-Louis Cauchy (1789–1857). Arthur Cayley (1821–1895) is credited with having been the first to introduce the common present-day notation of vertical bars enclosing a square matrix. For more interesting facts about the history of determinants, the reader is advised to read the article by Price (1947).

2.3.2. The Rank of a Matrix

Let $\mathbf{A} = (a_{ij})$ be a matrix of order $m \times n$. Let $\mathbf{u}'_1, \mathbf{u}'_2, \ldots, \mathbf{u}'_m$ denote the row vectors of $\mathbf{A}$, and let $\mathbf{v}_1, \mathbf{v}_2, \ldots, \mathbf{v}_n$ denote its column vectors. Consider the linear spans of the row and column vectors, namely, $V_1 = L(\mathbf{u}'_1, \mathbf{u}'_2, \ldots, \mathbf{u}'_m), V_2 = L(\mathbf{v}_1, \mathbf{v}_2, \ldots, \mathbf{v}_n)$, respectively.

Theorem 2.3.1. The vector spaces V_1 and V_2 have the same dimension.

Proof. See Lancaster (1969, Theorem 1.15.1), or Searle (1982, Section 6.6). □

Thus, for any matrix $\mathbf{A}$, the number of linearly independent rows is the same as the number of linearly independent columns.

Definition 2.3.7. The rank of a matrix $\mathbf{A}$ is the number of its linearly independent rows (or columns). The rank of $\mathbf{A}$ is denoted by $r(\mathbf{A})$. □

Theorem 2.3.2. If a matrix $\mathbf{A}$ has a nonzero minor of order r, and if all minors of order $r + 1$ and higher (if they exist) are zero, then $\mathbf{A}$ has rank r.

Proof. See Lancaster (1969, Lemma 1, Section 1.15). □

For example, if $\mathbf{A}$ is the matrix

$$\mathbf{A} = \begin{bmatrix} 2 & 3 & -1 \\ 0 & 1 & 2 \\ 2 & 4 & 1 \end{bmatrix},$$

then $r(\mathbf{A}) = 2$. This is because $\det(\mathbf{A}) = 0$ and at least one minor of order 2 is different from zero.

There are several properties associated with the rank of a matrix. Some of these properties are the following:

1. $r(\mathbf{A}) = r(\mathbf{A}')$.
2. The rank of $\mathbf{A}$ is unchanged if $\mathbf{A}$ is multiplied by a nonsingular matrix. Thus if $\mathbf{A}$ is an $m \times n$ matrix and $\mathbf{P}$ is an $n \times n$ nonsingular matrix, then $r(\mathbf{A}) = r(\mathbf{AP})$.
3. $r(\mathbf{A}) = r(\mathbf{AA}') = r(\mathbf{A}'\mathbf{A})$.
4. If the matrix $\mathbf{A}$ is partitioned as $\mathbf{A} = [\mathbf{A}_1 : \mathbf{A}_2]$, where $\mathbf{A}_1$ and $\mathbf{A}_2$ are submatrices of the same order, then $r(\mathbf{A}_1 + \mathbf{A}_2) \leq r(\mathbf{A}) \leq r(\mathbf{A}_1) + r(\mathbf{A}_2)$. More generally, if the matrices $\mathbf{A}_1, \mathbf{A}_2, \ldots, \mathbf{A}_k$ are of the same order and if $\mathbf{A}$ is partitioned as $\mathbf{A} = [\mathbf{A}_1 : \mathbf{A}_2 : \cdots : \mathbf{A}_k]$, then

$$
r\left(\sum_{i=1}^{k} \mathbf{A}_i \right) \leq r(\mathbf{A}) \leq \sum_{i=1}^{k} r(\mathbf{A}_i).
$$

5. If the product $\mathbf{AB}$ is defined, then $r(\mathbf{A}) + r(\mathbf{B}) - n \leq r(\mathbf{AB}) \leq \min\{r(\mathbf{A}), r(\mathbf{B})\}$, where n is the number of columns of $\mathbf{A}$ (or the number of rows of $\mathbf{B}$).
6. $r(\mathbf{A} \otimes \mathbf{B}) = r(\mathbf{A})r(\mathbf{B})$.
7. $r(\mathbf{A} \oplus \mathbf{B}) = r(\mathbf{A}) + r(\mathbf{B})$.

Definition 2.3.8.   Let $\mathbf{A}$ be a matrix of order $m \times n$ and rank r. Then we have the following:

1. $\mathbf{A}$ is said to have a full row rank if $r = m < n$.
2. $\mathbf{A}$ is said to have a full column rank if $r = n < m$.
3. $\mathbf{A}$ is of full rank if $r = m = n$. In this case, $\det(\mathbf{A}) \neq 0$, that is, $\mathbf{A}$ is a nonsingular matrix.   □

2.3.3. The Inverse of a Matrix

Let $\mathbf{A} = (a_{ij})$ be a nonsingular matrix of order $n \times n$. The inverse of $\mathbf{A}$, denoted by $\mathbf{A}^{-1}$, is an $n \times n$ matrix that satisfies the condition $\mathbf{AA}^{-1} = \mathbf{A}^{-1}\mathbf{A} = \mathbf{I}_n$.

The inverse of $\mathbf{A}$ can be computed as follows: Let c_{ij} be the cofactor of a_{ij} (see Definition 2.3.6). Define the matrix $\mathbf{C}$ as $\mathbf{C} = (c_{ij})$. The transpose of $\mathbf{C}$ is called the adjugate or adjoint of $\mathbf{A}$ and is denoted by adj $\mathbf{A}$. The inverse of $\mathbf{A}$ is then given by

$$
\mathbf{A}^{-1} = \frac{\text{adj } \mathbf{A}}{\det(\mathbf{A})}.
$$

It can be verified that

$$A\left[\frac{\operatorname{adj}A}{\det(A)}\right] = \left[\frac{\operatorname{adj}A}{\det(A)}\right]A = I_n.$$

For example, if A is the matrix

$$A = \begin{bmatrix} 2 & 0 & 1 \\ -3 & 2 & 0 \\ 2 & 1 & 1 \end{bmatrix},$$

then $\det(A) = -3$, and

$$\operatorname{adj}A = \begin{bmatrix} 2 & 1 & -2 \\ 3 & 0 & -3 \\ -7 & -2 & 4 \end{bmatrix}.$$

Hence,

$$A^{-1} = \begin{bmatrix} -\frac{2}{3} & -\frac{1}{3} & \frac{2}{3} \\ -1 & 0 & 1 \\ \frac{7}{3} & \frac{2}{3} & -\frac{4}{3} \end{bmatrix}.$$

Some properties of the inverse operation are given below:

1. $(AB)^{-1} = B^{-1}A^{-1}$.
2. $(A')^{-1} = (A^{-1})'$.
3. $\det(A^{-1}) = 1/\det(A)$.
4. $(A^{-1})^{-1} = A$.
5. $(A \otimes B)^{-1} = A^{-1} \otimes B^{-1}$.
6. $(A \oplus B)^{-1} = A^{-1} \oplus B^{-1}$.
7. If A is partitioned as

$$A = \begin{bmatrix} A_{11} & A_{12} \\ A_{21} & A_{22} \end{bmatrix},$$

where A_{ij} is of order $n_i \times n_j$ $(i, j = 1, 2)$, then

$$\det(A) = \begin{cases} \det(A_{11}) \cdot \det(A_{22} - A_{21}A_{11}^{-1}A_{12}) & \text{if } A_{11} \text{ is nonsingular,} \\ \det(A_{22}) \cdot \det(A_{11} - A_{12}A_{22}^{-1}A_{21}) & \text{if } A_{22} \text{ is nonsingular.} \end{cases}$$

The inverse of $\mathbf{A}$ is partitioned as

$$\mathbf{A}^{-1} = \begin{bmatrix} \mathbf{B}_{11} & \mathbf{B}_{12} \\ \mathbf{B}_{21} & \mathbf{B}_{22} \end{bmatrix},$$

where

$$\mathbf{B}_{11} = \left(\mathbf{A}_{11} - \mathbf{A}_{12}\mathbf{A}_{22}^{-1}\mathbf{A}_{21}\right)^{-1},$$

$$\mathbf{B}_{12} = -\mathbf{B}_{11}\mathbf{A}_{12}\mathbf{A}_{22}^{-1},$$

$$\mathbf{B}_{21} = -\mathbf{A}_{22}^{-1}\mathbf{A}_{21}\mathbf{B}_{11},$$

$$\mathbf{B}_{22} = \mathbf{A}_{22}^{-1} + \mathbf{A}_{22}^{-1}\mathbf{A}_{21}\mathbf{B}_{11}\mathbf{A}_{12}\mathbf{A}_{22}^{-1}.$$

2.3.4. Generalized Inverse of a Matrix

This inverse represents a more general concept than the one discussed in the previous section. Let $\mathbf{A}$ be a matrix of order $m \times n$. Then, a generalized inverse of $\mathbf{A}$, denoted by $\mathbf{A}^-$, is a matrix of order $n \times m$ that satisfies the condition

$$\mathbf{A}\mathbf{A}^-\mathbf{A} = \mathbf{A}. \tag{2.4}$$

Note that $\mathbf{A}^-$ is defined even if $\mathbf{A}$ is not a square matrix. If $\mathbf{A}$ is a square matrix, it does not have to be nonsingular. Furthermore, condition (2.4) can be satisfied by infinitely many matrices (see, for example, Searle, 1982, Chapter 8). If $\mathbf{A}$ is nonsingular, then (2.4) is satisfied by only $\mathbf{A}^{-1}$. Thus $\mathbf{A}^{-1}$ is a special case of $\mathbf{A}^-$.

Theorem 2.3.3.

1. If $\mathbf{A}$ is a symmetric matrix, then $\mathbf{A}^-$ can be chosen to be symmetric.
2. $\mathbf{A}(\mathbf{A}'\mathbf{A})^-\mathbf{A}'\mathbf{A} = \mathbf{A}$ for any matrix $\mathbf{A}$.
3. $\mathbf{A}(\mathbf{A}'\mathbf{A})^-\mathbf{A}'$ is invariant to the choice of a generalized inverse of $\mathbf{A}'\mathbf{A}$.

Proof. See Searle (1982, pages 221–222). ☐

2.3.5. Eigenvalues and Eigenvectors of a Matrix

Let $\mathbf{A}$ be a square matrix of order $n \times n$. By definition, a scalar λ is said to be an eigenvalue (or characteristic root) of $\mathbf{A}$ if $\mathbf{A} - \lambda\mathbf{I}_n$ is a singular matrix, that is,

$$\det(\mathbf{A} - \lambda\mathbf{I}_n) = 0. \tag{2.5}$$

Thus an eigenvalue of $\mathbf{A}$ satisfies a polynomial equation of degree n called the characteristic equation of $\mathbf{A}$. If λ is a multiple solution (or root) of equation (2.5), that is, (2.5) has several roots, say m, that are equal to λ, then λ is said to be an eigenvalue of multiplicity m.

Since $r(\mathbf{A} - \lambda \mathbf{I}_n) < n$ by the fact that $\mathbf{A} - \lambda \mathbf{I}_n$ is singular, the columns of $\mathbf{A} - \lambda \mathbf{I}_n$ must be linearly related. Hence, there exists a nonzero vector $\mathbf{v}$ such that

$$(\mathbf{A} - \lambda \mathbf{I}_n)\mathbf{v} = \mathbf{0}, \tag{2.6}$$

or equivalently,

$$\mathbf{A}\mathbf{v} = \lambda \mathbf{v}. \tag{2.7}$$

A vector satisfying (2.7) is called an eigenvector (or a characteristic vector) corresponding to the eigenvalue λ. From (2.7) we note that the linear transformation of $\mathbf{v}$ by the matrix $\mathbf{A}$ is a scalar multiple of $\mathbf{v}$.

The following theorems describe certain properties associated with eigenvalues and eigenvectors. The proofs of these theorems can be found in standard matrix algebra books (see the annotated bibliography).

Theorem 2.3.4. A square matrix $\mathbf{A}$ is singular if and only if at least one of its eigenvalues is equal to zero. In particular, if $\mathbf{A}$ is symmetric, then its rank is equal to the number of its nonzero eigenvalues.

Theorem 2.3.5. The eigenvalues of a symmetric matrix are real.

Theorem 2.3.6. Let $\mathbf{A}$ be a square matrix, and let $\lambda_1, \lambda_2, \ldots, \lambda_k$ denote its distinct eigenvalues. If $\mathbf{v}_1, \mathbf{v}_2, \ldots, \mathbf{v}_k$ are eigenvectors of $\mathbf{A}$ corresponding to $\lambda_1, \lambda_2, \ldots, \lambda_k$, respectively, then $\mathbf{v}_1, \mathbf{v}_2, \ldots, \mathbf{v}_k$ are linearly independent. In particular, if $\mathbf{A}$ is symmetric, then $\mathbf{v}_1, \mathbf{v}_2, \ldots, \mathbf{v}_k$ are orthogonal to one another, that is, $\mathbf{v}_i' \mathbf{v}_j = 0$ for $i \neq j$ $(i, j = 1, 2, \ldots, k)$.

Theorem 2.3.7. Let $\mathbf{A}$ and $\mathbf{B}$ be two matrices of orders $m \times m$ and $n \times n$, respectively. Let $\lambda_1, \lambda_2, \ldots, \lambda_m$ be the eigenvalues of $\mathbf{A}$, and $v_1, v_2, \ldots, v_n$ be the eigenvalues of $\mathbf{B}$. Then we have the following:

1. The eigenvalues of $\mathbf{A} \otimes \mathbf{B}$ are of the form $\lambda_i v_j$ $(i = 1, 2, \ldots, m;\ j = 1, 2, \ldots, n)$.
2. The eigenvalues of $\mathbf{A} \oplus \mathbf{B}$ are $\lambda_1, \lambda_2, \ldots, \lambda_m;\ v_1, v_2, \ldots, v_n$.

Theorem 2.3.8. Let $\lambda_1, \lambda_2, \ldots, \lambda_n$ be the eigenvalues of a matrix $\mathbf{A}$ of order $n \times n$. Then the following hold:

1. $\mathrm{tr}(\mathbf{A}) = \sum_{i=1}^{n} \lambda_i$.
2. $\det(\mathbf{A}) = \prod_{i=1}^{n} \lambda_i$.

Theorem 2.3.9.   Let **A** and **B** be two matrices of orders $m \times n$ and $n \times m$ ($n \geq m$), respectively. The nonzero eigenvalues of **BA** are the same as those of **AB**.

2.3.6. Some Special Matrices

1. The vector $\mathbf{1}_n$ is a column vector of ones of order $n \times 1$.
2. The matrix $\mathbf{J}_n$ is a matrix of ones of order $n \times n$.
3. *Idempotent Matrix.* A square matrix **A** for which $\mathbf{A}^2 = \mathbf{A}$ is called an idempotent matrix. For example, the matrix $\mathbf{A} = \mathbf{I}_n - (1/n)\mathbf{J}_n$ is idempotent of order $n \times n$. The eigenvalues of an idempotent matrix are equal to zeros and ones. It follows from Theorem 2.3.8 that the rank of an idempotent matrix, which is the same as the number of eigenvalues that are equal to 1, is also equal to its trace. Idempotent matrices are used in many applications in statistics (see Section 2.4).
4. *Orthogonal Matrix.* A square matrix **A** is orthogonal if $\mathbf{A}'\mathbf{A} = \mathbf{I}$. From this definition it follows that (i) **A** is orthogonal if and only if $\mathbf{A}' = \mathbf{A}^{-1}$; (ii) $|\det(\mathbf{A})| = 1$. A special orthogonal matrix is the Householder matrix, which is a symmetric matrix of the form

$$\mathbf{H} = \mathbf{I} - 2\mathbf{u}\mathbf{u}'/\mathbf{u}'\mathbf{u},$$

where **u** is a nonzero vector. Orthogonal matrices occur in many applications of matrix algebra and play an important role in statistics, as will be seen in Section 2.4.

2.3.7. The Diagonalization of a Matrix

Theorem 2.3.10 (The Spectral Decomposition Theorem).   Let **A** be a symmetric matrix of order $n \times n$. There exists an orthogonal matrix **P** such that $\mathbf{A} = \mathbf{P}\Lambda\mathbf{P}'$, where $\Lambda = \text{Diag}(\lambda_1, \lambda_2, \ldots, \lambda_n)$ is a diagonal matrix whose diagonal elements are the eigenvalues of **A**. The columns of **P** are the corresponding orthonormal eigenvectors of **A**.

Proof. See Basilevsky (1983, Theorem 5.8, page 200).   □

If **P** is partitioned as $\mathbf{P} = [\mathbf{p}_1 : \mathbf{p}_2 : \cdots : \mathbf{p}_n]$, where $\mathbf{p}_i$ is an eigenvector of **A** with eigenvalue λ_i ($i = 1, 2, \ldots, n$), then **A** can be written as

$$\mathbf{A} = \sum_{i=1}^{n} \lambda_i \mathbf{p}_i \mathbf{p}_i'.$$

For example, if

$$A = \begin{bmatrix} 1 & 0 & -2 \\ 0 & 0 & 0 \\ -2 & 0 & 4 \end{bmatrix},$$

then A has two distinct eigenvalues, $\lambda_1 = 0$ of multiplicity 2 and $\lambda_2 = 5$. For $\lambda_1 = 0$ we have two orthonormal eigenvectors, $p_1 = (2, 0, 1)'/\sqrt{5}$ and $p_2 = (0, 1, 0)'$. Note that p_1 and p_2 span the kernel (null space) of the linear transformation represented by A. For $\lambda_2 = 5$ we have the normal eigenvector $p_3 = (1, 0, -2)'/\sqrt{5}$, which is orthogonal to both p_1 and p_2. Hence, P and Λ in Theorem 2.3.10 for the matrix A are

$$P = \begin{bmatrix} \dfrac{2}{\sqrt{5}} & 0 & \dfrac{1}{\sqrt{5}} \\ 0 & 1 & 0 \\ \dfrac{1}{\sqrt{5}} & 0 & \dfrac{-2}{\sqrt{5}} \end{bmatrix},$$

$$\Lambda = \text{Diag}(0, 0, 5).$$

The next theorem gives a more general form of the spectral decomposition theorem.

Theorem 2.3.11 (The Singular-Value Decomposition Theorem). Let A be a matrix of order $m \times n$ ($m \leq n$) and rank r. There exist orthogonal matrices P and Q such that $A = P[D : 0]Q'$, where $D = \text{Diag}(\lambda_1, \lambda_2, \ldots, \lambda_m)$ is a diagonal matrix with nonnegative diagonal elements called the singular values of A, and 0 is a zero matrix of order $m \times (n - m)$. The diagonal elements of D are the square roots of the eigenvalues of AA'.

Proof. See, for example, Searle (1982, pages 316–317). ☐

2.3.8. Quadratic Forms

Let $A = (a_{ij})$ be a symmetric matrix of order $n \times n$, and let $x = (x_1, x_2, \ldots, x_n)'$ be a column vector of order $n \times 1$. The function

$$q(x) = x'Ax$$

$$= \sum_{i=1}^{n} \sum_{j=1}^{n} a_{ij} x_i x_j$$

is called a quadratic form in x.

A quadratic form $\mathbf{x'Ax}$ is said to be the following:

1. Positive definite if $\mathbf{x'Ax} > 0$ for all $\mathbf{x} \neq \mathbf{0}$ and is zero only if $\mathbf{x} = \mathbf{0}$.
2. Positive semidefinite if $\mathbf{x'Ax} \geq 0$ for all $\mathbf{x}$ and $\mathbf{x'Ax} = 0$ for at least one nonzero value of $\mathbf{x}$.
3. Nonnegative definite if $\mathbf{A}$ is either positive definite or positive semidefinite.

Theorem 2.3.12. Let $\mathbf{A} = (a_{ij})$ be a symmetric matrix of order $n \times n$. Then $\mathbf{A}$ is positive definite if and only if either of the following two conditions is satisfied:

1. The eigenvalues of $\mathbf{A}$ are all positive.
2. The leading principal minors of $\mathbf{A}$ are all positive, that is,

$$a_{11} > 0, \qquad \det\left(\begin{bmatrix} a_{11} & a_{12} \\ a_{21} & a_{22} \end{bmatrix}\right) > 0, \ldots, \qquad \det(\mathbf{A}) > 0.$$

Proof. The proof of part 1 follows directly from the spectral decomposition theorem. For the proof of part 2, see Lancaster (1969, Theorem 2.14.4). □

Theorem 2.3.13. Let $\mathbf{A} = (a_{ij})$ be a symmetric matrix of order $n \times n$. Then $\mathbf{A}$ is positive semidefinite if and only if its eigenvalues are nonnegative with at least one of them equal to zero.

Proof. See Basilevsky (1983, Theorem 5.10, page 203). □

2.3.9. The Simultaneous Diagonalization of Matrices

By simultaneous diagonalization we mean finding a matrix, say $\mathbf{Q}$, that can reduce several square matrices to a diagonal form. In many situations there may be a need to diagonalize several matrices simultaneously. This occurs frequently in statistics, particularly in analysis of variance.

The proofs of the following theorems can be found in Graybill (1983, Chapter 12).

Theorem 2.3.14. Let $\mathbf{A}$ and $\mathbf{B}$ be symmetric matrices of order $n \times n$.

1. If $\mathbf{A}$ is positive definite, then there exists a nonsingular matrix $\mathbf{Q}$ such that $\mathbf{Q'AQ} = \mathbf{I}_n$ and $\mathbf{Q'BQ} = \mathbf{D}$, where $\mathbf{D}$ is a diagonal matrix whose diagonal elements are the roots of the polynomial equation $\det(\mathbf{B} - \lambda\mathbf{A}) = 0$.

2. If **A** and **B** are positive semidefinite, then there exists a nonsingular matrix **Q** such that

$$Q'AQ = D_1,$$

$$Q'BQ = D_2,$$

where D_1 and D_2 are diagonal matrices (for a detailed proof of this result, see Newcomb, 1960).

Theorem 2.3.15. Let $A_1, A_2, \ldots, A_k$ be symmetric matrices of order $n \times n$. Then there exists an orthogonal matrix **P** such that

$$A_i = P\Lambda_i P', \qquad i = 1, 2, \ldots, k,$$

where Λ_i is a diagonal matrix, if and only if $A_i A_j = A_j A_i$ for all $i \neq j$ $(i, j = 1, 2, \ldots, k)$.

2.3.10. Bounds on Eigenvalues

Let **A** be a symmetric matrix of order $n \times n$. We denote the ith eigenvalue of **A** by $e_i(A)$, $i = 1, 2, \ldots, n$. The smallest and largest eigenvalues of **A** are denoted by $e_{\min}(A)$ and $e_{\max}(A)$, respectively.

Theorem 2.3.16. $e_{\min}(A) \leq x'Ax/x'x \leq e_{\max}(A)$.

Proof. This follows directly from the spectral decomposition theorem. □

The ratio $x'Ax/x'x$ is called Rayleigh's quotient for **A**. The lower and upper bounds in Theorem 2.3.16 can be achieved by choosing **x** to be an eigenvector associated with $e_{\min}(A)$ and $e_{\max}(A)$, respectively. Thus Theorem 2.3.16 implies that

$$\inf_{x \neq 0} \left[\frac{x'Ax}{x'x} \right] = e_{\min}(A), \qquad (2.8)$$

$$\sup_{x \neq 0} \left[\frac{x'Ax}{x'x} \right] = e_{\max}(A). \qquad (2.9)$$

Theorem 2.3.17. If **A** is a symmetric matrix and **B** is a positive definite matrix, both of order $n \times n$, then

$$e_{\min}(B^{-1}A) \leq \frac{x'Ax}{x'Bx} \leq e_{\max}(B^{-1}A)$$

Proof. The proof is left to the reader. □

Note that the above lower and upper bounds are equal to the infimum and supremum, respectively, of the ratio $\mathbf{x}'\mathbf{A}\mathbf{x}/\mathbf{x}'\mathbf{B}\mathbf{x}$ for $\mathbf{x} \neq \mathbf{0}$.

Theorem 2.3.18. If $\mathbf{A}$ is a positive semidefinite matrix and $\mathbf{B}$ is a positive definite matrix, both of order $n \times n$, then for any i $(i = 1, 2, \ldots, n)$,

$$e_i(\mathbf{A})e_{\min}(\mathbf{B}) \leq e_i(\mathbf{AB}) \leq e_i(\mathbf{A})e_{\max}(\mathbf{B}). \tag{2.10}$$

Furthermore, if $\mathbf{A}$ is positive definite, then for any i $(i = 1, 2, \ldots, n)$,

$$\frac{e_i^2(\mathbf{AB})}{e_{\max}(\mathbf{A})e_{\max}(\mathbf{B})} \leq e_i(\mathbf{A})e_i(\mathbf{B}) \leq \frac{e_i^2(\mathbf{AB})}{e_{\min}(\mathbf{A})e_{\min}(\mathbf{B})}$$

Proof. See Anderson and Gupta (1963, Corollary 2.2.1).   □

A special case of the double inequality in (2.10) is

$$e_{\min}(\mathbf{A})e_{\min}(\mathbf{B}) \leq e_i(\mathbf{AB}) \leq e_{\max}(\mathbf{A})e_{\max}(\mathbf{B}),$$

for all i $(i = 1, 2, \ldots, n)$.

Theorem 2.3.19. Let $\mathbf{A}$ and $\mathbf{B}$ be symmetric matrices of order $n \times n$. Then, the following hold:

1. $e_i(\mathbf{A}) \leq e_i(\mathbf{A} + \mathbf{B})$, $i = 1, 2, \ldots, n$, if $\mathbf{B}$ is nonnegative definite.
2. $e_i(\mathbf{A}) < e_i(\mathbf{A} + \mathbf{B})$, $i = 1, 2, \ldots, n$, if $\mathbf{B}$ is positive definite.

Proof. See Bellman (1970, Theorem 3, page 117).   □

Theorem 2.3.20 (Schur's Theorem). Let $\mathbf{A} = (a_{ij})$ be a symmetric matrix of order $n \times n$, and let $\|\mathbf{A}\|_2$ denote its Euclidean norm, defined as

$$\|\mathbf{A}\|_2 = \left(\sum_{i=1}^{n} \sum_{j=1}^{n} a_{ij}^2 \right)^{1/2}.$$

Then

$$\sum_{i=1}^{n} e_i^2(\mathbf{A}) = \|\mathbf{A}\|_2^2.$$

Proof. See Lancaster (1969, Theorem 7.3.1).   □

Since $\|\mathbf{A}\|_2 \leq n \max_{i,j} |a_{ij}|$, then from Theorem 2.3.20 we conclude that

$$|e_{\max}(\mathbf{A})| \leq n \max_{i,j} |a_{ij}|.$$

Theorem 2.3.21. Let $\mathbf{A}$ be a symmetric matrix of order $n \times n$, and let m and s be defined as

$$m = \frac{\text{tr}(\mathbf{A})}{n}, \qquad s = \left(\frac{\text{tr}(\mathbf{A}^2)}{n} - m^2 \right)^{1/2}.$$

Then

$$m - s(n-1)^{1/2} \leq e_{\min}(\mathbf{A}) \leq m - \frac{s}{(n-1)^{1/2}},$$

$$m + \frac{s}{(n-1)^{1/2}} \leq e_{\max}(\mathbf{A}) \leq m + s(n-1)^{1/2},$$

$$e_{\max}(\mathbf{A}) - e_{\min}(\mathbf{A}) \leq s(2n)^{1/2}.$$

Proof. See Wolkowicz and Styan (1980, Theorems 2.1 and 2.5). □

2.4. APPLICATIONS OF MATRICES IN STATISTICS

The use of matrix algebra is quite prevalent in statistics. In fact, in the areas of experimental design, linear models, and multivariate analysis, matrix algebra is considered the most frequently used branch of mathematics. Applications of matrices in these areas are well documented in several books, for example, Basilevsky (1983), Graybill (1983), Magnus and Neudecker (1988), and Searle (1982). We shall therefore not attempt to duplicate the material given in these books.

Let us consider the following applications:

2.4.1. The Analysis of the Balanced Mixed Model

In analysis of variance, a linear model associated with a given experimental situation is said to be balanced if the numbers of observations in the subclasses of the data are the same. For example, the two-way crossed-classification model with interaction,

$$y_{ijk} = \mu + \alpha_i + \beta_j + (\alpha\beta)_{ij} + \epsilon_{ijk}, \qquad (2.11)$$

$i = 1, 2, \ldots, a$; $j = 1, 2, \ldots, b$; $k = 1, 2, \ldots, n$, is balanced, since there are n observations for each combination of i and j. Here, α_i and β_j represent the main effects of the factors under consideration, $(\alpha\beta)_{ij}$ denotes the interaction effect, and ϵ_{ijk} is a random error term. Model (2.11) can be written in vector form as

$$\mathbf{y} = \mathbf{H}_0 \boldsymbol{\tau}_0 + \mathbf{H}_1 \boldsymbol{\tau}_1 + \mathbf{H}_2 \boldsymbol{\tau}_2 + \mathbf{H}_3 \boldsymbol{\tau}_3 + \mathbf{H}_4 \boldsymbol{\tau}_4, \qquad (2.12)$$

where $\mathbf{y}$ is the vector of observations, $\tau_0 = \mu$, $\tau_1 = (\alpha_1, \alpha_2, \ldots, \alpha_a)'$, $\tau_2 = (\beta_1, \beta_2, \ldots, \beta_b)'$, $\tau_3 = [(\alpha\beta)_{11}, (\alpha\beta)_{12}, \ldots, (\alpha\beta)_{ab}]'$, and $\tau_4 = (\epsilon_{111}, \epsilon_{112}, \ldots, \epsilon_{abn})'$. The matrices $\mathbf{H}_i$ ($i = 0, 1, 2, 3, 4$) can be expressed as direct products of the form

$$\mathbf{H}_0 = \mathbf{1}_a \otimes \mathbf{1}_b \otimes \mathbf{1}_n,$$

$$\mathbf{H}_1 = \mathbf{I}_a \otimes \mathbf{1}_b \otimes \mathbf{1}_n,$$

$$\mathbf{H}_2 = \mathbf{1}_a \otimes \mathbf{I}_b \otimes \mathbf{1}_n,$$

$$\mathbf{H}_3 = \mathbf{I}_a \otimes \mathbf{I}_b \otimes \mathbf{1}_n,$$

$$\mathbf{H}_4 = \mathbf{I}_a \otimes \mathbf{I}_b \otimes \mathbf{I}_n.$$

In general, any balanced linear model can be written in vector form as

$$\mathbf{y} = \sum_{l=0}^{\nu} \mathbf{H}_l \tau_l, \tag{2.13}$$

where $\mathbf{H}_l$ ($l = 0, 1, \ldots, \nu$) is a direct product of identity matrices and vectors of ones (see Khuri, 1982). If $\tau_0, \tau_1, \ldots, \tau_\theta$ ($\theta < \nu - 1$) are fixed unknown parameter vectors (fixed effects), and $\tau_{\theta+1}, \tau_{\theta+2}, \ldots, \tau_\nu$ are random vectors (random effects), then model (2.11) is called a balanced mixed model. Furthermore, if we assume that the random effects are independent and have the normal distributions $N(\mathbf{0}, \sigma_l^2 \mathbf{I}_{c_l})$, where c_l is the number of columns of $\mathbf{H}_l$, $l = \theta+1, \theta+2, \ldots, \nu$, then, because model (2.11) is balanced, its statistical analysis becomes very simple. Here, the σ_l^2's are called the model's variance components. A balanced mixed model can be written as

$$\mathbf{y} = \mathbf{Xg} + \mathbf{Zh} \tag{2.14}$$

where $\mathbf{Xg} = \sum_{l=0}^{\theta} \mathbf{H}_l \tau_l$ is the fixed portion of the model, and $\mathbf{Zh} = \sum_{l=\theta+1}^{\nu} \mathbf{H}_l \tau_l$ is its random portion. The variance–covariance matrix of $\mathbf{y}$ is given by

$$\Sigma = \sum_{l=\theta+1}^{\nu} \mathbf{A}_l \sigma_l^2,$$

where $\mathbf{A}_l = \mathbf{H}_l \mathbf{H}_l'$ ($l = \theta+1, \theta+2, \ldots, \nu$). Note that $\mathbf{A}_l \mathbf{A}_p = \mathbf{A}_p \mathbf{A}_l$ for all $l \neq p$. Hence, the matrices $\mathbf{A}_l$ can be diagonalized simultaneously (see Theorem 2.3.15).

If $\mathbf{y}'\mathbf{Ay}$ is a quadratic form in $\mathbf{y}$, then $\mathbf{y}'\mathbf{Ay}$ is distributed as a noncentral chi-squared variate $\chi_m'^2(\eta)$ if and only if $\mathbf{A}\Sigma$ is idempotent of rank m, where η is the noncentrality parameter and is given by $\eta = \mathbf{g}'\mathbf{X}'\mathbf{AXg}$ (see Searle, 1971, Section 2.5).

The total sum of squares, $\mathbf{y}'\mathbf{y}$, can be uniquely partitioned as

$$\mathbf{y}'\mathbf{y} = \sum_{l=0}^{\nu} \mathbf{y}'\mathbf{P}_l \mathbf{y},$$

where the $\mathbf{P}_l$'s are idempotent matrices such that $\mathbf{P}_l\mathbf{P}_s = 0$ for all $l \neq s$ (see Khuri, 1982). The quadratic form $\mathbf{y}'\mathbf{P}_l\mathbf{y}$ $(l = 0, 1, \ldots, \nu)$ is positive semidefinite and represents the sum of squares for the lth effect in model (2.13).

Theorem 2.4.1. Consider the balanced mixed model (2.14), where the random effects are assumed to be independently and normally distributed with zero means and variance–covariance matrices $\sigma_l^2\mathbf{I}_{c_l}$ $(l = \theta + 1, \theta + 2, \ldots, \nu)$. Then we have the following:

1. $\mathbf{y}'\mathbf{P}_0\mathbf{y}, \mathbf{y}'\mathbf{P}_1\mathbf{y}, \ldots, \mathbf{y}'\mathbf{P}_\nu\mathbf{y}$ are statistically independent.
2. $\mathbf{y}'\mathbf{P}_l\mathbf{y}/\delta_l$ is distributed as a noncentral chi-squared variate with degrees of freedom equal to the rank of $\mathbf{P}_l$ and noncentrality parameter given by $\eta_l = \mathbf{g}'\mathbf{X}'\mathbf{P}_l\mathbf{X}\mathbf{g}/\delta_l$ for $l = 0, 1, \ldots, \theta$, where δ_l is a particular linear combination of the variance components $\sigma_{\theta+1}^2, \sigma_{\theta+2}^2, \ldots, \sigma_\nu^2$. However, for $l = \theta + 1, \theta + 2, \ldots, \nu$, that is, for the random effects, $\mathbf{y}'\mathbf{P}_l\mathbf{y}/\delta_l$ is distributed as a central chi-squared variate with m_l degrees of freedom, where $m_l = r(\mathbf{P}_l)$.

Proof. See Theorem 4.1 in Khuri (1982). □

Theorem 2.4.1 provides the basis for a complete analysis of any balanced mixed model, as it can be used to obtain exact tests for testing the significance of the fixed effects and the variance components.

A linear function $\mathbf{a}'\mathbf{g}$, of $\mathbf{g}$ in model (2.14), is estimable if there exists a linear function, $\mathbf{c}'\mathbf{y}$, of the observations such that $E(\mathbf{c}'\mathbf{y}) = \mathbf{a}'\mathbf{g}$. In Searle (1971, Section 5.4) it is shown that $\mathbf{a}'\mathbf{g}$ is estimable if and only if $\mathbf{a}'$ belongs to the linear span of the rows of $\mathbf{X}$. In Khuri (1984) we have the following theorem:

Theorem 2.4.2. Consider the balanced mixed model in (2.14). Then we have the following:

1. $r(\mathbf{P}_l\mathbf{X}) = r(\mathbf{P}_l)$, $l = 0, 1, \ldots, \theta$.
2. $r(\mathbf{X}) = \sum_{l=0}^{\theta} r(\mathbf{P}_l\mathbf{X})$.
3. $\mathbf{P}_0\mathbf{X}\mathbf{g}, \mathbf{P}_1\mathbf{X}\mathbf{g}, \ldots, \mathbf{P}_\theta\mathbf{X}\mathbf{g}$ are linearly independent and span the space of all estimable linear functions of $\mathbf{g}$.

Theorem 2.4.2 is useful in identifying a basis of estimable linear functions of the fixed effects in model (2.14).

2.4.2. The Singular-Value Decomposition

The singular-value decomposition of a matrix is far more useful, both in statistics and in matrix algebra, then is commonly realized. For example, it plays a significant role in regression analysis. Let us consider the linear model

$$\mathbf{y} = \mathbf{X}\boldsymbol{\beta} + \boldsymbol{\epsilon}, \tag{2.15}$$

where $\mathbf{y}$ is a vector of n observations, $\mathbf{X}$ is an $n \times p$ $(n \geq p)$ matrix consisting of known constants, $\boldsymbol{\beta}$ is an unknown parameter vector, and $\boldsymbol{\epsilon}$ is a random error vector. Using Theorem 2.3.11, the matrix $\mathbf{X}'$ can be expressed as

$$\mathbf{X}' = \mathbf{P}[\mathbf{D}:\mathbf{0}]\mathbf{Q}', \tag{2.16}$$

where $\mathbf{P}$ and $\mathbf{Q}$ are orthogonal matrices of orders $p \times p$ and $n \times n$, respectively, and $\mathbf{D}$ is a diagonal matrix of order $p \times p$ consisting of nonnegative diagonal elements. These are the singular values of $\mathbf{X}$ (or of $\mathbf{X}'$) and are the positive square roots of the eigenvalues of $\mathbf{X}'\mathbf{X}$. From (2.16) we get

$$\mathbf{X} = \mathbf{Q}\begin{bmatrix} \mathbf{D} \\ \mathbf{0}' \end{bmatrix}\mathbf{P}'. \tag{2.17}$$

If the columns of $\mathbf{X}$ are linearly related, then they are said to be multicollinear. In this case, $\mathbf{X}$ has rank r $(<p)$, and the columns of $\mathbf{X}$ belong to a vector subspace of dimension r. At least one of the eigenvalues of $\mathbf{X}'\mathbf{X}$, and hence at least one of the singular values of $\mathbf{X}$, will be equal to zero. In practice, such exact multicollinearities rarely occur in statistical applications. Rather, the columns of $\mathbf{X}$ may be "nearly" linearly related. In this case, the rank of $\mathbf{X}$ is p, but some of the singular values of $\mathbf{X}$ will be "near zero." We shall use the term multicollinearity in a broader sense to describe the latter situation. It is also common to use the term "ill conditioning" to refer to the same situation.

The presence of multicollinearities in $\mathbf{X}$ can have adverse effects on the least-squares estimate, $\hat{\boldsymbol{\beta}}$, of $\boldsymbol{\beta}$ in (2.15). This can be easily seen from the fact that $\hat{\boldsymbol{\beta}} = (\mathbf{X}'\mathbf{X})^{-1}\mathbf{X}'\mathbf{y}$ and $\text{Var}(\hat{\boldsymbol{\beta}}) = (\mathbf{X}'\mathbf{X})^{-1}\sigma^2$, where σ^2 is the error variance. Large variances associated with the elements of $\hat{\boldsymbol{\beta}}$ can therefore be expected when the columns of $\mathbf{X}$ are multicollinear. This causes $\hat{\boldsymbol{\beta}}$ to become an unreliable estimate of $\boldsymbol{\beta}$. For a detailed study of multicollinearity and its effects, see Belsley, Kuh, and Welsch (1980, Chapter 3), Montgomery and Peck (1982, Chapter 8), and Myers (1990, Chapter 3).

The singular-value decomposition of $\mathbf{X}$ can provide useful information for detecting multicollinearity, as we shall now see. Let us suppose that the columns of $\mathbf{X}$ are multicollinear. Because of this, some of the singular values of $\mathbf{X}$, say p_2 $(<p)$ of them, will be "near zero." Let us partition $\mathbf{D}$ in (2.17) as

$$\mathbf{D} = \begin{bmatrix} \mathbf{D}_1 & \mathbf{0} \\ \mathbf{0} & \mathbf{D}_2 \end{bmatrix},$$

where $\mathbf{D}_1$ and $\mathbf{D}_2$ are of orders $p_1 \times p_1$ and $p_2 \times p_2$ ($p_1 = p - p_2$), respectively. The diagonal elements of D_2 consist of those singular values of $\mathbf{X}$ labeled as "near zero." Let us now write (2.17) as

$$\mathbf{XP} = \mathbf{Q} \begin{bmatrix} \mathbf{D}_1 & \mathbf{0} \\ \mathbf{0} & \mathbf{D}_2 \\ \mathbf{0} & \mathbf{0} \end{bmatrix}. \tag{2.18}$$

Let us next partition $\mathbf{P}$ and $\mathbf{Q}$ as $\mathbf{P} = [\mathbf{P}_1 : \mathbf{P}_2], \mathbf{Q} = [\mathbf{Q}_1 : \mathbf{Q}_2]$, where $\mathbf{P}_1$ and $\mathbf{P}_2$ have p_1 and p_2 columns, respectively, and $\mathbf{Q}_1$ and $\mathbf{Q}_2$ have p_1 and $n - p_1$ columns, respectively. From (2.18) we conclude that

$$\mathbf{XP}_1 = \mathbf{Q}_1 \mathbf{D}_1, \tag{2.19}$$

$$\mathbf{XP}_2 \approx \mathbf{0}, \tag{2.20}$$

where $\approx$ represents approximate equality. The matrix $\mathbf{XP}_2$ is "near zero" because of the smallness of the diagonal elements of $\mathbf{D}_2$.

We note from (2.20) that each column of $\mathbf{P}_2$ provides a "near"-linear relationship among the columns of $\mathbf{X}$. If (2.20) were an exact equality, then the columns of $\mathbf{P}_2$ would provide an orthonormal basis for the null space of $\mathbf{X}$.

We have mentioned that the presence of multicollinearity is indicated by the "smallness" of the singular values of $\mathbf{X}$. The problem now is to determine what "small" is. For this purpose it is common in statistics to use the condition number of $\mathbf{X}$, denoted by $\kappa(\mathbf{X})$. By definition

$$\kappa(\mathbf{X}) = \frac{\lambda_{max}}{\lambda_{min}},$$

where λ_{max} and λ_{min} are, respectively, the largest and smallest singular values of $\mathbf{X}$. Since the singular values of $\mathbf{X}$ are the positive square roots of the eigenvalues of $\mathbf{X'X}$, then $\kappa(\mathbf{X})$ can also be written as

$$\kappa(\mathbf{X}) = \sqrt{\frac{e_{max}(\mathbf{X'X})}{e_{min}(\mathbf{X'X})}}.$$

If $\kappa(\mathbf{X})$ is less than 10, then there is no serious problem with multicollinearity. Values of $\kappa(\mathbf{X})$ between 10 and 30 indicate moderate to strong multicollinearity, and if $\kappa > 30$, severe multicollinearity is implied.

More detailed discussions concerning the use of the singular-value decomposition in regression can be found in Mandel (1982). See also Lowerre (1982). Good (1969) described several applications of this decomposition in statistics and in matrix algebra.

2.4.3. Extrema of Quadratic Forms

In many statistical problems there is a need to find the extremum (maximum or minimum) of a quadratic form or a ratio of quadratic forms. Let us, for example, consider the following problem:

Let $X_1, X_2, \ldots, X_n$ be a collection of random vectors, all having the same number of elements. Suppose that these vectors are independently and identically distributed (i.i.d.) as $N(\mu, \Sigma)$, where both μ and Σ are unknown. Consider testing the hypothesis $H_0: \mu = \mu_0$ versus its alternative $H_a: \mu \neq \mu_0$, where μ_0 is some hypothesized value of μ. We need to develop a test statistic for testing H_0.

The multivariate hypothesis H_0 is true if and only if the univariate hypotheses

$$H_0(\lambda): \lambda'\mu = \lambda'\mu_0$$

are true for all $\lambda \neq 0$. A test statistic for testing $H_0(\lambda)$ is the following:

$$t(\lambda) = \frac{\lambda'(\overline{X} - \mu_0)\sqrt{n}}{\sqrt{\lambda'S\lambda}},$$

where $\overline{X} = \sum_{i=1}^{n} X_i/n$ and S is the sample variance–covariance matrix, which is an unbiased estimator of Σ, and is given by

$$S = \frac{1}{n-1} \sum_{i=1}^{n} (X_i - \overline{X})(X_i - \overline{X})'.$$

Large values of $t^2(\lambda)$ indicate falsehood of $H_0(\lambda)$. Since H_0 is rejected if and only if $H_0(\lambda)$ is rejected for at least one λ, then the condition to reject H_0 at the α-level is $\sup_{\lambda \neq 0}[t^2(\lambda)] > c_\alpha$, where c_α is the upper $100\alpha\%$ point of the distribution of $\sup_{\lambda \neq 0}[t^2(\lambda)]$. But

$$\sup_{\lambda \neq 0}\left[t^2(\lambda)\right] = \sup_{\lambda \neq 0} \frac{n|\lambda'(\overline{X} - \mu_0)|^2}{\lambda'S\lambda}$$

$$= n \sup_{\lambda \neq 0} \frac{\lambda'(\overline{X} - \mu_0)(\overline{X} - \mu_0)'\lambda}{\lambda'S\lambda}$$

$$= n\, e_{\max}\left[S^{-1}(\overline{X} - \mu_0)(\overline{X} - \mu_0)'\right],$$

by Theorem 2.3.17.

Now,

$$e_{\max}\left[S^{-1}(\overline{X} - \mu_0)(\overline{X} - \mu_0)'\right] = e_{\max}\left[(\overline{X} - \mu_0)'S^{-1}(\overline{X} - \mu_0)\right],$$

$$= (\overline{X} - \mu_0)'S^{-1}(\overline{X} - \mu_0).$$

by Theorem 2.3.9.

Hence,

$$\sup_{\boldsymbol{\lambda} \neq 0} \left[t^2(\boldsymbol{\lambda}) \right] = n(\overline{\mathbf{X}} - \boldsymbol{\mu}_0)' \mathbf{S}^{-1} (\overline{\mathbf{X}} - \boldsymbol{\mu}_0)$$

is the test statistic for the multivariate hypothesis H_0. This is called Hotelling's T^2-statistic. Its critical values are obtained in terms of the critical values of the F-distribution (see, for example, Morrison, 1967, Chapter 4).

Another example of using the extremum of a ratio of quadratic forms is in the determination of the canonical correlation coefficient between two random vectors (see Exercise 2.26). The article by Bush and Olkin (1959) lists several similar statistical applications.

2.4.4. The Parameterization of Orthogonal Matrices

Orthogonal matrices are used frequently in statistics, especially in linear models and multivariate analysis (see, for example, Graybill, 1961, Chapter 11; James, 1954).

The n^2 elements of an $n \times n$ orthogonal matrix $\mathbf{Q}$ are subject to $n(n + 1)/2$ constraints because $\mathbf{Q}'\mathbf{Q} = \mathbf{I}_n$. These elements can therefore be represented by $n^2 - n(n + 1)/2 = n(n - 1)/2$ independent parameters. The need for such a representation arises in several situations. For example, in the design of experiments, there may be a need to search for an orthogonal matrix that satisfies a certain optimality criterion. Using the independent parameters of an orthogonal matrix can facilitate this search. Khuri and Myers (1981) followed this approach in their construction of a response surface design that is robust to nonnormality of the error distribution associated with the response function. Another example is the generation of random orthogonal matrices for carrying out simulation experiments. This was used by Heiberger, Velleman, and Ypelaar (1983) to construct test data with special properties for multivariate linear models. Anderson, Olkin, and Underhill (1987) proposed a procedure to generate random orthogonal matrices.

Methods to parameterize an orthogonal matrix were reviewed in Khuri and Good (1989). One such method is to use the relationship between an orthogonal matrix and a skew-symmetric matrix. If $\mathbf{Q}$ is an orthogonal matrix with determinant equal to one, then it can be written in the form

$$\mathbf{Q} = e^{\mathbf{T}},$$

where $\mathbf{T}$ is a skew-symmetric matrix (see, for example, Gantmacher, 1959). The elements of $\mathbf{T}$ above its main diagonal can be used to parameterize $\mathbf{Q}$. This exponential mapping is defined by the infinite series

$$e^{\mathbf{T}} = \mathbf{I} + \mathbf{T} + \frac{\mathbf{T}^2}{2!} + \frac{\mathbf{T}^3}{3!} + \cdots.$$

The exponential parameterization was used in a theorem concerning the asymptotic joint density function of the eigenvalues of the sample variance–covariance matrix (Muirhead, 1982, page 394).

Another parameterization of $\mathbf{Q}$ is given by

$$\mathbf{Q} = (\mathbf{I} - \mathbf{U})(\mathbf{I} + \mathbf{U})^{-1},$$

where $\mathbf{U}$ is a skew-symmetric matrix. This relationship is valid provided that $\mathbf{Q}$ does not have the eigenvalue -1. Otherwise, $\mathbf{Q}$ can be written as

$$\mathbf{Q} = \mathbf{L}(\mathbf{I} - \mathbf{U})(\mathbf{I} + \mathbf{U})^{-1},$$

where $\mathbf{L}$ is a diagonal matrix in which each element on the diagonal is either 1 or -1. Arthur Cayley (1821–1895) is credited with having introduced the relationship between $\mathbf{Q}$ and $\mathbf{U}$.

Finally, the recent article by Olkin (1990) illustrates the strong interplay between statistics and linear algebra. The author listed several areas of statistics with a strong linear algebra component.

FURTHER READING AND ANNOTATED BIBLIOGRAPHY

Anderson, T. W., and S. D. Gupta (1963). "Some inequalities on characteristic roots of matrices," *Biometrika*, **50**, 522–524.

Anderson, T. W., I. Olkin, and L. G. Underhill (1987). "Generation of random orthogonal matrices." *SIAM J. Sci. Statist. Comput.*, **8**, 625–629.

Basilevsky, A. (1983). *Applied Matrix Algebra in the Statistical Sciences*. North-Holland, New York. (This book addresses topics in matrix algebra that are useful in both applied and theoretical branches of the statistical sciences.)

Bellman, R. (1970). *Introduction to Matrix Analysis*, 2nd ed. McGraw-Hill, New York. (An excellent reference book on matrix algebra. The minimum–maximum characterization of eigenvalues is discussed in Chap. 7. Kronecker products are studied in Chap. 12. Some applications of matrices to stochastic processes and probability theory are given in Chap. 14.)

Belsley, D. A., E. Kuh, and R. E. Welsch (1980). *Regression Diagnostics*. Wiley, New York. (This is a good reference for learning about multicollinearity in linear statistical models that was discussed in Section 2.4.2. Examples are provided based on actual econometric data.)

Bush, K. A., and I. Olkin (1959). "Extrema of quadratic forms with applications to statistics." *Biometrika*, **46**, 483–486.

Gantmacher, F. R. (1959). *The Theory of Matrices*, Vols. I and II. Chelsea, New York. (These two volumes provide a rather more advanced study of matrix algebra than standard introductory texts. Methods to parameterize an orthogonal matrix, which were mentioned in Section 2.4.4, are discussed in Vol. I.)

Golub, G. H., and C. F. Van Loan (1983). *Matrix Computations*. Johns Hopkins University Press, Baltimore, Maryland.

Good, I. J. (1969). "Some applications of the singular decomposition of a matrix." *Technometrics*, **11**, 823–831.

Graybill, F. A. (1961). *An Introduction to Linear Statistical Models*, Vol. I. McGraw-Hill, New York. (This is considered a classic textbook in experimental statistics. It is concerned with the mathematical treatment, using matrix algebra, of linear statistical models.)

Graybill, F. A. (1983). *Matrices with Applications in Statistics*, 2nd ed. Wadsworth, Belmont, California. (This frequently referenced textbook contains a great number of theorems in matrix algebra, and describes many properties of matrices that are pertinent to linear model and mathematical statistics.)

Healy, M. J. R. (1986). *Matrices for Statistics*. Clarendon Press, Oxford, England. (This is a short book that provides a brief coverage of some basic concepts in matrix algebra. Some applications in statistics are also mentioned.)

Heiberger, R. M., P. F. Velleman, and M. A. Ypelaar (1983). "Generating test data with independently controllable features for multivariate general linear forms." *J. Amer. Statist. Assoc.*, **78**, 585–595.

Henderson, H. V., F. Pukelsheim, and S. R. Searle (1983). "On the history of the Kronecker product." *Linear and Multilinear Algebra*, **14**, 113–120.

Henderson, H. V., and S. R. Searle (1981). "The vec-permutation matrix, the vec operator and Kronecker products: A review." *Linear and Multilinear Algebra*, **9**, 271–288.

Hoerl, A. E., and R. W. Kennard (1970). "Ridge regression: Applications to nonorthogonal problems." *Technometrics*, **12**, 69–82.

James, A. T. (1954). "Normal multivariate analysis and the orthogonal group." *Ann. Math. Statist.*, **25**, 40–75.

Khuri, A. I. (1982). "Direct products: A powerful tool for the analysis of balanced data." *Comm. Statist. Theory Methods*, **11**, 2903–2920.

Khuri, A. I. (1984). "Interval estimation of fixed effects and of functions of variance components in balanced mixed models." *Sankhyā, Series B*, **46**, 10–28. (Section 5 in this article gives a procedure for the construction of exact simultaneous confidence intervals on estimable linear functions of the fixed effects in a balanced mixed model.)

Khuri, A. I., and I. J. Good (1989). "The parameterization of orthogonal matrices: A review mainly for statisticians." *South African Statist. J.*, **23**, 231–250.

Khuri, A. I., and R. H. Myers (1981). "Design related robustness of tests in regression models." *Comm. Statist. Theory Methods*, **10**, 223–235.

Lancaster, P. (1969). *Theory of Matrices*. Academic Press, New York. (This book is written primarily for students of applied mathematics, engineering, or science who want to acquire a good knowledge of the theory of matrices. Chap. 7 has an interesting discussion concerning the behavior of matrix eigenvalues under perturbation of the elements of the matrix.)

Lowerre, J. M. (1982). "An introduction to modern matrix methods and statistics." *Amer. Statist.*, **36**, 113–115. (An application of the singular-value decomposition is given in Section 2 of this article.)

Magnus, J. R., and H. Neudecker (1988). *Matrix Differential Calculus with Applications in Statistics and Econometrics*. Wiley, New York. (This book consists of six parts. Part one deals with the basics of matrix algebra. The remaining parts are devoted to the development of matrix differential calculus and its applications to statistics and econometrics. Part four has a chapter on inequalities concerning eigenvalues that pertains to Section 2.3.10 in this chapter.)

Mandel, J. (1982). "Use of the singular-value decomposition in regression analysis." *Amer. Statist.*, **36**, 15–24.

Marcus, M., and H. Minc (1988). *Introduction to Linear Algebra*. Dover, New York. (This book presents an introduction to the fundamental concepts of linear algebra and matrix theory.)

Marsaglia, G., and G. P. H. Styan (1974). "Equalities and inequalities for ranks of matrices." *Linear and Multilinear Algebra*, **2**, 269–292. (This is an interesting collection of results on ranks of matrices. It includes a wide variety of equalities and inequalities for ranks of products, of sums, and of partitioned matrices.)

May, W. G. (1970). *Linear Algebra*. Scott, Foresman and Company, Glenview, Illinois.

Montgomery, D. C., and E. A. Peck (1982). *Introduction to Linear Regression Analysis*. Wiley, New York. (Chap. 8 in this book has an interesting discussion concerning multicollinearity. It includes the sources of multicollinearity, its harmful effects in regression, available diagnostics, and a survey of remedial measures. This chapter provides useful additional information to the material in Section 2.4.2.)

Morrison, D. F. (1967). *Multivariate Statistical Methods*. McGraw-Hill, New York. (This book can serve as an introductory text to multivariate analysis.)

Muirhead, R. J. (1982). *Aspects of Multivariate Statistical Theory*. Wiley, New York. (This book is designed as a text for a graduate-level course in multivariate analysis.)

Myers, R. H. (1990). *Classical and Modern Regression with Applications*, 2nd ed. PWS-Kent, Boston. (Chap. 8 in this book should be useful reading concerning multicollinearity and its effects.)

Newcomb, R. W. (1960). "On the simultaneous diagonalization of two semidefinite matrices." *Quart. Appl. Math.*, **19**, 144–146.

Olkin, I. (1990). "Interface between statistics and linear algebra." In *Matrix Theory and Applications*, Vol. 40, C. R. Johnson, ed., American Mathematical Society, Providence, Rhode Island, pp. 233–256.

Price, G. B. (1947). "Some identities in the theory of determinants." *Amer. Math. Monthly*, **54**, 75–90. (Section 10 in this article gives some history of the theory of determinants.)

Rogers, G. S. (1984). "Kronecker products in ANOVA—a first step." *Amer. Statist.*, **38**, 197–202.

Searle, S. R. (1971). *Linear Models*. Wiley, New York.

Searle, S. R. (1982). *Matrix Algebra Useful for Statistics*. Wiley, New York. (This is a useful book introducing matrix algebra in a manner that is helpful in the statistical analysis of data and in statistics in general. Chaps. 13, 14, and 15 present applications of matrices in regression and linear models.)

Seber, G. A. F. (1984). *Multivariate Observations*. Wiley, New York. (This is a good reference on applied multivariate analysis that is suited for a graduate-level course.)

Smith, D. E. (1958). *History of Mathematics*. Vol. I. Dover, New York. (This interesting book contains, among other things, some history concerning the development of the theory of determinants and matrices.)

Wolkowicz, H., and G. P. H. Styan (1980). "Bounds for eigenvalues using traces." *Linear Algebra Appl.*, **29**, 471–506.

EXERCISES

In Mathematics

2.1. Show that a set of $n \times 1$ vectors, $\mathbf{u}_1, \mathbf{u}_2, \ldots, \mathbf{u}_m$, is always linearly dependent if $m > n$.

2.2. Let W be a vector subspace of V such that $W = L(\mathbf{u}_1, \mathbf{u}_2, \ldots, \mathbf{u}_n)$, where the $\mathbf{u}_i$'s $(i = 1, 2, \ldots, n)$ are linearly independent. If $\mathbf{v}$ is any vector in V that is not in W, then the vectors $\mathbf{u}_1, \mathbf{u}_2, \ldots, \mathbf{u}_n, \mathbf{v}$ are linearly independent.

2.3. Prove Theorem 2.1.3.

2.4. Prove part 1 of Theorem 2.2.1.

2.5. Let $T: U \to V$ be a linear transformation. Show that T is one-to-one if and only if whenever $\mathbf{u}_1, \mathbf{u}_2, \ldots, \mathbf{u}_n$ are linearly independent in U, then $T(\mathbf{u}_1), T(\mathbf{u}_2), \ldots, T(\mathbf{u}_n)$ are linearly independent in V.

2.6. Let $T: R_n \to R_m$ be represented by an $n \times m$ matrix of rank ρ.
  (a) Show that $\dim[T(R_n)] = \rho$.
  (b) Show that if $n \leq m$ and $\rho = n$, then T is one-to-one.

2.7. Show that $\text{tr}(\mathbf{A}'\mathbf{A}) = 0$ if and only if $\mathbf{A} = \mathbf{0}$.

2.8. Let $\mathbf{A}$ be a symmetric positive semidefinite matrix of order $n \times n$. Show that $\mathbf{v}'\mathbf{A}\mathbf{v} = 0$ if and only if $\mathbf{A}\mathbf{v} = \mathbf{0}$.

2.9. The matrices $\mathbf{A}$ and $\mathbf{B}$ are symmetric and positive semidefinite of order $n \times n$ such that $\mathbf{A}\mathbf{B} = \mathbf{B}\mathbf{A}$. Show that $\mathbf{A}\mathbf{B}$ is positive semidefinite.

2.10. If $\mathbf{A}$ is a symmetric $n \times n$ matrix, and $\mathbf{B}$ is an $n \times n$ skew-symmetric matrix, then show that $\text{tr}(\mathbf{A}\mathbf{B}) = 0$.

2.11. Suppose that $\text{tr}(\mathbf{P}\mathbf{A}) = 0$ for every skew-symmetric matrix $\mathbf{P}$. Show that the matrix $\mathbf{A}$ is symmetric.

2.12. Let A be an $n \times n$ matrix and C be a nonsingular matrix of order $n \times n$. Show that A, $C^{-1}AC$, and CAC^{-1} have the same set of eigenvalues.

2.13. Let A be an $n \times n$ symmetric matrix, and let λ be an eigenvalue of A of multiplicity k. Then $A - \lambda I_n$ has rank $n - k$.

2.14. Let A be a nonsingular matrix of order $n \times n$, and let c and d be $n \times 1$ vectors. If $d'A^{-1}c \neq -1$, then

$$(A + cd')^{-1} = A^{-1} - \frac{(A^{-1}c)(d'A^{-1})}{1 + d'A^{-1}c}.$$

This is known as the Sherman-Morrison formula.

2.15. Show that if A and $I_k + V'A^{-1}U$ are nonsingular, then

$$(A + UV')^{-1} = A^{-1} - A^{-1}U(I_k + V'A^{-1}U)^{-1}V'A^{-1},$$

where A is of order $n \times n$, and U and V are of order $n \times k$. This result is known as the Sherman-Morrison-Woodbury formula and is a generalization of the result in Exercise 2.14.

2.16. Prove Theorem 2.3.17.

2.17. Let A and B be $n \times n$ idempotent matrices. Show that $A - B$ is idempotent if and only if $AB = BA = B$.

2.18. Let A be an orthogonal matrix. What can be said about the eigenvalues of A?

2.19. Let A be a symmetric matrix of order $n \times n$, and let L be a matrix of order $n \times m$. Show that

$$e_{\min}(A)\operatorname{tr}(L'L) \leq \operatorname{tr}(L'AL) \leq e_{\max}(A)\operatorname{tr}(L'L)$$

2.20. Let A be a nonnegative definite matrix of order $n \times n$, and let L be a matrix of order $n \times m$. Show that
(a) $e_{\min}(L'AL) \geq e_{\min}(A)e_{\min}(L'L)$,
(b) $e_{\max}(L'AL) \leq e_{\max}(A)e_{\max}(L'L)$.

2.21. Let A and B be $n \times n$ symmetric matrices with A nonnegative definite. Show that

$$e_{\min}(B)\operatorname{tr}(A) \leq \operatorname{tr}(AB) \leq e_{\max}(B)\operatorname{tr}(A).$$

2.22. Let A^- be a g-inverse of A. Show that

(a) A^-A is idempotent,

(b) $r(A^-) \geq r(A)$,

(c) $r(A) = r(A^-A)$.

In Statistics

2.23. Let $y = (y_1, y_2, \ldots, y_n)'$ be a normal random vector $N(0, \sigma^2 I_n)$. Let $\bar{y}$ and s^2 be the sample mean and sample variance given by

$$\bar{y} = \frac{1}{n} \sum_{i=1}^{n} y_i,$$

$$s^2 = \frac{1}{n-1} \left[\sum_{i=1}^{n} y_i^2 - \frac{(\sum_{i=1}^{n} y_i)^2}{n} \right].$$

(a) Show that A is an idempotent matrix of rank $n - 1$, where A is an $n \times n$ matrix such that $y'Ay = (n - 1)s^2$.

(b) What distribution does $(n - 1)s^2/\sigma^2$ have?

(c) Show that $\bar{y}$ and Ay are uncorrelated; then conclude that $\bar{y}$ and s^2 are statistically independent.

2.24. Consider the one-way classification model

$$y_{ij} = \mu + \alpha_i + \epsilon_{ij}, \quad i = 1, 2, \ldots, a; \quad j = 1, 2, \ldots, n_i,$$

where μ and α_i $(i = 1, 2, \ldots, a)$ are unknown parameters and ϵ_{ij} is a random error with a zero mean. Show that

(a) $\alpha_i - \alpha_{i'}$ is an estimable linear function for all $i \neq i'$ $(i, i' = 1, 2, \ldots, a)$,

(b) μ is nonestimable.

2.25. Consider the linear model

$$y = X\beta + \epsilon,$$

where X is a known matrix of order $n \times p$ and rank r $(\leq p)$, β is an unknown parameter vector, and ϵ is a random error vector such that $E(\epsilon) = 0$ and $\text{Var}(\epsilon) = \sigma^2 I_n$.

(a) Show that $X(X'X)^- X'$ is an idempotent matrix.

(b) Let $l'y$ be an unbiased linear estimator of $\lambda'\beta$. Show that

$$\text{Var}(\lambda'\hat{\beta}) \leq \text{Var}(l'y),$$

where $\lambda'\hat{\beta} = \lambda'(X'X)^- X'y$.

The result given in part (b) is known as the *Gauss–Markov theorem*.

2.26. Consider the linear model in Exercise 2.25, and suppose that $r(\mathbf{X}) = p$. Hoerl and Kennard (1970) introduced an estimator of $\boldsymbol{\beta}$ called the ridge estimator $\boldsymbol{\beta}^*$:

$$\boldsymbol{\beta}^* = \left(\mathbf{X}'\mathbf{X} + k\mathbf{I}_p\right)^{-1}\mathbf{X}'\mathbf{y},$$

where k is a "small" fixed number. For an appropriate value of k, $\boldsymbol{\beta}^*$ provides improved accuracy in the estimation of $\boldsymbol{\beta}$ over the least-squares estimator $\hat{\boldsymbol{\beta}} = (\mathbf{X}'\mathbf{X})^{-1}\mathbf{X}'\mathbf{y}$. Let $\mathbf{X}'\mathbf{X} = \mathbf{P}\boldsymbol{\Lambda}\mathbf{P}'$ be the spectral decomposition of $\mathbf{X}'\mathbf{X}$. Show that $\boldsymbol{\beta}^* = \mathbf{P}\mathbf{D}\mathbf{P}'\hat{\boldsymbol{\beta}}$, where $\mathbf{D}$ is a diagonal matrix whose ith diagonal element is $\lambda_i/(\lambda_i + k)$, $i = 1, 2, \ldots, p$, and where $\lambda_1, \lambda_2, \ldots, \lambda_p$ are the diagonal elements of $\boldsymbol{\Lambda}$.

2.27. Consider the ratio

$$\rho^2 = \frac{(\mathbf{x}'\mathbf{A}\mathbf{y})^2}{(\mathbf{x}'\mathbf{B}_1\mathbf{x})(\mathbf{y}'\mathbf{B}_2\mathbf{y})},$$

where $\mathbf{A}$ is a matrix of order $m \times n$ and $\mathbf{B}_1, \mathbf{B}_2$ are positive definite of orders $m \times m$ and $n \times n$, respectively. Show that

$$\sup_{\mathbf{x},\mathbf{y}} \rho^2 = e_{\max}\left(\mathbf{B}_1^{-1}\mathbf{A}\mathbf{B}_2^{-1}\mathbf{A}'\right).$$

[*Hint:* Define $\mathbf{C}_1$ and $\mathbf{C}_2$ as symmetric nonsingular matrices such that $\mathbf{C}_1^2 = \mathbf{B}_1, \mathbf{C}_2^2 = \mathbf{B}_2$. Let $\mathbf{C}_1\mathbf{x} = \mathbf{u}, \mathbf{C}_2\mathbf{y} = \mathbf{v}$. Then ρ^2 can be written as

$$\rho^2 = \frac{\left(\mathbf{u}'\mathbf{C}_1^{-1}\mathbf{A}\mathbf{C}_2^{-1}\mathbf{v}\right)^2}{(\mathbf{u}'\mathbf{u})(\mathbf{v}'\mathbf{v})} = \left(\boldsymbol{\nu}'\mathbf{C}_1^{-1}\mathbf{A}\mathbf{C}_2^{-1}\boldsymbol{\tau}\right)^2,$$

where $\boldsymbol{\nu} = \mathbf{u}/(\mathbf{u}'\mathbf{u})^{1/2}, \boldsymbol{\tau} = \mathbf{v}/(\mathbf{v}'\mathbf{v})^{1/2}$ are unit vectors. Verify the result of this problem after noting that ρ^2 is now the square of a dot product.]

Note: This exercise has the following application in multivariate analysis: Let $\mathbf{z}_1$ and $\mathbf{z}_2$ be random vectors with zero means and variance–covariance matrices $\boldsymbol{\Sigma}_{11}, \boldsymbol{\Sigma}_{22}$, respectively. Let $\boldsymbol{\Sigma}_{12}$ be the covariance matrix of $\mathbf{z}_1$ and $\mathbf{z}_2$. On choosing $\mathbf{A} = \boldsymbol{\Sigma}_{12}, \mathbf{B}_1 = \boldsymbol{\Sigma}_{11}, \mathbf{B}_2 = \boldsymbol{\Sigma}_{22}$, the positive square root of the supremum of ρ^2 is called the *canonical correlation coefficient* between $\mathbf{z}_1$ and $\mathbf{z}_2$. It is a measure of the linear association between $\mathbf{z}_1$ and $\mathbf{z}_2$ (see, for example, Seber, 1984, Section 5.7).

CHAPTER 3

Limits and Continuity of Functions

The notions of limits and continuity of functions lie at the kernel of calculus. The general concept of continuity is very old in mathematics. It had its inception long ago in ancient Greece. We owe to Aristotle (384–322 B.C.) the first known definition of continuity: "A thing is continuous when of any two successive parts the limits at which they touch are one and the same and are, as the word implies, held together" (see Smith, 1958, page 93). Our present definitions of limits and continuity of functions, however, are substantially those given by Augustin-Louis Cauchy (1789–1857).

In this chapter we introduce the concepts of limits and continuity of real-valued functions, and study some of their properties. The domains of definition of the functions will be subsets of R, the set of real numbers. A typical subset of R will be denoted by D.

3.1. LIMITS OF A FUNCTION

Before defining the notion of a limit of a function, let us understand what is meant by the notation $x \to a$, where a and x are elements in R. If a is finite, then $x \to a$ means that x can have values that belong to a neighborhood $N_r(a)$ of a (see Definition 1.6.1) for any $r > 0$, but $x \neq a$, that is, $0 < |x - a| < r$. Such a neighborhood is called a deleted neighborhood of a, that is, a neighborhood from which the point a has been removed. If a is infinite ($-\infty$ or $+\infty$), then $x \to a$ indicates that $|x|$ can get larger and larger without any constraint on the extent of its increase. Thus $|x|$ can have values greater than any positive number. In either case, whether a is finite or infinite, we say that x tends to a or approaches a.

Let us now study the behavior of a function $f(x)$ as $x \to a$.

Definition 3.1.1. Suppose that the function $f(x)$ is defined in a deleted neighborhood of a point $a \in R$. Then $f(x)$ is said to have a limit L as $x \to a$

57

if for every $\epsilon > 0$ there exists a $\delta > 0$ such that

$$|f(x) - L| < \epsilon \tag{3.1}$$

for all x for which

$$0 < |x - a| < \delta. \tag{3.2}$$

In this case, we write $f(x) \to L$ as $x \to a$, which is equivalent to saying that $\lim_{x \to a} f(x) = L$. Less formally, we say that $f(x) \to L$ as $x \to a$ if, however small the positive number ϵ might be, $f(x)$ differs from L by less than ϵ for values of x sufficiently close to a. $\square$

NOTE 3.1.1.   When $f(x)$ has a limit as $x \to a$, it is considered to be finite. If this is not the case, then $f(x)$ is said to have an infinite limit ($-\infty$ or $+\infty$) as $x \to a$. This limit exists only in the extended real number system, which consists of the real number system combined with the two symbols, $-\infty$ and $+\infty$. In this case, for every positive number M there exists a $\delta > 0$ such that $|f(x)| > M$ if $0 < |x - a| < \delta$. If a is infinite and L is finite, then $f(x) \to L$ as $x \to a$ if for any $\epsilon > 0$ there exists a positive number N such that inequality (3.1) is satisfied for all x for which $|x| > N$. In case both a and L are infinite, then $f(x) \to L$ as $x \to a$ if for any $B > 0$ there exists a positive number A such that $|f(x)| > B$ if $|x| > A$.

NOTE 3.1.2.   If $f(x)$ has a limit L as $x \to a$, then L must be unique. To show this, suppose that L_1 and L_2 are two limits of $f(x)$ as $x \to a$. Then, for any $\epsilon > 0$ there exist $\delta_1 > 0$, $\delta_2 > 0$ such that

$$|f(x) - L_1| < \frac{\epsilon}{2}, \qquad \text{if } 0 < |x - a| < \delta_1,$$

$$|f(x) - L_2| < \frac{\epsilon}{2}, \qquad \text{if } 0 < |x - a| < \delta_2.$$

Hence, if $\delta = \min(\delta_1, \delta_2)$, then

$$|L_1 - L_2| = |L_1 - f(x) + f(x) - L_2|$$
$$\leq |f(x) - L_1| + |f(x) - L_2|$$
$$< \epsilon$$

for all x for which $0 < |x - a| < \delta$. Since $|L_1 - L_2|$ is smaller than ϵ, which is an arbitrary positive number, we must have $L_1 = L_2$ (why?).

NOTE 3.1.3.   The limit of $f(x)$ as described in Definition 3.1.1 is actually called a two-sided limit. This is because x can approach a from either side. There are, however, cases where $f(x)$ can have a limit only when x approaches a from one side. Such a limit is called a one-sided limit.

By definition, if $f(x)$ has a limit as x approaches a from the left, symbolically written as $x \to a^-$, then $f(x)$ has a left-sided limit, which we denote by L^-. In this case we write

$$\lim_{x \to a^-} f(x) = L^-.$$

If, however, $f(x)$ has a limit as x approaches a from the right, symbolically written as $x \to a^+$, then $f(x)$ has a right-sided limit, denoted by L^+, that is,

$$\lim_{x \to a^+} f(x) = L^+.$$

From the above definition it follows that $f(x)$ has a left-sided limit L^- as $x \to a^-$ if for every $\epsilon > 0$ there exists a $\delta > 0$ such that

$$|f(x) - L^-| < \epsilon$$

for all x for which $0 < a - x < \delta$. Similarly, $f(x)$ has a right-sided limit L^+ as $x \to a^+$ if for every $\epsilon > 0$ there exists a $\delta > 0$ such that

$$|f(x) - L^+| < \epsilon$$

for all x for which $0 < x - a < \delta$.

Obviously, if $f(x)$ has a two-sided limit L as $x \to a$, then L^- and L^+ both exist and are equal to L. Vice versa, if $L^- = L^+$, then $f(x)$ has a two-sided limit L as $x \to a$, where L is the common value of L^- and L^+ (why?). We can then state that $\lim_{x \to a} f(x) = L$ if and only if

$$\lim_{x \to a^-} f(x) = \lim_{x \to a^+} f(x) = L.$$

Thus to determine if $f(x)$ has a limit as $x \to a$, we first need to find out if it has a left-sided limit L^- and a right-sided limit L^+ as $x \to a$. If this is the case and $L^- = L^+ = L$, then $f(x)$ has a limit L as $x \to a$.

Throughout the remainder of the book, we shall drop the characterization "two-sided" when making a reference to a two-sided limit L of $f(x)$. Instead, we shall simply state that L is the limit of $f(x)$.

EXAMPLE 3.1.1. Consider the function

$$f(x) = \begin{cases} (x-1)/(x^2 - 1), & x \neq -1, 1, \\ 4, & x = 1. \end{cases}$$

This function is defined everywhere except at $x = -1$. Let us find its limit as $x \to a$, where $a \in R$. We note that

$$\lim_{x \to a} f(x) = \lim_{x \to a} \frac{x - 1}{x^2 - 1} = \lim_{x \to a} \frac{1}{x + 1}.$$

This is true even if $a = 1$, because $x \neq a$ as $x \to a$. We now claim that if $a \neq -1$, then

$$\lim_{x \to a} \frac{1}{x+1} = \frac{1}{a+1}.$$

To prove this claim, we need to find a $\delta > 0$ such that for any $\epsilon > 0$,

$$\left| \frac{1}{x+1} - \frac{1}{a+1} \right| < \epsilon \tag{3.3}$$

if $0 < |x - a| < \delta$. Let us therefore consider the following two cases:

CASE 1. $a > -1$. In this case we have

$$\left| \frac{1}{x+1} - \frac{1}{a+1} \right| = \frac{|x-a|}{|x+1|\,|a+1|}. \tag{3.4}$$

If $|x - a| < \delta$, then

$$a - \delta + 1 < x + 1 < a + \delta + 1. \tag{3.5}$$

Since $a + 1 > 0$, we can choose $\delta > 0$ such that $a - \delta + 1 > 0$, that is, $\delta < a + 1$. From (3.4) and (3.5) we then get

$$\left| \frac{1}{x+1} - \frac{1}{a+1} \right| < \frac{\delta}{(a+1)(a-\delta+1)}.$$

Let us constrain δ even further by requiring that

$$\frac{\delta}{(a+1)(a-\delta+1)} < \epsilon.$$

This is accomplished by choosing $\delta > 0$ so that

$$\delta < \frac{(a+1)^2 \epsilon}{1 + (a+1)\epsilon}.$$

Since

$$\frac{(a+1)^2 \epsilon}{1 + (a+1)\epsilon} < a + 1,$$

inequality (3.3) will be satisfied by all x for which $|x - a| < \delta$, where

$$0 < \delta < \frac{(a+1)^2 \epsilon}{1 + (a+1)\epsilon}. \tag{3.6}$$

CASE 2. $a < -1$. Here, we choose $\delta > 0$ such that $a + \delta + 1 < 0$, that is, $\delta < -(a + 1)$. From (3.5) we conclude that

$$|x + 1| > -(a + \delta + 1).$$

Hence, from (3.4) we get

$$\left| \frac{1}{x + 1} - \frac{1}{a + 1} \right| < \frac{\delta}{(a + 1)(a + \delta + 1)}.$$

As before, we further constrain δ by requiring that it satisfy the inequality

$$\frac{\delta}{(a + 1)(a + \delta + 1)} < \epsilon,$$

or equivalently, the inequality

$$\delta < \frac{(a + 1)^2 \epsilon}{1 - (a + 1)\epsilon}.$$

Note that

$$\frac{(a + 1)^2 \epsilon}{1 - (a + 1)\epsilon} < -(a + 1).$$

Consequently, inequality (3.3) can be satisfied by choosing δ such that

$$0 < \delta < \frac{(a + 1)^2 \epsilon}{1 - (a + 1)\epsilon}. \tag{3.7}$$

Cases 1 and 2 can be combined by rewriting (3.6) and (3.7) using the single double inequality

$$0 < \delta < \frac{|a + 1|^2 \epsilon}{1 + |a + 1|\epsilon}.$$

If $a = -1$, then no limit exists as $x \to a$. This is because

$$\lim_{x \to -1} f(x) = \lim_{x \to -1} \frac{1}{x + 1}.$$

If $x \to -1^-$, then

$$\lim_{x \to -1^-} \frac{1}{x+1} = -\infty,$$

and, as $x \to -1^+$,

$$\lim_{x \to -1^+} \frac{1}{x+1} = \infty.$$

Since the left-sided and right-sided limits are not equal, no limit exists as $x \to -1$.

EXAMPLE 3.1.2.   Let $f(x)$ be defined as

$$f(x) = \begin{cases} 1 + \sqrt{x}, & x \geq 0, \\ x, & x < 0. \end{cases}$$

This function has no limit as $x \to 0$, since

$$\lim_{x \to 0^-} f(x) = \lim_{x \to 0^-} x = 0,$$

$$\lim_{x \to 0^+} f(x) = \lim_{x \to 0^+} (1 + \sqrt{x}) = 1.$$

However, for any $a \neq 0$, $\lim_{x \to a} f(x)$ exists.

EXAMPLE 3.1.3.   Let $f(x)$ be given by

$$f(x) = \begin{cases} x \cos x, & x \neq 0, \\ 0, & x = 0. \end{cases}$$

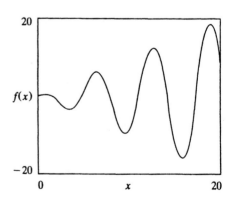

Figure 3.1. The graph of the function $f(x)$.

Then $\lim_{x \to 0} f(x) = 0$. This is true because $|f(x)| \leq |x|$ in any deleted neighborhood of $a = 0$. As $x \to \infty$, $f(x)$ oscillates unboundedly, since

$$-x \leq x \cos x \leq x.$$

Thus $f(x)$ has no limit as $x \to \infty$. A similar conclusion can be reached when $x \to -\infty$ (see Figure 3.1).

3.2. SOME PROPERTIES ASSOCIATED WITH LIMITS OF FUNCTIONS

The following theorems give some fundamental properties associated with function limits.

Theorem 3.2.1. Let $f(x)$ and $g(x)$ be real-valued functions defined on $D \subset R$. Suppose that $\lim_{x \to a} f(x) = L$ and $\lim_{x \to a} g(x) = M$. Then

1. $\lim_{x \to a}[f(x) + g(x)] = L + M$,
2. $\lim_{x \to a}[f(x)g(x)] = LM$,
3. $\lim_{x \to a}[1/g(x)] = 1/M$ if $M \neq 0$,
4. $\lim_{x \to a}[f(x)/g(x)] = L/M$ if $M \neq 0$.

Proof. We shall only prove parts 2 and 3. The proof of part 1 is straightforward, and part 4 results from applying parts 2 and 3.

Proof of Part 2. Consider the following three cases:

CASE 1. Both L and M are finite. Let $\epsilon > 0$ be given and let $\tau > 0$ be such that

$$\tau(\tau + |L| + |M|) < \epsilon. \tag{3.8}$$

This inequality is satisfied by all values of τ for which

$$0 < \tau < \frac{-(|L| + |M|) + \sqrt{(|L| + |M|)^2 + 4\epsilon}}{2}.$$

Now, there exist $\delta_1 > \delta_2 > 0$ such that

$$|f(x) - L| < \tau \quad \text{if } 0 < |x - a| < \delta_1,$$

$$|g(x) - M| < \tau \quad \text{if } 0 < |x - a| < \delta_2.$$

Then, for any x such that $0 < |x - a| < \delta$ where $\delta = \min(\delta_1, \delta_2)$,

$$\begin{aligned}
|f(x)g(x) - LM| &= |M[f(x) - L] + f(x)[g(x) - M]| \\
&\leq \tau|M| + \tau|f(x)| \\
&\leq \tau|M| + \tau[|L| + |f(x) - L|] \\
&\leq \tau(\tau + |L| + |M|) \\
&< \epsilon,
\end{aligned}$$

which proves part 2.

CASE 2. One of L and M is finite and the other is infinite. Without any loss of generality, we assume that L is finite and $M = \infty$. Let us also assume that $L \neq 0$, since $0 \cdot \infty$ is indeterminate. Let $A > 0$ be given. There exists a $\delta_1 > 0$ such that $|f(x)| > |L|/2$ if $0 < |x - a| < \delta_1$ (why?). Also, there exists a $\delta_2 > 0$ such that $|g(x)| > 2A/|L|$ if $0 < |x - a| < \delta_2$. Let $\delta = \min(\delta_1, \delta_2)$. If $0 < |x - a| < \delta$, then

$$\begin{aligned}
|f(x)g(x)| &= |f(x)||g(x)| \\
&< \frac{|L|}{2}\frac{2}{|L|}A = A.
\end{aligned}$$

This means that $\lim_{x \to a} f(x)g(x) = \infty$, which proves part 2.

CASE 3. Both L and M are infinite. Suppose that $L = \infty$, $M = \infty$. In this case, for a given $B > 0$ there exist $\kappa_1 > 0$, $\kappa_2 > 0$ such that

$$\begin{aligned}
|f(x)| &> \sqrt{B} \qquad \text{if } 0 < |x - a| < \kappa_1, \\
|g(x)| &> \sqrt{B} \qquad \text{if } 0 < |x - a| < \kappa_2.
\end{aligned}$$

Then,

$$|f(x)g(x)| > B, \quad \text{if } 0 < |x - a| < \kappa,$$

where $\kappa = \min(\kappa_1, \kappa_2)$. This implies that $\lim_{x \to a} f(x)g(x) = \infty$, which proves part 2.

Proof of Part 3. Let $\epsilon > 0$ be given. If $M \neq 0$, then there exists a $\lambda_1 > 0$ such that $|g(x)| > |M|/2$ if $0 < |x - a| < \lambda_1$. Also, there exists a $\lambda_2 > 0$ such that $|g(x) - M| < \epsilon M^2/2$ if $0 < |x - a| < \lambda_2$. Then,

$$\begin{aligned}
\left|\frac{1}{g(x)} - \frac{1}{M}\right| &= \frac{|g(x) - M|}{|g(x)||M|} \\
&< \frac{2|g(x) - M|}{|M|^2} \\
&< \epsilon,
\end{aligned}$$

if $0 < |x - a| < \lambda$, where $\lambda = \min(\lambda_1, \lambda_2)$. $\qquad \square$

Theorem 3.2.1 is also true if L and M are one-sided limits of $f(x)$ and $g(x)$, respectively.

Theorem 3.2.2.   If $f(x) \le g(x)$, then $\lim_{x \to a} f(x) \le \lim_{x \to a} g(x)$.

Proof. Let $\lim_{x \to a} f(x) = L$, $\lim_{x \to a} g(x) = M$. Suppose that $L - M > 0$. By Theorem 3.2.1, $L - M$ is the limit of the function $h(x) = f(x) - g(x)$. Therefore, there exists a $\delta > 0$ such that

$$|h(x) - (L - M)| < \frac{L - M}{2}, \qquad (3.9)$$

if $0 < |x - a| < \delta$. Inequality (3.9) implies that $h(x) > (L - M)/2 > 0$, which is not possible, since, by assumption, $h(x) = f(x) - g(x) \le 0$. We must then have $L - M \le 0$.   □

3.3. THE o, O NOTATION

These symbols provide a convenient way to describe the limiting behavior of a function $f(x)$ as x tends to a certain limit.

Let $f(x)$ and $g(x)$ be two functions defined on $D \subset R$. The function $g(x)$ is positive and usually has a simple form such as 1, x, or $1/x$. Suppose that there exists a positive number K such that

$$|f(x)| \le Kg(x)$$

for all $x \in E$, where $E \subset D$. Then, $f(x)$ is said to be of an order of magnitude not exceeding that of $g(x)$. This fact is denoted by writing

$$f(x) = O(g(x))$$

for all $x \in E$. In particular, if $g(x) = 1$, then $f(x)$ is necessarily a bounded function on E. For example,

$$\cos x = O(1) \qquad \text{for all } x,$$
$$x = O(x^2) \qquad \text{for large values of } x,$$
$$x^2 + 2x = O(x^2) \qquad \text{for large values of } x,$$
$$\sin x = O(|x|) \qquad \text{for all } x.$$

The last relationship is true because

$$\left| \frac{\sin x}{x} \right| \le 1$$

for all values of x, where x is measured in radians.

Let us now suppose that the relationship between $f(x)$ and $g(x)$ is such that

$$\lim_{x \to a} \frac{f(x)}{g(x)} = 0.$$

Then we say that $f(x)$ is of a smaller order of magnitude than $g(x)$ in a deleted neighborhood of a. This fact is denoted by writing

$$f(x) = o(g(x)) \qquad \text{as } x \to a,$$

which is equivalent to saying that $f(x)$ tends to zero more rapidly than $g(x)$ as $x \to a$. The o symbol can also be used when x tends to infinity. In this case we write

$$f(x) = o(g(x)) \qquad \text{for } x > A,$$

where A is some positive number. For example,

$$x^2 = o(x) \qquad \text{as } x \to 0,$$
$$\tan x^3 = o(x^2) \qquad \text{as } x \to 0,$$
$$\sqrt{x} = o(x) \qquad \text{as } x \to \infty.$$

If $f(x)$ and $g(x)$ are any two functions such that

$$\frac{f(x)}{g(x)} \to 1 \qquad \text{as } x \to a,$$

then $f(x)$ and $g(x)$ are said to be asymptotically equal, written symbolically $f(x) \sim g(x)$, as $x \to a$. For example,

$$x^2 \sim x^2 + 3x + 1 \qquad \text{as } x \to \infty,$$
$$\sin x \sim x \qquad \text{as } x \to 0.$$

On the basis of the above definitions, the following properties can be deduced:

1. $O(f(x) + g(x)) = O(f(x)) + O(g(x))$.
2. $O(f(x)g(x)) = O(f(x))O(g(x))$.
3. $o(f(x)g(x)) = O(f(x))o(g(x))$.
4. If $f(x) \sim g(x)$ as $x \to a$, then $f(x) = g(x) + o(g(x))$ as $x \to a$.

3.4. CONTINUOUS FUNCTIONS

A function $f(x)$ may have a limit L as $x \to a$. This limit, however, may not be equal to the value of the function at $x = a$. In fact, the function may not even be defined at this point. If $f(x)$ is defined at $x = a$ and $L = f(a)$, then $f(x)$ is said to be continuous at $x = a$.

Definition 3.4.1. Let $f: D \to R$, where $D \subset R$, and let $a \in D$. Then $f(x)$ is continuous at $x = a$ if for every $\epsilon > 0$ there exists a $\delta > 0$ such that

$$|f(x) - f(a)| < \epsilon$$

for all $x \in D$ for which $|x - a| < \delta$.

It is important here to note that in order for $f(x)$ to be continuous at $x = a$, it is necessary that it be defined at $x = a$ as well as at all other points inside a neighborhood $N_r(a)$ of the point a for some $r > 0$. Thus to show continuity of $f(x)$ at $x = a$, the following conditions must be verified:

1. $f(x)$ is defined at all points inside a neighborhood of the point a.
2. $f(x)$ has a limit from the left and a limit from the right as $x \to a$, and that these two limits are equal to L.
3. The value of $f(x)$ at $x = a$ is equal to L.

For convenience, we shall denote the left-sided and right-sided limits of $f(x)$ as $x \to a$ by $f(a^-)$ and $f(a^+)$, respectively.

If any of the above conditions is violated, then $f(x)$ is said to be discontinuous at $x = a$. There are two kinds of discontinuity. □

Definition 3.4.2. A function $f: D \to R$ has a discontinuity of the first kind at $x = a$ if $f(a^-)$ and $f(a^+)$ exist, but at least one of them is different from $f(a)$. The function $f(x)$ has a discontinuity of the second kind at the same point if at least one of $f(a^-)$ and $f(a^+)$ does not exist. □

Definition 3.4.3. A function $f: D \to R$ is continuous on $E \subset D$ if it is continuous at every point of E.

For example, the function

$$f(x) = \begin{cases} \dfrac{x-1}{x^2-1}, & x \ge 0,\ x \neq 1, \\ \dfrac{1}{2}, & x = 1, \end{cases}$$

is defined for all $x \ge 0$ and is continuous at $x = 1$. This is true because, as was shown in Example 3.1.1,

$$\lim_{x \to 1} f(x) = \lim_{x \to 1} \frac{1}{x+1} = \frac{1}{2},$$

which is equal to the value of the function at $x = 1$. Furthermore, $f(x)$ is continuous at all other points of its domain. Note that if $f(1)$ were different from $\frac{1}{2}$, then $f(x)$ would have a discontinuity of the first kind at $x = 1$.

Let us now consider the function

$$f(x) = \begin{cases} x + 1, & x > 0, \\ 0, & x = 0, \\ x - 1, & x < 0. \end{cases}$$

This function is continuous everywhere except at $x = 0$, since it has no limit as $x \to 0$ by the fact that $f(0^-) = -1$ and $f(0^+) = 1$. The discontinuity at this point is therefore of the first kind.

An example of a discontinuity of the second kind is given by the function

$$f(x) = \begin{cases} \cos\dfrac{1}{x}, & x \neq 0, \\ 0, & x = 0, \end{cases}$$

which has a discontinuity of the second kind at $x = 0$, since neither $f(0^-)$ nor $f(0^+)$ exists.

Definition 3.4.4. A function $f: D \to R$ is left-continuous at $x = a$ if $\lim_{x \to a^-} f(x) = f(a)$. It is right-continuous at $x = a$ if $\lim_{x \to a^+} f(x) = f(a)$.

□

Obviously, a left-continuous or a right-continuous function is not necessarily continuous. In order for $f(x)$ to be continuous at $x = a$ it is necessary and sufficient that $f(x)$ be both left-continuous and right-continuous at this point.

For example, the function

$$f(x) = \begin{cases} x - 1, & x \leq 0, \\ 1, & x > 0 \end{cases}$$

is left-continuous at $x = 0$, since $f(0^-) = -1 = f(0)$. If $f(x)$ were defined so that $f(x) = x - 1$ for $x < 0$ and $f(x) = 1$ for $x \geq 0$, then it would be right-continuous at $x = 0$.

Definition 3.4.5. The function $f: D \to R$ is uniformly continuous on $E \subset D$ if for every $\epsilon > 0$ there exists a $\delta > 0$ such that

$$|f(x_1) - f(x_2)| < \epsilon \tag{3.10}$$

for all $x_1, x_2 \in E$ for which $|x_1 - x_2| < \delta$. □

This definition appears to be identical to the definition of continuity. That is not exactly the case. Uniform continuity is always associated with a set such as E in Definition 3.4.5, whereas continuity can be defined at a single point a. Furthermore, inequality (3.10) is true for all pairs of points $x_1, x_2 \in E$ such that $|x_1 - x_2| < \delta$. Hence, δ depends only on ϵ, not on the particular locations of x_1, x_2. On the other hand, in the definition of continuity (Definition 3.4.1) δ depends on ϵ as well as on the location of the point where continuity is considered. In other words, δ can change from one point to another for the same given $\epsilon > 0$. If, however, for a given $\epsilon > 0$, the same δ can be used with all points in some set $E \subset D$, then $f(x)$ is uniformly continuous on E. For this reason, whenever $f(x)$ is uniformly continuous on E, δ can be described as being "portable," which means it can be used everywhere inside E provided that $\epsilon > 0$ remains unchanged.

Obviously, if $f(x)$ is uniformly continuous on E, then it is continuous there. The converse, however, is not true. For example, consider the function $f: (0, 1) \to R$ given by $f(x) = 1/x$. Here, $f(x)$ is continuous at all points of $E = (0, 1)$, but is not uniformly continuous there. To demonstrate this fact, let us first show that $f(x)$ is continuous on E. Let $\epsilon > 0$ be given and let $a \in E$. Since $a > 0$, there exists a $\delta_1 > 0$ such that the neighborhood $N_{\delta_1}(a)$ is a subset of E. This can be accomplished by choosing δ_1 such that $0 < \delta_1 < a$. Now, for all $x \in N_{\delta_1}(a)$,

$$\left| \frac{1}{x} - \frac{1}{a} \right| = \frac{|x - a|}{ax} < \frac{\delta_1}{a(a - \delta_1)}.$$

Let $\delta_2 > 0$ be such that for the given $\epsilon > 0$,

$$\frac{\delta_2}{a(a - \delta_2)} < \epsilon,$$

which can be satisfied by requiring that

$$0 < \delta_2 < \frac{a^2 \epsilon}{1 + a\epsilon}.$$

Since

$$\frac{a^2 \epsilon}{1 + a\epsilon} < a,$$

then

$$\left| \frac{1}{x} - \frac{1}{a} \right| < \epsilon \qquad (3.11)$$

if $|x - a| < \delta$, where

$$\delta < \min\left(a, \frac{a^2\epsilon}{1 + a\epsilon}\right) = \frac{a^2\epsilon}{1 + a\epsilon}.$$

It follows that $f(x) = 1/x$ is continuous at every point of E. We note here the dependence of δ on both ϵ and a.

Let us now demonstrate that $f(x) = 1/x$ is not uniformly continuous on E. Define G to be the set

$$G = \left\{\frac{a^2\epsilon}{1 + a\epsilon}\,\middle|\, a \in E\right\}.$$

In order for $f(x) = 1/x$ to be uniformly continuous on E, the infimum of G must be positive. If this is possible, then for a given $\epsilon > 0$, (3.11) will be satisfied by all x for which

$$|x - a| < \inf(G),$$

and for all $a \in (0, 1)$. However, this cannot happen, since $\inf(G) = 0$. Thus it is not possible to find a single δ for which (3.11) will work for all $a \in (0, 1)$.

Let us now try another function defined on the same set $E = (0, 1)$, namely, the function $f(x) = x^2$. In this case, for a given $\epsilon > 0$, let $\delta > 0$ be such that

$$\delta^2 + 2\delta a - \epsilon < 0. \tag{3.12}$$

Then, for any $a \in E$, if $|x - a| < \delta$ we get

$$\begin{aligned}
|x^2 - a^2| &= |x - a|\,|x + a| \\
&= |x - a|\,|x - a + 2a| \\
&\le \delta(\delta + 2a) < \epsilon. \tag{3.13}
\end{aligned}$$

It is easy to see that this inequality is satisfied by all $\delta > 0$ for which

$$0 < \delta < -a + \sqrt{a^2 + \epsilon}. \tag{3.14}$$

If H is the set

$$H = \left\{-a + \sqrt{a^2 + \epsilon}\,\middle|\, a \in E\right\},$$

then it can be verified that $\inf(H) = -1 + \sqrt{1 + \epsilon}$. Hence, by choosing δ such that

$$\delta \le \inf(H),$$

inequality (3.14), and hence (3.13), will be satisfied for all $a \in E$. The function $f(x) = x^2$ is therefore uniformly continuous on E.

The above examples demonstrate that continuity and uniform continuity on a set E are not always equivalent. They are, however, equivalent under certain conditions on E. This will be illustrated in Theorem 3.4.6 in the next subsection.

3.4.1. Some Properties of Continuous Functions

Continuous functions have some interesting properties, some of which are given in the following theorems:

Theorem 3.4.1. Let $f(x)$ and $g(x)$ be two continuous functions defined on a set $D \subset R$. Then:

1. $f(x) + g(x)$ and $f(x)g(x)$ are continuous on D.
2. $af(x)$ is continuous on D, where a is a constant.
3. $f(x)/g(x)$ is continuous on D provided that $g(x) \neq 0$ on D.

Proof. The proof is left as an exercise. $\square$

Theorem 3.4.2. Suppose that $f: D \to R$ is continuous on D, and $g: f(D) \to R$ is continuous on $f(D)$, the image of D under f. Then the composite function $h: D \to R$ defined as $h(x) = g[f(x)]$ is continuous on D.

Proof. Let $\epsilon > 0$ be given, and let $a \in D$. Since g is continuous at $f(a)$, there exists a $\delta' > 0$ such that $|g[f(x)] - g[f(a)]| < \epsilon$ if $|f(x) - f(a)| < \delta'$. Since $f(x)$ is continuous at $x = a$, there exists a $\delta > 0$ such that $|f(x) - f(a)| < \delta'$ if $|x - a| < \delta$. It follows that by taking $|x - a| < \delta$ we must have $|h(x) - h(a)| < \epsilon$. $\square$

Theorem 3.4.3. If $f(x)$ is continuous at $x = a$ and $f(a) > 0$, then there exists a neighborhood $N_\delta(a)$ in which $f(x) > 0$.

Proof. Since $f(x)$ is continuous at $x = a$, there exists a $\delta > 0$ such that

$$|f(x) - f(a)| < \tfrac{1}{2}f(a),$$

if $|x - a| < \delta$. This implies that

$$f(x) > \tfrac{1}{2}f(a) > 0$$

for all $x \in N_\delta(a)$. $\square$

Theorem 3.4.4 (The Intermediate-Value Theorem). Let $f: D \to R$ be continuous, and let $[a, b]$ be a closed interval contained in D. Suppose that

$f(a) < f(b)$. If λ is a number such that $f(a) < \lambda < f(b)$, then there exists a point c, where $a < c < b$, such that $\lambda = f(c)$.

Proof. Let $g: D \to R$ be defined as $g(x) = f(x) - \lambda$. This function is continuous and is such that $g(a) < 0, g(b) > 0$. Consider the set

$$S = \{x \in [a, b] | g(x) < 0\}.$$

This set is nonempty, since $a \in S$ and is bounded from above by b. Hence, by Theorem 1.5.1 the least upper bound of S exists. Let $c = \text{lub}(S)$. Since $S \subset [a, b]$, then $c \in [a, b]$.

Now, for every positive integer n, there exists a point $x_n \in S$ such that

$$c - \frac{1}{n} < x_n \le c.$$

Otherwise, if $x \le c - 1/n$ for all $x \in S$, then $c - 1/n$ will be an upper bound of S, contrary to the definition of c. Consequently, $\lim_{n \to \infty} x_n = c$. Since $g(x)$ is continuous on $[a, b]$, then

$$g(c) = \lim_{x_n \to c} g(x_n) \le 0, \qquad (3.15)$$

by Theorem 3.2.2 and the fact that $g(x_n) < 0$. From (3.15) we conclude that $c < b$, since $g(b) > 0$.

Let us suppose that $g(c) < 0$. Then, by Theorem 3.4.3, there exists a neighborhood $N_\delta(c)$, for some $\delta > 0$, such that $g(x) < 0$ for all $x \in N_\delta(c) \cap [a, b]$. Consequently, there exists a point $x_0 \in [a, b]$ such that $c < x_0 < c + \delta$ and $g(x_0) < 0$. This means that x_0 belongs to S and is greater than c, a contradiction. Therefore, by inequality (3.15) we must have $g(c) = 0$, that is, $f(c) = \lambda$. We note that $c > a$, since $c \ge a$, but $c \ne a$. This last is true because if $a = c$, then $g(c) < 0$, a contradiction. This completes the proof of the theorem.   □

The direct implication of the intermediate-value theorem is that a continuous function possesses the property of assuming at least once every value between any two distinct values taken inside its domain.

Theorem 3.4.5. Suppose that $f: D \to R$ is continuous and that D is closed and bounded. Then $f(x)$ is bounded in D.

Proof. Let a be the greatest lower bound of D, which exists because D is bounded. Since D is closed, then $a \in D$ (why?). Furthermore, since $f(x)$

is continuous, then for a given $\epsilon > 0$ there exists a $\delta_1 > 0$ such that

$$f(a) - \epsilon < f(x) < f(a) + \epsilon$$

if $|x - a| < \delta_1$. The function $f(x)$ is therefore bounded in $N_{\delta_1}(a)$. Define $\mathscr{A}$ to be the set

$$\mathscr{A} = \{ x \in D | f(x) \text{ is bounded} \}.$$

This set is nonempty and bounded, and $N_{\delta_1}(a) \cap D \subset \mathscr{A}$. We need to show that $D - \mathscr{A}$ is an empty set.

As before, the least upper bound of $\mathscr{A}$ exists (since it is bounded) and belongs to D (since D is closed). Let $c = \text{lub}(\mathscr{A})$. By the continuity of $f(x)$, there exists a neighborhood $N_{\delta_2}(c)$ in which $f(x)$ is bounded for some $\delta_2 > 0$. If $D - \mathscr{A}$ is nonempty, then $N_{\delta_2}(c) \cap (D - \mathscr{A})$ is also nonempty [if $N_{\delta_2}(c) \subset \mathscr{A}$, then $c + (\delta_2/2) \in \mathscr{A}$, a contradiction]. Let $x_0 \in N_{\delta_2}(c) \cap (D - \mathscr{A})$. Then, on one hand, $f(x_0)$ is not bounded, since $x_0 \in (D - \mathscr{A})$. On the other hand, $f(x_0)$ must be bounded, since $x_0 \in N_{\delta_2}(c)$. This contradiction leads us to conclude that $D - \mathscr{A}$ must be empty and that $f(x)$ is bounded in D. $\qquad \square$

Corollary 3.4.1. If $f: D \to R$ is continuous, where D is closed and bounded, then $f(x)$ achieves its infimum and supremum at least once in D, that is, there exists $\xi, \eta \in D$ such that

$$f(\xi) \leq f(x) \qquad \text{for all } x \in D,$$
$$f(\eta) \geq f(x) \qquad \text{for all } x \in D.$$

Equivalently,

$$f(\xi) = \inf_{x \in D} f(x),$$
$$f(\eta) = \sup_{x \in D} f(x).$$

Proof. By Theorem 3.4.5, $f(D)$ is a bounded set. Hence, its least upper bound exists. Let $M = \text{lub } f(D)$, which is the same as $\sup_{x \in D} f(x)$. If there exists no point x in D for which $f(x) = M$, then $M - f(x) > 0$ for all $x \in D$. Consequently, $1/[M - f(x)]$ is continuous on D by Theorem 3.4.1, and is hence bounded there by Theorem 3.4.5.

Now, if $\delta > 0$ is any given positive number, we can find a value x for which $f(x) > M - \delta$, or

$$\frac{1}{M - f(x)} > \frac{1}{\delta}.$$

This implies that $1/[M - f(x)]$ is not bounded, a contradiction. Therefore, there must exist a point $\eta \in D$ at which $f(\eta) = M$.

The proof concerning the existence of a point $\xi \in D$ such that $f(\xi) = \inf_{x \in D} f(x)$ is similar. $\square$

The requirement that D be closed in Corollary 3.4.1 is essential. For example, the function $f(x) = 2x - 1$, which is defined on $D = \{x \mid 0 < x < 1\}$, cannot achieve its infimum, namely -1, in D. For if there exists a $\xi \in D$ such that $f(\xi) \leq 2x - 1$ for all $x \in D$, then there exists a $\delta > 0$ such that $0 < \xi - \delta$. Hence,

$$f(\xi - \delta) = 2\xi - 2\delta - 1 < f(\xi),$$

a contradiction.

Theorem 3.4.6. Let $f: D \rightarrow R$ be continuous on D. If D is closed and bounded, then f is uniformly continuous on D.

Proof. Suppose that f is not uniformly continuous. Then, by using the logical negation of the statement concerning uniform continuity in Definition 3.4.5, we may conclude that there exists an $\epsilon > 0$ such that for every $\delta > 0$, we can find $x_1, x_2 \in D$ with $|x_1 - x_2| < \delta$ for which $|f(x_1) - f(x_2)| \geq \epsilon$. On this basis, by choosing $\delta = 1$, we can find $u_1, v_1 \in D$ with $|u_1 - v_1| < 1$ for which $|f(u_1) - f(v_1)| \geq \epsilon$. Similarly, we can find $u_2, v_2 \in D$ with $|u_2 - v_2| < \frac{1}{2}$ for which $|f(u_2) - f(v_2)| \geq \epsilon$. By continuing in this process we can find $u_n, v_n \in D$ with $|u_n - v_n| < 1/n$ for which $|f(u_n) - f(v_n)| \geq \epsilon$, $n = 3, 4, \ldots$.

Now, let S be the set

$$S = \{u_n \mid n = 1, 2, \ldots\}$$

This set is bounded, since $S \subset D$. Hence, its least upper bound exists. Let $c = \text{lub}(S)$. Since D is closed, then $c \in D$. Thus, as in the proof of Theorem 3.4.4, we can find points $u_{n_1}, u_{n_2}, \ldots, u_{n_k}, \ldots$ in S such that $\lim_{k \rightarrow \infty} u_{n_k} = c$. Since $f(x)$ is continuous, there exists a $\delta' > 0$ such that

$$|f(x) - f(c)| < \frac{\epsilon}{2},$$

if $|x - c| < \delta'$ for any given $\epsilon > 0$. Let us next choose k large enough such that if $n_k > N$, where N is some large positive number, then

$$|u_{n_k} - c| < \frac{\delta'}{2} \quad \text{and} \quad \frac{1}{n_k} < \frac{\delta'}{2}. \tag{3.16}$$

Since $|u_{n_k} - v_{n_k}| < 1/n_k$, then

$$|v_{n_k} - c| \leq |u_{n_k} - v_{n_k}| + |u_{n_k} - c|$$

$$< \frac{1}{n_k} + \frac{\delta'}{2} < \delta' \tag{3.17}$$

for $n_k > N$. From (3.16) and (3.17) and the continuity of $f(x)$ we conclude that

$$\left|f(u_{n_k}) - f(c)\right| < \frac{\epsilon}{2} \quad \text{and} \quad \left|f(v_{n_k}) - f(c)\right| < \frac{\epsilon}{2}.$$

Thus,

$$\left|f(u_{n_k}) - f(v_{n_k})\right| \leq \left|f(u_{n_k}) - f(c)\right| + \left|f(v_{n_k}) - f(c)\right|$$

$$< \epsilon. \tag{3.18}$$

However, as was seen earlier,

$$\left|f(u_n) - f(v_n)\right| \geq \epsilon,$$

hence,

$$\left|f(u_{n_k}) - f(v_{n_k})\right| \geq \epsilon,$$

which contradicts (3.18). This leads us to assert that $f(x)$ is uniformly continuous on D. □

3.4.2. Lipschitz Continuous Functions

Lipschitz continuity is a specialized form of uniform continuity.

Definition 3.4.6. The function $f: D \to R$ is said to satisfy the Lipschitz condition on a set $E \subset D$ if there exist constants, K and α, where $K > 0$ and $0 < \alpha \leq 1$ such that

$$\left|f(x_1) - f(x_2)\right| \leq K|x_1 - x_2|^\alpha$$

for all $x_1, x_2 \in E$. □

Notationally, whenever $f(x)$ satisfies the Lipschitz condition with constants K and α on a set E, we say that it is Lip(K, α) on E. In this case, $f(x)$ is called a Lipschitz continuous function. It is easy to see that a Lipschitz continuous function on E is also uniformly continuous there.

As an example of a Lipschitz continuous function, consider $f(x) = \sqrt{x}$, where $x \geq 0$. We claim that $\sqrt{x}$ is Lip($1, \frac{1}{2}$) on its domain. To show this, we first write

$$\left|\sqrt{x_1} - \sqrt{x_2}\right| \leq \sqrt{x_1} + \sqrt{x_2}.$$

Hence,

$$\left| \sqrt{x_1} - \sqrt{x_2} \right|^2 \leq |x_1 - x_2|.$$

Thus,

$$\left| \sqrt{x_1} - \sqrt{x_2} \right| \leq |x_1 - x_2|^{1/2},$$

which proves our claim.

3.5. INVERSE FUNCTIONS

From Chapter 1 we recall that one of the basic characteristics of a function $y = f(x)$ is that two values of y are equal if they correspond to the same value of x. If we were to reverse the roles of x and y so that two values of x are equal whenever they correspond to the same value of y, then x becomes a function of y. Such a function is called the inverse function of f and is denoted by f^{-1}. We conclude that the inverse of $f: D \to R$ exists if and only if f is one-to-one.

Definition 3.5.1.   Let $f: D \to R$. If there exists a function $f^{-1}: f(D) \to D$ such that $f^{-1}[f(x)] = x$ and all $x \in D$ and $f[f^{-1}(y)] = y$ for all $y \in f(D)$, then f^{-1} is called the inverse function of f.   □

Definition 3.5.2.   Let $f: D \to R$. Then, f is said to be monotone increasing [decreasing] on D if whenever $x_1, x_2 \in D$ are such that $x_1 < x_2$, then $f(x_1) \leq f(x_2)$ $[f(x_1) \geq f(x_2)]$. The function f is strictly monotone increasing [decreasing] on D if $f(x_1) < f(x_2)$ $[f(x_1) > f(x_2)]$ whenever $x_1 < x_2$.
          □

If f is either monotone increasing or monotone decreasing on D, then it is called a monotone function on D. In particular, if it is either strictly monotone increasing or strictly monotone decreasing, then $f(x)$ is strictly monotone on D.

Strictly monotone functions have the property that their inverse functions exist. This will be shown in the next theorem.

Theorem 3.5.1.   Let $f: D \to R$ be strictly monotone increasing (or decreasing) on D. Then, there exists a unique inverse function f^{-1}, which is strictly monotone increasing (or decreasing) on $f(D)$.

Proof. Let us suppose that f is strictly monotone increasing on D. To show that f^{-1} exists as a strictly monotone increasing function on $f(D)$.

Suppose that $x_1, x_2 \in D$ are such that $f(x_1) = f(x_2) = y$. If $x_1 \neq x_2$, then $x_1 < x_2$ or $x_2 < x_1$. Since f is strictly monotone increasing, then $f(x_1) < f(x_2)$ or $f(x_2) < f(x_1)$. In either case, $f(x_1) \neq f(x_2)$, which contradicts the assumption that $f(x_1) = f(x_2)$. Hence, $x_1 = x_2$, that is, f is one-to-one and has therefore a unique inverse f^{-1}.

The inverse f^{-1} is strictly monotone increasing on $f(D)$. To show this, suppose that $f(x_1) < f(x_2)$. Then, $x_1 < x_2$. If not, we must have $x_1 \geq x_2$. In this case, $f(x_1) = f(x_2)$ when $x_1 = x_2$, or $f(x_1) > f(x_2)$ when $x_1 > x_2$, since f is strictly monotone increasing. However, this is contrary to the assumption that $f(x_1) < f(x_2)$. Thus $x_1 < x_2$ and f^{-1} is strictly monotone increasing.

The proof of Theorem 3.5.1 when "increasing" is replaced with "decreasing" is similar. $\square$

Theorem 3.5.2.   Suppose that $f: D \to R$ is continuous and strictly monotone increasing (decreasing) on $[a, b] \subset D$. Then, f^{-1} is continuous and strictly monotone increasing (decreasing) on $f([a, b])$.

Proof. By Theorem 3.5.1 we only need to show the continuity of f^{-1}. Suppose that f is strictly monotone increasing. The proof when f is strictly monotone decreasing is similar.

Since f is continuous on a closed and bounded interval, then by Corollary 3.4.1 it must achieve its infimum and supremum on $[a, b]$. Furthermore, because f is strictly monotone increasing, its infimum and supremum must be attained at only a and b, respectively. Thus

$$f([a, b]) = [f(a), f(b)].$$

Let $d \in [f(a), f(b)]$. There exists a unique value c, $a \leq c \leq b$, such that $f(c) = d$. For any $\epsilon > 0$, let τ be defined as

$$\tau = \min[f(c) - f(c - \epsilon), f(c + \epsilon) - f(c)].$$

Then there exists a δ, $0 < \delta < \tau$, such that all the x's in $[a, b]$ that satisfy the inequality

$$|f(x) - d| < \delta$$

must also satisfy the inequality

$$|x - c| < \epsilon.$$

This is true because

$$f(c) - f(c) + f(c - \epsilon) < d - \delta < f(x) < d + \delta$$
$$< f(c) + f(c + \epsilon) - f(c),$$

that is,

$$f(c - \epsilon) < f(x) < f(c + \epsilon).$$

Using the fact that f^{-1} is strictly monotone increasing (by Theorem 3.5.1), we conclude that

$$c - \epsilon < x < c + \epsilon,$$

that is, $|x - c| < \epsilon$. It follows that $x = f^{-1}(y)$ is continuous on $[f(a), f(b)]$.
□

Note that in general if $y = f(x)$, the equation $y - f(x) = 0$ may not produce a unique solution for x in terms of y. If, however, the domain of f can be partitioned into subdomains on each of which f is strictly monotone, then f can have an inverse on each of these subdomains.

EXAMPLE 3.5.1.   Consider the function $f: R \to R$ defined by $y = f(x) = x^3$. It is easy to see that f is strictly monotone increasing for all $x \in R$. It therefore has a unique inverse given by $f^{-1}(y) = y^{1/3}$.

EXAMPLE 3.5.2.   Let $f: [-1, 1] \to R$ be such that $y = f(x) = x^5 - x$. From Figure 3.2 it can be seen that f is strictly monotone increasing on $D_1 = [-1, -5^{-1/4}]$ and $D_2 = [5^{-1/4}, 1]$, but is strictly monotone decreasing on $D_3 = [-5^{-1/4}, 5^{-1/4}]$. This function has therefore three inverses, one on each of D_1, D_2, and D_3. By Theorem 3.5.2, all three inverses are continuous.

EXAMPLE 3.5.3.   Let $f: R \to [-1, 1]$ be the function $y = f(x) = \sin x$, where x is measured in radians. There is no unique inverse on R, since the sine function is not strictly monotone there. If, however, we restrict the domain of f to $D = [-\pi/2, \pi/2]$, then f is strictly monotone increasing there and has the unique inverse $f^{-1}(y) = \text{Arcsin } y$ (see Example 1.3.4). The inverse of f on $[\pi/2, 3\pi/2]$ is given by $f^{-1}(y) = \pi - \text{Arcsin } y$. We can similarly find the inverse of f on $[3\pi/2, 5\pi/2]$, $[5\pi/2, 7\pi/2]$, etc.

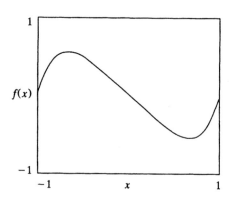

Figure 3.2. The graph of the function $f(x) = x^5 - x$.

3.6. CONVEX FUNCTIONS

Convex functions are frequently used in operations research. They also happen to be continuous, as will be shown in this section. The natural domains for such functions are convex sets.

Definition 3.6.1. A set $D \subset R$ is convex if $\lambda x_1 + (1 - \lambda)x_2 \in D$ whenever x_1, x_2 belong to D and $0 \leq \lambda \leq 1$. Geometrically, a convex set contains the line segment connecting any two of its points. The same definition actually applies to convex sets in R^n, the n-dimensional Euclidean space ($n \geq 2$). For example, each of the following sets is convex:

1. Any interval in R.
2. Any sphere in R^3, and in general, any hypersphere in R^n, $n \geq 4$.
3. The set $\{(x, y) \in R^2| \ |x| + |y| \leq 1\}$. See Figure 3.3.   □

Definition 3.6.2. A function $f: D \to R$ is convex if

$$f[\lambda x_1 + (1 - \lambda)x_2] \leq \lambda f(x_1) + (1 - \lambda)f(x_2) \qquad (3.19)$$

for all $x_1, x_2 \in D$ and any λ such that $0 \leq \lambda \leq 1$. The function f is strictly convex if inequality (3.19) is strict for $x_1 \neq x_2$.

Geometrically, inequality (3.19) means that if P and Q are any two points on the graph of $y = f(x)$, then the portion of the graph between P and Q lies below the chord PQ (see Figure 3.4). Examples of convex functions include $f(x) = x^2$ on R, $f(x) = \sin x$ on $[\pi, 2\pi]$, $f(x) = e^x$ on R, $f(x) = -\log x$ for $x > 0$, to name just a few.   □

Definition 3.6.3. A function $f: D \to R$ is concave if $-f$ is convex.   □

We note that if $f: [a, b] \to R$ is convex and the values of f at a and b are finite, then $f(x)$ is bounded from above on $[a, b]$ by $M = \max\{f(a), f(b)\}$. This is true because if $x \in [a, b]$, then $x = \lambda a + (1 - \lambda)b$ for some $\lambda \in [0, 1]$,

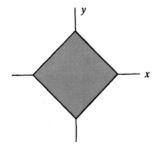

Figure 3.3. The set $\{(x, y) \in R^2| \ |x| + |y| \leq 1\}$.

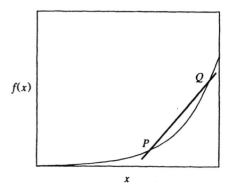

Figure 3.4. The graph of a convex function.

since $[a, b]$ is a convex set. Hence,

$$f(x) \le \lambda f(a) + (1 - \lambda) f(b)$$
$$\le \lambda M + (1 - \lambda) M = M.$$

The function $f(x)$ is also bounded from below. To show this, we first note that any $x \in [a, b]$ can be written as

$$x = \frac{a + b}{2} + t,$$

where

$$a - \frac{a + b}{2} \le t \le b - \frac{a + b}{2}.$$

Now,

$$f\left(\frac{a + b}{2}\right) \le \frac{1}{2} f\left(\frac{a + b}{2} + t\right) + \frac{1}{2} f\left(\frac{a + b}{2} - t\right), \qquad (3.20)$$

since if $(a + b)/2 + t$ belongs to $[a, b]$, then so does $(a + b)/2 - t$. From (3.20) we then have

$$f\left(\frac{a + b}{2} + t\right) \ge 2f\left(\frac{a + b}{2}\right) - f\left(\frac{a + b}{2} - t\right).$$

Since

$$f\left(\frac{a + b}{2} - t\right) \le M,$$

then

$$f\left(\frac{a + b}{2} + t\right) \ge 2f\left(\frac{a + b}{2}\right) - M,$$

that is, $f(x) \geq m$ for all $x \in [a, b]$, where

$$m = 2f\left(\frac{a+b}{2}\right) - M.$$

Another interesting property of convex functions is given in Theorem 3.6.1.

Theorem 3.6.1. Let $f: D \to R$ be a convex function, where D is an open interval. Then f is Lip$(K, 1)$ on any closed interval $[a, b]$ contained in D, that is,

$$|f(x_1) - f(x_2)| \leq K|x_1 - x_2| \qquad (3.21)$$

for all $x_1, x_2 \in [a, b]$.

Proof. Consider the closed interval $[a - \epsilon, b + \epsilon]$, where $\epsilon > 0$ is chosen so that this interval is contained in D. Let m' and M' be, respectively, the lower and upper bounds of f (as was seen earlier) on $[a - \epsilon, b + \epsilon]$. Let x_1, x_2 be any two distinct points in $[a, b]$. Define z_1 and λ as

$$z_1 = x_2 + \frac{\epsilon(x_2 - x_1)}{|x_1 - x_2|},$$

$$\lambda = \frac{|x_1 - x_2|}{\epsilon + |x_1 - x_2|}.$$

Then $z_1 \in [a - \epsilon, b + \epsilon]$. This is true because $(x_1 - x_1)/|x_1 - x_2|$ is either equal to 1 or to -1. Since $x_2 \in [a, b]$, then

$$a - \epsilon \leq x_2 - \epsilon \leq x_2 + \frac{\epsilon(x_2 - x_1)}{|x_1 - x_2|} \leq x_2 + \epsilon \leq b + \epsilon.$$

Furthermore, it can be verified that

$$x_2 = \lambda z_1 + (1 - \lambda) x_1.$$

We then have

$$f(x_2) \leq \lambda f(z_1) + (1 - \lambda) f(x_1) = \lambda[f(z_1) - f(x_1)] + f(x_1).$$

Thus,

$$f(x_2) - f(x_1) \leq \lambda[f(z_1) - f(x_1)]$$
$$\leq \lambda[M' - m']$$
$$\leq \frac{|x_1 - x_2|}{\epsilon}(M' - m') = K|x_1 - x_2|, \qquad (3.22)$$

where $K = (M' - m')/\epsilon$. Since inequality (3.22) is true for any $x_1, x_2 \in [a, b]$, we must also have

$$f(x_1) - f(x_2) \leq K|x_1 - x_2|. \tag{3.23}$$

From inequalities (3.22) and (3.23) we conclude that

$$|f(x_1) - f(x_2)| \leq K|x_1 - x_2|$$

for any $x_1, x_2 \in [a, b]$, which shows that $f(x)$ is Lip$(K, 1)$ on $[a, b]$. $\square$

Using Theorem 3.6.1 it is easy to prove the following corollary:

Corollary 3.6.1. Let $f: D \to R$ be a convex function, where D is an open interval. If $[a, b]$ is any closed interval contained in D, then $f(x)$ is uniformly continuous on $[a, b]$ and is therefore continuous on D.

Note that if $f(x)$ is convex on (a, b), then it does not have to be continuous at the end points of the interval. It is easy to see, for example, that the function $f: [-1, 1] \to R$ defined as

$$f(x) = \begin{cases} x^2, & -1 < x < 1, \\ 2, & x = 1, -1 \end{cases}$$

is convex on $(-1, 1)$, but is discontinuous at $x = -1, 1$.

3.7. CONTINUOUS AND CONVEX FUNCTIONS IN STATISTICS

The most vivid examples of continuous functions in statistics are perhaps the cumulative distribution functions of continuous random variables. If X is a continuous random variable, then its cumulative distribution function

$$F(x) = P(X \leq x)$$

is continuous on R. In this case,

$$P(X = a) = \lim_{n \to \infty} \left[F\left(a + \frac{1}{n}\right) - F\left(a - \frac{1}{n}\right) \right] = 0,$$

that is, the distribution of X assigns a zero probability to any single value. This is a basic characteristic of continuous random variables.

Examples of continuous distributions include the beta, Cauchy, chi-squared, exponential, gamma, Laplace, logistic, lognormal, normal, t, uniform, and the Weibull distributions. Most of these distributions are described in introductory statistics books. A detailed account of their properties and uses is given in the two books by Johnson and Kotz (1970a, 1970b).

It is interesting to note that if X is any random variable (not necessarily continuous), then its cumulative distribution function, $F(x)$, is right-continuous on R (see, for example, Harris, 1966, page 55). This function is also monotone increasing on R. If $F(x)$ is strictly monotone increasing, then by Theorem 3.5.1 it has a unique inverse $F^{-1}(y)$. In this case, if Y has the uniform distribution over the open interval $(0, 1)$, then the random variable $F^{-1}(Y)$ has the cumulative distribution function $F(x)$. To show this, consider $X = F^{-1}(Y)$. Then,

$$P(X \leq x) = P\left[F^{-1}(Y) \leq x\right]$$
$$= P[Y \leq F(x)]$$
$$= F(x).$$

This result has an interesting application in sampling. If $Y_1, Y_2, \ldots, Y_n$ form an independent random sample from the uniform distribution $U(0, 1)$, then $F^{-1}(Y_1), F^{-1}(Y_2), \ldots, F^{-1}(Y_n)$ will form an independent sample from a distribution with the cumulative distribution function $F(x)$. In other words, samples from any distribution can be generated through sampling from the uniform distribution $U(0, 1)$. This result forms the cornerstone of Monte Carlo simulation in statistics. Such a method provides an artificial way of collecting "data." There are situations where the actual taking of a physical sample is either impossible or too expensive. In such situations, useful information can often be derived from simulated sampling. Monte Carlo simulation is also used in the study of the relative performance of test statistics and parameter estimators when the data come from certain specified parent distributions.

Another example of the use of continuous functions in statistics is in limit theory. For example, it is known that if $\{X_n\}_{n=1}^{\infty}$ is a sequence of random variables that converges in probability to c, and if $g(x)$ is a continuous function at $x = c$, then the random variable $g(X_n)$ converges in probability to $g(c)$ as $n \to \infty$. By definition, a sequence of random variables $\{X_n\}_{n=1}^{\infty}$ converges in probability to a constant c if for a given $\epsilon > 0$,

$$\lim_{n \to \infty} P(|X_n - c| \geq \epsilon) = 0.$$

In particular, if $\{X_n\}_{n=1}^{\infty}$ is a sequence of estimators of a parameter c, then X_n is said to be a consistent estimator of c if X_n converges in probability to c. For example, the sample mean

$$\bar{X}_n = \frac{1}{n} \sum_{i=1}^{n} X_i$$

of a sample of size n from a population with a finite mean μ is a consistent estimator of μ according to the law of large numbers (see, for example, Lindgren, 1976, page 155). Other types of convergence in statistics can be found in standard mathematical statistics books (see the annotated bibliography).

Convex functions also play an important role in statistics, as can be seen from the following examples.

If $f(x)$ is a convex function and X is a random variable with a finite mean $\mu = E(X)$, then

$$E[f(X)] \geq f[E(X)].$$

Equality holds if and only if X is constant with probability 1. This inequality is known as *Jensen's inequality*. If f is strictly convex, the inequality is strict unless X is constant with probability 1. A proof of Jensen's inequality is given in Section 6.7.4. See also Lehmann (1983, page 50).

Jensen's inequality has useful applications in statistics. For example, it can be used to show that if $x_1, x_2, \ldots, x_n$ are n positive scalars, then their arithmetic mean is greater than or equal to their geometric mean, which is equal to $(\Pi_{i=1}^{n} x_i)^{1/n}$. This can be shown as follows:

Consider the convex function $f(x) = -\log x$. Let X be a discrete random variable that takes the values $x_1, x_2, \ldots, x_n$ with probabilities equal to $1/n$, that is,

$$P(X = x) = \begin{cases} \dfrac{1}{n}, & x = x_1, x_2, \ldots, x_n \\ 0 & \text{otherwise.} \end{cases}$$

Then, by Jensen's inequality,

$$E(-\log X) \geq -\log E(X). \tag{3.24}$$

However,

$$E(-\log X) = -\frac{1}{n} \sum_{i=1}^{n} \log x_i, \tag{3.25}$$

and

$$-\log E(X) = -\log\left(\frac{1}{n} \sum_{i=1}^{n} x_i\right). \tag{3.26}$$

By using (3.25) and (3.26) in (3.24) we get

$$\frac{1}{n} \sum_{i=1}^{n} \log x_i \le \log\left(\frac{1}{n} \sum_{i=1}^{n} x_i\right),$$

or

$$\log\left[\left(\prod_{i=1}^{n} x_i\right)^{1/n}\right] \le \log\left(\frac{1}{n} \sum_{i=1}^{n} x_i\right).$$

Since the logarithmic function is monotone increasing, we conclude that

$$\left(\prod_{i=1}^{n} x_i\right)^{1/n} \le \frac{1}{n} \sum_{i=1}^{n} x_i.$$

Jensen's inequality can also be used to show that the arithmetic mean is greater than or equal to the harmonic mean. This assertion can be shown as follows:

Consider the function $f(x) = x^{-1}$, which is convex for $x > 0$. If X is a random variable with $P(X > 0) = 1$, then by Jensen's inequality,

$$E\left(\frac{1}{X}\right) \ge \frac{1}{E(X)}. \tag{3.27}$$

In particular, if X has the discrete distribution described earlier, then

$$E\left(\frac{1}{X}\right) = \frac{1}{n} \sum_{i=1}^{n} \frac{1}{x_i}$$

and

$$E(X) = \frac{1}{n} \sum_{i=1}^{n} x_i.$$

By substitution in (3.27) we get

$$\frac{1}{n} \sum_{i=1}^{n} \frac{1}{x_i} \ge \left(\frac{1}{n} \sum_{i=1}^{n} x_i\right)^{-1},$$

or

$$\frac{1}{n} \sum_{i=1}^{n} x_i \ge \left(\frac{1}{n} \sum_{i=1}^{n} \frac{1}{x_i}\right)^{-1}. \tag{3.28}$$

The quantity on the right of inequality (3.28) is the harmonic mean of $x_1, x_2, \ldots, x_n$.

Another example of the use of convex functions in statistics is in the general theory of estimation. Let $X_1, X_2, \ldots, X_n$ be a random sample of size n from a population whose distribution depends on an unknown parameter θ. Let $\omega(X_1, X_2, \ldots, X_n)$ be an estimator of θ. By definition, the loss function $L[\theta, \omega(X_1, X_2, \ldots, X_n)]$ is a nonnegative function that measures the loss incurred when θ is estimated by $\omega(X_1, X_2, \ldots, X_n)$. The expected value (mean) of the loss function is called the risk function, denoted by $R(\theta, \omega)$, that is,

$$ R(\theta, \omega) = E\{L[\theta, \omega(X_1, X_2, \ldots, X_n)]\}. $$

The loss function is taken to be a convex function of θ. Examples of loss functions include the squared error loss,

$$ L[\theta, \omega(X_1, X_2, \ldots, X_n)] = [\theta - \omega(X_1, X_2, \ldots, X_n)]^2, $$

and the absolute error loss,

$$ L[\theta, \omega(X_1, X_2, \ldots, X_n)] = |\theta - \omega(X_1, X_2, \ldots, X_n)|. $$

The first loss function is strictly convex, whereas the second is convex, but not strictly convex.

The goodness of an estimator of θ is judged on the basis of its risk function, assuming that a certain loss function has been selected. The smaller the risk, the more desirable the estimator. An estimator $\omega^*(X_1, X_2, \ldots, X_n)$ is said to be admissible if there is no other estimator $\omega(X_1, X_2, \ldots, X_n)$ of θ such that

$$ R(\theta, \omega) \leq R(\theta, \omega^*) $$

for all $\theta \in \Omega$ (Ω is the parameter space), with strict inequality for at least one θ. An estimator $\omega_0(X_1, X_2, \ldots, X_n)$ is said to be a minimax estimator if it minimizes the supremum (with respect to θ) of the risk function, that is,

$$ \sup_{\theta \in \Omega} R(\theta, \omega_0) \leq \sup_{\theta \in \Omega} R(\theta, \omega), $$

where $\omega(X_1, X_2, \ldots, X_n)$ is any other estimator of θ. It should be noted that a minimax estimator may not be admissible.

EXAMPLE 3.7.1.   Let $X_1, X_2, \ldots, X_{20}$ be a random sample of size 20 from the normal distribution $N(\theta, 1), -\infty < \theta < \infty$. Let $\omega_1(X_1, X_2, \ldots, X_{20}) = \overline{X}_{20}$ be the sample mean, and let $\omega_2(X_1, X_2, \ldots, X_{20}) = 0$. Then, using a squared error loss function,

$$R(\theta, \omega_1) = E\left[\left(\overline{X}_{20} - \theta\right)^2\right] = \text{Var}\left(\overline{X}_{20}\right) = \tfrac{1}{20},$$

$$R(\theta, \omega_2) = E\left[(0 - \theta)^2\right] = \theta^2.$$

In this case,

$$\sup_{\theta} \left[R(\theta, \omega_1)\right] = \tfrac{1}{20},$$

whereas

$$\sup_{\theta} \left[R(\theta, \omega_2)\right] = \infty.$$

Thus $\omega_1 = \overline{X}_{20}$ is a better estimator than $\omega_2 = 0$. It can be shown that $\overline{X}_{20}$ is the minimax estimator of θ with respect to a squared error loss function. Note, however, that $\overline{X}_{20}$ is not an admissible estimator, since

$$R(\theta, \omega_1) \leq R(\theta, \omega_2)$$

for $\theta \geq 20^{-1/2}$ or $\theta \leq -20^{-1/2}$. However, for $-20^{-1/2} < \theta < 20^{-1/2}$,

$$R(\theta, \omega_2) < R(\theta, \omega_1).$$

FURTHER READING AND ANNOTATED BIBLIOGRAPHY

Corwin, L. J., and R. H. Szczarba (1982). *Multivariable Calculus.* Dekker, New York.

Fisz, M. (1963). *Probability Theory and Mathematical Statistics,* 3rd ed. Wiley, New York. (Some continuous distributions are described in Chap. 5. Limit theorems concerning sequences of random variables are discussed in Chap. 6.)

Fulks, W. (1978). *Advanced Calculus,* 3rd ed. Wiley, New York.

Hardy, G. H. (1955). *4 Course of Pure Mathematics,* 10th ed. The University Press, Cambridge, England. (Section 89 of Chap. 4 gives definitions concerning the *o* and *O* symbols introduced in Section 3.3 of this chapter.)

Harris, B. (1966). *Theory of Probability.* Addison-Wesley, Reading, Massachusetts. (Some continuous distributions are given in Section 3.5.)

Henle, J. M., and E. M. Kleinberg (1979). *Infinitesimal Calculus.* The MIT Press, Cambridge, Massachusetts.

Hillier, F. S., and G. J. Lieberman (1967). *Introduction to Operations Research.* Holden-Day, San Francisco. (Convex sets and functions are described in Appendix 1.)

Hogg, R. V., and A. T. Craig (1965). *Introduction to Mathematical Statistics,* 2nd ed. Macmillan, New York. (Loss and risk functions are discussed in Section 9.3.)

Hyslop, J. M. (1954). *Infinite Series,* 5th ed. Oliver and Boyd, Edinburgh, England. (Chap. 1 gives definitions and summaries of results concerning the *o, O* notation.)

Johnson, N. L., and S. Kotz (1970a). *Continuous Univariate Distributions—1*. Houghton Mifflin, Boston.

Johnson, N. L., and S. Kotz (1970b). *Continuous Univariate Distributions—2*. Houghton Mifflin, Boston.

Lehmann, E. L. (1983). *Theory of Point Estimation*. Wiley, New York. (Section 1.6 discusses convex functions and their uses as loss functions.)

Lindgren, B. W. (1976). *Statistical Theory*, 3rd ed. Macmillan, New York. (The concepts of loss and utility, or negative loss, used in statistical decision theory are discussed in Chap. 8.)

Randles, R. H., and D. A. Wolfe (1979). *Introduction to the Theory of Nonparametric Statistics*. Wiley, New York. (Some mathematical statistics results, including Jensen's inequality, are given in the Appendix.)

Rao, C. R. (1973). *Linear Statistical Inference and Its Applications*, 2nd ed. Wiley, New York.

Roberts, A. W., and D. E. Varberg (1973). *Convex Functions*. Academic Press, New York. (This is a handy reference book that contains all the central facts about convex functions.)

Roussas, G. G. (1973). *A First Course in Mathematical Statistics*. Addison-Wesley, Reading, Massachusetts. (Chap. 12 defines and provides discussions concerning admissible and minimax estimators.)

Rudin, W. (1964). *Principles of Mathematical Analysis*, 2nd ed. McGraw-Hill, New York. (Limits of functions and some properties of continuous functions are given in Chap. 4.)

Sagan, H. (1974). *Advanced Calculus*. Houghton Mifflin, Boston.

Smith, D. E. (1958). *History of Mathematics*, Vol. 1. Dover, New York.

EXERCISES

In Mathematics

3.1. Determine if the following limits exist:

(a)
$$\lim_{x \to 1} \frac{x^5 - 1}{x - 1},$$

(b)
$$\lim_{x \to 0} x \sin \frac{1}{x},$$

(c)
$$\lim_{x \to 0} \left(\sin \frac{1}{x} \right) \Big/ \left(\sin \frac{1}{x} \right),$$

(d) $\lim_{x \to 0} f(x)$, where $f(x) = \begin{cases} \dfrac{x^2 - 1}{x - 1}, & x > 0, \\[2mm] \dfrac{x^3 - 1}{2(x - 1)}, & x < 0. \end{cases}$

3.2. Show that
   (a) $\tan x^3 = o(x^2)$ as $x \to 0$.
   (b) $x = o(\sqrt{x})$ as $x \to 0$,
   (c) $O(1) = o(x)$ as $x \to \infty$,
   (d) $f(x)g(x) = x^{-1} + O(1)$ as $x \to 0$, where $f(x) = x + o(x^2)$, $g(x) = x^{-2} + O(x^{-1})$.

3.3. Determine where the following functions are continuous, and indicate the points of discontinuity (if any):

   (a) $f(x) = \begin{cases} x \sin(1/x), & x \neq 0, \\ 0, & x = 0, \end{cases}$

   (b) $f(x) = \begin{cases} [(x - 1)/(2 - x)]^{1/2}, & x \neq 2, \\ 1, & x = 2, \end{cases}$

   (c) $f(x) = \begin{cases} x^{-m/n}, & x \neq 0, \\ 0, & x = 0, \end{cases}$

   where m and n are positive integers,

   (d) $f(x) = \dfrac{x^4 - 2x^2 + 3}{x^3 - 1}, \qquad x \neq 1.$

3.4. Show that $f(x)$ is continuous at $x = a$ if and only if it is both left-continuous and right-continuous at $x = a$.

3.5. Use Definition 3.4.1 to show that the function

$$f(x) = x^2 - 1$$

is continuous at any point $a \in R$.

3.6. For what values of x is

$$f(x) = \lim_{n \to \infty} \frac{3nx}{1 - nx}$$

continuous?

3.7. Consider the function

$$f(x) = \frac{x - |x|}{x}, \qquad -1 < x < 1, \qquad x \neq 0.$$

Can $f(x)$ be defined at $x = 0$ so that it will be continuous there?

3.8. Let $f(x)$ be defined for all $x \in R$ and continuous at $x = 0$. Furthermore,

$$f(a + b) = f(a) + f(b),$$

for all a, b in R. Show that $f(x)$ is uniformly continuous everywhere in R.

3.9. Let $f(x)$ be defined as

$$f(x) = \begin{cases} 2x - 1, & 0 \leq x \leq 1, \\ x^3 - 5x^2 + 5, & 1 \leq x \leq 2. \end{cases}$$

Determine if $f(x)$ is uniformly continuous on $[0, 2]$.

3.10. Show that $f(x) = \cos x$ is uniformly continuous on R.

3.11. Prove Theorem 3.4.1.

3.12. A function $f: D \to R$ is called upper semicontinuous at $a \in D$ if for a given $\epsilon > 0$ there exists a $\delta > 0$ such that

$$f(x) < f(a) + \epsilon$$

for all $x \in N_\delta(a) \cap D$. If the above inequality is replaced with

$$f(x) > f(a) - \epsilon,$$

then $f(x)$ is said to be lower semicontinuous.
   Show that if D is closed and bounded, then
(a) $f(x)$ is bounded from above on D if $f(x)$ is upper semicontinuous.
(b) $f(x)$ is bounded from below on D if $f(x)$ is lower semicontinuous.

3.13. Let $f: [a, b] \to R$ be continuous such that $f(x) = 0$ for every rational number in $[a, b]$. Show that $f(x) = 0$ for every x in $[a, b]$.

3.14. For what values of x does the function

$$f(x) = 3 + |x - 1| + |x + 1|$$

have a unique inverse?

3.15. Let $f: R \rightarrow R$ be defined as

$$f(x) = \begin{cases} x, & x \le 1, \\ 2x - 1, & x > 1. \end{cases}$$

Find the inverse function f^{-1},

3.16. Let $f(x) = 2x^2 - 8x + 8$. Find the inverse of $f(x)$ for
  (a) $x \le -2$,
  (b) $x > 2$.

3.17. Suppose that $f: [a, b] \rightarrow R$ is a convex function. Show that for a given $\epsilon > 0$ there exists a $\delta > 0$ such that

$$\sum_{i=1}^{n} |f(a_i) - f(b_i)| < \epsilon$$

for every finite, pairwise disjoint family of open subintervals $\{(a_i, b_i)\}_{i=1}^{n}$ of $[a, b]$ for which $\sum_{i=1}^{n}(b_i - a_i) < \delta$.
  Note: A function satisfying this property is said to be absolutely continuous on $[a, b]$.

3.18. Let $f: [a, b] \rightarrow R$ be a convex function. Show that if $a_1, a_2, \ldots, a_n$ are positive numbers and $x_1, x_2, \ldots, x_n$ are points in $[a, b]$, then

$$f\left(\frac{\sum_{i=1}^{n} a_i x_i}{A} \right) \le \frac{\sum_{i=1}^{n} a_i f(x_i)}{A},$$

where $A = \sum_{i=1}^{n} a_i$.

3.19. Let $f(x)$ be continuous on $D \subset R$. Let S be the set of all $x \in D$ such that $f(x) = 0$. Show that S is a closed set.

3.20. Let $f(x)$ be a convex function on $D \subset R$. Show that $\exp[f(x)]$ is also convex on D.

In Statistics

3.21. Let X be a continuous random variable with the cumulative distribution function

$$F(x) = 1 - e^{-x/\theta}, \qquad x > 0.$$

This is known as the exponential distribution. Its mean and variance are $\mu = \theta$, $\sigma^2 = \theta^2$, respectively. Generate a random sample of five observations from an exponential distribution with mean 2.

[*Hint:* Select a ten-digit number from the table of random numbers, for example, 8389611097. Divide it by 10^{10} to obtain the decimal number 0.8389611097. This number can be regarded as an observation from the uniform distribution $U(0, 1)$. Now, solve the equation $F(x) = 0.8389611097$. The resulting value of x is considered as an observation from the prescribed exponential distribution. Repeat this process four more times, each time selecting a new decimal number from the table of random numbers.]

3.22. Verify Jensen's inequality in each of the following two cases:
(a) X is normally distributed and $f(x) = |x|$.
(b) X has the exponential distribution and $f(x) = e^{-x}$.

3.23. Use the definition of convergence in probability to verify that if the sequence of random variables $\{X_n\}_{n=1}^{\infty}$ converges in probability to zero, then so does the sequence $\{X_n^2\}_{n=1}^{\infty}$.

3.24. Show that

$$E(X^2) \geq [E(|X|)]^2.$$

[*Hint:* Let $Y = |X|$. Apply Jensen's inequality to Y with $f(x) = x^2$.] Deduce that if X has a mean μ and a variance σ^2, then

$$E(|X - \mu|) \leq \sigma.$$

3.25. Consider the exponential distribution described in Exercise 3.21. Let $X_1, X_2, \ldots, X_n$ be a sample of size n from this distribution. Consider the following estimators of θ:
(a) $\omega_1(X_1, X_2, \ldots, X_n) = \overline{X}_n$, the sample mean.
(b) $\omega_2(X_1, X_2, \ldots, X_n) = \overline{X}_n + 1$,
(c) $\omega_3(X_1, X_2, \ldots, X_n) = X_n$.
Determine the risk function corresponding to a squared error loss function for each one of these estimators. Which estimator has the smallest risk for all values of θ?

CHAPTER 4

Differentiation

Differentiation originated in connection with the problems of drawing tangents to curves and of finding maxima and minima of functions. Pierre de Fermat (1601–1665), the founder of the modern theory of numbers, is credited with having put forth the main ideas on which differential calculus is based.

In this chapter, we shall introduce the notion of differentiation and study its applications in the determination of maxima and minima of functions. We shall restrict our attention to real-valued functions defined on R, the set of real numbers. The study of differentiation in connection with multivariable functions, that is, functions defined on R^n ($n \geq 1$), will be considered in Chapter 7.

4.1. THE DERIVATIVE OF A FUNCTION

The notion of differentiation was motivated by the need to find the tangent to a curve at a given point. Fermat's approach to this problem was inspired by a geometric reasoning. His method uses the idea of a tangent as the limiting position of a secant when two of its points of intersection with the curve tend to coincide. This has lead to the modern notation associated with the derivative of a function, which we now introduce.

Definition 4.1.1. Let $f(x)$ be a function defined in a neighborhood $N_r(x_0)$ of a point x_0. Consider the ratio

$$\phi(h) = \frac{f(x_0 + h) - f(x_0)}{h}, \tag{4.1}$$

where h is a nonzero increment of x_0 such that $-r < h < r$. If $\phi(h)$ has a limit as $h \to 0$, then this limit is called the derivative of $f(x)$ at x_0 and is denoted by $f'(x_0)$. It is also common to use the notation

$$\frac{df(x)}{dx}\bigg|_{x=x_0} = f'(x_0).$$

We thus have

$$f'(x_0) = \lim_{h \to 0} \frac{f(x_0 + h) - f(x_0)}{h}. \tag{4.2}$$

By putting $x = x_0 + h$, formula (4.2) can be written as

$$f'(x_0) = \lim_{x \to x_0} \frac{f(x) - f(x_0)}{x - x_0}.$$

If $f'(x_0)$ exists, then $f(x)$ is said to be differentiable at $x = x_0$. Geometrically, $f'(x_0)$ is the slope of the tangent to the graph of the function $y = f(x)$ at the point (x_0, y_0), where $y_0 = f(x_0)$. If $f(x)$ has a derivative at every point of a set D, then $f(x)$ is said to be differentiable on D.

It is important to note that in order for $f'(x_0)$ to exist, the left-sided and right-sided limits of $\phi(h)$ in formula (4.1) must exist and be equal as $h \to 0$, or as x approaches x_0 from either side. It is possible to consider only one-sided derivatives at $x = x_0$. These occur when $\phi(h)$ has just a one-sided limit as $h \to 0$. We shall not, however, concern ourselves with such derivatives in this chapter.   □

Functions that are differentiable at a point must necessarily be continuous there. This will be shown in the next theorem.

Theorem 4.1.1.   Let $f(x)$ be defined in a neighborhood of a point x_0. If $f(x)$ has a derivative at x_0, then it must be continuous at x_0.

Proof. From Definition 4.1.1 we can write

$$f(x_0 + h) - f(x_0) = h\phi(h).$$

If the derivative of $f(x)$ exists at x_0, then $\phi(h) \to f'(x_0)$ as $h \to 0$. It follows from Theorem 3.2.1(2) that

$$f(x_0 + h) - f(x_0) \to 0$$

as $h \to 0$. Thus for a given $\epsilon > 0$ there exists a $\delta > 0$ such that

$$|f(x_0 + h) - f(x_0)| < \epsilon$$

if $|h| < \delta$. This indicates that $f(x)$ is continuous at x_0.   □

It should be noted that even though continuity is a necessary condition for differentiability, it is not a sufficient condition, as can be seen from the following example: Let $f(x)$ be defined as

$$f(x) = \begin{cases} x \sin \dfrac{1}{x}, & x \neq 0, \\ 0, & x = 0. \end{cases}$$

This function is continuous at $x = 0$, since $f(0) = \lim_{x \to 0} f(x) = 0$ by the fact that

$$\left| x \sin \frac{1}{x} \right| \leq |x|$$

for all x. However, $f(x)$ is not differentiable at $x = 0$. This is because when $x = 0$,

$$\phi(h) = \frac{f(h) - f(0)}{h}$$

$$= \frac{h \sin \dfrac{1}{h} - 0}{h}, \qquad \text{since } h \neq 0,$$

$$= \sin \frac{1}{h},$$

which does not have a limit as $h \to 0$. Hence, $f'(0)$ does not exist.

If $f(x)$ is differentiable on a set D, then $f'(x)$ is a function defined on D. In the event $f'(x)$ itself is differentiable on D, then its derivative is called the second derivative of $f(x)$ and is denoted by $f''(x)$. It is also common to use the notation

$$\frac{d^2 f(x)}{dx^2} = f''(x).$$

By the same token, we can define the nth ($n \geq 2$) derivative of $f(x)$ as the derivative of the $(n - 1)$st derivative of $f(x)$. We denote this derivative by

$$\frac{d^n f(x)}{dx^n} = f^{(n)}(x), \qquad n = 2, 3, \dots.$$

We shall now discuss some rules that pertain to differentiation. The reader is expected to know how to differentiate certain elementary functions such as polynomial, exponential, and trigonometric functions.

Theorem 4.1.2. Let $f(x)$ and $g(x)$ be defined and differentiable on a set D. Then

1. $[\alpha f(x) + \beta g(x)]' = \alpha f'(x) + \beta g'(x)$, where α and β are constants.
2. $[f(x)g(x)]' = f'(x)g(x) + f(x)g'(x)$.
3. $[f(x)/g(x)]' = [f'(x)g(x) - f(x)g'(x)]/g^2(x)$ if $g(x) \neq 0$.

Proof. The proof of (1) is straightforward. To prove (2) we write

$$\lim_{h \to 0} \frac{f(x+h)g(x+h) - f(x)g(x)}{h}$$

$$= \lim_{h \to 0} \frac{[f(x+h) - f(x)]g(x+h) + f(x)[g(x+h) - g(x)]}{h}$$

$$= \lim_{h \to 0} g(x+h) \lim_{h \to 0} \frac{f(x+h) - f(x)}{h}$$

$$+ f(x) \lim_{h \to 0} \frac{g(x+h) - g(x)}{h}.$$

However, $\lim_{h \to 0} g(x+h) = g(x)$, since $g(x)$ is continuous (because it is differentiable). Hence,

$$\lim_{h \to 0} \frac{f(x+h)g(x+h) - f(x)g(x)}{h} = g(x)f'(x) + f(x)g'(x).$$

Now, to prove (3) we write

$$\lim_{h \to 0} \frac{f(x+h)/g(x+h) - f(x)/g(x)}{h}$$

$$= \lim_{h \to 0} \frac{g(x)f(x+h) - f(x)g(x+h)}{hg(x)g(x+h)}$$

$$= \lim_{h \to 0} \frac{g(x)[f(x+h) - f(x)] - f(x)[g(x+h) - g(x)]}{hg(x)g(x+h)}$$

$$= \frac{\lim_{h \to 0}\{g(x)[f(x+h) - f(x)]/h - f(x)[g(x+h) - g(x)]/h\}}{g(x)\lim_{h \to 0} g(x+h)}$$

$$= \frac{g(x)f'(x) - f(x)g'(x)}{g^2(x)}. \qquad \square$$

Theorem 4.1.3 (The Chain Rule).   Let $f: D_1 \to R$ and $g: D_2 \to R$ be two functions. Suppose that $f(D_1) \subset D_2$. If $f(x)$ is differentiable on D_1 and $g(x)$ is differentiable on D_2, then the composite function $h(x) = g[f(x)]$ is differentiable on D_1, and

$$\frac{dg[f(x)]}{dx} = \frac{dg[f(x)]}{df(x)} \frac{df(x)}{dx}.$$

Proof. Let $z = f(x)$ and $t = f(x + h)$. By the fact that $g(z)$ is differentiable we can write

$$g[f(x+h)] - g[f(x)] = g(t) - g(z)$$
$$= (t - z)g'(z) + o(t - z), \qquad (4.3)$$

where, if we recall, the o-notation was introduced in Section 3.3. We then have

$$\frac{g[f(x+h)] - g[f(x)]}{h} = \frac{t-z}{h}g'(z) + \frac{o(t-z)}{t-z} \cdot \frac{t-z}{h}. \qquad (4.4)$$

As $h \to 0$, $t \to z$, and hence

$$\lim_{h \to 0} \frac{t-z}{h} = \lim_{h \to 0} \frac{f(x+h) - f(x)}{h} = \frac{df(x)}{dx}.$$

Now, by taking the limits of both sides of (4.4) as $h \to 0$ and noting that

$$\lim_{h \to 0} \frac{o(t-z)}{t-z} = \lim_{t \to z} \frac{o(t-z)}{t-z} = 0,$$

we conclude that

$$\frac{dg[f(x)]}{dx} = \frac{df(x)}{dx} \frac{dg[f(x)]}{df(x)}. \qquad \square$$

NOTE 4.1.1.   We recall that $f(x)$ must be continuous in order for $f'(x)$ to exist. However, if $f'(x)$ exists, it does not have to be continuous. Care should be exercised when showing that $f'(x)$ is continuous. For example, let us consider the function

$$f(x) = \begin{cases} x^2 \sin \dfrac{1}{x}, & x \neq 0, \\ 0, & x = 0. \end{cases}$$

Suppose that it is desired to show that $f'(x)$ exists, and if so, to determine if it is continuous. To do so, let us first find out if $f'(x)$ exists at $x = 0$:

$$f'(0) = \lim_{h \to 0} \frac{f(h) - f(0)}{h}$$

$$= \lim_{h \to 0} \frac{h^2 \sin \dfrac{1}{h}}{h}$$

$$= \lim_{h \to 0} h \sin \frac{1}{h} = 0.$$

Thus the derivative of $f(x)$ exists at $x = 0$ and is equal to zero. For $x \neq 0$, it is clear that the derivative of $f(x)$ exists. By applying Theorem 4.1.2 and using our knowledge of the derivatives of elementary functions, $f'(x)$ can be written as

$$f'(x) = \begin{cases} 2x \sin \dfrac{1}{x} - \cos \dfrac{1}{x}, & x \neq 0, \\ 0, & x = 0. \end{cases}$$

We note that $f'(x)$ exists for all x, but is not continuous at $x = 0$, since

$$\lim_{x \to 0} f'(x) = \lim_{x \to 0} \left(2x \sin \frac{1}{x} - \cos \frac{1}{x} \right)$$

does not exist, because $\cos(1/x)$ has no limit as $x \to 0$. However, for any nonzero value of x, $f'(x)$ is continuous.

If $f(x)$ is a convex function, then we have the following interesting result, whose proof can be found in Roberts and Varberg (1973, Theorem C, page 7):

Theorem 4.1.4. If $f: (a, b) \to R$ is convex on the open interval (a, b), then the set S where $f'(x)$ fails to exist is either finite or countable. Moreover, $f'(x)$ is continuous on $(a, b) - S$, the complement of S with respect to (a, b).

For example, the function $f(x) = |x|$ is convex on R. Its derivatives does not exist at $x = 0$ (why?), but is continuous everywhere else.

The sign of $f'(x)$ provides information about the behavior of $f(x)$ in a neighborhood of x. More specifically, we have the following theorem:

Theorem 4.1.5. Let $f: D \to R$, where D is an open set. Suppose that $f'(x)$ is positive at a point $x_0 \in D$. Then there is a neighborhood $N_\delta(x_0) \subset D$ such that for each x in this neighborhood, $f(x) > f(x_0)$ if $x > x_0$, and $f(x) < f(x_0)$ if $x < x_0$.

Proof. Let $\epsilon = f'(x_0)/2$. Then, there exists a $\delta > 0$ such that

$$f'(x_0) - \epsilon < \frac{f(x) - f(x_0)}{x - x_0} < f'(x_0) + \epsilon$$

if $|x - x_0| < \delta$. Hence, if $x > x_0$,

$$f(x) - f(x_0) > \frac{(x - x_0)f'(x_0)}{2},$$

which shows that $f(x) > f(x_0)$ since $f'(x_0) > 0$. Furthermore, since

$$\frac{f(x) - f(x_0)}{x - x_0} > 0,$$

then $f(x) < f(x_0)$ if $x < x_0$. $\quad\square$

If $f'(x_0) < 0$, it can be similarly shown that $f(x) < f(x_0)$ if $x > x_0$, and $f(x) > f(x_0)$ if $x < x_0$.

4.2. THE MEAN VALUE THEOREM

This is one of the most important theorems in differential calculus. It is also known as the theorem of the mean. Before proving the mean value theorem, let us prove a special case of it known as *Rolle's theorem.*

Theorem 4.2.1 (Rolle's Theorem). Let $f(x)$ be continuous on the closed interval $[a, b]$ and differentiable on the open interval (a, b). If $f(a) = f(b)$, then there exists a point c, $a < c < b$, such that $f'(c) = 0$.

Proof. Let d denote the common value of $f(a)$ and $f(b)$. Define $h(x) = f(x) - d$. Then $h(a) = h(b) = 0$. If $h(x)$ is also zero on (a, b), then $h'(x) = 0$ for $a < x < b$ and the theorem is proved. Let us therefore assume that $h(x) \neq 0$ for some $x \in (a, b)$. Since $h(x)$ is continuous on $[a, b]$ [because $f(x)$ is], then by Corollary 3.4.1 it must achieve its supremum M at a point ξ in $[a, b]$, and its infimum m at a point η in $[a, b]$. If $h(x) > 0$ for some $x \in (a, b)$, then we must obviously have $a < \xi < b$, because $h(x)$ vanishes at both end points. We now claim that $h'(\xi) = 0$. If $h'(\xi) > 0$ or < 0, then by Theorem 4.1.5, there exists a point x_1 in a neighborhood $N_\delta(\xi) \subset (a, b)$ at which $h(x_1) > h(\xi)$, a contradiction, since $h(\xi) = M$. Thus $h'(\xi) = 0$, which implies that $f'(\xi) = 0$, since $h'(x) = f'(x)$ for all $x \in (a, b)$. We can similarly arrive at the conclusion that $f'(\eta) = 0$ if $h(x) < 0$ for some $x \in (a, b)$. In this case, if $h'(\eta) \neq 0$, then by Theorem 4.1.5 there exists a point x_2 in a neighborhood $N_{\delta_2}(\eta) \subset (a, b)$ at which $h(x_2) < h(\eta) = m$, a contradiction, since m is the infimum of $h(x)$ over $[a, b]$.

Thus in both cases, whether $h(x) > 0$ or < 0 for some $x \in (a, b)$, we must have a point c, $a < c < b$, such that $f'(c) = 0$.   □

Rolle's theorem has the following geometric interpretation: If $f(x)$ satisfies the conditions of Theorem 4.2.1, then the graph of $y = f(x)$ must have a tangent line that is parallel to the x-axis at some point c between a and b. Note that there can be several points like c inside (a, b). For example, the function $y = x^3 - 5x^2 + 3x - 1$ satisfies the conditions of Rolle's theorem on the interval $[a, b]$, where $a = 0$ and $b = (5 + \sqrt{13})/2$. In this case, $f(a) = f(b) = -1$, and $f'(x) = 3x^2 - 10x + 3$ vanishes at $x = \frac{1}{3}$ and $x = 3$.

Theorem 4.2.2 (The Mean Value Theorem). If $f(x)$ is continuous on the closed interval $[a, b]$ and differentiable on the open interval (a, b), then there exists a point c, $a < c < b$, such that

$$f(b) = f(a) + (b - a)f'(c).$$

Proof. Consider the function

$$\Phi(x) = f(x) - f(a) - A(x - a),$$

where

$$A = \frac{f(b) - f(a)}{b - a}.$$

The function $\Phi(x)$ is continuous on $[a, b]$ and is differentiable on (a, b), since $\Phi'(x) = f'(x) - A$. Furthermore, $\Phi(a) = \Phi(b) = 0$. It follows from Rolle's theorem that there exists a point c, $a < c < b$, such that $\Phi'(c) = 0$. Thus,

$$f'(c) = \frac{f(b) - f(a)}{b - a},$$

which proves the theorem.   □

The mean value theorem has also a nice geometric interpretation. If the graph of the function $y = f(x)$ has a tangent line at each point of its length between two points P_1 and P_2 (see Figure 4.1), then there must be a point Q between P_1 and P_2 at which the tangent line is parallel to the secant line through P_1 and P_2. Note that there can be several points on the curve between P_1 and P_2 that have the same property as Q, as can be seen from Figure 4.1.

The mean value theorem is useful in the derivation of several interesting results, as will be seen in the remainder of this chapter.

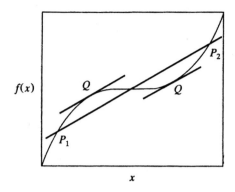

Figure 4.1. Tangent lines parallel to the secant line.

Corollary 4.2.1. If $f(x)$ has a derivative $f'(x)$ that is nonnegative (nonpositive) on an interval (a, b), then $f(x)$ is monotone increasing (decreasing) on (a, b). If $f'(x)$ is positive (negative) on (a, b), then $f(x)$ is strictly monotone increasing (decreasing) there.

Proof. Let x_1 and x_2 be two points in (a, b) such that $x_1 < x_2$. By the mean value theorem, there exists a point x_0, $x_1 < x_0 < x_2$, such that

$$f(x_2) = f(x_1) + (x_2 - x_1)f'(x_0).$$

If $f'(x_0) \geq 0$, then $f(x_2) \geq f(x_1)$ and $f(x)$ is monotone increasing. Similarly, if $f'(x_0) \leq 0$, then $f(x_2) \leq f(x_1)$ and $f(x)$ is monotone decreasing. If, however, $f'(x) > 0$, or $f'(x) < 0$ on (a, b), then strict monotonicity follows over (a, b). □

Theorem 4.2.3. If $f(x)$ is monotone increasing [decreasing] on an interval (a, b), and if $f(x)$ is differentiable there, then $f'(x) \geq 0 \, [f'(x) \leq 0]$ on (a, b).

Proof. Let $x_0 \in (a, b)$. There exists a neighborhood $N_r(x_0) \subset (a, b)$. Then, for any $x \in N_r(x_0)$ such that $x \neq x_0$, the ratio

$$\frac{f(x) - f(x_0)}{x - x_0}$$

is nonnegative. This is true because $f(x) \geq f(x_0)$ if $x > x_0$ and $f(x) \leq f(x_0)$ if $x < x_0$. By taking the limit of this ratio as $x \to x_0$, we claim that $f'(x_0) \geq 0$. To prove this claim, suppose that $f'(x_0) < 0$. Then there exists a $\delta > 0$ such that

$$\left| \frac{f(x) - f(x_0)}{x - x_0} - f'(x_0) \right| < -\frac{1}{2}f'(x_0)$$

if $|x - x_0| < \delta$. It follows that

$$\frac{f(x) - f(x_0)}{x - x_0} < \frac{1}{2} f'(x_0) < 0.$$

Thus $f(x) < f(x_0)$ if $x > x_0$, which is a contradiction. Hence, $f'(x_0) \geq 0$. A similar argument can be used to show that $f'(x_0) \leq 0$ when $f(x)$ is monotone decreasing. $\square$

Note that strict monotonicity on a set D does not necessarily imply that $f'(x) > 0$, or $f'(x) < 0$, for all x in D. For example, the function $f(x) = x^3$ is strictly monotone increasing for all x, but $f'(0) = 0$.

We recall from Theorem 3.5.1 that strict monotonicity of $f(x)$ is a sufficient condition for the existence of the inverse function f^{-1}. The next theorem shows that under certain conditions, f^{-1} is a differentiable function.

Theorem 4.2.4. Suppose that $f(x)$ is strictly monotone increasing (or decreasing) and continuous on an interval $[a, b]$. If $f'(x)$ exists and is different from zero at $x_0 \in (a, b)$, then the inverse function f^{-1} is differentiable at $y_0 = f(x_0)$ and its derivative is given by

$$\left. \frac{df^{-1}(y)}{dy} \right|_{y=y_0} = \frac{1}{f'(x_0)}.$$

Proof. By Theorem 3.5.2, $f^{-1}(y)$ exists and is continuous. Let $x_0 \in (a, b)$, and let $N_r(x_0) \subset (a, b)$ for some $r > 0$. Then, for any $x \in N_r(x_0)$,

$$\frac{f^{-1}(y) - f^{-1}(y_0)}{y - y_0} = \frac{x - x_0}{f(x) - f(x_0)}$$

$$= \frac{1}{[f(x) - f(x_0)]/(x - x_0)}, \qquad (4.5)$$

where $y = f(x)$. Now, since both f and f^{-1} are continuous, then $x \to x_0$ if and only if $y \to y_0$. By taking the limits of all sides in formula (4.5), we conclude that the derivative of f^{-1} at y_0 exists and is equal to

$$\left. \frac{df^{-1}(y)}{dy} \right|_{y=y_0} = \frac{1}{f'(x_0)}. \qquad \square$$

The following theorem gives a more general version of the mean value theorem. It is due to Augustin-Louis Cauchy and has an important application in calculating certain limits, as will be seen later in this chapter.

Theorem 4.2.5 (Cauchy's Mean Value Theorem).   If $f(x)$ and $g(x)$ are continuous on the closed interval $[a, b]$ and differentiable on the open interval (a, b), then there exists a point c, $a < c < b$, such that

$$[f(b) - f(a)]g'(c) = [g(b) - g(a)]f'(c),$$

Proof. The proof is based on using Rolle's theorem in a manner similar to that of Theorem 4.2.2. Define the function $\psi(x)$ as

$$\psi(x) = [f(b) - f(x)][g(b) - g(a)] - [f(b) - f(a)][g(b) - g(x)].$$

This function is continuous on $[a, b]$ and is differentiable on (a, b), since

$$\psi'(x) = -f'(x)[g(b) - g(a)] + g'(x)[f(b) - f(a)].$$

Furthermore, $\psi(a) = \psi(b) = 0$. Thus by Rolle's theorem, there exists a point c, $a < c < b$, such that $\psi'(c) = 0$, that is,

$$-f'(c)[g(b) - g(a)] + g'(c)[f(b) - f(a)] = 0. \tag{4.6}$$

In particular, if $g(b) - g(a) \neq 0$ and $f'(x)$ and $g'(x)$ do not vanish at the same point in (a, b), then formula (4.6) an be written as

$$\frac{f'(c)}{g'(c)} = \frac{f(b) - f(a)}{g(b) - g(a)}. \qquad \square$$

An immediate application of this theorem is a very popular method in calculating the limits of certain ratios of functions. This method is attributed to Guillaume Francois Marquis de l'Hospital (1661–1704) and is known as *l'Hospital's rule*. It deals with the limit of the ratio $f(x)/g(x)$ as $x \to a$ when both the numerator and the denominator tend simultaneously to zero or to infinity as $x \to a$. In either case, we have what is called an indeterminate ratio caused by having $0/0$ or ∞/∞ as $x \to a$.

Theorem 4.2.6 (l'Hospital's Rule).   Let $f(x)$ and $g(x)$ be continuous on the closed interval $[a, b]$ and differentiable on the open interval (a, b). Suppose that we have the following:

1. $g(x)$ and $g'(x)$ are not zero at any point inside (a, b).
2. $\lim_{x \to a^+} f'(x)/g'(x)$ exists.
3. $f(x) \to 0$ and $g(x) \to 0$ as $x \to a^+$, or $f(x) \to \infty$ and $g(x) \to \infty$ as $x \to a^+$.

Then,

$$\lim_{x \to a^+} \frac{f(x)}{g(x)} = \lim_{x \to a^+} \frac{f'(x)}{g'(x)}.$$

Proof. For the sake of simplicity, we shall drop the $+$ sign from a^+ and simply write $x \to a$ when x approaches a from the right. Let us consider the following cases:

CASE 1. $f(x) \to 0$ and $g(x) \to 0$ as $x \to a$, where a is finite. Let $x \in (a, b)$. By applying Cauchy's mean value theorem on the interval $[a, x]$ we get

$$\frac{f(x)}{g(x)} = \frac{f(x) - f(a)}{g(x) - g(a)} = \frac{f'(c)}{g'(c)}.$$

where $a < c < x$. Note that $f(a) = g(a) = 0$, since $f(x)$ and $g(x)$ are continuous and their limits are equal to zero when $x \to a$. Now, as $x \to a, c \to a$; hence

$$\lim_{x \to a} \frac{f(x)}{g(x)} = \lim_{c \to a} \frac{f'(c)}{g'(c)} = \lim_{x \to a} \frac{f'(x)}{g'(x)}.$$

CASE 2. $f(x) \to 0$ and $g(x) \to 0$ as $x \to \infty$. Let $z = 1/x$. As $x \to \infty, z \to 0$. Then

$$\lim_{x \to \infty} \frac{f(x)}{g(x)} = \lim_{z \to 0} \frac{f_1(z)}{g_1(z)}, \tag{4.7}$$

where $f_1(z) = f(1/z)$ and $g_1(z) = g(1/z)$. These functions are continuous since $f(x)$ and $g(x)$ are, and $z \neq 0$ as $z \to 0$ (see Theorem 3.4.2). Here, we find it necessary to set $f_1(0) = g_1(0) = 0$ so that $f_1(z)$ and $g_1(z)$ will be continuous at $z = 0$, since their limits are equal to zero. This is equivalent to defining $f(x)$ and $g(x)$ to be zero at infinity in the extended real number system. Furthermore, by the chain rule of Theorem 4.1.3 we have

$$f_1'(z) = f'(x)\left(-\frac{1}{z^2}\right),$$

$$g_1'(z) = g'(x)\left(-\frac{1}{z^2}\right).$$

If we now apply Case 1, we get

$$\lim_{z \to 0} \frac{f_1(z)}{g_1(z)} = \lim_{z \to 0} \frac{f_1'(z)}{g_1'(z)}$$

$$= \lim_{x \to \infty} \frac{f'(x)}{g'(x)}. \tag{4.8}$$

From (4.7) and (4.8) we then conclude that

$$\lim_{x \to \infty} \frac{f(x)}{g(x)} = \lim_{x \to \infty} \frac{f'(x)}{g'(x)}.$$

CASE 3. $f(x) \to \infty$ and $g(x) \to \infty$ as $x \to a$, where a is finite. Let $\lim_{x \to a}[f'(x)/g'(x)] = L$. Then for a given $\epsilon > 0$ there exists a $\delta > 0$ such that $a + \delta < b$ and

$$\left| \frac{f'(x)}{g'(x)} - L \right| < \epsilon, \tag{4.9}$$

if $a < x < a + \delta$. By applying Cauchy's mean value theorem on the interval $[x, a + \delta]$ we get

$$\frac{f(x) - f(a + \delta)}{g(x) - g(a + \delta)} = \frac{f'(d)}{g'(d)},$$

where $x < d < a + \delta$. From inequality (4.9) we then have

$$\left| \frac{f(x) - f(a + \delta)}{g(x) - g(a + \delta)} - L \right| < \epsilon$$

for all x such that $a < x < a + \delta$. It follows that

$$
\begin{aligned}
L &= \lim_{x \to a} \frac{f(x) - f(a + \delta)}{g(x) - g(a + \delta)} \\
&= \lim_{x \to a} \frac{f(x)}{g(x)} \lim_{x \to a} \frac{1 - f(a + \delta)/f(x)}{1 - g(a + \delta)/g(x)} \\
&= \lim_{x \to a} \frac{f(x)}{g(x)}
\end{aligned}
$$

since both $f(x)$ and $g(x)$ tend to ∞ as $x \to a$.

CASE 4. $f(x) \to \infty$ and $g(x) \to \infty$ as $x \to \infty$. This can be easily shown by using the techniques applied in Cases 2 and 3.

CASE 5. $\lim_{x \to a} f'(x)/g'(x) = \infty$, where a is finite or infinite. Let us consider the ratio $g(x)/f(x)$. We have that

$$\lim_{x \to a} \frac{g'(x)}{f'(x)} = 0.$$

Hence,

$$\lim_{x \to a} \frac{g(x)}{f(x)} = \lim_{x \to a} \frac{g'(x)}{f'(x)} = 0.$$

If A is any positive number, then there exists a $\delta > 0$ such that

$$\frac{g(x)}{f(x)} < \frac{1}{A},$$

if $a < x < a + \delta$. Thus for such values of x,

$$\frac{f(x)}{g(x)} > A,$$

which implies that

$$\lim_{x \to a} \frac{f(x)}{g(x)} = \infty.$$

When applying l'Hospital's rule, the ratio $f'(x)/g'(x)$ may assume the indeterminate from $0/0$ or ∞/∞ as $x \to a$. In this case, higher-order derivatives of $f(x)$ and $g(x)$, assuming such derivatives exist, will be needed. In general, if the first $n - 1$ ($n \geq 1$) derivatives of $f(x)$ and $g(x)$ tend simultaneously to zero or to ∞ as $x \to a$, and if the nth derivatives, $f^{(n)}(x)$ and $g^{(n)}(x)$, exist and satisfy the same conditions as those imposed on $f'(x)$ and $g'(x)$ in Theorem 4.2.6, then

$$\lim_{x \to a} \frac{f(x)}{g(x)} = \lim_{x \to a} \frac{f^{(n)}(x)}{g^{(n)}(x)}. \qquad \square$$

A Historical Note

According to Eves (1976, page 342), in 1696 the Marquis de l'Hospital assembled the lecture notes of his teacher Johann Bernoulli (1667–1748) into the world's first textbook on calculus. In this book, the so-called l'Hospital's rule is found. It is perhaps more accurate to refer to this rule as the Bernoulli-l'Hospital rule. Note that the name l'Hospital follows the old French spelling and the letter s is not to be pronounced. In modern French this name is spelled as l'Hôpital.

EXAMPLE 4.2.1. $\lim_{x \to 0} \dfrac{\sin x}{x} = \lim_{x \to 0} \dfrac{\cos x}{1} = 1.$

This is a well-known limit. It implies that $\sin x$ and x are asymptotically equal, that is, $\sin x \sim x$ as $x \to 0$ (see Section 3.3).

EXAMPLE 4.2.2. $\lim_{x \to 0} \dfrac{1 - \cos x}{x^2} = \lim_{x \to 0} \dfrac{\sin x}{2x} = \lim_{x \to 0} \dfrac{\cos x}{2} = \dfrac{1}{2}.$

We note here that l'Hospital's rule was applied twice before reaching the limit $\frac{1}{2}$.

EXAMPLE 4.2.3. $\lim\limits_{x \to \infty} \dfrac{a^x}{x}$, where $a > 1$.

This is of the form ∞/∞ as $x \to \infty$. Since $a^x = e^{x \log a}$, then

$$\lim_{x \to \infty} \frac{a^x}{x} = \lim_{x \to \infty} \frac{e^{x \log a}(\log a)}{1} = \infty.$$

This is also a well-known limit. On the basis of this result it can be shown that (see Exercise 4.12) the following hold:

1. $\lim\limits_{x \to \infty} \dfrac{a^x}{x^m} = \infty$, where $a > 1, m > 0$.

2. $\lim\limits_{x \to \infty} \dfrac{\log x}{x^m} = 0$, where $m > 0$.

EXAMPLE 4.2.4. $\lim_{x \to 0^+} x^x$.

This is of the form 0^0 as $x \to 0^+$, which is indeterminate. It can be reduced to the form $0/0$ or ∞/∞ so that l'Hospital's rule can apply. To do so we write x^x as

$$x^x = e^{x \log x}.$$

However,

$$x \log x = \frac{\log x}{1/x},$$

which is of the form $-\infty/\infty$ as $x \to 0^+$. By l'Hospital's rule we then have

$$\lim_{x \to 0^+} (x \log x) = \lim_{x \to 0^+} \frac{1/x}{-1/x^2}$$

$$= \lim_{x \to 0^+} (-x)$$

$$= 0.$$

It follows that

$$\lim_{x \to 0^+} x^x = \lim_{x \to 0^+} e^{x \log x}$$

$$= \exp\left[\lim_{x \to 0^+} (x \log x)\right]$$

$$= 1.$$

EXAMPLE 4.2.5. $\lim\limits_{x \to \infty} x \log\left(\dfrac{x+1}{x-1}\right)$.

This is of the form $\infty \times 0$ as $x \to \infty$, which is indeterminate. But

$$x \log\left(\frac{x+1}{x-1}\right) = \frac{\log\left(\dfrac{x+1}{x-1}\right)}{1/x}$$

is of the form $0/0$ as $x \to \infty$. Hence,

$$\lim_{x \to \infty} x \log\left(\frac{x+1}{x-1}\right) = \lim_{x \to \infty} \frac{\dfrac{-2}{(x-1)(x+1)}}{-1/x^2}$$

$$= \lim_{x \to \infty} \frac{2}{(1-1/x)(1+1/x)}$$

$$= 2.$$

We can see from the foregoing examples that the use of l'Hospital's rule can facilitate the process of finding the limit of the ratio $f(x)/g(x)$ as $x \to a$. In many cases, it is easier to work with $f'(x)/g'(x)$ than with the original ratio. Many other indeterminate forms can also be resolved by l'Hospital's rule by first reducing them to the form $0/0$ or ∞/∞ as was shown in Examples 4.2.4 and 4.2.5.

It is important here to remember that the application of l'Hospital's rule requires that the limit of $f'(x)/g'(x)$ exist as a finite number or be equal to infinity in the extended real number system as $x \to a$. If this is not the case, then it does not necessarily follow that the limit of $f(x)/g(x)$ does not exist. For example, consider $f(x) = x^2 \sin(1/x)$ and $g(x) = x$. Here, $f(x)/g(x)$ tends to zero as $x \to 0$, as was seen earlier in this chapter. However, the ratio

$$\frac{f'(x)}{g'(x)} = 2x \sin\frac{1}{x} - \cos\frac{1}{x}$$

has no limit as $x \to 0$, since it oscillates inside a small neighborhood of the origin.

4.3. TAYLOR'S THEOREM

This theorem is also known as the general mean value theorem, since it is considered as an extension of the mean value theorem. It was formulated by the English mathematician Brook Taylor (1685–1731) in 1712 and has since become a very important theorem in calculus. Taylor used his theorem to expand functions into infinite series. However, full recognition of the importance of Taylor's expansion was not realized until 1755, when Leonhard Euler (1707–1783) applied it in his differential calculus, and still later, when Joseph Louis Lagrange (1736–1813) used it as the foundation of his theory of functions.

Theorem 4.3.1 (Taylor's Theorem).   If the $(n-1)$st $(n \geq 1)$ derivative of $f(x)$, namely $f^{(n-1)}(x)$, is continuous on the closed interval $[a, b]$ and the nth derivative $f^{(n)}(x)$ exists on the open interval (a, b), then for each $x \in [a, b]$

we have

$$f(x) = f(a) + (x - a)f'(a) + \frac{(x-a)^2}{2!}f''(a)$$

$$+ \cdots + \frac{(x-a)^{n-1}}{(n-1)!}f^{(n-1)}(a) + \frac{(x-a)^n}{n!}f^{(n)}(\xi),$$

where $a < \xi < x$.

Proof. The method to prove this theorem is very similar to the one used for Theorem 4.2.2. For a fixed x in $[a, b]$ let the function $\psi_n(t)$ be defined as

$$\psi_n(t) = g_n(t) - \left(\frac{x-t}{x-a}\right)^n g_n(a),$$

where $a \le t \le b$ and

$$g_n(t) = f(x) - f(t) - (x - t)f'(t)$$

$$- \frac{(x-t)^2}{2!}f''(t) - \cdots - \frac{(x-t)^{n-1}}{(n-1)!}f^{(n-1)}(t). \qquad (4.10)$$

The function $\psi_n(t)$ has the following properties:

1. $\psi_n(a) = 0$ and $\psi_n(x) = 0$.
2. $\psi_n(t)$ is a continuous function of t on $[a, b]$.
3. The derivative of $\psi_n(t)$ with respect to t exists on (a, b). This derivative is equal to

$$\psi_n'(t) = g_n'(t) + \frac{n(x-t)^{n-1}}{(x-a)^n}g_n(a)$$

$$= -f'(t) + f'(t) - (x - t)f''(t) + (x - t)f''(t)$$

$$- \cdots + \frac{(x-t)^{n-2}}{(n-2)!}f^{(n-1)}(t)$$

$$- \frac{(x-t)^{n-1}}{(n-1)!}f^{(n)}(t) + \frac{n(x-t)^{n-1}}{(x-a)^n}g_n(a)$$

$$= -\frac{(x-t)^{n-1}}{(n-1)!}f^{(n)}(t) + \frac{n(x-t)^{n-1}}{(x-a)^n}g_n(a).$$

By applying Rolle's theorem to $\psi_n(t)$ on the interval $[a, x]$ we can assert that there exists a value ξ, $a < \xi < x$, such that $\psi_n'(\xi) = 0$, that is,

$$-\frac{(x-\xi)^{n-1}}{(n-1)!}f^{(n)}(\xi) + \frac{n(x-\xi)^{n-1}}{(x-a)^n}g_n(a) = 0,$$

or

$$g_n(a) = \frac{(x-a)^n}{n!}f^{(n)}(\xi). \tag{4.11}$$

Using formula (4.10) in (4.11), we finally get

$$f(x) = f(a) + (x-a)f'(a) + \frac{(x-a)^2}{2!}f''(a)$$

$$+ \cdots + \frac{(x-a)^{n-1}}{(n-1)!}f^{(n-1)}(a) + \frac{(x-a)^n}{n!}f^{(n)}(\xi). \tag{4.12}$$

This is known as *Taylor's formula*. It can also be expressed as

$$f(a+h) = f(a) + hf'(a) + \frac{h^2}{2!}f''(a)$$

$$+ \cdots + \frac{h^{n-1}}{(n-1)!}f^{(n-1)}(a) + \frac{h^n}{n!}f^{(n)}(a + \theta_n h), \tag{4.13}$$

where $h = x - a$ and $0 < \theta_n < 1$. □

In particular, if $f(x)$ has derivatives of all orders in some neighborhood $N_r(a)$ of the point a, formula (4.12) can provide a series expansion of $f(x)$ for $x \in N_r(a)$ as $n \to \infty$. The last term in formula (4.12), or formula (4.13), is called the remainder of Taylor's series and is denoted by R_n. Thus,

$$R_n = \frac{(x-a)^n}{n!}f^{(n)}(\xi)$$

$$= \frac{h^n}{n!}f^{(n)}(a + \theta_n h).$$

If $R_n \to 0$ as $n \to \infty$, then

$$f(x) = f(a) + \sum_{n=1}^{\infty} \frac{(x-a)^n}{n!} f^{(n)}(a), \qquad (4.14)$$

or

$$f(a+h) = f(a) + \sum_{n=1}^{\infty} \frac{h^n}{n!} f^{(n)}(a). \qquad (4.15)$$

This results in what is known as *Taylor's series*. Thus the validity of Taylor's series is contingent on having $R_n \to 0$ as $n \to \infty$, and on having derivatives of all orders for $f(x)$ in $N_r(a)$. The existence of these derivatives alone is not sufficient to guarantee a valid expansion.

A special form of Taylor's series is *Maclaurin's series*, which results when $a = 0$. Formula (4.14) then reduces to

$$f(x) = f(0) + \sum_{n=1}^{\infty} \frac{x^n}{n!} f^{(n)}(0). \qquad (4.16)$$

In this case, the remainder takes the form

$$R_n = \frac{x^n}{n!} f^{(n)}(\theta_n x). \qquad (4.17)$$

The sum of the first n terms in Maclaurin's series provides an approximation for the value of $f(x)$. The size of the remainder determines how close the sum is to $f(x)$. Since the remainder depends on θ_n, which lies in the interval $(0,1)$, an upper bound on R_n that is free of θ_n will therefore be needed to assess the accuracy of the approximation. For example, let us consider the function $f(x) = \cos x$. In this case,

$$f^{(n)}(x) = \cos\left(x + \frac{n\pi}{2}\right), \qquad n = 1, 2, \ldots,$$

and

$$f^{(n)}(0) = \cos\left(\frac{n\pi}{2}\right)$$

$$= \begin{cases} 0, & n \text{ odd}, \\ (-1)^{n/2}, & n \text{ even}. \end{cases}$$

Formula (4.16) becomes

$$\cos x = 1 + \sum_{n=1}^{\infty} (-1)^n \frac{x^{2n}}{(2n)!}$$

$$= 1 - \frac{x^2}{2!} + \frac{x^4}{4!} - \cdots + (-1)^n \frac{x^{2n}}{(2n)!} + R_{2n+1},$$

where from formula (4.17) R_{2n+1} is

$$R_{2n+1} = \frac{x^{2n+1}}{(2n+1)!} \cos \left[\theta_{2n+1} x + \frac{(2n+1)\pi}{2} \right].$$

An upper bound on $|R_{2n+1}|$ is then given by

$$|R_{2n+1}| \le \frac{|x|^{2n+1}}{(2n+1)!}.$$

Therefore, the error of approximating $\cos x$ with the sum

$$s_{2n} = 1 - \frac{x^2}{2!} + \frac{x^4}{4!} - \cdots + (-1)^n \frac{x^{2n}}{(2n)!}$$

does not exceed $|x|^{2n+1}/(2n+1)!$, where x is measured in radians. For example, if $x = \pi/3$ and $n = 3$, the sum

$$s_6 = 1 - \frac{x^2}{2} + \frac{x^4}{4!} - \frac{x^6}{6!} = 0.49996$$

approximates $\cos(\pi/3)$ with an error not exceeding

$$\frac{|x|^7}{7!} = 0.00027.$$

The true value of $\cos(\pi/3)$ is 0.5.

4.4. MAXIMA AND MINIMA OF A FUNCTION

In this section we consider the problem of finding the extreme values of a function $y = f(x)$ whose derivative $f'(x)$ exists in any open set inside its domain of definition.

Definition 4.4.1. A function $f: D \to R$ has a local (or relative) maximum at a point $x_0 \in D$ if there exists a $\delta > 0$ such that $f(x) \leq f(x_0)$ for all $x \in N_\delta(x_0) \cap D$. The function f has a local (or relative) minimum at x_0 if $f(x) \geq f(x_0)$ for all $x \in N_\delta(x_0) \cap D$. Local maxima and minima are referred to as local optima (or local extrema). $\square$

Definition 4.4.2. A function $f: D \to R$ has an absolute maximum (minimum) over D if there exists a point $x^* \in D$ such that $f(x) \leq f(x^*)[f(x) \geq f(x^*)]$ for all $x \in D$. Absolute maxima and minima are called absolute optima (or extrema). $\square$

The determination of local optima of $f(x)$ can be greatly facilitated if $f(x)$ is differentiable.

Theorem 4.4.1. Let $f(x)$ be differentiable on the open interval (a, b). If $f(x)$ has a local maximum, or a local minimum, at a point x_0 in (a, b), then $f'(x_0) = 0$.

Proof. Suppose that $f(x)$ has a local maximum at x_0. Then, $f(x) \leq f(x_0)$ for all x in some neighborhood $N_\delta(x_0) \subset (a, b)$. It follows that

$$\frac{f(x) - f(x_0)}{x - x_0} \begin{cases} \leq 0 & \text{if } x > x_0, \\ \geq 0 & \text{if } x < x_0, \end{cases} \tag{4.18}$$

for all x in $N_\delta(x_0)$. As $x \to x_0^+$, the ratio in (4.18) will have a nonpositive limit, and if $x \to x_0^-$, the ratio will have a nonnegative limit. Since $f'(x_0)$ exists, these two limits must be equal and equal to $f'(x_0)$ as $x \to x_0$. We therefore conclude that $f'(x_0) = 0$. The proof when $f(x)$ has a local minimum is similar. $\square$

It is important here to note that $f'(x_0) = 0$ is a necessary condition for a differentiable function to have a local optimum at x_0. It is not, however, a sufficient condition. That is, if $f'(x_0) = 0$, then it is not necessarily true that x_0 is a point of local optimum. For example, the function $f(x) = x^3$ has a zero derivative at the origin, but $f(x)$ does not have a local optimum there (why not?). In general, a value x_0 for which $f'(x_0) = 0$ is called a stationary value for the function. Thus a stationary value does not necessarily correspond to a local optimum.

We should also note that Theorem 4.4.1 assumes that $f(x)$ is differentiable in a neighborhood of x_0. If this condition is not fulfilled, the theorem ceases to be true. The existence of $f'(x_0)$ is not prerequisite for $f(x)$ to have a local optimum at x_0. In fact, $f(x)$ can have a local optimum at x_0 even if $f'(x_0)$ does not exist. For example, $f(x) = |x|$ has a local minimum at $x = 0$, but $f'(0)$ does not exist.

We recall from Corollary 3.4.1 that if $f(x)$ is continuous on $[a, b]$, then it must achieve its absolute optima at some points inside $[a, b]$. These points can be interior points, that is, points that belong to the open interval (a, b), or they can be end (boundary) points. In particular, if $f'(x)$ exists on (a, b), to determine the locations of the absolute optima we must solve the equation $f'(x) = 0$ and then compare the values of $f(x)$ at the roots of this equation with $f(a)$ and $f(b)$. The largest (smallest) of these values is the absolute maximum (minimum). In the event $f'(x) \neq 0$ on (a, b), then $f(x)$ must achieve its absolute optimum at an end point.

4.4.1. A Sufficient Condition for a Local Optimum

We shall make use of Taylor's expansion to come up with a sufficient condition for $f(x)$ to have a local optimum at $x = x_0$.

Suppose that $f(x)$ has n derivatives in a neighborhood $N_\delta(x_0)$ such that $f'(x_0) = f''(x_0) = \cdots = f^{(n-1)}(x_0) = 0$, but $f^{(n)}(x_0) \neq 0$. Then by Taylor's theorem we have

$$ f(x) = f(x_0) + \frac{h^n}{n!} f^{(n)}(x_0 + \theta_n h) $$

for any x in $N_\delta(x_0)$, where $h = x - x_0$ and $0 < \theta_n < 1$. Furthermore, if we assume that $f^{(n)}(x)$ is continuous at x_0, then

$$ f^{(n)}(x_0 + \theta_n h) = f^{(n)}(x_0) + o(1), $$

where, as we recall from Section 3.3, $o(1) \to 0$ as $h \to 0$. We can therefore write

$$ f(x) - f(x_0) = \frac{h^n}{n!} f^{(n)}(x_0) + o(h^n). \tag{4.19} $$

In order for $f(x)$ to have a local optimum at x_0, $f(x) - f(x_0)$ must have the same sign (positive or negative) for small values of h inside a neighborhood of 0. But, from (4.19), the sign of $f(x) - f(x_0)$ is determined by the sign of $h^n f^{(n)}(x_0)$. We can then conclude that if n is even, then a local optimum is achieved at x_0. In this case, a local maximum occurs at x_0 if $f^{(n)}(x_0) < 0$, whereas $f^{(n)}(x_0) > 0$ indicates that x_0 is a point of local minimum. If, however, n is odd, then x_0 is not a point of local optimum, since $f(x) - f(x_0)$ changes sign around x_0. In this case, the point on the graph of $y = f(x)$ whose abscissa is x_0 is called a *saddle point*.

In particular, if $f'(x_0) = 0$ and $f''(x_0) \neq 0$, then x_0 is a point of local optimum. When $f''(x_0) < 0$, $f(x)$ has a local maximum at x_0, and when $f''(x_0) > 0$, $f(x)$ has a local minimum at x_0.

EXAMPLE 4.4.1. Let $f(x) = 2x^3 - 3x^2 - 12x + 6$. Then $f'(x) = 6x^2 - 6x - 12 = 0$ at $x = -1, 2$, and

$$f''(x) = \begin{cases} 12x - 6 = -18 & \text{at } x = -1, \\ 18 & \text{at } x = 2. \end{cases}$$

We have then a local maximum at $x = -1$ and a local minimum at $x = 2$.

EXAMPLE 4.4.2. $f(x) = x^4 - 1$. In this case,

$$f'(x) = 4x^3 = 0 \quad \text{at } x = 0,$$
$$f''(x) = 12x^2 = 0 \quad \text{at } x = 0,$$
$$f'''(x) = 24x = 0 \quad \text{at } x = 0,$$
$$f^{(4)}(x) = 24.$$

Then, $x = 0$ is a point of local minimum.

EXAMPLE 4.4.3. Consider $f(x) = (x + 5)^2(x^3 - 10)$. We have

$$f'(x) = 5(x + 5)(x - 1)(x + 2)^2,$$
$$f''(x) = 10(x + 2)(2x^2 + 8x - 1),$$
$$f'''(x) = 10(6x^2 + 24x + 15).$$

Here, $f'(x) = 0$ at $x = -5, -2$, and 1. At $x = -5$ there is a local maximum, since $f''(-5) = -270 < 0$. At $x = 1$ we have a local minimum, since $f''(1) = 270 > 0$. However, at $x = -2$ a saddle point occurs, since $f''(-2) = 0$ and $f'''(-2) = -90 \neq 0$.

EXAMPLE 4.4.4. $f(x) = (2x + 1)/(x + 4)$, $0 \leq x \leq 5$. Then

$$f'(x) = 7/(x + 4)^2.$$

In this case, $f'(x)$ does not vanish anywhere in $(0, 5)$. Thus $f(x)$ has no local maxima or local minima in that open interval. Being continuous on $[0, 5]$, $f(x)$ must achieve its absolute optima at the end points. Since $f'(x) > 0$, $f(x)$ is strictly monotone increasing on $[0, 5]$ by Corollary 4.2.1. Its absolute minimum and absolute maximum are therefore attained at $x = 0$ and $x = 5$, respectively.

4.5. APPLICATIONS IN STATISTICS

Differential calculus has many applications in statistics. Let us consider some of these applications.

4.5.1. Functions of Random Variables

Let Y be a continuous random variable whose cumulative distribution function is $F(y) = P(Y \leq y)$. If $F(y)$ is differentiable for all y, then its derivative $F'(y)$ is called the density function of Y and is denoted by $f(y)$. Continuous random variables for which $f(y)$ exists are said to be absolutely continuous.

Let Y be an absolutely continuous random variable, and let W be another random variable which can be expressed as a function of Y of the form $W = \psi(Y)$. Suppose that this function is strictly monotone and differentiable over its domain. By Theorem 3.5.1, ψ has a unique inverse ψ^{-1}, which is also differentiable by Theorem 4.2.4. Let $G(w)$ denote the cumulative distribution function of W.

If ψ is strictly monotone increasing, then

$$G(w) = P(W \leq w) = P\left[Y \leq \psi^{-1}(w)\right] = F\left[\psi^{-1}(w)\right].$$

If it is strictly monotone decreasing, then

$$G(w) = P(W \leq w) = P\left[Y \geq \psi^{-1}(w)\right] = 1 - F\left[\psi^{-1}(w)\right].$$

By differentiating $G(w)$ using the chain rule we obtain the density function $g(w)$ for W, namely,

$$g(w) = \frac{dF\left[\psi^{-1}(w)\right]}{d\psi^{-1}(w)} \frac{d\psi^{-1}(w)}{dw}$$

$$= f\left[\psi^{-1}(w)\right] \frac{d\psi^{-1}(w)}{dw} \tag{4.20}$$

if ψ is strictly monotone increasing, and

$$g(w) = -\frac{dF\left[\psi^{-1}(w)\right]}{d\psi^{-1}(w)} \frac{d\psi^{-1}(w)}{dw}$$

$$= -f\left[\psi^{-1}(w)\right] \frac{d\psi^{-1}(w)}{dw} \tag{4.21}$$

if ψ is strictly monotone decreasing. By combining (4.20) and (4.21) we obtain

$$g(w) = f\left[\psi^{-1}(w)\right] \left| \frac{d\psi^{-1}(w)}{dw} \right|. \tag{4.22}$$

For example, suppose that Y has the uniform distribution $U(0, 1)$ whose density function is

$$f(y) = \begin{cases} 1, & 0 < y < 1, \\ 0 & \text{elsewhere}. \end{cases}$$

Let $W = -\log Y$. Using formula (4.22), the density function of W is given by

$$g(w) = \begin{cases} e^{-w}, & 0 < w < \infty, \\ 0 & \text{elsewhere}. \end{cases}$$

The Mean and Variance of $W = \psi(Y)$

The mean and variance of the random variable W can be obtained by using its density function:

$$E(W) = \int_{-\infty}^{\infty} wg(w)\, dw,$$

$$\mathrm{Var}(W) = E[W - E(W)]^2$$

$$= \int_{-\infty}^{\infty} [w - E(W)]^2 g(w)\, dw.$$

In some cases, however, the exact distribution of Y may not be known, or $g(w)$ may be a complicated function to integrate. In such cases, approximate expressions for the mean and variance of W can be obtained by applying Taylor's expansion around the mean of Y, μ. If we assume that $\psi''(y)$ exists, then

$$\psi(y) = \psi(\mu) + (y - \mu)\psi'(\mu) + o(y - \mu).$$

If $o(y - \mu)$ is small enough, first-order approximations of $E(W)$ and $\mathrm{Var}(W)$ can be obtained, namely,

$$E(W) \approx \psi(\mu), \quad \text{since } E(Y - \mu) = 0; \quad \mathrm{Var}(W) \approx \sigma^2 [\psi'(\mu)]^2, \quad (4.23)$$

where $\sigma^2 = \mathrm{Var}(Y)$, and the symbol $\approx$ denotes approximate equality. If $o(y - \mu)$ is not small enough, then higher-order approximations can be utilized provided that certain derivatives of $\psi(y)$ exist. For example, if $\psi'''(y)$ exists, then

$$\psi(y) = \psi(\mu) + (y - \mu)\psi'(\mu) + \tfrac{1}{2}(y - \mu)^2 \psi''(\mu) + o\left[(y - \mu)^2\right].$$

In this case, if $o[(y - \mu)^2]$ is small enough, then second-order approximations can be obtained for $E(W)$ and $\mathrm{Var}(W)$ of the form

$$E(W) \approx \psi(\mu) + \tfrac{1}{2}\sigma^2 \psi''(\mu), \quad \text{since } E(Y - \mu)^2 = \sigma^2,$$

$$\mathrm{Var}(W) \approx E\{Q(Y) - E[Q(Y)]\}^2,$$

where

$$Q(Y) = \psi(\mu) + (Y - \mu)\psi'(\mu) + \tfrac{1}{2}(Y - \mu)^2\psi''(\mu).$$

Thus,

$$\mathrm{Var}(W) \approx \sigma^2[\psi'(\mu)]^2 + \tfrac{1}{4}[\psi''(\mu)]^2\,\mathrm{Var}\big[(Y - \mu)^2\big]$$
$$+ \psi'(\mu)\psi''(\mu)\,E\big[(Y - \mu)^3\big].$$

Variance Stabilizing Transformations

One of the basic assumptions of regression and analysis of variance is the constancy of the variance σ^2 of a response variable Y on which experimental data are obtained. This assumption is often referred to as the assumption of homoscedasticity. There are situations, however, in which σ^2 is not constant for all the data. When this happens, Y is said to be heteroscedastic. Heteroscedasticity can cause problems and difficulties in connection with the statistical analysis of the data (for a survey of the problems of heteroscedasticity, see Judge et al., 1980).

Some situations that lead to heteroscedasticity are (see Wetherill et al., 1986, page 200):

  i. *The Use of Averaged Data.* In many experimental situations, the data used in a regression program consist of averages of samples that are different in size. This happens sometimes in survey analysis.

  ii. *Variances Depending on the Explanatory Variables.* The variance of an observation can sometimes depend on the explanatory (or input) variables in the hypothesized model, as is the case with some econometric models. For example, if the response variable is household expenditure and one explanatory variable is household income, then the variance of the observations may be a function of household income.

  iii. *Variances Depending on the Mean Response.* The response variable Y may have a distribution whose variance is a function of its mean, that is, $\sigma^2 = h(\mu)$, where μ is the mean of Y. The Poisson distribution, for example, has the property that $\sigma^2 = \mu$. Thus as μ changes (as a function of some explanatory variables), then so will σ^2. The following example illustrates this situation (see Chatterjee and Price, 1977, page 39): Let Y be the number of accidents, and x be the speed of operating a lathe in a machine shop. Suppose that a linear relationship is assumed between Y and x of the form

$$Y = \beta_0 + \beta_1 x + \epsilon,$$

where ϵ is a random error with a zero mean. Here, Y has the Poisson distribution with mean $\mu = \beta_0 + \beta_1 x$. The variance of Y, being equal to μ, will not be constant, since it depends on x.

Heteroscedasticity due to dependence on the mean response can be removed, or at least reduced, by a suitable transformation of the response variable Y. So let us suppose that $\sigma^2 = h(\mu)$. Let $W = \psi(Y)$. We need to find a proper transformation ψ that causes W to have almost the constant variance property. If this can be accomplished, then ψ is referred to as a variance stabilizing transformation.

If the first-order approximation of $\mathrm{Var}(W)$ by Taylor's expansion is adequate, then by formula (4.23) we can select ψ so that

$$h(\mu)[\psi'(\mu)]^2 = c, \qquad (4.24)$$

where c is a constant. Without loss of generality, let $c = 1$. A solution of (4.24) is given by

$$\psi(\mu) = \int \frac{d\mu}{\sqrt{h(\mu)}}.$$

Thus if $W = \psi(Y)$, then $\mathrm{Var}(W)$ will have a variance approximately equal to one. For example, if $h(\mu) = \mu$, as is the case with the Poisson distribution, then

$$\psi(\mu) = \int \frac{d\mu}{\sqrt{\mu}} = 2\sqrt{\mu}.$$

Hence, $W = 2\sqrt{Y}$ will have a variance approximately equal to one. (In this case, it is more common to use the transformation $W = \sqrt{Y}$ which has a variance approximately equal to 0.25). Thus in the earlier example regarding the relationship between the number of accidents and the speed of operating a lathe, we need to regress $\sqrt{Y}$ against x in order to ensure approximate homosecdasticity.

The relationship (if any) between σ^2 and μ may be determined by theoretical considerations based on a knowledge of the type of data used—for example, Poisson data. In practice, however, such knowledge may not be known a priori. In this case, the appropriate transformation is selected empirically on the basis of residual analysis of the data. See, for example, Box and Draper (1987, Chapter 8), Montgomery and Peck (1982, Chapter 3). If possible, a transformation is selected to correct nonnormality (if the original data are not believed to be normally distributed) as well as heteroscedasticity. In this respect, a useful family of transformations introduced by Box and Cox (1964) can be used. These authors considered the power family of transformations defined by

$$\psi(Y) = \begin{cases} (Y^\lambda - 1)/\lambda, & \lambda \neq 0, \\ \log Y, & \lambda = 0. \end{cases}$$

This family may only be applied when Y has positive values. Furthermore, since by l'Hospital's rule

$$\lim_{\lambda \to 0} \frac{Y^{\lambda} - 1}{\lambda} = \lim_{\lambda \to 0} Y^{\lambda} \log Y = \log Y,$$

the Box–Cox transformation is a continuous function of λ. An estimate of λ can be obtained from the data using the method of maximum likelihood (see Montgomery and Peck, 1982, Section 3.7.1; Box and Draper, 1987, Section 8.4).

Asymptotic Distributions

The asymptotic distributions of functions of random variables are of special interest in statistical limit theory. By definition, a sequence of random variables $\{Y_n\}_{n=1}^{\infty}$ converges in distribution to the random variable Y if

$$\lim_{n \to \infty} F_n(y) = F(y)$$

at each point y where $F(y)$ is continuous, where $F_n(y)$ is the cumulative distribution function of Y_n ($n = 1, 2, \dots$) and $F(y)$ is the cumulative distribution function of Y (see Section 5.3 concerning sequences of functions). This form of convergence is denoted by writing

$$Y_n \xrightarrow{d} Y.$$

An illustration of convergence in distribution is provided by the well-known *central limit theorem*. It states that if $\{Y_n\}_{n=1}^{\infty}$ is a sequence of independent and identically distributed random variables with common mean and variance, μ and σ^2, respectively, that are both finite, and if $\bar{Y}_n = \sum_{i=1}^{n} Y_i / n$ is the sample mean of a sample size n, then as $n \to \infty$,

$$\frac{\bar{Y}_n - \mu}{\sigma/\sqrt{n}} \xrightarrow{d} Z,$$

where Z has the standard normal distribution $N(0, 1)$.

An extension of the central limit theorem that includes functions of random variables is given by the following theorem:

Theorem 4.5.1. Let $\{Y_n\}_{n=1}^{\infty}$ be a sequence of independent and identically distributed random variables with mean μ and variance σ^2 (both finite), and let $\bar{Y}_n$ be the sample mean of a sample of size n. If $\psi(y)$ is a function whose derivative $\psi'(y)$ exists and is continuous in a neighborhood of μ such that

$\psi'(\mu) \neq 0$, then as $n \to \infty$,

$$\frac{\psi(\bar{Y}_n) - \psi(\mu)}{\sigma|\psi'(\mu)|/\sqrt{n}} \xrightarrow{d} Z.$$

Proof. See Wilks (1962, page 259). ☐

On the basis of Theorem 4.5.1 we can assert that when n is large enough, $\psi(\bar{Y}_n)$ is approximately distributed as a normal variate with a mean $\psi(\mu)$ and a standard deviation $(\sigma/\sqrt{n})|\psi'(\mu)|$. For example, if $\psi(y) = y^2$, then as $n \to \infty$,

$$\frac{\bar{Y}_n^2 - \mu^2}{2|\mu|\sigma/\sqrt{n}} \xrightarrow{d} Z.$$

4.5.2. Approximating Response Functions

Perhaps the most prevalent use of Taylor's expansion in statistics is in the area of linear models. Let Y denote a response variable, such as the yield of a product, whose mean $\mu(x)$ is believed to depend on an explanatory (or input) variable x such as temperature or pressure. The true relationship between μ and x is usually unknown. However, if $\mu(x)$ is considered to have derivatives of all orders, then it is possible to approximate its values by using low-order terms of a Taylor's series over a limited range of interest. In this case, $\mu(x)$ can be represented approximately by a polynomial of degree d (≥ 1) of the form

$$\mu(x) = \beta_0 + \sum_{j=1}^{d} \beta_j x^j,$$

where $\beta_0, \beta_1, \ldots, \beta_d$ are unknown parameters. Estimates of these parameters are obtained by running n ($\geq d + 1$) experiments in which n observations, $y_1, y_2, \ldots, y_n$, on Y are obtained for specified values of x. This leads us to the linear model

$$y_i = \beta_0 + \sum_{j=1}^{d} \beta_j x_i^j + \epsilon_i, \qquad i = 1, 2, \ldots, n, \tag{4.25}$$

where ϵ_i is a random error. The method of least squares can then be used to estimate the unknown parameters in (4.25). The adequacy of model (4.25) to represent the true mean response $\mu(x)$ can be checked using the given data provided that replicated observations are available at some points inside the region of interest. For more details concerning the adequacy of fit of linear models and the method of least squares, see, for example, Box and Draper (1987, Chapters 2 and 3) and Khuri and Cornell (1996, Chapter 2).

4.5.3. The Poisson Process

A random phenomenon that arises through a process which continues in time (or space) in a manner controlled by the laws of probability is called a stochastic process. A particular example of such a process is the Poisson process, which is associated with the number of events that take place over a period of time—for example, the arrival of customers at a service counter, or the arrival of α-rays, emitted from a radioactive source, at a Geiger counter.

Define $p_n(t)$ as the probability of n arrivals during a time interval of length t. For a Poisson process, the following postulates are assumed to hold:

1. The probability of exactly one arrival during a small time interval of length h is approximately proportional to h, that is,

$$p_1(h) = \lambda h + o(h)$$

   as $h \to 0$, where λ is a constant.

2. The probability of more than one arrival during a small time interval of length h is negligible, that is,

$$\sum_{n>1} p_n(h) = o(h)$$

   as $h \to 0$.

3. The probability of an arrival occurring during a small time interval $(t, t+h)$ does not depend on what happened prior to t. This means that the events defined according to the number of arrivals occurring during nonoverlapping time intervals are independent.

On the basis of the above postulates, an expression for $p_n(t)$ can be found as follows: For $n \geq 1$ and for small h we have approximately

$$p_n(t+h) = p_n(t)p_0(h) + p_{n-1}(t)p_1(h)$$

$$= p_n(t)[1 - \lambda h + o(h)] + p_{n-1}(t)[\lambda h + o(h)], \quad (4.26)$$

since the probability of no arrivals during the time interval $(t, t+h)$ is approximately equal to $1 - p_1(h)$. For $n = 0$ we have

$$p_0(t+h) = p_0(t)p_0(h)$$

$$= p_0(t)[1 - \lambda h + o(h)]. \quad (4.27)$$

From (4.26) and (4.27) we then get for $n \geq 1$,

$$\frac{p_n(t+h) - p_n(t)}{h} = p_n(t)\left[-\lambda + \frac{o(h)}{h}\right] + p_{n-1}(t)\left[\lambda + \frac{o(h)}{h}\right],$$

and for $n = 0$,

$$\frac{p_0(t+h) - p_0(t)}{h} = p_0(t)\left[-\lambda + \frac{o(h)}{h}\right].$$

By taking the limit as $h \to 0$ we obtain the derivatives

$$p_n'(t) = -\lambda p_n(t) + \lambda p_{n-1}(t), \qquad n \geq 1 \qquad (4.28)$$

$$p_0'(t) = -\lambda p_0(t). \qquad (4.29)$$

From (4.29) the solution for $p_0(t)$ is given by

$$p_0(t) = e^{-\lambda t}, \qquad (4.30)$$

since $p_0(t) = 1$ when $t = 0$ (that is, initially there were no arrivals). By substituting (4.30) in (4.28) when $n = 1$ we get

$$p_1'(t) = -\lambda p_1(t) + \lambda e^{-\lambda t}. \qquad (4.31)$$

If we now multiply the two sides of (4.31) by $e^{\lambda t}$ we obtain

$$e^{\lambda t} p_1'(t) + \lambda p_1(t) e^{\lambda t} = \lambda,$$

or

$$\left[e^{\lambda t} p_1(t)\right]' = \lambda.$$

Hence,

$$e^{\lambda t} p_1(t) = \lambda t + c,$$

where c is a constant. This constant must be equal to zero, since $p_1(0) = 0$. We then have

$$p_1(t) = \lambda t e^{-\lambda t}.$$

By continuing in this process and using equation (4.28) we can find $p_2(t)$, then $p_3(t), \ldots$, etc. In general, it can be shown that

$$p_n(t) = \frac{e^{-\lambda t}(\lambda t)^n}{n!}, \qquad n = 0, 1, 2, \ldots . \qquad (4.32)$$

In particular, if $t = 1$, then formula (4.32) gives the probability of n arrivals during one unit of time, namely,

$$p_n(1) = \frac{e^{-\lambda}\lambda^n}{n!}, \qquad n = 0, 1, \ldots .$$

This gives the probability mass function of a Poisson random variable with mean λ.

4.5.4. Minimizing the Sum of Absolute Deviations

Consider a data set consisting of n observations $y_1, y_2, \ldots, y_n$. For an arbitrary real number a, let $D(a)$ denote the sum of absolute deviations of the data from a, that is,

$$D(a) = \sum_{i=1}^{n} |y_i - a|.$$

For a given a, $D(a)$ represents a measure of spread, or variation, for the data set. Since the value of $D(a)$ varies with a, it may be of interest to determine its minimum. We now show that $D(a)$ is minimized when $a = \mu^*$, where μ^* denotes the median of the data set. By definition, μ^* is a value that falls in the middle when the observations are arranged in order of magnitude. It is a measure of location like the mean. If we write $y_{(1)} \leq y_{(2)} \leq \cdots \leq y_{(n)}$ for the ordered y_i's, then when n is odd, μ^* is the unique value $y_{(n/2+1/2)}$; whereas when n is even, μ^* is any value such that $y_{(n/2)} \leq \mu^* \leq y_{(n/2+1)}$. In the latter case, μ^* is sometimes chosen as the middle of the interval.

There are several ways to show that μ^* minimizes $D(a)$. The following simple proof is due to Blyth (1990):

On the interval $y_{(k)} < a < y_{(k+1)}$, $k = 1, 2, \ldots, n - 1$, we have

$$D(a) = \sum_{i=1}^{n} |y_{(i)} - a|$$

$$= \sum_{i=1}^{k} (a - y_{(i)}) + \sum_{i=k+1}^{n} (y_{(i)} - a)$$

$$= ka - \sum_{i=1}^{k} y_{(i)} + \sum_{i=k+1}^{n} y_{(i)} - (n - k)a.$$

The function $D(a)$ is continuous for all a and is differentiable everywhere except at $y_1, y_2, \ldots, y_n$. For $a \neq y_i$ ($i = 1, 2, \ldots, n$), the derivative $D'(a)$ is given by

$$D'(a) = 2(k - n/2).$$

If $k \neq n/2$, then $D'(a) \neq 0$ on $(y_{(k)}, y_{(k+1)})$, and by Corollary 4.2.1, $D(a)$ must be strictly monotone on $[y_{(k)}, y_{(k+1)}]$, $k = 1, 2, \ldots, n - 1$.

Now, when n is odd, $D(a)$ is strictly monotone decreasing for $a \leq y_{(n/2+1/2)}$, because $D'(a) < 0$ over $(y_{(k)}, y_{(k+1)})$ for $k < n/2$. It is strictly monotone increasing for $a \geq y_{(n/2+1/2)}$, because $D'(a) > 0$ over $(y_{(k)}, y_{(k+1)})$ for $k > n/2$. Hence, $\mu^* = y_{(n/2+1/2)}$ is a point of absolute minimum for $D(a)$. Furthermore, when n is even, $D(a)$ is strictly monotone decreasing for $a \leq y_{(n/2)}$, because $D'(a) < 0$ over $(y_{(k)}, y_{(k+1)})$ for $k < n/2$. Also, $D(a)$ is constant over $(y_{(n/2)}, y_{(n/2+1)})$, because $D'(a) = 0$ for $k = n/2$, and is strictly monotone increasing for $a \geq y_{(n/2+1)}$, because $D'(a) > 0$ over $(y_{(k)}, y_{(k+1)})$ for $k > n/2$. This indicates that $D(a)$ achieves its absolute minimum at any point μ^* such that $y_{(n/2)} \leq \mu^* \leq y_{(n/2+1)}$, which completes the proof.

FURTHER READING AND ANNOTATED BIBLIOGRAPHY

Apostol, T. M. (1964). *Mathematical Analysis*. Addison-Wesley, Reading, Massachusetts. (Chap. 5 discusses differentiation of functions of one variable.)

Blyth, C. R. (1990). "Minimizing the sum of absolute deviations." *Amer. Statist.*, **44**, 329.

Box, G. E. P., and D. R. Cox (1964). "An analysis of transformations." *J. Roy. Statist. Soc. Ser. B*, **26**, 211–243.

Box, G. E. P., and N. R. Draper (1987). *Empirical Model-Building and Response Surfaces*. Wiley, New York. (Chap. 2 introduces the idea of approximating the mean of a response variable using low-order polynomials; Chap. 3 discusses the method of least squares for fitting empirical models; the use of transformations, including those for stabilizing variances, is described in Chap. 8.)

Box, G. E. P., W. G. Hunter, and J. S. Hunter (1978). *Statistics for Experimenters*. Wiley, New York. (Various transformations are listed in Chap. 7, which include the Box–Cox and variance stabilizing transformations.)

Buck, R. C. (1956). *Advanced Calculus*, McGraw-Hill, New York. (Chap. 2 discusses the mean value theorem and l'Hospital's rule.)

Chatterjee, S., and B. Price (1977). *Regression Analysis by Example*. Wiley, New York. (Chap. 2 includes a discussion concerning variance stabilizing transformations, in addition to detection and removal of the effects of heteroscedasticity in regression analysis.)

Cooke, W. P. (1988). "L'Hôpital's rule in a Poisson derivation." *Amer. Math. Monthly*, **95**, 253–254.

Eggermont, P. P. B. (1988). "Noncentral difference quotients and the derivative." *Amer. Math. Monthly*, **95**, 551–553.

Eves, H. (1976). *An Introduction to the History of Mathematics*, 4th ed. Holt, Rinehart and Winston, New York.

Fulks, W. (1978). *Advanced Calculus*, 3rd ed. Wiley, New York. (Differentiation is the subject of Chap. 4.)

Georgiev, A. A. (1984). "Kernel estimates of functions and their derivatives with applications," *Statist. Probab. Lett.*, **2**, 45–50.

Hardy, G. H. (1955). *A Course of Pure Mathematics*, 10th ed. The University Press, Cambridge, England. (Chap. 6 covers differentiation and provides some interesting examples.)

Hogg, R. V., and A. T. Craig (1965). *Introduction to Mathematical Statistics*, 2nd ed. Macmillan, New York. (Chap. 4 discusses distributions of functions of random variables.)

James, A. T., and R. A. J. Conyers (1985). "Estimation of a derivative by a difference quotient: Its application to hepatocyte lactate metabolism." *Biometrics*, **41**, 467–476.

Judge, G. G., W. E. Griffiths, R. C. Hill, and T. C. Lee (1980). *The Theory and Practice of Econometrics*. Wiley, New York.

Khuri, A. I., and J. A. Cornell (1996). *Response Surfaces*, 2nd ed. Dekker, New York. (Chaps. 1 and 2 discuss the polynomial representation of a response surface and the method of least squares.)

Lindgren, B. W. (1976). *Statistical Theory*, 3rd ed. Macmillan, New York. (Section 3.2 discusses the development of the Poisson process.)

Menon, V. V., B. Prasad, and R. S. Singh (1984). "Non-parametric recursive estimates of a probability density function and its derivatives." *J. Statist. Plann. Inference*, **9**, 73–82.

Montgomery, D. C., and E. A. Peck (1982). *Introduction to Linear Regression Analysis*. Wiley, New York. (Chap. 3 presents several methods useful for checking the validity of the basic regression assumptions. Several variance stabilizing transformations are also listed.)

Parzen, E. (1962). *Stochastic Processes*. Holden-Day, San Francisco. (Chap. 1 introduces the definition of stochastic processes including the Poisson process.)

Roberts, A. W., and D. E. Varberg (1973). *Convex Functions*. Academic Press, New York. (Chap. 1 discusses a characterization of convex functions using derivatives; Chap. 5 discusses maxima and minima of differentiable functions.)

Roussas, G. G. (1973). *A First Course in Mathematical Statistics*. Addison-Wesley, Reading, Massachusetts. (Chap. 3 discusses absolutely continuous random variables.)

Rudin, W. (1964). *Principles of Mathematical Analysis*. 2nd ed. McGraw-Hill, New York. (Differentiation is discussed in Chap. 5.)

Sagan, H. (1974). *Advanced Calculus*. Houghton Mifflin, Boston. (Chap. 3 discusses differentiation.)

Wetherill, G. B., P. Duncombe, M. Kenward, J. Köllerström, S. R. Paul, and B. J. Vowden (1986). *Regression Analysis with Applications*. Chapman and Hall, London, England. (Section 9.2 discusses the sources of heteroscedasticity in regression analysis.)

Wilks, S. S. (1962). *Mathematical Statistics*. Wiley, New York. (Chap. 9 considers limit theorems including asymptotic distributions of functions of the sample mean.)

EXERCISES

In Mathematics

4.1. Let $f(x)$ be defined in a neighborhood of the origin. Show that if $f'(0)$ exists, then

$$\lim_{h \to 0} \frac{f(h) - f(-h)}{2h} = f'(0).$$

Give a counterexample to show that the converse is not true in general, that is, if the above limit exists, then it is not necessary that $f'(0)$ exists.

4.2. Let $f(x)$ and $g(x)$ have derivatives up to order n on $[a, b]$. Let $h(x) = f(x)g(x)$. Show that

$$h^{(n)}(x) = \sum_{k=0}^{n} \binom{n}{k} f^{(k)}(x) g^{(n-k)}(x).$$

(This is known as Leibniz's formula.)

4.3. Suppose that $f(x)$ has a derivative at a point x_0, $a < x_0 < b$. Show that there exists a neighborhood $N_\delta(x_0)$ and a positive number A such that

$$|f(x) - f(x_0)| < A|x - x_0|$$

for all $x \in N_\delta(x_0)$, $x \neq x_0$.

4.4. Suppose that $f(x)$ is differentiable on $(0, \infty)$ and $f'(x) \to 0$ as $x \to \infty$. Let $g(x) = f(x + 1) - f(x)$. Prove that $g(x) \to 0$ as $x \to \infty$.

4.5. Let the function $f(x)$ be defined as

$$f(x) = \begin{cases} x^3 - 2x, & x \geq 1, \\ ax^2 - bx + 1, & x < 1. \end{cases}$$

For what values of a and b does $f(x)$ have a continuous derivative?

4.6. Suppose that $f(x)$ is twice differentiable on $(0, \infty)$. Let m_0, m_1, m_2 be the least upper bounds of $|f(x)|$, $|f'(x)|$, and $|f''(x)|$, respectively, on $(0, \infty)$.

(a) Show that

$$|f'(x)| \le \frac{m_0}{h} + hm_2$$

for all x in $(0, \infty)$ and for every $h > 0$.

(b) Deduce from (a) that

$$m_1^2 \le 4m_0 m_2.$$

4.7. Suppose that $\lim_{x \to x_0} f'(x)$ exists. Does it follow that $f(x)$ is differentiable at x_0? Give a proof to show that the statement is correct or produce a counterexample to show that it is false.

4.8. Show that $D(a) = \sum_{i=1}^{n} |y_i - a|$ has no derivatives with respect to a at $y_1, y_2, \ldots, y_n$.

4.9. Suppose that the function $f(x)$ is such that $f'(x)$ and $f''(x)$ are continuous in a neighborhood of the origin and satisfies $f(0) = 0$. Show that

$$\lim_{x \to 0} \frac{d}{dx} \left[\frac{f(x)}{x} \right] = \frac{1}{2} f''(0).$$

4.10. Show that if $f'(x)$ exists and is bounded for all x, then $f(x)$ is uniformly continuous on R, the set of real numbers.

4.11. Suppose that $g: R \to R$ and that $|g'(x)| < M$ for all $x \in R$, where M is a positive constant. Define $f(x) = x + cg(x)$, where c is a positive constant. Show that it is possible to choose c small enough so that f is a one-to-one function.

4.12. Suppose that $f(x)$ is continuous on $[0, \infty)$, $f'(x)$ exists on $(0, \infty)$, $f(0) = 0$, and $f'(x)$ is monotone increasing on $(0, \infty)$. Show that $g(x)$ is monotone increasing on $(0, \infty)$ where $g(x) = f(x)/x$.

4.13. Show that if $a > 1$ and $m > 0$, then
(a) $\lim_{x \to \infty} (a^x/x^m) = \infty$,
(b) $\lim_{x \to \infty} [(\log x)/x^m] = 0$.

4.14. Apply l'Hospital's rule to find the limit

$$\lim_{x \to \infty} \left(1 + \frac{1}{x} \right)^x.$$

4.15. (a) Find $\lim_{x \to 0^+}(\sin x)^x$.

(b) Find $\lim_{x \to 0^+}(e^{-1/x}/x)$.

4.16. Show that

$$\lim_{x \to 0}\left[1 + ax + o(x)\right]^{1/x} = e^a,$$

where a is a constant and $o(x)$ is any function whose order of magnitude is less than that of x as $x \to 0$.

4.17. Consider the functions $f(x) = 4x^3 + 6x^2 - 10x + 2$ and $g(x) = 3x^4 + 4x^3 - 5x^2 + 1$. Show that

$$\frac{f'(x)}{g'(x)} \neq \frac{f(1) - f(0)}{g(1) - g(0)}$$

for any $x \in (0, 1)$. Does this contradict Cauchy's mean value theorem?

4.18. Suppose that $f(x)$ is differentiable for $a \leq x \leq b$. If $f'(a) < f'(b)$ and γ is a number such that $f'(a) < \gamma < f'(b)$, show that there exists a ξ, $a < \xi < b$, for which $f'(\xi) = \gamma$ [a similar result holds if $f'(a) > f'(b)$]. [*Hint:* Consider the function $g(x) = f(x) - \gamma(x - a)$. Show that $g(x)$ has a minimum at ξ.]

4.19. Suppose that $f(x)$ is differentiable on (a, b). Let $x_1, x_2, \ldots, x_n$ be in (a,b), and let $\lambda_1, \lambda_2, \ldots, \lambda_n$ be positive numbers such that $\sum_{i=1}^n \lambda_i = 1$. Show that there exists a point c in (a, b) such that

$$\sum_{i=1}^n \lambda_i f'(x_i) = f'(c).$$

[*Note:* This is a generalization of the result in Exercise 4.18.]

4.20. Let $x_1, x_2, \ldots, x_n$ and $y_1, y_2, \ldots, y_n$ be in (a, b) such that $x_i < y_i$ $(i = 1, 2, \ldots, n)$. Show that if $f(x)$ is differentiable on (a, b), then there exists a point c in (a, b) such that

$$\sum_{i=1}^n [f(y_i) - f(x_i)] = f'(c) \sum_{i=1}^n (y_i - x_i).$$

4.21. Give a Maclaurin's series expansion of the function $f(x) = \log(1 + x)$.

4.22. Discuss the maxima and minima of the function $f(x) = (x^4 + 3)/(x^2 + 2)$.

4.23. Determine if $f(x) = e^{3x}/x$ has an absolute minimum on $(0, \infty)$.

4.24. For what values of a and b is the function

$$f(x) = \frac{1}{x^2 + ax + b}$$

bounded on the interval $[-1, 1]$? Find the absolute maximum on that interval.

In Statistics

4.25. Let Y be a continuous random variable whose cumulative distribution function, $F(y)$, is strictly monotone. Let $G(y)$ be another strictly monotone, continuous cumulative distribution function. Show that the cumulative distribution function of the random variable $G^{-1}[F(Y)]$ is $G(y)$.

4.26. Let Y have the cumulative distribution function

$$F(y) = \begin{cases} 1 - e^{-y}, & y \geq 0, \\ 0, & y < 0. \end{cases}$$

Find the density function of $W = \sqrt{Y}$.

4.27. Let Y be normally distributed with mean 1 and variance 0.04. Let $W = Y^3$.
   (a) Find the density function of W.
   (b) Find the exact mean and variance of W.
   (c) Find approximate values for the mean and variance of W using Taylor's expansion, and compare the results with those of (b).

4.28. Let Z be normally distributed with mean 0 and variance 1. Let $Y = Z^2$. Find the density function of Y.
   [*Note:* The function $\psi(z) = z^2$ is not strictly monotone for all z.]

4.29. Let X be a random variable that denotes the age at failure of a component. The failure rate is defined as the probability of failure in a finite interval of time, say of length h, given the age of the component, say x. This failure rate is therefore equal to

$$P(x \leq X \leq x + h \mid X \geq x).$$

Consider the following limit:

$$\lim_{h \to 0} \frac{1}{h} P(x \leq X \leq x + h \mid X \geq x).$$

If this limit exists, then it is called the hazard rate, or instantaneous failure rate.

(a) Give an expression for the failure rate in terms of $F(x)$, the cumulative distribution function of X.

(b) Suppose that X has the exponential distribution with the cumulative distribution function

$$F(x) = 1 - e^{-x/\sigma}, \qquad x \ge 0,$$

where σ is a positive constant. Show that X has a constant hazard rate.

(c) Show that any random variable with a constant hazard rate must have the exponential distribution.

4.30. Consider a Poisson process with parameter λ over the interval $(0, t)$. Divide this interval into n equal subintervals of length $h = t/n$. We consider that we have a "success" in a given subinterval if one arrival occurs in that subinterval. If there are no arrivals, then we consider that we have a "failure." Let Y_n denote the number of "successes" in the n subintervals of length h. Then we have approximately

$$P(Y_n = r) = \binom{n}{r} p_n^r (1 - p_n)^{n-r}, \qquad r = 0, 1, \ldots, n,$$

where p_n is approximately equal to $\lambda h = \lambda t/n$. Show that

$$\lim_{n \to \infty} P(Y_n = r) = \frac{e^{-\lambda t}(\lambda t)^r}{r!}.$$

CHAPTER 5

Infinite Sequences and Series

The study of the theory of infinite sequences and series is an integral part of advanced calculus. All limiting processes, such as differentiation and integration, can be investigated on the basis of this theory.

The first example of an infinite series is attributed to Archimedes, who showed that the sum

$$1 + \frac{1}{4} + \cdots + \frac{1}{4^n}$$

was less than $\frac{4}{3}$ for any value of n. However, it was not until the nineteenth century that the theory of infinite series was firmly established by Augustin-Louis Cauchy (1789–1857).

In this chapter we shall study the theory of infinite sequences and series, and investigate their convergence. Unless otherwise stated, the terms of all sequences and series considered in this chapter are real-valued.

5.1. INFINITE SEQUENCES

In Chapter 1 we introduced the general concept of a function. An infinite sequence is a particular function $f: J^+ \to R$ defined on the set of all positive integers. For a given $n \in J^+$, the value of this function, namely $f(n)$, is called the nth term of the infinite sequence and is denoted by a_n. The sequence itself is denoted by the symbol $\{a_n\}_{n=1}^{\infty}$. In some cases, the integer with which the infinite sequence begins is different from one. For example, it may be equal to zero or to some other integer. For the sake of simplicity, an infinite sequence will be referred to as just a sequence.

Since a sequence is a function, then, in particular, the sequence $\{a_n\}_{n=1}^{\infty}$ can have the following properties:

1. It is bounded if there exists a constant $K > 0$ such that $|a_n| \le K$ for all n.

132

2. It is monotone increasing if $a_n \le a_{n+1}$ for all n, and is monotone decreasing if $a_n \ge a_{n+1}$ for all n.

3. It converges to a finite number c if $\lim_{n \to \infty} a_n = c$, that is, for a given $\epsilon > 0$ there exists an integer N such that

$$|a_n - c| < \epsilon \qquad \text{if } n > N.$$

In this case, c is called the limit of the sequence and this fact is denoted by writing $a_n \to c$ as $n \to \infty$. If the sequence does not converge to a finite limit, then it is said to be divergent.

4. It is said to oscillate if it does not converge to a finite limit, nor to $+\infty$ or $-\infty$ as $n \to \infty$.

EXAMPLE 5.1.1. Let $a_n = (n^2 + 2n)/(2n^2 + 3)$. Then $a_n \to \frac{1}{2}$ as $n \to \infty$, since

$$\lim_{n \to \infty} a_n = \lim_{n \to \infty} \frac{1 + 2/n}{2 + 3/n^2}$$
$$= \frac{1}{2}.$$

EXAMPLE 5.1.2. Consider $a_n = \sqrt{n+1} - \sqrt{n}$. This sequence converges to zero, since

$$a_n = \frac{(\sqrt{n+1} - \sqrt{n})(\sqrt{n+1} + \sqrt{n})}{\sqrt{n+1} + \sqrt{n}}$$
$$= \frac{1}{\sqrt{n+1} + \sqrt{n}}.$$

Hence, $a_n \to 0$ as $n \to \infty$.

EXAMPLE 5.1.3. Suppose that $a_n = 2^n/n^3$. Here, the sequence is divergent, since by Example 4.2.3,

$$\lim_{n \to \infty} \frac{2^n}{n^3} = \infty.$$

EXAMPLE 5.1.4. Let $a_n = (-1)^n$. This sequence oscillates, since it is equal to 1 when n is even and to -1 when n is odd.

Theorem 5.1.1. Every convergent sequence is bounded.

Proof. Suppose that $\{a_n\}_{n=1}^{\infty}$ converges to c. Then, there exists an integer N such that

$$|a_n - c| < 1 \qquad \text{if } n > N.$$

For such values of n, we have

$$|a_n| < \max(|c - 1|, |c + 1|).$$

It follows that

$$|a_n| < K$$

for all n, where

$$K = \max(|a_1| + 1, |a_2| + 1, \ldots, |a_N| + 1, |c - 1|, |c + 1|). \qquad \square$$

The converse of Theorem 5.1.1 is not necessarily true. That is, if a sequence is bounded, then it does not have to be convergent. As a counterexample, consider the sequence given in Example 5.1.4. This sequence is bounded, but is not convergent. To guarantee converge of a bounded sequence we obviously need an additional condition.

Theorem 5.1.2. Every bounded monotone (increasing or decreasing) sequence converges.

Proof. Suppose that $\{a_n\}_{n=1}^{\infty}$ is a bounded and monotone increasing sequence (the proof is similar if the sequence is monotone decreasing). Since the sequence is bounded, it must be bounded from above and hence has a least upper bound c (see Theorem 1.5.1). Thus $a_n \leq c$ for all n. Furthermore, for any given $\epsilon > 0$ there exists an integer N such that

$$c - \epsilon < a_N \leq c;$$

otherwise $c - \epsilon$ would be an upper bound of $\{a_n\}_{n=1}^{\infty}$. Now, because the sequence is monotone increasing,

$$c - \epsilon < a_N \leq a_{N+1} \leq a_{N+2} \leq \cdots \leq c,$$

that is,

$$c - \epsilon < a_n \leq c \qquad \text{for } n \geq N.$$

We can write

$$c - \epsilon < a_n < c + \epsilon,$$

or equivalently,

$$|a_n - c| < \epsilon \qquad \text{if } n \geq N.$$

This indicates that $\{a_n\}_{n=1}^{\infty}$ converges to c. $\qquad \square$

Using Theorem 5.1.2 it is easy to prove the following corollary.

Corollary 5.1.1.

1. If $\{a_n\}_{n=1}^{\infty}$ is bounded from above and is monotone increasing, then $\{a_n\}_{n=1}^{\infty}$ converges to $c = \sup_{n \geq 1} a_n$.
2. If $\{a_n\}_{n=1}^{\infty}$ is bounded from below and is monotone decreasing, then $\{a_n\}_{n=1}^{\infty}$ converges to $d = \inf_{n \geq 1} a_n$.

EXAMPLE 5.1.5.   Consider the sequence $\{a_n\}_{n=1}^{\infty}$, where $a_1 = \sqrt{2}$ and $a_{n+1} = \sqrt{2 + \sqrt{a_n}}$ for $n \geq 1$. This sequence is bounded, since $a_n < 2$ for all n, as can be easily shown using mathematical induction: We have $a_1 = \sqrt{2} < 2$. If $a_n < 2$, then $a_{n+1} < \sqrt{2 + \sqrt{2}} < 2$. Furthermore, the sequence is monotone increasing, since $a_n \leq a_{n+1}$ for $n = 1, 2, \ldots$, which can also be shown by mathematical induction. Hence, by Theorem 5.1.2 $\{a_n\}_{n=1}^{\infty}$ must converge. To find its limit, we note that

$$\lim_{n \to \infty} a_{n+1} = \lim_{n \to \infty} \sqrt{2 + \sqrt{a_n}}$$

$$= \sqrt{2 + \sqrt{\lim_{n \to \infty} a_n}}\ .$$

If c denotes the limit of a_n as $n \to \infty$, then

$$c = \sqrt{2 + \sqrt{c}}\ .$$

By solving this equation under the condition $c \geq \sqrt{2}$ we find that the only solution is $c = 1.831$.

Definition 5.1.1.   Consider the sequence $\{a_n\}_{n=1}^{\infty}$. An infinite collection of its terms, picked out in a manner that preserves the original order of the terms of the sequence, is called a subsequence of $\{a_n\}_{n=1}^{\infty}$. More formally, any sequence of the form $\{b_n\}_{n=1}^{\infty}$, where $b_n = a_{k_n}$ such that $k_1 < k_2 < \cdots < k_n < \cdots$ is a subsequence of $\{a_n\}_{n=1}^{\infty}$. Note that $k_n \geq n$ for $n \geq 1$.   □

Theorem 5.1.3.   A sequence $\{a_n\}_{n=1}^{\infty}$ converges to c if and only if every subsequence of $\{a_n\}_{n=1}^{\infty}$ converges to c.

Proof. The proof is left to the reader.   □

It should be noted that if a sequence diverges, then it does not necessarily follow that every one of its subsequences must diverge. A sequence may fail to converge, yet several of its subsequences converge. For example, the sequence whose nth term is $a_n = (-1)^n$ is divergent, as was seen earlier. However, the two subsequences $\{b_n\}_{n=1}^{\infty}$ and $\{c_n\}_{n=1}^{\infty}$, where $b_n = a_{2n} = 1$ and $c_n = a_{2n-1} = -1$ ($n = 1, 2, \ldots$), are both convergent.

We have noted earlier that a bounded sequence may not converge. It is possible, however, that one of its subsequences is convergent. This is shown in the next theorem.

Theorem 5.1.4. Every bounded sequence has a convergent subsequence.

Proof. Suppose that $\{a_n\}_{n=1}^{\infty}$ is a bounded sequence. Without loss of generality we can consider that the number of distinct terms of the sequence is infinite. (If this is not the case, then there exists an infinite subsequence of $\{a_n\}_{n=1}^{\infty}$ that consists of terms that are equal. Obviously, such a subsequence converges.) Let G denote the set consisting of all terms of the sequence. Then G is a bounded infinite set. By Theorem 1.6.2, G must have a limit point, say c. Also, by Theorem 1.6.1, every neighborhood of c must contain infinitely many points of G. It follows that we can find integers $k_1 < k_2 < k_3 < \cdots$ such that

$$|a_{k_n} - c| < \frac{1}{n} \qquad \text{for } n = 1, 2, \ldots.$$

Thus for a given $\epsilon > 0$ there exists an integer $N > 1/\epsilon$ such that $|a_{k_n} - c| < \epsilon$ if $n > N$. This indicates that the subsequence $\{a_{k_n}\}_{n=1}^{\infty}$ converges to c.
□

We conclude from Theorem 5.1.4 that a bounded sequence can have several convergent subsequences. The limit of each of these subsequences is called a subsequential limit. Let E denote the set of all subsequential limits of $\{a_n\}_{n=1}^{\infty}$. This set is bounded, since the sequence is bounded (why?).

Definition 5.1.2. Let $\{a_n\}_{n=1}^{\infty}$ be a bounded sequence, and let E be the set of all its subsequential limits. Then the least upper bound of E is called the upper limit of $\{a_n\}_{n=1}^{\infty}$ and is denoted by $\limsup_{n \to \infty} a_n$. Similarly, the greatest lower bound of E is called the lower limit of $\{a_n\}_{n=1}^{\infty}$ and is denoted by $\liminf_{n \to \infty} a_n$. For example, the sequence $\{a_n\}_{n=1}^{\infty}$, where $a_n = (-1)^n[1 + (1/n)]$, has two subsequential limits, namely -1 and 1. Thus $E = \{-1, 1\}$, and $\limsup_{n \to \infty} a_n = 1$, $\liminf_{n \to \infty} a_n = -1$. □

Theorem 5.1.5. The sequence $\{a_n\}_{n=1}^{\infty}$ converges to c if any only if

$$\liminf_{n \to \infty} a_n = \limsup_{n \to \infty} a_n = c.$$

Proof. The proof is left to the reader. □

Theorem 5.1.5 implies that when a sequence converges, the set of all its subsequential limits consists of a single element, namely the limit of the sequence.

5.1.1. The Cauchy Criterion

We have seen earlier that the definition of convergence of a sequence $\{a_n\}_{n=1}^{\infty}$ requires finding the limit of a_n as $n \to \infty$. In some cases, such a limit may be difficult to figure out. For example, consider the sequence whose nth term is

$$a_n = 1 - \frac{1}{3} + \frac{1}{5} - \frac{1}{7} + \cdots + \frac{(-1)^{n-1}}{2n-1}, \qquad n = 1, 2, \ldots. \qquad (5.1)$$

It is not easy to calculate the limit of a_n in order to find out if the sequence converges. Fortunately, however, there is another convergence criterion for sequences, known as the *Cauchy criterion* after Augustin-Louis Cauchy (it was known earlier to Bernhard Bolzano, 1781–1848, a Czechoslovakian priest whose mathematical work was undeservedly overlooked by his lay and clerical contemporaries; see Boyer, 1968, page 566).

Theorem 5.1.6 (The Cauchy Criterion).   The sequence $\{a_n\}_{n=1}^{\infty}$ converges if and only if it satisfies the following condition, known as the ϵ-condition: For each $\epsilon > 0$ there is an integer N such that

$$|a_m - a_n| < \epsilon \qquad \text{for all } m > N, n > N.$$

Proof. Necessity: If the sequence converges, then it must satisfy the ϵ-condition. Let $\epsilon > 0$ be given. Since the sequence $\{a_n\}_{n=1}^{\infty}$ converges, then there exists a number c and an integer N such that

$$|a_n - c| < \frac{\epsilon}{2} \qquad \text{if } n > N.$$

Hence, for $m > N$, $n > N$ we must have

$$|a_m - a_n| = |a_m - c + c - a_n|$$
$$\leq |a_m - c| + |a_n - c| < \epsilon.$$

Sufficiency: If the sequence satisfies the ϵ-condition, then it must converge. If the ϵ-condition is satisfied, then there is an integer N such that for any given $\epsilon > 0$,

$$|a_n - a_{N+1}| < \epsilon$$

for all values of $n \geq N + 1$. Thus for such values of n,

$$a_{N+1} - \epsilon < a_n < a_{N+1} + \epsilon. \qquad (5.2)$$

The sequence $\{a_n\}_{n=1}^{\infty}$ is therefore bounded, since from the double inequality (5.2) we can assert that

$$|a_n| < \max(|a_1| + 1, |a_2| + 1, \ldots, |a_N| + 1, |a_{N+1} - \epsilon|, |a_{N+1} + \epsilon|)$$

for all n. By Theorem 5.1.4, $\{a_n\}_{n=1}^{\infty}$ has a convergent subsequence $\{a_{k_n}\}_{n=1}^{\infty}$. Let c be the limit of this subsequence. If we invoke again the ϵ-condition, we can find an integer N' such that

$$|a_m - a_{k_n}| < \epsilon' \qquad \text{if } m > N', k_n \geq n \geq N',$$

where $\epsilon' < \epsilon$. By fixing m and letting $k_n \to \infty$ we get

$$|a_m - c| \leq \epsilon' < \epsilon \qquad \text{if } m > N'.$$

This indicates that the sequence $\{a_n\}_{n=1}^{\infty}$ is convergent and has c as its limit. $\square$

Definition 5.1.3. A sequence $\{a_n\}_{n=1}^{\infty}$ that satisfies the ϵ-condition of the Cauchy criterion is said to be a Cauchy sequence. $\square$

EXAMPLE 5.1.6. With the help of the Cauchy criterion it is now possible to show that the sequence $\{a_n\}_{n=1}^{\infty}$ whose nth term is defined by formula (5.1) is a Cauchy sequence and is therefore convergent. To do so, let $m > n$. Then,

$$a_m - a_n = (-1)^n \left[\frac{1}{2n+1} - \frac{1}{2n+3} + \cdots + \frac{(-1)^{p-1}}{2n+2p-1} \right], \qquad (5.3)$$

where $p = m - n$. We claim that the quantity inside brackets in formula (5.3) is positive. This can be shown by grouping successive terms in pairs. Thus if p is even, the quantity is equal to

$$\left(\frac{1}{2n+1} - \frac{1}{2n+3} \right) + \left(\frac{1}{2n+5} - \frac{1}{2n+7} \right) + \cdots$$
$$+ \left(\frac{1}{2n+2p-3} - \frac{1}{2n+2p-1} \right),$$

which is positive, since the difference inside each parenthesis is positive. If $p = 1$, the quantity is obviously positive, since it is then equal to $1/(2n+1)$. If $p \geq 3$ is an odd integer, the quantity can be written as

$$\left(\frac{1}{2n+1} - \frac{1}{2n+3} \right) + \left(\frac{1}{2n+5} - \frac{1}{2n+7} \right) + \cdots$$
$$+ \left(\frac{1}{2n+2p-5} - \frac{1}{2n+2p-3} \right) + \frac{1}{2n+2p-1},$$

which is also positive. Hence, for any p,

$$|a_m - a_n| = \frac{1}{2n+1} - \frac{1}{2n+3} + \cdots + \frac{(-1)^{p-1}}{2n+2p-1}. \qquad (5.4)$$

We now claim that

$$|a_m - a_n| < \frac{1}{2n+1}.$$

To prove this claim, let us again consider two cases. If p is even, then

$$|a_m - a_n| = \frac{1}{2n+1} - \left(\frac{1}{2n+3} - \frac{1}{2n+5} \right) - \cdots$$
$$- \left(\frac{1}{2n+2p-5} - \frac{1}{2n+2p-3} \right) - \frac{1}{2n+2p-1}$$
$$< \frac{1}{2n+1}, \qquad (5.5)$$

since all the quantities inside parentheses in (5.5) are positive. If p is odd, then

$$|a_m - a_n| = \frac{1}{2n+1} - \left(\frac{1}{2n+3} - \frac{1}{2n+5} \right) - \cdots$$
$$- \left(\frac{1}{2n+2p-3} - \frac{1}{2n+2p-1} \right) < \frac{1}{2n+1},$$

which proves our claim. On the basis of this result we can assert that for a given $\epsilon > 0$,

$$|a_m - a_n| < \epsilon \qquad \text{if } m > n > N,$$

where N is such that

$$\frac{1}{2N+1} < \epsilon,$$

or equivalently,

$$N > \frac{1}{2\epsilon} - \frac{1}{2}.$$

This shows that $\{a_n\}_{n=1}^{\infty}$ is a Cauchy sequence.

EXAMPLE 5.1.7.   Consider the sequence $\{a_n\}_{n=1}^{\infty}$, where

$$a_n = (-1)^n \left(1 + \frac{1}{n}\right).$$

We have seen earlier that $\liminf_{n \to \infty} a_n = -1$ and $\limsup_{n \to \infty} a_n = 1$. Thus by Theorem 5.1.5 this sequence is not convergent. We can arrive at the same conclusion using the Cauchy criterion by showing that the ϵ-condition is not satisfied. This occurs whenever we can find an $\epsilon > 0$ such that for however N may be chosen,

$$|a_m - a_n| \geq \epsilon$$

for some $m > N$, $n > N$. In our example, if N is any positive integer, then the inequality

$$|a_m - a_n| \geq 2 \tag{5.6}$$

can be satisfied by choosing $m = \nu$ and $n = \nu + 1$, where ν is an odd integer greater than N.

5.2. INFINITE SERIES

Let $\{a_n\}_{n=1}^{\infty}$ be a given sequence. Consider the symbolic expression

$$\sum_{n=1}^{\infty} a_n = a_1 + a_2 + \cdots + a_n + \cdots . \tag{5.7}$$

By definition, this expression is called an infinite series, or just a series for simplicity, and a_n is referred to as the nth term of the series. The finite sum

$$s_n = \sum_{i=1}^{n} a_i, \qquad n = 1, 2, \ldots,$$

is called the nth partial sum of the series.

Definition 5.2.1.   Consider the series $\sum_{n=1}^{\infty} a_n$. Let s_n be its nth partial sum ($n = 1, 2, \ldots$).

1. The series is said to be convergent if the sequence $\{s_n\}_{n=1}^{\infty}$ converges. In this case, if $\lim_{n \to \infty} s_n = s$, where s is finite, then we say that the series converges to s, or that s is the sum of the series. Symbolically, this is

expressed by writing

$$s = \sum_{n=1}^{\infty} a_n.$$

2. If s_n does not tend to a finite limit, then the series is said to be divergent. □

Definition 5.2.1 formulates convergence of a series in terms of convergence of the associated sequence of its partial sums. By applying the Cauchy criterion (Theorem 5.1.6) to the latter sequence, we arrive at the following condition of convergence for a series:

Theorem 5.2.1. The series $\sum_{n=1}^{\infty} a_n$, converges if and only if for a given $\epsilon > 0$ there is an integer N such that

$$\left| \sum_{i=m+1}^{n} a_i \right| < \epsilon \qquad \text{for all } n > m > N. \tag{5.8}$$

Inequality (5.8) follows from applying Theorem 5.1.6 to the sequence $\{s_n\}_{n=1}^{\infty}$ of partial sums of the series and noting that

$$|s_n - s_m| = \left| \sum_{i=m+1}^{n} a_i \right| \qquad \text{for } n > m.$$

In particular, if $n = m + 1$, then inequality (5.8) becomes

$$|a_{m+1}| < \epsilon \tag{5.9}$$

for all $m > N$. This implies that $\lim_{m \to \infty} a_{m+1} = 0$, and hence $\lim_{n \to \infty} a_n = 0$. We therefore conclude the following result:

RESULT 5.2.1. If $\sum_{n=1}^{\infty} a_n$ is a convergent series, then $\lim_{n \to \infty} a_n = 0$.

It is important here to note that the convergence of the nth term of a series to zero as $n \to \infty$ is a necessary condition for the convergence of the series. It is not, however, a sufficient condition, that is, if $\lim_{n \to \infty} a_n = 0$, then it does not follow that $\sum_{n=1}^{\infty} a_n$ converges. For example, as we shall see later, the series $\sum_{n=1}^{\infty} (1/n)$ is divergent, and its nth term goes to zero as $n \to \infty$. It is true, however, that if $\lim_{n \to \infty} a_n \neq 0$, then $\sum_{n=1}^{\infty} a_n$ is divergent. This follows from applying the law of contraposition to the necessary condition of convergence. We conclude the following:

1. If $a_n \to 0$ as $n \to \infty$, then no conclusion can be reached regarding convergence or divergence of $\sum_{n=1}^{\infty} a_n$.

2. If $a_n \nrightarrow 0$ as $n \to \infty$, then $\sum_{n=1}^{\infty} a_n$ is divergent. For example, the series $\sum_{n=1}^{\infty} [n/(n+1)]$ is divergent, since

$$\lim_{n \to \infty} \frac{n}{n+1} = 1 \neq 0.$$

EXAMPLE 5.2.1.   One of the simplest series is the geometric series, $\sum_{n=1}^{\infty} a^n$. This series is divergent if $|a| \geq 1$, since $\lim_{n \to \infty} a^n \neq 0$. It is convergent if $|a| < 1$ by the Cauchy criterion: Let $n > m$. Then

$$s_n - s_m = a^{m+1} + a^{m+2} + \cdots + a^n. \tag{5.10}$$

By multiplying the two sides of (5.10) by a, we get

$$a(s_n - s_m) = a^{m+2} + a^{m+3} + \cdots + a^{n+1}. \tag{5.11}$$

If we now subtract (5.11) from (5.10), we obtain

$$s_n - s_m = \frac{a^{m+1} - a^{n+1}}{1 - a}. \tag{5.12}$$

Since $|a| < 1$, we can find an integer N such that for $m > N$, $n > N$,

$$|a|^{m+1} < \frac{\epsilon(1-a)}{2},$$

$$|a|^{n+1} < \frac{\epsilon(1-a)}{2}.$$

Hence, for a given $\epsilon > 0$,

$$|s_n - s_m| < \epsilon \qquad \text{if } n > m > N.$$

Formula (5.12) can actually be used to find the sum of the geometric series when $|a| < 1$. Let $m = 1$. By taking the limits of both sides of (5.12) as $n \to \infty$ we get

$$\lim_{n \to \infty} s_n = s_1 + \frac{a^2}{1-a}, \qquad \text{since } \lim_{n \to \infty} a^{n+1} = 0,$$

$$= a + \frac{a^2}{1-a}$$

$$= \frac{a}{1-a}.$$

EXAMPLE 5.2.2.   Consider the series $\sum_{n=1}^{\infty}(1/n!)$. This series converges by the Cauchy criterion. To show this, we first note that

$$n! = n(n-1)(n-2) \times \cdots \times 3 \times 2 \times 1$$

$$\geq 2^{n-1} \quad \text{for } n = 1, 2, \ldots .$$

Hence, for $n > m$,

$$|s_n - s_m| = \frac{1}{(m+1)!} + \frac{1}{(m+2)!} + \cdots + \frac{1}{n!}$$

$$\leq \frac{1}{2^m} + \frac{1}{2^{m+1}} + \cdots + \frac{1}{2^{n-1}}$$

$$= 2 \sum_{i=m+1}^{n} \frac{1}{2^i}.$$

This is a partial sum of a convergent geometric series with $a = \frac{1}{2} < 1$ [see formula (5.10)]. Consequently, $|s_n - s_m|$ can be made smaller than any given $\epsilon > 0$ by choosing m and n large enough.

Theorem 5.2.2.   If $\sum_{n=1}^{\infty} a_n$ and $\sum_{n=1}^{\infty} b_n$ are two convergent series, and if c is a constant, then the following series are also convergent:

1. $\sum_{n=1}^{\infty}(ca_n) = c\sum_{n=1}^{\infty} a_n$.
2. $\sum_{n=1}^{\infty}(a_n + b_n) = \sum_{n=1}^{\infty} a_n + \sum_{n=1}^{\infty} b_n$.

Proof. The proof is left to the reader.   □

Definition 5.2.2.   The series $\sum_{n=1}^{\infty} a_n$ is absolutely convergent if $\sum_{n=1}^{\infty}|a_n|$ is convergent.   □

For example, the series $\sum_{n=1}^{\infty}[(-1)^n/n!]$ is absolutely convergent, since $\sum_{n=1}^{\infty}(1/n!)$ is convergent, as was seen in Example 5.2.2.

Theorem 5.2.3.   Every absolutely convergent series is convergent.

Proof. Consider the series $\sum_{n=1}^{\infty} a_n$, and suppose that $\sum_{n=1}^{\infty}|a_n|$ is convergent. We have that

$$\left| \sum_{i=m+1}^{n} a_i \right| \leq \sum_{i=m+1}^{n} |a_i|. \tag{5.13}$$

By applying the Cauchy criterion to $\sum_{n=1}^{\infty} |a_n|$ we can find an integer N such that for a given $\epsilon > 0$,

$$\sum_{i=m+1}^{n} |a_i| < \epsilon \qquad \text{if } n > m > N. \tag{5.14}$$

From (5.13) and (5.14) we conclude that $\sum_{n=1}^{\infty} a_n$ satisfies the Cauchy criterion and is therefore convergent by Theorem 5.2.1. $\qquad\square$

Note that it is possible that $\sum_{n=1}^{\infty} a_n$ is convergent while $\sum_{n=1}^{\infty} |a_n|$ is divergent. In this case, the series $\sum_{n=1}^{\infty} a_n$ is said to be conditionally convergent. Examples of this kind of series will be seen later.

In the next section we shall discuss convergence of series whose terms are positive.

5.2.1. Tests of Convergence for Series of Positive Terms

Suppose that the terms of the series $\sum_{n=1}^{\infty} a_n$ are such that $a_n > 0$ for $n > K$, where K is a constant. Without loss of generality we shall consider that $K = 1$. Such a series is called a series of positive terms.

Series of positive terms are interesting because the study of their convergence is comparatively simple and can be used in the determination of convergence of more general series whose terms are not necessarily positive. It is easy to see that a series of positive terms diverges if and only if its sum is $+\infty$.

In what follows we shall introduce techniques that simplify the process of determining whether or not a given series of positive terms is convergent. We refer to these techniques as tests of convergence. The advantage of these tests is that they are in general easier to apply than the Cauchy criterion. This is because evaluating or obtaining inequalities involving the expression $\sum_{i=m+1}^{n} a_i$ in Theorem 5.2.1 can be somewhat difficult. The tests of convergence, however, have the disadvantage that they can sometime fail to determine convergence or divergence, as we shall soon find out. It should be remembered that these tests apply only to series of positive terms.

The Comparison Test

This test is based on the following theorem:

Theorem 5.2.4. Let $\sum_{n=1}^{\infty} a_n$ and $\sum_{n=1}^{\infty} b_n$ be two series of positive terms such that $a_n \leq b_n$ for $n > N_0$, where N_0 is a fixed integer.

   i. If $\sum_{n=1}^{\infty} b_n$ converges, then so does $\sum_{n=1}^{\infty} a_n$.
   ii. If $\sum_{n=1}^{\infty} a_n$ is divergent, then $\sum_{n=1}^{\infty} b_n$ is divergent too.

Proof. We have that

$$\sum_{i=m+1}^{n} a_i \le \sum_{i=m+1}^{n} b_i \quad \text{for } n > m > N_0. \tag{5.15}$$

If $\sum_{n=1}^{\infty} b_n$ is convergent, then for a given $\epsilon > 0$ there exists an integer N_1 such that

$$\sum_{i=m+1}^{n} b_i < \epsilon \quad \text{for } n > m > N_1. \tag{5.16}$$

From (5.15) and (5.16) it follows that if $n > m > N$, where $N = \max(N_0, N_1)$, then

$$\sum_{i=m+1}^{n} a_i < \epsilon,$$

which proves (i).

The proof of (ii) follows from applying the law of contraposition to (i).
□

To determine convergence or divergence of $\sum_{n=1}^{\infty} a_n$ we thus need to have in our repertoire a collection of series of positive terms whose behavior (with regard to convergence or divergence) is known. These series can then be compared against $\sum_{n=1}^{\infty} a_n$. For this purpose, the following series can be useful:

a. $\sum_{n=1}^{\infty} 1/n$. This is a divergent series called the harmonic series.
b. $\sum_{n=1}^{\infty} 1/n^k$. This is divergent if $k < 1$ and is convergent if $k > 1$.

To prove that the harmonic series is divergent, let us consider its nth partial sum, namely,

$$s_n = \sum_{i=1}^{n} \frac{1}{i}.$$

Let $A > 0$ be an arbitrary positive number. Choose n large enough so that $n > 2^m$, where $m > 2A$. Then for such values of n,

$$s_n > \left(1 + \frac{1}{2}\right) + \left(\frac{1}{3} + \frac{1}{4}\right) + \left(\frac{1}{5} + \frac{1}{6} + \frac{1}{7} + \frac{1}{8}\right) + \cdots$$

$$+ \left(\frac{1}{2^{m-1}+1} + \cdots + \frac{1}{2^m}\right)$$

$$> \frac{1}{2} + \frac{2}{4} + \frac{4}{8} + \cdots + \frac{2^{m-1}}{2^m} = \frac{m}{2} > A. \tag{5.17}$$

Since A is arbitrary and s_n is a monotone increasing function of n, inequality (5.17) implies that $s_n \to \infty$ as $n \to \infty$. This proves divergence of the harmonic series.

Let us next consider the series in (b). If $k < 1$, then $1/n^k > 1/n$ and $\sum_{n=1}^{\infty}(1/n^k)$ must be divergent by Theorem 5.2.4(ii). Suppose now that $k > 1$. Consider the nth partial sum of the series, namely,

$$s_n' = \sum_{i=1}^{n} \frac{1}{i^k}.$$

Then, by choosing m large enough so that $2^m > n$ we get

$$s_n' \le \sum_{i=1}^{2^m - 1} \frac{1}{i^k}$$

$$= 1 + \left(\frac{1}{2^k} + \frac{1}{3^k} \right) + \left(\frac{1}{4^k} + \frac{1}{5^k} + \frac{1}{6^k} + \frac{1}{7^k} \right) + \cdots$$

$$+ \left[\frac{1}{\left(2^{m-1}\right)^k} + \cdots + \frac{1}{\left(2^m - 1\right)^k} \right]$$

$$\le 1 + \left(\frac{1}{2^k} + \frac{1}{2^k} \right) + \left(\frac{1}{4^k} + \frac{1}{4^k} + \frac{1}{4^k} + \frac{1}{4^k} \right) + \cdots$$

$$+ \left[\frac{1}{\left(2^{m-1}\right)^k} + \cdots + \frac{1}{\left(2^{m-1}\right)^k} \right]$$

$$= 1 + \frac{2}{2^k} + \frac{4}{4^k} + \cdots + \frac{2^{m-1}}{\left(2^{m-1}\right)^k}$$

$$= \sum_{i=1}^{m} a^{i-1}, \tag{5.18}$$

where $a = 1/2^{k-1}$. But the right-hand side of (5.18) represents the mth partial sum of a convergent geometric series (since $a < 1$). Hence, as $m \to \infty$, the right-hand side of (5.18) converges to

$$\sum_{i=1}^{\infty} a^{i-1} = \frac{1}{1-a} \qquad \text{(see Example 5.2.1)}.$$

Thus the sequence $\{s_n'\}_{n=1}^{\infty}$ is bounded. Since it is also monotone increasing, it must be convergent (see Theorem 5.1.2). This proves convergence of the series $\sum_{n=1}^{\infty}(1/n^k)$ for $k > 1$.

Another version of the comparison test in Theorem 5.2.4 that is easier to implement is given by the following theorem:

Theorem 5.2.5. Let $\sum_{n=1}^{\infty} a_n$ and $\sum_{n=1}^{\infty} b_n$ be two series of positive terms. If there exists a positive constant l such that a_n and lb_n are asymptotically equal, $a_n \sim lb_n$ as $n \to \infty$ (see Section 3.3), that is,

$$\lim_{n \to \infty} \frac{a_n}{b_n} = l,$$

then the two series are either both convergent or both divergent.

Proof. There exists an integer N such that

$$\left| \frac{a_n}{b_n} - l \right| < \frac{l}{2} \qquad \text{if } n > N,$$

or equivalently,

$$\frac{l}{2} < \frac{a_n}{b_n} < \frac{3l}{2} \qquad \text{whenever } n > N.$$

If $\sum_{n=1}^{\infty} a_n$ is convergent, then $\sum_{n=1}^{\infty} b_n$ is convergent by a combination of Theorem 5.2.2(1) and 5.2.4(i), since $b_n < (2/l)a_n$. Similarly, if $\sum_{n=1}^{\infty} b_n$ converges, then so does $\sum_{n=1}^{\infty} a_n$, since $a_n < (3l/2)b_n$. If $\sum_{n=1}^{\infty} a_n$ is divergent, then $\sum_{n=1}^{\infty} b_n$ is divergent too by a combination of Theorems 5.2.2(1) and 5.2.4(ii), since $b_n > (2/3l)a_n$. Finally, $\sum_{n=1}^{\infty} a_n$ diverges if the same is true of $\sum_{n=1}^{\infty} b_n$, since $a_n > (l/2)b_n$. □

EXAMPLE 5.2.3. The series $\sum_{n=1}^{\infty} (n + 2)/(n^3 + 2n + 1)$ is convergent, since

$$\frac{n + 2}{n^3 + 2n + 1} \sim \frac{1}{n^2} \qquad \text{as } n \to \infty,$$

which is the nth term of a convergent series [recall that $\sum_{n=1}^{\infty} (1/n^k)$ is convergent if $k > 1$].

EXAMPLE 5.2.4. $\sum_{n=1}^{\infty} 1/\sqrt{n(n + 1)}$ is divergent, because

$$\frac{1}{\sqrt{n(n + 1)}} \sim \frac{1}{n} \qquad \text{as } n \to \infty,$$

which is the nth term of the divergent harmonic series.

The Ratio or d'Alembert's Test

This test is usually attributed to the French mathematician Jean Baptiste d'Alembert (1717–1783), but is also known as Cauchy's ratio test after Augustin-Louis Cauchy (1789–1857).

Theorem 5.2.6. Let $\sum_{n=1}^{\infty} a_n$ be a series of positive terms. Then the following hold:

1. The series converges if $\lim \sup_{n \to \infty}(a_{n+1}/a_n) < 1$ (see Definition 5.1.2).
2. The series diverges if $\lim \inf_{n \to \infty}(a_{n+1}/a_n) > 1$ (see Definition 5.1.2).
3. If $\lim \inf_{n \to \infty}(a_{n+1}/a_n) \le 1 \le \lim \sup_{n \to \infty}(a_{n+1}/a_n)$, no conclusion can be made regarding convergence or divergence of the series (that is, the ratio test fails).

In particular, if $\lim_{n \to \infty}(a_{n+1}/a_n) = r$ exists, then the following hold:

1. The series converges if $r < 1$.
2. The series diverges if $r > 1$.
3. The test fails if $r = 1$.

Proof. Let $p = \lim \inf_{n \to \infty}(a_{n+1}/a_n)$, $q = \lim \sup_{n \to \infty}(a_{n+1}/a_n)$.

1. If $q < 1$, then by the definition of the upper limit (Definition 5.1.2), there exists an integer N such that

$$\frac{a_{n+1}}{a_n} < q' \qquad \text{for } n \ge N, \tag{5.19}$$

where q' is chosen such that $q < q' < 1$. (If $a_{n+1}/a_n \ge q'$ for infinitely many values of n, then the sequence $\{a_{n+1}/a_n\}_{n=1}^{\infty}$ has a subsequential limit greater than or equal to q', which exceeds q. This contradicts the definition of q.) From (5.19) we then get

$$a_{N+1} < a_N q'$$

$$a_{N+2} < a_{N+1} q' < a_N q'^2,$$

$$\vdots$$

$$a_{N+m} < a_{N+m-1} q' < a_N q'^m,$$

where $m \ge 1$. Thus for $n > N$,

$$a_n < a_N q'^{(n-N)} = a_N (q')^{-N} q'^n.$$

Hence, the series converges by comparison with the convergent geometric series $\sum_{n=1}^{\infty} q'^n$, since $q' < 1$.

2. If $p > 1$, then in an analogous manner we can find an integer N such that

$$\frac{a_{n+1}}{a_n} > p' \qquad \text{for } n \geq N, \tag{5.20}$$

where p' is chosen such that $p > p' > 1$. But this implies that a_n cannot tend to zero as $n \to \infty$, and the series is therefore divergent by Result 5.2.1.

3. If $p \leq 1 \leq q$, then we can demonstrate by using an example that the ratio test is inconclusive: Consider the two series $\sum_{n=1}^{\infty}(1/n)$, $\sum_{n=1}^{\infty}(1/n^2)$. For both series, $p = q = 1$ and hence $p \leq 1 \leq q$, since $\lim_{n \to \infty}(a_{n+1}/a_n) = 1$. But the first series is divergent while the second is convergent, as was seen earlier. $\qquad \square$

EXAMPLE 5.2.5.   Consider the same series as in Example 5.2.2. This series was shown to be convergent by the Cauchy criterion. Let us now apply the ratio test. In this case,

$$\lim_{n \to \infty} \frac{a_{n+1}}{a_n} = \lim_{n \to \infty} \frac{1}{(n+1)!} \Big/ \frac{1}{n!}$$

$$= \lim_{n \to \infty} \frac{1}{n+1} = 0 < 1,$$

which indicates convergence by Theorem 5.2.6(1).

Nurcombe (1979) stated and proved the following extension of the ratio test:

Theorem 5.2.7.   Let $\sum_{n=1}^{\infty} a_n$ be a series of positive terms, and k be a fixed positive integer.

1. If $\lim_{n \to \infty}(a_{n+k}/a_n) < 1$, then the series converges.
2. If $\lim_{n \to \infty}(a_{n+k}/a_n) > 1$, then the series diverges.

This test reduces to the ratio test when $k = 1$.

The Root or Cauchy's Test

This is a more powerful test than the ratio test. It is based on the following theorem:

Theorem 5.2.8. Let $\sum_{n=1}^{\infty} a_n$ be a series of positive terms. Let $\limsup_{n \to \infty} a_n^{1/n} = \rho$. Then we have the following:

1. The series converges if $\rho < 1$.
2. The series diverges if $\rho > 1$.
3. The test is inconclusive if $\rho = 1$.

In particular, if $\lim_{n \to \infty} a_n^{1/n} = \tau$ exists, then we have the following:

1. The series converges if $\tau < 1$.
2. The series diverges if $\tau > 1$.
3. The test is inconclusive if $\tau = 1$.

Proof.

1. As in Theorem 5.2.6(1), if $\rho < 1$, then there is an integer N such that

$$a_n^{1/n} < \rho' \qquad \text{for } n \geq N,$$

where ρ' is chosen such that $\rho < \rho' < 1$. Thus

$$a_n < \rho'^n \qquad \text{for } n \geq N.$$

The series is therefore convergent by comparison with the convergent geometric series $\sum_{n=1}^{\infty} \rho'^n$, since $\rho' < 1$.

2. Suppose that $\rho > 1$. Let $\epsilon > 0$ be such that $\epsilon < \rho - 1$. Then

$$a_n^{1/n} > \rho - \epsilon > 1$$

for infinitely many values of n (why?). Thus for such values of n,

$$a_n > (\rho - \epsilon)^n,$$

which implies that a_n cannot tend to zero as $n \to \infty$ and the series is therefore divergent by Result 5.2.1.

3. Consider again the two series $\sum_{n=1}^{\infty}(1/n), \sum_{n=1}^{\infty}(1/n^2)$. In both cases $\rho = 1$ (see Exercise 5.18). The test therefore fails, since the first series is divergent and the second is convergent. $\square$

NOTE 5.2.1. We have mentioned earlier that the root test is more powerful than the ratio test. By this we mean that whenever the ratio test shows convergence or divergence, then so does the root test; whenever the root test is inconclusive, the ratio test is inconclusive too. However, there are situations where the ratio test fails, but the root test doe not (see Example 5.2.6). This fact is based on the following theorem:

Theorem 5.2.9. If $a_n > 0$, then

$$\liminf_{n \to \infty} \frac{a_{n+1}}{a_n} \le \liminf_{n \to \infty} a_n^{1/n} \le \limsup_{n \to \infty} a_n^{1/n} \le \limsup_{n \to \infty} \frac{a_{n+1}}{a_n}.$$

Proof. It is sufficient to prove the two inequalities

$$\limsup_{n \to \infty} a_n^{1/n} \le \limsup_{n \to \infty} \frac{a_{n+1}}{a_n}, \tag{5.21}$$

$$\liminf_{n \to \infty} \frac{a_{n+1}}{a_n} \le \liminf_{n \to \infty} a_n^{1/n}. \tag{5.22}$$

Inequality (5.21): Let $q = \limsup_{n \to \infty}(a_{n+1}/a_n)$. If $q = \infty$, then there is nothing to prove. Let us therefore consider that q is finite. If we choose q' such that $q < q'$, then as in the proof of Theorem 5.2.6(1), we can find an integer N such that

$$a_n < a_N(q')^{-N} q'^n \quad \text{for } n > N.$$

Hence,

$$a_n^{1/n} < \left[a_N(q')^{-N} \right]^{1/n} q'. \tag{5.23}$$

As $n \to \infty$, the limit of the right-hand side of inequality (5.23) is q'. It follows that

$$\limsup_{n \to \infty} a_n^{1/n} \le q'. \tag{5.24}$$

Since (5.24) is true for any $q' > q$, then we must also have

$$\limsup_{n \to \infty} a_n^{1/n} \le q.$$

Inequality (5.22): Let $p = \liminf_{n \to \infty}(a_{n+1}/a_n)$. We can consider p to be finite (if $p = \infty$, then $q = \infty$ and the proof of the theorem will be complete; if $p = -\infty$, then there is nothing to prove). Let p' be chosen such that $p' < p$. As in the proof of Theorem 5.2.6(2), we can find an integer N such that

$$\frac{a_{n+1}}{a_n} > p' \qquad \text{for } n \ge N. \tag{5.25}$$

From (5.25) it is easy to show that

$$a_n > a_N(p')^{-N} p'^n \qquad \text{for } n \ge N.$$

Hence, for such values of n,

$$a_n^{1/n} > \left[a_N(p')^{-N}\right]^{1/n} p'.$$

Consequently,

$$\liminf_{n \to \infty} a_n^{1/n} \geq p'. \tag{5.26}$$

Since (5.26) is true for any $p' < p$, then

$$\liminf_{n \to \infty} a_n^{1/n} \geq p.$$

From Theorem 5.2.9 we can easily see that whenever $q < 1$, then $\limsup_{n \to \infty} a_n^{1/n} < 1$; whenever $p > 1$, then $\limsup_{n \to \infty} a_n^{1/n} > 1$. In both cases, if convergence or divergence of the series is resolved by the ratio test, then it can also be resolved by the root test. If, however, the root test fails (when $\limsup_{n \to \infty} a_n^{1/n} = 1$), then the ratio test fails too by Theorem 5.2.6(3). On the other hand, it is possible for the ratio test to be inconclusive whereas the root test is not. This occurs when

$$\liminf_{n \to \infty} \frac{a_{n+1}}{a_n} \leq \liminf_{n \to \infty} a_n^{1/n} \leq \limsup_{n \to \infty} a_n^{1/n} < 1 \leq \limsup_{n \to \infty} \frac{a_{n+1}}{a_n}. \qquad \square$$

EXAMPLE 5.2.6.   Consider the series $\sum_{n=1}^{\infty}(a^n + b^n)$, where $0 < a < b < 1$. This can be written as $\sum_{n=1}^{\infty} c_n$, where for $n \geq 1$,

$$c_n = \begin{cases} a^{(n+1)/2} & \text{if } n \text{ is odd,} \\ b^{n/2} & \text{if } n \text{ is even.} \end{cases}$$

Now,

$$\frac{c_{n+1}}{c_n} = \begin{cases} (b/a)^{(n+1)/2} & \text{if } n \text{ is odd,} \\ a(a/b)^{n/2} & \text{if } n \text{ is even,} \end{cases}$$

$$c_n^{1/n} = \begin{cases} a^{(n+1)/(2n)} & \text{if } n \text{ is odd,} \\ b^{1/2} & \text{if } n \text{ is even.} \end{cases}$$

As $n \to \infty$, c_{n+1}/c_n has two limits, namely 0 and ∞; $c_n^{1/n}$ has two limits, $a^{1/2}$ and $b^{1/2}$. Thus

$$\liminf_{n \to \infty} \frac{c_{n+1}}{c_n} = 0,$$

$$\limsup_{n \to \infty} \frac{c_{n+1}}{c_n} = \infty,$$

$$\limsup_{n \to \infty} c_n^{1/n} = b^{1/2} < 1.$$

Since $0 \le 1 \le \infty$, we can clearly see that the ratio test is inconclusive, whereas the root test indicates that the series is convergent.

Maclaurin's (or Cauchy's) Integeral Test

This test was introduced by Colin Maclaurin (1698–1746) and then rediscovered by Cauchy. The description and proof of this test will be given in Chapter 6.

Cauchy's Condensation Test

Let us consider the following theorem:

Theorem 5.2.10. Let $\sum_{n=1}^{\infty} a_n$ be a series of positive terms, where a_n is a monotone decreasing function of n ($= 1, 2, \ldots$). Then $\sum_{n=1}^{\infty} a_n$ converges or diverges if and only if the same is true of the series $\sum_{n=1}^{\infty} 2^n a_{2^n}$.

Proof. Let s_n and t_m be the nth and mth partial sums, respectively, of $\sum_{n=1}^{\infty} a_n$ and $\sum_{n=1}^{\infty} 2^n a_{2^n}$. If m is such that $n < 2^m$, then

$$s_n \le a_1 + (a_2 + a_3) + (a_4 + a_5 + a_6 + a_7)$$
$$+ \cdots + (a_{2^m} + a_{2^m+1} + \cdots + a_{2^m+2^m-1})$$
$$\le a_1 + 2a_2 + 4a_4 + \cdots + 2^m a_{2^m} = t_m. \tag{5.27}$$

Furthermore, if $n > 2^m$, then

$$s_n \ge a_1 + a_2 + (a_3 + a_4) + \cdots + (a_{2^{m-1}+1} + \cdots + a_{2^m})$$
$$\ge \frac{a_1}{2} + a_2 + 2a_4 + \cdots + 2^{m-1} a_{2^m} = \frac{t_m}{2}. \tag{5.28}$$

If $\sum_{n=1}^{\infty} 2^n a_{2^n}$ diverges, then $t_m \to \infty$ as $m \to \infty$. Hence, from (5.28), $s_n \to \infty$ as $n \to \infty$, and the series $\sum_{n=1}^{\infty} a_n$ is also divergent.

Now, if $\sum_{n=1}^{\infty} 2^n a_{2^n}$ converges, then the sequence $\{t_m\}_{m=1}^{\infty}$ is bounded. From (5.27), the sequence $\{s_n\}_{n=1}^{\infty}$ is also bounded. It follows that $\sum_{n=1}^{\infty} a_n$ is a convergent series (see Exercise 5.13).   □

EXAMPLE 5.2.7.   Consider again the series $\sum_{n=1}^{\infty}(1/n^k)$. We have already seen that this series converges if $k > 1$ and diverges if $k \leq 1$. Let us now apply Cauchy's condensation test. In this case,

$$\sum_{n=1}^{\infty} 2^n a_{2^n} = \sum_{n=1}^{\infty} 2^n \frac{1}{2^{nk}} = \sum_{n=1}^{\infty} \frac{1}{2^{n(k-1)}}$$

is a geometric series $\sum_{n=1}^{\infty} b^n$, where $b = 1/2^{k-1}$. If $k \leq 1$, then $b \geq 1$ and the series diverges. If $k > 1$, then $b < 1$ and the series converges. It is interesting to note that in this example, both the ratio and the root tests fail.

The following tests enable us to handle situations where the ratio test fails. These tests are particular cases on a general test called Kummer's test.

Kummer's Test

This test is named after the German mathematician Ernst Eduard Kummer (1810–1893).

Theorem 5.2.11.   Let $\sum_{n=1}^{\infty} a_n$ and $\sum_{n=1}^{\infty} b_n$ be two series of positive terms. Suppose that the series $\sum_{n=1}^{\infty} b_n$ is divergent. Let

$$\lim_{n \to \infty} \left(\frac{1}{b_n} \frac{a_n}{a_{n+1}} - \frac{1}{b_{n+1}} \right) = \lambda,$$

Then $\sum_{n=1}^{\infty} a_n$ converges if $\lambda > 0$ and diverges if $\lambda < 0$.

Proof. Suppose that $\lambda > 0$. We can find an integer N such that for $n > N$,

$$\frac{1}{b_n} \frac{a_n}{a_{n+1}} - \frac{1}{b_{n+1}} > \frac{\lambda}{2}. \tag{5.29}$$

Inequality (5.29) can also be written as

$$a_{n+1} < \frac{2}{\lambda} \left(\frac{a_n}{b_n} - \frac{a_{n+1}}{b_{n+1}} \right). \tag{5.30}$$

If s_n is the nth partial sum of $\sum_{n=1}^{\infty} a_n$, then from (5.30) and for $n > N$,

$$s_{n+1} < s_{N+1} + \frac{2}{\lambda} \sum_{i=N+2}^{n+1} \left(\frac{a_{i-1}}{b_{i-1}} - \frac{a_i}{b_i} \right),$$

that is,

$$s_{n+1} < s_{N+1} + \frac{2}{\lambda} \left(\frac{a_{N+1}}{b_{N+1}} - \frac{a_{n+1}}{b_{n+1}} \right),$$

$$s_{n+1} < s_{N+1} + \frac{2}{\lambda} \frac{a_{N+1}}{b_{N+1}} \qquad \text{for } n > N. \tag{5.31}$$

Inequality (5.31) indicates that the sequence $\{s_n\}_{n=1}^{\infty}$ is bounded. Hence, the series $\sum_{n=1}^{\infty} a_n$ is convergent (see Exercise 5.13).

Now, let us suppose that $\lambda < 0$. We can find an integer N such that

$$\frac{1}{b_n} \frac{a_n}{a_{n+1}} - \frac{1}{b_{n+1}} < 0 \qquad \text{for } n > N.$$

Thus for such values of n,

$$a_{n+1} > \frac{a_n}{b_n} b_{n+1}. \tag{5.32}$$

It is easy to verify that because of (5.32),

$$a_n > \frac{a_{N+1}}{b_{N+1}} b_n \tag{5.33}$$

for $n \geq N + 2$. Since $\sum_{n=1}^{\infty} b_n$ is divergent, then from (5.33) and the use of the comparison test we conclude that $\sum_{n=1}^{\infty} a_n$ is divergent too. $\square$

Two particular cases of Kummer's test are Raabe's test and Gauss's test.

Raabe's Test

This test was established in 1832 by J. L. Raabe.

Theorem 5.2.12. Suppose that $\sum_{n=1}^{\infty} a_n$ is a series of positive terms and that

$$\frac{a_n}{a_{n+1}} = 1 + \frac{\tau}{n} + o\left(\frac{1}{n} \right) \qquad \text{as } n \to \infty.$$

Then $\sum_{n=1}^{\infty} a_n$ converges if $\tau > 1$ and diverges if $\tau < 1$.

Proof. We have that

$$\frac{a_n}{a_{n+1}} = 1 + \frac{\tau}{n} + o\left(\frac{1}{n}\right).$$

This means that

$$n\left(\frac{a_n}{a_{n+1}} - 1 - \frac{\tau}{n}\right) \to 0 \tag{5.34}$$

as $n \to \infty$. Equivalently, (5.34) can be expressed as

$$\lim_{n \to \infty} \left(\frac{na_n}{a_{n+1}} - n - 1\right) = \tau - 1. \tag{5.35}$$

Let $b_n = 1/n$ in (5.35). This is the nth term of a divergent series. If we now apply Kummer's test, we conclude that the series $\sum_{n=1}^{\infty} a_n$ converges if $\tau - 1 > 0$ and diverges if $\tau - 1 < 0$. $\square$

Gauss's Test

This test is named after Carl Friedrich Gauss (1777–1855). It provides a slight improvement over Raabe's test in that it usually enables us to handle the case $\tau = 1$. For such a value of τ, Raabe's test is inconclusive.

Theorem 5.2.13. Let $\sum_{n=1}^{\infty} a_n$ be a series of positive terms. Suppose that

$$\frac{a_n}{a_{n+1}} = 1 + \frac{\theta}{n} + O\left(\frac{1}{n^{\delta+1}}\right), \qquad \delta > 0.$$

Then $\sum_{n=1}^{\infty} a_n$ converges if $\theta > 1$ and diverges if $\theta \leq 1$.

Proof. Since

$$O\left(\frac{1}{n^{\delta+1}}\right) = o\left(\frac{1}{n}\right),$$

then by Raabe's test, $\sum_{n=1}^{\infty} a_n$ converges if $\theta > 1$ and diverges if $\theta < 1$. Let us therefore consider $\theta = 1$. We have

$$\frac{a_n}{a_{n+1}} = 1 + \frac{1}{n} + O\left(\frac{1}{n^{\delta+1}}\right).$$

Put $b_n = 1/(n \log n)$, and consider

$$\lim_{n \to \infty} \left(\frac{1}{b_n} \frac{a_n}{a_{n+1}} - \frac{1}{b_{n+1}} \right)$$

$$= \lim_{n \to \infty} \left\{ n \log n \left[1 + \frac{1}{n} + O\left(\frac{1}{n^{\delta+1}} \right) \right] - (n+1)\log(n+1) \right\}$$

$$= \lim_{n \to \infty} \left[(n+1)\log\frac{n}{n+1} + (n \log n)O\left(\frac{1}{n^{\delta+1}} \right) \right] = -1.$$

This is true because

$$\lim_{n \to \infty} \left[(n+1)\log\frac{n}{n+1} \right] = -1 \qquad \text{(by l'Hospital's rule)}$$

and

$$\lim_{n \to \infty} (n \log n)O\left(\frac{1}{n^{\delta+1}} \right) = 0 \qquad [\text{see Example } 4.2.3(2)].$$

Since $\sum_{n=1}^{\infty}[1/(n \log n)]$ is a divergent series (this can be shown by using Cauchy's condensation test), then by Kummer's test, the series $\sum_{n=1}^{\infty} a_n$ is divergent. □

EXAMPLE 5.2.8. Gauss established his test in order to determine the convergence of the so-called hypergeometric series. He managed to do so in an article published in 1812. This series is of the form $1 + \sum_{n=1}^{\infty} a_n$, where

$$a_n = \frac{\alpha(\alpha+1)(\alpha+2)\cdots(\alpha+n-1)\beta(\beta+1)(\beta+2)\cdots(\beta+n-1)}{n!\gamma(\gamma+1)(\gamma+2)\cdots(\gamma+n-1)},$$

$$n = 1,2,\ldots,$$

where α, β, γ are real numbers, and none of them is zero or a negative integer. We have

$$\frac{a_n}{a_{n+1}} = \frac{(n+1)(n+\gamma)}{(n+\alpha)(n+\beta)} = \frac{n^2 + (\gamma+1)n + \gamma}{n^2 + (\alpha+\beta)n + \alpha\beta}$$

$$= 1 + \frac{\gamma+1-\alpha-\beta}{n} + O\left(\frac{1}{n^2} \right).$$

In this case, $\theta = \gamma+1-\alpha-\beta$ and $\delta = 1$. By Gauss's test, this series is convergent if $\theta > 1$, or $\gamma > \alpha + \beta$, and is divergent if $\theta \leq 1$, or $\gamma \leq \alpha + \beta$.

5.2.2. Series of Positive and Negative Terms

Consider the series $\sum_{n=1}^{\infty} a_n$, where a_n may be positive or negative for $n \geq 1$. The convergence of this general series can be determined by the Cauchy criterion (Theorem 5.1.6). However, it is more convenient to consider the series $\sum_{n=1}^{\infty} |a_n|$ of absolute values, to which the tests of convergence in Section 5.2.1 can be applied. We recall from Definition 5.2.2 that if the latter series converges, then the series $\sum_{n=1}^{\infty} a_n$ is absolutely convergent. This is a stronger type of convergence than the one given in Definition 5.2.1, since by Theorem 5.2.3 convergence of $\sum_{n=1}^{\infty} |a_n|$ implies convergence of $\sum_{n=1}^{\infty} a_n$. The converse, however, is not necessarily true, that is, convergence of $\sum_{n=1}^{\infty} a_n$ does not necessarily imply convergence of $\sum_{n=1}^{\infty} |a_n|$. For example, consider the series

$$\sum_{n=1}^{\infty} \frac{(-1)^{n-1}}{2n-1} = 1 - \frac{1}{3} + \frac{1}{5} - \frac{1}{7} + \cdots. \tag{5.36}$$

This series is convergent by the result of Example 5.1.6. It is not, however, absolutely convergent, since $\sum_{n=1}^{\infty} [1/(2n-1)]$ is divergent by comparison with the harmonic series $\sum_{n=1}^{\infty} (1/n)$, which is divergent. We recall that a series such as (5.36) that converges, but not absolutely, is called a conditionally convergent series.

The series in (5.36) belongs to a special class of series known as alternating series.

Definition 5.2.3. The series $\sum_{n=1}^{\infty} (-1)^{n-1} a_n$, where $a_n > 0$ for $n \geq 1$, is called an alternating series. □

The following theorem, which was established by Gottfried Wilhelm Leibniz (1646–1716), can be used to determine convergence of alternating series:

Theorem 5.2.14. Let $\sum_{n=1}^{\infty} (-1)^{n-1} a_n$ be an alternating series such that the sequence $\{a_n\}_{n=1}^{\infty}$ is monotone decreasing and converges to zero as $n \to \infty$. Then the series is convergent.

Proof. Let s_n be the nth partial sum of the series, and let m be an integer such that $m < n$. Then

$$s_n - s_m = \sum_{i=m+1}^{n} (-1)^{i-1} a_i$$

$$= (-1)^m \left[a_{m+1} - a_{m+2} + \cdots + (-1)^{n-m-1} a_n \right]. \tag{5.37}$$

Since $\{a_n\}_{n=1}^{\infty}$ is monotone decreasing, it is easy to show that the quantity inside brackets in (5.37) is nonnegative. Hence,

$$|s_n - s_m| = a_{m+1} - a_{m+2} + \cdots + (-1)^{n-m-1} a_n.$$

Now, if $n - m$ is odd, then

$$|s_n - s_m| = a_{m+1} - (a_{m+2} - a_{m+3}) - \cdots - (a_{n-1} - a_n)$$
$$\leq a_{m+1}.$$

If $n - m$ is even, then

$$|s_n - s_m| = a_{m+1} - (a_{m+2} - a_{m+3}) - \cdots - (a_{n-2} - a_{n-1}) - a_n$$
$$\leq a_{m+1}.$$

Thus in both cases

$$|s_n - s_m| \leq a_{m+1}.$$

Since the sequence $\{a_n\}_{n=1}^{\infty}$ converges to zero, then for a given $\epsilon > 0$ there exists an integer N such that for $m \geq N$, $a_{m+1} < \epsilon$. Consequently,

$$|s_n - s_m| < \epsilon \qquad \text{if } n > m \geq N.$$

By Theorem 5.2.1, the alternating series is convergent. $\square$

EXAMPLE 5.2.9. The series given by formula (5.36) was shown earlier to be convergent. This result can now be easily verified with the help of Theorem 5.2.14.

EXAMPLE 5.2.10. The series $\sum_{n=1}^{\infty}(-1)^n/n^k$ is absolutely convergent if $k > 1$, is conditionally convergent if $0 < k \leq 1$, and is divergent if $k \leq 0$ (since the nth term does not go to zero).

EXAMPLE 5.2.11. The series $\sum_{n=2}^{\infty}(-1)^n/(\sqrt{n} \log n)$ is conditionally convergent, since it converges by Theorem 5.2.14, but the series of absolute values diverges by Cauchy's condensation test (Theorem 5.2.10).

5.2.3. Rearrangement of Series

One of the main differences between infinite series and finite series is that whereas the latter are amenable to the laws of algebra, the former are not necessarily so. In particular, if the order of terms of an infinite series is altered, its sum (assuming it converges) may, in general, change; or worse, the altered series may even diverge. Before discussing this rather disturbing phenomenon, let us consider the following definition:

Definition 5.2.4. Let J^+ denote the set of positive integers and $\sum_{n=1}^{\infty} a_n$ be a given series. Then a second series such as $\sum_{n=1}^{\infty} b_n$ is said to be a rearrangement of $\sum_{n=1}^{\infty} a_n$ if there exists a one-to-one and onto function $f: J^+ \to J^+$ such that $b_n = a_{f(n)}$ for $n \geq 1$.

For example, the series

$$1 + \tfrac{1}{3} - \tfrac{1}{2} + \tfrac{1}{5} + \tfrac{1}{7} - \tfrac{1}{4} + \cdots, \tag{5.38}$$

where two positive terms are followed by one negative term, is a rearrangement of the alternating harmonic series

$$1 - \tfrac{1}{2} + \tfrac{1}{3} - \tfrac{1}{4} + \tfrac{1}{5} - \cdots. \tag{5.39}$$

The series in (5.39) is conditionally convergent, as is the series in (5.38). However, the two series have different sums (see Exercise 5.21).   □

Fortunately, for absolutely convergent series we have the following theorem:

Theorem 5.2.15. If the series $\sum_{n=1}^{\infty} a_n$ is absolutely convergent, then any rearrangement of it remains absolutely convergent and has the same sum.

Proof. Suppose that $\sum_{n=1}^{\infty} a_n$ is absolutely convergent and that $\sum_{n=1}^{\infty} b_n$ is a rearrangement of it. By Theorem 5.2.1, for a given $\epsilon > 0$, there exists an integer N such that for all $n > m > N$,

$$\sum_{i=m+1}^{n} |a_i| < \frac{\epsilon}{2}.$$

We then have

$$\sum_{k=1}^{\infty} |a_{m+k}| \leq \frac{\epsilon}{2} \qquad \text{if } m > N.$$

Now, let us choose an integer M large enough so that

$$\{1, 2, \ldots, N+1\} \subset \{f(1), f(2), \ldots, f(M)\}.$$

It follows that if $n > M$, then $f(n) \geq N+2$. Consequently, for $n > m > M$,

$$\sum_{i=m+1}^{n} |b_i| = \sum_{i=m+1}^{n} |a_{f(i)}|$$

$$\leq \sum_{k=1}^{\infty} |a_{N+k+1}| \leq \frac{\epsilon}{2}.$$

This implies that the series $\sum_{n=1}^{\infty}|b_n|$ satisfies the Cauchy criterion of Theorem 5.2.1. Therefore, $\sum_{n=1}^{\infty}b_n$ is absolutely convergent.

We now show that the two series have the same sum. Let $s = \sum_{n=1}^{\infty}a_n$, and s_n be its nth partial sum. Then, for a given $\epsilon > 0$ there exists an integer N large enough so that

$$|s_{N+1} - s| < \frac{\epsilon}{2}.$$

If t_n is the nth partial sum of $\sum_{n=1}^{\infty}b_n$, then

$$|t_n - s| \leq |t_n - s_{N+1}| + |s_{N+1} - s|.$$

By choosing M large enough as was done earlier, and by taking $n > M$, we get

$$|t_n - s_{N+1}| = \left| \sum_{i=1}^{n} b_i - \sum_{i=1}^{N+1} a_i \right|$$

$$= \left| \sum_{i=1}^{n} a_{f(i)} - \sum_{i=1}^{N+1} a_i \right|$$

$$\leq \sum_{k=1}^{\infty} |a_{N+k+1}| \leq \frac{\epsilon}{2},$$

since if $n > M$,

$$\{a_1, a_2, \ldots, a_{N+1}\} \subset \{a_{f(1)}, a_{f(2)}, \ldots, a_{f(n)}\}.$$

Hence, for $n > M$,

$$|t_n - s| < \epsilon,$$

which shows that the sum of the series $\sum_{n=1}^{\infty}b_n$ is s.   □

Unlike absolutely convergent series, those that are conditionally convergent are susceptible to rearrangements of their terms. To demonstrate this, let us consider the following alternating series:

$$\sum_{n=1}^{\infty} a_n = \sum_{n=1}^{\infty} \frac{(-1)^{n-1}}{\sqrt{n}}.$$

This series is conditionally convergent, since it is convergent by Theorem 5.2.14 while $\sum_{n=1}^{\infty}(1/\sqrt{n})$ is divergent. Let us consider the following rearrangement:

$$\sum_{n=1}^{\infty} b_n = 1 + \frac{1}{\sqrt{3}} - \frac{1}{\sqrt{2}} + \frac{1}{\sqrt{5}} + \frac{1}{\sqrt{7}} - \frac{1}{\sqrt{4}} + \cdots \qquad (5.40)$$

in which two positive terms are followed by one that is negative. Let s'_{3n} denote the sum of the first $3n$ terms of (5.40). Then

$$
s'_{3n} = \left(1 + \frac{1}{\sqrt{3}} - \frac{1}{\sqrt{2}}\right) + \left(\frac{1}{\sqrt{5}} + \frac{1}{\sqrt{7}} - \frac{1}{\sqrt{4}}\right) + \cdots
$$
$$
+ \left(\frac{1}{\sqrt{4n-3}} + \frac{1}{\sqrt{4n-1}} - \frac{1}{\sqrt{2n}}\right)
$$
$$
= \left(1 - \frac{1}{\sqrt{2}}\right) + \left(\frac{1}{\sqrt{3}} - \frac{1}{\sqrt{4}}\right) + \cdots + \left(\frac{1}{\sqrt{2n-1}} - \frac{1}{\sqrt{2n}}\right)
$$
$$
+ \frac{1}{\sqrt{2n+1}} + \frac{1}{\sqrt{2n+3}} + \cdots + \frac{1}{\sqrt{4n-3}} + \frac{1}{\sqrt{4n-1}}
$$
$$
= s_{2n} + \frac{1}{\sqrt{2n+1}} + \frac{1}{\sqrt{2n+3}} + \cdots + \frac{1}{\sqrt{4n-1}},
$$

where s_{2n} is the sum of the first $2n$ terms of the original series. We note that

$$
s'_{3n} > s_{2n} + \frac{n}{\sqrt{4n-1}}. \tag{5.41}
$$

If s is the sum of the original series, then $\lim_{n \to \infty} s_{2n} = s$ in (5.41). But, since

$$
\lim_{n \to \infty} \frac{n}{\sqrt{4n-1}} = \infty,
$$

the sequence $\{s'_{3n}\}_{n=1}^{\infty}$ is not convergent, which implies that the series in (5.40) is divergent. This clearly shows that a rearrangement of a conditionally convergent series can change its character. This rather unsettling characteristic of conditionally convergent series is depicted in the following theorem due to Georg Riemann (1826–1866):

Theorem 5.2.16. A conditionally convergent series can always be rearranged so as to converge to any given number s, or to diverge to $+\infty$ or to $-\infty$.

Proof. The proof can be found in several books, for example, Apostol (1964, page 368), Fulks (1978, page 489), Knopp (1951, page 318), and Rudin (1964, page 67). □

5.2.4. Multiplication of Series

Suppose that $\sum_{n=1}^{\infty} a_n$ and $\sum_{n=1}^{\infty} b_n$ are two series. We recall from Theorem 5.2.2 that if these series are convergent, then their sum is a convergent series obtained by adding the two series term by term. The product of these two series, however, requires a more delicate operation. There are several ways to define this product. We shall consider the so-called Cauchy's product.

Definition 5.2.5. Let $\sum_{n=0}^{\infty} a_n$ and $\sum_{n=0}^{\infty} b_n$ be two series in which the summation index starts at zero instead of one. Cauchy's product of these two series is the series $\sum_{n=0}^{\infty} c_n$, where

$$c_n = \sum_{k=0}^{n} a_k b_{n-k}, \qquad n = 0, 1, 2, \ldots,$$

that is,

$$\sum_{n=0}^{\infty} c_n = a_0 b_0 + (a_0 b_1 + a_1 b_0) + (a_0 b_2 + a_1 b_1 + a_2 b_0) + \cdots.$$

Other products could have been defined by simply adopting different arrangements of the terms that make up the series $\sum_{n=0}^{\infty} c_n$. □

The question now is: under what condition will Cauchy's product of two series converge? The answer to this question is given in the next theorem.

Theorem 5.2.17. Let $\sum_{n=0}^{\infty} c_n$ be Cauchy's product of $\sum_{n=0}^{\infty} a_n$ and $\sum_{n=0}^{\infty} b_n$. Suppose that these two series are convergent and have sums equal to s and t, respectively.

1. If at least one of $\sum_{n=0}^{\infty} a_n$ and $\sum_{n=0}^{\infty} b_n$ converges absolutely, then $\sum_{n=0}^{\infty} c_n$ converges and its sum is equal to st (this result is known as Mertens's theorem).
2. If both series are absolutely convergent, then $\sum_{n=0}^{\infty} c_n$ converges absolutely to the product st (this result is due to Cauchy).

Proof.

1. Suppose that $\sum_{n=0}^{\infty} a_n$ is the series that converges absolutely. Let s_n, t_n, and u_n denote the partial sums $\sum_{i=0}^{n} a_i$, $\sum_{i=0}^{n} b_i$, and $\sum_{i=0}^{n} c_i$, respectively. We need to show that $u_n \to st$ as $n \to \infty$. We have that

$$u_n = a_0 b_0 + (a_0 b_1 + a_1 b_0) + \cdots + (a_0 b_n + a_1 b_{n-1} + \cdots + a_n b_0)$$

$$= a_0 t_n + a_1 t_{n-1} + \cdots + a_n t_0. \tag{5.42}$$

Let β_n denote the remainder of the series $\sum_{n=0}^{\infty} b_n$ with respect to t_n, that is, $\beta_n = t - t_n$ ($n = 0, 1, 2, \ldots$). By making the proper substitution in

(5.42) we get

$$u_n = a_0(t - \beta_n) + a_1(t - \beta_{n-1}) + \cdots + a_n(t - \beta_0)$$
$$= ts_n - (a_0 \beta_n + a_1 \beta_{n-1} + \cdots + a_n \beta_0). \qquad (5.43)$$

Since $s_n \to s$ as $n \to \infty$, the proof of (1) will be complete if we can show that the sum inside parentheses in (5.43) goes to zero as $n \to \infty$. We now proceed to show that this is the case.

Let $\epsilon > 0$ be given. Since the sequence $\{\beta_n\}_{n=0}^\infty$ converges to zero, there exists an integer N such that

$$|\beta_n| < \epsilon \qquad \text{if } n > N.$$

Hence,

$$|a_0 \beta_n + a_1 \beta_{n-1} + \cdots + a_n \beta_0|$$
$$\leq |a_n \beta_0 + a_{n-1} \beta_1 + \cdots + a_{n-N} \beta_N|$$
$$+ |a_{n-N-1} \beta_{N+1} + a_{n-N-2} \beta_{N+2} + \cdots + a_0 \beta_n|$$
$$< |a_n \beta_0 + a_{n-1} \beta_1 + \cdots + a_{n-N} \beta_N| + \epsilon \sum_{i=0}^{n-N-1} |a_i|$$
$$< B \sum_{i=n-N}^{n} |a_i| + \epsilon s^*, \qquad (5.44)$$

where $B = \max(|\beta_0|, |\beta_1|, \ldots, |\beta_N|)$ and s^* is the sum of the series $\sum_{n=0}^\infty |a_n| (\sum_{n=0}^\infty a_n$ is absolutely convergent). Furthermore, because of this and by the Cauchy criterion we can find an integer M such that

$$\sum_{i=n-N}^{n} |a_i| < \epsilon \qquad \text{if } n - N > M + 1.$$

Thus when $n > N + M + 1$ we get from inequality (5.44)

$$|a_0 \beta_n + a_1 \beta_{n-1} + \cdots + a_n \beta_0| \leq \epsilon(B + s^*).$$

Since ϵ can be arbitrarily small, we conclude that

$$\lim_{n \to \infty} (a_0 \beta_n + a_1 \beta_{n-1} + \cdots + a_n \beta_0) = 0.$$

2. Let v_n denote the nth partial sum of $\sum_{i=0}^{\infty}|c_i|$. Then

$$v_n = |a_0 b_0| + |a_0 b_1 + a_1 b_0| + \cdots + |a_0 b_n + a_1 b_{n-1} + \cdots + a_n b_0|$$

$$\leq |a_0||b_0| + |a_0||b_1| + |a_1||b_0|$$

$$+ \cdots + |a_0||b_n| + |a_1||b_{n-1}| + \cdots + |a_n||b_0|$$

$$= |a_0|t_n^* + |a_1|t_{n-1}^* + \cdots + |a_n|t_0^*,$$

where $t_k^* = \sum_{i=0}^{k}|b_i|$, $k = 0, 1, 2, \ldots, n$. Thus,

$$v_n \leq (|a_0| + |a_1| + \cdots + |a_n|)t_n^*$$

$$\leq s^* t^* \qquad \text{for all } n,$$

where t^* is the sum of the series $\sum_{n=0}^{\infty}|b_n|$, which is convergent by assumption. We conclude that the sequence $\{v_n\}_{n=0}^{\infty}$ is bounded. Since $v_n \geq 0$, then by Exercise 5.12 this sequence is convergent, and therefore $\sum_{n=0}^{\infty} c_n$ converges absolutely. By part (1), the sum of this series is st. $\qquad \square$

It should be noted that absolute convergence of at least one of $\sum_{n=0}^{\infty} a_n$ and $\sum_{n=0}^{\infty} b_n$ is an essential condition for the validity of part (1) of Theorem 5.2.17. If this condition is not satisfied, then $\sum_{n=0}^{\infty} c_n$ may not converge. For example, consider the series $\sum_{n=0}^{\infty} a_n, \sum_{n=0}^{\infty} b_n$, where

$$a_n = b_n = \frac{(-1)^n}{\sqrt{n+1}}, \qquad n = 0, 1, \ldots .$$

These two series are convergent by Theorem 5.2.14. They are not, however, absolutely convergent, and their Cauchy's product is divergent (see Exercise 5.22).

5.3. SEQUENCES AND SERIES OF FUNCTIONS

All the sequences and series considered thus far in this chapter had constant terms. We now extend our study to sequences and series whose terms are functions of x.

Definition 5.3.1. Let $\{f_n(x)\}_{n=1}^{\infty}$ be a sequence of functions defined on a set $D \subset R$.

1. If there exists a function $f(x)$ defined on D such that for every x in D,

$$\lim_{n \to \infty} f_n(x) = f(x),$$

   then the sequence $\{f_n(x)\}_{n=1}^{\infty}$ is said to converge to $f(x)$ on D. Thus for a given $\epsilon > 0$ there exists an integer N such that $|f_n(x) - f(x)| < \epsilon$ if $n > N$. In general, N depends on ϵ as well as on x.

2. If $\sum_{n=1}^{\infty} f_n(x)$ converges for every x in D to $s(x)$, then $s(x)$ is said to be the sum of the series. In this case, for a given $\epsilon > 0$ there exists an integer N such that

$$|s_n(x) - s(x)| < \epsilon \qquad \text{if } n > N,$$

   where $s_n(x)$ is the nth partial sum of the series $\sum_{n=1}^{\infty} f_n(x)$. The integer N depends on ϵ and, in general, on x also.

3. In particular, if N in (1) depends on ϵ but not on $x \in D$, then the sequence $\{f_n(x)\}_{n=1}^{\infty}$ is said to converge uniformly to $f(x)$ on D. Similarly, if N in (2) depends on ϵ, but not on $x \in D$, then the series $\sum_{n=1}^{\infty} f_n(x)$ converges uniformly to $s(x)$ on D.   □

The Cauchy criterion for sequences (Theorem 5.1.6) and its application to series (Theorem 5.2.1) apply to sequences and series of functions. In case of uniform convergence, the integer N described in this criterion depends only on ϵ.

Theorem 5.3.1. Let $\{f_n(x)\}_{n=1}^{\infty}$ be a sequence of functions defined on $D \subset R$ and converging to $f(x)$. Define the number λ_n as

$$\lambda_n = \sup_{x \in D} |f_n(x) - f(x)|.$$

Then the sequence converges uniformly to $f(x)$ on D if and only if $\lambda_n \to 0$ as $n \to \infty$.

Proof. Sufficiency: Suppose that $\lambda_n \to 0$ as $n \to \infty$. To show that $f_n(x) \to f(x)$ uniformly on D. Let $\epsilon > 0$ be given. Then there exists an integer N such that for $n > N$, $\lambda_n < \epsilon$. Hence, for such values of n,

$$|f_n(x) - f(x)| \le \lambda_n < \epsilon$$

for all $x \in D$. Since N depends only on ϵ, the sequence $\{f_n(x)\}_{n=1}^{\infty}$ converges uniformly to $f(x)$ on D.

Necessity: Suppose that $f_n(x) \to f(x)$ uniformly on D. To show that $\lambda_n \to 0$. Let $\epsilon > 0$ be given. There exists an integer N that depends only on ϵ such that for $n > N$,

$$|f_n(x) - f(x)| < \frac{\epsilon}{2}$$

for all $x \in D$. It follows that

$$\lambda_n = \sup_{x \in D} |f_n(x) - f(x)| \le \frac{\epsilon}{2}.$$

Thus $\lambda_n \to 0$ as $n \to \infty$. $\qquad\square$

Theorem 5.3.1 can be applied to convergent series of functions by replacing $f_n(x)$ and $f(x)$ with $s_n(x)$ and $s(x)$, respectively, where $s_n(x)$ is the nth partial sum of the series and $s(x)$ is its sum.

EXAMPLE 5.3.1. Let $f_n(x) = \sin(2\pi x/n)$, $0 \le x \le 1$. Then $f_n(x) \to 0$ as $n \to \infty$. Furthermore,

$$\left| \sin\left(\frac{2\pi x}{n} \right) \right| \le \frac{2\pi x}{n} \le \frac{2\pi}{n}.$$

In this case,

$$\lambda_n = \sup_{0 \le x \le 1} \left| \sin\left(\frac{2\pi x}{n} \right) \right| \le \frac{2\pi}{n}.$$

Thus $\lambda_n \to 0$ as $n \to \infty$, and the sequence $\{f_n(x)\}_{n=1}^{\infty}$ converges uniformly to $f(x) = 0$ on $[0, 1]$.

The next theorem provides a simple test for uniform convergence of series of functions. It is due to Karl Weierstrass (1815–1897).

Theorem 5.3.2 (Weierstrass's M-test). Let $\sum_{n=1}^{\infty} f_n(x)$ be a series of functions defined on $D \subset R$. If there exists a sequence $\{M_n\}_{n=1}^{\infty}$ of constants such that

$$|f_n(x)| \le M_n, \qquad n = 1, 2,, \dots,$$

for all $x \in D$, and if $\sum_{n=1}^{\infty} M_n$ converges, then $\sum_{n=1}^{\infty} f_n(x)$ converges uniformly on D.

Proof. Let $\epsilon > 0$ be given. By the Cauchy criterion (Theorem 5.2.1), there exists an integer N such that

$$\sum_{i=m+1}^{n} M_i < \epsilon$$

for all $n > m > N$. Hence, for all such values of m, n, and for all $x \in D$,

$$\left| \sum_{i=m+1}^{n} f_i(x) \right| \leq \sum_{i=m+1}^{n} |f_i(x)|$$

$$\leq \sum_{i=m+1}^{n} M_i < \epsilon.$$

This implies that $\sum_{n=1}^{\infty} f_n(x)$ converges uniformly on D by the Cauchy criterion. $\square$

We note that Weierstrass's M-test is easier to apply than Theorem 5.3.1, since it does not require specifying the sum $s(x)$ of the series.

EXAMPLE 5.3.2.   Let us investigate convergence of the sequence $\{f_n(x)\}_{n=1}^{\infty}$, where $f_n(x)$ is defined as

$$f_n(x) = \begin{cases} 2x + \dfrac{1}{n}, & 0 \leq x < 1, \\[2mm] \exp\left(\dfrac{x}{n}\right), & 1 \leq x < 2, \\[2mm] 1 - \dfrac{1}{n}, & x \geq 2. \end{cases}$$

This sequence converges to

$$f(x) = \begin{cases} 2x, & 0 \leq x < 1, \\ 1, & x \geq 1. \end{cases}$$

Now,

$$|f_n(x) - f(x)| = \begin{cases} 1/n, & 0 \leq x < 1, \\ \exp(x/n) - 1, & 1 \leq x < 2, \\ 1/n, & x \geq 2. \end{cases}$$

However, for $1 \leq x < 2$,

$$\exp(x/n) - 1 < \exp(2/n) - 1.$$

Furthermore, by Maclaurin's series expansion,

$$\exp\left(\frac{2}{n}\right) - 1 = \sum_{k=1}^{\infty} \frac{(2/n)^k}{k!} > \frac{2}{n} > \frac{1}{n}.$$

Thus,

$$\sup_{0 \le x < \infty} |f_n(x) - f(x)| = \exp(2/n) - 1,$$

which tends to zero as $n \to \infty$. Therefore, the sequence $\{f_n(x)\}_{n=1}^{\infty}$ converges uniformly to $f(x)$ on $[0, \infty)$.

EXAMPLE 5.3.3. Consider the series $\sum_{n=1}^{\infty} f_n(x)$, where

$$f_n(x) = \frac{x^n}{n^3 + nx^n}, \qquad 0 \le x \le 1.$$

The function $f_n(x)$ is monotone increasing with respect to x. It follows that for $0 \le x \le 1$,

$$|f_n(x)| = f_n(x) \le \frac{1}{n^3 + n}, \qquad n = 1, 2, \dots.$$

But the series $\sum_{n=1}^{\infty} [1/(n^3 + n)]$ is convergent. Hence, $\sum_{n=1}^{\infty} f_n(x)$ is uniformly convergent on $[0, 1]$ by Weierstrass's M-test.

5.3.1. Properties of Uniformly Convergent Sequences and Series

Sequences and series of functions that are uniformly convergent have several interesting properties. We shall study some of these properties in this section.

Theorem 5.3.3. Let $\{f_n(x)\}_{n=1}^{\infty}$ be uniformly convergent to $f(x)$ on a set D. If for each n, $f_n(x)$ has a limit τ_n as $x \to x_0$, where x_0 is a limit point of D, then the sequence $\{\tau_n\}_{n=1}^{\infty}$ converges to $\tau_0 = \lim_{x \to x_0} f(x)$. This is equivalent to stating that

$$\lim_{n \to \infty} \left[\lim_{x \to x_0} f_n(x) \right] = \lim_{x \to x_0} \left[\lim_{n \to \infty} f_n(x) \right].$$

Proof. Let us first show that $\{\tau_n\}_{n=1}^{\infty}$ is a convergent sequence. By the Cauchy criterion (Theorem 5.1.6), there exists an integer N such that for a given $\epsilon > 0$,

$$|f_m(x) - f_n(x)| < \frac{\epsilon}{2} \qquad \text{for all } m > N, n > N. \qquad (5.45)$$

The integer N depends only on ϵ, and inequality (5.45) is true for all $x \in D$, since the sequence is uniformly convergent. By taking the limit as $x \to x_0$ in (5.45) we get

$$|\tau_m - \tau_n| \le \frac{\epsilon}{2} \qquad \text{if } m > N, n > N,$$

which indicates that $\{\tau_n\}_{n=1}^{\infty}$ is a Cauchy sequence and is therefore convergent. Let $\tau_0 = \lim_{n \to \infty} \tau_n$. We now need to show that $f(x)$ has a limit and that this limit is equal to τ_0. Let $\epsilon > 0$ be given. There exists an integer N_1 such that for $n > N_1$,

$$|f(x) - f_n(x)| < \frac{\epsilon}{4}$$

for all $x \in D$, by the uniform convergence of the sequence. Furthermore, there exists an integer N_2 such that

$$|\tau_n - \tau_0| < \frac{\epsilon}{4} \qquad \text{if } n > N_2.$$

Thus for $n > \max(N_1, N_2)$,

$$|f(x) - f_n(x)| + |\tau_n - \tau_0| < \frac{\epsilon}{2}$$

for all $x \in D$. Then

$$|f(x) - \tau_0| \leq |f(x) - f_n(x)| + |f_n(x) - \tau_n| + |\tau_n - \tau_0|$$

$$< |f_n(x) - \tau_n| + \frac{\epsilon}{2} \qquad\qquad\qquad (5.46)$$

if $n > \max(N_1, N_2)$ for all $x \in D$. By taking the limit as $x \to x_0$ in (5.46) we get

$$\left| \lim_{x \to x_0} f(x) - \tau_0 \right| \leq \frac{\epsilon}{2} \qquad\qquad\qquad (5.47)$$

by the fact that

$$\lim_{x \to x_0} |f_n(x) - \tau_n| = 0 \qquad \text{for } n = 1, 2, \dots .$$

Since ϵ is arbitrarily small, inequality (5.47) implies that

$$\lim_{x \to x_0} f(x) = \tau_0. \qquad \square$$

Corollary 5.3.1. Let $\{f_n(x)\}_{n=1}^{\infty}$ be a sequence of continuous functions that converges uniformly to $f(x)$ on a set D. Then $f(x)$ is continuous on D.

Proof. The proof follows directly from Theorem 5.3.3, since $\tau_n = f_n(x_0)$ for $n \geq 1$ and $\tau_0 = \lim_{n \to \infty} \tau_n = \lim_{n \to \infty} f_n(x_0) = f(x_0)$. $\qquad \square$

Corollary 5.3.2. Let $\sum_{n=1}^{\infty} f_n(x)$ be a series of functions that converges uniformly to $s(x)$ on a set D. If for each n, $f_n(x)$ has a limit τ_n as $x \to x_0$, then the series $\sum_{n=1}^{\infty} \tau_n$ converges and has a sum equal to $s_0 = \lim_{x \to x_0} s(x)$, that is,

$$\lim_{x \to x_0} \sum_{n=1}^{\infty} f_n(x) = \sum_{n=1}^{\infty} \lim_{x \to x_0} f_n(x).$$

Proof. The proof is left to the reader. $\square$

By combining Corollaries 5.3.1 and 5.3.2 we conclude the following corollary:

Corollary 5.3.3. Let $\sum_{n=1}^{\infty} f_n(x)$ be a series of continuous functions that converges uniformly to $s(x)$ on a set D. Then $s(x)$ is continuous on D.

EXAMPLE 5.3.4.   Let $f_n(x) = x^2/(1+x^2)^{n-1}$ be defined on $[1, \infty)$ for $n \geq 1$. Let $s_n(x)$ be the nth partial sum of the series $\sum_{n=1}^{\infty} f_n(x)$. Then,

$$s_n(x) = \sum_{k=1}^{n} \frac{x^2}{(1+x^2)^{k-1}} = x^2 \left[\frac{1 - \dfrac{1}{(1+x^2)^n}}{1 - \dfrac{1}{1+x^2}} \right],$$

by using the fact that the sum of the finite geometric series $\sum_{k=1}^{n} a^{k-1}$ is

$$\sum_{k=1}^{n} a^{k-1} = \frac{1-a^n}{1-a}. \tag{5.48}$$

Since $1/(1+x^2) < 1$ for $x \geq 1$, then as $n \to \infty$,

$$s_n(x) \to \frac{x^2}{1 - \dfrac{1}{1+x^2}} = 1 + x^2.$$

Thus,

$$\sum_{n=1}^{\infty} \frac{x^2}{(1+x^2)^{n-1}} = 1 + x^2.$$

Now, let $x_0 = 1$, then

$$\sum_{n=1}^{\infty} \lim_{x \to 1} \frac{x^2}{(1+x^2)^{n-1}} = \sum_{n=1}^{\infty} \frac{1}{2^{n-1}} = \frac{1}{1 - \frac{1}{2}} = 2$$

$$= \lim_{x \to 1} (1 + x^2),$$

which results from applying formula (5.48) with $a = \frac{1}{2}$ and then letting $n \to \infty$. This provides a verification to Corollary 5.3.2. Note that the series $\sum_{n=1}^{\infty} f_n(x)$ is uniformly convergent by Weierstrass's M-test (why?).

Corollaries 5.3.2 and 5.3.3 clearly show that the properties of the function $f_n(x)$ carry over to the sum $s(x)$ of the series $\sum_{n=1}^{\infty} f_n(x)$ when the series is uniformly convergent.

Another property that $s(x)$ shares with the $f_n(x)$'s is given by the following theorem:

Theorem 5.3.4. Let $\sum_{n=1}^{\infty} f_n(x)$ be a series of functions, where $f_n(x)$ is differentiable on $[a, b]$ for $n \geq 1$. Suppose that $\sum_{n=1}^{\infty} f_n(x)$ converges at least at one point $x_0 \in [a, b]$ and that $\sum_{n=1}^{\infty} f_n'(x)$ converges uniformly on $[a, b]$. Then we have the following:

1. $\sum_{n=1}^{\infty} f_n(x)$ converges uniformly to $s(x)$ on $[a, b]$.
2. $s'(x) = \sum_{n=1}^{\infty} f_n'(x)$, that is, the derivative of $s(x)$ is obtained by a term-by-term differentiation of the series $\sum_{n=1}^{\infty} f_n(x)$.

Proof.

1. Let $x \neq x_0$ be a point in $[a, b]$. By the mean value theorem (Theorem 4.2.2), there exists a point ξ_n between x and x_0 such that for $n \geq 1$,

$$f_n(x) - f_n(x_0) = (x - x_0) f_n'(\xi_n). \qquad (5.49)$$

Since $\sum_{n=1}^{\infty} f_n'(x)$ is uniformly convergent on $[a, b]$, then by the Cauchy criterion, there exists an integer N such that

$$\left| \sum_{i=m+1}^{n} f_i'(x) \right| < \frac{\epsilon}{b-a}$$

for all $n > m > N$ and for any $x \in [a, b]$. From (5.49) we get

$$\left| \sum_{i=m+1}^{n} [f_i(x) - f_i(x_0)] \right| = |x - x_0| \left| \sum_{i=m+1}^{n} f_i'(\xi_i) \right|$$

$$< \frac{\epsilon}{b-a} |x - x_0|$$

$$< \epsilon$$

for all $n > m > N$ and for any $x \in [a, b]$. This shows that

$$\sum_{n=1}^{\infty} [f_n(x) - f_n(x_0)]$$

is uniformly convergent on D. Consequently,

$$\sum_{n=1}^{\infty} f_n(x) = \sum_{n=1}^{\infty} [f_n(x) - f_n(x_0)] + s(x_0)$$

is uniformly convergent to $s(x)$ on D, where $s(x_0)$ is the sum of the series $\sum_{n=1}^{\infty} f_n(x_0)$, which was assumed to be convergent.

2. Let $\phi_n(h)$ denote the ratio

$$\phi_n(h) = \frac{f_n(x + h) - f_n(x)}{h}, \qquad n = 1, 2, \ldots,$$

where both x and $x + h$ belong to $[a, b]$. By invoking the mean value theorem again, $\phi_n(h)$ can be written as

$$\phi_n(h) = f_n'(x + \theta_n h), \qquad n = 1, 2, \ldots,$$

where $0 < \theta_n < 1$. Furthermore, by the uniform convergence of $\sum_{n=1}^{\infty} f_n'(x)$ we can deduce that $\sum_{n=1}^{\infty} \phi_n(h)$ is also uniformly convergent on $[-r, r]$ for some $r > 0$. But

$$\sum_{n=1}^{\infty} \phi_n(h) = \sum_{n=1}^{\infty} \frac{f_n(x + h) - f_n(x)}{h}$$

$$= \frac{s(x + h) - s(x)}{h}, \qquad (5.50)$$

where $s(x)$ is the sum of the series $\sum_{n=1}^{\infty} f_n(x)$. Let us now apply Corollary 5.3.2 to $\sum_{n=1}^{\infty} \phi_n(h)$. We get

$$\lim_{h \to 0} \sum_{n=1}^{\infty} \phi_n(h) = \sum_{n=1}^{\infty} \lim_{h \to 0} \phi_n(h). \qquad (5.51)$$

From (5.50) and (5.51) we then have

$$\lim_{h \to 0} \frac{s(x + h) - s(x)}{h} = \sum_{n=1}^{\infty} f_n'(x).$$

Thus,

$$s'(x) = \sum_{n=1}^{\infty} f_n'(x). \qquad \square$$

5.4. POWER SERIES

A power series is a special case of the series of functions discussed in Section 5.3. It is of the form $\sum_{n=0}^{\infty} a_n x^n$, where the a_n's are constants. We have already encountered such series in connection with Taylor's and Maclaurin's series in Section 4.3.

Obviously, just as with any series of functions, the convergence of a power series depends on the values of x. By definition, if there exists a number $\rho > 0$ such that $\sum_{n=0}^{\infty} a_n x^n$ is convergent if $|x| < \rho$ and is divergent if $|x| > \rho$, then ρ is said to be the radius of convergence of the series, and the interval $(-\rho, \rho)$ is called the interval of convergence. The set of all values of x for which the power series converges is called its region of convergence.

The definition of the radius of convergence implies that $\sum_{n=0}^{\infty} a_n x^n$ is absolutely convergent within its interval of convergence. This is shown in the next theorem.

Theorem 5.4.1. Let ρ be the radius of convergence of $\sum_{n=0}^{\infty} a_n x^n$. Suppose that $\rho > 0$. Then $\sum_{n=0}^{\infty} a_n x^n$ converges absolutely for all x inside the interval $(-\rho, \rho)$.

Proof. Let x be such that $|x| < \rho$. There exists a point $x_0 \in (-\rho, \rho)$ such that $|x| < |x_0|$. Then, $\sum_{n=0}^{\infty} a_n x_0^n$ is a convergent series. By Result 5.2.1, $a_n x_0^n \to 0$ as $n \to \infty$, and hence $\{a_n x_0^n\}_{n=0}^{\infty}$ is a bounded sequence by Theorem 5.1.1. Thus

$$|a_n x_0^n| < K \qquad \text{for all } n.$$

Now,

$$|a_n x^n| = \left| a_n \left(\frac{x}{x_0} \right)^n x_0^n \right|$$

$$< K \eta^n,$$

where

$$\eta = \left| \frac{x}{x_0} \right| < 1.$$

Since the geometric series $\sum_{n=0}^{\infty} \eta^n$ is convergent, then by the comparison test (see Theorem 5.2.4), the series $\sum_{n=0}^{\infty} |a_n x^n|$ is convergent. □

To determine the radius of convergence we shall rely on some of the tests of convergence given in Section 5.2.1.

Theorem 5.4.2. Let $\sum_{n=0}^{\infty} a_n x^n$ be a power series. Suppose that

$$\lim_{n \to \infty} \left| \frac{a_{n+1}}{a_n} \right| = p.$$

Then the radius of convergence of the power series is

$$\rho = \begin{cases} 1/p, & 0 < p < \infty, \\ 0, & p = \infty, \\ \infty, & p = 0. \end{cases}$$

Proof. The proof follows from applying the ratio test given in Theorem 5.2.6 to the series $\sum_{n=0}^{\infty} |a_n x^n|$: We have that if

$$\lim_{n \to \infty} \left| \frac{a_{n+1} x^{n+1}}{a_n x^n} \right| < 1,$$

then $\sum_{n=0}^{\infty} a_n x^n$ is absolutely convergent. This inequality can be written as

$$p|x| < 1. \tag{5.52}$$

If $0 < p < \infty$, then absolute convergence occurs if $|x| < 1/p$ and the series diverges when $|x| > 1/p$. Thus $\rho = 1/p$. If $p = \infty$, the series diverges whenever $x \neq 0$. In this case, $\rho = 0$. If $p = 0$, then (5.52) holds for any value of x, that is, $\rho = \infty$. $\square$

Theorem 5.4.3. Let $\sum_{n=0}^{\infty} a_n x^n$ be a power series. Suppose that

$$\limsup_{n \to \infty} |a_n|^{1/n} = q.$$

Then,

$$\rho = \begin{cases} 1/q, & 0 < q < \infty, \\ 0, & q = \infty, \\ \infty, & q = 0. \end{cases}$$

Proof. This result follows from applying the root test in Theorem 5.2.8 to the series $\sum_{n=0}^{\infty} |a_n x^n|$. Details of the proof are similar to those given in Theorem 5.4.2. $\square$

The determination of the region of convergence of $\sum_{n=0}^{\infty} a_n x^n$ depends on the value of ρ. We know that the series converges if $|x| < \rho$ and diverges if $|x| > \rho$. The convergence of the series at $x = \rho$ and $x = -\rho$ has to be determined separately. Thus the region of convergence can be $(-\rho, \rho)$, $[-\rho, \rho)$, $(-\rho, \rho]$, or $[-\rho, \rho]$.

EXAMPLE 5.4.1. Consider the geometric series $\sum_{n=0}^{\infty} x^n$. By applying either Theorem 5.4.2 or Theorem 5.4.3, it is easy to show that $\rho = 1$. The series diverges if $x = 1$ or -1. Thus the region of convergence is $(-1, 1)$. The sum of this series can be obtained from formula (5.48) by letting n go to infinity. Thus

$$\sum_{n=0}^{\infty} x^n = \frac{1}{1-x}, \qquad -1 < x < 1. \tag{5.53}$$

EXAMPLE 5.4.2. Consider the series $\sum_{n=0}^{\infty}(x^n/n!)$. Here,

$$\lim_{n \to \infty} \left| \frac{a_{n+1}}{a_n} \right| = \lim_{n \to \infty} \frac{n!}{(n+1)!}$$

$$= \lim_{n \to \infty} \frac{1}{n+1} = 0.$$

Thus $\rho = \infty$, and the series converges absolutely for any value of x. This particular series is Maclaurin's expansion of e^x, that is,

$$e^x = \sum_{n=0}^{\infty} \frac{x^n}{n!}.$$

EXAMPLE 5.4.3. Suppose we have the series $\sum_{n=1}^{\infty}(x^n/n)$. Then

$$\lim_{n \to \infty} \left| \frac{a_{n+1}}{a_n} \right| = \lim_{n \to \infty} \frac{n}{n+1} = 1,$$

and $\rho = 1$. When $x = 1$ we get the harmonic series, which is divergent. When $x = -1$ we get the alternating harmonic series, which is convergent by Theorem 5.2.14. Thus the region of convergence is $[-1, 1)$.

In addition to being absolutely convergent within its interval of convergence, a power series is also uniformly convergent there. This is shown in the next theorem.

Theorem 5.4.4. Let $\sum_{n=0}^{\infty} a_n x^n$ be a power series with a radius of convergence $\rho \, (> 0)$. Then we have the following:

1. The series converges uniformly on the interval $[-r, r]$, where $r < \rho$.
2. If $s(x) = \sum_{n=0}^{\infty} a_n x^n$, then $s(x)$ (i) is continuous on $[-r, r]$; (ii) is differentiable on $[-r, r]$ and has derivative

$$s'(x) = \sum_{n=1}^{\infty} n a_n x^{n-1}, \qquad -r \le x \le r;$$

and (iii) has derivatives of all orders on $[-r, r]$ and

$$\frac{d^k s(x)}{dx^k} = \sum_{n=k}^{\infty} \frac{a_n n!}{(n-k)!} x^{n-k}, \qquad k = 1, 2, \ldots, \; -r \le x \le r.$$

Proof.

1. If $|x| \le r$, then $|a_n x^n| \le |a_n| r^n$ for $n \ge 0$. Since $\sum_{n=0}^{\infty} |a_n| r^n$ is convergent by Theorem 5.4.1, then by the Weierstrass M-test (Theorem 5.3.2), $\sum_{n=0}^{\infty} a_n x^n$ is uniformly convergent on $[-r, r]$.

2. (i) Continuity of $s(x)$ follows directly from Corollary 5.3.3. (ii) To show this result, we first note that the two series $\sum_{n=0}^{\infty} a_n x^n$ and $\sum_{n=1}^{\infty} n a_n x^{n-1}$ have the same radius of convergence. This is true by Theorem 5.4.3 and the fact that

$$\limsup_{n \to \infty} |n a_n|^{1/n} = \limsup_{n \to \infty} |a_n|^{1/n},$$

since $\lim_{n \to \infty} n^{1/n} = 1$ as $n \to \infty$. We can then assert that $\sum_{n=1}^{\infty} n a_n x^{n-1}$ is uniformly convergent on $[-r, r]$. By Theorem 5.3.4, $s(x)$ is differentiable on $[-r, r]$, and its derivative is obtained by a term-by-term differentiation of $\sum_{n=0}^{\infty} a_n x^n$. (iii) This follows from part (ii) by repeated differentiation of $s(x)$. □

Under a certain condition, the interval on which the power series converges uniformly can include the end points of the interval of convergence. This is discussed in the next theorem.

Theorem 5.4.5. Let $\sum_{n=0}^{\infty} a_n x^n$ be a power series with a finite nonzero radius of convergence ρ. If $\sum_{n=0}^{\infty} a_n \rho^n$ is absolutely convergent, then the power series is uniformly convergent on $[-\rho, \rho]$.

Proof. The proof is similar to that of part 1 of Theorem 5.4.4. In this case, for $|x| \le \rho$, $|a_n x^n| \le |a_n| \rho^n$. Since $\sum_{n=0}^{\infty} |a_n| \rho^n$ is convergent, then $\sum_{n=0}^{\infty} a_n x^n$ is uniformly convergent on $[-\rho, \rho]$ by the Weierstrass M-test. □

EXAMPLE 5.4.4. Consider the geometric series of Example 5.4.1. This series is uniformly convergent on $[-r, r]$, where $r < 1$. Furthermore, by differentiating the two sides of (5.53) we get

$$\sum_{n=1}^{\infty} n x^{n-1} = \frac{1}{(1-x)^2}, \qquad -1 < x < 1.$$

This provides a series expansion of $1/(1-x)^2$ within the interval $(-1, 1)$. By repeated differentiation it is easy to show that for $-1 < x < 1$,

$$\frac{1}{(1-x)^k} = \sum_{n=0}^{\infty} \binom{n+k-1}{n} x^n, \qquad k = 1, 2, \dots .$$

The radius of convergence of this series is $\rho = 1$, the same as for the original series.

EXAMPLE 5.4.5. Suppose we have the series

$$\sum_{n=1}^{\infty} \frac{2^n}{2n^2 + n} \left(\frac{x}{1-x} \right)^n ,$$

which can be written as

$$\sum_{n=1}^{\infty} \frac{z^n}{2n^2 + n} ,$$

where $z = 2x/(1-x)$. This is a power series in z. By Theorem 5.4.2, the radius of convergence of this series is $\rho = 1$. We note that when $z = 1$ the series $\sum_{n=1}^{\infty} [1/(2n^2 + n)]$ is absolutely convergent. Thus by Theorem 5.4.5, the given series is uniformly convergent for $|z| \leq 1$, that is, for values of x satisfying

$$-\frac{1}{2} \leq \frac{x}{1-x} \leq \frac{1}{2} ,$$

or equivalently,

$$-1 \leq x \leq \tfrac{1}{3} .$$

5.5. SEQUENCES AND SERIES OF MATRICES

In Section 5.3 we considered sequences and series whose terms were scalar functions of x rather than being constant as was done in Sections 5.1 and 5.2. In this section we consider yet another extension, in which the terms of the series are matrices rather than scalars. We shall provide a brief discussion of this extension. The interested reader can find a more detailed study of this topic in Gantmacher (1959), Lancaster (1969), and Graybill (1983). As in Chapter 2, all matrix elements considered here are real.

For the purpose of our study of sequences and series of matrices we first need to introduce the norm of a matrix.

Definition 5.5.1. Let **A** be a matrix of order $m \times n$. A norm of **A**, denoted by $\|\mathbf{A}\|$, is a real-valued function of **A** with the following properties:

1. $\|\mathbf{A}\| \geq 0$, and $\|\mathbf{A}\| = 0$ if and only if $\mathbf{A} = \mathbf{0}$.
2. $\|c\mathbf{A}\| = |c| \|\mathbf{A}\|$, where c is a scalar.
3. $\|\mathbf{A} + \mathbf{B}\| \leq \|\mathbf{A}\| + \|\mathbf{B}\|$, where **B** is any matrix of order $m \times n$.
4. $\|\mathbf{AC}\| \leq \|\mathbf{A}\| \|\mathbf{C}\|$, where **C** is any matrix for which the product **AC** is defined.   □

If $\mathbf{A} = (a_{ij})$, then examples of matrix norms that satisfy properties 1, 2, 3, and 4 include the following:

1. The Euclidean norm, $\|\mathbf{A}\|_2 = (\sum_{i=1}^m \sum_{j=1}^n a_{ij}^2)^{1/2}$.
2. The spectral norm, $\|\mathbf{A}\|_s = [e_{\max}(\mathbf{A}'\mathbf{A})]^{1/2}$, where $e_{\max}(\mathbf{A}'\mathbf{A})$ is the largest eigenvalue of $\mathbf{A}'\mathbf{A}$.

Definition 5.5.2. Let $\mathbf{A}_k = (a_{ijk})$ be matrices of orders $m \times n$ for $k \geq 1$. The sequence $\{\mathbf{A}_k\}_{k=1}^\infty$ is said to converge to the $m \times n$ matrix $\mathbf{A} = (a_{ij})$ if $\lim_{k \to \infty} a_{ijk} = a_{ij}$ for $i = 1, 2, \ldots, m$; $j = 1, 2, \ldots, n$.   □

For example, the sequence of matrices

$$\mathbf{A}_k = \begin{bmatrix} \dfrac{1}{k} & \dfrac{1}{k^2} - 1 & \dfrac{k-1}{k+1} \\[2mm] 2 & 0 & \dfrac{k^2}{2-k^2} \end{bmatrix}, \qquad k = 1, 2, \ldots,$$

converges to

$$\mathbf{A} = \begin{bmatrix} 0 & -1 & 1 \\ 2 & 0 & -1 \end{bmatrix}$$

as $k \to \infty$. The sequence

$$\mathbf{A}_k = \begin{bmatrix} \dfrac{1}{k} & k^2 - 2 \\[2mm] 1 & \dfrac{k}{1+k} \end{bmatrix}, \qquad k = 1, 2, \ldots,$$

does not converge, since $k^2 - 2$ goes to infinite as $k \to \infty$.

From Definition 5.5.2 it is easy to see that $\{A_k\}_{k=1}^{\infty}$ converges to A if and only if

$$\lim_{k \to \infty} \|A_k - A\| = 0,$$

where $\|\cdot\|$ is any matrix norm.

Definition 5.5.3. Let $\{A_k\}_{k=1}^{\infty}$ be a sequence of matrices of order $m \times n$. Then $\sum_{k=1}^{\infty} A_k$ is called an infinite series (or just a series) of matrices. This series is said to converge to the $m \times n$ matrix $S = (s_{ij})$ if and only if the series $\sum_{k=1}^{\infty} a_{ijk}$ converges for all $i = 1, 2, \ldots, m$; $j = 1, 2, \ldots, n$, where a_{ijk} is the (i, j)th element of A_k, and

$$\sum_{k=1}^{\infty} a_{ijk} = s_{ij}, \qquad i = 1, 2, \ldots, m; j = 1, 2, \ldots, n. \tag{5.54}$$

The series $\sum_{k=1}^{\infty} A_k$ is divergent if at least one of the series in (5.54) is divergent. □

From Definition 5.5.3 and Result 5.2.1 we conclude that $\sum_{k=1}^{\infty} A_k$ diverges if $\lim_{k \to \infty} a_{ijk} \neq 0$ for at least one pair (i, j), that is, if $\lim_{k \to \infty} A_k \neq 0$.

A particular type of infinite series of matrices is the power series $\sum_{k=0}^{\infty} \alpha_k A^k$, where A is a square matrix, α_k is a scalar ($k = 0, 1, \ldots$), and A^0 is by definition the identity matrix I. For example, the power series

$$I + A + \frac{1}{2!}A^2 + \frac{1}{3!}A^3 + \cdots + \frac{1}{k!}A^k + \cdots$$

represents an expansion of the exponential matrix function $\exp(A)$ (see Gantmacher, 1959).

Theorem 5.5.1. Let A be an $n \times n$ matrix. Then $\lim_{k \to \infty} A^k = 0$ if $\|A\| < 1$, where $\|\cdot\|$ is any matrix norm.

Proof. From property 4 in Definition 5.5.1 we can write

$$\|A^k\| \leq \|A\|^k, \qquad k = 1, 2, \ldots .$$

Since $\|A\| < 1$, then $\lim_{k \to \infty} \|A^k\| = 0$, which implies that $\lim_{k \to \infty} A^k = 0$ (why?). □

Theorem 5.5.2. Let A be a symmetric matrix of order $n \times n$ such that $|\lambda_i| < 1$ for $i = 1, 2, \ldots, n$, where λ_i is the ith eigenvalue of A (all the eigenvalues of A are real by Theorem 2.3.5). Then $\sum_{k=0}^{\infty} A^k$ converges to $(I - A)^{-1}$.

Proof. By the spectral decomposition theorem (Theorem 2.3.10) there exists an orthogonal matrix $\mathbf{P}$ such that $\mathbf{A} = \mathbf{P}\boldsymbol{\Lambda}\mathbf{P}'$, where $\boldsymbol{\Lambda}$ is a diagonal matrix whose diagonal elements are the eigenvalues of $\mathbf{A}$. Then

$$\mathbf{A}^k = \mathbf{P}\boldsymbol{\Lambda}^k\mathbf{P}', \qquad k = 0, 1, 2, \ldots .$$

Since $|\lambda_i| < 1$ for all i, then $\boldsymbol{\Lambda}^k \to \mathbf{0}$ and hence $\mathbf{A}^k \to \mathbf{0}$ as $k \to \infty$. Furthermore, the matrix $\mathbf{I} - \mathbf{A}$ is nonsingular, since

$$\mathbf{I} - \mathbf{A} = \mathbf{P}(\mathbf{I} - \boldsymbol{\Lambda})\mathbf{P}'$$

and all the diagonal elements of $\mathbf{I} - \boldsymbol{\Lambda}$ are positive.

Now, for any nonnegative integer k we have the following identity:

$$(\mathbf{I} - \mathbf{A})(\mathbf{I} + \mathbf{A} + \mathbf{A}^2 + \cdots + \mathbf{A}^k) = \mathbf{I} - \mathbf{A}^{k+1}.$$

Hence,

$$\mathbf{I} + \mathbf{A} + \cdots + \mathbf{A}^k = (\mathbf{I} - \mathbf{A})^{-1}(\mathbf{I} - \mathbf{A}^{k+1}).$$

By letting k go to infinity we get

$$\sum_{k=0}^{\infty} \mathbf{A}^k = (\mathbf{I} - \mathbf{A})^{-1},$$

since $\lim_{k \to \infty} \mathbf{A}^{k+1} = \mathbf{0}$.　　　□

Theorem 5.5.3. Let $\mathbf{A}$ be a symmetric $n \times n$ matrix and λ be any eigenvalue of $\mathbf{A}$. Then $|\lambda| \leq \|\mathbf{A}\|$, where $\|\mathbf{A}\|$ is any matrix norm of $\mathbf{A}$.

Proof. We have that $\mathbf{A}\mathbf{v} = \lambda\mathbf{v}$, where $\mathbf{v}$ is an eigenvector of $\mathbf{A}$ for the eigenvalue λ. If $\|\mathbf{A}\|$ is any matrix norm of $\mathbf{A}$, then

$$\|\lambda\mathbf{v}\| = |\lambda| \|\mathbf{v}\| = \|\mathbf{A}\mathbf{v}\| \leq \|\mathbf{A}\| \|\mathbf{v}\|.$$

Since $\mathbf{v} \neq \mathbf{0}$, we conclude that

$$|\lambda| \leq \|\mathbf{A}\|.　　　　□$$

Corollary 5.5.1. Let $\mathbf{A}$ be a symmetric matrix of order $n \times n$ such that $\|\mathbf{A}\| < 1$, where $\|\mathbf{A}\|$ is any matrix norm of $\mathbf{A}$. Then $\sum_{k=0}^{\infty} \mathbf{A}^k$ converges to $(\mathbf{I} - \mathbf{A})^{-1}$.

Proof. This result follows from Theorem 5.5.2, since for $i = 1, 2, \ldots, n$, $|\lambda_i| \leq \|\mathbf{A}\| < 1$.　　　□

5.6. APPLICATIONS IN STATISTICS

Sequences and series have many useful applications in statistics. Some of these applications will be discussed in this section.

5.6.1. Moments of a Discrete Distribution

Perhaps one of the most visible applications of infinite series in statistics is in the study of the distribution of a discrete random variable that can assume a countable number of values. Under certain conditions, this distribution can be completely determined by its moments. By definition, the moments of a distribution are a set of descriptive constants that are useful for measuring its properties.

Let X be a discrete random variable that takes on the values $x_0, x_1, \ldots, x_n, \ldots$, with probabilities $p(n)$, $n \geq 0$. Then, by definition, the kth central moment of X, denoted by μ_k, is

$$\mu_k = E\big[(X - \mu)^k\big] = \sum_{n=0}^{\infty} (x_n - \mu)^k p(n), \qquad k = 1, 2, \ldots,$$

where $\mu = E(X) = \sum_{n=0}^{\infty} x_n p(n)$ is the mean of X. We note that $\mu_2 = \sigma^2$ is the variance of X. The kth noncentral moment of X is given by the series

$$\mu'_k = E(X^k) = \sum_{n=0}^{\infty} x_n^k p(n), \qquad k = 1, 2, \ldots . \qquad (5.55)$$

We note that $\mu'_1 = \mu$. If, for some integer N, $|x_n| \geq 1$ for $n > N$, and if the series in (5.55) converges absolutely, then so does the series for μ'_j ($j = 1, 2, \ldots, k - 1$). This follows from applying the comparison test:

$$|x_n|^j p(n) \leq |x_n|^k p(n) \qquad \text{if } j < k \text{ and } n > N.$$

Examples of discrete random variables with a countable number of values include the Poisson (see Section 4.5.3) and the negative binomial. The latter random variable represents the number n of failures before the rth success when independent trials are performed, each of which has two probability outcomes, success or failure, with a constant probability p of success on each trial. Its probability mass function is therefore of the form

$$p(n) = \binom{n + r - 1}{n} p^r (1 - p)^n, \qquad n = 0, 1, 2, \ldots .$$

By contrast, the Poisson random variable has the probability mass function

$$p(n) = \frac{e^{-\lambda}\lambda^n}{n!}, \qquad n = 0,1,2,\ldots,$$

where λ is the mean of X. We can verify that λ is the mean by writing

$$\mu = \sum_{n=0}^{\infty} n \frac{e^{-\lambda}\lambda^n}{n!}$$

$$= \lambda e^{-\lambda} \sum_{n=1}^{\infty} \frac{\lambda^{n-1}}{(n-1)!}$$

$$= \lambda e^{-\lambda} \sum_{n=0}^{\infty} \frac{\lambda^n}{n!}$$

$$= \lambda e^{-\lambda}(e^{\lambda}), \qquad \text{by Maclaurin's expansion of } e^{\lambda}$$

$$= \lambda.$$

The second noncentral moment of the Poisson distribution is

$$\mu_2' = \sum_{n=0}^{\infty} n^2 \frac{e^{-\lambda}\lambda^n}{n!}$$

$$= e^{-\lambda}\lambda \sum_{n=1}^{\infty} n \frac{\lambda^{n-1}}{(n-1)!}$$

$$= e^{-\lambda}\lambda \sum_{n=1}^{\infty} (n-1+1) \frac{\lambda^{n-1}}{(n-1)!}$$

$$= e^{-\lambda}\lambda \left[\lambda \sum_{n=2}^{\infty} \frac{\lambda^{n-2}}{(n-2)!} + \sum_{n=1}^{\infty} \frac{\lambda^{n-1}}{(n-1)!} \right]$$

$$= e^{-\lambda}\lambda [\lambda e^{\lambda} + e^{\lambda}]$$

$$= \lambda^2 + \lambda.$$

In general, the kth noncentral moment of the Poisson distribution is given by the series

$$\mu_k' = \sum_{n=0}^{\infty} n^k \frac{e^{-\lambda}\lambda^n}{n!}, \qquad k = 1,2,\ldots,$$

which converges for any k. This can be shown, for example, by the ratio test

$$\lim_{n \to \infty} \frac{a_{n+1}}{a_n} = \lim_{n \to \infty} \left(\frac{n+1}{n} \right)^k \frac{\lambda}{n+1}$$

$$= 0 < 1.$$

Thus all the noncentral moments of the Poisson distribution exist.
Similarly, for the negative binomial distribution we have

$$\mu = \sum_{n=0}^{\infty} n \binom{n+r-1}{n} p^r (1-p)^n$$

$$= \frac{r(1-p)}{p} \qquad \text{(why?)}, \tag{5.56}$$

$$\mu_2' = \sum_{n=0}^{\infty} n^2 \binom{n+r-1}{n} p^r (1-p)^n$$

$$= \frac{r(1-p)(1+r-rp)}{p^2} \qquad \text{(why?)} \tag{5.57}$$

and the kth noncentral moment,

$$\mu_k' = \sum_{n=0}^{\infty} n^k \binom{n+r-1}{n} p^r (1-p)^n, \qquad k = 1, 2, \ldots, \tag{5.58}$$

exists for any k, since, by the ratio test,

$$\lim_{n \to \infty} \frac{a_{n+1}}{a_n} = \lim_{n \to \infty} \frac{\left(\dfrac{n+1}{n} \right)^k \left(\dfrac{n+r}{n+1} \right)}{\left(\dfrac{n+r-1}{n} \right)} (1-p)$$

$$= (1-p) \lim_{n \to \infty} \left(\frac{n+1}{n} \right)^k \left(\frac{n+r}{n+1} \right)$$

$$= 1 - p < 1,$$

which proves convergence of the series in (5.58).

A very important inequality that concerns the mean μ and variance σ^2 of any random variable X (not just the discrete ones) is Chebyshev's inequality, namely,

$$P(|X - \mu| \geq r\sigma) \leq \frac{1}{r^2},$$

or equivalently,

$$P(|X - \mu| < r\sigma) \geq 1 - \frac{1}{r^2}, \tag{5.59}$$

where r is any positive number (see, for example, Lindgren, 1976, Section 2.3.2). The importance of this inequality stems from the fact that it is independent of the exact distribution of X and connects the variance of X with the distribution of its values. For example, inequality (5.59) states that at least $(1 - 1/r^2) \times 100\%$ of the values of X fall within $r\sigma$ from its mean, where $\sigma = \sqrt{\sigma^2}$ is the standard deviation of X.

Chebyshev's inequality is a special case of a more general inequality called Markov's inequality. If b is a nonzero constant and $h(x)$ is a nonnegative function, then

$$P[h(X) \geq b^2] \leq \frac{1}{b^2} E[h(X)],$$

provided that $E[h(X)]$ exists. Chebyshev's inequality follows from Markov's inequality by choosing $h(X) = (X - \mu)^2$.

Another important result that concerns the moments of a distribution is given by the following theorem, regarding what is known as the *Stieltjes moment problem*, which also applies to any random variable:

Theorem 5.6.1. Suppose that the moments μ'_k $(k = 1, 2, \ldots)$ of a random variable X exist, and the series

$$\sum_{k=1}^{\infty} \frac{\mu'_k}{k!} \tau^k \tag{5.60}$$

is absolutely convergent for some $\tau > 0$. Then these moments uniquely determine the cumulative distribution function $F(x)$ of X.

Proof. See, for example, Fisz (1963, Theorem 3.2.1).   □

In particular, if

$$|\mu'_k| \leq M^k, \qquad k = 1, 2, \ldots,$$

for some constant M, then the series in (5.60) converges absolutely for any $\tau > 0$ by the comparison test. This is true because the series $\sum_{k=1}^{\infty} (M^k/k!)\tau^k$ converges (for example, by the ratio test) for any value of τ.

It should be noted that absolute convergence of the series in (5.60) is a sufficient condition for the unique determination of $F(x)$, but is not a necessary condition. This is shown in Rao (1973, page 106). Furthermore, if some moments of X fail to exist, then the remaining moments that do exist cannot determine $F(x)$ uniquely. The following counterexample is given in Fisz (1963, page 74):

Let X be a discrete random variable that takes on the values $x_n = 2^n/n^2$, $n \geq 1$, with probabilities $p(n) = 1/2^n$, $n \geq 1$. Then

$$\mu = E(X) = \sum_{n=1}^{\infty} \frac{1}{n^2},$$

which exists, because the series is convergent. However, μ'_2 does not exist, because

$$\mu'_2 = E(X^2) = \sum_{n=1}^{\infty} \frac{2^n}{n^4}$$

and this series is divergent, since $2^n/n^4 \to \infty$ as $n \to \infty$.

Now, let Y be another discrete random variable that takes on the value zero with probability $\frac{1}{2}$ and the values $y_n = 2^{n+1}/n^2$, $n \geq 1$, with probabilities $q(n) = 1/2^{n+1}$, $n \geq 1$. Then,

$$E(Y) = \sum_{n=1}^{\infty} \frac{1}{n^2} = E(X).$$

The second noncentral moment of Y does not exist, since

$$\mu'_2 = E(Y^2) = \sum_{n=1}^{\infty} \frac{2^{n+1}}{n^4},$$

and this series is divergent.

Since μ'_2 does not exist for both X and Y, none of their noncentral moments of order $k > 2$ exist either, as can be seen from applying the comparison test. Thus X and Y have the same first noncentral moments, but do not have noncentral moments of any order greater than 1. These two random variables have obviously different distributions.

5.6.2. Moment and Probability Generating Functions

Let X be a discrete random variable that takes on the values $x_0, x_1, x_2, \ldots$ with probabilities $p(n)$, $n \geq 0$.

The Moment Generating Function of X

This function is defined as

$$\phi(t) = E(e^{tX}) = \sum_{n=0}^{\infty} e^{tx_n} p(n) \qquad (5.61)$$

provided that the series converges. In particular, if $x_n = n$ for $n \geq 0$, then

$$\phi(t) = \sum_{n=0}^{\infty} e^{tn} p(n), \tag{5.62}$$

which is a power series in e^t. If ρ is the radius of convergence for this series, then by Theorem 5.4.4, $\phi(t)$ is a continuous function of t and has derivatives of all orders inside its interval of convergence. Since

$$\frac{d^k \phi(t)}{dt^k}\bigg|_{t=0} = E(X^k) = \mu'_k, \qquad k = 1, 2, \ldots, \tag{5.63}$$

$\phi(t)$, when it exists, can be used to obtain all noncentral moments of X, which can completely determine the distribution of X by Theorem 5.6.1.

From (5.63), by using Maclaurin's expansion of $\phi(t)$, we can obtain an expression for this function as a power series in t:

$$\phi(t) = \phi(0) + \sum_{n=1}^{\infty} \frac{t^n}{n!} \phi^{(n)}(0)$$

$$= 1 + \sum_{n=1}^{\infty} \frac{\mu'_n}{n!} t^n. \tag{5.64}$$

Let us now go back to the series in (5.62). If

$$\lim_{n \to \infty} \frac{p(n+1)}{p(n)} = p,$$

then by Theorem 5.4.2, the radius of convergence ρ is

$$\rho = \begin{cases} 1/p, & 0 < p < \infty, \\ 0, & p = \infty, \\ \infty, & p = 0. \end{cases}$$

Alternatively, if $\limsup_{n \to \infty} [p(n)]^{1/n} = q$, then

$$\rho = \begin{cases} 1/q, & 0 < q < \infty, \\ 0, & q = \infty, \\ \infty, & q = 0. \end{cases}$$

For example, for the Poisson distribution, where

$$p(n) = \frac{e^{-\lambda}\lambda^n}{n!}, \qquad n = 0, 1, 2, \ldots,$$

we have $\lim_{n \to \infty}[p(n+1)/p(n)] = \lim_{n \to \infty}[\lambda/(n+1)] = 0$. Hence, $\rho = \infty$, that is,

$$\phi(t) = \sum_{n=0}^{\infty} \frac{e^{-\lambda}\lambda^n}{n!} e^{tn}$$

converges uniformly for any value of t for which $e^t < \infty$, that is, $-\infty < t < \infty$. As a matter of fact, a closed-form expression for $\phi(t)$ can be found, since

$$\phi(t) = e^{-\lambda} \sum_{n=0}^{\infty} \frac{(\lambda e^t)^n}{n!}$$

$$= e^{-\lambda} \exp(\lambda e^t)$$

$$= \exp(\lambda e^t - \lambda) \qquad \text{for all } t. \tag{5.65}$$

The kth noncentral moment of X is then given by

$$\mu'_k = \left. \frac{d^k\phi(t)}{dt^k} \right|_{t=0} = \left. \frac{d^k(\lambda e^t - \lambda)}{dt^k} \right|_{t=0}.$$

In particular, the first two noncentral moments are

$$\mu'_1 = \mu = \lambda,$$

$$\mu'_2 = \lambda + \lambda^2.$$

This confirms our earlier finding concerning these two moments.

It should be noted that formula (5.63) is valid provided that there exists a $\delta > 0$ such that the neighborhood $N_\delta(0)$ is contained inside the interval of convergence. For example, let X have the probability mass function

$$p(n) = \frac{6}{\pi^2 n^2}, \qquad n = 1, 2, \ldots .$$

Then

$$\lim_{n \to \infty} \frac{p(n+1)}{p(n)} = 1.$$

Hence, by Theorem 5.4.4, the series

$$\phi(t) = \sum_{n=1}^{\infty} \frac{6}{\pi^2 n^2} e^{tn}$$

converges uniformly for values of t satisfying $e^t \leq r$, where $r \leq 1$, or equivalently, for $t \leq \log r \leq 0$. If, however, $t > 0$, then the series diverges. Thus there does not exist a neighborhood $N_\delta(0)$ that is contained inside the interval of convergence for any $\delta > 0$. Consequently, formula (5.63) does not hold in this case.

From the moment generating function we can derive a series of constants that play a role similar to that of the moments. These constants are called cumulants. They have properties that are, in certain circumstances, more useful than those of the moments. Cumulants were originally defined and studied by Thiele (1903).

By definition, the cumulants of X, denoted by $\kappa_1, \kappa_2, \ldots, \kappa_n, \ldots$ are constants that satisfy the following identity in t:

$$
\exp\left(\kappa_1 t + \frac{\kappa_2 t^2}{2!} + \cdots + \frac{\kappa_n t^n}{n!} + \cdots \right)
$$

$$
= 1 + \mu_1' t + \frac{\mu_2'}{2!} t^2 + \cdots + \frac{\mu_n'}{n!} t^n + \cdots . \tag{5.66}
$$

Using formula (5.64), this identity can be written as

$$
\sum_{n=1}^{\infty} \frac{\kappa_n}{n!} t^n = \log \phi(t), \tag{5.67}
$$

provided that $\phi(t)$ exists and is positive. By definition, the natural logarithm of the moment generating function of X is called the cumulant generating function.

Formula (5.66) can be used to express the noncentral moments in terms of the cumulants, and vice versa. Kendall and Stuart (1977, Section 3.14) give a general relationship that can be used for this purpose. For example,

$$
\kappa_1 = \mu_1',
$$

$$
\kappa_2 = \mu_2' - \mu_1'^2,
$$

$$
\kappa_3 = \mu_3' - 3\mu_1' \mu_2' + 2\mu_1'^3.
$$

The cumulants have an interesting property in that they are, except for κ_1, invariant to any constant shift c in X. That is, for $n = 2, 3, \ldots, \kappa_n$ is not changed if X is replaced by $X + c$. This follows from noting that

$$
E\left[e^{(X+c)t} \right] = e^{ct} \phi(t),
$$

which is the moment generating function of $X + c$. But

$$
\log\left[e^{ct} \phi(t) \right] = ct + \log \phi(t).
$$

By comparison with (5.67) we can then conclude that except for κ_1, the cumulants of $X + c$ are the same as those of X. This contrasts sharply with the noncentral moments of X, which are not invariant to such a shift.

Another advantage of using cumulants is that they can be employed to obtain approximate expressions for the percentile points of the distribution of X (see Section 9.5.1).

EXAMPLE 5.6.1. Let X be a Poisson random variable whose moment generating function is given by formula (5.65). By applying (5.67) we get

$$\sum_{n=1}^{\infty} \frac{\kappa_n}{n!} t^n = \log\left[\exp(\lambda e^t - \lambda)\right]$$

$$= \lambda e^t - \lambda$$

$$= \lambda \sum_{n=1}^{\infty} \frac{t^n}{n!}.$$

Here, we have made use of Maclaurin's expansion of e^t. This series converges for any value of t. It follows that $\kappa_n = \lambda$ for $n = 1, 2, \ldots$.

The Probability Generating Function

This is similar to the moment generating function. It is defined as

$$\psi(t) = E(t^X) = \sum_{n=0}^{\infty} t^{x_n} p(n).$$

In particular, if $x_n = n$ for $n \geq 0$, then

$$\psi(t) = \sum_{n=0}^{\infty} t^n p(n), \tag{5.68}$$

which is a power series in t. Within its interval of convergence, this series represents a continuous function with derivatives of all orders. We note that $\psi(0) = p(0)$ and that

$$\frac{1}{k!} \frac{d^k \psi(t)}{dt^k} \bigg|_{t=0} = p(k), \qquad k = 1, 2, \ldots . \tag{5.69}$$

Thus, the entire probability distribution of X is completely determined by $\psi(t)$.

The probability generating function is also useful in determining the moments of X. This is accomplished by using the relation

$$\frac{d^k \psi(t)}{dt^k}\bigg|_{t=1} = \sum_{n=k}^{\infty} n(n-1) \cdots (n-k+1) p(n)$$

$$= E[X(X-1) \cdots (X-k+1)]. \tag{5.70}$$

The quantity on the right-hand side of (5.70) is called the kth factorial moment of X, which we denote by θ_k. The noncentral moments of X can be derived from the θ_k's. For example,

$$\mu'_1 = \theta_1,$$

$$\mu'_2 = \theta_2 + \theta_1,$$

$$\mu'_3 = \theta_3 + 3\theta_2 + \theta_1,$$

$$\mu'_4 = \theta_4 + 6\theta_3 + 7\theta_2 + \theta_1.$$

Obviously, formula (5.70) is valid provided that $t = 1$ belongs to the interval of convergence of the series in (5.68).

If a closed-form expression is available for the moment generating function, then a corresponding expression can be obtained for $\psi(t)$ by replacing e^t with t. For example, from formula (5.65), the probability generating function for the Poisson distribution is given by $\psi(t) = \exp(\lambda t - \lambda)$.

5.6.3. Some Limit Theorems

In Section 3.7 we defined convergence in probability of a sequence of random variables. In Section 4.5.1 convergence in distribution of the same sequence was introduced. In this section we introduce yet another type of convergence.

Definition 5.6.1. A sequence $\{X_n\}_{n=1}^{\infty}$ of random variables converges in quadratic mean to a random variable X if

$$\lim_{n \to \infty} E(X_n - X)^2 = 0.$$

This convergence is written symbolically as $X_n \xrightarrow{q.m.} X$. ☐

Convergence in quadratic mean implies convergence in probability. This follows directly from applying Markov's inequality: If $X_n \xrightarrow{q.m.} X$, then for any $\epsilon > 0$,

$$P(|X_n - X| \geq \epsilon) \leq \frac{1}{\epsilon^2} E(X_n - X)^2 \to 0$$

as $n \to \infty$. This shows that the sequence $\{X_n\}_{n=1}^{\infty}$ converges in probability to X.

5.6.3.1. The Weak Law of Large Numbers (Khinchine's Theorem)

Let $\{X_i\}_{i=1}^{\infty}$ be a sequence of independent and identically distributed random variables with a finite mean μ. Then $\bar{X}_n$ converges in probability to μ as $n \to \infty$, where $\bar{X}_n = (1/n)\sum_{i=1}^{n} X_i$ is the sample mean of a sample of size n.

Proof. See, for example, Lindgren (1976, Section 2.5.1) or Rao (1973, Section 2c.3). $\square$

Definition 5.6.2. A sequence $\{X_n\}_{n=1}^{\infty}$ of random variables converges strongly, or almost surely, to a random variable X, written symbolically as $X_n \xrightarrow{\text{a.s.}} X$, if for any $\epsilon > 0$,

$$\lim_{N \to \infty} P\left(\sup_{n \geq N} |X_n - X| > \epsilon \right) = 0. \qquad \square$$

Theorem 5.6.2. Let $\{X_n\}_{n=1}^{\infty}$ be a sequence of random variables. Then we have the following:

1. If $X_n \xrightarrow{\text{a.s.}} c$, where c is constant, then X_n converges in probability to c.
2. If $X_n \xrightarrow{\text{q.m.}} c$, and the series $\sum_{n=1}^{\infty} E(X_n - c)^2$ converges, then $X_n \xrightarrow{\text{a.s.}} c$.

5.6.3.2. The Strong Law of Large Numbers (Kolmogorov's Theorem)

Let $\{X_n\}_{n=1}^{\infty}$ be a sequence of independent random variables such that $E(X_n) = \mu_n$ and $\text{Var}(X_n) = \sigma_n^2$, $n = 1, 2, \ldots$. If the series $\sum_{n=1}^{\infty} \sigma_n^2/n^2$ converges, then $\bar{X}_n \xrightarrow{\text{a.s.}} \bar{\mu}_n$, where $\bar{\mu}_n = (1/n)\sum_{i=1}^{n} \mu_i$.

Proof. See Rao (1973, Section 2c.3). $\square$

5.6.3.3. The Continuity Theorem for Probability Generating Functions

See Feller (1968, page 280).

Suppose that for every $k \geq 1$, the sequence $\{p_k(n)\}_{n=0}^{\infty}$ represents a discrete probability distribution. Let $\psi_k(t) = \sum_{n=0}^{\infty} t^n p_k(n)$ be the corresponding probability generating function ($k = 1, 2, \ldots$). In order for a limit

$$q_n = \lim_{k \to \infty} p_k(n)$$

to exist for every $n = 0, 1, \ldots$, it is necessary and sufficient that the limit

$$\psi(t) = \lim_{k \to \infty} \psi_k(t)$$

exist for every t in the open interval $(0, 1)$. In this case,

$$\psi(t) = \sum_{n=0}^{\infty} t^n q_n.$$

This theorem implies that a sequence of discrete probability distributions converges if and only if the corresponding probability generating functions converge. It is important here to point out that the q_n's may not form a discrete probability distribution (because they may not sum to 1). The function $\psi(t)$ may not therefore be a probability generating function.

5.6.4. Power Series and Logarithmic Series Distributions

The power series distribution, which was introduced by Kosambi (1949), represents a family of discrete distributions, such as the binomial, Poisson, and negative binomial. Its probability mass function is given by

$$p(n) = \frac{a_n \theta^n}{f(\theta)}, \qquad n = 0, 1, 2, \ldots,$$

where $a_n \geq 0$, $\theta > 0$, and $f(\theta)$ is the function

$$f(\theta) = \sum_{n=0}^{\infty} a_n \theta^n. \tag{5.71}$$

This function is defined provided that θ falls inside the interval of convergence of the series in (5.71).

For example, for the Poisson distribution, $\theta = \lambda$, where λ is the mean, $a_n = 1/n!$ for $n = 0, 1, 2, \ldots$, and $f(\theta) = e^\lambda$. For the negative binomial, $\theta = 1 - p$ and $a_n = \binom{n + r - 1}{n}$, $n = 0, 1, 2, \ldots$, where $n = $ number of failures, $r = $ number of successes, and $p = $ probability of success on each trial, and thus

$$f(\theta) = \sum_{n=0}^{\infty} \binom{n + r - 1}{n} (1 - p)^n = \frac{1}{p^r}.$$

A special case of the power series distribution is the logarithmic series distribution. It was first introduced by Fisher, Corbet, and Williams (1943) while studying abundance and diversity for insect trap data. The probability mass function for this distribution is

$$p(n) = -\frac{\theta^n}{n \log(1 - \theta)}, \qquad n = 1, 2, \ldots,$$

where $0 < \theta < 1$.

The logarithmic series distribution is useful in the analysis of various kinds of data. A description of some of its applications can be found, for example, in Johnson and Kotz (1969, Chapter 7).

5.6.5. Poisson Approximation to Power Series Distributions

See Pérez-Abreu (1991).

The Poisson distribution can provide an approximation to the distribution of the sum of random variables having power series distributions. This is based on the following theorem:

Theorem 5.6.3. For each $k \geq 1$, let $X_1, X_2, \ldots, X_k$ be independent non-negative integer-valued random variables with a common power series distribution

$$p_k(n) = a_n \theta_k^n / f(\theta_k), \qquad n = 0, 1, \ldots,$$

where $a_n \geq 0$ $(n = 0, 1, \ldots)$ are independent of k and

$$f(\theta_k) = \sum_{n=0}^{\infty} a_n \theta_k^n, \qquad \theta_k > 0.$$

Let $a_0 > 0$, $\lambda > 0$ be fixed and $S_k = \sum_{i=1}^{k} X_i$. If $k\theta_k \to \lambda$ as $k \to \infty$, then

$$\lim_{k \to \infty} P(S_k = n) = e^{-\lambda_0} \lambda_0^n / n!, \qquad n = 0, 1, \ldots,$$

where $\lambda_0 = \lambda a_1 / a_0$.

Proof. See Pérez-Abreu (1991, page 43).   □

By using this theorem we can obtain the well-known Poisson approximation to the binomial and the negative binomial distributions as shown below.

EXAMPLE 5.6.2 (The Binomial Distribution). For each $k \geq 1$, let $X_1, \ldots, X_k$ be a sequence of independent Bernoulli random variables with success probability p_k. Let $S_k = \sum_{i=1}^{k} X_i$. Suppose that $kp_k \to \lambda > 0$ as $k \to \infty$. Then, for each $n = 0, 1, \ldots,$

$$\lim_{k \to \infty} P(S_k = n) = \lim_{k \to \infty} \binom{k}{n} p_k^n (1 - p_k)^{k-n}$$

$$= e^{-\lambda} \lambda^n / n!.$$

This follows from the fact that the Bernoulli distribution with success probability p_k is a power series distribution with $\theta_k = p_k / (1 - p_k)$ and $f(\theta_k) =$

$1 + \theta_k$. Since $a_0 = a_1 = 1$, and $k\theta_k \to \lambda$ as $k \to \infty$, we get from applying Theorem 5.6.3 that

$$\lim_{k \to \infty} P(S_k = n) = e^{-\lambda}\lambda^n/n!.$$

EXAMPLE 5.6.3 (The Negative Binomial Distribution). We recall that a random variable Y has the negative binomial distribution if it represents the number of failures n (in repeated trials) before the kth success ($k \geq 1$). Let p_k denote the probability of success on a single trial. Let $X_1, X_2, \ldots, X_k$ be random variables defined as

X_1 = number of failures occurring before the 1st success,

X_2 = number of failures occurring between the 1st success
      and the 2nd success,

$\vdots$

X_k = number of failures occurring between the $(k - 1)$st
      success and the kth success.

Such random variables have what is known as the geometric distribution. It is a special case of the negative binomial distribution with $k = 1$. The common probability distribution of the X_i's is

$$P(X_i = n) = p_k(1 - p_k)^n, \qquad n = 0, 1 \ldots; \quad i = 1, 2, \ldots, k.$$

This is a power series distribution with $a_n = 1$ ($n = 0, 1, \ldots$), $\theta_k = 1 - p_k$, and

$$f(\theta_k) = \sum_{n=0}^{\infty} (1 - p_k)^n = \frac{1}{1 - (1 - p_k)} = \frac{1}{p_k}.$$

It is easy to see that $X_1, X_2, \ldots, X_k$ are independent and that $Y = S_k = \sum_{i=1}^{k} X_i$.

Let us now assume that $k(1 - p_k) \to \lambda > 0$ as $k \to \infty$. Then from Theorem 5.6.3 we obtain the following result:

$$\lim_{k \to \infty} P(S_k = n) = e^{-\lambda}\lambda^n/n!, \qquad n = 0, 1 \ldots.$$

5.6.6. A Ridge Regression Application

Consider the linear model

$$\mathbf{y} = \mathbf{X}\boldsymbol{\beta} + \boldsymbol{\epsilon},$$

where $\mathbf{y}$ is a vector of n response values, $\mathbf{X}$ is an $n \times p$ matrix of rank p, $\boldsymbol{\beta}$ is a vector of p unknown parameters, and $\boldsymbol{\epsilon}$ is a random error vector such that $E(\boldsymbol{\epsilon}) = \mathbf{0}$ and $\text{Var}(\boldsymbol{\epsilon}) = \sigma^2 \mathbf{I}_n$. All variables in this model are corrected for their means and scaled to unit length, so that $\mathbf{X}'\mathbf{X}$ and $\mathbf{X}'\mathbf{y}$ are in correlation form.

We recall from Section 2.4.2 that if the columns of $\mathbf{X}$ are multicollinear, then the least-squares estimator of $\boldsymbol{\beta}$, namely $\hat{\boldsymbol{\beta}} = (\mathbf{X}'\mathbf{X})^{-1}\mathbf{X}'\mathbf{y}$, is an unreliable estimator due to large variances associated with its elements. There are several methods that can be used to combat multicollinearity. A review of such methods can be found in Ofir and Khuri (1986). Ridge regression is one of the most popular of these methods. This method, which was developed by Hoerl and Kennard (1970a, b), is based on adding a positive constant k to the diagonal elements of $\mathbf{X}'\mathbf{X}$. This leads to a biased estimator $\boldsymbol{\beta}^*$ of $\boldsymbol{\beta}$ called the ridge regression estimator and is given by

$$\boldsymbol{\beta}^* = (\mathbf{X}'\mathbf{X} + k\mathbf{I}_n)^{-1}\mathbf{X}'\mathbf{y}.$$

The elements of $\boldsymbol{\beta}^*$ can have substantially smaller variances than the corresponding elements of $\hat{\boldsymbol{\beta}}$ (see, for example, Montgomery and Peck, 1982, Section 8.5.3).

Draper and Herzberg (1987) showed that the ridge regression residual sum of squares can be represented as a power series in k. More specifically, consider the vector of predicted responses,

$$\hat{\mathbf{y}}_k = \mathbf{X}\boldsymbol{\beta}^*$$
$$= \mathbf{X}(\mathbf{X}'\mathbf{X} + k\mathbf{I}_n)^{-1}\mathbf{X}'\mathbf{y}, \tag{5.72}$$

which is based on using $\boldsymbol{\beta}^*$. Formula (5.72) can be written as

$$\hat{\mathbf{y}}_k = \mathbf{X}(\mathbf{X}'\mathbf{X})^{-1}\left[\mathbf{I}_n + k(\mathbf{X}'\mathbf{X})^{-1}\right]^{-1}\mathbf{X}'\mathbf{y}. \tag{5.73}$$

From Theorem 5.5.2, if all the eigenvalues of $k(\mathbf{X}'\mathbf{X})^{-1}$ are less than one in absolute value, then

$$\left[\mathbf{I}_n + k(\mathbf{X}'\mathbf{X})^{-1}\right]^{-1} = \sum_{i=0}^{\infty}(-1)^i k^i (\mathbf{X}'\mathbf{X})^{-i}. \tag{5.74}$$

From (5.73) and (5.74) we get

$$\hat{\mathbf{y}}_k = \left(\mathbf{H}_1 - k\mathbf{H}_2 + k^2\mathbf{H}_3 - k^3\mathbf{H}_4 + \cdots\right)\mathbf{y},$$

where $\mathbf{H}_i = \mathbf{X}(\mathbf{X}'\mathbf{X})^{-i}\mathbf{X}'$, $i \geq 1$. Thus the ridge regression residual sum of squares, which is the sum of squares of deviations of the elements of $\mathbf{y}$ from the corresponding elements of $\hat{\mathbf{y}}_k$, is

$$(\mathbf{y} - \hat{\mathbf{y}}_k)'(\mathbf{y} - \hat{\mathbf{y}}_k) = \mathbf{y}'\mathbf{Q}\mathbf{y},$$

where

$$\mathbf{Q} = \left(\mathbf{I}_n - \mathbf{H}_1 + k\mathbf{H}_2 - k^2\mathbf{H}_3 + k^3\mathbf{H}_4 - \cdots \right)^2.$$

It can be shown (see Exercise 5.32) that

$$\mathbf{y}'\mathbf{Q}\mathbf{y} = SS_E + \sum_{i=3}^{\infty} (i-2)(-k)^{i-1} S_i, \qquad (5.75)$$

where SS_E is the usual least-squares residual sum of squares, which can be obtained when $k = 0$, that is,

$$SS_E = \mathbf{y}'\left[\mathbf{I}_n - \mathbf{X}(\mathbf{X}'\mathbf{X})^{-1}\mathbf{X}' \right]\mathbf{y},$$

and $S_i = \mathbf{y}'\mathbf{H}_i\mathbf{y}$, $i \geq 3$. The terms to the right of (5.75), other than SS_E, are bias sums of squares induced by the presence of a nonzero k. Draper and Herzberg (1987) demonstrated by means of an example that the series in (5.75) may diverge or else converge very slowly, depending on the value of k.

FURTHER READING AND ANNOTATED BIBLIOGRAPHY

Apostol, T. M. (1964). *Mathematical Analysis*. Addison-Wesley, Reading, Massachusetts. (Infinite series are discussed in Chap. 12.)

Boyer, C. B. (1968). *A History of Mathematics*. Wiley, New York.

Draper, N. R., and A. M. Herzberg (1987). "A ridge-regression sidelight." *Amer. Statist.*, **41**, 282–283.

Draper, N. R., and H. Smith (1981). *Applied Regression Analysis*, 2nd ed. Wiley, New York. (Chap. 6 discusses ridge regression in addition to the various statistical procedures for selecting variables in a regression model.)

Feller, W. (1968). *An Introduction to Probability Theory and Its Applications*, Vol. I, 3rd ed. Wiley, New York.

Fisher, R. A., and E. A. Cornish (1960). "The percentile points of distribution having known cumulants." *Technometrics*, **2**, 209–225.

Fisher, R. A., A. S. Corbet, and C. B. Williams (1943). "The relation between the number of species and the number of individuals in a random sample of an animal population." *J. Anim. Ecology*, **12**, 42–58.

Fisz, M. (1963). *Probability Theory and Mathematical Statistics*, 3rd ed. Wiley, New York. (Chap. 5 deals almost exclusively with limit distributions for sums of independent random variables.)

Fulks, W. (1978). *Advanced Calculus*, 3rd ed. Wiley, New York. (Chap. 2 discusses limits of sequences; Chap. 13 deals with infinite series of constant terms; Chap. 14 deals with sequences and series of functions; Chap. 15 provides a study of power series.)

Gantmacher, F. R. (1959). *The Theory of Matrices*, Vol. I. Chelsea, New York.

Graybill, F. A. (1983). *Matrices with Applications in Statistics*, 2nd ed. Wadsworth, Belmont, California. (Chap. 5 includes a section on sequences and series of matrices.)

Hirschman, I. I., Jr. (1962). *Infinite Series*. Holt, Rinehart and Winston, New York. (This book is designed to be used in applied courses beyond the advanced calculus level. It emphasizes applications of the theory of infinite series.)

Hoerl, A. E., and R. W. Kennard (1970a). "Ridge regression: Biased estimation for non-orthogonal problems." *Technometrics*, **12**, 55–67.

Hoerl, A. E., and R. W. Kennard (1970b). "Ridge regression: Applications to non-orthogonal problems." *Technometrics*, **12**, 69–82; Correction. **12**, 723.

Hogg, R. V., and A. T. Craig (1965). *Introduction to Mathematical Statistics*, 2nd ed. Macmillan, New York.

Hyslop, J. M. (1954). *Infinite Series*, 5th ed. Oliver and Boyd, Edinburgh. (This book presents a concise treatment of the theory of infinite series. It provides the basic elements of this theory in a clear and easy-to-follow manner.)

Johnson, N. L., and S. Kotz (1969). *Discrete Distributions*. Houghton Mifflin, Boston. (Chaps. 1 and 2 contain discussions concerning moments, cumulants, generating functions, and power series distributions.)

Kendall, M., and A. Stuart (1977). *The Advanced Theory of Statistics*, Vol. 1, 4th ed. Macmillan, New York. (Moments, cumulants, and moment generating functions are discussed in Chap. 3.)

Knopp, K. (1951). *Theory and Application of Infinite Series*. Blackie and Son, London. (This reference book provides a detailed and comprehensive study of the theory of infinite series. It contains many interesting examples.)

Kosambi, D. D. (1949). "Characteristic properties of series distributions." *Proc. Nat. Inst. Sci. India*, **15**, 109–113.

Lancaster, P. (1969). *Theory of Matrices*. Academic Press, New York. (Chap. 5 discusses functions of matrices in addition to sequences and series involving matrix terms.)

Lindgren, B. W. (1976). *Statistical Theory*, 3rd ed. Macmillan, New York. (Chap. 2 contains a section on moments of a distribution and a proof of Markov's inequality.)

Montgomery, D. C., and E. A. Peck (1982). *Introduction to Linear Regression Analysis*. Wiley, New York. (Chap. 8 discusses the effect of multicollinearity and the methods for dealing with it including ridge regression.)

Nurcombe, J. R. (1979). "A sequence of convergence tests." *Amer. Math. Monthly*, **86**, 679–681.

Ofir, C., and A. I. Khuri (1986). "Multicollinearity in marketing models: Diagnostics and remedial measures." *Internat. J. Res. Market.*, **3**, 181–205. (This is a review article that surveys the problem of multicollinearity in linear models and the various remedial measures for dealing with it.)

Pérez-Abreu, V. (1991). "Poisson approximation to power series distributions." *Amer. Statist.*, **45**, 42–45.

Pye, W. C., and P. G. Webster (1989). "A note on Raabe's test extended." *Math. Comput. Ed.*, **23**, 125–128.

Rao, C. R. (1973). *Linear Statistical Inference and Its Applications*, 2nd ed. Wiley, New York. (Chap. 2 contains a section on limit theorems in statistics, including the weak and strong laws of large numbers.)

Rudin, W. (1964). *Principles of Mathematical Analysis*, 2nd ed. McGraw-Hill, New York. (Sequences and series of scalar constants are discussed in Chap. 3; sequences and series of functions are studied in Chap. 7.)

Thiele, T. N. (1903). *Theory of Observations*. Layton, London. Reprinted in *Ann. Math. Statist.*, **2**, 165–307 (1931).

Wilks, S. S. (1962). *Mathematical Statistics*. Wiley, New York. (Chap. 4 discusses different types of convergence of random variables; Chap. 5 presents several results concerning the moments of a distribution.)

Withers, C. S. (1984). "Asymptotic expansions for distributions and quantiles with power series cumulants." *J. Roy. Statist. Soc. Ser. B*, **46**, 389–396.

EXERCISES

In Mathematics

5.1. Suppose that $\{a_n\}_{n=1}^{\infty}$ is a bounded sequence of positive terms.
   (a) Define $b_n = \max\{a_1, a_2, \ldots, a_n\}$, $n = 1, 2, \ldots$. Show that the sequence $\{b_n\}_{n=1}^{\infty}$ converges, and identify its limit.
   (b) Suppose further that $a_n \to c$ as $n \to \infty$, where $c > 0$. Show that $c_n \to c$, where $\{c_n\}_{n=1}^{\infty}$ is the sequence of geometric means, $c_n = (\prod_{i=1}^{n} a_i)^{1/n}$.

5.2. Suppose that $\{a_n\}_{n=1}^{\infty}$ and $\{b_n\}_{n=1}^{\infty}$ are any two Cauchy sequences. Let $d_n = |a_n - b_n|$, $n = 1, 2, \ldots$. Show that the sequence $\{d_n\}_{n=1}^{\infty}$ converges.

5.3. Prove Theorem 5.1.3.

5.4. Show that if $\{a_n\}_{n=1}^{\infty}$ is a bounded sequence, then the set E of all its subsequential limits is also bounded.

5.5. Suppose that $a_n \to c$ as $n \to \infty$ and that $\{a_i\}_{i=1}^{\infty}$ is a sequence of positive terms for which $\sum_{i=1}^{n} \alpha_i \to \infty$ as $n \to \infty$.
   (a) Show that

$$\frac{\sum_{i=1}^{n} \alpha_i a_i}{\sum_{i=1}^{n} \alpha_i} \to c \qquad \text{as } n \to \infty.$$

   In particular, if $\alpha_i = 1$ for all i, then

$$\frac{1}{n} \sum_{i=1}^{n} a_i \to c \quad \text{as } n \to \infty.$$

   (b) Show that the converse of the special case in (a) does not always hold by giving a counterexample of a sequence $\{a_n\}_{n=1}^{\infty}$ that does not converge, yet $(\sum_{i=1}^{n} a_i)/n$ converges as $n \to \infty$.

5.6. Let $\{a_n\}_{n=1}^{\infty}$ be a sequence of positive terms such that

$$\frac{a_{n+1}}{a_n} \to b \qquad \text{as } n \to \infty,$$

where $0 < b < 1$. Show that there exist constants c and r such that $0 < r < 1$ and $c > 0$ for which $a_n < cr^n$ for sufficiently large values of n.

5.7. Suppose that we have the sequence $\{a_n\}_{n=1}^{\infty}$, where $a_1 = 1$ and

$$a_{n+1} = \frac{a_n(3b + a_n^2)}{3a_n^2 + b}, \qquad b > 0, \quad n = 1, 2, \dots.$$

Show that the sequence converges, and find its limit.

5.8. Show that the sequence $\{a_n\}_{n=1}^{\infty}$ converges, and find its limit, where $a_1 = 1$ and

$$a_{n+1} = (2 + a_n)^{1/2}, \qquad n = 1, 2, \dots.$$

5.9. Let $\{a_n\}_{n=1}^{\infty}$ be a sequence and $s_n = \sum_{i=1}^{n} a_i$.
 (a) Show that $\limsup_{n \to \infty} (s_n/n) \leq \limsup_{n \to \infty} a_n$.
 (b) If s_n/n converges as $n \to \infty$, then show that $a_n/n \to 0$ as $n \to \infty$.

5.10. Show that the sequence $\{a_n\}_{n=1}^{\infty}$, where $a_n = \sum_{i=1}^{n}(1/i)$, is not a Cauchy sequence and is therefore divergent.

5.11. Suppose that the sequence $\{a_n\}_{n=1}^{\infty}$ satisfies the following condition: There is an r, $0 < r < 1$, such that

$$|a_{n+1} - a_n| < br^n, \qquad n = 1, 2, \dots,$$

where b is a positive constant. Show that this sequence converges.

5.12. Show that if $a_n \geq 0$ for all n, then $\sum_{n=1}^{\infty} a_n$ converges if and only if $\{s_n\}_{n=1}^{\infty}$ is a bounded sequence, where s_n is the nth partial sum of the series.

5.13. Show that the series $\sum_{n=1}^{\infty}[1/(3n - 1)(3n + 2)]$ converges to $\frac{1}{6}$.

5.14. Show that the series $\sum_{n=1}^{\infty}(n^{1/n} - 1)^p$ is divergent for $p \leq 1$.

5.15. Let $\sum_{n=1}^{\infty} a_n$ be a divergent series of positive terms.
 (a) If $\{a_n\}_{n=1}^{\infty}$ is a bounded sequence, then show that $\sum_{n=1}^{\infty}[a_n/(1 + a_n)]$ diverges.

(b) Show that (a) is true even if $\{a_n\}_{n=1}^\infty$ is not a bounded sequence.

5.16. Let $\sum_{n=1}^\infty a_n$ be a divergent series of positive terms. Show that

$$\frac{a_n}{s_n^2} \le \frac{1}{s_{n-1}} - \frac{1}{s_n}, \qquad n = 2, 3, \ldots,$$

where s_n is the nth partial sum of the series; then deduce that $\sum_{n=1}^\infty (a_n/s_n^2)$ converges.

5.17. Let $\sum_{n=1}^\infty a_n$ be a convergent series of positive terms. Let $r_n = \sum_{i=n}^\infty a_i$. Show that for $m < n$,

$$\sum_{i=m}^n \frac{a_i}{r_i} > 1 - \frac{r_n}{r_m},$$

and deduce that $\sum_{n=1}^\infty (a_n/r_n)$ diverges.

5.18. Given the two series $\sum_{n=1}^\infty (1/n), \sum_{n=1}^\infty (1/n^2)$. Show that

$$\lim_{n \to \infty} \left(\frac{1}{n} \right)^{1/n} = 1,$$

$$\lim_{n \to \infty} \left(\frac{1}{n^2} \right)^{1/n} = 1.$$

5.19. Test for convergence of the series $\sum_{n=1}^\infty a_n$, where

(a) $\qquad a_n = (n^{1/n} - 1)^n$,

(b) $\qquad a_n = \dfrac{\log(1+n)}{\log(1 + e^{n^2})}$,

(c) $\qquad a_n = \dfrac{1 \times 3 \times 5 \times \cdots \times (2n-1)}{2 \times 4 \times 6 \times \cdots \times 2n} \cdot \dfrac{1}{2n+1}$,

(d) $\qquad a_n = \sqrt{n + \sqrt{n}} - \sqrt{n}$,

(e) $\qquad a_n = \dfrac{(-1)^n 4^n}{n^n}$,

(f) $\qquad a_n = \sin\left[\left(n + \frac{1}{n} \right) \pi \right]$.

5.20. Determine the values of x for which each of the following series converges uniformly:

(a)
$$\sum_{n=1}^{\infty} \frac{n+2}{3^n} x^{2n},$$

(b)
$$\sum_{n=1}^{\infty} \frac{10^n}{n} x^n,$$

(c)
$$\sum_{n=1}^{\infty} (n+1)^2 x^n,$$

(d)
$$\sum_{n=1}^{\infty} \frac{\cos(nx)}{n(n^2+1)}.$$

5.21. Consider the series

$$\sum_{n=1}^{\infty} a_n = \sum_{n=1}^{\infty} \frac{(-1)^{n-1}}{n}.$$

Let $\sum_{n=1}^{\infty} b_n$ be a certain rearrangement of $\sum_{n=1}^{\infty} a_n$ given by

$$1 + \tfrac{1}{3} - \tfrac{1}{2} + \tfrac{1}{5} + \tfrac{1}{7} - \tfrac{1}{4} + \tfrac{1}{9} + \tfrac{1}{11} - \tfrac{1}{6} + \cdots,$$

where two positive terms are followed by one negative. Show that the sum of the original series is less than $\frac{10}{12}$, whereas that of the rearranged series (which is convergent) exceeds $\frac{11}{12}$.

5.22. Consider Cauchy's product of $\sum_{n=0}^{\infty} a_n$ with itself, where

$$a_n = \frac{(-1)^n}{\sqrt{n+1}}, \qquad n = 0, 1, 2, \ldots.$$

Show that this product is divergent. [*Hint:* Show that the nth term of this product does not go to zero as $n \to \infty$.]

5.23. Consider the sequence of functions $\{f_n(x)\}_{n=1}^{\infty}$, where for $n = 1, 2, \ldots$

$$f_n(x) = \frac{nx}{1+nx^2}, \qquad x \geq 0.$$

Find the limit of this sequence, and determine whether or not the convergence is uniform on $[0, \infty)$.

5.24. Consider the series $\sum_{n=1}^{\infty}(1/n^x)$.

   (a) Show that this series converges uniformly on $[1 + \delta, \infty)$, where δ is any positive number. [*Note:* The function represented by this series is known as *Riemann's ζ-function* and is denoted by $\zeta(x)$.]

   (b) Is $\zeta(x)$ differentiable on $[1 + \delta, \infty)$? If so, give a series expansion for $\zeta'(x)$.

In Statistics

5.25. Prove formulas (5.56) and (5.57).

5.26. Find a series expansion for the moment generating function of the negative binomial distribution. For what values of t does this series converge uniformly? In this case, can formula (5.63) be applied to obtain an expression for the kth noncentral moment $(k = 1, 2, ...)$ of this distribution? Why or why not?

5.27. Find the first three cumulants of the negative binomial distribution.

5.28. Show that the moments μ'_n $(n = 1, 2, ...)$ of a random variable X determine the cumulative distribution functions of X uniquely if

$$\limsup_{n \to \infty} \frac{|\mu'_n|^{1/n}}{n} \quad \text{is finite}.$$

[*Hint:* Use the fact that $n! \sim \sqrt{2\pi} n^{n+1/2} e^{-n}$ as $n \to \infty$.]

5.29. Find the moment generating function of the logarithmic series distribution, and deduce that the mean and variance of this distribution are given by

$$\mu = \alpha\theta/(1 - \theta),$$

$$\sigma^2 = \mu\left(\frac{1}{1 - \theta} - \mu\right),$$

where $\alpha = -1/\log(1 - \theta)$.

5.30. Let $\{X_n\}_{n=1}^{\infty}$ be a sequence of binomial random variables where the probability mass function of X_n $(n = 1, 2, ...)$ is given by

$$p_n(k) = \binom{n}{k} p^k (1 - p)^{n-k}, \qquad k = 0, 1, 2, ..., n,$$

where $0 < p < 1$. Further, let the random variable Y_n be defined as

$$Y_n = \frac{X_n}{n} - p.$$

(a) Show that $E(X_n) = np$ and $\text{Var}(X_n) = np(1-p)$.
(b) Apply Chebyshev's inequality to show that

$$P(|Y_n| \geq \epsilon) \leq \frac{p(1-p)}{n\epsilon^2},$$

where $\epsilon > 0$.
(c) Deduce from (b) that Y_n converges in probability to zero. [*Note:* This result is known as *Bernoulli's law of large numbers.*]

5.31. Let $X_1, X_2, \ldots, X_n$ be a sequence of independent Bernoulli random variables with success probability p_n. Let $S_n = \sum_{i=1}^{n} X_i$. Suppose that $np_n \to \mu > 0$ as $n \to \infty$.

(a) Give an expression for $\phi_n(t)$, the moment generating function of S_n.
(b) Show that

$$\lim_{n \to \infty} \phi_n(t) = \exp(\mu e^t - \mu),$$

which is the moment generating function of a Poisson distribution with mean μ.

5.32. Prove formula (5.75).

CHAPTER 6

Integration

The origin of integral calculus can be traced back to the ancient Greeks. They were motivated by the need to measure the length of a curve, the area of a surface, or the volume of a solid. Archimedes used techniques very similar to actual integration to determine the length of a segment of a curve. Democritus (410 B.C.) had the insight to consider that a cone was made up of infinitely many plane cross sections parallel to the base.

The theory of integration received very little stimulus after Archimedes's remarkable achievements. It was not until the beginning of the seventeenth century that the interest in Archimedes's ideas began to develop. Johann Kepler (1571–1630) was the first among European mathematicians to develop the ideas of infinitesimals in connection with integration. The use of the term "integral" is due to the Swiss mathematician Johann Bernoulli (1667–1748).

In the present chapter we shall study integration of real-valued functions of a single variable x according to the concepts put forth by the German mathematician Georg Friedrich Riemann (1826–1866). He was the first to establish a rigorous analytical foundation for integration, based on the older geometric approach.

6.1. SOME BASIC DEFINITIONS

Let $f(x)$ be a function defined and bounded on a finite interval $[a, b]$. Suppose that this interval is partitioned into a finite number of subintervals by a set of points $P = \{x_0, x_1, \dots, x_n\}$ such that $a = x_0 < x_1 < x_2 < \cdots < x_n = b$. This set is called a partition of $[a, b]$. Let $\Delta x_i = x_i - x_{i-1}$ $(i = 1, 2, \dots, n)$, and Δ_p be the largest of $\Delta x_1, \Delta x_2, \dots, \Delta x_n$. This value is called the norm of P. Consider the sum

$$S(P, f) = \sum_{i=1}^{n} f(t_i) \Delta x_i,$$

where t_i is a point in the subinterval $[x_{i-1}, x_i]$, $i = 1, 2, \dots, n$.

205

The function $f(x)$ is said to be Riemann integrable on $[a, b]$ if a number A exists with the following property: For any given $\epsilon > 0$ there exists a number $\delta > 0$ such that

$$|A - S(P, f)| < \epsilon$$

for any partition P of $[a, b]$ with a norm $\Delta_p < \delta$, and for any choice of the point t_i in $[x_{i-1}, x_i]$, $i = 1, 2, \ldots, n$. The number A is called the Riemann integral of $f(x)$ on $[a, b]$ and is denoted by $\int_a^b f(x)\, dx$. The integration symbol $\int$ was first used by the German mathematician Gottfried Wilhelm Leibniz (1646–1716) to represent a sum (it was derived from the first letter of the Latin word summa, which means a sum).

6.2. THE EXISTENCE OF THE RIEMANN INTEGRAL

In order to investigate the existence of the Riemann integral, we shall need the following theorem:

Theorem 6.2.1. Let $f(x)$ be a bounded function on a finite interval, $[a, b]$. For every partition $P = \{x_0, x_1, \ldots, x_n\}$ of $[a, b]$, let m_i and M_i be, respectively, the infimum and supremum of $f(x)$ on $[x_{i-1}, x_i]$, $i = 1, 2, \ldots, n$. If, for a given $\epsilon > 0$, there exists a $\delta > 0$ such that

$$US_P(f) - LS_P(f) < \epsilon \qquad\qquad (6.1)$$

whenever $\Delta_p < \delta$, where Δ_p is the norm of P, and

$$LS_P(f) = \sum_{i=1}^n m_i \Delta x_i,$$

$$US_P(f) = \sum_{i=1}^n M_i \Delta x_i,$$

then $f(x)$ is Riemann integrable on $[a, b]$. Conversely, if $f(x)$ is Riemann integrable, then inequality (6.1) holds for any partition P such that $\Delta_p < \delta$. [The sums, $LS_P(f)$ and $US_P(f)$, are called the lower sum and upper sum, respectively, of $f(x)$ with respect to the partition P.]

In order to prove Theorem 6.2.1 we need the following lemmas:

Lemma 6.2.1. Let P and P' be two partitions of $[a, b]$ such that $P' \supset P$ (P' is called a refinement of P and is constructed by adding partition points between those that belong to P). Then

$$US_{P'}(f) \le US_P(f),$$
$$LS_{P'}(f) \ge LS_P(f).$$

Proof. Let $P = \{x_0, x_1, \ldots, x_n\}$. By the nature of the partition P', the ith subinterval $\Delta x_i = x_i - x_{i-1}$ is divided into k_i parts $\Delta_{x_i}^{(1)}, \Delta_{x_i}^{(2)}, \ldots, \Delta_{x_i}^{(k_i)}$, where $k_i \geq 1$, $i = 1, 2, \ldots, n$. If $m_i^{(j)}$ and $M_i^{(j)}$ denote, respectively, the infimum and supremum of $f(x)$ on $\Delta_{x_i}^{(j)}$, then $m_i \leq m_i^{(j)} \leq M_i^{(j)} \leq M_i$ for $j = 1, 2, \ldots, k_i$; $i = 1, 2, \ldots, n$, where m_i and M_i are the infimum and supremum of $f(x)$ on $[x_{i-1}, x_i]$, respectively. It follows that

$$LS_P(f) = \sum_{i=1}^{n} m_i \Delta x_i \leq \sum_{i=1}^{n} \sum_{j=1}^{k_i} m_i^{(j)} \Delta_{x_i}^{(j)} = LS_{P'}(f)$$

$$US_{P'}(f) = \sum_{i=1}^{n} \sum_{j=1}^{k_i} M_i^{(j)} \Delta_{x_i}^{(j)} \leq \sum_{i=1}^{n} M_i \Delta x_i = US_P(f). \qquad \square$$

Lemma 6.2.2. Let P and P' be any two partitions of $[a, b]$. Then $LS_P(f) \leq US_{P'}(f)$.

Proof. Let $P'' = P \cup P'$. The partition P'' is a refinement of both P and P'. Then, by Lemma 6.2.1,

$$LS_P(f) \leq LS_{P''}(f) \leq US_{P''}(f) \leq US_{P'}(f). \qquad \square$$

Proof of Theorem 6.2.1

Let $\epsilon > 0$ be given. Suppose that inequality (6.1) holds for any partition P whose norm Δ_p is less than δ. Let $S(P, f) = \sum_{i=1}^{n} f(t_i) \Delta x_i$, where t_i is a point in $[x_{i-1}, x_i]$, $i = 1, 2, \ldots, n$. By the definition of $LS_P(f)$ and $US_P(f)$ we can write

$$LS_P(f) \leq S(P, f) \leq US_P(f). \qquad (6.2)$$

Let m and M be the infimum and supremum, respectively, of $f(x)$ on $[a, b]$; then

$$m(b - a) \leq LS_P(f) \leq US_P(f) \leq M(b - a). \qquad (6.3)$$

Let us consider two sets of lower and upper sums of $f(x)$ with respect to partitions $P, P', P'', \ldots$ such that $P \subset P' \subset P'' \subset \cdots$. Then, by Lemma 6.2.1, the set of upper sums is decreasing, and the set of lower sums is increasing. Furthermore, because of (6.3), the set of upper sums is bounded from below by $m(b - a)$, and the set of lower sums is bounded from above by $M(b - a)$. Hence, the infimum of $US_P(f)$ and the supremum of $LS_P(f)$ with respect to P do exist (see Theorem 1.5.1).

From Lemma 6.2.2 it is easy to deduce that

$$\sup_P LS_P(f) \leq \inf_P US_P(f).$$

Now, suppose that for the given $\epsilon > 0$ there exists a $\delta > 0$ such that

$$US_P(f) - LS_P(f) < \epsilon \tag{6.4}$$

for any partition whose norm Δ_p is less than δ. We have that

$$LS_P(f) \leq \sup_P LS_P(f) \leq \inf_P US_P(f) \leq US_P(f). \tag{6.5}$$

Hence,

$$\inf_P US_P(f) - \sup_P LS_P(f) < \epsilon.$$

Since $\epsilon > 0$ is arbitrary, we conclude that if (6.1) is satisfied, then

$$\inf_P US_P(f) = \sup_P LS_P(f). \tag{6.6}$$

Furthermore, from (6.2), (6.4), and (6.5) we obtain

$$|S(P, f) - A| < \epsilon,$$

where A is the common value of $\inf_P US_P(f)$ and $\sup_P LS_P(f)$. This proves that A is the Riemann integral of $f(x)$ on $[a, b]$.

Let us now show that the converse of the theorem is true, that is, if $f(x)$ is Riemann integrable on $[a, b]$, then inequality (6.1) holds.

If $f(x)$ is Riemann integrable, then for a given $\epsilon > 0$ there exists a $\delta > 0$ such that

$$\left| \sum_{i=1}^{n} f(t_i) \Delta x_i - A \right| < \frac{\epsilon}{3} \tag{6.7}$$

and

$$\left| \sum_{i=1}^{n} f(t_i') \Delta x_i - A \right| < \frac{\epsilon}{3} \tag{6.8}$$

for any partition $P = \{x_0, x_1, \ldots, x_n\}$ of $[a, b]$ with a norm $\Delta_p < \delta$, and any choices of t_i, t_i' in $[x_{i-1}, x_i]$, $i = 1, 2, \ldots, n$, where $A = \int_a^b f(x)\,dx$. From (6.7) and (6.8) we then obtain

$$\left| \sum_{i=1}^{n} [f(t_i) - f(t_i')] \Delta x_i \right| < \frac{2\epsilon}{3}.$$

Now, $M_i - m_i$ is the supremum of $f(x) - f(x')$ for x, x' in $[x_{i-1}, x_i]$, $i = 1, 2, \ldots, n$. It follows that for a given $\eta > 0$ we can choose t_i, t_i' in $[x_{i-1}, x_i]$ so that

$$f(t_i) - f(t_i') > M_i - m_i - \eta, \qquad i = 1, 2, \ldots, n,$$

for otherwise $M_i - m_i - \eta$ would be an upper bound for $f(x) - f(x')$ for all x, x' in $[x_{i-1}, x_i]$, which is a contradiction. In particular, if $\eta = \epsilon/[3(b - a)]$, then we can find t_i, t_i' in $[x_{i-1}, x_i]$ such that

$$US_P(f) - LS_P(f) = \sum_{i=1}^{n} (M_i - m_i) \Delta x_i$$

$$< \sum_{i=1}^{n} [f(t_i) - f(t_i')] \Delta x_i + \eta(b - a)$$

$$< \epsilon.$$

This proves the validity of inequality (6.1).   □

Corollary 6.2.1. Let $f(x)$ be a bounded function on $[a, b]$. Then $f(x)$ is Riemann integrable on $[a, b]$ if and only if $\inf_P US_P(f) = \sup_P LS_P(f)$, where $LS_P(f)$ and $US_P(f)$ are, respectively, the lower and upper sums of $f(x)$ with respect to a partition P of $[a, b]$.

Proof. See Exercise 6.1.   □

EXAMPLE 6.2.1.   Let $f(x): [0, 1] \to R$ be the function $f(x) = x^2$. Then, $f(x)$ is Riemann integrable on $[0, 1]$. To show this, let $P = \{x_0, x_1, \ldots, x_n\}$ be any partition of $[0, 1]$, where $x_0 = 0$, $x_n = 1$. Then

$$LS_P(f) = \sum_{i=1}^{n} x_{i-1}^2 \Delta x_i,$$

$$US_P(f) = \sum_{i=1}^{n} x_i^2 \Delta x_i.$$

Hence,

$$US_P(f) - LS_P(f) = \sum_{i=1}^{n} \left(x_i^2 - x_{i-1}^2 \right) \Delta x_i$$

$$\leq \Delta_p \sum_{i=1}^{n} \left(x_i^2 - x_{i-1}^2 \right),$$

where Δ_p is the norm of P. But

$$\sum_{i=1}^{n} \left(x_i^2 - x_{i-1}^2 \right) = x_n^2 - x_0^2 = 1.$$

Thus

$$US_P(f) - LS_P(f) \leq \Delta_p.$$

It follows that for a given $\epsilon > 0$ we can choose $\delta = \epsilon$ such that for any partition P whose norm Δ_p is less than δ,

$$US_P(f) - LS_P(f) < \epsilon.$$

By Theorem 6.2.1, $f(x) = x^2$ is Riemann integrable on $[0, 1]$.

EXAMPLE 6.2.2.   Consider the function $f(x)$: $[0, 1] \to R$ such that $f(x) = 0$ if x a rational number and $f(x) = 1$ if x is irrational. Since every subinterval of $[0, 1]$ contains both rational and irrational numbers, then for any partition $P = \{x_0, x_1, \ldots, x_n\}$ of $[0, 1]$ we have

$$US_P(f) = \sum_{i=1}^{n} M_i \Delta x_i = \sum_{i=1}^{n} \Delta x_i = 1,$$

$$LS_P(f) = \sum_{i=1}^{n} m_i \Delta x_i = \sum_{i=1}^{n} 0 \Delta x_i = 0.$$

It follows that

$$\inf_P US_P(f) = 1 \qquad \text{and} \qquad \sup_P LS_P(f) = 0.$$

By Corollary 6.2.1, $f(x)$ is not Riemann integrable on $[0, 1]$.

6.3. SOME CLASSES OF FUNCTIONS THAT ARE RIEMANN INTEGRABLE

There are certain classes of functions that are Riemann integrable. Identifying a given function as a member of such a class can facilitate the determination of its Riemann integrability. Some of these classes of functions include: (i) continuous functions; (ii) monotone functions; (iii) functions of bounded variation.

Theorem 6.3.1.   If $f(x)$ is continuous on $[a, b]$, then it is Riemann integrable there.

Proof.   Since $f(x)$ is continuous on a closed and bounded interval, then by Theorem 3.4.6 it must be uniformly continuous on $[a, b]$. Consequently, for a given $\epsilon > 0$ there exists a $\delta > 0$ that depends only on ϵ such that for any x_1, x_2 in $[a, b]$ we have

$$|f(x_1) - f(x_2)| < \frac{\epsilon}{b - a}$$

if $|x_1 - x_2| < \delta$. Let $P = \{x_0, x_1, \ldots, x_n\}$ be a partition of P with a norm $\Delta_p < \delta$. Then

$$US_P(f) - LS_P(f) = \sum_{i=1}^{n} (M_i - m_i)\Delta x_i,$$

where m_i and M_i are, respectively, the infimum and supremum of $f(x)$ on $[x_{i-1}, x_i]$, $i = 1, 2, \ldots, n$. By Corollary 3.4.1 there exist points ξ_i, η_i in $[x_{i-1}, x_i]$ such that $m_i = f(\xi_i), M_i = f(\eta_i), i = 1, 2, \ldots, n$. Since $|\eta_i - \xi_i| \leq \Delta_p < \delta$ for $i = 1, 2, \ldots, n$, then

$$US_P(f) - LS_P(f) = \sum_{i=1}^{n} [f(\eta_i) - f(\xi_i)]\Delta x_i$$

$$< \frac{\epsilon}{b-a} \sum_{i=1}^{n} \Delta x_i = \epsilon.$$

By Theorem 6.2.1 we conclude that $f(x)$ is Riemann integrable on $[a, b]$.
□

It should be noted that continuity is a sufficient condition for Riemann integrability, but is not a necessary one. A function $f(x)$ can have discontinuities in $[a, b]$ and still remains Riemann integrable on $[a, b]$. For example, consider the function

$$f(x) = \begin{cases} -1, & -1 \leq x < 0, \\ 1, & 0 \leq x \leq 1. \end{cases}$$

This function is discontinuous at $x = 0$. However, it is Riemann integrable on $[-1, 1]$. To show this, let $\epsilon > 0$ be given, and let $P = \{x_0, x_1, \ldots, x_n\}$ be a partition of $[-1, 1]$ such that $\Delta_p < \epsilon/2$. By the nature of this function, $f(x_i) - f(x_{i-1}) \geq 0$, and the infimum and supremum of $f(x)$ on $[x_{i-1}, x_i]$ are equal to $f(x_{i-1})$ and $f(x_i)$, respectively, $i = 1, 2, \ldots, n$. Hence,

$$US_P(f) - LS_P(f) = \sum_{i=1}^{n} (M_i - m_i)\Delta x_i$$

$$= \sum_{i=1}^{n} [f(x_i) - f(x_{i-1})]\Delta x_i$$

$$< \Delta_p \sum_{i=1}^{n} [f(x_i) - f(x_{i-1})] = \Delta_p[f(1) - f(-1)]$$

$$< \frac{\epsilon}{2}[f(1) - f(-1)] = \epsilon.$$

The function $f(x)$ is therefore Riemann integrable on $[-1, 1]$ by Theorem 6.2.1.

On the basis of this example it is now easy to prove the following theorem:

Theorem 6.3.2. If $f(x)$ is monotone increasing (or monotone decreasing) on $[a, b]$, then it is Riemann integrable there.

Theorem 6.3.2 can be used to construct a function that has a countable number of discontinuities in $[a, b]$ and is also Riemann integrable (see Exercise 6.2).

6.3.1. Functions of Bounded Variation

Let $f(x)$ be defined on $[a, b]$. This function is said to be of bounded variation on $[a, b]$ if there exists a number $M > 0$ such that for any partition $P = \{x_0, x_1, \ldots, x_n\}$ of $[a, b]$ we have

$$\sum_{i=1}^{n} |\Delta f_i| \leq M,$$

where $\Delta f_i = f(x_i) - f(x_{i-1})$, $i = 1, 2, \ldots, n$.

Any function that is monotone increasing (or decreasing) on $[a, b]$ is also of bounded variation there. To show this, let $f(x)$ be monotone increasing on $[a, b]$. Then

$$\sum_{i=1}^{n} |\Delta f_i| = \sum_{i=1}^{n} [f(x_i) - f(x_{i-1})] = f(b) - f(a).$$

Hence, if M is any number greater than or equal to $f(b) - f(a)$, then $\sum_{i=1}^{n} |\Delta f_i| \leq M$ for any partition P of $[a, b]$.

Another example of a function of bounded variation is given in the next theorem.

Theorem 6.3.3. If $f(x)$ is continuous on $[a, b]$ and its derivative $f'(x)$ exists and is bounded on (a, b), then $f(x)$ is of bounded variation on $[a, b]$.

Proof. Let $P = \{x_0, x_1, \ldots, x_n\}$ be a partition of $[a, b]$. By applying the mean value theorem (Theorem 4.2.2) on each $[x_{i-1}, x_i]$, $i = 1, 2, \ldots, n$, we obtain

$$\sum_{i=1}^{n} |\Delta f_i| = \sum_{i=1}^{n} |f'(\xi_i) \Delta x_i|$$

$$\leq K \sum_{i=1}^{n} (x_i - x_{i-1}) = K(b - a),$$

where $x_{i-1} < \xi_i < x_i$, $i = 1, 2, \ldots, n$, and $K > 0$ is such that $|f'(x)| \leq K$ on (a, b).   □

It should be noted that any function of bounded variation on $[a, b]$ is also bounded there. This is true because if $a < x < b$, then $P = \{a, x, b\}$ is a partition of $[a, b]$. Hence,

$$|f(x) - f(a)| + |f(b) - f(x)| \leq M.$$

for some positive number M. This implies that $|f(x)|$ is bounded on $[a, b]$ since

$$|f(x)| \leq \tfrac{1}{2}\big[|f(x) - f(a)| + |f(x) - f(b)| + |f(a) + f(b)|\big]$$
$$\leq \tfrac{1}{2}\big[M + |f(a) + f(b)|\big].$$

The converse of this result, however, is not necessarily true, that is, if $f(x)$ is bounded, then it may not be of bounded variation. For example, the function

$$f(x) = \begin{cases} x \cos\left(\dfrac{\pi}{2x}\right), & 0 < x \leq 1, \\ 0, & x = 0 \end{cases}$$

is bounded on $[0, 1]$, but is not of bounded variation there. It can be shown that for the partition

$$P = \left\{0, \frac{1}{2n}, \frac{1}{2n-1}, \ldots, \frac{1}{3}, \frac{1}{2}, 1\right\},$$

$\sum_{i=1}^{2n} |\Delta f_i| \to \infty$ as $n \to \infty$ and hence cannot be bounded by a constant M for all n (see Exercise 6.4).

Theorem 6.3.4. If $f(x)$ is of bounded variation on $[a, b]$, then it is Riemann integrable there.

Proof. Let $\epsilon > 0$ be given, and let $P = \{x_0, x_1, \ldots, x_n\}$ be a partition of $[a, b]$. Then

$$US_P(f) - LS_P(f) = \sum_{i=1}^{n} (M_i - m_i)\,\Delta x_i, \tag{6.9}$$

where m_i and M_i are the infimum and supremum of $f(x)$ on $[x_{i-1}, x_i]$, respectively, $i = 1, 2, \ldots, n$. By the properties of m_i and M_i, there exist ξ_i

and η_i in $[x_{i-1}, x_i]$ such that for $i = 1, 2, \ldots, n$,

$$m_i \le f(\xi_i) < m_i + \epsilon',$$

$$M_i - \epsilon' < f(\eta_i) \le M_i,$$

where ϵ' is a small positive number to be determined later. It follows that

$$M_i - m_i - 2\epsilon' < f(\eta_i) - f(\xi_i) \le M_i - m_i, \qquad i = 1, 2, \ldots, n.$$

Hence,

$$M_i - m_i < 2\epsilon' + f(\eta_i) - f(\xi_i)$$

$$\le 2\epsilon' + |f(\xi_i) - f(\eta_i)|, \qquad i = 1, 2, \ldots, n.$$

From formula (6.9) we obtain

$$US_P(f) - LS_P(f) < 2\epsilon' \sum_{i=1}^{n} \Delta x_i + \sum_{i=1}^{n} |f(\xi_i) - f(\eta_i)| \Delta x_i. \quad (6.10)$$

Now, if Δ_p is the norm of P, then

$$\sum_{i=1}^{n} |f(\xi_i) - f(\eta_i)| \Delta x_i \le \Delta_p \sum_{i=1}^{n} |f(\xi_i) - f(\eta_i)|$$

$$\le \Delta_p \sum_{i=1}^{m} |f(z_i) - f(z_{i-1})|, \quad (6.11)$$

where $\{z_0, z_1, \ldots, z_m\}$ is a partition Q of $[a, b]$, which consists of the points $x_0, x_1, \ldots, x_n$ as well as the points $\xi_1, \eta_1, \xi_2, \eta_2, \ldots, \xi_n, \eta_n$, that is, Q is a refinement of P obtained by adding the ξ_i's and η_i's $(i = 1, 2, \ldots, n)$. Since $f(x)$ is of bounded variation on $[a, b]$, there exists a number $M > 0$ such that

$$\sum_{i=1}^{m} |f(z_i) - f(z_{i-1})| \le M. \quad (6.12)$$

From (6.10), (6.11), and (6.12) it follows that

$$US_P(f) - LS_P(f) < 2\epsilon'(b - a) + M\Delta_p. \quad (6.13)$$

Let us now select the partition P such that $\Delta_p < \delta$, where $M\delta < \epsilon/2$. If we also choose ϵ' such that $2\epsilon'(b - a) < \epsilon/2$, then from (6.13) we obtain $US_P(f) - LS_P(f) < \epsilon$. The function $f(x)$ is therefore Riemann integrable on $[a, b]$ by Theorem 6.2.1. $\square$

6.4. PROPERTIES OF THE RIEMANN INTEGRAL

The Riemann integral has several properties that are useful at both the theoretical and practical levels. Most of these properties are fairly simple and striaghtforward. We shall therefore not prove every one of them in this section.

Theorem 6.4.1. If $f(x)$ and $g(x)$ are Riemann integrable on $[a, b]$ and if c_1 and c_2 are constants, then $c_1 f(x) + c_2 g(x)$ is Riemann integrable on $[a, b]$, and

$$\int_a^b [c_1 f(x) + c_2 g(x)]\, dx = c_1 \int_a^b f(x)\, dx + c_2 \int_a^b g(x)\, dx.$$

Theorem 6.4.2. If $f(x)$ is Riemann integrable on $[a, b]$, and $m \le f(x) \le M$ for all x in $[a, b]$, then

$$m(b - a) \le \int_a^b f(x)\, dx \le M(b - a).$$

Theorem 6.4.3. If $f(x)$ and $g(x)$ are Riemann integrable on $[a, b]$, and if $f(x) \le g(x)$ for all x in $[a, b]$, then

$$\int_a^b f(x)\, dx \le \int_a^b g(x)\, dx.$$

Theorem 6.4.4. If $f(x)$ is Riemann integrable on $[a, b]$ and if $a < c < b$, then

$$\int_a^b f(x)\, dx = \int_a^c f(x)\, dx + \int_c^b f(x)\, dx.$$

Theorem 6.4.5. If $f(x)$ is Riemann integrable on $[a, b]$, then so is $|f(x)|$ and

$$\left| \int_a^b f(x)\, dx \right| \le \int_a^b |f(x)|\, dx.$$

Proof. Let $P = \{x_0, x_1, \ldots, x_n\}$ be a partition of $[a, b]$. Let m_i and M_i be the infimum and supremum of $f(x)$, respectively, on $[x_{i-1}, x_i]$; and let m_i', M_i' be the same for $|f(x)|$. We claim that

$$M_i - m_i \ge M_i' - m_i', \qquad i = 1, 2, \ldots, n.$$

It is obvious that $M_i - m_i = M_i' - m_i'$ if $f(x)$ is either nonnegative or nonpositive for all x in $[x_{i-1}, x_i]$, $i = 1, 2, \ldots, n$. Let us therefore suppose that $f(x)$ is negative on D_i^- and nonnegative on D_i^+, where D_i^- and D_i^+ are such that $D_i^- \cup D_i^+ = D_i = [x_{i-1}, x_i]$ for $i = 1, 2, \ldots, n$. We than have

$$M_i - m_i = \sup_{D_i^+} f(x) - \inf_{D_i^-} f(x)$$

$$= \sup_{D_i^+} |f(x)| - \inf_{D_i^-} (-|f(x)|)$$

$$= \sup_{D_i^+} |f(x)| + \sup_{D_i^-} |f(x)|$$

$$\geq \sup_{D_i} |f(x)| = M_i',$$

since $\sup_{D_i} |f(x)| = \max\{\sup_{D_i^+} |f(x)|, \sup_{D_i^-} |f(x)|\}$. Hence, $M_i - m_i \geq M_i' \geq M_i' - m_i'$ for $i = 1, 2, \ldots, n$, which proves our claim.

$$US_P(|f|) - LS_P(|f|) = \sum_{i=1}^{n} (M_i' - m_i') \, \Delta x_i \leq \sum_{i=1}^{n} (M_i - m_i) \, \Delta x_i.$$

Hence,

$$US_P(|f|) - LS_P(|f|) \leq US_P(f) - LS_P(f). \tag{6.14}$$

Since $f(x)$ is Riemann integrable, the right-hand side of inequality (6.14) can be made smaller than any given $\epsilon > 0$ by a proper choice of the norm Δ_p of P. It follows that $|f(x)|$ is Riemann integrable on $[a, b]$ by Theorem 6.2.1. Furthermore, since $\mp f(x) \leq |f(x)|$ for all x in $[a, b]$, then $\int_a^b \mp f(x) \, dx \leq \int_a^b |f(x)|$ by Theorem 6.4.3, that is,

$$\mp \int_a^b f(x) \, dx \leq \int_a^b |f(x)| \, dx.$$

Thus, $|\int_a^b f(x) \, dx| \leq \int_a^b |f(x)| \, dx$.   □

Corollary 6.4.1. If $f(x)$ is Riemann integrable on $[a, b]$, then so is $f^2(x)$.

Proof. Using the same notation as in the proof of Theorem 6.4.5, we have that $m_i'^2$ and $M_i'^2$ are, respectively, the infimum and supremum of $f^2(x)$ on $[x_{i-1}, x_i]$ for $i = 1, 2, \ldots, n$. Now,

$$M_i'^2 - m_i'^2 = (M_i' - m_i')(M_i' + m_i')$$

$$\leq 2M'(M_i' - m_i')$$

$$\leq 2M'(M_i - m_i), \qquad i = 1, 2, \ldots, n, \tag{6.15}$$

where M' is the supremum of $|f(x)|$ on $[a, b]$. The Riemann integrability of $f^2(x)$ now follows from inequality (6.15) by the Riemann integrability of $f(x)$.

$\square$

Corollary 6.4.2. If $f(x)$ and $g(x)$ are Riemann integrable on $[a, b]$, then so is their product $f(x)g(x)$.

Proof. This follows directly from the identity

$$4f(x)g(x) = [f(x) + g(x)]^2 - [f(x) - g(x)]^2, \qquad (6.16)$$

and the fact that the squares of $f(x) + g(x)$ and $f(x) - g(x)$ are Riemann integrable on $[a, b]$ by Theorem 6.4.1 and Corollary 6.4.1. $\square$

Theorem 6.4.6 (The Mean Value Theorem for Integrals). If $f(x)$ is continuous on $[a, b]$, then there exists a point $c \in [a, b]$ such that

$$\int_a^b f(x)\, dx = (b - a)f(c).$$

Proof. By Theorem 6.4.2 we have

$$m \le \frac{1}{b - a} \int_a^b f(x)\, dx \le M,$$

where m and M are, respectively, the infimum and supremum of $f(x)$ on $[a, b]$. Since $f(x)$ is continuous, then by Corollary 3.4.1 it must attain the values m and M at some points inside $[a, b]$. Furthermore, by the intermediate-value theorem (Theorem 3.4.4), $f(x)$ assumes every value between m and M. Hence, there is a point $c \in [a, b]$ such that

$$f(c) = \frac{1}{b - a} \int_a^b f(x)\, dx. \qquad \square$$

Definition 6.4.1. Let $f(x)$ be Riemann integrable on $[a, b]$. The function

$$F(x) = \int_a^x f(t)\, dt, \qquad a \le x \le b,$$

is called an indefinite integral of $f(x)$. $\square$

Theorem 6.4.7. If $f(x)$ is Riemann integrable on $[a, b]$, then $F(x) = \int_a^x f(t)\, dt$ is uniformly continuous on $[a, b]$.

Proof. Let x_1, x_2 be in $[a, b]$, $x_1 < x_2$. Then,

$$|F(x_2) - F(x_1)| = \left| \int_a^{x_2} f(t)\, dt - \int_a^{x_1} f(t)\, dt \right|$$

$$= \left| \int_{x_1}^{x_2} f(t)\, dt \right|, \qquad \text{by Theorem 6.4.4}$$

$$\leq \int_{x_1}^{x_2} |f(t)|\, dt, \qquad \text{by Theorem 6.4.5}$$

$$\leq M'(x_2 - x_1),$$

where M' is the supremum of $|f(x)|$ on $[a, b]$. Thus if $\epsilon > 0$ is given, then $|F(x_2) - F(x_1)| < \epsilon$ provided that $|x_1 - x_2| < \epsilon/M'$. This proves uniform continuity of $F(x)$ on $[a, b]$. ☐

The next theorem presents a practical way for evaluating the Riemann integral on $[a, b]$.

Theorem 6.4.8. Suppose that $f(x)$ is continuous on $[a, b]$. Let $F(x) = \int_a^x f(t)\, dt$. Then we have the following:

    i. $dF(x)/dx = f(x)$, $a \leq x \leq b$.
    ii. $\int_a^b f(x)\, dx = G(b) - G(a)$, where $G(x) = F(x) + c$, and c is an arbitrary constant.

Proof. We have

$$\frac{dF(x)}{dx} = \frac{d}{dx} \int_a^x f(t)\, dt = \lim_{h \to 0} \frac{1}{h} \left[\int_a^{x+h} f(t)\, dt - \int_a^x f(t)\, dt \right]$$

$$= \lim_{h \to 0} \frac{1}{h} \int_x^{x+h} f(t)\, dt, \qquad \text{by Theorem 6.4.4}$$

$$= \lim_{h \to 0} f(x + \theta h),$$

by Theorem 6.4.6, where $0 \leq \theta \leq 1$. Hence,

$$\frac{dF(x)}{dx} = \lim_{h \to 0} f(x + \theta h) = f(x)$$

by the continuity of $f(x)$. This result indicates that an indefinite integral of $f(x)$ is any function whose derivative is equal to $f(x)$. It is therefore unique up to a constant. Thus both $F(x)$ and $F(x) + c$, where c is an arbitrary constant, are considered to be indefinite integrals.

To prove the second part of the theorem, let $G(x)$ be defined on $[a, b]$ as

$$G(x) = F(x) + c = \int_a^x f(t)\, dt + c,$$

that is, $G(x)$ is an indefinite integral of $f(x)$. If $x = a$, then $G(a) = c$, since $F(a) = 0$. Also, if $x = b$, then $G(b) = F(b) + c = \int_a^b f(t)\, dt + G(a)$. It follows that

$$\int_a^b f(t)\, dt = G(b) - G(a). \qquad \square$$

This result is known as the *fundamental theorem of calculus*. It is generally attributed to Isaac Barrow (1630–1677), who was the first to realize that differentiation and integration are inverse operations. One advantage of this theorem is that it provides a practical way to evaluate the integral of $f(x)$ on $[a, b]$.

6.4.1. Change of Variables in Riemann Integration

There are situations in which the variable x in a Riemann integral is a function of some other variable, say u. In this case, it may be of interest to determine how the integral can be expressed and evaluated under the given transformation. One advantage of this change of variable is the possibility of simplifying the actual evaluation of the integral, provided that the transformation is properly chosen.

Theorem 6.4.9. Let $f(x)$ be continuous on $[\alpha, \beta]$, and let $x = g(u)$ be a function whose derivative $g'(u)$ exists and is continuous on $[c, d]$. Suppose that the range of g is contained inside $[\alpha, \beta]$. If a, b are points in $[\alpha, \beta]$ such that $a = g(c)$ and $b = g(d)$, then

$$\int_a^b f(x)\, dx = \int_c^d f[g(u)] g'(u)\, du.$$

Proof. Let $F(x) = \int_a^x f(t)\, dt$. By Theorem 6.4.8, $F'(x) = f(x)$. Let $G(u)$ be defined as

$$G(u) = \int_c^u f[g(t)] g'(t)\, dt.$$

Since f, g, and g' are continuous, then by Theorem 6.4.8 we have

$$\frac{dG(u)}{du} = f[g(u)] g'(u). \qquad (6.17)$$

However, according to the chain rule (Theorem 4.1.3),

$$\frac{dF[g(u)]}{du} = \frac{dF[g(u)]}{dg(u)} \frac{dg(u)}{du}$$

$$= f[g(u)]g'(u). \tag{6.18}$$

From formulas (6.17) and (6.18) we conclude that

$$G(u) - F[g(u)] = \lambda, \tag{6.19}$$

where λ is a constant. If a and b are points in $[\alpha, \beta]$ such that $a = g(c), b = g(d)$, then when $u = c$, we have $G(c) = 0$ and $\lambda = -F[g(c)] = -F(a) = 0$. Furthermore, when $u = d, G(d) = \int_c^d f[g(t)]g'(t) dt$. From (6.19) we then obtain

$$G(d) = \int_c^d f[g(t)]g'(t) \, dt = F[g(d)] + \lambda$$

$$= F(b)$$

$$= \int_a^b f(x) \, dx.$$

For example, consider the integral $\int_1^2 (2t^2 - 1)^{1/2} t \, dt$. Let $x = 2t^2 - 1$. Then $dx = 4t \, dt$, and by Theorem 6.4.9,

$$\int_1^2 (2t^2 - 1)^{1/2} t \, dt = \frac{1}{4} \int_1^7 x^{1/2} \, dx.$$

An indefinite integral of $x^{1/2}$ is given by $\frac{2}{3}x^{3/2}$. Hence,

$$\int_1^2 (2t^2 - 1)^{1/2} t \, dt = \frac{1}{4}\left(\frac{2}{3}\right)(7^{3/2} - 1) = \frac{1}{6}(7^{3/2} - 1). \qquad \square$$

6.5. IMPROPER RIEMANN INTEGRALS

In our study of the Riemann integral we have only considered integrals of functions that are bounded on a finite interval $[a, b]$. We now extend the scope of Riemann integration to include situations where the integrand can become unbounded at one or more points inside the range of integration, which can also be infinite. In such situations, the Riemann integral is called an improper integral.

There are two kinds of improper integrals. If $f(x)$ is Riemann integrable on $[a, b]$ for any $b > a$, then $\int_a^\infty f(x) \, dx$ is called an improper integral of the first kind, where the range of integration is infinite. If, however, $f(x)$

becomes infinite at a finite number of points inside the range of integration, then the integral $\int_a^b f(x)\,dx$ is said to be improper of the second kind.

Definition 6.5.1.   Let $F(z) = \int_a^z f(x)\,dx$. Suppose that $F(z)$ exists for any value of z greater than a. If $F(z)$ has a finite limit L as $z \to \infty$, then the improper integral $\int_a^\infty f(x)\,dx$ is said to converge to L. In this case, L represents the Riemann integral of $f(x)$ on $[a, \infty)$ and we write

$$L = \int_a^\infty f(x)\,dx.$$

On the other hand, if $L = \pm\infty$, then the improper integral $\int_a^\infty f(x)\,dx$ is said to diverge. By the same token, we can define the integral $\int_{-\infty}^a f(x)\,dx$ as the limit, if it exists, of $\int_{-z}^a f(x)\,dx$ as $z \to \infty$. Also, $\int_{-\infty}^\infty f(x)\,dx$ is defined as

$$\int_{-\infty}^\infty f(x)\,dx = \lim_{u \to \infty} \int_{-u}^a f(x)\,dx + \lim_{z \to \infty} \int_a^z f(x)\,dx,$$

where a is any finite number, provided that both limits exist.

The convergence of $\int_a^\infty f(x)\,dx$ can be determined by using the Cauchy criterion in a manner similar to the one used in the study of convergence of sequences (see Section 5.1.1).

Theorem 6.5.1.   The improper integral $\int_a^\infty f(x)\,dx$ converges if and only if for a given $\epsilon > 0$ there exists a z_0 such that

$$\left| \int_{z_1}^{z_2} f(x)\,dx \right| < \epsilon, \tag{6.20}$$

whenever z_1 and z_2 exceed z_0.

Proof.   If $F(z) = \int_a^z f(x)\,dx$ has a limit L as $z \to \infty$, then for a given $\epsilon > 0$ there exists z_0 such that for $z > z_0$.

$$|F(z) - L| < \frac{\epsilon}{2}.$$

Now, if both z_1 and z_2 exceed z_0, then

$$\left| \int_{z_1}^{z_2} f(x)\,dx \right| = |F(z_2) - F(z_1)|$$

$$\leq |F(z_2) - L| + |F(z_1) - L| < \epsilon.$$

Vice versa, if condition (6.20) is satisfied, then we need to show that $F(z)$ has a limit as $z \to \infty$. Let us therefore define the sequence $\{g_n\}_{n=1}^\infty$, where g_n is given by

$$g_n = \int_a^{a+n} f(x)\, dx, \qquad n = 1, 2, \dots .$$

It follows that for any $\epsilon > 0$,

$$|g_n - g_m| = \left| \int_{a+m}^{a+n} f(x)\, dx \right| < \epsilon,$$

if m and n are large enough. This implies that $\{g_n\}_{n=1}^{\infty}$ is a Cauchy sequence; hence it converges by Theorem 5.1.6. Let $g = \lim_{n \to \infty} g_n$. To show that $\lim_{z \to \infty} F(z) = g$, let us write

$$|F(z) - g| = |F(z) - g_n + g_n - g|$$
$$\leq |F(z) - g_n| + |g_n - g|. \qquad (6.21)$$

Suppose $\epsilon > 0$ is given. There exists an integer N_1 such that $|g_n - g| < \epsilon/2$ if $n > N_1$. Also, there exists an integer N_2 such that

$$|F(z) - g_n| = \left| \int_{a+n}^{z} f(x)\, dx \right| < \frac{\epsilon}{2} \qquad (6.22)$$

if $z > a + n > N_2$. Thus by choosing $z > a + n$, where $n > \max(N_1, N_2 - a)$, we get from inequalities (6.21) and (6.22)

$$|F(z) - g| < \epsilon.$$

This completes the proof.   □

Definition 6.5.2. If the improper integral $\int_a^\infty |f(x)|\, dx$ is convergent, then the integral $\int_a^\infty f(x)\, dx$ is said to be absolutely convergent. If $\int_a^\infty f(x)\, dx$ is convergent but not absolutely, then it is said to be conditionally convergent.   □

It is easy to show that an improper integral is convergent if it converges absolutely.

As with the case of series of positive terms, there are comparison tests that can be used to test for convergence of improper integrals of the first kind of nonnegative functions. These tests are described in the following theorems.

Theorem 6.5.2. Let $f(x)$ be a nonnegative function that is Riemann integrable on $[a, b]$ for every $b \geq a$. Suppose that there exists a function $g(x)$ such that $f(x) \leq g(x)$ for $x \geq a$. If $\int_a^\infty g(x)\, dx$ converges, then so does $\int_a^\infty f(x)\, dx$

and we have

$$\int_a^\infty f(x)\, dx \leq \int_a^\infty g(x)\, dx.$$

Proof. See Exercise 6.7.   □

Theorem 6.5.3. Let $f(x)$ and $g(x)$ be nonnegative functions that are Riemann integrable on $[a, b]$ for every $b \geq a$. If

$$\lim_{x \to \infty} \frac{f(x)}{g(x)} = k,$$

where k is a positive constant, then $\int_a^\infty f(x)\, dx$ and $\int_a^\infty g(x)\, dx$ are either both convergent or both divergent.

Proof. See Exercise 6.8.   □

EXAMPLE 6.5.1.   Consider the integral $\int_1^\infty e^{-x} x^2\, dx$. We have that $e^x = 1 + \sum_{n=1}^\infty (x^n/n!)$. Hence, for $x \geq 1$, $e^x > x^p/p!$, where p is any positive integer. If p is chosen such that $p - 2 \geq 2$, then

$$e^{-x} x^2 < \frac{p!}{x^{p-2}} \leq \frac{p!}{x^2}.$$

However, $\int_1^\infty (dx/x^2) = [-1/x]_1^\infty = 1$. Therefore, by Theorem 6.5.2, the integral of $e^{-x} x^2$ on $[1, \infty)$ is convergent.

EXAMPLE 6.5.2.   The integral $\int_0^\infty [(\sin x)/(x + 1)^2]\, dx$ is absolutely convergent, since

$$\frac{|\sin x|}{(x + 1)^2} \leq \frac{1}{(x + 1)^2}$$

and

$$\int_0^\infty \frac{dx}{(x + 1)^2} = \left[-\frac{1}{x + 1} \right]_0^\infty = 1.$$

EXAMPLE 6.5.3.   The integral $\int_0^\infty (\sin x/x)\, dx$ is conditionally convergent. We first show that $\int_0^\infty (\sin x/x)\, dx$ is convergent. We have that

$$\int_0^\infty \frac{\sin x}{x}\, dx = \int_0^1 \frac{\sin x}{x}\, dx + \int_1^\infty \frac{\sin x}{x}\, dx. \tag{6.23}$$

By Exercise 6.3, $(\sin x)/x$ is Riemann integrable on $[0, 1]$, since it is continuous there except at $x = 0$, which is a discontinuity of the first kind (see Definition 3.4.2). As for the second integral in (6.23), we have for $z_2 > z_1 > 1$,

$$\int_{z_1}^{z_2} \frac{\sin x}{x}\, dx = \left[-\frac{\cos x}{x} \right]_{z_1}^{z_2} - \int_{z_1}^{z_2} \frac{\cos x}{x^2}\, dx$$

$$= \frac{\cos z_1}{z_1} - \frac{\cos z_2}{z_2} - \int_{z_1}^{z_2} \frac{\cos x}{x^2}\, dx.$$

Thus

$$\left| \int_{z_1}^{z_2} \frac{\sin x}{x}\, dx \right| \le \frac{1}{z_1} + \frac{1}{z_2} + \int_{z_1}^{z_2} \frac{dx}{x^2} = \frac{2}{z_1}.$$

Since $2/z_1$ can be made arbitrarily small by choosing z_1 large enough, then by Theorem 6.5.1, $\int_1^\infty (\sin x/x)\, dx$ is convergent and so is $\int_0^\infty (\sin x/x)\, dx$.

It remains to show that $\int_0^\infty (\sin x/x)\, dx$ is not absolutely convergent. This follows from the fact that (see Exercise 6.10)

$$\lim_{n \to \infty} \int_0^{n\pi} \left| \frac{\sin x}{x} \right| dx = \infty.$$

Convergence of improper integrals of the first kind can be used to determine convergence of series of positive terms (see Section 5.2.1). This is based on the next theorem.

Theorem 6.5.4 (Maclaurin's Integral Test). Let $\sum_{n=1}^\infty a_n$ be a series of positive terms such that $a_{n+1} \le a_n$ for $n \ge 1$. Let $f(x)$ be a positive nonincreasing function defined on $[1, \infty)$ such that $f(n) = a_n$, $n = 1, 2, \ldots$, and $f(x) \to 0$ as $x \to \infty$. Then, $\sum_{n=1}^\infty a_n$ converges if and only if the improper integral $\int_1^\infty f(x)\, dx$ converges.

Proof. If $n \ge 1$ and $n \le x \le n + 1$, then

$$a_n = f(n) \ge f(x) \ge f(n + 1) = a_{n+1}.$$

By Theorem 6.4.2 we have for $n \ge 1$

$$a_n \ge \int_n^{n+1} f(x)\, dx \ge a_{n+1}. \tag{6.24}$$

If $s_n = \sum_{k=1}^n a_k$ is the nth partial sum of the series, then from inequality (6.24) we obtain

$$s_n \ge \int_1^{n+1} f(x)\, dx \ge s_{n+1} - a_1. \tag{6.25}$$

If the series $\sum_{n=1}^{\infty} a_n$ converges to the sum s, then $s \geq s_n$ for all n. Consequently, the sequence whose nth term is $F(n + 1) = \int_1^{n+1} f(x)\,dx$ is monotone increasing and is bounded by s; hence it must have a limit. Therefore, the integral $\int_1^{\infty} f(x)\,dx$ converges.

Now, let us suppose that $\int_1^{\infty} f(x)\,dx$ is convergent and is equal to L. Then from inequality (6.25) we obtain

$$ s_{n+1} \leq a_1 + \int_1^{n+1} f(x)\,dx \leq a_1 + L, \qquad n \geq 1, \qquad (6.26) $$

since $f(x)$ is positive. Inequality (6.26) indicates that the monotone increasing sequence $\{s_n\}_{n=1}^{\infty}$ is bounded hence it has a limit, which is the sum of the series.   □

Theorem 6.5.4 provides a test of convergence for a series of positive terms. Of course, the usefulness of this test depends on how easy it is to integrate the function $f(x)$.

As an example of using the integral test, consider the harmonic series $\sum_{n=1}^{\infty}(1/n)$. If $f(x)$ is defined as $f(x) = 1/x$, $x \geq 1$, then $F(x) = \int_1^x f(t)\,dt = \log x$. Since $F(x)$ goes to infinity as $x \to \infty$, the harmonic series must therefore be divergent, as was shown in Chapter 5. On the other hand, the series $\sum_{n=1}^{\infty}(1/n^2)$ is convergent, since $F(x) = \int_1^x(dt/t^2) = 1 - 1/x$, which converges to 1 as $x \to \infty$.

6.5.1. Improper Riemann Integrals of the Second Kind

Let us now consider integrals of the form $\int_a^b f(x)\,dx$ where $[a, b]$ is a finite interval and the integrand becomes infinite at a finite number of points inside $[a, b]$. Such integrals are called improper integrals of the second kind. Suppose, for example, that $f(x) \to \infty$ as $x \to a^+$. Then $\int_a^b f(x)\,dx$ is said to converge if the limit

$$ \lim_{\epsilon \to 0^+} \int_{a+\epsilon}^b f(x)\,dx $$

exists and is finite. Similarly, if $f(x) \to \infty$ as $x \to b^-$, then $\int_a^b f(x)\,dx$ is convergent if the limit

$$ \lim_{\epsilon \to 0^+} \int_a^{b-\epsilon} f(x)\,dx $$

exists. Furthermore, if $f(x) \to \infty$ as $x \to c$, where $a < c < b$, then $\int_a^b f(x)\,dx$ is the sum of $\int_a^c f(x)\,dx$ and $\int_c^b f(x)\,dx$ provided that both integrals converge. By definition, if $f(x) \to \infty$ as $x \to x_0$, where $x_0 \in [a, b]$, then x_0 is said to be a singularity of $f(x)$.

The following theorems can help in determining convergence of integrals of the second kind. They are similar to Theorems 6.5.1, 6.5.2, and 6.5.3. Their proofs will therefore be omitted.

Theorem 6.5.5. If $f(x) \to \infty$ as $x \to a^+$, then $\int_a^b f(x) \, dx$ converges if and only if for a given $\epsilon > 0$ there exists a z_0 such that

$$\left| \int_{z_1}^{z_2} f(x) \, dx \right| < \epsilon,$$

where z_1 and z_2 are any two numbers such that $a < z_1 < z_2 < z_0 < b$.

Theorem 6.5.6. Let $f(x)$ be a nonnegative function such that $\int_c^b f(x) \, dx$ exists for every c in $(a, b]$. If there exists a function $g(x)$ such that $f(x) \le g(x)$ for all x in $(a, b]$, and if $\int_c^b g(x) \, dx$ converges as $c \to a^+$, then so does $\int_c^b f(x) \, dx$ and we have

$$\int_a^b f(x) \, dx \le \int_a^b g(x) \, dx.$$

Theorem 6.5.7. Let $f(x)$ and $g(x)$ be nonnegative functions that are Riemann integrable on $[c, b]$ for every c such that $a < c \le b$. If

$$\lim_{x \to a^+} \frac{f(x)}{g(x)} = k,$$

where k is a positive constant, then $\int_a^b f(x) \, dx$ and $\int_a^b g(x) \, dx$ are either both convergent or both divergent.

Definition 6.5.3. Let $\int_a^b f(x) \, dx$ be an improper integral of the second kind. If $\int_a^b |f(x)| \, dx$ converges, then $\int_a^b f(x) \, dx$ is said to converge absolutely. If, however, $\int_a^b f(x) \, dx$ is convergent, but not absolutely, then it is said to be conditionally convergent. □

Theorem 6.5.8. If $\int_a^b |f(x)| \, dx$ converges, then so does $\int_a^b f(x) \, dx$.

EXAMPLE 6.5.4.   Consider the integral $\int_0^1 e^{-x} x^{n-1} \, dx$, where $n > 0$. If $0 < n < 1$, then the integral is improper of the second kind, since $x^{n-1} \to \infty$ as $x \to 0^+$. Thus, $x = 0$ is a singularity of the integrand. Since

$$\lim_{x \to 0^+} \frac{e^{-x} x^{n-1}}{x^{n-1}} = 1,$$

then the behavior of $\int_0^1 e^{-x} x^{n-1} \, dx$ with regard to convergence or divergence is the same as that of $\int_0^1 x^{n-1} \, dx$. But $\int_0^1 x^{n-1} \, dx = (1/n)[x^n]_0^1 = 1/n$ is convergent, and so is $\int_0^1 e^{-x} x^{n-1} \, dx$.

EXAMPLE 6.5.5. $\int_0^1 (\sin x / x^2)\, dx$. The integrand has a singularity at $x = 0$. Let $g(x) = 1/x$. Then, $(\sin x)/[x^2 g(x)] \to 1$ as $x \to 0^+$. But $\int_0^1 (dx/x) = [\log x]_0^1$ is divergent, since $\log x \to -\infty$ as $x \to 0^+$. Therefore, $\int_0^1 (\sin x / x^2)\, dx$ is divergent.

EXAMPLE 6.5.6. Consider the integral $\int_0^2 (x^2 - 3x + 1)/[x(x-1)^2]\, dx$. Here, the integrand has two singularities, namely $x = 0$ and $x = 1$, inside $[0, 2]$. We can therefore write

$$\int_0^2 \frac{x^2 - 3x + 1}{x(x-1)^2}\, dx = \lim_{t \to 0^+} \int_t^{1/2} \frac{x^2 - 3x + 1}{x(x-1)^2}\, dx$$

$$+ \lim_{u \to 1^-} \int_{1/2}^u \frac{x^2 - 3x + 1}{x(x-1)^2}\, dx$$

$$+ \lim_{v \to 1^+} \int_v^2 \frac{x^2 - 3x + 1}{x(x-1)^2}\, dx.$$

We note that

$$\frac{x^2 - 3x + 1}{x(x-1)^2} = \frac{1}{x} - \frac{1}{(x-1)^2}.$$

Hence,

$$\int_0^2 \frac{x^2 - 3x + 1}{x(x-1)^2}\, dx = \lim_{t \to 0^+} \left[\log x + \frac{1}{x-1} \right]_t^{1/2}$$

$$+ \lim_{u \to 1^-} \left[\log x + \frac{1}{x-1} \right]_{1/2}^u$$

$$+ \lim_{v \to 1^+} \left[\log x + \frac{1}{x-1} \right]_v^2.$$

None of the above limits exists as a finite number. This integral is therefore divergent.

6.6. CONVERGENCE OF A SEQUENCE OF RIEMANN INTEGRALS

In the present section we confine our attention to the limiting behavior of integrals of a sequence of functions $\{f_n(x)\}_{n=1}^\infty$.

Theorem 6.6.1. Suppose that $f_n(x)$ is Riemann integrable on $[a, b]$ for $n \geq 1$. If $f_n(x)$ converges uniformly to $f(x)$ on $[a, b]$ as $n \to \infty$, then $f(x)$ is Riemann integrable on $[a, b]$ and

$$\lim_{n \to \infty} \int_a^b f_n(x) \, dx = \int_a^b f(x) \, dx.$$

Proof. Let us first show that $f(x)$ is Riemann integrable on $[a, b]$. Let $\epsilon > 0$ be given. Since $f_n(x)$ converges uniformly to $f(x)$, then there exists an integer n_0 that depends only on ϵ such that

$$|f_n(x) - f(x)| < \frac{\epsilon}{3(b - a)} \tag{6.27}$$

if $n > n_0$ for all $x \in [a, b]$. Let $n_1 > n_0$. Since $f_{n_1}(x)$ is Riemann integrable on $[a, b]$, then by Theorem 6.2.1 there exists a $\delta > 0$ such that

$$US_P(f_{n_1}) - LS_P(f_{n_1}) < \frac{\epsilon}{3} \tag{6.28}$$

for any partition P of $[a, b]$ with a norm $\Delta_p < \delta$. Now, from inequality (6.27) we have

$$f(x) < f_{n_1}(x) + \frac{\epsilon}{3(b - a)},$$

$$f(x) > f_{n_1}(x) - \frac{\epsilon}{3(b - a)}.$$

We conclude that

$$US_P(f) \leq US_P(f_{n_1}) + \frac{\epsilon}{3}, \tag{6.29}$$

$$LS_P(f) \geq LS_P(f_{n_1}) - \frac{\epsilon}{3}. \tag{6.30}$$

From inequalities (6.28), (6.29), and (6.30) it follows that if $\Delta_p < \delta$, then

$$US_P(f) - LS_P(f) \leq US_P(f_{n_1}) - LS_P(f_{n_1}) + \frac{2\epsilon}{3} < \epsilon. \tag{6.31}$$

Inequality (6.31) shows that $f(x)$ is Riemann integrable on $[a, b]$, again by Theorem 6.2.1.

Let us now show that

$$\lim_{n \to \infty} \int_a^b f_n(x)\, dx = \int_a^b f(x)\, dx. \tag{6.32}$$

From inequality (6.27) we have for $n > n_0$,

$$\left| \int_a^b f_n(x)\, dx - \int_a^b f(x)\, dx \right| \leq \int_a^b |f_n(x) - f(x)|\, dx$$

$$< \frac{\epsilon}{3},$$

and the result follows, since ϵ is an arbitrary positive number.   □

6.7. SOME FUNDAMENTAL INEQUALITIES

In this section we consider certain well-known inequalities for the Riemann integral.

6.7.1. The Cauchy–Schwarz Inequality

Theorem 6.7.1. Suppose that $f(x)$ and $g(x)$ are such that $f^2(x)$ and $g^2(x)$ are Riemann integrable on $[a, b]$. Then

$$\left[\int_a^b f(x) g(x)\, dx \right]^2 \leq \left[\int_a^b f^2(x)\, dx \right] \left[\int_a^b g^2(x)\, dx \right]. \tag{6.33}$$

The limits of integration may be finite or infinite.

Proof. Let c_1 and c_2 be constants, not both zero. Without loss of generality, let us assume that $c_2 \neq 0$. Then

$$\int_a^b \left[c_1 f(x) + c_2 g(x) \right]^2 dx \geq 0.$$

Thus the quadratic form

$$c_1^2 \int_a^b f^2(x)\, dx + 2c_1 c_2 \int_a^b f(x) g(x)\, dx + c_2^2 \int_a^b g^2(x)\, dx$$

is nonnegative for all c_1 and c_2. It follows that its discriminant, namely,

$$c_2^2 \left[\int_a^b f(x) g(x)\, dx \right]^2 - c_2^2 \left[\int_a^b f^2(x)\, dx \right] \left[\int_a^b g^2(x)\, dx \right]$$

must be nonpositive, that is,

$$\left[\int_a^b f(x) g(x)\, dx \right]^2 \le \left[\int_a^b f^2(x)\, dx \right] \left[\int_a^b g^2(x)\, dx \right].$$

It is easy to see that if $f(x)$ and $g(x)$ are linearly related [that is, there exist constants τ_1 and τ_2, not both zero, such that $\tau_1 f(x) + \tau_2 g(x) = 0$], then inequality (6.33) becomes an equality.   □

6.7.2. Hölder's Inequality

This is a generalization of the Cauchy–Schwarz inequality due to Otto Hölder (1859–1937). To prove Hölder's inequality we need the following lemmas:

Lemma 6.7.1. Let $a_1, a_1, \ldots, a_n$; $\lambda_1, \lambda_2, \ldots, \lambda_n$ be nonnegative numbers such that $\sum_{i=1}^n \lambda_i = 1$. Then

$$\prod_{i=1}^n a_i^{\lambda_i} \le \sum_{i=1}^n \lambda_i a_i. \tag{6.34}$$

The right-hand side of inequality (6.34) is a weighted arithmetic mean of the a_i's, and the left-hand side is a weighted geometric mean.

Proof. This lemma is an extension of a result given in Section 3.7 concerning the properties of convex functions (see Exercise 6.19).   □

Lemma 6.7.2. Suppose that $f_1(x), f_2(x), \ldots, f_n(x)$ are nonnegative and Riemann integrable on $[a, b]$. If $\lambda_1, \lambda_2, \ldots, \lambda_n$ are nonnegative numbers such that $\sum_{i=1}^n \lambda_i = 1$, then

$$\int_a^b \left[\prod_{i=1}^n f_i^{\lambda_i}(x) \right] dx \le \prod_{i=1}^n \left[\int_a^b f_i(x)\, dx \right]^{\lambda_i}. \tag{6.35}$$

Proof. Without loss of generality, let us assume that $\int_a^b f_i(x)\, dx > 0$ for $i = 1, 2, \ldots, n$ [inequality (6.35) is obviously true if at least one $f_i(x)$ is identically equal to zero]. By Lemma 6.7.1 we have

$$\frac{\int_a^b \left[\prod_{i=1}^n f_i^{\lambda_i}(x) \right] dx}{\prod_{i=1}^n \left[\int_a^b f_i(x)\, dx \right]^{\lambda_i}}$$

$$= \int_a^b \left[\frac{f_1(x)}{\int_a^b f_1(x)\, dx} \right]^{\lambda_1} \left[\frac{f_2(x)}{\int_a^b f_2(x)\, dx} \right]^{\lambda_2} \cdots \left[\frac{f_n(x)}{\int_a^b f_n(x)\, dx} \right]^{\lambda_n} dx$$

$$\leq \int_a^b \sum_{i=1}^n \frac{\lambda_i f_i(x)}{\int_a^b f_i(x)\, dx}\, dx = \sum_{i=1}^n \lambda_i = 1.$$

Hence, inequality (6.35) follows.   □

Theorem 6.7.2 (Hölder's Inequality).   Let p and q be two positive numbers such that $1/p + 1/q = 1$. If $|f(x)|^p$ and $|g(x)|^q$ are Riemann integrable on $[a, b]$, then

$$\left| \int_a^b f(x) g(x)\, dx \right| \leq \left[\int_a^b |f(x)|^p\, dx \right]^{1/p} \left[\int_a^b |g(x)|^q\, dx \right]^{1/q}.$$

Proof. Define the functions

$$u(x) = |f(x)|^p, \qquad v(x) = |g(x)|^q.$$

Then, by Lemma 6.7.2,

$$\int_a^b u(x)^{1/p} v(x)^{1/q}\, dx \leq \left[\int_a^b u(x)\, dx \right]^{1/p} \left[\int_a^b v(x)\, dx \right]^{1/q},$$

that is,

$$\int_a^b |f(x)|\,|g(x)|\, dx \leq \left[\int_a^b |f(x)|^p\, dx \right]^{1/p} \left[\int_a^b |g(x)|^q\, dx \right]^{1/q}. \quad (6.36)$$

The theorem follows from inequality (6.36) and the fact that

$$\left| \int_a^b f(x) g(x)\, dx \right| \leq \int_a^b |f(x)|\,|g(x)|\, dx.$$

We note that the Cauchy-Schwarz inequality can be deduced from Theorem 6.7.2 by taking $p = q = 2$.   □

6.7.3. Minkowski's Inequality

The following inequality is due to Hermann Minkowski (1864–1909).

Theorem 6.7.3. Suppose that $f(x)$ and $g(x)$ are functions such that $|f(x)|^p$ and $|g(x)|^p$ are Riemann integrable on $[a, b]$, where $1 \le p < \infty$. Then

$$\left[\int_a^b |f(x) + g(x)|^p \, dx \right]^{1/p} \le \left[\int_a^b |f(x)|^p \, dx \right]^{1/p} + \left[\int_a^b |g(x)|^p \, dx \right]^{1/p}.$$

Proof. The theorem is obviously true if $p = 1$ by the triangle inequality. We therefore assume that $p > 1$. Let q be a positive number such that $1/p + 1/q = 1$. Hence, $p = p(1/p + 1/q) = 1 + p/q$. Let us now write

$$|f(x) + g(x)|^p = |f(x) + g(x)||f(x) + g(x)|^{p/q}$$

$$\le |f(x)||f(x) + g(x)|^{p/q} + |g(x)||f(x) + g(x)|^{p/q}.$$

$$(6.37)$$

By applying Hölder's inequality to the two terms on the right-hand side of inequality (6.37) we obtain

$$\int_a^b |f(x)||f(x) + g(x)|^{p/q} \, dx$$

$$\le \left[\int_a^b |f(x)|^p \, dx \right]^{1/p} \left[\int_a^b |f(x) + g(x)|^p \, dx \right]^{1/q}, \qquad (6.38)$$

$$\int_a^b |g(x)||f(x) + g(x)|^{p/q} \, dx$$

$$\le \left[\int_a^b |g(x)|^p \, dx \right]^{1/p} \left[\int_a^b |f(x) + g(x)|^p \, dx \right]^{1/q}. \qquad (6.39)$$

From inequalities (6.37), (6.38), and (6.39) we conclude that

$$\int_a^b |f(x) + g(x)|^p \, dx \le \left[\int_a^b |f(x) + g(x)|^p \, dx \right]^{1/q}$$

$$\times \left\{ \left[\int_a^b |f(x)|^p \, dx \right]^{1/p} + \left[\int_a^b |g(x)|^p \, dx \right]^{1/p} \right\}.$$

$$(6.40)$$

Since $1 - 1/q = 1/p$, inequality (6.40) can be written as

$$\left[\int_a^b |f(x) + g(x)|^p \, dx\right]^{1/p} \le \left[\int_a^b |f(x)|^p \, dx\right]^{1/p} + \left[\int_a^b |g(x)|^p \, dx\right]^{1/p}.$$

Minkowski's inequality can be extended to integrals involving more than two functions. It can be shown (see Exercise 6.20) that if $|f_i(x)|^p$ is Riemann integrable on $[a, b]$ for $i = 1, 2, \ldots, n$, then

$$\left[\int_a^b \left|\sum_{i=1}^n f_i(x)\right|^p \, dx\right]^{1/p} \le \sum_{i=1}^n \left[\int_a^b |f_i(x)|^p \, dx\right]^{1/p}. \qquad \square$$

6.7.4. Jensen's Inequality

Theorem 6.7.4. Let X be a random variable with a finite expected value, $\mu = E(X)$. If $\phi(x)$ is a twice differentiable convex function, then

$$E[\phi(X)] \ge \phi[E(X)].$$

Proof. Since $\phi(x)$ is convex and $\phi''(x)$ exists, then we must have $\phi''(x) \ge 0$. By applying the mean value theorem (Theorem 4.2.2) around μ we obtain

$$\phi(X) = \phi(\mu) + (X - \mu)\phi'(c),$$

where c is between μ and X. If $X - \mu > 0$, then $c > \mu$ and hence $\phi'(c) \ge \phi'(\mu)$, since $\phi''(x)$ is nonnegative. Thus,

$$\phi(X) - \phi(\mu) = (X - \mu)\phi'(c) \ge (X - \mu)\phi'(\mu). \qquad (6.41)$$

On the other hand, if $X - \mu < 0$, then $c < \mu$ and $\phi'(c) \le \phi'(\mu)$. Hence,

$$\phi(X) - \phi(\mu) = (X - \mu)\phi'(c) \ge (X - \mu)\phi'(\mu). \qquad (6.42)$$

From inequalities (6.41) and (6.42) we conclude that

$$E[\phi(X) - \phi(\mu)] \ge \phi'(\mu)E(X - \mu) = 0,$$

which implies that

$$E[\phi(X)] \ge \phi(\mu),$$

since $E[\phi(\mu)] = \phi(\mu)$. $\qquad \square$

6.8. RIEMANN–STIELTJES INTEGRAL

In this section we consider a more general integral, namely the Riemann–Stieltjes integral. The concept on which this integral is based can be attributed to a combination of ideas by Georg Friedrich Riemann (1826–1866) and the Dutch mathematician Thomas Joannes Stieltjes (1856–1894).

The Riemann-Stieltjes integral involves two functions $f(x)$ and $g(x)$, both defined on the interval $[a, b]$, and is denoted by $\int_a^b f(x)\, dg(x)$. In particular, if $g(x) = x$ we obtain the Riemann integral $\int_a^b f(x)\, dx$. Thus the Riemann integral is a special case of the Riemann-Stieltjes integral.

The definition of the Riemann-Stieltjes integral of $f(x)$ with respect to $g(x)$ on $[a, b]$ is similar to that of the Riemann integral. If $f(x)$ is bounded on $[a, b]$, if $g(x)$ is monotone increasing on $[a, b]$, and if $P = \{x_0, x_1, \ldots, x_n\}$ is a partition of $[a, b]$, then as in Section 6.2, we define the sums

$$LS_P(f, g) = \sum_{i=1}^{n} m_i \, \Delta g_i,$$

$$US_P(f, g) = \sum_{i=1}^{n} M_i \, \Delta g_i,$$

where m_i and M_i are, respectively, the infimum and supremum of $f(x)$ on $[x_{i-1}, x_i]$, $\Delta g_i = g(x_i) - g(x_{i-1})$, $i = 1, 2, \ldots, n$. If for a given $\epsilon > 0$ there exists a $\delta > 0$ such that

$$US_P(f, g) - LS_P(f, g) < \epsilon \tag{6.43}$$

whenever $\Delta_p < \delta$, where Δ_p is the norm of P, then $f(x)$ is said to be Riemann–Stieltjes integrable with respect to $g(x)$ on $[a, b]$. In this case,

$$\int_a^b f(x)\, dg(x) = \inf_P US_P(f, g) = \sup_P LS_P(f, g).$$

Condition (6.43) is both necessary and sufficient for the existence of the Riemann–Stieltjes integral.

Equivalently, suppose that for a given partition $P = \{x_0, x_1, \ldots, x_n\}$ we define the sum

$$S(P, f, g) = \sum_{i=1}^{n} f(t_i) \, \Delta g_i, \tag{6.44}$$

where t_i is a point in the interval $[x_{i-1}, x_i]$, $i = 1, 2, \ldots, n$. Then $f(x)$ is Riemann–Stieltjes integrable with respect to $g(x)$ on $[a, b]$ if for any $\epsilon > 0$

there exists a $\delta > 0$ such that

$$\left| S(P, f, g) - \int_a^b f(x) \, dg(x) \right| < \epsilon \tag{6.45}$$

for any partition P of $[a, b]$ with a norm $\Delta_p < \delta$, and for any choice of the point t_i in $[x_{i-1}, x_i]$, $i = 1, 2, \ldots, n$.

Theorems concerning the Riemann–Stieltjes integral are very similar to those seen earlier concerning the Riemann integral. In particular, we have the following theorems:

Theorem 6.8.1. If $f(x)$ is continuous on $[a, b]$, then $f(x)$ is Riemann–Stieltjes integrable on $[a, b]$.

Proof. See Exercise 6.21.   □

Theorem 6.8.2. If $f(x)$ is monotone increasing (or monotone decreasing) on $[a, b]$, and $g(x)$ is continuous on $[a, b]$, then $f(x)$ is Riemann–Stieltjes integrable with respect to $g(x)$ on $[a, b]$.

Proof. See Exercise 6.22.   □

The next theorem shows that under certain conditions, the Riemann–Stieltjes integral reduces to the Riemann integral.

Theorem 6.8.3. Suppose that $f(x)$ is Riemann–Stieltjes integrable with respect to $g(x)$ on $[a, b]$, where $g(x)$ has a continuous derivative $g'(x)$ on $[a, b]$. Then

$$\int_a^b f(x) \, dg(x) = \int_a^b f(x) g'(x) \, dx.$$

Proof. Let $P = \{x_0, x_1, \ldots, x_n\}$ be a partition of $[a, b]$. Consider the sum

$$S(P, h) = \sum_{i=1}^n h(t_i) \, \Delta x_i, \tag{6.46}$$

where $h(x) = f(x) g'(x)$ and $x_{i-1} \le t_i \le x_i$, $i = 1, 2, \ldots, n$. Let us also consider the sum

$$S(P, f, g) = \sum_{i=1}^n f(t_i) \, \Delta g_i, \tag{6.47}$$

If we apply the mean value theorem (Theorem 4.2.2) to $g(x)$, we obtain

$$\Delta g_i = g(x_i) - g(x_{i-1}) = g'(z_i)\,\Delta x_i, \qquad i = 1, 2, \dots, n, \qquad (6.48)$$

where $x_{i-1} < z_i < x_i$, $i = 1, 2, \dots, n$. From (6.46), (6.47), and (6.48) we can then write

$$S(P, f, g) - S(P, h) = \sum_{i=1}^{n} f(t_i)[g'(z_i) - g'(t_i)]\,\Delta x_i. \qquad (6.49)$$

Since $f(x)$ is bounded on $[a, b]$ and $g'(x)$ is uniformly continuous on $[a, b]$ by Theorem 3.4.6, then for a given $\epsilon > 0$ there exists a $\delta_1 > 0$, which depends only on ϵ, such that

$$|g'(z_i) - g'(t_i)| < \frac{\epsilon}{2M(b-a)},$$

if $|z_i - t_i| < \delta_1$, where $M > 0$ is such that $|f(x)| \leq M$ on $[a, b]$. From (6.49) it follows that if the partition P has a norm $\Delta_p < \delta_1$, then

$$|S(P, f, g) - S(P, h)| < \frac{\epsilon}{2}. \qquad (6.50)$$

Now, since $f(x)$ is Riemann–Stieltjes integrable with respect to $g(x)$ on $[a, b]$, then by definition, for the given $\epsilon > 0$ there exists a $\delta_2 > 0$ such that

$$\left| S(P, f, g) - \int_a^b f(x)\,dg(x) \right| < \frac{\epsilon}{2}, \qquad (6.51)$$

if the norm Δ_p of P is less than δ_2. We conclude from (6.50) and (6.51) that if the norm of P is less than $\min(\delta_1, \delta_2)$, then

$$\left| S(P, h) - \int_a^b f(x)\,dg(x) \right| < \epsilon.$$

Since ϵ is arbitrary, this inequality implies that $\int_a^b f(x)\,dg(x)$ is in fact the Riemann integral $\int_a^b h(x)\,dx = \int_a^b f(x)g'(x)\,dx$. $\qquad\square$

Using Theorem 6.8.3, it is easy to see that if, for example, $f(x) = 1$ and $g(x) = x^2$, then $\int_a^b f(x)\,dg(x) = \int_a^b f(x)g'(x)\,dx = \int_a^b 2x\,dx = b^2 - a^2$.

It should be noted that Theorems 6.8.1 and 6.8.2 provide sufficient conditions for the existence of $\int_a^b f(x)\,dg(x)$. It is possible, however, for the Riemann–Stieltjes integral to exist even if $g(x)$ is a discontinuous function. For example, consider the function $g(x) = \gamma I(x - c)$, where γ is a nonzero constant, $a < c < b$, and $I(x - c)$ is such that

$$I(x - c) = \begin{cases} 0, & x < c, \\ 1, & x \geq c. \end{cases}$$

The quantity γ represents what is called a jump at $x = c$. If $f(x)$ is bounded on $[a, b]$ and is continuous at $x = c$, then

$$\int_a^b f(x)\, dg(x) = \gamma f(c). \tag{6.52}$$

To show the validity of formula (6.52), let $P = \{x_0, x_1, \ldots, x_n\}$ be any partition of $[a, b]$. Then, $\Delta g_i = g(x_i) - g(x_{i-1})$ will be zero as long as $x_i < c$ or $x_{i-1} \geq c$. Suppose, therefore, that there exists a k, $1 \leq k \leq n$, such that $x_{k-1} < c \leq x_k$. In this case, the sum $S(P, f, g)$ in formula (6.44) takes the form

$$S(P, f, g) = \sum_{i=1}^n f(t_i)\, \Delta g_i = \gamma f(t_k).$$

It follows that

$$|S(P, f, g) - \gamma f(c)| = |\gamma| |f(t_k) - f(c)|. \tag{6.53}$$

Now, let $\epsilon > 0$ be given. Since $f(x)$ is continuous at $x = c$, then there exists a $\delta > 0$ such that

$$|f(t_k) - f(c)| < \frac{\epsilon}{|\gamma|},$$

if $|t_k - c| < \delta$. Thus if the norm Δ_p of P is chosen so that $\Delta_p < \delta$, then

$$|S(P, f, g) - \gamma f(c)| < \epsilon. \tag{6.54}$$

Equality (6.52) follows from comparing inequalities (6.45) and (6.54).

It is now easy to show that if

$$g(x) = \begin{cases} \lambda, & a \leq x < b, \\ \lambda', & x = b, \end{cases}$$

and if $f(x)$ is continuous at $x = b$, then

$$\int_a^b f(x)\, dg(x) = (\lambda' - \lambda) f(b). \tag{6.55}$$

The previous examples represent special cases of a class of functions defined on $[a, b]$ called step functions. These functions are constant on $[a, b]$

except for a finite number of jump discontinuities. We can generalize formula (6.55) to this class of functions as can be seen in the next theorem.

Theorem 6.8.4. Let $g(x)$ be a step function defined on $[a, b]$ with jump discontinuities at $x = c_1, c_2, \ldots, c_n$, and $a < c_1 < c_2 < \cdots < c_n = b$, such that

$$g(x) = \begin{cases} \lambda_1, & a \leq x < c_1, \\ \lambda_2, & c_1 \leq x < c_2, \\ \vdots & \vdots \\ \lambda_n, & c_{n-1} \leq x < c_n, \\ \lambda_{n+1}, & x = c_n. \end{cases}$$

If $f(x)$ is bounded on $[a, b]$ and is continuous at $x = c_1, c_2, \ldots, c_n$, then

$$\int_a^b f(x) \, dg(x) = \sum_{i=1}^n (\lambda_{i+1} - \lambda_i) f(c_i). \tag{6.56}$$

Proof. The proof can be easily obtained by first writing the integral in formula (6.56) as

$$\int_a^b f(x) \, dg(x) = \int_a^{c_1} f(x) \, dg(x) + \int_{c_1}^{c_2} f(x) \, dg(x)$$
$$+ \cdots + \int_{c_{n-1}}^{c_n} f(x) \, dg(x). \tag{6.57}$$

If we now apply formula (6.55) to each integral in (6.57) we obtain

$$\int_a^{c_1} f(x) \, dg(x) = (\lambda_2 - \lambda_1) f(c_1),$$

$$\int_{c_1}^{c_2} f(x) \, dg(x) = (\lambda_3 - \lambda_2) f(c_2),$$

$$\vdots$$

$$\int_{c_{n-1}}^{c_n} f(x) \, dg(x) = (\lambda_{n+1} - \lambda_n) f(c_n).$$

By adding up all these integrals we obtain formula (6.56). □

EXAMPLE 6.8.1. One example of a step function is the greatest-integer function $[x]$, which is defined as the greatest integer less than or equal to x. If $f(x)$ is bounded on $[0, n]$ and is continuous at $x = 1, 2, \ldots, n$, where n is a positive integer, then by Theorem 6.8.4 we can write

$$\int_0^n f(x)\, d[x] = \sum_{i=1}^n f(i). \tag{6.58}$$

It follows that every finite sum of the form $\sum_{i=1}^n a_i$ can be expressed as a Riemann–Stieltjes integral with respect to $[x]$ of a function $f(x)$ continuous on $[0, n]$ such that $f(i) = a_i$, $i = 1, 2, \ldots, n$. The Riemann–Stieltjes integral has therefore the distinct advantage of making finite sums expressible as integrals.

6.9. APPLICATIONS IN STATISTICS

Riemann integration plays an important role in statistical distribution theory. Perhaps the most prevalent use of the Riemann integral is in the study of the distributions of continuous random variables.

We recall from Section 4.5.1 that a continuous random variable X with a cumulative distribution function $F(x) = P(X \leq x)$ is absolutely continuous if $F(x)$ is differentiable. In this case, there exists a function $f(x)$ called the density function of X such that $F'(x) = f(x)$, that is,

$$F(x) = \int_{-\infty}^x f(t)\, dt. \tag{6.59}$$

In general, if X is a continuous random variable, it need not be absolutely continuous. It is true, however, that most common distributions that are continuous are also absolutely continuous.

The probability distribution of an absolutely continuous random variable is completely determined by its density function. For example, from (6.59) it follows that

$$P(a < X < b) = F(b) - F(a) = \int_a^b f(x)\, dx. \tag{6.60}$$

Note that the value of this probability remains unchanged if one or both of the end points of the interval $[a, b]$ are included. This is true because the probability assigned to these individual points is zero when X has a continuous distribution. The mean μ and variance σ^2 of X are given by

$$\mu = E(X) = \int_{-\infty}^{\infty} x f(x)\, dx,$$

$$\sigma^2 = \text{Var}(X) = \int_{-\infty}^{\infty} (x - \mu)^2 f(x)\, dx.$$

In general, the kth central moment of X, denoted by μ_k $(k = 1, 2, \ldots)$, is defined as

$$\mu_k = E\left[(X - \mu)^k\right] = \int_{-\infty}^{\infty} (x - \mu)^k f(x)\, dx. \tag{6.61}$$

We note that $\sigma^2 = \mu_2$. Similarly, the kth noncentral moment of X, denoted by μ'_k $(k = 1, 2, \ldots)$, is defined as

$$\mu'_k = E(X^k). \tag{6.62}$$

The first noncentral moment of X is its mean μ, while its first central moment is equal to zero.

We note that if the domain of the density function $f(x)$ is infinite, then μ_k and μ'_k are improper integrals of the first kind. Therefore, they may or may not exist. If $\int_{-\infty}^{\infty} |x|^k f(x)\, dx$ exists, then so does μ'_k (see Definition 6.5.2 concerning absolute convergence of improper integrals). The latter integral is called the kth absolute moment and is denoted by ν_k. If ν_k exists, then the noncentral moments of order j for $j \le k$ exist also. This follows because of the inequality

$$|x^j| \le |x|^k + 1 \qquad \text{if } j \le k, \tag{6.63}$$

which is true because $|x|^j \le |x|^k$ if $|x| \ge 1$ and $|x|^j < 1$ if $|x| < 1$. Hence, $|x^j| \le |x|^k + 1$ for all x. Consequently, from (6.63) we obtain the inequality

$$\nu_j \le \nu_k + 1, \qquad j \le k,$$

which implies that μ'_j exists for $j \le k$. Since the central moment μ_j in formula (6.61) is expressible in terms of noncentral moments of order j or smaller, the existence of ν_j also implies the existence of μ_j.

EXAMPLE 6.9.1.   Consider a random variable X with the density function

$$f(x) = \frac{1}{\pi(1 + x^2)}, \qquad -\infty < x < \infty.$$

Such a random variable has the so-called Cauchy distribution. Its mean μ does not exist. This follows from the fact that in order of μ to exist, the two limits in the following formula must exist:

$$\mu = \frac{1}{\pi} \lim_{a \to \infty} \int_{-a}^{0} \frac{x\, dx}{1 + x^2} + \frac{1}{\pi} \lim_{b \to \infty} \int_{0}^{b} \frac{x\, dx}{1 + x^2}. \tag{6.64}$$

But, $\int_{-a}^{0} x\,dx/(1+x^2) = -\frac{1}{2}\log(1+a^2) \to -\infty$ as $a \to \infty$, and $\int_{0}^{b} x\,dx/(1+x^2)$ $= \frac{1}{2}\log(1+b^2) \to \infty$ as $b \to \infty$. The integral $\int_{-\infty}^{\infty} x\,dx/(1+x^2)$ is therefore divergent, and hence μ does not exist.

It should be noted that it would be incorrect to state that

$$\mu = \frac{1}{\pi} \lim_{a \to \infty} \int_{-a}^{a} \frac{x\,dx}{1+x^2}, \tag{6.65}$$

which is equal to zero. This is because the limits in (6.64) must exist for any a and b that tend to infinity. The limit in formula (6.65) requires that $a = b$. Such a limit is therefore considered as a subsequential limit.

The higher-order moments of the Cauchy distribution do not exist either. It is easy to verify that

$$\mu_k' = \frac{1}{\pi} \int_{-\infty}^{\infty} \frac{x^k\,dx}{1+x^2} \tag{6.66}$$

is divergent for $k \geq 1$.

EXAMPLE 6.9.2.   Consider a random variable X that has the logistic distribution with the density function

$$f(x) = \frac{e^x}{(1+e^x)^2}, \qquad -\infty < x < \infty.$$

The mean of X is

$$\mu = \int_{-\infty}^{\infty} \frac{xe^x}{(1+e^x)^2}\,dx$$

$$= \int_{0}^{1} \log\left(\frac{u}{1-u}\right)du, \tag{6.67}$$

where $u = e^x/(1+e^x)$. We recognize the integral in (6.67) as being an improper integral of the second kind with singularities at $u = 0$ and $u = 1$. We therefore write

$$\mu = \lim_{a \to 0^+} \int_{a}^{1/2} \log\left(\frac{u}{1-u}\right)du + \lim_{b \to 1^-} \int_{1/2}^{b} \log\left(\frac{u}{1-u}\right)du. \tag{6.68}$$

Thus,

$$\mu = \lim_{a \to 0^+} \left[u \log u + (1-u)\log(1-u) \right]_a^{1/2}$$

$$+ \lim_{b \to 1^-} \left[u \log u + (1-u)\log(1-u) \right]_{1/2}^b$$

$$= \lim_{b \to 1^-} \left[b \log b + (1-b)\log(1-b) \right]$$

$$- \lim_{a \to 0^+} \left[a \log a + (1-a)\log(1-a) \right]. \tag{6.69}$$

By applying l'Hospital's rule (Theorem 4.2.6) we find that

$$\lim_{b \to 1^-} (1-b)\log(1-b) = \lim_{a \to 0^+} a \log a = 0.$$

We thus have

$$\mu = \lim_{b \to 1^-} (b \log b) - \lim_{a \to 0^+} \left[(1-a)\log(1-a) \right] = 0.$$

The variance σ^2 of X can be shown to be equal to $\pi^2/3$ (see Exercise 6.24).

6.9.1. The Existence of the First Negative Moment of a Continuous Distribution

Let X be a continuous random variable with a density function $f(x)$. By definition, the first negative moment of X is $E(X^{-1})$. The existence of such a moment will be explored in this section.

The need to evaluate a first negative moment can arise in many practical applications. Here are some examples.

EXAMPLE 6.9.3.   Let $\mathscr{P}$ be a population with a mean μ and a variance σ^2. The coefficient of variation is a measure of variation in the population per unit mean and is equal to $\sigma/|\mu|$, assuming that $\mu \neq 0$. An estimate of this ratio is $s/|\bar{y}|$, where s and $\bar{y}$ are, respectively, the sample standard deviation and sample mean of a sample randomly chosen from $\mathscr{P}$. If the population is normally distributed, then $\bar{y}$ is also normally distributed and is statistically independent of s. In this case, $E(s/|\bar{y}|) = E(s)E(1/|\bar{y}|)$. The question now is whether $E(1/|\bar{y}|)$ exists or not.

EXAMPLE 6.9.4 (Calibration or Inverse Regression).   Consider the simple linear regression model

$$E(y) = \beta_0 + \beta_1 x. \tag{6.70}$$

In most regression situations, the interest is in predicting the response y for a given value of x. For this purpose we use the prediction equation $\hat{y} = \hat{\beta}_0 + \hat{\beta}_1 x$, where $\hat{\beta}_0$ and $\hat{\beta}_1$ are the least-squares estimators of β_0 and β_1, respectively. These are obtained from the data set $\{(x_1, y_1), (x_2, y_2), \ldots, (x_n, y_n)\}$ that results from running n experiments in which y is measured for specified settings of x. There are other situations, however, where the interest is in predicting the value of x, say x_0, that corresponds to an observed value of y, say y_0. This is an inverse regression problem known as the calibration problem (see Graybill, 1976, Section 8.5; Montgomery and Peck, 1982, Section 9.7).

For example, in calibrating a new type of thermometer, n readings, $y_1, y_2, \ldots, y_n$, are taken at predetermined known temperature values, $x_1, x_2, \ldots, x_n$ (these values are known by using a standard temperature gauge). Suppose that the relationship between the x_i's and the y_i's is well represented by the model in (6.70). If a new reading y_0 is observed using the new thermometer, then it is of interest to estimate the correct temperature x_0 (that is, the temperature on the standard gauge corresponding to the observed temperature reading y_0).

In another calibration problem, the date of delivery of a pregnant woman can be estimated by the size y of the head of her unborn child, which can be determined by a special electronic device (sonogram). If the relationship between y and the number of days x left until delivery is well represented by model (6.70), then for a measured value of y, say y_0, it is possible to estimate x_0, the corresponding value of x.

In general, from model (6.70) we have $E(y_0) = \beta_0 + \beta_1 x_0$. If $\beta_1 \neq 0$, we can solve for x_0 and obtain

$$x_0 = \frac{E(y_0) - \beta_0}{\beta_1}.$$

Hence, to estimate x_0 we use

$$\hat{x}_0 = \frac{y_0 - \hat{\beta}_0}{\hat{\beta}_1} = \bar{x} + \frac{y_0 - \bar{y}}{\hat{\beta}_1},$$

since $\hat{\beta}_0 = \bar{y} - \hat{\beta}_1 \bar{x}$, where $\bar{x} = (1/n)\sum_{i=1}^{n} x_i$, $\bar{y} = (1/n)\sum_{i=1}^{n} y_i$. If the response y is normally distributed with a variance σ^2, then $\bar{y}$ and $\hat{\beta}_1$ are statistically independent. Since y_0 is also statistically independent of $\hat{\beta}_1$ (y_0 does not belong to the data set used to estimate β_1), then the expected value of $\hat{x}_0$ is given by

$$E(\hat{x}_0) = \bar{x} + E\left(\frac{y_0 - \bar{y}}{\hat{\beta}_1}\right)$$

$$= \bar{x} + E(y_0 - \bar{y}) E\left(\frac{1}{\hat{\beta}_1}\right).$$

Here again it is of interest to know if $E(1/\hat{\beta}_1)$ exists.

Now, suppose that the density function $f(x)$ of the continuous random variable X is defined on $(0, \infty)$. Let us also assume that $f(x)$ is continuous. The expected value of X^{-1} is

$$E(X^{-1}) = \int_0^\infty \frac{f(x)}{x}\,dx. \tag{6.71}$$

This is an improper integral with a singularity at $x = 0$. In particular, if $f(0) > 0$, then $E(X^{-1})$ does not exist, because

$$\lim_{x \to 0^+} \frac{f(x)/x}{1/x} = f(0) > 0.$$

By Theorem 6.5.7, the integrals $\int_0^\infty (f(x)/x)\,dx$ and $\int_0^\infty (dx/x)$ are of the same kind. Since the latter is divergent, then so is the former. Note that if $f(x)$ is defined on $(-\infty, \infty)$ and $f(0) > 0$, then $E(X^{-1})$ does not exist either. In this case,

$$\int_{-\infty}^\infty \frac{f(x)}{x}\,dx = \int_{-\infty}^0 \frac{f(x)}{x}\,dx + \int_0^\infty \frac{f(x)}{x}\,dx$$

$$= -\int_{-\infty}^0 \frac{f(x)}{|x|}\,dx + \int_0^\infty \frac{f(x)}{x}\,dx.$$

Both integrals on the right-hand side are divergent.

A sufficient condition for the existence of $E(X^{-1})$ is given by the following theorem [see Piegorsch and Casella (1985)]:

Theorem 6.9.1. Let $f(x)$ be a continuous density function for a random variable X defined on $(0, \infty)$. If

$$\lim_{x \to 0^+} \frac{f(x)}{x^\alpha} < \infty \qquad \text{for some } \alpha > 0, \tag{6.72}$$

then $E(X^{-1})$ exists.

Proof. Since the limit of $f(x)/x^\alpha$ is finite as $x \to 0^+$, there exist finite constants M and $\delta > 0$ such that $f(x)/x^\alpha < M$ if $0 < x < \delta$. Hence,

$$\int_0^\delta \frac{f(x)}{x}\,dx < M \int_0^\delta x^{\alpha - 1}\,dx = \frac{M\delta^\alpha}{\alpha}.$$

Thus,

$$E(X^{-1}) = \int_0^\infty \frac{f(x)}{x}\,dx = \int_0^\delta \frac{f(x)}{x}\,dx + \int_\delta^\infty \frac{f(x)}{x}\,dx$$

$$< \frac{M\delta^\alpha}{\alpha} + \frac{1}{\delta}\int_\delta^\infty f(x)\,dx \le \frac{M\delta^\alpha}{\alpha} + \frac{1}{\delta} < \infty. \qquad \square$$

It should be noted that the condition of Theorem 6.9.1 is not a necessary one. Piegorsch and Casella (1985) give an example of a family of density functions that all violate condition (6.72), with some members having a finite first negative moment and others not having one (see Exercise 6.25).

Corollary 6.9.1. Let $f(x)$ be a continuous density function for a random variable X defined on $(0,\infty)$ such that $f(0) = 0$. If $f'(0)$ exists and is finite, then $E(X^{-1})$ exists.

Proof. We have that

$$f'(0) = \lim_{x\to 0^+} \frac{f(x) - f(0)}{x} = \lim_{x\to 0^+} \frac{f(x)}{x}.$$

By applying Theorem 6.9.1 with $\alpha = 1$ we conclude that $E(X^{-1})$ exists. $\square$

EXAMPLE 6.9.5. Let X be a normal random variable with a mean μ and a variance σ^2. Its density function is given by

$$f(x) = \frac{1}{\sqrt{2\pi\sigma^2}} \exp\left[-\frac{1}{2\sigma^2}(x-\mu)^2\right], \qquad -\infty < x < \infty.$$

In this example, $f(0) > 0$. Hence, $E(X^{-1})$ does not exist. Consequently, $E(1/|\bar{y}|)$ in Example 6.9.3 does not exist if the population $\mathscr{P}$ is normally distributed, since the density function of $|\bar{y}|$ is positive at zero. Also, in Example 6.9.4, $E(1/\hat{\beta}_1)$ does not exist, because $\hat{\beta}_1$ is normally distributed if the response y satisfies the assumption of normality.

EXAMPLE 6.9.6. Let X be a continuous random variable with the density function

$$f(x) = \frac{1}{\Gamma(n/2)2^{n/2}} x^{n/2-1} e^{-x/2}, \qquad 0 < x < \infty,$$

where n is a positive integer and $\Gamma(n/2)$ is the value of the gamma function, $\int_0^\infty e^{-x} x^{n/2-1}\, dx$. This is the density function of a chi-squared random variable with n degrees of freedom.

Let us consider the limit

$$\lim_{x \to 0^+} \frac{f(x)}{x^\alpha} = \frac{1}{\Gamma(n/2) 2^{n/2}} \lim_{x \to 0^+} x^{n/2-\alpha-1}\, e^{-x/2}$$

for $\alpha > 0$. This limit exists and is equal to zero if $n/2 - \alpha - 1 > 0$, that is, if $n > 2(1 + \alpha) > 2$. Thus by Theorem 6.9.1, $E(X^{-1})$ exists if the number of degrees of freedom exceeds 2.

More recently, Khuri and Casella (2002) presented several extensions and generalizations of the results in Piegorsch and Casella (1985), including a necessary and sufficient condition for the existence of $E(X^{-1})$.

6.9.2. Transformation of Continuous Random Variables

Let Y be a continuous random variable with a density function $f(y)$. Let $W = \psi(Y)$, where $\psi(y)$ is a function whose derivative exists and is continuous on a set A. Suppose that $\psi'(y) \neq 0$ for all $y \in A$, that is, $\psi(y)$ is strictly monotone. We recall from Section 4.5.1 that the density function $g(w)$ of W is given by

$$g(w) = f\left[\psi^{-1}(w)\right] \left| \frac{d\psi^{-1}(w)}{dw} \right|, \qquad w \in B, \tag{6.73}$$

where B is the image of A under ψ and $y = \psi^{-1}(w)$ is the inverse function of $w = \psi(y)$. This result can be easily obtained by applying the change of variables technique in Section 6.4.1. This is done as follows: We have that for any w_1 and w_2 such that $w_1 \leq w_2$,

$$P(w_1 \leq W \leq w_2) = \int_{w_1}^{w_2} g(w)\, dw. \tag{6.74}$$

If $\psi'(y) > 0$, then $\psi(y)$ is strictly monotone increasing. Hence,

$$P(w_1 \leq W \leq w_2) = P(y_1 \leq Y \leq y_2), \tag{6.75}$$

where y_1 and y_2 are such that $w_1 = \psi(y_1), w_2 = \psi(y_2)$. But

$$P(y_1 \leq Y \leq y_2) = \int_{y_1}^{y_2} f(y)\, dy. \tag{6.76}$$

Let us now apply the change of variables $y = \psi^{-1}(w)$ to the integral in (6.76).

By Theorem 6.4.9 we have

$$\int_{y_1}^{y_2} f(y)\, dy = \int_{w_1}^{w_2} f\left[\psi^{-1}(w)\right] \frac{d\psi^{-1}(w)}{dw}\, dw. \tag{6.77}$$

From (6.75) and (6.76) we obtain

$$P(w_1 \le W \le w_2) = \int_{w_1}^{w_2} f\left[\psi^{-1}(w)\right] \frac{d\psi^{-1}(w)}{dw}\, dw. \tag{6.78}$$

On the other hand, if $\psi'(y) < 0$, then $P(w_1 \le W \le w_2) = P(y_2 \le Y \le y_1)$. Consequently,

$$\begin{aligned}
P(w_1 \le W \le w_2) &= \int_{y_2}^{y_1} f(y)\, dy \\
&= \int_{w_2}^{w_1} f\left[\psi^{-1}(w)\right] \frac{d\psi^{-1}(w)}{dw}\, dw \\
&= -\int_{w_1}^{w_2} f\left[\psi^{-1}(w)\right] \frac{d\psi^{-1}(w)}{dw}\, dw.
\end{aligned} \tag{6.79}$$

By combining (6.78) and (6.79) we obtain

$$P(w_1 \le W \le w_2) = \int_{w_1}^{w_2} f\left[\psi^{-1}(w)\right] \left| \frac{d\psi^{-1}(w)}{dw} \right| dw. \tag{6.80}$$

Formula (6.73) now follows from comparing (6.74) and (6.80).

The Case Where $w = \psi(y)$ Has No Unique Inverse

Formula (6.73) requires that $w = \psi(y)$ has a unique inverse, the existence of which is guaranteed by the nonvanishing of the derivative $\psi'(y)$. Let us now consider the following extension: The function $\psi(y)$ is continuously differentiable, but its derivative can vanish at a finite number of points inside its domain. We assume that the domain of $\psi(y)$ can be partitioned into a finite number, say n, of disjoint subdomains, denoted by $I_1, I_2, \ldots, I_n$, on each of which $\psi(y)$ is strictly monotone (decreasing or increasing). Hence, on each I_i $(i = 1, 2, \ldots, n)$, $\psi(y)$ has a unique inverse. Let ψ_i denote the restriction of the function ψ to I_i, that is, $\psi_i(y)$ has a unique inverse, $y = \psi_i^{-1}(w)$, $i = 1, 2, \ldots, n$. Since $I_1, I_2, \ldots, I_n$ are disjoint, for any w_1 and w_2 such that $w_1 \le w_2$ we have

$$P(w_1 \le W \le w_2) = \sum_{i=1}^{n} P\left[Y \in \psi_i^{-1}(w_1, w_2)\right], \tag{6.81}$$

where $\psi_i^{-1}(w_1, w_2)$ is the inverse image of $[w_1, w_2]$, which is a subset of I_i, $i = 1, 2, \ldots, n$. Now, on the ith subdomain we have

$$
P\left[Y \in \psi_i^{-1}(w_1, w_2)\right] = \int_{\psi_i^{-1}(w_1, w_2)} f(y)\, dy
$$

$$
= \int_{T_i} f\left[\psi_i^{-1}(w)\right] \left| \frac{d\psi_i^{-1}(w)}{dw} \right| dw, \qquad i = 1, 2, \ldots, n,
$$

$$(6.82)$$

where T_i is the image of $\psi_i^{-1}(w_1, w_2)$ under ψ_i. Formula (6.82) follows from applying formula (6.80) to the function $\psi_i(y)$, $i = 1, 2, \ldots, n$. Note that $T_i = \psi_i[\psi_i^{-1}(w_1, w_2)] = [w_1, w_2] \cap \psi_i(I_i)$, $i = 1, 2, \ldots, n$ (why?). Thus T_i is a subset of both $[w_1, w_2]$ and $\psi_i(I_i)$. We can therefore write the integral in (6.82) as

$$
\int_{T_i} f\left[\psi_i^{-1}(w)\right] \left| \frac{d\psi_i^{-1}(w)}{dw} \right| dw = \int_{w_1}^{w_2} \delta_i(w) f\left[\psi_i^{-1}(w)\right] \left| \frac{d\psi_i^{-1}(w)}{dw} \right| dw, \quad (6.83)
$$

where $\delta_i(w) = 1$ if $w \in \psi_i(I_i)$ and $\delta_i(w) = 0$ otherwise, $i = 1, 2, \ldots, n$. Using (6.82) and (6.83) in formula (6.81), we obtain

$$
P(w_1 \le W \le w_2) = \sum_{i=1}^{n} \int_{w_1}^{w_2} \delta_i(w) f\left[\psi_i^{-1}(w)\right] \left| \frac{d\psi_i^{-1}(w)}{dw} \right| dw
$$

$$
= \int_{w_1}^{w_2} \sum_{i=1}^{n} \delta_i(w) f\left[\psi_i^{-1}(w)\right] \left| \frac{d\psi_i^{-1}(w)}{dw} \right| dw,
$$

from which we deduce that the density function of W is given by

$$
g(w) = \sum_{i=1}^{n} \delta_i(w) f\left[\psi_i^{-1}(w)\right] \left| \frac{d\psi_i^{-1}(w)}{dw} \right|. \qquad (6.84)
$$

EXAMPLE 6.9.7. Let Y have the standard normal distribution with the density function

$$
f(y) = \frac{1}{\sqrt{2\pi}} \exp\left(-\frac{y^2}{2}\right), \qquad -\infty < y < \infty.
$$

Define the random variable W as $W = Y^2$. In this case, the function $w = y^2$

has two inverse functions on $(-\infty, \infty)$, namely,

$$
y = \begin{cases} -\sqrt{w}, & y \leq 0, \\ \sqrt{w}, & y > 0. \end{cases}
$$

Thus $I_1 = (-\infty, 0]$, $I_2 = (0, \infty)$, and $\psi_1(I_1) = [0, \infty)$, $\psi_2(I_2) = (0, \infty)$. Hence, $\delta_1(w) = 1$, $\delta_2(w) = 1$ if $w \in (0, \infty)$. By applying formula (6.84) we then get

$$
\begin{aligned}
g(w) &= \frac{1}{\sqrt{2\pi}} e^{-w/2} \left| \frac{-1}{2\sqrt{w}} \right| + \frac{1}{\sqrt{2\pi}} e^{-w/2} \left| \frac{1}{2\sqrt{w}} \right| \\
&= \frac{1}{\sqrt{2\pi}} \frac{e^{-w/2}}{\sqrt{w}}, \qquad w > 0.
\end{aligned}
$$

This represents the density function of a chi-squared random variable with one degree of freedom.

6.9.3. The Riemann–Stieltjes Integral Representation of the Expected Value

Let X be a random variable with a cumulative distribution function $F(x)$. Suppose that $h(x)$ is Riemann–Stieltjes integrable with respect to $F(x)$ on $(-\infty, \infty)$. Then the expected value of $h(X)$ is defined as

$$
E[h(X)] = \int_{-\infty}^{\infty} h(x) \, dF(x). \tag{6.85}
$$

Formula (6.85) provides a unified representation of expected values for both discrete and continuous random variables.

If X is a continuous random variable with a density function $f(x)$, that is, $F'(x) = f(x)$, then

$$
E[h(X)] = \int_{-\infty}^{\infty} h(x) f(x) \, dx.
$$

If, however, X has a discrete distribution with a probability mass function $p(x)$ and takes the values $c_1, c_2, \ldots, c_n$, then

$$
E[h(X)] = \sum_{i=1}^{n} h(c_i) p(c_i). \tag{6.86}
$$

Formula (6.86) follows from applying Theorem 6.8.4. Here, $F(x)$ is a step function with jump discontinuities at $c_1, c_2, \ldots, c_n$ such that

$$F(x) = \begin{cases} 0, & -\infty < x < c_1 \\ p(c_1), & c_1 \leq x < c_2, \\ p(c_1) + p(c_2), & c_2 \leq x < c_3, \\ \vdots & \\ \sum_{i=1}^{n-1} p(c_i), & c_{n-1} \leq x < c_n, \\ \sum_{i=1}^{n} p(c_i) = 1, & c_n \leq x < \infty. \end{cases}$$

Thus, by formula (6.56) we obtain

$$\int_{-\infty}^{\infty} h(x)\, dF(x) = \sum_{i=1}^{n} p(c_i) h(c_i). \tag{6.87}$$

For example, suppose that X has the discrete uniform distribution with $p(x) = 1/n$ for $x = c_1, c_2, \ldots, c_n$. Its cumulative distribution function $F(c)$ can be expressed as

$$F(x) = P[X \leq x] = \frac{1}{n} \sum_{i=1}^{n} I(x - c_i),$$

where $I(x - c_i)$ is equal to zero if $x < c_i$ and is equal to one if $x \geq c_i$ $(i = 1, 2, \ldots, n)$. The expected value of $h(X)$ is

$$E[h(X)] = \frac{1}{n} \sum_{i=1}^{n} h(c_i).$$

EXAMPLE 6.9.8. The moment generating function $\phi(t)$ of a random variable X with a cumulative distribution function $F(x)$ is defined as the expected value of $h_t(X) = e^{tX}$, that is,

$$\phi(t) = E(e^{tX}) = \int_{-\infty}^{\infty} e^{tx}\, dF(x),$$

where t is a scalar. If X is a discrete random variable with a probability mass function $p(x)$ and takes the values $c_1, c_2, \ldots, c_n, \ldots$, then by letting n go to infinity in (6.87) we obtain (see also Section 5.6.2)

$$\phi(t) = \sum_{i=1}^{\infty} p(c_i) e^{tc_i}.$$

The moment generating function of a continuous random variable with a density function $f(x)$ is

$$\phi(t) = \int_{-\infty}^{\infty} e^{tx} f(x)\, dx. \tag{6.88}$$

The convergence of the integral in (6.88) depends on the choice of the scalar t. For example, for the gamma distribution $G(\alpha, \beta)$ with the density function

$$f(x) = \frac{x^{\alpha-1} e^{-x/\beta}}{\Gamma(\alpha)\beta^{\alpha}}, \qquad \alpha > 0, \quad \beta > 0, \quad 0 < x < \infty,$$

$\phi(t)$ is of the form

$$\phi(t) = \int_{0}^{\infty} \frac{e^{tx} x^{\alpha-1} e^{-x/\beta}}{\Gamma(\alpha)\beta^{\alpha}}\, dx$$

$$= \int_{0}^{\infty} \frac{x^{\alpha-1} \exp[-x(1-\beta t)/\beta]}{\Gamma(\alpha)\beta^{\alpha}}\, dx.$$

If we set $y = x(1 - \beta t)/\beta$, we obtain

$$\phi(t) = \int_{0}^{\infty} \frac{\beta}{(1-\beta t)\Gamma(\alpha)\beta^{\alpha}} \left(\frac{\beta y}{1-\beta t}\right)^{\alpha-1} e^{-y}\, dy.$$

Thus,

$$\phi(t) = \frac{1}{(1-\beta t)^{\alpha}} \int_{0}^{\infty} \frac{y^{\alpha-1} e^{-y}}{\Gamma(\alpha)}\, dy$$

$$= (1 - \beta t)^{-\alpha},$$

since $\int_{0}^{\infty} e^{-y} y^{\alpha-1}\, dy = \Gamma(\alpha)$ by the definition of the gamma function. We note that $\phi(t)$ exists for all values of α provided that $1 - \beta t > 0$, that is, $t < 1/\beta$.

6.9.4. Chebyshev's Inequality

In Section 5.6.1 there was a mention of Chebyshev's inequality. Using the Riemann–Stieltjes integral representation of the expected value, it is now possible to provide a proof for this important inequality.

Theorem 6.9.2. Let X be a random variable (discrete or continuous) with a mean μ and a variance σ^2. Then, for any positive constant r,

$$P(|X - \mu| \geq r\sigma) \leq \frac{1}{r^2}.$$

Proof. By definition, σ^2 is the expected value of $h(X) = (X - \mu)^2$. Thus,

$$\sigma^2 = \int_{-\infty}^{\infty} (x - \mu)^2 \, dF(x), \tag{6.89}$$

where $F(x)$ is the cumulative distribution function of X. Let us now partition $(-\infty, \infty)$ into three disjoint intervals: $(-\infty, \mu - r\sigma]$, $(\mu - r\sigma, \mu + r\sigma)$, $[\mu + r\sigma, \infty)$. The integral in (6.89) can therefore be written as

$$\sigma^2 = \int_{-\infty}^{\mu - r\sigma} (x - \mu)^2 \, dF(x) + \int_{\mu - r\sigma}^{\mu + r\sigma} (x - \mu)^2 \, dF(x)$$

$$+ \int_{\mu + r\sigma}^{\infty} (x - \mu)^2 \, dF(x)$$

$$\geq \int_{-\infty}^{\mu - r\sigma} (x - \mu)^2 \, dF(x) + \int_{\mu + r\sigma}^{\infty} (x - \mu)^2 \, dF(x). \tag{6.90}$$

We note that in the first integral in (6.90), $x \leq \mu - r\sigma$, so that $x - \mu \leq -r\sigma$. Hence, $(x - \mu)^2 \geq r^2\sigma^2$. Also, in the second integral, $x - \mu \geq r\sigma$. Hence, $(x - \mu)^2 \geq r^2\sigma^2$. Consequently,

$$\int_{-\infty}^{\mu - r\sigma} (x - \mu)^2 \, dF(x) \geq r^2\sigma^2 \int_{-\infty}^{\mu - r\sigma} dF(x) = r^2\sigma^2 P(X \leq \mu - r\sigma),$$

$$\int_{\mu + r\sigma}^{\infty} (x - \mu)^2 \, dF(x) \geq r^2\sigma^2 \int_{\mu + r\sigma}^{\infty} dF(x) = r^2\sigma^2 P(X \geq \mu + r\sigma).$$

From inequality (6.90) we then have

$$\sigma^2 \geq r^2\sigma^2 \left[P(X - \mu \leq -r\sigma) + P(X - \mu \geq r\sigma) \right]$$

$$= r^2\sigma^2 P(|X - \mu| \geq r\sigma),$$

which implies that

$$P(|X - \mu| \geq r\sigma) \leq \frac{1}{r^2}. \qquad \square$$

FURTHER READING AND ANNOTATED BIBLIOGRAPHY

DeCani, J. S., and R. A. Stine (1986). "A note on deriving the information matrix for a logistic distribution." *Amer. Statist.*, **40**, 220–222. (This article uses calculus techniques, such as integration and l'Hospital's rule, in determining the mean and variance of the logistic distribution as was seen in Example 6.9.2.)

Fulks, W. (1978). *Advanced Calculus*, 3rd ed. Wiley, New York. (Chap. 5 discusses the Riemann integral; Chap. 16 provides a study of improper integrals.)

Graybill, F. A. (1976). *Theory and Application of the Linear Model*. Duxbury Press, North Scituate, Massachusetts. (Section 8.5 discusses the calibration problem for a simple linear regression model as was seen in Example 6.9.4.)

Hardy, G. H., J. E. Littlewood, and G. Pólya (1952). *Inequalities*, 2nd ed. Cambridge University Press, Cambridge, England. (This is a classic and often referenced book on inequalities. Chap. 6 is relevant to the present chapter.)

Hartig, D. (1991). "L'Hôpital's rule via integration." *Amer. Math. Monthly*, **98**, 156–157.

Khuri, A. I., and G. Casella (2002). "The existence of the first negative moment revisited." *Amer. Statist.*, **56**, 44–47. (This article demonstrates the utility of the comparison test given in Theorem 6.5.3 in showing the existence of the first negative moment of a continuous random variable.)

Lindgren, B. W. (1976). *Statistical Theory*, 3rd ed. Macmillan, New York. (Section 2.2.2 gives the Riemann–Stieltjes integral representation of the expected value of a random variable as was seen in Section 6.9.3.)

Montgomery, D. C., and E. A. Peck (1982). *Introduction to Linear Regression Analysis*. Wiley, New York. (The calibration problem for a simple linear regression model is discussed in Section 9.7.)

Moran, P. A. P. (1968). *An Introduction to Probability Theory*. Clarendon Press, Oxford, England. (Section 5.9 defines moments of a random variable using the Riemann–Stieltjes integral representation of the expected value; Section 5.10 discusses a number of inequalities pertaining to these moments.)

Piegorsch, W. W., and G. Casella (1985). "The existence of the first negative moment." *Amer. Statist.*, **39**, 60–62. (This article gives a sufficient condition for the existence of the first negative moment of a continuous random variable as was seen in Section 6.9.1.)

Roussas, G. G. (1973). *A First Course in Mathematical Statistics*. Addison-Wesley, Reading, Massachusetts. (Chap. 9 is concerned with transformations of continuous random variables as was seen in Section 6.9.2. See, in particular, Theorems 2 and 3 in this chapter.)

Taylor, A. E., and W. R. Mann (1972). *Advanced Calculus*, 2nd ed. Wiley, New York. (Chap. 18 discusses the Riemann integral as well as the Riemann–Stieltjes integral; improper integrals are studied in Chap. 22.)

Wilks, S. S. (1962). *Mathematical Statistics*. Wiley, New York. (Chap. 3 uses the Riemann–Stieltjes integral to define expected values and moments of random variables; functions of random variables are discussed in Section 2.8.)

EXERCISES

In Mathematics

6.1. Let $f(x)$ be a bounded function defined on the interval $[a, b]$. Let P be a partition of $[a, b]$. Show that $f(x)$ is Riemann integrable on $[a, b]$ if and only if

$$\inf_{P} US_{P}(f) = \sup_{P} LS_{P}(f) = \int_{a}^{b} f(x) \, dx.$$

6.2. Construct a function that has a countable number of discontinuities in $[0, 1]$ and is Riemann integrable on $[0, 1]$.

6.3. Show that if $f(x)$ is continuous on $[a, b]$ except for a finite number of discontinuities of the first kind (see Definition 3.4.2), then $f(x)$ is Riemann integrable on $[a, b]$.

6.4. Show that the function

$$f(x) = \begin{cases} x \cos(\pi/2x), & 0 < x \le 1, \\ 0, & x = 0, \end{cases}$$

is not of bounded variation on $[0, 1]$.

6.5. Let $f(x)$ and $g(x)$ have continuous derivatives with $g'(x) > 0$. Suppose that $\lim_{x \to \infty} f(x) = \infty$, $\lim_{x \to \infty} g(x) = \infty$, and $\lim_{x \to \infty} f'(x)/g'(x) = L$, where L is finite.

(a) Show that for a given $\epsilon > 0$ there exists a constant $M > 0$ such that for $x > M$,

$$|f'(x) - Lg'(x)| < \epsilon g'(x).$$

Hence, if λ_1 and λ_2 are such that $M < \lambda_1 < \lambda_2$, then

$$\left| \int_{\lambda_1}^{\lambda_2} [f'(x) - Lg'(x)] \, dx \right| < \int_{\lambda_1}^{\lambda_2} \epsilon g'(x) \, dx.$$

(b) Deduce from (a) that

$$\left| \frac{f(\lambda_2)}{g(\lambda_2)} - L \right| < \epsilon + \frac{|f(\lambda_1)|}{g(\lambda_2)} + |L| \frac{g(\lambda_1)}{g(\lambda_2)}.$$

(c) Make use of (b) to show that for a sufficiently large λ_2,

$$\left| \frac{f(\lambda_2)}{g(\lambda_2)} - L \right| < 3\epsilon,$$

and hence $\lim_{x \to \infty} f(x)/g(x) = L$.
[*Note:* This problem verifies l'Hospital's rule for the ∞/∞ indeterminate form by using integration properties without relying on Cauchy's mean value theorem as in Section 4.2 (see Hartig, 1991)].

6.6. Show that if $f(x)$ is continuous on $[a, b]$, and if $g(x)$ is a nonnegative Riemann integrable function on $[a, b]$ such that $\int_a^b g(x) \, dx > 0$, then there exists a constant c, $a \leq c \leq b$, such that

$$\frac{\int_a^b f(x) g(x) \, dx}{\int_a^b g(x) \, dx} = f(c).$$

6.7. Prove Theorem 6.5.2.

6.8. Prove Theorem 6.5.3.

6.9. Suppose that $f(x)$ is a positive monotone decreasing function defined on $[a, \infty)$ such that $f(x) \to 0$ as $x \to \infty$. Show that if $f(x)$ is Riemann–Stieltjes integrable with respect to $g(x)$ on $[a, b]$ for every $b \geq a$, where $g(x)$ is bounded on $[a, \infty)$, then the integral $\int_a^\infty f(x) \, dg(x)$ is convergent.

6.10. Show that $\lim_{n \to \infty} \int_0^{n\pi} |(\sin x)/x| \, dx = \infty$, where n is a positive integer. [*Hint:* Show first that

$$\int_0^{n\pi} \left| \frac{\sin x}{x} \right| dx = \int_0^{\pi} \sin x \left[\frac{1}{x} + \frac{1}{x + \pi} + \cdots + \frac{1}{x + (n-1)\pi} \right] dx.]$$

6.11. Apply Maclaurin's integral test to determine convergence or divergence of the following series:

(**a**)
$$\sum_{n=1}^{\infty} \frac{\log n}{n\sqrt{n}},$$

(**b**)
$$\sum_{n=1}^{\infty} \frac{n + 4}{2n^3 + 1},$$

(**c**)
$$\sum_{n=1}^{\infty} \frac{1}{\sqrt{n + 1} - 1}.$$

6.12. Consider the sequence $\{f_n(x)\}_{n=1}^{\infty}$, where $f_n(x) = nx/(1 + nx^2)$, $x \geq 0$. Find the limit of $\int_1^2 f_n(x) \, dx$ as $n \to \infty$.

6.13. Consider the improper integral $\int_0^1 x^{m-1} (1 - x)^{n-1} \, dx$.

(a) Show that the integral converges if $m > 0$, $n > 0$. In this case, the function $B(m, n)$ defined as

$$B(m, n) = \int_0^1 x^{m-1}(1 - x)^{n-1} \, dx$$

is called the beta function.

(b) Show that

$$B(m, n) = 2 \int_0^{\pi/2} \sin^{2m-1} \theta \cos^{2n-1} \theta \, d\theta$$

(c) Show that

$$B(m, n) = \int_0^\infty \frac{x^{m-1}}{(1+x)^{m+n}} \, dx = \int_0^\infty \frac{x^{n-1}}{(1+x)^{m+n}} \, dx$$

(d) Show that

$$B(m, n) = \int_0^1 \frac{x^{m-1} + x^{n-1}}{(1+x)^{m+n}} \, dx.$$

6.14. Determine whether each of the following integrals is convergent or divergent:

(a)
$$\int_0^\infty \frac{dx}{\sqrt{1 + x^3}},$$

(b)
$$\int_0^\infty \frac{dx}{(1 + x^3)^{1/3}},$$

(c)
$$\int_0^1 \frac{dx}{(1 - x^3)^{1/3}},$$

(d)
$$\int_0^\infty \frac{dx}{\sqrt{x}(1 + 2x)}.$$

6.15. Let $f_1(x)$ and $f_2(x)$ be bounded on $[a, b]$, and $g(x)$ be monotone increasing on $[a, b]$. If $f_1(x)$ and $f_2(x)$ are Riemann–Stieltjes integrable with respect to $g(x)$ on $[a, b]$, then show that $f_1(x)f_2(x)$ is also Riemann–Stieltjes integrable with respect to $g(x)$ on $[a, b]$.

6.16. Let $f(x)$ be a function whose first n derivatives are continuous on $[a, b]$, and let

$$h_n(x) = f(b) - f(x) - (b - x)f'(x) - \cdots - \frac{(b - x)^{n-1}}{(n - 1)!} f^{(n-1)}(x).$$

Show that

$$h_n(a) = \frac{1}{(n - 1)!} \int_a^b (b - x)^{n-1} f^{(n)}(x) \, dx$$

and hence

$$f(b) = f(a) + (b - a)f'(a) + \cdots + \frac{(b - a)^{n-1}}{(n - 1)!} f^{(n-1)}(a)$$

$$+ \frac{1}{(n - 1)!} \int_a^b (b - x)^{n-1} f^{(n)}(x) \, dx.$$

This represents Taylor's expansion of $f(x)$ around $x = a$ (see Section 4.3) with a remainder R_n given by

$$R_n = \frac{1}{(n - 1)!} \int_a^b (b - x)^{n-1} f^{(n)}(x) \, dx.$$

[*Note:* This form of Taylor's theorem has the advantage of providing an exact formula for R_n, which does not involve an undetermined number θ_n as was seen in Section 4.3.]

6.17. Suppose that $f(x)$ is monotone and its derivative $f'(x)$ is Riemann integrable on $[a, b]$. Let $g(x)$ be continuous on $[a, b]$. Show that there exists a number c, $a \leq c \leq b$, such that

$$\int_a^b f(x)g(x) \, dx = f(a) \int_a^c g(x) \, dx + f(b) \int_c^b g(x) \, dx.$$

6.18. Deduce from Exercise 6.17 that for any $b > a > 0$,

$$\left| \int_a^b \frac{\sin x}{x} \, dx \right| \leq \frac{4}{a}.$$

6.19. Prove Lemma 6.7.1.

6.20. Show that if $f_1(x), f_2(x), \ldots, f_n(x)$ are such that $|f_i(x)|^p$ is Riemann integrable on $[a, b]$ for $i = 1, 2, \ldots, n$, where $1 \leq p < \infty$, then

$$\left[\int_a^b \left| \sum_{i=1}^n f_i(x) \right|^p dx \right]^{1/p} \leq \sum_{i=1}^n \left[\int_a^b |f_i(x)|^p dx \right]^{1/p}.$$

6.21. Prove Theorem 6.8.1.

6.22. Prove Theorem 6.8.2.

In Statistics

6.23. Show that the integral in formula (6.66) is divergent for $k \geq 1$.

6.24. Consider the random variable X that has the logistic distribution described in Example 6.9.2. Show that $\mathrm{Var}(X) = \pi^2/3$.

6.25. Let $\{f_n(x)\}_{n=1}^\infty$ be a family density functions defined by

$$f_n(x) = \frac{|\log^n x|^{-1}}{\int_0^\lambda |\log^n t|^{-1} dt}, \qquad 0 < x < \lambda,$$

where $\lambda \in (0, 1)$.
(a) Show that condition (6.72) of Theorem 6.9.1 is not satisfied by any $f_n(x)$, $n \geq 1$.
(b) Show that when $n = 1$, $E(X^{-1})$ does not exist, where X is a random variable with the density function $f_1(x)$.
(c) Show that for $n > 1$, $E(X_n^{-1})$ exists, where X_n is a random variable with the density function $f_n(x)$.

6.26. Let X be a random variable with a continuous density function $f(x)$ on $(0, \infty)$. Suppose that $f(x)$ is bounded near zero. Then $E(X^{-\alpha})$ exits, where $\alpha \in (0, 1)$.

6.27. Let X be a random variable with a continuous density function $f(x)$ on $(0, \infty)$. If $\lim_{x \to 0^+}(f(x)/x^\alpha)$ is equal to a positive constant k for some $\alpha > 0$, then $E[X^{-(1+\alpha)}]$ does not exist.

6.28. The random variable Y has the t-distributions with n degrees of freedom. Its density function is given by

$$f(y) = \frac{\Gamma\left(\dfrac{n+1}{2}\right)}{\sqrt{n\pi}\,\Gamma\left(\dfrac{n}{2}\right)}\left(1 + \frac{y^2}{n}\right)^{-(n+1)/2}, \qquad -\infty < y < \infty,$$

where $\Gamma(m)$ is the gamma function $\int_0^\infty e^{-x} x^{m-1}\, dx$, $m > 0$. Find the density function of $W = |Y|$.

6.29. Let X be a random variable with a mean μ and a variance σ^2.

(a) Show that Chebyshev's inequality can be expressed as

$$P(|X - \mu| \geq r) \leq \frac{\sigma^2}{r^2},$$

where r is any positive constant.

(b) Let $\{X_n\}_{n=1}^\infty$ be a sequence of independent and identically distributed random variables. If the common mean and variance of the X_i's are μ and σ^2, respectively, then show that

$$P(|\bar{X}_n - \mu| \geq r) \leq \frac{\sigma^2}{nr^2},$$

where $\bar{X}_n = (1/n)\sum_{i=1}^n X_i$ and r is any positive constant.

(c) Deduce from (b) that $\bar{X}_n$ converges in probability to μ as $n \to \infty$, that is, for every $\epsilon > 0$,

$$P(|\bar{X}_n - \mu| \geq \epsilon) \to 0 \qquad \text{as } n \to \infty.$$

6.30. Let X be a random variable with a cumulative distribution function $F(x)$. Let μ'_k be its kth noncentral moment,

$$\mu'_k = E(X^k) = \int_{-\infty}^\infty x^k\, dF(x).$$

Let ν_k be the kth absolute moment of X,

$$\nu_k = E(|X|^k) = \int_{-\infty}^\infty |x|^k\, dF(x).$$

Suppose that ν_k exists for $k = 1, 2, \ldots, n$.

(a) Show that $v_k^2 \le v_{k-1} v_{k+1}$, $k = 1, 2, \ldots, n-1$. [*Hint:* For any u and v,

$$0 \le \int_{-\infty}^{\infty} \left[u|x|^{(k-1)/2} + v|x|^{(k+1)/2} \right]^2 dF(x)$$

$$= u^2 v_{k-1} + 2uv v_k + v^2 v_{k+1} .]$$

(b) Deduce from (a) that

$$v_1 \le v_2^{1/2} \le v_3^{1/3} \le \cdots \le v_n^{1/n} .$$

CHAPTER 7

Multidimensional Calculus

In the previous chapters we have mainly dealt with real-valued functions of a single variable x. In this chapter we extend the notions of limits, continuity, differentiation, and integration to multivariable functions, that is, functions of several variables. These functions can be real-valued or possibly vector-valued. More specifically, if R^n denotes the n-dimensional Euclidean space, $n \geq 1$, then we shall in general consider functions defined on a set $D \subset R^n$ and have values in R^m, $m \geq 1$. Such functions are represented symbolically as $\mathbf{f}: D \to R^m$, where for $\mathbf{x} = (x_1, x_2, \ldots, x_n)' \in D$,

$$\mathbf{f}(\mathbf{x}) = [f_1(\mathbf{x}), f_2(\mathbf{x}), \ldots, f_m(\mathbf{x})]'$$

and $f_i(\mathbf{x})$ is a real-valued function of $x_1, x_2, \ldots, x_n$ $(i = 1, 2, \ldots, m)$.

Even though the basic framework of the methodology in this chapter is general and applies in any number of dimensions, most of the examples are associated with two- or three-dimensional spaces. At this stage, it would be helpful to review the basic concepts given in Chapters 1 and 2. This can facilitate the understanding of the methodology and its development in a multidimensional environment.

7.1. SOME BASIC DEFINITIONS

Some of the concepts described in Chapter 1 pertained to one-dimensional Euclidean spaces. In this section we extend these concepts to higher-dimensional Euclidean spaces.

Any point $\mathbf{x}$ in R^n can be represented as a column vector of the form $(x_1, x_2, \ldots, x_n)'$, where x_i is the ith element of $\mathbf{x}$ $(i = 1, 2, \ldots, n)$. The Euclidean norm of $\mathbf{x}$ was defined in Chapter 2 (see Definition 2.1.4) as $\|\mathbf{x}\|_2 = (\sum_{i=1}^n x_i^2)^{1/2}$. For simplicity we shall drop the subindex 2 and denote this norm by $\|\mathbf{x}\|$.

Let $\mathbf{x}_0 \in R^n$. A neighborhood $N_r(\mathbf{x}_0)$ of $\mathbf{x}_0$ is a set of points in R^n that lie within some distance, say r, from $\mathbf{x}_0$, that is,

$$N_r(\mathbf{x}_0) = \{\mathbf{x} \in R^n | \|\mathbf{x} - \mathbf{x}_0\| < r\}.$$

If $\mathbf{x}_0$ is deleted from $N_r(\mathbf{x}_0)$, we obtain the so-called deleted neighborhood of $\mathbf{x}_0$, which we denote by $N_r^d(\mathbf{x}_0)$.

A point $\mathbf{x}_0$ in R^n is a limit point of a set $A \subset R^n$ if every neighborhood of $\mathbf{x}_0$ contains an element $\mathbf{x}$ of A such that $\mathbf{x} \neq \mathbf{x}_0$, that is, every deleted neighborhood of $\mathbf{x}_0$ contains points of A.

A set $A \subset R^n$ is closed if every limit point of A belongs to A.

A point $\mathbf{x}_0$ in R^n is an interior of a set $A \subset R^n$ if there exists an $r > 0$ such that $N_r(\mathbf{x}_0) \subset A$.

A set $A \subset R^n$ is open if for every point $\mathbf{x}$ in A there exists a neighborhood $N_r(\mathbf{x})$ that is contained in A. Thus A is open if it consists entirely of interior points.

A point $p \in R^n$ is a boundary point of a set $A \subset R^n$ if every neighborhood of p contains points of A as well as points of $\overline{A}$, the complement of A with respect to R^n. The set of all boundary points of A is called its boundary and is denoted by $Br(A)$.

A set $A \subset R^n$ is bounded if there exists an $r > 0$ such that $\|\mathbf{x}\| \leq r$ for all $\mathbf{x}$ in A.

Let $\mathbf{g} \colon J^+ \to R^n$ be a vector-valued function defined on the set of all positive integers. Let $\mathbf{g}(i) = \mathbf{a}_i$, $i \geq 1$. Then $\{\mathbf{a}_i\}_{i=1}^{\infty}$ represents a sequence of points in R^n. By a subsequence of $\{\mathbf{a}_i\}_{i=1}^{\infty}$ we mean a sequence $\{\mathbf{a}_{k_i}\}_{i=1}^{\infty}$ such that $k_1 < k_2 < \cdots < k_i < \cdots$ and $k_i \geq i$ for $i \geq 1$ (see Definition 5.1.1).

A sequence $\{\mathbf{a}_i\}_{i=1}^{\infty}$ converges to a point $c \in R^n$ if for a given $\epsilon > 0$ there exists an integer N such that $\|\mathbf{a}_i - \mathbf{c}\| < \epsilon$ whenever $i > N$. This is written symbolically as $\lim_{i \to \infty} \mathbf{a}_i = \mathbf{c}$, or $\mathbf{a}_i \to \mathbf{c}$ as $i \to \infty$.

A sequence $\{\mathbf{a}_i\}_{i=1}^{\infty}$ is bounded if there exists a number $K > 0$ such that $\|\mathbf{a}_i\| \leq K$ for all i.

7.2. LIMITS OF A MULTIVARIABLE FUNCTION

We recall from Chapter 3 that for a function of a single variable x, its limit at a point is considered when x approaches the point from two directions, left and right. Here, for a function of several variables, say $x_1, x_2, \ldots, x_n$, its limit at a point $\mathbf{a} = (a_1, a_2, \ldots, a_n)'$ is considered when $\mathbf{x} = (x_1, x_2, \ldots, x_n)'$ approaches $\mathbf{a}$ in any possible way. Thus when $n > 1$ there are infinitely many ways in which $\mathbf{x}$ can approach $\mathbf{a}$.

Definition 7.2.1. Let $\mathbf{f} \colon D \to R^m$, where $D \subset R^n$. Then $\mathbf{f}(\mathbf{x})$ is said to have a limit $\mathbf{L} = (L_1, L_2, \ldots, L_m)'$ as $\mathbf{x}$ approaches $\mathbf{a}$, written symbolically as $\mathbf{x} \to \mathbf{a}$, where $\mathbf{a}$ is a limit point of D, if for a given $\epsilon > 0$ there exists a $\delta > 0$ such that $\|\mathbf{f}(\mathbf{x}) - \mathbf{L}\| < \epsilon$ for all $\mathbf{x}$ in $D \cap N_\delta^d(\mathbf{a})$, where $N_\delta^d(\mathbf{a})$ is a deleted neighborhood of $\mathbf{a}$ of radius δ. If it exists, this limit is written symbolically as $\lim_{\mathbf{x} \to \mathbf{a}} \mathbf{f}(\mathbf{x}) = \mathbf{L}$.   □

Note that whenever a limit of $\mathbf{f}(\mathbf{x})$ exists, its value must be the same no matter how $\mathbf{x}$ approaches $\mathbf{a}$. It is important here to understand the meaning of "$\mathbf{x}$ approaches $\mathbf{a}$." By this we do not necessarily mean that $\mathbf{x}$ moves along a straight line leading into $\mathbf{a}$. Rather, we mean that $\mathbf{x}$ moves closer and closer to $\mathbf{a}$ along any curve that goes through $\mathbf{a}$.

EXAMPLE 7.2.1.   Consider the behavior of the function

$$f(x_1, x_2) = \frac{x_1^3 - x_2^3}{x_1^2 + x_2^2}$$

as $\mathbf{x} = (x_1, x_2)' \to \mathbf{0}$, where $\mathbf{0} = (0,0)'$. This function is defined everywhere in R^2 except at $\mathbf{0}$. It is convenient here to represent the point $\mathbf{x}$ using polar coordinates, r and θ, such that $x_1 = r \cos \theta, x_2 = r \sin \theta, r > 0,\ 0 \le \theta \le 2\pi$. We then have

$$f(x_1, x_2) = \frac{r^3 \cos^3 \theta - r^3 \sin^3 \theta}{r^2 \cos^2 \theta + r^2 \sin^2 \theta}$$

$$= r(\cos^3 \theta - \sin^3 \theta).$$

Since $\mathbf{x} \to \mathbf{0}$ if and only if $r \to 0$, $\lim_{\mathbf{x} \to \mathbf{0}} f(x_1, x_2) = 0$ no matter how $\mathbf{x}$ approaches $\mathbf{0}$.

EXAMPLE 7.2.2.   Consider the function

$$f(x_1, x_2) = \frac{x_1 x_2}{x_1^2 + x_2^2}.$$

Using polar coordinates again, we obtain

$$f(x_1, x_2) = \cos \theta \sin \theta,$$

which depends on θ, but not on r. Since θ can have infinitely many values, $f(x_1, x_2)$ cannot be made close to any one constant L no matter how small r is. Thus the limit of this function does not exist as $\mathbf{x} \to \mathbf{0}$.

EXAMPLE 7.2.3.   Let $f(x_1, x_2)$ be defined as

$$f(x_1, x_2) = \frac{x_2(x_1^2 + x_2^2)}{x_2^2 + (x_1^2 + x_2^2)^2}.$$

This function is defined everywhere in R^2 except at $(0,0)'$. On the line $x_1 = 0$, $f(0, x_2) = x_2^3/(x_2^2 + x_2^4)$, which goes to zero as $x_2 \to 0$. When $x_2 = 0$, $f(x_1, 0) = 0$ for $x_1 \neq 0$; hence, $f(x_1, 0) \to 0$ as $x_1 \to 0$. Furthermore, for any other straight line $x_2 = tx_1$ $(t \neq 0)$ through the origin we have

$$f(x_1, tx_1) = \frac{tx_1(x_1^2 + t^2 x_1^2)}{t^2 x_1^2 + (x_1^2 + t^2 x_1^2)^2},$$

$$= \frac{tx_1(1 + t^2)}{t^2 + x_1^2(1 + t^2)^2}, \qquad x_1 \neq 0,$$

which has a limit equal to zero as $x_1 \to 0$. We conclude that the limit of $f(x_1, x_2)$ is zero as $\mathbf{x} \to \mathbf{0}$ along any straight line through the origin. However, $f(x_1, x_2)$ does not have a limit as $\mathbf{x} \to \mathbf{0}$. For example, along the circle $x_2 = x_1^2 + x_2^2$ that passes through the origin,

$$f(x_1, x_2) = \frac{(x_1^2 + x_2^2)^2}{(x_1^2 + x_2^2)^2 + (x_1^2 + x_2^2)^2} = \frac{1}{2}, \qquad x_1^2 + x_2^2 \neq 0.$$

Hence, $f(x_1, x_2) \to \frac{1}{2} \neq 0$.

This example demonstrates that a function may not have a limit as $\mathbf{x} \to \mathbf{a}$ even though its limit exists for approaches toward $\mathbf{a}$ along straight lines.

7.3. CONTINUITY OF A MULTIVARIABLE FUNCTION

The notion of continuity for a function of several variables is much the same as that for a function of a single variable.

Definition 7.3.1.   Let $\mathbf{f} \colon D \to R^m$, where $D \subset R^n$, and let $\mathbf{a} \in D$. Then $\mathbf{f}(\mathbf{x})$ is continuous at $\mathbf{a}$ if

$$\lim_{\mathbf{x} \to \mathbf{a}} \mathbf{f}(\mathbf{x}) = \mathbf{f}(\mathbf{a}),$$

where $\mathbf{x}$ remains in D as it approaches $\mathbf{a}$. This is equivalent to stating that for a given $\epsilon > 0$ there exits a $\delta > 0$ such that

$$\|\mathbf{f}(\mathbf{x}) - \mathbf{f}(\mathbf{a})\| < \epsilon$$

for all $\mathbf{x} \in D \cap N_\delta(\mathbf{a})$.

If $\mathbf{f}(\mathbf{x})$ is continuous at every point $\mathbf{x}$ in D, then it is said to be continuous in D. In particular, if $\mathbf{f}(\mathbf{x})$ is continuous in D and if δ (in the definition of continuity) depends only on ϵ (that is, δ is the same for all points in D for the given ϵ), then $\mathbf{f}(\mathbf{x})$ is said to be uniformly continuous in D. $\qquad\square$

We now present several theorems that provide some important properties of multivariable continuous functions. These theorems are analogous to those given in Chapter 3. Let us first consider the following lemmas (the proofs are left to the reader):

Lemma 7.3.1. Every bounded sequence in R^n has a convergent subsequence.

This lemma is analogous to Theorem 5.1.4.

Lemma 7.3.2. Suppose that $f, g: D \to R$ are real-valued continuous functions, where $D \subset R^n$. Then we have the following:

1. $f + g$, $f - g$, and fg are continuous in D.
2. $|f|$ is continuous in D.
3. $1/f$ is continuous in D provided that $f(\mathbf{x}) \neq 0$ for all $\mathbf{x}$ in D.

This lemma is analogous to Theorem 3.4.1.

Lemma 7.3.3. Suppose that $\mathbf{f}: D \to R^m$ is continuous, where $D \subset R^n$, and that $\mathbf{g}: G \to R^v$ is also continuous, where $G \subset R^m$ is the image of D under $\mathbf{f}$. Then the composite function $\mathbf{g} \circ \mathbf{f}: D \to R^v$, defined as $\mathbf{g} \circ \mathbf{f}(\mathbf{x}) = \mathbf{g}[\mathbf{f}(\mathbf{x})]$, is continuous in D.

This lemma is analogous to Theorem 3.4.2.

Theorem 7.3.1. Let $f: D \to R$ be a real-valued continuous function defined on a closed and bounded set $D \subset R^n$. Then there exist points $\mathbf{p}$ and $\mathbf{q}$ in D for which

$$f(\mathbf{p}) = \sup_{x \in D} f(\mathbf{x}), \qquad\qquad (7.1)$$

$$f(\mathbf{q}) = \inf_{x \in D} f(\mathbf{x}). \qquad\qquad (7.2)$$

Thus $f(\mathbf{x})$ attains each of its infimum and supremum at least once in D.

Proof. Let us first show that $f(\mathbf{x})$ is bounded in D. We shall prove this by contradiction. Suppose that $f(\mathbf{x})$ is not bounded in D. Then we can find a sequence of points $\{\mathbf{p}_i\}_{i=1}^\infty$ in D such that $|f(\mathbf{p}_i)| \geq i$ for $i \geq 1$ and hence

$|f(\mathbf{p}_i)| \to \infty$ as $i \to \infty$. Since the terms of this sequence are elements in a bounded set, $\{\mathbf{p}_i\}_{i=1}^{\infty}$ must be a bounded sequence. By Lemma 7.3.1, this sequence has a convergent subsequence $\{\mathbf{p}_{k_i}\}_{i=1}^{\infty}$. Let $\mathbf{p}_0$ be the limit of this subsequence, which is also a limit point of D; hence, it belongs to D, since D is closed. Now, on one hand, $|f(\mathbf{p}_{k_i})| \to |f(\mathbf{p}_0)|$ as $i \to \infty$, by the continuity of $f(\mathbf{x})$ and hence of $|f(\mathbf{x})|$ [see Lemma 7.3.2(2)]. On the other hand, $|f(\mathbf{p}_{k_i})| \to \infty$. This contradiction shows that $f(\mathbf{x})$ must be bounded in D. Consequently, the infimum and supremum of $f(\mathbf{x})$ in D are finite.

Suppose now equality (7.1) does not hold for any $\mathbf{p} \in D$. Then, $M - f(\mathbf{x}) > 0$ for all $\mathbf{x} \in D$, where $M = \sup_{\mathbf{x} \in D} f(\mathbf{x})$. Consequently, $[M - f(\mathbf{x})]^{-1}$ is positive and continuous in D by Lemma 7.3.2(3) and is therefore bounded by the first half of this proof. However, if $\delta > 0$ is any given positive number, then, by the definition of M, we can find a point $\mathbf{x}_\delta$ in D for which $f(\mathbf{x}_\delta) > M - \delta$, or

$$\frac{1}{M - f(\mathbf{x}_\delta)} > \frac{1}{\delta}.$$

This implies that $[M - f(\mathbf{x})]^{-1}$ is not bounded, a contradiction, which proves equality (7.1). The proof of equality (7.2) is similar. $\qquad \square$

Theorem 7.3.2. Suppose that D is a closed and bounded set in R^n. If $\mathbf{f}: D \to R^m$ is continuous, then it is uniformly continuous in D.

Proof. We shall prove this theorem by contradiction. Suppose that $\mathbf{f}$ is not uniformly continuous in D. Then there exists an $\epsilon > 0$ such that for every $\delta > 0$ we can find $\mathbf{a}$ and $\mathbf{b}$ in D such that $\|\mathbf{a} - \mathbf{b}\| < \delta$, but $\|\mathbf{f}(\mathbf{a}) - \mathbf{f}(\mathbf{b})\| \geq \epsilon$. Let us choose $\delta = 1/i$, $i \geq 1$. We can therefore find two sequences $\{\mathbf{a}_i\}_{i=1}^{\infty}, \{\mathbf{b}_i\}_{i=1}^{\infty}$ with $\mathbf{a}_i, \mathbf{b}_i \in D$ such that $\|\mathbf{a}_i - \mathbf{b}_i\| < 1/i$, and

$$\|\mathbf{f}(\mathbf{a}_i) - \mathbf{f}(\mathbf{b}_i)\| \geq \epsilon \qquad (7.3)$$

for $i \geq 1$. Now, the sequence $\{\mathbf{a}_i\}_{i=1}^{\infty}$ is bounded. Hence, by Lemma 7.3.1, it has a convergent subsequence $\{\mathbf{a}_{k_i}\}_{i=1}^{\infty}$ whose limit, denoted by $\mathbf{a}_0$, is in D, since D is closed. Also, since $\mathbf{f}$ is continuous at $\mathbf{a}_0$, we can find a $\lambda > 0$ such that $\|\mathbf{f}(\mathbf{x}) - \mathbf{f}(\mathbf{a}_0)\| < \epsilon/2$ if $\|\mathbf{x} - \mathbf{a}_0\| < \lambda$, where $\mathbf{x} \in D$. By the convergence of $\{\mathbf{a}_{k_i}\}_{i=1}^{\infty}$ to $\mathbf{a}_0$, we can choose k_i large enough so that

$$\frac{1}{k_i} < \frac{\lambda}{2} \qquad (7.4)$$

and

$$\|\mathbf{a}_{k_i} - \mathbf{a}_0\| < \frac{\lambda}{2}. \qquad (7.5)$$

From (7.5) it follows that

$$\|\mathbf{f}(\mathbf{a}_{k_i}) - \mathbf{f}(\mathbf{a}_0)\| < \frac{\epsilon}{2}. \tag{7.6}$$

Furthermore, since $\|\mathbf{a}_{k_i} - \mathbf{b}_{k_i}\| < 1/k_i$, we can write

$$\|\mathbf{b}_{k_i} - \mathbf{a}_0\| \le \|\mathbf{a}_{k_i} - \mathbf{a}_0\| + \|\mathbf{a}_{k_i} - \mathbf{b}_{k_i}\|$$

$$< \frac{\lambda}{2} + \frac{1}{k_i} < \lambda.$$

Hence, by the continuity of $\mathbf{f}$ at $\mathbf{a}_0$,

$$\|\mathbf{f}(\mathbf{b}_{k_i}) - \mathbf{f}(\mathbf{a}_0)\| < \frac{\epsilon}{2}. \tag{7.7}$$

From inequalities (7.6) and (7.7) we conclude that whenever k_i satisfies inequalities (7.4) and (7.5),

$$\|\mathbf{f}(\mathbf{a}_{k_i}) - \mathbf{f}(\mathbf{b}_{k_i})\| \le \|\mathbf{f}(\mathbf{a}_{k_i}) - \mathbf{f}(\mathbf{a}_0)\| + \|\mathbf{f}(\mathbf{b}_{k_i}) - \mathbf{f}(\mathbf{a}_0)\| < \epsilon,$$

which contradicts inequality (7.3). This leads us to assert that $\mathbf{f}$ is uniformly continuous in D. $\square$

7.4. DERIVATIVES OF A MULTIVARIABLE FUNCTION

In this section we generalize the concept of differentiation given in Chapter 4 to a multivariable function $\mathbf{f}: D \to R^m$, where $D \subset R^n$.

Let $\mathbf{a} = (a_1, a_2, \ldots, a_n)'$ be an interior point of D. Suppose that the limit

$$\lim_{h_i \to 0} \frac{\mathbf{f}(a_1, a_2, \ldots, a_i + h_i, \ldots, a_n) - \mathbf{f}(a_1, a_2, \ldots, a_i, \ldots, a_n)}{h_i}$$

exists; then $\mathbf{f}$ is said to have a partial derivative with respect to x_i at $\mathbf{a}$. This derivative is denoted by $\partial \mathbf{f}(\mathbf{a})/\partial x_i$, or just $\mathbf{f}_{x_i}(\mathbf{a})$, $i = 1, 2, \ldots, n$. Hence, partial differentiation with respect to x_i is done in the usual fashion while treating all the remaining variables as constants. For example, if $f: R^3 \to R$ is defined as $f(x_1, x_2, x_3) = x_1 x_2^2 + x_2 x_3^3$, then at any point $\mathbf{x} \in R^3$ we have

$$\frac{\partial f(\mathbf{x})}{\partial x_1} = x_2^2,$$

$$\frac{\partial f(\mathbf{x})}{\partial x_2} = 2 x_1 x_2 + x_3^3,$$

$$\frac{\partial f(\mathbf{x})}{\partial x_3} = 3 x_2 x_3^2.$$

In general, if f_j is the jth element of $\mathbf{f}$ $(j = 1, 2, \ldots, m)$, then the terms $\partial f_j(\mathbf{x})/\partial x_i$, for $i = 1, 2, \ldots, n$; $j = 1, 2, \ldots, m$, constitute an $m \times n$ matrix called the Jacobian matrix (after Carl Gustav Jacobi, 1804–1851) of $\mathbf{f}$ at $\mathbf{x}$ and is denoted by $\mathbf{J_f}(\mathbf{x})$. If $m = n$, the determinant of $\mathbf{J_f}(\mathbf{x})$ is called the Jacobian determinant; it is sometimes represented as

$$\det[\mathbf{J_f}(\mathbf{x})] = \frac{\partial(f_1, f_2, \ldots, f_n)}{\partial(x_1, x_2, \ldots, x_n)}. \tag{7.8}$$

For example, if $\mathbf{f}: R^3 \to R^2$ is such that

$$\mathbf{f}(x_1, x_2, x_3) = \left(x_1^2 \cos x_2, \; x_2^2 + x_3^2 e^{x_1} \right)',$$

then

$$\mathbf{J_f}(x_1, x_2, x_3) = \begin{bmatrix} 2 x_1 \cos x_2 & -x_1^2 \sin x_2 & 0 \\ x_3^2 e^{x_1} & 2 x_2 & 2 x_3 e^{x_1} \end{bmatrix}.$$

Higher-order partial derivatives of $\mathbf{f}$ are defined similarly. For example, the second-order partial derivative of $\mathbf{f}$ with respect to x_i at $\mathbf{a}$ is defined as

$$\lim_{h_i \to 0} \frac{\mathbf{f}_{x_i}(a_1, a_2, \ldots, a_i + h_i, \ldots, a_n) - \mathbf{f}_{x_i}(a_1, a_2, \ldots, a_i, \ldots, a_n)}{h_i}$$

and is denoted by $\partial^2 \mathbf{f}(\mathbf{a})/\partial x_i^2$, or $\mathbf{f}_{x_i x_i}(\mathbf{a})$. Also, the second-order partial derivative of $\mathbf{f}$ with respect to x_i and x_j, $i \neq j$, at $\mathbf{a}$ is given by

$$\lim_{h_j \to 0} \frac{\mathbf{f}_{x_i}(a_1, a_2, \ldots, a_j + h_j, \ldots, a_n) - \mathbf{f}_{x_i}(a_1, a_2, \ldots, a_j, \ldots, a_n)}{h_j}$$

and is denoted by $\partial^2 \mathbf{f}(\mathbf{a})/\partial x_j \, \partial x_i$, or $\mathbf{f}_{x_j x_i}(\mathbf{a})$, $i \neq j$.

Under certain conditions, the order in which differentiation with respect to x_i and x_j takes place is irrelevant, that is, $\mathbf{f}_{x_i x_j}(\mathbf{a})$ is identical to $\mathbf{f}_{x_j x_i}(\mathbf{a})$, $i \neq j$. This property is known as the commutative property of partial differentiation and is proved in the next theorem.

Theorem 7.4.1. Let $\mathbf{f} \colon D \to R^m$, where $D \subset R^n$, and let $\mathbf{a}$ be an interior point of D. Suppose that in a neighborhood of $\mathbf{a}$ the following conditions are satisfied:

1. $\partial \mathbf{f}(\mathbf{x})/\partial x_i$ and $\partial \mathbf{f}(\mathbf{x})/\partial x_j$ exist and are finite ($i, j = 1, 2, \ldots, n$, $i \neq j$).
2. Of the derivatives $\partial^2 \mathbf{f}(\mathbf{x})/\partial x_i \, \partial x_j, \partial^2 \mathbf{f}(\mathbf{x})/\partial x_j \, \partial x_i$ one exists and is continuous.

Then

$$\frac{\partial^2 \mathbf{f}(\mathbf{a})}{\partial x_i \, \partial x_j} = \frac{\partial^2 \mathbf{f}(\mathbf{a})}{\partial x_j \, \partial x_i}.$$

Proof. Let us suppose that $\partial^2 \mathbf{f}(\mathbf{x})/\partial x_j \, \partial x_i$ exists and is continuous in a neighborhood of $\mathbf{a}$. Without loss of generality we assume that $i < j$. If $\partial^2 \mathbf{f}(\mathbf{a})/\partial x_i \, \partial x_j$ exists, then it must be equal to the limit

$$\lim_{h_i \to 0} \frac{\mathbf{f}_{x_j}(a_1, a_2, \ldots, a_i + h_i, \ldots, a_n) - \mathbf{f}_{x_j}(a_1, a_2, \ldots, a_i, \ldots, a_n)}{h_i},$$

that is,

$$\lim_{h_i \to 0} \frac{1}{h_i} \left[\lim_{h_j \to 0} \frac{1}{h_j} \left\{ \mathbf{f}(a_1, a_2, \ldots, a_i + h_i, \ldots, a_j + h_j, \ldots, a_n) \right.\right.$$

$$\left. - \mathbf{f}(a_1, a_2, \ldots, a_i + h_i, \ldots, a_j, \ldots, a_n) \right\}$$

$$- \lim_{h_j \to 0} \frac{1}{h_j} \left\{ \mathbf{f}(a_1, a_2, \ldots, a_i, \ldots, a_j + h_j, \ldots, a_n) \right.$$

$$\left.\left. - \mathbf{f}(a_1, a_2, \ldots, a_i, \ldots, a_j, \ldots, a_n) \right\} \right]. \quad (7.9)$$

Let us denote $\mathbf{f}(x_1, x_2, \ldots, x_j + h_j, \ldots, x_n) - \mathbf{f}(x_1, x_2, \ldots, x_j, \ldots, x_n)$ by $\boldsymbol{\psi}(x_1, x_2, \ldots, x_n)$. Then the double limit in (7.9) can be written as

$$\lim_{h_i \to 0} \lim_{h_j \to 0} \frac{1}{h_i h_j} \left[\boldsymbol{\psi}(a_1, a_2, \ldots, a_i + h_i, \ldots, a_j, \ldots, a_n) \right.$$

$$\left. - \boldsymbol{\psi}(a_1, a_2, \ldots, a_i, \ldots, a_j, \ldots, a_n) \right]$$

$$= \lim_{h_i \to 0} \lim_{h_j \to 0} \frac{1}{h_j} \left[\frac{\partial \boldsymbol{\psi}(a_1, a_2, \ldots, a_i + \theta_i h_i, \ldots, a_j, \ldots, a_n)}{\partial x_i} \right], \quad (7.10)$$

where $0 < \theta_i < 1$. In formula (7.10) we have applied the mean value theorem (Theorem 4.2.2) to ψ as if it were a function of the single variable x_i (since $\partial \mathbf{f}/\partial x_i$, and hence $\partial \psi/\partial x_i$ exists in a neighborhood of $\mathbf{a}$). The right-hand side of (7.10) can then be written as

$$\lim_{h_i \to 0} \lim_{h_j \to 0} \frac{1}{h_j} \left[\frac{\partial \mathbf{f}(a_1, a_2, \ldots, a_i + \theta_i h_i, \ldots, a_j + h_j, \ldots, a_n)}{\partial x_i} \right.$$

$$\left. - \frac{\partial \mathbf{f}(a_1, a_2, \ldots, a_i + \theta_i h_i, \ldots, a_j, \ldots, a_n)}{\partial x_i} \right]$$

$$= \lim_{h_i \to 0} \lim_{h_j \to 0} \frac{\partial}{\partial x_j} \left[\frac{\partial \mathbf{f}(a_1, a_2, \ldots, a_i + \theta_i h_i, \ldots, a_j + \theta_j h_j, \ldots, a_n)}{\partial x_i} \right],$$

$$(7.11)$$

where $0 < \theta_j < 1$. In formula (7.11) we have again made use of the mean value theorem, since $\partial^2 \mathbf{f}(\mathbf{x})/\partial x_j \, \partial x_i$ exists in the given neighborhood around $\mathbf{a}$. Furthermore, since $\partial^2 \mathbf{f}(\mathbf{x})/\partial x_j \, \partial x_i$ is continuous in this neighborhood, the double limit in (7.11) is equal to $\partial^2 \mathbf{f}(\mathbf{a})/\partial x_j \, \partial x_i$. This establishes the assertion that the two second-order partial derivatives of $\mathbf{f}$ are equal. □

EXAMPLE 7.4.1. Consider the function $f: R^3 \to R$, where $f(x_1, x_2, x_3) = x_1 e^{x_2} + x_2 \cos x_1$. Then

$$\frac{\partial f(x_1, x_2, x_3)}{\partial x_1} = e^{x_2} - x_2 \sin x_1,$$

$$\frac{\partial f(x_1, x_2, x_3)}{\partial x_2} = x_1 e^{x_2} + \cos x_1,$$

$$\frac{\partial^2 f(x_1, x_2, x_3)}{\partial x_2 \, \partial x_1} = e^{x_2} - \sin x_1,$$

$$\frac{\partial^2 f(x_1, x_2, x_3)}{\partial x_1 \, \partial x_2} = e^{x_2} - \sin x_1.$$

7.4.1. The Total Derivative

Let $f(\mathbf{x})$ be a real-valued function defined on a set $D \subset R^n$, where $\mathbf{x} = (x_1, x_2, \ldots, x_n)'$. Suppose that $x_1, x_2, \ldots, x_n$ are functions of a single variable t. Then f is a function of t. The ordinary derivative of f with respect to t, namely df/dt, is called the total derivative of f.

Let us now assume that for the values of t under consideration dx_i/dt exists for $i = 1, 2, \ldots, n$ and that $\partial f(\mathbf{x})/\partial x_i$ exists and is continuous in the interior of D for $i = 1, 2, \ldots, n$. Under these considerations, the total derivative of f is given by

$$\frac{df}{dt} = \sum_{i=1}^{n} \frac{\partial f(\mathbf{x})}{\partial x_i} \frac{dx_i}{dt}. \tag{7.12}$$

To show this we proceed as follows: Let $\Delta x_1, \Delta x_2, \ldots, \Delta x_n$ be increments of $x_1, x_2, \ldots, x_n$ that correspond to an increment Δt of t. In turn, f will have the increment Δf. We then have

$$\Delta f = f(x_1 + \Delta x_1, x_2 + \Delta x_2, \ldots, x_n + \Delta x_n) - f(x_1, x_2, \ldots, x_n).$$

This can be written as

$$\begin{aligned}
\Delta f = &\left[f(x_1 + \Delta x_1, x_2 + \Delta x_2, \ldots, x_n + \Delta x_n) \right.\\
&\left. - f(x_1, x_2 + \Delta x_2, \ldots, x_n + \Delta x_n) \right]\\
&+ \left[f(x_1, x_2 + \Delta x_2, \ldots, x_n + \Delta x_n) \right.\\
&\left. - f(x_1, x_2, x_3 + \Delta x_3, \ldots, x_n + \Delta x_n) \right]\\
&+ \left[f(x_1, x_2, x_3 + \Delta x_3, \ldots, x_n + \Delta x_n) \right.\\
&\left. - f(x_1, x_2, x_3, x_4 + \Delta x_4, \ldots, x_n + \Delta x_n) \right]\\
&+ \cdots + \left[f(x_1, x_2, \ldots, x_{n-1}, x_n + \Delta x_n) - f(x_1, x_2, \ldots, x_n) \right].
\end{aligned}$$

By applying the mean value theorem to the difference in each bracket we obtain

$$\begin{aligned}
\Delta f = &\Delta x_1 \frac{\partial f(x_1 + \theta_1 \Delta x_1, x_2 + \Delta x_2, \ldots, x_n + \Delta x_n)}{\partial x_1}\\
&+ \Delta x_2 \frac{\partial f(x_1, x_2 + \theta_2 \Delta x_2, x_3 + \Delta x_3, \ldots, x_n + \Delta x_n)}{\partial x_2}\\
&+ \Delta x_3 \frac{\partial f(x_1, x_2, x_3 + \theta_3 \Delta x_3, x_4 + \Delta x_4, \ldots, x_n + \Delta x_n)}{\partial x_3}\\
&+ \cdots + \Delta x_n \frac{\partial f(x_1, x_2, \ldots, x_{n-1}, x_n + \theta_n \Delta x_n)}{\partial x_n},
\end{aligned}$$

where $0 < \theta_i < 1$ for $i = 1, 2, \ldots, n$. Hence,

$$\frac{\Delta f}{\Delta t} = \frac{\Delta x_1}{\Delta t} \frac{\partial f(x_1 + \theta_1 \Delta x_1, x_2 + \Delta x_2, \ldots, x_n + \Delta x_n)}{\partial x_1}$$

$$+ \frac{\Delta x_2}{\Delta t} \frac{\partial f(x_1, x_2 + \theta_2 \Delta x_2, x_3 + \Delta x_3, \ldots, x_n + \Delta x_n)}{\partial x_2}$$

$$+ \frac{\Delta x_3}{\Delta t} \frac{\partial f(x_1, x_2, x_3 + \theta_3 \Delta x_3, x_4 + \Delta x_4, \ldots, x_n + \Delta x_n)}{\partial x_3}$$

$$+ \cdots + \frac{\Delta x_n}{\Delta t} \frac{\partial f(x_1, x_2, \ldots, x_{n-1}, x_n + \theta_n \Delta x_n)}{\partial x_n}. \qquad (7.13)$$

As $\Delta t \to 0$, $\Delta x_i / \Delta t \to dx_i / dt$, and the partial derivatives in (7.13), being continuous, tend to $\partial f(\mathbf{x}) / \partial x_i$ for $i = 1, 2, \ldots, n$. Thus $\Delta f / \Delta t$ tends to the right-hand side of formula (7.12).

For example, consider the function $f(x_1, x_2) = x_1^2 - x_2^3$, where $x_1 = e^t \cos t$, $x_2 = \cos t + \sin t$. Then,

$$\frac{df}{dt} = 2x_1(e^t \cos t - e^t \sin t) - 3x_2^2(-\sin t + \cos t)$$

$$= 2e^t \cos t(e^t \cos t - e^t \sin t) - 3(\cos t + \sin t)^2(-\sin t + \cos t)$$

$$= (\cos t - \sin t)(2e^{2t} \cos t - 6 \sin t \cos t - 3).$$

Of course, the same result could have been obtained by expressing f directly as a function of t via x_1 and x_2 and then differentiating it with respect to t.

We can generalize formula (7.12) by assuming that each of $x_1, x_2, \ldots, x_n$ is a function of several variables including the variable t. In this case, we need to consider the partial derivative $\partial f / \partial t$, which can be similarly shown to have the value

$$\frac{\partial f}{\partial t} = \sum_{i=1}^{n} \frac{\partial f(\mathbf{x})}{\partial x_i} \frac{\partial x_i}{\partial t}. \qquad (7.14)$$

In general, the expression

$$df = \sum_{i=1}^{n} \frac{\partial f(\mathbf{x})}{\partial x_i} dx_i \qquad (7.15)$$

is called the total differential of f at $\mathbf{x}$.

EXAMPLE 7.4.2. Consider the equation $f(x_1, x_2) = 0$, which in general represents a relation between x_1 and x_2. It may or may not define x_2 as a function of x_1. In this case, x_2 is said to be an implicit function of x_1. If x_2 can be obtained as a function of x_1, then we write $x_2 = g(x_1)$. Consequently, $f[x_1, g(x_1)]$ will be identically equal to zero. Hence, $f[x_1, g(x_1)]$, being a function of one variable x_1, will have a total derivative identically equal to zero. By applying formula (7.12) with $t = x_1$ we obtain

$$\frac{df}{dx_1} = \frac{\partial f}{\partial x_1} + \frac{\partial f}{\partial x_2}\frac{dx_2}{dx_1} \equiv 0.$$

If $\partial f/\partial x_2 \neq 0$, then the derivative of x_2 is given by

$$\frac{dx_2}{dx_1} = \frac{-\partial f/\partial x_1}{\partial f/\partial x_2}. \tag{7.16}$$

In particular, if $f(x_1, x_2) = 0$ is of the form $x_1 - h(x_2) = 0$, and if this equation can be solved uniquely for x_2 in terms of x_1, then x_2 represents the inverse function of h, that is, $x_2 = h^{-1}(x_1)$. Thus according to formula (7.16),

$$\frac{dh^{-1}}{dx_1} = \frac{1}{dh/dx_2}.$$

This agrees with the formula for the derivative of the inverse function given in Theorem 4.2.4.

7.4.2. Directional Derivatives

Let $\mathbf{f}\colon D \to R^m$, where $D \subset R^n$, and let $\mathbf{v}$ be a unit vector in R^n (that is, a vector whose length is equal to one), which represents a certain direction in the n-dimensional Euclidean space. By definition, the directional derivative of $\mathbf{f}$ at a point $\mathbf{x}$ is the interior of D in the direction of $\mathbf{v}$ is given by the limit

$$\lim_{h \to 0} \frac{\mathbf{f}(\mathbf{x} + h\mathbf{v}) - \mathbf{f}(\mathbf{x})}{h},$$

if it exists. In particular, if $\mathbf{v} = \mathbf{e}_i$, the unit vector in the direction of the ith coordinate axis, then the directional derivative of $\mathbf{f}$ in the direction of $\mathbf{v}$ is just the partial derivative of $\mathbf{f}$ with respect to x_i $(i = 1, 2, \ldots, n)$.

Lemma 7.4.1. Let $\mathbf{f}\colon D \to R^m$, where $D \subset R^n$. If the partial derivatives $\partial f_j/\partial x_i$ exist at a point $\mathbf{x} = (x_1, x_2, \ldots, x_n)'$ in the interior of D for $i = 1, 2, \ldots, n$; $j = 1, 2, \ldots, m$, where f_j is the jth element of $\mathbf{f}$, then the directional derivative of $\mathbf{f}$ at $\mathbf{x}$ in the direction of a unit vector $\mathbf{v}$ exists and is equal to $\mathbf{J}_\mathbf{f}(\mathbf{x})\mathbf{v}$, where $\mathbf{J}_\mathbf{f}(\mathbf{x})$ is the $m \times n$ Jacobian of $\mathbf{f}$ at $\mathbf{x}$.

Proof. Let us first consider the directional derivative of f_j in the direction of **v**. To do so, we rotate the coordinate axes so that **v** coincides with the direction of the ξ_1-axis, where $\xi_1, \xi_2, \ldots, \xi_n$ are the resulting new coordinates. By the well-known relations for rotation of axes in analytic geometry of n dimensions we have

$$x_i = \xi_1 v_i + \sum_{l=2}^{n} \xi_l \lambda_{li}, \qquad i = 1, 2, \ldots, n, \tag{7.17}$$

where v_i is the ith element of **v** $(i = 1, 2, \ldots, n)$ and λ_{li} is the ith element of λ_l, the unit vector in the direction of the ξ_l-axis $(l = 2, 3, \ldots, n)$.

Now, the directional derivative of f_j in the direction of **v** can be obtained by first expressing f_j as a function of $\xi_1, \xi_2, \ldots, \xi_n$ using the relations (7.17) and then differentiating it with respect to ξ_1. By formula (7.14), this is equal to

$$\frac{\partial f_j}{\partial \xi_1} = \sum_{i=1}^{n} \frac{\partial f_j}{\partial x_i} \frac{\partial x_i}{\partial \xi_1}$$

$$= \sum_{i=1}^{n} \frac{\partial f_j}{\partial x_i} v_i, \qquad j = 1, 2, \ldots, m. \tag{7.18}$$

From formula (7.18) we conclude that the directional derivative of $\mathbf{f} = (f_1, f_2, \ldots, f_m)'$ in the direction of **v** is equal to $\mathbf{J_f}(\mathbf{x})\mathbf{v}$.   □

EXAMPLE 7.4.3.   Let **f**: $R^3 \to R^2$ be defined as

$$\mathbf{f}(x_1, x_2, x_3) = \begin{bmatrix} x_1^2 + x_2^2 + x_3^2 \\ x_1^2 - x_1 x_2 + x_3^2 \end{bmatrix}.$$

The directional derivative of **f** at $\mathbf{x} = (1, 2, 1)'$ in the direction of $\mathbf{v} = (1/\sqrt{2}, -1/\sqrt{2}, 0)'$ is

$$\mathbf{J_f}(\mathbf{x})\mathbf{v} = \begin{bmatrix} 2x_1 & 2x_2 & 2x_3 \\ 2x_1 - x_2 & -x_1 & 2x_3 \end{bmatrix}_{(1,2,1)} \begin{bmatrix} \dfrac{1}{\sqrt{2}} \\ -\dfrac{1}{\sqrt{2}} \\ 0 \end{bmatrix}$$

$$= \begin{bmatrix} 2 & 4 & 2 \\ 0 & -1 & 2 \end{bmatrix} \begin{bmatrix} \dfrac{1}{\sqrt{2}} \\ -\dfrac{1}{\sqrt{2}} \\ 0 \end{bmatrix} = \begin{bmatrix} \dfrac{-2}{\sqrt{2}} \\ \dfrac{1}{\sqrt{2}} \end{bmatrix}.$$

Definition 7.4.1. Let $f: D \to R$, where $D \subset R^n$. If the partial derivatives $\partial f / \partial x_i$ $(i = 1, 2, \ldots, n)$ exist at a point $\mathbf{x} = (x_1, x_2, \ldots, x_n)'$ in the interior of D, then the vector $(\partial f / \partial x_1, \partial f / \partial x_2, \ldots, \partial f / \partial x_n)'$ is called the gradient of f at $\mathbf{x}$ and is denoted by $\nabla f(\mathbf{x})$. $\square$

Using Definition 7.4.1, the directional derivative of f at $\mathbf{x}$ in the direction of a unit vector $\mathbf{v}$ can be expressed as $\nabla f(\mathbf{x})'\mathbf{v}$, where $\nabla f(\mathbf{x})'$ denotes the transpose of $\nabla f(\mathbf{x})$.

The Geometric Meaning of the Gradient

Let $f: D \to R$, where $D \subset R^n$. Suppose that the partial derivatives of f exist at a point $\mathbf{x} = (x_1, x_2, \ldots, x_n)'$ in the interior of D. Let C denote a smooth curve that lies on the surface of $f(\mathbf{x}) = c_0$, where c_0 is a constant, and passes through the point $\mathbf{x}$. This curve can be represented by the equations $x_1 = g_1(t), x_2 = g_2(t), \ldots, x_n = g_n(t)$, where $a \le t \le b$. By formula (7.12), the total derivative of f with respect to t at $\mathbf{x}$ is

$$\frac{df}{dt} = \sum_{i=1}^{n} \frac{\partial f(\mathbf{x})}{\partial x_i} \frac{dg_i}{dt}. \tag{7.19}$$

The vector $\boldsymbol{\lambda} = (dg_1/dt, dg_2/dt, \ldots, dg_n/dt)'$ is tangent to C at $\mathbf{x}$. Thus from (7.19) we obtain

$$\frac{df}{dt} = \nabla f(\mathbf{x})'\boldsymbol{\lambda}. \tag{7.20}$$

Now, since $f[g_1(t), g_2(t), \ldots, g_n(t)] = c_0$ along C, then $df/dt = 0$ and hence $\nabla f(\mathbf{x})'\boldsymbol{\lambda} = 0$. This indicates that the gradient vector is orthogonal to $\boldsymbol{\lambda}$, and hence to C, at $\mathbf{x} \in D$. Since this result is true for any smooth curve through $\mathbf{x}$, we conclude that the gradient vector $\nabla f(\mathbf{x})$ is orthogonal to the surface of $f(\mathbf{x}) = c_0$ at $\mathbf{x}$.

Definition 7.4.2. Let $f: D \to R$, where $D \subset R^n$. Then $\nabla f: D \to R^n$. The Jacobian matrix of $\nabla f(\mathbf{x})$ is called the Hessian matrix of f and is denoted by $\mathbf{H}_f(\mathbf{x})$. Thus $\mathbf{H}_f(\mathbf{x}) = \mathbf{J}_{\nabla f}(\mathbf{x})$, that is,

$$\mathbf{H}_f(\mathbf{x}) = \begin{bmatrix} \dfrac{\partial^2 f(\mathbf{x})}{\partial x_1^2} & \dfrac{\partial^2 f(\mathbf{x})}{\partial x_2 \, \partial x_1} & \cdots & \dfrac{\partial^2 f(\mathbf{x})}{\partial x_n \, \partial x_1} \\ \vdots & \vdots & & \vdots \\ \dfrac{\partial^2 f(\mathbf{x})}{\partial x_1 \, \partial x_n} & \dfrac{\partial^2 f(\mathbf{x})}{\partial x_2 \, \partial x_n} & \cdots & \dfrac{\partial^2 f(\mathbf{x})}{\partial x_n^2} \end{bmatrix}. \tag{7.21}$$

The determinant of $\mathbf{H}_f(\mathbf{x})$ is called the *Hessian determinant*. If the conditions of Theorem 7.4.1 regarding the commutative property of partial differentiation are valid, then $\mathbf{H}_f(\mathbf{x})$ is a symmetric matrix. As we shall see in Section 7.7, the Hessian matrix plays an important role in the identification of maxima and minima of a multivariable function.   □

7.4.3. Differentiation of Composite Functions

Let $\mathbf{f}$: $D_1 \to R^m$, where $D_1 \subset R^n$, and let $\mathbf{g}$: $D_2 \to R^p$, where $D_2 \subset R^m$. Let $\mathbf{x}_0$ be an interior point of D_1 and $\mathbf{f}(\mathbf{x}_0)$ be an interior point of D_2. If the $m \times n$ Jacobian matrix $\mathbf{J}_f(\mathbf{x}_0)$ and the $p \times m$ Jacobian matrix $\mathbf{J}_g[\mathbf{f}(\mathbf{x}_0)]$ both exist, then the $p \times n$ Jacobian matrix $\mathbf{J}_h(\mathbf{x}_0)$ for the composite function $\mathbf{h} = \mathbf{g} \circ \mathbf{f}$ exists and is given by

$$\mathbf{J}_h(\mathbf{x}_0) = \mathbf{J}_g[\mathbf{f}(\mathbf{x}_0)]\mathbf{J}_f(\mathbf{x}_0). \tag{7.22}$$

To prove formula (7.22), let us consider the (k, i)th element of $\mathbf{J}_h(\mathbf{x}_0)$, namely $\partial h_k(\mathbf{x}_0)/\partial x_i$, where $h_k(\mathbf{x}_0) = g_k[\mathbf{f}(\mathbf{x}_0)]$ is the kth element of $\mathbf{h}(\mathbf{x}_0) = \mathbf{g}[\mathbf{f}(\mathbf{x}_0)]$, $i = 1, 2, \ldots, n$; $k = 1, 2, \ldots, p$, By applying formula (7.14) we obtain

$$\frac{\partial h_k(\mathbf{x}_0)}{\partial x_i} = \sum_{j=1}^{m} \frac{\partial g_k[\mathbf{f}(\mathbf{x}_0)]}{\partial f_j} \frac{\partial f_j(\mathbf{x}_0)}{\partial x_i}, \qquad i = 1, 2, \ldots, n; k = 1, 2, \ldots, p,$$

$$\tag{7.23}$$

where $f_j(\mathbf{x}_0)$ is the jth element of $\mathbf{f}(\mathbf{x}_0)$, $j = 1, 2, \ldots, m$. But $\partial g_k[\mathbf{f}(\mathbf{x}_0)]/\partial f_j$ is the (k, j)th element of $\mathbf{J}_g[\mathbf{f}(\mathbf{x}_0)]$, and $\partial f_j(\mathbf{x}_0)/\partial x_i$ is the (j, i)th element of $\mathbf{J}_f(\mathbf{x}_0)$, $i = 1, 2, \ldots, n$; $j = 1, 2, \ldots, m$; $k = 1, 2, \ldots, p$. Hence, formula (7.22) follows from formula (7.23) and the rule of matrix multiplication.

In particular, if $m = n = p$, then from formula (7.22), the determinant of $\mathbf{J}_h(\mathbf{x}_0)$ is given by

$$\det[\mathbf{J}_h(\mathbf{x}_0)] = \det[\mathbf{J}_g(\mathbf{f}(\mathbf{x}_0))]\det[\mathbf{J}_f(\mathbf{x}_0)]. \tag{7.24}$$

Using the notation in formula (7.8), formula (7.24) can be expressed as

$$\frac{\partial(h_1, h_2, \ldots, h_n)}{\partial(x_1, x_2, \ldots, x_n)} = \frac{\partial(g_1, g_2, \ldots, g_n)}{\partial(f_1, f_2, \ldots, f_n)} \frac{\partial(f_1, f_2, \ldots, f_n)}{\partial(x_1, x_2, \ldots, x_n)}. \tag{7.25}$$

EXAMPLE 7.4.4.   Let $\mathbf{f}$: $R^2 \to R^3$ be given by

$$\mathbf{f}(x_1, x_2) = \begin{bmatrix} x_1^2 - x_2 \cos x_1 \\ x_1 x_2 \\ x_1^3 + x_2^3 \end{bmatrix}.$$

Let $g: R^3 \to R$ be defined as

$$g(\xi_1, \xi_2, \xi_3) = \xi_1 - \xi_2^2 + \xi_3,$$

where

$$\xi_1 = x_1^2 - x_2 \cos x_1,$$
$$\xi_2 = x_1 x_2,$$
$$\xi_3 = x_1^3 + x_2^3.$$

In this case,

$$\mathbf{J_f}(\mathbf{x}) = \begin{bmatrix} 2x_1 + x_2 \sin x_1 & -\cos x_1 \\ x_2 & x_1 \\ 3x_1^2 & 3x_2^2 \end{bmatrix},$$

$$\mathbf{J_g}[\mathbf{f}(\mathbf{x})] = (1, -2\xi_2, 1).$$

Hence, by formula (7.22),

$$\mathbf{J_h}(\mathbf{x}) = (1, -2\xi_2, 1) \begin{bmatrix} 2x_1 + x_2 \sin x_1 & -\cos x_1 \\ x_2 & x_1 \\ 3x_1^2 & 3x_2^2 \end{bmatrix}$$

$$= (2x_1 + x_2 \sin x_1 - 2x_1 x_2^2 + 3x_1^2, -\cos x_1 - 2x_1^2 x_2 + 3x_2^2).$$

7.5. TAYLOR'S THEOREM FOR A MULTIVARIABLE FUNCTION

We shall now consider a multidimensional analogue of Taylor's theorem, which was discussed in Section 4.3 for a single-variable function.

Let us first introduce the following notation: Let $\mathbf{x} = (x_1, x_2, \ldots, x_n)'$. Then $\mathbf{x}'\nabla$ denotes a first-order differential operator of the form

$$\mathbf{x}'\nabla = \sum_{i=1}^{n} x_i \frac{\partial}{\partial x_i}.$$

The symbol ∇, called the del operator, was used earlier to define the gradient vector. If m is a positive integer, then $(\mathbf{x}'\nabla)^m$ denotes an mth-order differential operator. For example, for $m = n = 2$,

$$(\mathbf{x}'\nabla)^2 = \left(x_1 \frac{\partial}{\partial x_1} + x_2 \frac{\partial}{\partial x_2} \right)^2$$

$$= x_1^2 \frac{\partial^2}{\partial x_1^2} + 2x_1 x_2 \frac{\partial^2}{\partial x_1 \, \partial x_2} + x_2^2 \frac{\partial^2}{\partial x_2^2}.$$

Thus $(\mathbf{x}'\boldsymbol{\nabla})^2$ is obtained by squaring $x_1\,\partial/\partial x_1 + x_2\,\partial/\partial x_2$ in the usual fashion, except that the squares of $\partial/\partial x_1$ and $\partial/\partial x_2$ are replaced by $\partial^2/\partial x_1^2$ and $\partial^2/\partial x_2^2$, respectively, and the product of $\partial/\partial x_1$ and $\partial/\partial x_2$ is replaced by $\partial^2/\partial x_1\,\partial x_2$ (here we are assuming that the commutative property of partial differentiation holds once these differential operators are applied to a real-valued function). In general, $(\mathbf{x}'\boldsymbol{\nabla})^m$ is obtained by a multinomial expansion of degree m of the form

$$
(\mathbf{x}'\boldsymbol{\nabla})^m = \sum_{k_1,k_2,\dots,k_n} \binom{m}{k_1,k_2,\dots,k_n} x_1^{k_1} x_2^{k_2} \cdots x_n^{k_n} \frac{\partial^m}{\partial x_1^{k_1}\,\partial x_2^{k_2}\,\cdots\,\partial x_n^{k_n}},
$$

where the sum is taken over all n-tuples $(k_1,k_2,\dots,k_n)$ for which $\sum_{i=1}^n k_i = m$, and

$$
\binom{m}{k_1,k_2,\dots,k_n} = \frac{m!}{k_1!\,k_2!\cdots k_n!}.
$$

If a real-valued function $f(\mathbf{x})$ has partial derivatives through order m, then an application of the differential operator $(\mathbf{x}'\boldsymbol{\nabla})^m$ to $f(\mathbf{x})$ results in

$$
(\mathbf{x}'\boldsymbol{\nabla})^m f(\mathbf{x}) = \sum_{k_1,k_2,\dots,k_n} \binom{m}{k_1,k_2,\dots,k_n} x_1^{k_1} x_2^{k_2} \cdots
$$

$$
\times x_n^{k_n} \frac{\partial^m f(\mathbf{x})}{\partial x_1^{k_1}\,\partial x_2^{k_2}\,\cdots\,\partial x_n^{k_n}}. \tag{7.26}
$$

The notation $(\mathbf{x}'\boldsymbol{\nabla})^m f(\mathbf{x}_0)$ indicates that $(\mathbf{x}'\boldsymbol{\nabla})^m f(\mathbf{x})$ is evaluated at $\mathbf{x}_0$.

Theorem 7.5.1. Let $f: D \to R$, where $D \subset R^n$, and let $N_\delta(\mathbf{x}_0)$ be a neighborhood of $\mathbf{x}_0 \in D$ such that $N_\delta(\mathbf{x}_0) \subset D$. If f and all its partial derivatives of order $\leq r$ exist and are continuous in $N_\delta(\mathbf{x}_0)$, then for any $\mathbf{x} \in N_\delta(\mathbf{x}_0)$,

$$
f(\mathbf{x}) = f(\mathbf{x}_0) + \sum_{i=1}^{r-1} \frac{\big[(\mathbf{x}-\mathbf{x}_0)'\boldsymbol{\nabla}\big]^i f(\mathbf{x}_0)}{i!} + \frac{\big[(\mathbf{x}-\mathbf{x}_0)'\boldsymbol{\nabla}\big]^r f(\mathbf{z}_0)}{r!}, \tag{7.27}
$$

where $\mathbf{z}_0$ is a point on the line segment from $\mathbf{x}_0$ to $\mathbf{x}$.

Proof. Let $\mathbf{h} = \mathbf{x} - \mathbf{x}_0$. Let the function $\phi(t)$ be defined as $\phi(t) = f(\mathbf{x}_0 + t\mathbf{h})$, where $0 \leq t \leq 1$. If $t = 0$, then $\phi(0) = f(\mathbf{x}_0)$ and $\phi(1) = f(\mathbf{x}_0 + \mathbf{h}) = f(\mathbf{x})$, if

$t = 1$. Now, by formula (7.12),

$$\frac{d\phi(t)}{dt} = \sum_{i=1}^{n} h_i \frac{\partial f(\mathbf{x})}{\partial x_i} \bigg|_{\mathbf{x} = \mathbf{x}_0 + t\mathbf{h}}$$

$$= (\mathbf{h}'\nabla) f(\mathbf{x}_0 + t\mathbf{h}),$$

where h_i is the ith element of $\mathbf{h}$ ($i = 1, 2, \ldots, n$). Furthermore, the derivative of order m of $\phi(t)$ is

$$\frac{d^m\phi(t)}{dt^m} = (\mathbf{h}'\nabla)^m f(\mathbf{x}_0 + t\mathbf{h}), \qquad 1 \le m \le r.$$

Since the partial derivatives of f through order r are continuous, then the same order derivatives of $\phi(t)$ are also continuous on $[0, 1]$ and

$$\frac{d^m\phi(t)}{dt^m} \bigg|_{t=0} = (\mathbf{h}'\nabla)^m f(\mathbf{x}_0), \qquad 1 \le m \le r.$$

If we now apply Taylor's theorem in Section 4.3 to the single-variable function $\phi(t)$, we obtain

$$\phi(t) = \phi(0) + \sum_{i=1}^{r-1} \frac{t^i}{i!} \frac{d^i\phi(t)}{dt^i} \bigg|_{t=0} + \frac{t^r}{r!} \frac{d^r\phi(t)}{dt^r} \bigg|_{t=\xi}, \qquad (7.28)$$

where $0 < \xi < t$. By setting $t = 1$ in formula (7.28), we obtain

$$f(\mathbf{x}) = f(\mathbf{x}_0) + \sum_{i=1}^{r-1} \frac{[(\mathbf{x} - \mathbf{x}_0)'\nabla]^i f(\mathbf{x}_0)}{i!} + \frac{[(\mathbf{x} - \mathbf{x}_0)'\nabla]^r f(\mathbf{z}_0)}{r!},$$

where $\mathbf{z}_0 = \mathbf{x}_0 + \xi\mathbf{h}$. Since $0 < \xi < 1$, the point $\mathbf{z}_0$ lies on the line segment between $\mathbf{x}_0$ and $\mathbf{x}$. $\quad\square$

In particular, if $f(\mathbf{x})$ has partial derivatives of all orders in $N_\delta(\mathbf{x}_0)$, then we have the series expansion

$$f(\mathbf{x}) = f(\mathbf{x}_0) + \sum_{i=1}^{\infty} \frac{[(\mathbf{x} - \mathbf{x}_0)'\nabla]^i f(\mathbf{x}_0)}{i!}. \qquad (7.29)$$

In this case, the last term in formula (7.27) serves as a remainder of Taylor's series.

EXAMPLE 7.5.1. Consider the function $f: R^2 \to R$ defined as $f(x_1, x_2) = x_1 x_2 + x_1^2 + e^{x_1} \cos x_2$. This function has partial derivatives of all orders. Thus in a neighborhood of $\mathbf{x}_0 = (0, 0)'$ we can write

$$f(x_1, x_2) = 1 + (\mathbf{x}'\nabla)f(0,0) + \frac{1}{2!}(\mathbf{x}'\nabla)^2 f(0,0) + \frac{1}{3!}(\mathbf{x}'\nabla)^3 f(\xi x_1, \xi x_2),$$

$$0 < \xi < 1.$$

It can be verified that

$$(\mathbf{x}'\nabla)f(0,0) = x_1,$$

$$(\mathbf{x}'\nabla)^2 f(0,0) = 3x_1^2 + 2x_1 x_2 - x_2^2,$$

$$(\mathbf{x}'\nabla)^3 f(\xi x_1, \xi x_2) = x_1^3 e^{\xi x_1} \cos(\xi x_2) - 3x_1^2 x_2 e^{\xi x_1} \sin(\xi x_2)$$

$$- 3x_1 x_2^2 e^{\xi x_1} \cos(\xi x_2) + x_2^3 e^{\xi x_1} \sin(\xi x_2).$$

Hence,

$$f(x_1, x_2)$$

$$= 1 + x_1 + \frac{1}{2!}(3x_1^2 + 2x_1 x_2 - x_2^2)$$

$$+ \frac{1}{3!}\{[x_1^3 - 3x_1 x_2^2] e^{\xi x_1} \cos(\xi x_2) + [x_2^3 - 3x_1^2 x_2] e^{\xi x_1} \sin(\xi x_2)\}.$$

The first three terms serve as a second-order approximation of $f(x_1, x_2)$, while the last term serves as a remainder.

7.6. INVERSE AND IMPLICIT FUNCTION THEOREMS

Consider the function $\mathbf{f}: D \to R^n$, where $D \subset R^n$. Let $\mathbf{y} = \mathbf{f}(\mathbf{x})$. The purpose of this section is to present conditions for the existence of an inverse function $\mathbf{f}^{-1}$ which expresses $\mathbf{x}$ as a function of $\mathbf{y}$. These conditions are given in the next theorem, whose proof can be found in Sagan (1974, page 371). See also Fulks (1978, page 346).

Theorem 7.6.1 (Inverse Function Theorem). Let $\mathbf{f}: D \to R^n$, where D is an open subset of R^n and $\mathbf{f}$ has continuous first-order partial derivatives in D. If for some $\mathbf{x}_0 \in D$, the $n \times n$ Jacobian matrix $\mathbf{J}_{\mathbf{f}}(\mathbf{x}_0)$ is nonsingular, that is,

$$\det[\mathbf{J_f}(\mathbf{x}_0)] = \frac{\partial(f_1, f_2, \ldots, f_n)}{\partial(x_1, x_2, \ldots, x_n)}\bigg|_{\mathbf{x}=\mathbf{x}_0} \neq 0,$$

where f_i is the ith element of $\mathbf{f}$ $(i = 1, 2, \ldots, n)$, then there exist an $\epsilon > 0$ and a $\delta > 0$ such that an inverse function $\mathbf{f}^{-1}$ exists in the neighborhood $N_\delta[\mathbf{f}(\mathbf{x}_0)]$ and takes values in the neighborhood $N_\epsilon(\mathbf{x}_0)$. Moreover, $\mathbf{f}^{-1}$ has continuous first-order partial derivatives in $N_\delta[\mathbf{f}(\mathbf{x}_0)]$, and its Jacobian matrix at $\mathbf{f}(\mathbf{x}_0)$ is the inverse of $\mathbf{J_f}(\mathbf{x}_0)$; hence,

$$\det\{\mathbf{J}_{\mathbf{f}^{-1}}[\mathbf{f}(\mathbf{x}_0)]\} = \frac{1}{\det[\mathbf{J_f}(\mathbf{x}_0)]}. \tag{7.30}$$

EXAMPLE 7.6.1.   Let $\mathbf{f}: R^3 \to R^3$ be given by

$$\mathbf{f}(x_1, x_2, x_3) = \begin{bmatrix} 2x_1 x_2 - x_2 \\ x_1^2 + x_2 + 2x_3^2 \\ x_1 x_2 + x_2 \end{bmatrix}.$$

Here,

$$\mathbf{J_f}(\mathbf{x}) = \begin{bmatrix} 2x_2 & 2x_1 - 1 & 0 \\ 2x_1 & 1 & 4x_3 \\ x_2 & x_1 + 1 & 0 \end{bmatrix},$$

and $\det[\mathbf{J_f}(\mathbf{x})] = -12x_2 x_3$. Hence, all $\mathbf{x} \in R^3$ at which $x_2 x_3 \neq 0$, $\mathbf{f}$ has an inverse function $\mathbf{f}^{-1}$. For example, if $D = \{(x_1, x_2, x_3) | x_2 > 0, x_3 > 0\}$, then $\mathbf{f}$ is invertible in D. From the equations

$$y_1 = 2x_1 x_2 - x_2,$$
$$y_2 = x_1^2 + x_2 + 2x_3^2,$$
$$y_3 = x_1 x_2 + x_2,$$

we obtain the inverse function $\mathbf{x} = \mathbf{f}^{-1}(\mathbf{y})$, where

$$x_1 = \frac{y_1 + y_3}{2y_3 - y_1},$$

$$x_2 = \frac{-y_1 + 2y_3}{3},$$

$$x_3 = \frac{1}{\sqrt{2}} \left[y_2 - \frac{(y_1 + y_3)^2}{(2y_3 - y_1)^2} - \frac{2y_3 - y_1}{3} \right]^{1/2}.$$

If, for example, we consider $\mathbf{x}_0 = (1, 1, 1)'$, then $\mathbf{y}_0 = \mathbf{f}(\mathbf{x}_0) = (1, 4, 2)'$, and $\det[\mathbf{J}_\mathbf{f}(\mathbf{x}_0)] = -12$. The Jacobian matrix of $\mathbf{f}^{-1}$ at $\mathbf{y}_0$ is

$$
\mathbf{J}_{\mathbf{f}^{-1}}(\mathbf{y}_0) = \begin{bmatrix} \frac{2}{3} & 0 & -\frac{1}{3} \\ -\frac{1}{3} & 0 & \frac{2}{3} \\ -\frac{1}{4} & \frac{1}{4} & 0 \end{bmatrix}.
$$

Its determinant is equal to

$$
\det\left[\mathbf{J}_{\mathbf{f}^{-1}}(\mathbf{y}_0)\right] = -\frac{1}{12}.
$$

We note that this is the reciprocal of $\det[\mathbf{J}_\mathbf{f}(\mathbf{x}_0)]$, as it should be according to formula (7.30).

The inverse function theorem can be viewed as providing a unique solution to a system of n equations given by $\mathbf{y} = \mathbf{f}(\mathbf{x})$. There are, however, situations in which $\mathbf{y}$ is not explicitly expressed as a function of $\mathbf{x}$. In general, we may have two vectors, $\mathbf{x}$ and $\mathbf{y}$, of orders $n \times 1$ and $m \times 1$, respectively, that satisfy the relation

$$
\mathbf{g}(\mathbf{x}, \mathbf{y}) = \mathbf{0}, \tag{7.31}
$$

where $\mathbf{g}: R^{m+n} \to R^n$. In this more general case, we have n equations involving $m + n$ variables, namely, the elements of $\mathbf{x}$ and those of $\mathbf{y}$. The question now is what conditions will allow us to solve equations (7.31) uniquely for $\mathbf{x}$ in terms of $\mathbf{y}$. The answer to this question is given in the next theorem, whose proof can be found in Fulks (1978, page 352).

Theorem 7.6.2 (Implicit Function Theorem). Let $\mathbf{g}: D \to R^n$, where D is an open subset of R^{m+n}, and $\mathbf{g}$ has continuous first-order partial derivatives in D. If there is a point $\mathbf{z}_0 \in D$, where $\mathbf{z}_0 = (\mathbf{x}_0', \mathbf{y}_0')'$ with $\mathbf{x}_0 \in R^n$, $\mathbf{y}_0 \in R^m$ such that $\mathbf{g}(\mathbf{z}_0) = \mathbf{0}$, and if at $\mathbf{z}_0$,

$$
\frac{\partial(g_1, g_2, \ldots, g_n)}{\partial(x_1, x_2, \ldots, x_n)} \neq 0,
$$

where g_i is the ith element of $\mathbf{g}$ $(i = 1, 2, \ldots, n)$, then there is a neighborhood $N_\delta(\mathbf{y}_0)$ of $\mathbf{y}_0$ in which the equation $\mathbf{g}(\mathbf{x}, \mathbf{y}) = \mathbf{0}$ can be solved uniquely for $\mathbf{x}$ as a continuously differentiable function of $\mathbf{y}$.

EXAMPLE 7.6.2. Let $\mathbf{g}: R^3 \to R^2$ be given by

$$\mathbf{g}(x_1, x_2, y) = \begin{bmatrix} x_1 + x_2 + y^2 - 18 \\ x_1 - x_1 x_2 + y - 4 \end{bmatrix}.$$

We have

$$\frac{\partial(g_1, g_2)}{\partial(x_1, x_2)} = \det\left(\begin{bmatrix} 1 & 1 \\ 1 - x_2 & -x_1 \end{bmatrix}\right) = x_2 - x_1 - 1.$$

Let $\mathbf{z} = (x_1, x_2, y)'$. At the point $\mathbf{z}_0 = (1, 1, 4)'$, for example, $\mathbf{g}(\mathbf{z}_0) = \mathbf{0}$ and $\partial(g_1, g_2)/\partial(x_1, x_2) = -1 \neq 0$. Hence, by Theorem 7.6.2, we can solve the equations

$$x_1 + x_2 + y^2 - 18 = 0, \tag{7.32}$$

$$x_1 - x_1 x_2 + y - 4 = 0 \tag{7.33}$$

uniquely in terms of y in some neighborhood of $y_0 = 4$. For example, if D in Theorem 7.6.2 is of the form

$$D = \{(x_1, x_2, y) | x_1 > 0, x_2 > 0, y < 4.06\},$$

then from equations (7.32) and (7.33) we obtain the solution

$$x_1 = \tfrac{1}{2}\left\{-(y^2 - 17) + \left[(y^2 - 17)^2 - 4y + 16\right]^{1/2}\right\},$$

$$x_2 = \tfrac{1}{2}\left\{19 - y^2 - \left[(y^2 - 17)^2 - 4y + 16\right]^{1/2}\right\}.$$

We note that the sign preceding the square root in the formula for x_1 was chosen as $+$, so that $x_1 = x_2 = 1$ when $y = 4$. It can be verified that $(y^2 - 17)^2 - 4y + 16$ is positive for $y < 4.06$.

7.7. OPTIMA OF A MULTIVARIABLE FUNCTION

Let $f(\mathbf{x})$ be a real-valued function defined on a set $D \subset R^n$. A point $\mathbf{x}_0 \in D$ is said to be a point of local maximum of f if there exists a neighborhood $N_\delta(\mathbf{x}_0) \subset D$ such that $f(\mathbf{x}) \leq f(\mathbf{x}_0)$ for all $\mathbf{x} \in N_\delta(\mathbf{x}_0)$. If $f(\mathbf{x}) \geq f(\mathbf{x}_0)$ for all $\mathbf{x} \in N_\delta(\mathbf{x}_0)$, then $\mathbf{x}_0$ is a point of local minimum. If one of these inequalities holds for all $\mathbf{x}$ in D, then $\mathbf{x}_0$ is called a point of absolute maximum, or a point of absolute minimum, respectively, of f in D. In either case, $\mathbf{x}_0$ is referred to as a point of optimum (or extremum), and the value of $f(\mathbf{x})$ at $\mathbf{x} = \mathbf{x}_0$ is called an optimum value of $f(\mathbf{x})$.

In this section we shall discuss conditions under which $f(\mathbf{x})$ attains local optima in D. Then, we shall investigate the determination of the optima of $f(\mathbf{x})$ over a constrained region of D.

As in the case of a single-variable function, if $f(\mathbf{x})$ has first-order partial derivatives at a point $\mathbf{x}_0$ in the interior of D, and if $\mathbf{x}_0$ is a point of local optimum, then $\partial f/\partial x_i = 0$ for $i = 1, 2, \ldots, n$ at $\mathbf{x}_0$. The proof of this fact is similar to that of Theorem 4.4.1. Thus the vanishing of the first-order partial derivatives of $f(\mathbf{x})$ at $\mathbf{x}_0$ is a necessary condition for a local optimum at $\mathbf{x}_0$, but is obviously not sufficient. The first-order partial derivatives can be zero without necessarily having a local optimum at $\mathbf{x}_0$.

In general, any point at which $\partial f/\partial x_i = 0$ for $i = 1, 2, \ldots, n$ is called a stationary point. It follows that any point of local optimum at which f has first-order partial derivatives is a stationary point, but not every stationary point is a point of local optimum. If no local optimum is attained at a stationary point $\mathbf{x}_0$, then $\mathbf{x}_0$ is called a saddle point. The following theorem gives the conditions needed to have a local optimum at a stationary point.

Theorem 7.7.1. Let $f: D \to R$, where $D \subset R^n$. Suppose that f has continuous second-order partial derivatives in D. If $\mathbf{x}_0$ is a stationary point of f, then at $\mathbf{x}_0$ f has the following:

i. A local minimum if $(\mathbf{h}'\nabla)^2 f(\mathbf{x}_0) > 0$ for all $\mathbf{h} = (h_1, h_2, \ldots, h_n)'$ in a neighborhood of $\mathbf{0}$, where the elements of $\mathbf{h}$ are not all equal to zero.

ii. A local maximum if $(\mathbf{h}'\nabla)^2 f(\mathbf{x}_0) < 0$, where $\mathbf{h}$ is the same as in (i).

iii. A saddle point if $(\mathbf{h}'\nabla)^2 f(\mathbf{x}_0)$ changes sign for values of $\mathbf{h}$ in a neighborhood of $\mathbf{0}$.

Proof. By applying Taylor's theorem to $f(\mathbf{x})$ in a neighborhood of $\mathbf{x}_0$ we obtain

$$f(\mathbf{x}_0 + \mathbf{h}) = f(\mathbf{x}_0) + (\mathbf{h}'\nabla)f(\mathbf{x}_0) + \frac{1}{2!}(\mathbf{h}'\nabla)^2 f(\mathbf{z}_0),$$

where $\mathbf{h}$ is a nonzero vector in a neighborhood of $\mathbf{0}$ and $\mathbf{z}_0$ is a point on the line segment from $\mathbf{x}_0$ to $\mathbf{x}_0 + \mathbf{h}$. Since $\mathbf{x}_0$ is a stationary point, then $(\mathbf{h}'\nabla)f(\mathbf{x}_0) = 0$. Hence,

$$f(\mathbf{x}_0 + \mathbf{h}) - f(\mathbf{x}_0) = \frac{1}{2!}(\mathbf{h}'\nabla)^2 f(\mathbf{z}_0).$$

Also, since the second-order partial derivatives of f are continuous at $\mathbf{x}_0$, then we can write

$$f(\mathbf{x}_0 + \mathbf{h}) - f(\mathbf{x}_0) = \frac{1}{2!}(\mathbf{h}'\nabla)^2 f(\mathbf{x}_0) + o(\|\mathbf{h}\|),$$

where $\|\mathbf{h}\| = (\mathbf{h}'\mathbf{h})^{1/2}$ and $o(\|\mathbf{h}\|) \to 0$ as $\mathbf{h} \to \mathbf{0}$. We note that for small values of $\|\mathbf{h}\|$, the sign of $f(\mathbf{x}_0 + \mathbf{h}) - f(\mathbf{x}_0)$ depends on the value of $(\mathbf{h}'\nabla)^2 f(\mathbf{x}_0)$. It follows that if

i. $(\mathbf{h}'\nabla)^2 f(\mathbf{x}_0) > 0$, then $f(\mathbf{x}_0 + \mathbf{h}) > f(\mathbf{x}_0)$ for all nonzero values of $\mathbf{h}$ in some neighborhood of $\mathbf{0}$. Thus $\mathbf{x}_0$ is a point of local minimum of f.

ii. $(\mathbf{h}'\nabla)^2 f(\mathbf{x}_0) < 0$, then $f(\mathbf{x}_0 + \mathbf{h}) < f(\mathbf{x}_0)$ for all nonzero values of $\mathbf{h}$ in some neighborhood of $\mathbf{0}$. In this case, $\mathbf{x}_0$ is a point of local maximum of f.

iii. $(\mathbf{h}'\nabla)^2 f(\mathbf{x}_0)$ changes sign inside a neighborhood of $\mathbf{0}$, then $\mathbf{x}_0$ is neither a point of local maximum nor a point of local minimum. Therefore, $\mathbf{x}_0$ must be a saddle point. $\square$

We note that $(\mathbf{h}'\nabla)^2 f(\mathbf{x}_0)$ can be written as a quadratic form of the form $\mathbf{h}'\mathbf{A}\mathbf{h}$, where $\mathbf{A} = \mathbf{H}_f(\mathbf{x}_0)$ is the $n \times n$ Hessian matrix of f evaluated at $\mathbf{x}_0$, that is,

$$
\mathbf{A} = \begin{bmatrix} f_{11} & f_{12} & \cdots & f_{1n} \\ f_{21} & f_{22} & \cdots & f_{2n} \\ \vdots & \vdots & & \vdots \\ f_{n1} & f_{n2} & \cdots & f_{nn} \end{bmatrix}, \tag{7.34}
$$

where for simplicity we have denoted $\partial^2 f(\mathbf{x}_0)/\partial x_i \, \partial x_j$ by f_{ij}, $i, j = 1, 2, \ldots, n$ [see formula (7.21)].

Corollary 7.7.1.   Let f be the same function as in Theorem 7.7.1, and let $\mathbf{A}$ be the matrix given by formula (7.34). If $\mathbf{x}_0$ is a stationary point of f, then at $\mathbf{x}_0$ f has the following:

i. A local minimum if $\mathbf{A}$ is positive definite, that is, the leading principal minors of $\mathbf{A}$ (see Definition 2.3.6) are all positive,

$$
f_{11} > 0, \quad \det\left(\begin{bmatrix} f_{11} & f_{12} \\ f_{21} & f_{22} \end{bmatrix}\right) > 0, \ldots, \qquad \det(\mathbf{A}) > 0. \tag{7.35}
$$

ii. A local maximum if $\mathbf{A}$ is negative definite, that is, the leading principal minors of $\mathbf{A}$ have alternating signs as follows:

$$
f_{11} < 0, \quad \det\left(\begin{bmatrix} f_{11} & f_{12} \\ f_{21} & f_{22} \end{bmatrix}\right) > 0, \ldots, (-1)^n \det(\mathbf{A}) > 0. \tag{7.36}
$$

iii. A saddle point if $\mathbf{A}$ is neither positive definite nor negative definite.

Proof.

i. By Theorem 7.7.1, f has a local minimum at $\mathbf{x}_0$ if $(\mathbf{h}'\nabla)^2 f(\mathbf{x}_0) = \mathbf{h}'\mathbf{A}\mathbf{h}$ is positive for all $\mathbf{h} \neq \mathbf{0}$, that is, if $\mathbf{A}$ is positive definite. By Theorem 2.3.12(2), $\mathbf{A}$ is positive definite if and only if its leading principal minors are all positive. The conditions stated in (7.35) are therefore sufficient for a local minimum at $\mathbf{x}_0$.

ii. $(\mathbf{h}'\nabla)^2 f(\mathbf{x}_0) < 0$ if and only if $\mathbf{A}$ is negative definite, or $-\mathbf{A}$ is positive definite. Now, a leading principal minor of order m $(= 1, 2, \ldots, n)$ of $-\mathbf{A}$ is equal to $(-1)^m$ multiplied by the corresponding leading principal minor of $\mathbf{A}$. This leads to conditions (7.36).

iii. If $\mathbf{A}$ is neither positive definite nor negative definite, then $(\mathbf{h}'\nabla)^2 f(\mathbf{x}_0)$ must change sign inside a neighborhood of $\mathbf{x}_0$. This makes $\mathbf{x}_0$ a saddle point.   □

A Special Case

If f is a function of only $n = 2$ variables, x_1 and x_2, then conditions (7.35) and (7.36) can be written as:

i. $f_{11} > 0, f_{11} f_{22} - f_{12}^2 > 0$ for a local minimum at $\mathbf{x}_0$.

ii. $f_{11} < 0, f_{11} f_{22} - f_{12}^2 > 0$ for a local maximum at $\mathbf{x}_0$.

If $f_{11} f_{22} - f_{12}^2 < 0$, then $\mathbf{x}_0$ is a saddle point, since in this case

$$\mathbf{h}'\mathbf{A}\mathbf{h} = h_1^2 \frac{\partial^2 f(\mathbf{x}_0)}{\partial x_1^2} + 2h_1 h_2 \frac{\partial^2 f(\mathbf{x}_0)}{\partial x_1 \partial x_2} + h_2^2 \frac{\partial^2 f(\mathbf{x}_0)}{\partial x_2^2}$$

$$= \frac{\partial^2 f(\mathbf{x}_0)}{\partial x_1^2} (h_1 - ah_2)(h_1 - bh_2),$$

where ah_2 and bh_2 are the real roots of the equation $\mathbf{h}'\mathbf{A}\mathbf{h} = 0$ with respect to h_1. Hence, $\mathbf{h}'\mathbf{A}\mathbf{h}$ changes sign in a neighborhood of $\mathbf{0}$.

If $f_{11} f_{22} - f_{12}^2 = 0$, then $\mathbf{h}'\mathbf{A}\mathbf{h}$ can be written as

$$\mathbf{h}'\mathbf{A}\mathbf{h} = \frac{\partial^2 f(\mathbf{x}_0)}{\partial x_1^2} \left[h_1 + h_2 \frac{\partial^2 f(\mathbf{x}_0)/\partial x_1 \partial x_2}{\partial^2 f(\mathbf{x}_0)/\partial x_1^2} \right]^2$$

provided that $\partial^2 f(\mathbf{x}_0)/\partial x_1^2 \neq 0$. Thus $\mathbf{h}'\mathbf{A}\mathbf{h}$ has the same sign as that of $\partial^2 f(\mathbf{x}_0)/\partial x_1^2$ except for those values of $\mathbf{h} = (h_1, h_2)'$ for which

$$h_1 + h_2 \frac{\partial^2 f(\mathbf{x}_0)/\partial x_1 \partial x_2}{\partial^2 f(\mathbf{x}_0)/\partial x_1^2} = 0,$$

in which case it is zero. In the event $\partial^2 f(\mathbf{x}_0)/\partial x_1^2 = 0$, then $\partial^2 f(\mathbf{x}_0)/\partial x_1 \partial x_2 = 0$, and $\mathbf{h}'\mathbf{Ah} = h_2^2 \partial^2 f(\mathbf{x}_0)/\partial x_2^2$, which has the same sign as that of $\partial^2 f(\mathbf{x}_0)/\partial x_2^2$, if it is different from zero, except for those values of $\mathbf{h} = (h_1, h_2)'$ for which $h_2 = 0$, where it is zero. It follows that when $f_{11} f_{22} - f_{12}^2 = 0$, $\mathbf{h}'\mathbf{Ah}$ has a constant sign for all $\mathbf{h}$ inside a neighborhood of $\mathbf{0}$. However, it can vanish for some nonzero values of $\mathbf{h}$. For such values of $\mathbf{h}$, the sign of $f(\mathbf{x}_0 + \mathbf{h}) - f(\mathbf{x}_0)$ depends on the signs of higher-order partial derivatives (higher than second order) of f at $\mathbf{x}_0$. These can be obtained from Taylor's expansion. In this case, no decision can be made regarding the nature of the stationary point until these higher-order partial derivatives (if they exist) have been investigated.

EXAMPLE 7.7.1.   Let $f: R^2 \to R$ be the function $f(x_1, x_2) = x_1^2 + 2x_2^2 - x_1$. Consider the equations

$$\frac{\partial f}{\partial x_1} = 2x_1 - 1 = 0,$$

$$\frac{\partial f}{\partial x_2} = 4x_2 = 0.$$

The only solution is $\mathbf{x}_0 = (0.5, 0)'$. The Hessian matrix is

$$\mathbf{A} = \begin{bmatrix} 2 & 0 \\ 0 & 4 \end{bmatrix},$$

which is positive definite, since $2 > 0$ and $\det(\mathbf{A}) = 8 > 0$. The point $\mathbf{x}_0$ is therefore a local minimum. Since it is the only one in R^2, it must also be the absolute minimum.

EXAMPLE 7.7.2.   Consider the function $f: R^3 \to R$, where

$$f(x_1, x_2, x_3) = \tfrac{1}{3}x_1^3 + 2x_2^2 + x_3^2 - 2x_1x_2 + 3x_1x_3 + x_2x_3$$
$$- 10x_1 + 4x_2 - 6x_3 + 1.$$

A stationary point must satisfy the equations

$$\frac{\partial f}{\partial x_1} = x_1^2 - 2x_2 + 3x_3 - 10 = 0, \tag{7.37}$$

$$\frac{\partial f}{\partial x_2} = -2x_1 + 4x_2 + x_3 + 4 = 0, \tag{7.38}$$

$$\frac{\partial f}{\partial x_3} = 3x_1 + x_2 + 2x_3 - 6 = 0. \tag{7.39}$$

From (7.38) and (7.39) we get

$$x_2 = x_1 - 2,$$
$$x_3 = 4 - 2x_1.$$

By substituting these expressions in equation (7.37) we obtain

$$x_1^2 - 8x_1 + 6 = 0.$$

This equation has two solutions, namely, $4 - \sqrt{10}$ and $4 + \sqrt{10}$. We therefore have two stationary points,

$$\mathbf{x}_0^{(1)} = \left(4 + \sqrt{10}, 2 + \sqrt{10}, -4 - 2\sqrt{10}\right)',$$
$$\mathbf{x}_0^{(2)} = \left(4 - \sqrt{10}, 2 - \sqrt{10}, -4 + 2\sqrt{10}\right)'.$$

Now, the Hessian matrix is

$$\mathbf{A} = \begin{bmatrix} 2x_1 & -2 & 3 \\ -2 & 4 & 1 \\ 3 & 1 & 2 \end{bmatrix}.$$

Its leading principal minors are $2x_1, 8x_1 - 4$, and $14x_1 - 56$. The last one is the determinant of $\mathbf{A}$. At $\mathbf{x}_0^{(1)}$ all three are positive. Therefore, $\mathbf{x}_0^{(1)}$ is a point of local minimum. At $\mathbf{x}_0^{(2)}$ the values of the leading principal minors are 1.675, 2.7018, and -44.272. In this case, $\mathbf{A}$ is neither positive definite over negative definite. Thus $\mathbf{x}_0^{(2)}$ is a saddle point.

7.8. THE METHOD OF LAGRANGE MULTIPLIERS

This method, which is due to Joseph Louis de Lagrange (1736–1813), is used to optimize a real-valued function $f(x_1, x_2, \ldots, x_n)$, where $x_1, x_2, \ldots, x_n$ are subject to m ($< n$) equality constraints of the form

$$
\begin{aligned}
g_1(x_1, x_2, \ldots, x_n) &= 0, \\
g_2(x_1, x_2, \ldots, x_n) &= 0, \\
&\vdots \\
g_m(x_1, x_2, \ldots, x_n) &= 0,
\end{aligned}
\tag{7.40}
$$

where $g_1, g_2, \ldots, g_m$ are differentiable functions.

The determination of the stationary points in this constrained optimization problem is done by first considering the function

$$F(\mathbf{x}) = f(\mathbf{x}) + \sum_{j=1}^{m} \lambda_j g_j(\mathbf{x}), \tag{7.41}$$

where $\mathbf{x} = (x_1, x_2, \ldots, x_n)'$ and $\lambda_1, \lambda_2, \ldots, \lambda_m$ are scalars called *Lagrange multipliers*. By differentiating (7.41) with respect to $x_1, x_2, \ldots, x_n$ and equating the partial derivatives to zero we obtain

$$\frac{\partial F}{\partial x_i} = \frac{\partial f}{\partial x_i} + \sum_{j=1}^{m} \lambda_j \frac{\partial g_j}{\partial x_i} = 0, \qquad i = 1, 2, \ldots, n. \tag{7.42}$$

Equations (7.40) and (7.42) consist of $m + n$ equations in $m + n$ unknowns, namely, $x_1, x_2, \ldots, x_n; \lambda_1, \lambda_2, \ldots, \lambda_m$. The solutions for $x_1, x_2, \ldots, x_n$ determine the locations of the stationary points. The following argument explains why this is the case:

Suppose that in equation (7.40) we can solve for m x_i's, for example, $x_1, x_2, \ldots, x_m$, in terms of the remaining $n - m$ variables. By Theorem 7.6.2, this is possible whenever

$$\frac{\partial(g_1, g_2, \ldots, g_m)}{\partial(x_1, x_2, \ldots, x_m)} \neq 0. \tag{7.43}$$

In this case, we can write

$$x_1 = h_1(x_{m+1}, x_{m+2}, \ldots, x_n),$$
$$x_2 = h_2(x_{m+1}, x_{m+2}, \ldots, x_n),$$
$$\vdots \tag{7.44}$$
$$x_m = h_m(x_{m+1}, x_{m+2}, \ldots, x_n).$$

Thus $f(\mathbf{x})$ is a function of only $n - m$ variables, namely, $x_{m+1}, x_{m+2}, \ldots, x_n$. If the partial derivatives of f with respect to these variables exist and if f has a local optimum, then these partial derivatives must necessarily vanish, that is,

$$\frac{\partial f}{\partial x_i} + \sum_{j=1}^{m} \frac{\partial f}{\partial h_j} \frac{\partial h_j}{\partial x_i} = 0, \qquad i = m + 1, m + 2, \ldots, n. \tag{7.45}$$

Now, if equations (7.44) are used to substitute $h_1, h_2, \ldots, h_m$ for $x_1, x_2, \ldots, x_m$, respectively, in equation (7.40), then we obtain the identities

$$g_1(h_1, h_2, \ldots, h_m, x_{m+1}, x_{m+2}, \ldots, x_n) \equiv 0,$$
$$g_2(h_1, h_2, \ldots, h_m, x_{m+1}, x_{m+2}, \ldots, x_n) \equiv 0,$$
$$\vdots$$
$$g_m(h_1, h_2, \ldots, h_m, x_{m+1}, x_{m+2}, \ldots, x_n) \equiv 0.$$

By differentiating these identities with respect to $x_{m+1}, x_{m+2}, \ldots, x_n$ we obtain

$$\frac{\partial g_k}{\partial x_i} + \sum_{j=1}^{m} \frac{\partial g_k}{\partial h_j} \frac{\partial h_j}{\partial x_i} = 0, \qquad i = m+1, m+2, \ldots, n; \; k = 1, 2, \ldots, m. \quad (7.46)$$

Let us now define the vectors

$$\boldsymbol{\delta}_k = \left(\frac{\partial g_k}{\partial x_{m+1}}, \frac{\partial g_k}{\partial x_{m+2}}, \ldots, \frac{\partial g_k}{\partial x_n} \right)', \qquad k = 1, 2, \ldots, m,$$

$$\boldsymbol{\gamma}_k = \left(\frac{\partial g_k}{\partial h_1}, \frac{\partial g_k}{\partial h_2}, \ldots, \frac{\partial g_k}{\partial h_m} \right)', \qquad k = 1, 2, \ldots, m,$$

$$\boldsymbol{\eta}_j = \left(\frac{\partial h_j}{\partial x_{m+1}}, \frac{\partial h_j}{\partial x_{m+2}}, \ldots, \frac{\partial h_j}{\partial x_n} \right)', \qquad j = 1, 2, \ldots, m,$$

$$\boldsymbol{\psi} = \left(\frac{\partial f}{\partial x_{m+1}}, \frac{\partial f}{\partial x_{m+2}}, \ldots, \frac{\partial f}{\partial x_n} \right)',$$

$$\boldsymbol{\tau} = \left(\frac{\partial f}{\partial h_1}, \frac{\partial f}{\partial h_2}, \ldots, \frac{\partial f}{\partial h_m} \right)'.$$

Equations (7.45) and (7.46) can then be written as

$$[\boldsymbol{\delta}_1 : \boldsymbol{\delta}_2 : \cdots : \boldsymbol{\delta}_m] + [\boldsymbol{\eta}_1 : \boldsymbol{\eta}_2 : \cdots : \boldsymbol{\eta}_m]\boldsymbol{\Gamma} = 0, \qquad (7.47)$$

$$\boldsymbol{\psi} + [\boldsymbol{\eta}_1 : \boldsymbol{\eta}_2 : \cdots : \boldsymbol{\eta}_m]\boldsymbol{\tau} = 0, \qquad (7.48)$$

where $\boldsymbol{\Gamma} = [\boldsymbol{\gamma}_1 : \boldsymbol{\gamma}_2 : \cdots : \boldsymbol{\gamma}_m]$, which is a nonsingular $m \times m$ matrix if condition (7.43) is valid. From equation (7.47) we have

$$[\boldsymbol{\eta}_1 : \boldsymbol{\eta}_2 : \cdots : \boldsymbol{\eta}_m] = -[\boldsymbol{\delta}_1 : \boldsymbol{\delta}_2 : \cdots : \boldsymbol{\delta}_m]\boldsymbol{\Gamma}^{-1}.$$

By making the proper substitution in equation (7.48) we obtain

$$\boldsymbol{\psi} + [\boldsymbol{\delta}_1 : \boldsymbol{\delta}_2 : \cdots : \boldsymbol{\delta}_m]\boldsymbol{\lambda} = 0, \qquad (7.49)$$

where

$$\boldsymbol{\lambda} = -\boldsymbol{\Gamma}^{-1}\boldsymbol{\tau}. \qquad (7.50)$$

Equations (7.49) can then be expressed as

$$\frac{\partial f}{\partial x_i} + \sum_{j=1}^{m} \lambda_j \frac{\partial g_j}{\partial x_i} = 0, \qquad i = m+1, m+2, \ldots, n. \tag{7.51}$$

From equation (7.50) we also have

$$\frac{\partial f}{\partial x_i} + \sum_{j=1}^{m} \lambda_j \frac{\partial g_j}{\partial x_i} = 0, \qquad i = 1, 2, \ldots, m. \tag{7.52}$$

Equations (7.51) and (7.52) can now be combined into a single vector equation of the form

$$\nabla f(\mathbf{x}) + \sum_{j=1}^{m} \lambda_j \nabla g_j = \mathbf{0},$$

which is the same as equation (7.42). We conclude that at a stationary point of f, the values of $x_1, x_2, \ldots, x_n$ and the corresponding values of $\lambda_1, \lambda_2, \ldots, \lambda_m$ must satisfy equations (7.40) and (7.42).

Sufficient Conditions for a Local Optimum in the Method of Lagrange Multipliers

Equations (7.42) are only necessary for a stationary point $\mathbf{x}_0$ to be a point of local optimum of f subject to the constraints given by equations (7.40). Sufficient conditions for a local optimum are given in Gillespie (1954, pages 97–98). The following is a reproduction of these conditions:

Let $\mathbf{x}_0$ be a stationary point of f whose coordinates satisfy equations (7.40) and (7.42), and let $\lambda_1, \lambda_2, \ldots, \lambda_m$ be the corresponding Lagrange multipliers. Let F_{ij} denote the second-order partial derivative of F in formula (7.41) with respect to x_i, and x_j, $i, j = 1, 2, \ldots, n$; $i \neq j$. Consider the $(m+n) \times (m+n)$ matrix

$$\mathbf{B}_1 = \begin{bmatrix} F_{11} & F_{12} & \cdots & F_{1n} & g_1^{(1)} & g_2^{(1)} & \cdots & g_m^{(1)} \\ F_{21} & F_{22} & \cdots & F_{2n} & g_1^{(2)} & g_2^{(2)} & \cdots & g_m^{(2)} \\ \vdots & \vdots & & \vdots & \vdots & \vdots & & \vdots \\ F_{n1} & F_{n2} & \cdots & F_{nn} & g_1^{(n)} & g_2^{(n)} & \cdots & g_m^{(n)} \\ g_1^{(1)} & g_1^{(2)} & \cdots & g_1^{(n)} & 0 & 0 & \cdots & 0 \\ g_2^{(1)} & g_2^{(2)} & \cdots & g_2^{(n)} & 0 & 0 & \cdots & 0 \\ \vdots & \vdots & & \vdots & \vdots & \vdots & & \vdots \\ g_m^{(1)} & g_m^{(2)} & \cdots & g_m^{(n)} & 0 & 0 & \cdots & 0 \end{bmatrix}, \tag{7.53}$$

where $g_j^{(i)} = \partial g_j / \partial x_i$, $i = 1, 2, \ldots, n$; $j = 1, 2, \ldots, m$. Let Δ_1 denote the determinant of $\mathbf{B}_1$. Furthermore, let $\Delta_2, \Delta_3, \ldots, \Delta_{n-m}$ denote a set of principal minors of $\mathbf{B}_1$ (see Definition 2.3.6), namely, the determinants of the principal submatrices $\mathbf{B}_2, \mathbf{B}_3, \ldots, \mathbf{B}_{n-m}$, where $\mathbf{B}_i$ is obtained by deleting the first $i - 1$ rows and the first $i - 1$ columns of $\mathbf{B}_1$ ($i = 2, 3, \ldots, n - m$). All the partial derivatives used in $\mathbf{B}_1, \mathbf{B}_2, \ldots, \mathbf{B}_{n-m}$ are evaluated at $\mathbf{x}_0$. Then sufficient conditions for $\mathbf{x}_0$ to be a point of local minimum of f are the following:

i. If m is even,

$$\Delta_1 > 0, \qquad \Delta_2 > 0, \ldots, \qquad \Delta_{n-m} > 0.$$

ii. If m is odd,

$$\Delta_1 < 0, \qquad \Delta_2 < 0, \ldots, \qquad \Delta_{n-m} < 0.$$

However, sufficient conditions for $\mathbf{x}_0$ to be a point of local maximum are the following:

i. If n is even,

$$\Delta_1 > 0, \qquad \Delta_2 < 0, \ldots, \qquad (-1)^{n-m} \Delta_{n-m} < 0.$$

ii. If n is odd,

$$\Delta_1 < 0, \qquad \Delta_2 > 0, \ldots, \qquad (-1)^{n-m} \Delta_{n-m} > 0.$$

EXAMPLE 7.8.1.   Let us find the minimum and maximum distances from the origin to the curve determined by the intersection of the plane $x_2 + x_3 = 0$ with the ellipsoid $x_1^2 + 2x_2^2 + x_3^2 + 2x_2 x_3 = 1$. Let $f(x_1, x_2, x_3)$ be the squared distance function from the origin, that is,

$$f(x_1, x_2, x_3) = x_1^2 + x_2^2 + x_3^2.$$

The equality constraints are

$$g_1(x_1, x_2, x_3) \equiv x_2 + x_3 = 0,$$
$$g_2(x_1, x_2, x_3) \equiv x_1^2 + 2x_2^2 + x_3^2 + 2x_2 x_3 - 1 = 0.$$

Then

$$F(x_1, x_2, x_3) = x_1^2 + x_2^2 + x_3^2 + \lambda_1(x_2 + x_3)$$
$$+ \lambda_2(x_1^2 + 2x_2^2 + x_3^2 + 2x_2 x_3 - 1),$$
$$\frac{\partial F}{\partial x_1} = 2x_1 + 2\lambda_2 x_1 = 0, \tag{7.54}$$

$$\frac{\partial F}{\partial x_2} = 2x_2 + \lambda_1 + 2\lambda_2(2x_2 + x_3) = 0, \tag{7.55}$$

$$\frac{\partial F}{\partial x_3} = 2x_3 + \lambda_1 + 2\lambda_2(x_2 + x_3) = 0. \tag{7.56}$$

Equations (7.54), (7.55), and (7.56) and the equality constraints are satisfied by the following sets of solutions:

I. $x_1 = 0, x_2 = 1, x_3 = -1, \lambda_1 = 2, \lambda_2 = -2$.
II. $x_1 = 0, x_2 = -1, x_3 = 1, \lambda_1 = -2, \lambda_2 = -2$.
III. $x_1 = 1, x_2 = 0, x_3 = 0, \lambda_1 = 0, \lambda_2 = -1$.
IV. $x_1 = -1, x_2 = 0, x_3 = 0, \lambda_1 = 0, \lambda_2 = -1$.

To determine if any of these four sets correspond to local maxima or minima, we need to examine the values of $\Delta_1, \Delta_2, \ldots, \Delta_{n-m}$. Here, the matrix $\mathbf{B}_1$ in formula (7.53) has the value

$$\mathbf{B}_1 = \begin{bmatrix} 2 + 2\lambda_2 & 0 & 0 & 0 & 2x_1 \\ 0 & 2 + 4\lambda_2 & 2\lambda_2 & 1 & 4x_2 + 2x_3 \\ 0 & 2\lambda_2 & 2 + 2\lambda_2 & 1 & 2x_2 + 2x_3 \\ 0 & 1 & 1 & 0 & 0 \\ 2x_1 & 4x_2 + 2x_3 & 2x_2 + 2x_3 & 0 & 0 \end{bmatrix}.$$

Since $n = 3$ and $m = 2$, only one Δ_i, namely, Δ_1, the determinant of $\mathbf{B}_1$, is needed. Furthermore, since m is even and n is odd, a sufficient condition for a local minimum is $\Delta_1 > 0$, and for a local maximum the condition is $\Delta_1 < 0$. It can be verified that $\Delta_1 = -8$ for solution sets I and II, and $\Delta_1 = 8$ for solution sets III and IV. We therefore have local maxima at the points $(0, 1, -1)$ and $(0, -1, 1)$ with a common maximum value $f_{\max} = 2$. We also have local minima at the points $(1, 0, 0)$ and $(-1, 0, 0)$ with a common minimum value $f_{\min} = 1$. Since these are the only local optima on the curve of intersection, we conclude that the minimum distance from the origin to this curve is 1 and the maximum distance is $\sqrt{2}$.

7.9. THE RIEMANN INTEGRAL OF A MULTIVARIABLE FUNCTION

In Chapter 6 we discussed the Riemann integral of a real-valued function of a single variable x. In this section we extend the concept of Riemann integration to real-valued functions of n variables, $x_1, x_2, \ldots, x_n$.

Definition 7.9.1.   The set of points in R^n whose coordinates satisfy the inequalities

$$a_i \leq x_i \leq b_i, \qquad i = 1, 2, \ldots, n, \tag{7.57}$$

where $a_i < b_i$, $i = 1, 2, \ldots, n$, form an n-dimensional cell denoted by $c_n(a, b)$. The content (or volume) of this cell is $\prod_{i=1}^{n}(b_i - a_i)$ and is denoted by $\mu[c_n(a,b)]$.

Suppose that P_i is a partition of the interval $[a_i, b_i]$, $i = 1, 2, \ldots, n$. The Cartesian product $P = \times_{i=1}^{n} P_i$ is a partition of $c_n(a, b)$ and consists of n-dimensional subcells of $c_n(a, b)$. We denote these subcells by $S_1, S_2, \ldots, S_\nu$. The content of S_i is denoted by $\mu(S_i)$, $i = 1, 2, \ldots, \nu$, where ν is the number of subcells.   □

We shall first define the Riemann integral of a real-valued function $f(\mathbf{x})$ on an n-dimensional cell; then we shall extend this definition to any bounded region in R^n.

7.9.1. The Riemann Integral on Cells

Let $f: D \rightarrow R$, where $D \subset R^n$. Suppose that $c_n(a, b)$ is an n-dimensional cell contained in D and that f is bounded on $c_n(a, b)$. Let P be a partition of $c_n(a, b)$ consisting of the subcells $S_1, S_2, \ldots, S_\nu$. Let m_i and M_i be, respectively, the infimum and supremum of f on S_i, $i = 1, 2, \ldots, \nu$. Consider the sums

$$LS_P(f) = \sum_{i=1}^{\nu} m_i \, \mu(S_i), \tag{7.58}$$

$$US_P(f) = \sum_{i=1}^{\nu} M_i \, \mu(S_i). \tag{7.59}$$

We note the similarity of these sums to the ones defined in Section 6.2. As before, we refer to $LS_P(f)$ and $US_P(f)$ as the lower and upper sums, respectively, of f with respect to the partition P.

The following theorem is an n-dimensional analogue of Theorem 6.2.1. The proof is left to the reader.

Theorem 7.9.1.   Let $f: D \rightarrow R$, where $D \subset R^n$. Suppose that f is bounded on $c_n(a, b) \subset D$. Then f is Riemann integrable on $c_n(a, b)$ if and only if for every $\epsilon > 0$ there exists a partition P of $c_n(a, b)$ such that

$$US_P(f) - LS_P(f) < \epsilon.$$

Definition 7.9.2.   Let P_1 and P_2 be two partitions of $c_n(a, b)$. Then P_2 is a refinement of P_1 if every point in P_1 is also a point in P_2, that is, $P_1 \subset P_2$.   □

Using this definition, it is possible to prove results similar to those of Lemmas 6.2.1 and 6.2.2. In particular, we have the following lemma:

Lemma 7.9.1. Let $f: D \to R$, where $D \subset R^n$. Suppose that f is bounded on $c_n(a, b) \subset D$. Then $\sup_P LS_p(f)$ and $\inf_P US_P(f)$ exist, and

$$\sup_P LS_P(f) \leq \inf_P US_P(f).$$

Definition 7.9.3. Let $f: c_n(a, b) \to R$ be a bounded function. Then f is Riemann integrable on $c_n(a, b)$ if and only if

$$\sup_P LS_P(f) = \inf_P US_P(f). \qquad (7.60)$$

Their common value is called the Riemann integral of f on $c_n(a, b)$ and is denoted by $\int_{c_n(a, b)} f(\mathbf{x}) \, d\mathbf{x}$. This is equivalent to the expression $\int_{a_1}^{b_1} \int_{a_2}^{b_2} \cdots \int_{a_n}^{b_n} f(x_1, x_2, \ldots, x_n) \, dx_1 \, dx_2 \cdots dx_n$. For example, for $n = 2, 3$ we have

$$\int_{c_2(a, b)} f(\mathbf{x}) \, d\mathbf{x} = \int_{a_1}^{b_1} \int_{a_2}^{b_2} f(x_1, x_2) \, dx_1 \, dx_2, \qquad (7.61)$$

$$\int_{c_3(a, b)} f(\mathbf{x}) \, d\mathbf{x} = \int_{a_1}^{b_1} \int_{a_2}^{b_2} \int_{a_3}^{b_3} f(x_1, x_2, x_3) \, dx_1 \, dx_2 \, dx_3. \qquad (7.62)$$

The integral in formula (7.61) is called a double Riemann integral, and the one in formula (7.62) is called a triple Riemann integral. In general, for $n \geq 2$, $\int_{c_n(a, b)} f(\mathbf{x}) \, d\mathbf{x}$ is called an n-tuple Riemann integral. $\qquad \square$

The integral $\int_{c_n(a, b)} f(\mathbf{x}) \, d\mathbf{x}$ has properties similar to those of a single-variable Riemann integral in Section 6.4. The following theorem is an extension of Theorem 6.3.1.

Theorem 7.9.2. If f is continuous on an n-dimensional cell $c_n(a, b)$, then it is Riemann integrable there.

7.9.2. Iterated Riemann Integrals on Cells

The definition of the n-tuple Riemann integral in Section 7.9.1 does not provide a practicable way to evaluate it. We now show that the evaluation of this integral can be obtained by performing n Riemann integrals each of which is carried out with respect to one variable. Let us first consider the double integral as in formula (7.61).

Lemma 7.9.2. Suppose that f is real-valued and continuous on $c_2(a, b)$. Define the function $g(x_2)$ as

$$g(x_2) = \int_{a_1}^{b_1} f(x_1, x_2) \, dx_1.$$

Then $g(x_2)$ is continuous on $[a_2, b_2]$.

Proof. Let $\epsilon > 0$ be given. Since f is continuous on $c_2(a, b)$, which is closed and bounded, then by Theorem 7.3.2, f is uniformly continuous on $c_2(a, b)$. We can therefore find a $\delta > 0$ such that

$$|f(\xi) - f(\eta)| < \frac{\epsilon}{b_1 - a_1}$$

if $\|\xi - \eta\| < \delta$, where $\xi = (x_1, x_2)'$, $\eta = (y_1, y_2)'$, and $x_1, y_1 \in [a_1, b_1]$, $x_2, y_2 \in [a_2, b_2]$. It follows that if $|y_2 - x_2| < \delta$, then

$$|g(y_2) - g(x_2)| = \left| \int_{a_1}^{b_1} [f(x_1, y_2) - f(x_1, x_2)] \, dx_1 \right|$$

$$\leq \int_{a_1}^{b_1} |f(x_1, y_2) - f(x_1, x_2)| \, dx_1$$

$$< \int_{a_1}^{b_1} \frac{\epsilon}{b_1 - a_1} \, dx_1, \tag{7.63}$$

since $\|(x_1, y_2)' - (x_1, x_2)'\| = |y_2 - x_2| < \delta$. From inequality (7.63) we conclude that

$$|g(y_2) - g(x_2)| < \epsilon$$

if $|y_2 - x_2| < \delta$. Hence, $g(x_2)$ is continuous on $[a_2, b_2]$. Consequently, from Theorem 6.3.1, $g(x_2)$ is Riemann integrable on $[a_2, b_2]$, that is, $\int_{a_2}^{b_2} g(x_2) \, dx_2$ exists. We call the integral

$$\int_{a_2}^{b_2} g(x_2) \, dx_2 = \int_{a_2}^{b_2} \left[\int_{a_1}^{b_1} f(x_1, x_2) \, dx_1 \right] dx_2 \tag{7.64}$$

an iterated integral of order 2.   □

The next theorem states that the iterated integral (7.64) is equal to the double integral $\int_{c_2(a, b)} f(\mathbf{x}) \, d\mathbf{x}$.

Theorem 7.9.3. If f is continuous on $c_2(a, b)$, then

$$\int_{c_2(a, b)} f(\mathbf{x}) \, d\mathbf{x} = \int_{a_2}^{b_2} \left[\int_{a_1}^{b_1} f(x_1, x_2) \, dx_1 \right] dx_2.$$

Proof. Exercise 7.22.   □

We note that the iterated integral in (7.64) was obtained by integrating first with respect to x_1, then with respect to x_2. This order of integration could have been reversed, that is, we could have integrated f with respect to x_2 and then with respect to x_1. The result would be the same in both cases. This is based on the following theorem due to Guido Fubini (1879–1943).

Theorem 7.9.4 (Fubini's Theorem).   If f is continuous on $c_2(a, b)$, then

$$\int_{c_2(a,b)} f(\mathbf{x})\, d\mathbf{x} = \int_{a_2}^{b_2}\left[\int_{a_1}^{b_1} f(x_1, x_2)\, dx_1\right] dx_2 = \int_{a_1}^{b_1}\left[\int_{a_2}^{b_2} f(x_1, x_2)\, dx_2\right] dx_1$$

Proof. See Corwin and Szczarba (1982, page 287). □

A generalization of this theorem to multiple integrals of order n is given by the next theorem [see Corwin and Szczarba (1982, Section 11.1)].

Theorem 7.9.5 (Generalized Fubini's Theorem).   If f is continuous on $c_n(a, b) = \{\mathbf{x} \mid a_i \le x_i \le b_i,\ i = 1, 2, \ldots, n\}$, then

$$\int_{c_n(a,b)} f(\mathbf{x})\, d\mathbf{x} = \int_{c_{n-1}^{(i)}(a,b)}\left[\int_{a_i}^{b_i} f(\mathbf{x})\, dx_i\right] d\mathbf{x}_{(i)}, \qquad i = 1, 2, \ldots, n,$$

where $d\mathbf{x}_{(i)} = dx_1\, dx_2 \cdots dx_{i-1}\, dx_{i+1} \cdots dx_n$ and $c_{n-1}^{(i)}(a, b)$ is an $(n-1)$-dimensional cell such that $a_1 \le x_1 \le b_1, a_2 \le x_2 \le b_2, \ldots, a_{i-1} \le x_{i-1} \le b_{i-1}, a_{i+1} \le x_{i+1} \le b_{i+1}, \ldots, a_n \le x_n \le b_n$.

7.9.3. Integration over General Sets

We now consider n-tuple Riemann integration over regions in R^n that are not necessarily cell shaped as in Section 7.9.1.

Let $f: D \to R$ be a bounded and continuous function, where D is a bounded region in R^n. There exists an n-dimensional cell $c_n(a, b)$ such that $D \subset c_n(a, b)$. Let $g: c_n(a, b) \to R$ be defined as

$$g(\mathbf{x}) = \begin{cases} f(\mathbf{x}), & \mathbf{x} \in D, \\ 0, & \mathbf{x} \notin D. \end{cases}$$

Then

$$\int_{c_n(a,b)} g(\mathbf{x})\, d\mathbf{x} = \int_D f(\mathbf{x})\, d\mathbf{x}. \tag{7.65}$$

The integral on the right-hand side of (7.65) is independent of the choice of $c_n(a, b)$ provided that it contains D. It should be noted that the function $g(\mathbf{x})$ may not be continuous on $Br(D)$, the boundary of D. This, however, should not affect the existence of the integral on the left-hand side of (7.65). The reason for this is given in Theorem 7.9.7. First, we need to define the so-called Jordan content of a set.

Definition 7.9.4. Let $D \subset R^n$ be a bounded set such that $D \subset c_n(a, b)$ for some n-dimensional cell. Let the function $\lambda_D: R^n \to R$ be defined as

$$\lambda_D(\mathbf{x}) = \begin{cases} 1, & \mathbf{x} \in D, \\ 0, & \mathbf{x} \notin D. \end{cases}$$

This is called the characteristic function of D. Suppose that

$$\sup_P LS_P(\lambda_D) = \inf_P US_P(\lambda_D), \tag{7.66}$$

where $LS_P(\lambda_D)$ and $US_P(\lambda_D)$ are, respectively, the lower and upper sums of $\lambda_D(\mathbf{x})$ with respect to a partition P of $c_n(a, b)$. Then, D is said to have an n-dimensional Jordan content denoted by $\mu_j(D)$, where $\mu_j(D)$ is equal to the common value of the terms in equality (7.66). In this case, D is said to be *Jordan measurable*. □

The proofs of the next two theorems can be found in Sagan (1974, Chapter 11).

Theorem 7.9.6. A bounded set $D \subset R^n$ is Jordan measurable if and only if its boundary $Br(D)$ has a Jordan content equal to zero.

Theorem 7.9.7. Let $f: D \to R$, where $D \subset R^n$ is bounded and Jordan measurable. If f is bounded and continuous in D except on a set that has a Jordan content equal to zero, then $\int_D f(\mathbf{x}) d\mathbf{x}$ exists.

It follows from Theorems 7.9.6 and 7.9.7 that the integral in equality (7.75) must exist even though $g(\mathbf{x})$ may not be continuous on the boundary $Br(D)$ of D, since $Br(D)$ has a Jordan content equal to zero.

EXAMPLE 7.9.1. Let $f(x_1, x_2) = x_1 x_2$ and D be the region

$$D = \{(x_1, x_2) | x_1^2 + x_2^2 \le 1, \, x_1 \ge 0, \, x_2 \ge 0\}.$$

It is easy to see that D is contained inside the two-dimensional cell

$$c_2(0, 1) = \{(x_1, x_2) | 0 \le x_1 \le 1, 0 \le x_2 \le 1\}.$$

Then

$$\int\int_D x_1 x_2 \, dx_1 \, dx_2 = \int_0^1 \left[\int_0^{(1-x_1^2)^{1/2}} x_1 x_2 \, dx_2 \right] dx_1.$$

We note that for a fixed x_1 in $[0, 1]$, the part of the line through $(x_1, 0)$ that lies inside D and is parallel to the x_2-axis is in fact the interval $0 \leq x_2 \leq (1 - x_1^2)^{1/2}$. For this reason, the limits of x_2 are 0 and $(1 - x_1^2)^{1/2}$. Consequently,

$$\int_0^1 \left[\int_0^{(1-x_1^2)^{1/2}} x_1 x_2 \, dx_2 \right] dx_1 = \int_0^1 x_1 \left[\int_0^{(1-x_1^2)^{1/2}} x_2 \, dx_2 \right] dx_1$$

$$= \tfrac{1}{2} \int_0^1 x_1 (1 - x_1^2) \, dx_1$$

$$= \tfrac{1}{8}.$$

In practice, it is not always necessary to make reference to $c_n(a, b)$ that encloses D in order to evaluate the integral on D. Rather, we only need to recognize that the limits of integration in the iterated Riemann integral depend in general on variables that have not yet been integrated out, as was seen in Example 7.9.1. Care should therefore be exercised in correctly identifying the limits of integration. By changing the order of integration (according to Fubini's theorem), it is possible to facilitate the evaluation of the integral.

EXAMPLE 7.9.2. Consider $\int \int_D e^{x_2^2} \, dx_1 \, dx_2$, where D is the region in the first quadrant bounded by $x_2 = 1$ and $x_1 = x_2$. In this example, it is easier to integrate first with respect to x_1 and then with respect to x_2. Thus

$$\int \int_D e^{x_2^2} \, dx_1 \, dx_2 = \int_0^1 \left[\int_0^{x_2} e^{x_2^2} \, dx_1 \right] dx_2$$

$$= \int_0^1 x_2 e^{x_2^2} \, dx_2$$

$$= \tfrac{1}{2}(e - 1).$$

EXAMPLE 7.9.3. Consider the integral $\int \int_D (x_1^2 + x_2^3) \, dx_1 \, dx_2$, where D is a region in the first quadrant bounded by $x_2 = x_1^2$ and $x_1 = x_2^4$. Hence,

$$\int \int_D (x_1^2 + x_2^3) \, dx_1 \, dx_2 = \int_0^1 \left[\int_{x_2^4}^{\sqrt{x_2}} (x_1^2 + x_2^3) \, dx_1 \right] dx_2$$

$$= \int_0^1 \left[\tfrac{1}{3} (x_2^{3/2} - x_2^{12}) + \left(\sqrt{x_2} - x_2^4 \right) x_2^3 \right] dx_2$$

$$= \tfrac{959}{4680}.$$

7.9.4. Change of Variables in n-Tuple Riemann Integrals

In this section we give an extension of the change of variables formula in Section 6.4.1 to n-tuple Riemann integrals.

Theorem 7.9.8. Suppose that D is a closed and bounded set in R^n. Let $f: D \to R$ be continuous. Suppose that $\mathbf{h}: D \to R^n$ is a one-to-one function with continuous first-order partial derivatives such that the Jacobian determinant,

$$\det[\mathbf{J}_{\mathbf{h}}(\mathbf{x})] = \frac{\partial(h_1, h_2, \ldots, h_n)}{\partial(x_1, x_2, \ldots, x_n)},$$

is different from zero for all $\mathbf{x}$ in D, where $\mathbf{x} = (x_1, x_2, \ldots, x_n)'$ and h_i is the ith element of $\mathbf{h}(i = 1, 2, \ldots, n)$. Then

$$\int_D f(\mathbf{x}) \, d\mathbf{x} = \int_{D'} f[\mathbf{g}(\mathbf{u})] |\det \mathbf{J}_{\mathbf{g}}(\mathbf{u})| \, d\mathbf{u}, \tag{7.67}$$

where $D' = \mathbf{h}(D), \mathbf{u} = \mathbf{h}(\mathbf{x})$, $\mathbf{g}$ is the inverse function of $\mathbf{h}$, and

$$\det \mathbf{J}_{\mathbf{g}}(\mathbf{u}) = \frac{\partial(g_1, g_2, \ldots, g_n)}{\partial(u_1, u_2, \ldots, u_n)}, \tag{7.68}$$

where g_i and u_i are, respectively, the ith elements of $\mathbf{g}$ and $\mathbf{u}$ $(i = 1, 2, \ldots, n)$.

Proof. See, for example, Corwin and Szczarba (1982, Theorem 6.2), or Sagan (1974, Theorem 115.1).　　□

EXAMPLE 7.9.4. Consider the integral $\int\int_D x_1 x_2^2 \, dx_1 \, dx_2$, where D is bounded by the four parabolas, $x_2^2 = x_1, x_2^2 = 3x_1, x_1^2 = x_2, x_1^2 = 4x_2$. Let $u_1 = x_2^2/x_1, u_2 = x_1^2/x_2$. The inverse transformation is given by

$$x_1 = \left(u_1 u_2^2\right)^{1/3}, \qquad x_2 = \left(u_1^2 u_2\right)^{1/3}.$$

From formula (7.68) we have

$$\frac{\partial(g_1, g_2)}{\partial(u_1, u_2)} = \frac{\partial(x_1, x_2)}{\partial(u_1, u_2)} = -\frac{1}{3}.$$

By applying formula (7.67) we obtain

$$\int\int_D x_1 x_2^2 \, dx_1 \, dx_2 = \frac{1}{3} \int\int_{D'} u_1^{5/3} u_2^{4/3} \, du_1 \, du_2,$$

where D' is a rectangular region in the $u_1 u_2$ space bounded by the lines $u_1 = 1, 3; u_2 = 1, 4$. Hence,

$$\int\int_D x_1 x_2^2 \, dx_1 \, dx_2 = \frac{1}{3} \int_1^3 u_1^{5/3} \, du_1 \int_1^4 u_2^{4/3} \, du_2$$

$$= \frac{3}{56}(3^{8/3} - 1)(4^{7/3} - 1).$$

7.10. DIFFERENTIATION UNDER THE INTEGRAL SIGN

Suppose that $f(x_1, x_2, \ldots, x_n)$ is a real-valued function defined on $D \subset R^n$. If some of the x_i's, for example, $x_{m+1}, x_{m+2}, \ldots, x_n$ $(n > m)$, are integrated out, we obtain a function that depends only on the remaining variables. In this section we discuss conditions under which the latter function is differentiable. For simplicity, we shall only consider functions of $n = 2$ variables.

Theorem 7.10.1. Let $f: D \to R$, where $D \subset R^2$ contains the two-dimensional cell $c_2(a, b) = \{(x_1, x_2) | a_1 \leq x_1 \leq b_1, a_2 \leq x_2 \leq b_2\}$. Suppose that f is continuous and has a continuous first-order partial derivative with respect to x_2 in D. Then, for $a_2 < x_2 < b_2$,

$$\frac{d}{dx_2} \int_{a_1}^{b_1} f(x_1, x_2) \, dx_1 = \int_{a_1}^{b_1} \frac{\partial f(x_1, x_2)}{\partial x_2} \, dx_1. \tag{7.69}$$

Proof. Let $h(x_2)$ be defined on $[a_2, b_2]$ as

$$h(x_2) = \int_{a_1}^{b_1} \frac{\partial f(x_1, x_2)}{\partial x_2} \, dx_1, \qquad a_2 \leq x_2 \leq b_2.$$

Since $\partial f / \partial x_2$ is continuous, then by Lemma 7.9.2, $h(x_2)$ is continuous on $[a_2, b_2]$. Now, let t be such that $a_2 < t < b_2$. By integrating $h(x_2)$ over the interval $[a_2, t]$ we obtain

$$\int_{a_2}^{t} h(x_2) \, dx_2 = \int_{a_2}^{t} \left[\int_{a_1}^{b_1} \frac{\partial f(x_1, x_2)}{\partial x_2} \, dx_1 \right] dx_2. \tag{7.70}$$

The order of integration in (7.70) can be reversed by Theorem 7.9.4. We than have

$$\int_{a_2}^{t} h(x_2) \, dx_2 = \int_{a_1}^{b_1} \left[\int_{a_2}^{t} \frac{\partial f(x_1, x_2)}{\partial x_2} \, dx_2 \right] dx_1$$

$$= \int_{a_1}^{b_1} [f(x_1, t) - f(x_1, a_2)] \, dx_1$$

$$= \int_{a_1}^{b_1} f(x_1, t) \, dx_1 - \int_{a_1}^{b_1} f(x_1, a_2) \, dx_1$$

$$= F(t) - F(a_2), \tag{7.71}$$

where $F(y) = \int_{a_1}^{b_1} f(x_1, y)\, dx_1$. If we now apply Theorem 6.4.8 and differentiate the two sides of (7.71) with respect to t we obtain $h(t) = F'(t)$, that is,

$$\int_{a_1}^{b_1} \frac{\partial f(x_1, t)}{\partial t}\, dx_1 = \frac{d}{dt} \int_{a_1}^{b_1} f(x_1, t)\, dx_1. \tag{7.72}$$

Formula (7.69) now follows from formula (7.72) on replacing t with x_2. $\square$

Theorem 7.10.2. Let f and D be the same as in Theorem 7.10.1. Furthermore, let $\lambda(x_2)$ and $\theta(x_2)$ be functions defined and having continuous derivatives on $[a_2, b_2]$ such that $a_1 \le \lambda(x_2) \le \theta(x_2) \le b_1$ for all x_2 in $[a_2, b_2]$. Then the function $G: [a_2, b_2] \to R$ defined by

$$G(x_2) = \int_{\lambda(x_2)}^{\theta(x_2)} f(x_1, x_2)\, dx_1$$

is differentiable for $a_2 < x_2 < b_2$, and

$$\frac{dG}{dx_2} = \int_{\lambda(x_2)}^{\theta(x_2)} \frac{\partial f(x_1, x_2)}{\partial x_2}\, dx_1 + \theta'(x_2) f[\theta(x_2), x_2] - \lambda'(x_2) f[\lambda(x_2), x_2].$$

Proof. Let us write $G(x_2)$ as $H(\lambda, \theta, x_2)$. Since both of λ and θ depend on x_2, then by applying the total derivative formula [see formula (7.12)] to H we obtain

$$\frac{dH}{dx_2} = \frac{\partial H}{\partial \lambda} \frac{d\lambda}{dx_2} + \frac{\partial H}{\partial \theta} \frac{d\theta}{dx_2} + \frac{\partial H}{\partial x_2}. \tag{7.73}$$

Now, by Theorem 6.4.8,

$$\frac{\partial H}{\partial \theta} = \frac{\partial}{\partial \theta} \int_\lambda^\theta f(x_1, x_2)\, dx_1 = f(\theta, x_2),$$

$$\frac{\partial H}{\partial \lambda} = \frac{\partial}{\partial \lambda} \int_\lambda^\theta f(x_1, x_2)\, dx_1$$

$$= -\frac{\partial}{\partial \lambda} \int_\theta^\lambda f(x_1, x_2)\, dx_1 = -f(\lambda, x_2).$$

Furthermore, by Theorem 7.10.1,

$$\frac{\partial H}{\partial x_2} = \frac{\partial}{\partial x_2} \int_\lambda^\theta f(x_1, x_2)\, dx_1 = \int_\lambda^\theta \frac{\partial f(x_1, x_2)}{\partial x_2}\, dx_1.$$

By making the proper substitution in formula (7.73) we finally conclude that

$$\frac{d}{dx_2} \int_{\lambda(x_2)}^{\theta(x_2)} f(x_1, x_2)\, dx_1 = \int_{\lambda(x_2)}^{\theta(x_2)} \frac{\partial f(x_1, x_2)}{\partial x_2}\, dx_1 + \theta'(x_2) f[\theta(x_2), x_2]$$

$$- \lambda'(x_2) f[\lambda(x_2), x_2]. \qquad \square$$

EXAMPLE 7.10.1.

$$\frac{d}{dx_2} \int_{x_2^2}^{\cos x_2} (x_1 x_2^2 - 1) e^{-x_1}\, dx_1 = \int_{x_2^2}^{\cos x_2} 2 x_1 x_2\, e^{-x_1}\, dx_1$$

$$- \sin x_2 (x_2^2 \cos x_2 - 1) e^{-\cos x_2}$$

$$- 2 x_2 (x_2^4 - 1) e^{-x_2^2}.$$

Theorems 7.10.1 and 7.10.2 can be used to evaluate certain integrals of the form $\int_a^b f(x)\, dx$. For example, consider the integral

$$I = \int_0^\pi x^2 \cos x\, dx.$$

Define the function

$$F(x_2) = \int_0^\pi \cos(x_1 x_2)\, dx_1,$$

where $x_2 \geq 1$. Then

$$F(x_2) = \frac{1}{x_2} \sin(x_1 x_2) \Big|_{x_1=0}^{x_1=\pi} = \frac{1}{x_2} \sin(\pi x_2).$$

If we now differentiate $F(x_2)$ two times, we obtain

$$F''(x_2) = \frac{2 \sin(\pi x_2) - 2\pi x_2 \cos(\pi x_2) - \pi^2 x_2^2 \sin(\pi x_2)}{x_2^3}.$$

Thus

$$\int_0^\pi x_1^2 \cos(x_1 x_2)\, dx_1 = \frac{2\pi x_2 \cos(\pi x_2) + \pi^2 x_2^2 \sin(\pi x_2) - 2 \sin(\pi x_2)}{x_2^3}.$$

By replacing x_2 with 1 we obtain

$$I = \int_0^\pi x_1^2 \cos x_1 \, dx_1 = -2\pi.$$

7.11. APPLICATIONS IN STATISTICS

Multidimensional calculus provides a theoretical framework for the study of multivariate distributions, that is, joint distributions of several random variables. It can also be used to estimate the parameters of a statistical model. We now provide details of some of these applications.

Let $\mathbf{X} = (X_1, X_2, \ldots, X_n)'$ be a random vector. The distribution of $\mathbf{X}$ is characterized by its cumulative distribution function, namely,

$$F(\mathbf{x}) = P(X_1 \leq x_1, X_2 \leq x_2, \ldots, X_n \leq x_n), \tag{7.74}$$

where $\mathbf{x} = (x_1, x_2, \ldots, x_n)'$. If $F(\mathbf{x})$ is continuous and has an nth-order mixed partial derivative with respect to $x_1, x_2, \ldots, x_n$, then the function

$$f(\mathbf{x}) = \frac{\partial^n F(\mathbf{x})}{\partial x_1 \, \partial x_2 \cdots \partial x_n},$$

is called the density function of $\mathbf{X}$. In this case, formula (7.74) can be written in the form

$$F(\mathbf{x}) = \int_{-\infty}^{x_1} \int_{-\infty}^{x_2} \cdots \int_{-\infty}^{x_n} f(\mathbf{z}) \, d\mathbf{z}.$$

where $\mathbf{z} = (z_1, z_2, \ldots, z_n)'$. If the random variable X_i $(i = 1, 2, \ldots, n)$ is considered separately, then its distribution function is called the ith marginal distribution of $\mathbf{X}$. Its density function $f_i(x_i)$, called the ith marginal density function, can be obtained by integrating out the remaining $n - 1$ variables from $f(\mathbf{x})$. For example, if $\mathbf{X} = (X_1, X_2)'$, then the marginal density function of X_1 is

$$f_1(x_1) = \int_{-\infty}^{\infty} f(x_1, x_2) \, dx_2.$$

Similarly, the marginal density function of X_2 is

$$f_2(x_2) = \int_{-\infty}^{\infty} f(x_1, x_2) \, dx_1.$$

In particular, if $X_1, X_2, \ldots, X_n$ are independent random variables, then the density function of $\mathbf{X} = (X_1, X_2, \ldots, X_n)'$ is the product of all the associated marginal density functions, that is, $f(\mathbf{x}) = \prod_{i=1}^{n} f_i(x_i)$.

If only $n - 2$ variables are integrated out from $f(\mathbf{x})$, we obtain the so-called bivariate density function of the remaining two variables. For example, if $\mathbf{X} = (X_1, X_2, X_3, X_4)'$, the bivariate density function of x_1 and x_2 is

$$f_{12}(x_1, x_2) = \int_{-\infty}^{\infty} \int_{-\infty}^{\infty} f(x_1, x_2, x_3, x_4) \, dx_3 \, dx_4.$$

Now, the mean of $\mathbf{X} = (X_1, X_2, \ldots, X_n)'$ is $\boldsymbol{\mu} = (\mu_1, \mu_2, \ldots, \mu_n)'$, where

$$\mu_i = \int_{-\infty}^{\infty} x_i f_i(x_i) \, dx_i, \qquad i = 1, 2, \ldots, n.$$

The variance–covariance matrix of $\mathbf{X}$ is the $n \times n$ matrix $\boldsymbol{\Sigma} = (\sigma_{ij})$, where

$$\sigma_{ij} = \int_{-\infty}^{\infty} \int_{-\infty}^{\infty} (x_i - \mu_i)(x_j - \mu_j) f_{ij}(x_i, x_j) \, dx_i \, dx_j,$$

where μ_i and μ_j are the means of X_i and X_j, respectively, and $f_{ij}(x_i, x_j)$ is the bivariate density function of X_i and X_j, $i \neq j$. If $i = j$, then σ_{ii} is the variance of X_i, where

$$\sigma_{ii} = \int_{-\infty}^{\infty} (x_i - \mu_i)^2 f_i(x_i) \, dx_i, \qquad i = 1, 2, \ldots, n.$$

7.11.1. Transformations of Random Vectors

In this section we consider a multivariate extension of formula (6.73) regarding the density function of a function of a single random variable. This is given in the next theorem.

Theorem 7.11.1. Let $\mathbf{X}$ be a random vector with a continuous density function $f(\mathbf{x})$. Let $\mathbf{g}: D \to R^n$, where D is an open subset of R^n such that $P(\mathbf{X} \in D) = 1$. Suppose that $\mathbf{g}$ satisfies the conditions of the inverse function theorem (Theorem 7.6.1), namely the following:

   i. $\mathbf{g}$ has continuous first-order partial derivatives in D.
  ii. The Jacobian matrix $\mathbf{J_g}(\mathbf{x})$ is nonsingular in D, that is,

$$\det[\mathbf{J_g}(\mathbf{x})] = \frac{\partial(g_1, g_2, \ldots, g_n)}{\partial(x_1, x_2, \ldots, x_n)} \neq 0$$

for all $\mathbf{x} \in D$, where g_i is the ith element of $\mathbf{g}$ $(i = 1, 2, \ldots, n)$.

Then the density function of $\mathbf{Y} = \mathbf{g}(\mathbf{X})$ is given by

$$h(\mathbf{y}) = f\left[\mathbf{g}^{-1}(\mathbf{y})\right]\left|\det\left[\mathbf{J}_{\mathbf{g}^{-1}}(\mathbf{y})\right]\right|,$$

where $\mathbf{g}^{-1}$ is the inverse function of $\mathbf{g}$.

Proof. By Theorem 7.6.1, the inverse function of $\mathbf{g}$ exists. Let us therefore write $\mathbf{X} = \mathbf{g}^{-1}(\mathbf{Y})$. Now, the cumulative distribution function of $\mathbf{Y}$ is

$$H(\mathbf{y}) = P\left[g_1(\mathbf{X}) \leq y_1, \quad g_2(\mathbf{X}) \leq y_2, \dots, \quad g_n(\mathbf{X}) \leq y_n\right]$$

$$= \int_{A_n} f(\mathbf{x})\, d\mathbf{x}, \tag{7.75}$$

where $A_n = \{\mathbf{x} \in D | g_i(\mathbf{x}) \leq y_i,\ i = 1, 2, \dots, n\}$. If we make the change of variable $\mathbf{w} = \mathbf{g}(\mathbf{x})$ in formula (7.75), then, by applying Theorem 7.9.8 with $\mathbf{g}^{-1}(\mathbf{w})$ used instead of $\mathbf{g}(\mathbf{u})$, we obtain

$$\int_{A_n} f(\mathbf{x})\, d\mathbf{x} = \int_{B_n} f\left[\mathbf{g}^{-1}(\mathbf{w})\right]\left|\det\left[\mathbf{J}_{\mathbf{g}^{-1}}(\mathbf{w})\right]\right| d\mathbf{w},$$

where $B_n = \mathbf{g}(A_n) = \{\mathbf{g}(\mathbf{x})| g_i(\mathbf{x}) \leq y_i,\ i = 1, 2, \dots, n\}$. Thus

$$H(\mathbf{y}) = \int_{-\infty}^{y_1}\int_{-\infty}^{y_2} \cdots \int_{-\infty}^{y_n} f\left[\mathbf{g}^{-1}(\mathbf{w})\right]\left|\det\left[\mathbf{J}_{\mathbf{g}^{-1}}(\mathbf{w})\right]\right| d\mathbf{w}.$$

It follows that the density function of $\mathbf{Y}$ is

$$h(\mathbf{y}) = f\left[\mathbf{g}^{-1}(\mathbf{y})\right]\left|\frac{\partial\left(g_1^{-1}, g_2^{-1}, \dots, g_n^{-1}\right)}{\partial(y_1, y_2, \dots, y_n)}\right|, \tag{7.76}$$

where g_i^{-1} is the ith element of $\mathbf{g}^{-1}$ $(i = 1, 2, \dots, n)$.   □

EXAMPLE 7.11.1.   Let $\mathbf{X} = (X_1, X_2)'$, where X_1 and X_2 are independent random variables that have the standard normal distribution. Here, the density function of $\mathbf{X}$ is the product of the density functions of X_1 and X_2. Thus

$$f(\mathbf{x}) = \frac{1}{2\pi}\exp\left[-\frac{1}{2}\left(x_1^2 + x_2^2\right)\right], \qquad -\infty < x_1,\ x_2 < \infty.$$

Let $\mathbf{Y} = (Y_1, Y_2)'$ be defined as

$$Y_1 = X_1 + X_2,$$
$$Y_2 = X_1 - 2X_2.$$

In this case, the set D in Theorem 7.11.1 is R^2, $g_1(\mathbf{x}) = x_1 + x_2$, $g_2(\mathbf{x}) = x_1 - 2x_2$, $g_1^{-1}(\mathbf{y}) = x_1 = \frac{1}{3}(2y_1 + y_2)$, $g_2^{-1}(\mathbf{y}) = x_2 = \frac{1}{3}(y_1 - y_2)$, and

$$\frac{\partial(g_1^{-1}, g_2^{-1})}{\partial(y_1, y_2)} = \det\begin{bmatrix} \frac{2}{3} & \frac{1}{3} \\ \frac{1}{3} & -\frac{1}{3} \end{bmatrix} = -\frac{1}{3}.$$

Hence, by formula (7.76), the density function of $\mathbf{y}$ is

$$h(\mathbf{y}) = \frac{1}{2\pi} \exp\left[-\frac{1}{2}\left(\frac{2y_1 + y_2}{3}\right)^2 - \frac{1}{2}\left(\frac{y_1 - y_2}{3}\right)^2 \right] \times \frac{1}{3}$$

$$= \frac{1}{6\pi} \exp\left[-\frac{1}{18}(5y_1^2 + 2y_1 y_2 + 2y_2^2) \right], \qquad -\infty < y_1, y_2 < \infty.$$

EXAMPLE 7.11.2. Suppose that it is desired to determine the density function of the random variable $V = X_1 + X_2$, where $X_1 \geq 0$, $X_2 \geq 0$, and $\mathbf{X} = (X_1, X_2)'$ has a continuous density function $f(x_1, x_2)$. This can be accomplished in two ways:

i. Let $Q(v)$ denote the cumulative distribution function of V and let $q(v)$ be its density function. Then

$$Q(v) = P(X_1 + X_2 \leq v)$$

$$= \int\int_A f(x_1, x_2)\, dx_1\, dx_2,$$

where $A = \{(x_1, x_2) | x_1 \geq 0,\ x_2 \geq 0,\ x_1 + x_2 \leq v\}$. We can write $Q(v)$ as

$$Q(v) = \int_0^v \left[\int_0^{v - x_2} f(x_1, x_2)\, dx_1 \right] dx_2.$$

If we now apply Theorem 7.10.2, we obtain

$$q(v) = \frac{dQ}{dv} = \int_0^v \frac{\partial}{\partial v}\left[\int_0^{v - x_2} f(x_1, x_2)\, dx_1 \right] dx_2$$

$$= \int_0^v f(v - x_2, x_2)\, dx_2. \tag{7.77}$$

ii. Consider the following transformation:

$$Y_1 = X_1 + X_2,$$

$$Y_2 = X_2.$$

Then

$$X_1 = Y_1 - Y_2,$$
$$X_2 = Y_2.$$

By Theorem 7.11.1, the density function of $\mathbf{Y} = (Y_1, Y_2)'$ is

$$h(y_1, y_2) = f(y_1 - y_2, y_2) \left| \frac{\partial(x_1, x_2)}{\partial(y_1, y_2)} \right|$$

$$= f(y_1 - y_2, y_2) \left| \det \begin{bmatrix} 1 & -1 \\ 0 & 1 \end{bmatrix} \right|$$

$$= f(y_1 - y_2, y_2), \quad y_1 \geq y_2 \geq 0.$$

By integrating y_2 out we obtain the marginal density function of $Y_1 = V$, namely,

$$q(v) = \int_0^v f(y_1 - y_2, y_2) \, dy_2 = \int_0^v f(v - x_2, x_2) \, dx_2.$$

This is identical to the density function given in formula (7.77).

7.11.2. Maximum Likelihood Estimation

Let $X_1, X_2, \ldots, X_n$ be a sample of size n from a population whose distribution depends on a set of p parameters, namely $\theta_1, \theta_2, \ldots, \theta_p$. We can regard this sample as forming a random vector $\mathbf{X} = (X_1, X_2, \ldots, X_n)'$. Suppose that $\mathbf{X}$ has the density function $f(\mathbf{x}, \boldsymbol{\theta})$, where $\mathbf{x} = (x_1, x_2, \ldots, x_n)'$ and $\boldsymbol{\theta} = (\theta_1, \theta_2, \ldots, \theta_p)'$. This density function is usually referred to as the likelihood function of $\mathbf{X}$; we denote it by $L(\mathbf{x}, \boldsymbol{\theta})$.

For a given sample, the maximum likelihood estimate of $\boldsymbol{\theta}$, denoted by $\hat{\boldsymbol{\theta}}$, is the value of $\boldsymbol{\theta}$ that maximizes $L(\mathbf{x}, \boldsymbol{\theta})$. If $L(\mathbf{x}, \boldsymbol{\theta})$ has partial derivatives with respect to $\theta_1, \theta_2, \ldots, \theta_p$, then $\hat{\boldsymbol{\theta}}$ is often obtained by solving the equations

$$\frac{\partial L(\mathbf{x}, \hat{\boldsymbol{\theta}})}{\partial \theta_i} = 0, \quad i = 1, 2, \ldots, p.$$

In most situations, it is more convenient to work with the natural logarithm of $L(\mathbf{x}, \boldsymbol{\theta})$; its maxima are attained at the same points as those of $L(\mathbf{x}, \boldsymbol{\theta})$. Thus $\hat{\boldsymbol{\theta}}$ satisfies the equation

$$\frac{\partial \log \left[L(\mathbf{x}, \hat{\boldsymbol{\theta}}) \right]}{\partial \theta_i} = 0, \quad i = 1, 2, \ldots, p. \tag{7.78}$$

Equations (7.78) are known as the *likelihood equations*.

EXAMPLE 7.11.3. Suppose that $X_1, X_2, \ldots, X_n$ form a sample of size n from a normal distribution with an unknown mean μ and a variance σ^2. Here, $\boldsymbol{\theta} = (\mu, \sigma^2)'$, and the likelihood function is given by

$$L(\mathbf{x}, \boldsymbol{\theta}) = \frac{1}{(2\pi\sigma^2)^{n/2}} \exp\left[-\frac{1}{2\sigma^2} \sum_{i=1}^{n} (x_i - \mu)^2 \right],$$

Let $L^*(\mathbf{x}, \boldsymbol{\theta}) = \log L(\mathbf{x}, \boldsymbol{\theta})$. Then

$$L^*(\mathbf{x}, \boldsymbol{\theta}) = -\frac{1}{2\sigma^2} \sum_{i=1}^{n} (x_i - \mu)^2 - \frac{n}{2} \log(2\pi\sigma^2).$$

The likelihood equations in formula (7.78) are of the form

$$\frac{\partial L^*}{\partial \mu} = \frac{1}{\sigma^2} \sum_{i=1}^{n} (x_i - \hat{\mu}) = 0 \tag{7.79}$$

$$\frac{\partial L^*}{\partial \sigma^2} = \frac{1}{2\hat{\sigma}^4} \sum_{i=1}^{n} (x_i - \hat{\mu})^2 - \frac{n}{2\hat{\sigma}^2} = 0. \tag{7.80}$$

Equations (7.79) and (7.80) can be written as

$$n(\bar{x} - \hat{\mu}) = 0, \tag{7.81}$$

$$\sum_{i=1}^{n} (x_i - \hat{\mu})^2 - n\hat{\sigma}^2 = 0, \tag{7.82}$$

where $\bar{x} = (1/n)\sum_{i=1}^{n} x_i$. If $n \geq 2$, then equations (7.81) and (7.82) have the solution

$$\hat{\mu} = \bar{x},$$

$$\hat{\sigma}^2 = \frac{1}{n} \sum_{i=1}^{n} (x_i - \bar{x})^2.$$

These are the maximum likelihood estimates of μ and σ^2, respectively.

It can be verified that $\hat{\mu}$ and $\hat{\sigma}^2$ are indeed the values of μ and σ^2 that maximize $L^*(\mathbf{x}, \boldsymbol{\theta})$. To show this, let us consider the Hessian matrix $\mathbf{A}$ of second-order partial derivatives of L^* (see formula 7.34),

$$\mathbf{A} = \begin{bmatrix} \dfrac{\partial^2 L^*}{\partial \mu^2} & \dfrac{\partial^2 L^*}{\partial \mu\, \partial \sigma^2} \\[2ex] \dfrac{\partial^2 L^*}{\partial \mu\, \partial \sigma^2} & \dfrac{\partial^2 L^*}{\partial \sigma^4} \end{bmatrix}.$$

Hence, for $\mu = \hat{\mu}$ and $\sigma^2 = \hat{\sigma}^2$,

$$\frac{\partial^2 L^*}{\partial \mu^2} = -\frac{n}{\hat{\sigma}^2},$$

$$\frac{\partial^2 L^*}{\partial \mu \, \partial \sigma^2} = -\frac{1}{\hat{\sigma}^4} \sum_{i=1}^{n} (x_i - \hat{\mu}) = 0,$$

$$\frac{\partial^2 L^*}{\partial \sigma^4} = -\frac{n}{2\hat{\sigma}^4}.$$

Thus $\partial^2 L^*/\partial \mu^2 < 0$ and $\det(\mathbf{A}) = n^2/2\hat{\sigma}^6 > 0$. Therefore, by Corollary 7.7.1, $(\hat{\mu}, \hat{\sigma}^2)$ is a point of local maximum of L^*. Since it is the only maximum, it must also be the absolute maximum.

Maximum likelihood estimators have interesting asymptotic properties. For more information on these properties, see, for example, Bickel and Doksum (1977, Section 4.4).

7.11.3. Comparison of Two Unbiased Estimators

Let X_1 and X_2 be two unbiased estimators of a parameter μ. Suppose that $\mathbf{X} = (X_1, X_2)'$ has the density function $f(x_1, x_2)$, $-\infty < x_1, x_2 < \infty$. To compare these estimators, we may consider the probability that one estimator, for example, X_1, is closer to μ than the other, X_2, that is,

$$p = P\big[|X_1 - \mu| < |X_2 - \mu|\big].$$

This probability can be expressed as

$$p = \int\int_D f(x_1, x_2) \, dx_1 \, dx_2, \tag{7.83}$$

where $D = \{(x_1, x_2)| \, |x_1 - \mu| < |x_2 - \mu|\}$. Let us now make the following change of variables using polar coordinates:

$$x_1 - \mu = r \cos \theta, \qquad x_2 - \mu = r \sin \theta.$$

By applying formula (7.67), the integral in (7.83) can be written as

$$p = \int\int_{D'} \tilde{g}(r, \theta) \left| \frac{\partial(x_1, x_2)}{\partial(r, \theta)} \right| dr \, d\theta$$

$$= \int\int_{D'} \tilde{g}(r, \theta) r \, dr \, d\theta,$$

where $\tilde{g}(r, \theta) = f(\mu + r \cos \theta, \mu + r \sin \theta)$ and

$$D'\left\{(r, \theta)|0 \le r < \infty, \frac{\pi}{4} \le \theta \le \frac{3\pi}{4}, \frac{5\pi}{4} \le \theta \le \frac{7\pi}{4}\right\}.$$

In particular, if **X** has the bivariate normal density, then

$$f(x_1, x_2) = \frac{1}{2\pi\sigma_1\sigma_2(1 - \rho^2)^{1/2}}$$

$$\times \exp\left\{-\frac{1}{2(1 - \rho^2)}\left[\frac{(x_1 - \mu)^2}{\sigma_1^2} - \frac{2\rho(x_1 - \mu)(x_2 - \mu)}{\sigma_1\sigma_2}\right.\right.$$

$$\left.\left. + \frac{(x_2 - \mu)^2}{\sigma_2^2}\right]\right\}, \qquad -\infty < x_1, x_2 < \infty,$$

and

$$\tilde{g}(r, \theta) = \frac{1}{2\pi\sigma_1\sigma_2(1 - \rho^2)^{1/2}}$$

$$\times \exp\left\{-\frac{r^2}{2(1 - \rho^2)}\left[\frac{\cos^2 \theta}{\sigma_1^2} - \frac{2\rho \cos \theta \sin \theta}{\sigma_1\sigma_2} + \frac{\sin^2 \theta}{\sigma_2^2}\right]\right\},$$

where σ_1^2 and σ_2^2 are the variances of X_1 and X_2, respectively, and ρ is their correlation coefficient. In this case,

$$p = 2\int_{\pi/4}^{3\pi/4}\left[\int_0^\infty \tilde{g}(r, \theta)r\,dr\right]d\theta.$$

It can be shown (see Lowerre, 1983) that

$$p = 1 - \frac{1}{\pi}\text{Arctan}\left[\frac{2\sigma_1\sigma_2(1 - \rho^2)^{1/2}}{\sigma_2^2 - \sigma_1^2}\right] \qquad (7.84)$$

if $\sigma_2 > \sigma_1$. A large value of p indicates that X_1 is closer to μ than X_2, which means that X_1 is a better estimator of μ than X_2.

7.11.4. Best Linear Unbiased Estimation

Let $X_1, X_2, \ldots, X_n$ be independent and identically distributed random variables with a common mean μ and a common variance σ^2. An estimator of the form $\hat{\phi} = \sum_{i=1}^{n} a_i X_i$, where the a_i's are constants, is said to be a linear estimator of μ. This estimator is unbiased if $E(\hat{\phi}) = \mu$, that is, if $\sum_{i=1}^{n} a_i = 1$, since $E(X_i) = \mu$ for $i = 1, 2, \ldots, n$. The variance of $\hat{\phi}$ is given by

$$\text{Var}(\hat{\phi}) = \sigma^2 \sum_{i=1}^{n} a_i^2.$$

The smaller the variance of $\hat{\phi}$, the more efficient $\hat{\phi}$ is as an estimator of μ. In particular, if $a_1, a_2, \ldots, a_n$ are chosen so that $\text{Var}(\hat{\phi})$ attains a minimum value, then $\hat{\phi}$ will have the smallest variance among all unbiased linear estimators of μ. In this case, $\hat{\phi}$ is called the *best linear unbiased estimator* (BLUE) of μ.

Thus to find the BLUE of μ we need to minimize the function $f = \sum_{i=1}^{n} a_i^2$ subject to the constraint $\sum_{i=1}^{n} a_i = 1$. This minimization problem can be solved using the method of Lagrange multipliers. Let us therefore write F [see formula (7.41)] as

$$F = \sum_{i=1}^{n} a_i^2 + \lambda \left(\sum_{i=1}^{n} a_i - 1 \right),$$

$$\frac{\partial F}{\partial a_i} = 2a_i + \lambda = 0, \qquad i = 1, 2, \ldots, n.$$

Hence, $a_i = -\lambda/2$ $(i = 1, 2, \ldots, n)$. Using the constraint $\sum_{i=1}^{n} a_i = 1$, we conclude that $\lambda = -2/n$. Thus $a_i = 1/n$, $i = 1, 2, \ldots, n$. To verify that this solution minimizes f, we need to consider the signs of $\Delta_1, \Delta_2, \ldots, \Delta_{n-1}$, where Δ_i is the determinant of $\mathbf{B}_i$ (see Section 7.8). Here, $\mathbf{B}_1$ is an $(n+1) \times (n+1)$ matrix of the form

$$\mathbf{B}_1 = \begin{bmatrix} 2\mathbf{I}_n & \mathbf{1}_n \\ \mathbf{1}'_n & 0 \end{bmatrix}.$$

It follows that

$$\Delta_1 = \det(\mathbf{B}_1) = -\frac{n 2^n}{2} < 0,$$

$$\Delta_2 = -\frac{(n-1) 2^{n-1}}{2} < 0,$$

$$\vdots$$

$$\Delta_{n-1} = -2^2 < 0.$$

Since the number of constraints, $m = 1$, is odd, then by the sufficient conditions described in Section 7.8 we must have a local minimum when $a_i = 1/n$, $i = 1, 2, \ldots, n$. Since this is the only local minimum in R^n, it must be the absolute minimum. Note that for such values of $a_1, a_2, \ldots, a_n$, $\hat{\phi}$ is the sample mean $\bar{X}_n$. We conclude that the sample mean is the most efficient (in terms of variance) unbiased linear estimator of μ.

7.11.5. Optimal Choice of Sample Sizes in Stratified Sampling

In stratified sampling, a finite population of N units is divided into r subpopulations, called strata, of sizes $N_1, N_2, \ldots, N_r$. From each stratum a random sample is drawn, and the drawn samples are obtained independently in the different strata. Let n_i be the size of the sample drawn from the ith stratum $(i = 1, 2, \ldots, r)$. Let y_{ij} denote the response value obtained from the jth unit within the ith stratum $(i = 1, 2, \ldots, r; \ j = 1, 2, \ldots, n_i)$. The population mean $\bar{Y}$ is

$$\bar{Y} = \frac{1}{N} \sum_{i=1}^{r} \sum_{j=1}^{N_i} y_{ij} = \frac{1}{N} \sum_{i=1}^{r} N_i \bar{Y}_i,$$

where $\bar{Y}_i$ is the true mean for the ith stratum $(i = 1, 2, \ldots, r)$. A stratified estimate of $\bar{Y}$ is $\bar{y}_{st}$ (st for stratified), where

$$\bar{y}_{st} = \frac{1}{N} \sum_{i=1}^{r} N_i \bar{y}_i,$$

in which $\bar{y}_i = (1/n_i)\sum_{j=1}^{n_i} y_{ij}$ is the mean of the sample from the ith stratum $(i = 1, 2, \ldots, r)$. If, in every stratum, $\bar{y}_i$ is unbiased for $\bar{Y}_i$, then $\bar{y}_{st}$ is an unbiased estimator of $\bar{Y}$. The variance of $\bar{y}_{st}$ is

$$\mathrm{Var}(\bar{y}_{st}) = \frac{1}{N^2} \sum_{i=1}^{r} N_i^2 \, \mathrm{Var}(\bar{y}_i).$$

Since $\bar{y}_i$ is the mean of a random sample from a finite population, then its variance is given by (see Cochran, 1963, page 22)

$$\mathrm{Var}(\bar{y}_i) = \frac{S_i^2}{n_i}(1 - f_i), \qquad i = 1, 2, \ldots, r,$$

where $f_i = n_i/N_i$, and

$$S_i^2 = \frac{1}{N_i - 1} \sum_{j=1}^{N_i} \left(y_{ij} - \bar{Y}_i \right)^2.$$

Hence,

$$\mathrm{Var}(\bar{y}_{\mathrm{st}}) = \sum_{i=1}^{r} \frac{1}{n_i} L_i^2 S_i^2 (1 - f_i),$$

where $L_i = N_i/N$ $(i = 1, 2, \ldots, r)$.

The sample sizes $n_1, n_2, \ldots, n_r$ can be chosen by the sampler in an optimal way, the optimality criterion being the minimization of $\mathrm{Var}(\bar{y}_{\mathrm{st}})$ for a specified cost of taking the samples. Here, the cost is defined by the formula

$$\mathrm{cost} = c_0 + \sum_{i=1}^{r} c_i n_i,$$

where c_i is the cost per unit in the ith stratum $(i = 1, 2, \ldots, r)$ and c_0 is the overhead cost. Thus the optimal choice of the sample sizes is reduced to finding the values of $n_1, n_2, \ldots, n_r$ that minimize $\sum_{i=1}^{r}(1/n_i)L_i^2 S_i^2 (1 - f_i)$ subject to the constraint

$$\sum_{i=1}^{r} c_i n_i = d - c_0, \tag{7.85}$$

where d is a constant. Using the method of Lagrange multipliers, we write

$$\begin{aligned}
F &= \sum_{i=1}^{r} \frac{1}{n_i} L_i^2 S_i^2 (1 - f_i) + \lambda \left(\sum_{i=1}^{r} c_i n_i + c_0 - d \right) \\
&= \sum_{i=1}^{r} \frac{1}{n_i} L_i^2 S_i^2 - \sum_{i=1}^{r} \frac{1}{N_i} L_i^2 S_i^2 + \lambda \left(\sum_{i=1}^{r} c_i n_i + c_0 - d \right).
\end{aligned}$$

Differentiating with respect to n_i $(i = 1, 2, \ldots, r)$, we obtain

$$\frac{\partial F}{\partial n_i} = - \frac{1}{n_i^2} L_i^2 S_i^2 + \lambda c_i = 0, \qquad i = 1, 2, \ldots, r,$$

Thus

$$n_i = (\lambda c_i)^{-1/2} L_i S_i, \qquad i = 1, 2, \ldots, r.$$

By substituting n_i in the equality constraint (7.85) we get

$$\sqrt{\lambda} = \frac{\sum_{i=1}^{r} \sqrt{c_i}\, L_i S_i}{d - c_0}.$$

Therefore,

$$n_i = \frac{(d - c_0) N_i S_i}{\sqrt{c_i} \, \Sigma_{j=1}^{r} \sqrt{c_j} \, N_j S_j}, \qquad i = 1, 2, \ldots, r. \tag{7.86}$$

It is easy to verify (using the sufficient conditions in Section 7.8) that the values of $n_1, n_2, \ldots, n_r$ given by equation (7.86) minimize $\text{Var}(\bar{y}_{\text{st}})$ under the constraint of equality (7.85). We conclude that $\text{Var}(\bar{y}_{\text{st}})$ is minimized when n_i is proportional to $(1/\sqrt{c_i}) N_i S_i$ $(i = 1, 2, \ldots, r)$. Consequently, n_i must be large if the corresponding stratum is large, if the cost of sampling per unit in that stratum is low, or if the variability within the stratum is large.

FURTHER READING AND ANNOTATED BIBLIOGRAPHY

Bickel, P. J., and K. A. Doksum (1977). *Mathematical Statistics*, Holden-Day, San Francisco. (Chap. 1 discusses distribution theory for transformation of random vectors.)

Brownlee, K. A. (1965). *Statistical Theory and Methodology*, 2nd ed. Wiley, New York. (See Section 9.8 with regard to the Behrens–Fisher test.)

Cochran, W. G. (1963). *Sampling Techniques*, 2nd ed. Wiley, New York. (This is a classic book on sampling theory as developed for use in sample surveys.)

Corwin, L. J., and R. H. Szczarba (1982). *Multivariate Calculus*. Marcel Dekker, New York. (This is a useful book that provides an introduction to multivariable calculus. The topics covered include continuity, differentiation, multiple integrals, line and surface integrals, differential forms, and infinite series.)

Fulks, W. (1978). *Advanced Calculus*, 3rd ed. Wiley, New York. (Chap. 8 discusses limits and continuity for a multivariable function; Chap. 10 covers the inverse function theorem; Chap. 11 discusses multiple integration.)

Gillespie, R. P. (1954). *Partial Differentiation*. Oliver and Boyd, Edinburgh, Scotland. (This concise book provides a brief introduction to multivariable calculus. It covers partial differentiation, Taylor's theorem, and maxima and minima of functions of several variables.)

Kaplan, W. (1991). *Advanced Calculus*, 4th ed. Addison-Wesley, Redwood City, California. (Topics pertaining to multivariable calculus are treated in several chapters including Chaps. 2, 3, 4, 5, and 6.)

Kaplan, W., and D. J. Lewis (1971). *Calculus and Linear Algebra*, Vol. II. Wiley, New York. (Chap. 12 gives a brief introduction to differential calculus of a multivariable function; Chap. 13 covers multiple integration.)

Lindgren, B. W. (1976). *Statistical Theory*, 3rd ed. Macmillan, New York. (Multivariate transformations are discussed in Chap. 10.)

Lowerre, J. M. (1983). "An integral of the bivariate normal and an application." *Amer. Statist.*, **37**, 235–236.

Rudin, W. (1964). *Principles of Mathematical Analysis*, 2nd ed. McGraw-Hill, New York. (Chap. 9 includes a study of multivariable functions.)

Sagan, H. (1974). *Advanced Calculus*. Houghton Mifflin, Boston. (Chap. 9 covers differential calculus of a multivariable function; Chap. 10 deals with the inverse function and implicit function theorems; Chap. 11 discusses multiple integration.)

Satterthwaite, F. E. (1946). "An approximate distribution of estimates of variance components." *Biometrics Bull.*, **2**, 110–114.

Taylor, A. E., and W. R. Mann (1972). *Advanced Calculus*, 2nd ed. Wiley, New York. (This book contains several chapters on multivariable calculus with many helpful exercises.)

Thibaudeau, Y., and G. P. H. Styan (1985). "Bounds for Chakrabarti's measure of imbalance in experimental design." In *Proceedings of the First International Tampere Seminar on Linear Statistical Models and Their Applications*, T. Pukkila and S. Puntanen, eds. University of Tampere, Tampere, Finland, pp. 323–347.

Wen, L. (2001). "A counterexample for the two-dimensional density function." *Amer. Math. Monthly*, **108**, 367–368.

EXERCISES

In Mathematics

7.1. Let $f(x_1, x_2)$ be a function defined on R^2 as

$$
f(x_1, x_2) = \begin{cases} \dfrac{|x_1|}{x_2^2} \exp\left(-\dfrac{|x_1|}{x_2^2}\right), & x_2 \neq 0, \\[3mm] 0, & x_2 = 0. \end{cases}
$$

(a) Show that $f(x_1, x_2)$ has a limit equal to zero as $\mathbf{x} = (x_1, x_2)' \to \mathbf{0}$ along any straight line through the origin.

(b) Show that $f(x_1, x_2)$ does not have a limit as $\mathbf{x} \to \mathbf{0}$.

7.2. Prove Lemma 7.3.1.

7.3. Prove Lemma 7.3.2.

7.4. Prove Lemma 7.3.3.

7.5. Consider the function

$$
f(x_1, x_2) = \begin{cases} \dfrac{x_1 x_2}{x_1^2 + x_2^2}, & (x_1, x_2) \neq (0, 0), \\[3mm] 0, & (x_1, x_2) = (0, 0). \end{cases}
$$

(a) Show that $f(x_1, x_2)$ is not continuous at the origin.

(b) Show that the partial derivatives of $f(x_1, x_2)$ with respect to x_1 and x_2 exist at the origin.

[*Note:* This exercise shows that a multivariable function does not have to be continuous at a point in order for its partial derivatives to exist at that point.]

7.6. The function $f(x_1, x_2, \ldots, x_k)$ is said to be homogeneous of degree n in $x_1, x_2, \ldots, x_k$ if for any nonzero scalar t,

$$f(tx_1, tx_2, \ldots, tx_k) = t^n f(x_1, x_2, \ldots, x_k)$$

for all $\mathbf{x} = (x_1, x_2, \ldots, x_k)'$ in the domain of f. Show that if $f(x_1, x_2, \ldots, x_k)$ is homogeneous of degree n, then

$$\sum_{i=1}^{k} x_i \frac{\partial f}{\partial x_i} = nf$$

[*Note:* This result is known as Euler's theorem for homogeneous functions.]

7.7. Consider the function

$$f(x_1, x_2) = \begin{cases} \dfrac{x_1^2 x_2}{x_1^4 + x_2^2}, & (x_1, x_2) \neq (0,0), \\ 0, & (x_1, x_2) = (0,0). \end{cases}$$

(a) Is f continuous at the origin? Why or why not?

(b) Show that f has a directional derivative in every direction at the origin.

7.8. Let S be a surface defined by the equation $f(\mathbf{x}) = c_0$, where $\mathbf{x} = (x_1, x_2, \ldots, x_k)'$ and c_0 is a constant. Let C denote a curve on S given by the equations $x_1 = g_1(t)$, $x_2 = g_2(t)$, $\ldots$, $x_k = g_k(t)$, where $g_1, g_2, \ldots, g_k$ are differentiable functions. Let s be the arc length of C measured from some fixed point in such a way that s increases with t. The curve can then be parameterized, using s instead of t, in the form $x_1 = h_1(s)$, $x_2 = h_2(s)$, $\ldots$, $x_k = h_k(s)$. Suppose that f has partial derivatives with respect to $x_1, x_2, \ldots, x_k$.

Show that the directional derivative of f at a point $\mathbf{x}$ on C in the direction of $\mathbf{v}$, where $\mathbf{v}$ is a unit tangent vector to C at $\mathbf{x}$ (in the direction of increasing s), is equal to df/ds.

7.9. Use Taylor's expansion in a neighborhood of the origin to obtain a second-order approximation for each of the following functions:

(a) $f(x_1, x_2) = \exp(x_2 \sin x_1)$.
(b) $f(x_1, x_2, x_3) = \sin(e^{x_1} + x_2^2 + x_3^3)$.
(c) $f(x_1, x_2) = \cos(x_1 x_2)$.

7.10. Suppose that $f(x_1, x_2)$ and $g(x_1, x_2)$ are continuously differentiable functions in a neighborhood of a point $\mathbf{x}_0 = (x_{10}, x_{20})'$. Consider the equation $u_1 = f(x_1, x_2)$. Suppose that $\partial f / \partial x_1 \neq 0$ at $\mathbf{x}_0$.
   (a) Show that

$$\frac{\partial x_1}{\partial x_2} = -\frac{\partial f}{\partial x_2} \Big/ \frac{\partial f}{\partial x_1}$$

   in a neighborhood of $\mathbf{x}_0$.
   (b) Suppose that in a neighborhood of $\mathbf{x}_0$,

$$\frac{\partial(f, g)}{\partial(x_1, x_2)} = 0.$$

Show that

$$\frac{\partial g}{\partial x_1} \frac{\partial x_1}{\partial x_2} + \frac{\partial g}{\partial x_2} = 0,$$

   that is, g is actually independent of x_2 in a neighborhood of $\mathbf{x}_0$.
   (c) Deduce from (b) that there exists a function $\phi: D \to R$, where $D \subset R$ is a neighborhood of $f(\mathbf{x}_0)$, such that

$$g(x_1, x_2) = \phi[f(x_1, x_2)]$$

   throughout a neighborhood of $\mathbf{x}_0$. In this case, the functions f and g are said to be functionally dependent.
   (d) Show that if f and g are functionally dependent, then

$$\frac{\partial(f, g)}{\partial(x_1, x_2)} = 0.$$

   [*Note:* From (b), (c), and (d) we conclude that f and g are functionally dependent on a set $\Delta \subset R^2$ if and only if $\partial(f, g)/\partial(x_1, x_2) = 0$ in Δ.]

7.11. Consider the equation

$$x_1 \frac{\partial u}{\partial x_1} + x_2 \frac{\partial u}{\partial x_2} + x_3 \frac{\partial u}{\partial x_3} = nu.$$

Let $\xi_1 = x_1/x_3$, $\xi_2 = x_2/x_3$, $\xi_3 = x_3$. Use this change of variables to show that the equation can be written as

$$\xi_3 \frac{\partial u}{\partial \xi_3} = nu.$$

Deduce that u is of the form

$$u = x_3^n F\left(\frac{x_1}{x_3}, \frac{x_2}{x_3}\right).$$

7.12. Let u_1 and u_2 be defined as

$$u_1 = x_1\left(1 - x_2^2\right)^{1/2} + x_2\left(1 - x_1^2\right)^{1/2},$$
$$u_2 = \left(1 - x_1^2\right)^{1/2}\left(1 - x_2^2\right)^{1/2} - x_1 x_2.$$

Show that u_1 and u_2 are functionally dependent.

7.13. Let $\mathbf{f}: R^3 \to R^3$ be defined as

$$\mathbf{u} = f(\mathbf{x}), \qquad \mathbf{x} = (x_1, x_2, x_3)', \qquad \mathbf{u} = (u_1, u_2, u_3)',$$

where $u_1 = x_1^3$, $u_2 = x_2^3$, $u_3 = x_3^3$.

(a) Show that the Jacobian matrix of $\mathbf{f}$ is not nonsingular in any subset $D \subset R^3$ that contains points on ay of the coordinate planes.

(b) Show that $\mathbf{f}$ has a unique inverse everywhere in R^3 including any subset D of the type described in (a).

[*Note:* This exercise shows that the nonvanishing of the Jacobian determinant in Theorem 7.6.1 (inverse function theorem) is a sufficient condition for the existence of an inverse function, but is not necessary.]

7.14. Consider the equations

$$g_1(x_1, x_2, y_1, y_2) = 0,$$
$$g_2(x_1, x_2, y_1, y_2) = 0,$$

where g_1 and g_2 are differentiable functions defined on a set $D \subset R^4$. Suppose that $\partial(g_1, g_2)/\partial(x_1, x_2) \neq 0$ in D. Show that

$$\frac{\partial x_1}{\partial y_1} = -\frac{\partial(g_1, g_2)}{\partial(y_1, x_2)} \bigg/ \frac{\partial(g_1, g_2)}{\partial(x_1, x_2)},$$

$$\frac{\partial x_2}{\partial y_1} = -\frac{\partial(g_1, g_2)}{\partial(x_1, y_1)} \bigg/ \frac{\partial(g_1, g_2)}{\partial(x_1, x_2)}.$$

7.15. Let $f(x_1, x_2, x_3) = 0$, $g(x_1, x_2, x_3) = 0$, where f and g are differentiable functions defined on a set $D \subset R^3$. Suppose that

$$\frac{\partial(f, g)}{\partial(x_2, x_3)} \neq 0, \qquad \frac{\partial(f, g)}{\partial(x_3, x_1)} \neq 0, \qquad \frac{\partial(f, g)}{\partial(x_1, x_2)} \neq 0$$

in D. Show that

$$\frac{dx_1}{\partial(f, g)/\partial(x_2, x_3)} = \frac{dx_2}{\partial(f, g)/\partial(x_3, x_1)} = \frac{dx_3}{\partial(f, g)/\partial(x_1, x_2)}.$$

7.16. Determine the stationary points of the following functions and check for local minima and maxima:

(a) $f = x_1^2 + x_2^2 + x_1 + x_2 + x_1 x_2$.

(b) $f = 2\alpha x_1^2 - x_1 x_2 + x_2^2 + x_1 - x_2 + 1$, where α is a scalar. Can α be chosen so that the stationary point is (i) a point of local minimum; (ii) a point of local maximum; (iii) a saddle point?

(c) $f = x_1^3 - 6x_1 x_2 + 3x_2^2 - 24x_1 + 4$.

(d) $f = x_1^4 + x_2^4 - 2(x_1 - x_2)^2$.

7.17. Consider the function

$$f = \frac{1 + p + \sum_{i=1}^m p_i}{\left(1 + p^2 + \sum_{i=1}^m p_i^2\right)^{1/2}},$$

which is defined on the region

$$C = \{(p_1, p_2, \ldots, p_m) \mid 0 < p \le p_i \le 1, i = 1, 2, \ldots, m\},$$

where p is a known constant. Show that

(a) $\partial f / \partial p_i$, for $i = 1, 2, \ldots, m$, vanish at exactly one point in C.

(b) The gradient vector $\nabla f = (\partial f/\partial p_1, \partial f/\partial p_2, \ldots, \partial f/\partial p_m)'$ does not vanish anywhere on the boundary of C.

(c) f attains its absolute maximum in the interior of C at the point $(p_1^o, p_2^o, \ldots, p_m^o)$, where

$$p_i^o = \frac{1 + p^2}{1 + p}, \qquad i = 1, 2, \ldots, m.$$

[*Note:* The function f was considered in an article by Thibaudeau and Styan (1985) concerning a measure of imbalance for experimental designs.]

7.18. Show that the function $f = (x_2 - x_1^2)(x_2 - 2x_1^2)$ does not have a local maximum or minimum at the origin, although it has a local minimum for $t = 0$ along every straight line given by the equations $x_1 = at, x_2 = bt$, where a and b are constants.

7.19. Find the optimal values of the function $f = x_1^2 + 12x_1x_2 + 2x_2^2$ subject to $4x_1^2 + x_2^2 = 25$. Determine the nature of the optima.

7.20. Find the minimum distance from the origin to the curve of intersection of the surfaces, $x_3(x_1 + x_2) = -2$ and $x_1x_2 = 1$.

7.21. Apply the method of Lagrange multipliers to show that

$$\left(x_1^2 x_2^2 x_3^2\right)^{1/3} \le \frac{1}{3}\left(x_1^2 + x_2^2 + x_3^2\right)$$

for all values of x_1, x_2, x_3.
[*Hint:* Find the maximum value of $f = x_1^2 x_2^2 x_3^2$ subject to $x_1^2 + x_2^2 + x_3^2 = c^2$, where c is a constant.]

7.22. Prove Theorem 7.9.3.

7.23. Evaluate the following integrals:
(a) $\int \int_D x_2 \sqrt{x_1} \, dx_1 \, dx_2$, where

$$D = \left\{(x_1, x_2) \mid x_1 > 0, x_2 > x_1^2, x_2 < 2 - x_1^2\right\}.$$

(b) $\int_0^1 [\int_0^{\sqrt{1-x_2^2}} (\frac{2}{3}x_1 + \frac{4}{3}x_2) \, dx_1] \, dx_2$.

7.24. Show that if $f(x_1, x_2)$ is continuous, then

$$\int_0^2 \left[\int_{x_1^2}^{4x_1 - x_1^2} f(x_1, x_2) \, dx_2\right] dx_1 = \int_0^4 \left[\int_{2 - \sqrt{4 - x_2}}^{\sqrt{x_2}} f(x_1, x_2) \, dx_1\right] dx_2.$$

7.25. Consider the integral

$$I = \int_0^1 \left[\int_{1 - x_1}^{1 - x_1^2} f(x_1, x_2) \, dx_2\right] dx_1.$$

(a) Write an equivalent expression for I by reversing the order of integration.
(b) If $g(x_1) = \int_{1 - x_1}^{1 - x_1^2} f(x_1, x_2) \, dx_2$, find dg/dx_1.

7.26. Evaluate $\int \int_D x_1 x_2 \, dx_1 \, dx_2$, where D is a region enclosed by the four parabolas $x_2^2 = x_1$, $x_2^2 = 2x_1$, $x_1^2 = x_2$, $x_1^2 = 2x_2$.
[*Hint:* Use a proper change of variables.]

7.27. Evaluate $\int\int\int_D(x_1^2+x_2^2)\,dx_1\,dx_2\,dx_3$, where D is a sphere of radius 1 centered at the origin.

[*Hint:* Make a change of variables using spherical polar coordinates of the form

$$x_1 = r \sin \theta \cos \phi,$$

$$x_2 = r \sin \theta \sin \phi,$$

$$x_3 = r \cos \theta,$$

$0 \le r \le 1, 0 \le \theta \le \pi, 0 \le \phi \le 2\pi.]$

7.28. Find the value of the integral

$$I = \int_0^{\sqrt{3}} \frac{dx}{\left(1+x^2\right)^3}.$$

[*Hint:* Consider the integral $\int_0^{\sqrt{3}} dx/(a+x^2)$, where $a > 0$.]

In Statistics

7.29. Suppose that the random vector $\mathbf{X} = (X_1, X_2)'$ has the density function

$$f(x_1, x_2) = \begin{cases} x_1 + x_2, & 0 < x_1 < 1, 0 < x_2 < 1, \\ 0 & \text{elsewhere}. \end{cases}$$

(a) Are the random variables X_1 and X_2 independent?
(b) Find the expected value of $X_1 X_2$.

7.30. Consider the density function $f(x_1, x_2)$ of $\mathbf{X} = (X_1, X_2)'$, where

$$f(x_1, x_2) = \begin{cases} 1, & -x_2 < x_1 < x_2, 0 < x_2 < 1, \\ 0 & \text{elsewhere}. \end{cases}$$

Show that X_1 and X_2 are uncorrelated random variables [that is, $E(X_1 X_2) = E(X_1)E(X_2)$], but are not independent.

7.31. The density function of $\mathbf{X} = (X_1, X_2)'$ is given by

$$f(x_1, x_2) = \begin{cases} \dfrac{1}{\Gamma(\alpha)\Gamma(\beta)} x_1^{\alpha-1} x_2^{\beta-1} e^{-x_1-x_2}, & 0 < x_1, x_2 < \infty, \\ 0 & \text{elsewhere}. \end{cases}$$

where $\alpha > 0$, $\beta > 0$, and $\Gamma(m)$ is the gamma function $\Gamma(m) = \int_0^\infty e^{-x} x^{m-1}\, dx$, $m > 0$. Suppose that Y_1 and Y_2 are random variables defined as

$$Y_1 = \frac{X_1}{X_1 + X_2},$$

$$Y_2 = X_1 + X_2.$$

(a) Find the joint density function of Y_1 and Y_2.
(b) Find the marginal densities of Y_1 and Y_2.
(c) Are Y_1 and Y_2 independent?

7.32. Suppose that $\mathbf{X} = (X_1, X_2)'$ has the density function

$$f(x_1, x_2) = \begin{cases} 10 x_1 x_2^2, & 0 < x_1 < x_2, 0 < x_2 < 1, \\ 0 & \text{elsewhere.} \end{cases}$$

Find the density function of $W = X_1 X_2$.

7.33. Find the density function of $W = (X_1^2 + X_2^2)^{1/2}$ given that $\mathbf{X} = (X_1, X_2)'$ has the density function

$$f(x_1, x_2) = \begin{cases} 4 x_1 x_2\, e^{-x_1^2 - x_2^2}, & x_1 > 0, \ x_2 > 0, \\ 0 & \text{elsewhere.} \end{cases}$$

7.34. Let $X_1, X_2, \ldots, X_n$ be independent random variables that have the exponential density $f(x) = e^{-x}$, $x > 0$. Let $Y_1, Y_2, \ldots, Y_n$ be n random variables defined as

$$Y_1 = X_1,$$

$$Y_2 = X_1 + X_2,$$

$$\vdots$$

$$Y_n = X_1 + X_2 + \cdots + X_n.$$

Find the density of $\mathbf{Y} = (Y_1, Y_2, \ldots, Y_n)'$, and then deduce the marginal density of Y_n.

7.35. Prove formula (7.84).

7.36. Let X_1 and X_2 be independent random variables such that $W_1 = (6/\sigma_1^2) X_1$ and $W_2 = (8/\sigma_2^2) X_2$ have the chi-squared distribution with six and eight degrees of freedom, respectively, where σ_1^2 and σ_2^2 are unknown parameters. Let $\theta = \frac{1}{7}\sigma_1^2 + \frac{1}{9}\sigma_2^2$. An unbiased estimator of θ is given by $\hat{\theta} = \frac{1}{7}X_1 + \frac{1}{9}X_2$, since X_1 and X_2 are unbiased estimators of σ_1^2 and σ_2^2, respectively.

Using Satterthwaite's approximation (see Satterthwaite, 1946), it can be shown that $\eta\hat{\theta}/\theta$ is approximately distributed as a chi-squared variate with η degrees of freedom, where η is given by

$$\eta = \frac{\left(\frac{1}{7}\sigma_1^2 + \frac{1}{9}\sigma_2^2\right)^2}{\frac{1}{6}\left(\frac{1}{7}\sigma_1^2\right)^2 + \frac{1}{8}\left(\frac{1}{9}\sigma_2^2\right)^2},$$

which can be written as

$$\eta = \frac{8(9 + 7\lambda)^2}{108 + 49\lambda^2},$$

where $\lambda = \sigma_2^2/\sigma_1^2$. It follows that the probability

$$p = P\left(\frac{\eta\hat{\theta}}{\chi_{0.025,\,\eta}^2} < \theta < \frac{\eta\hat{\theta}}{\chi_{0.975,\,\eta}^2}\right),$$

where $\chi_{\alpha,\,\eta}^2$ denotes the upper $100\alpha\%$ point of the chi-squared distribution with η degrees of freedom, is approximately equal to 0.95. Compute the exact value of p using double integration, given that $\lambda = 2$. Compare the result with the 0.95 value.

[*Notes:* (1) The density function of a chi-squared random variable with n degrees of freedom is given in Example 6.9.6. (2) In general, η is unknown. It can be estimated by $\hat{\eta}$ which results from replacing λ with $\hat{\lambda} = X_2/X_1$ in the formula for η. (3) The estimator $\hat{\theta}$ is used in the Behrens–Fisher test statistic for comparing the means of two populations with unknown variances σ_1^2 and σ_2^2, which are assumed to be unequal. If $\bar{Y}_1$ and $\bar{Y}_2$ are the means of two independent samples of sizes $n_1 = 7$ and $n_2 = 9$, respectively, randomly chosen from these populations, then θ is the variance of $\bar{Y}_1 - \bar{Y}_2$. In this case, X_1 and X_2 represent the corresponding sample variances. The Behrens–Fisher t-statistic is then given by

$$t = \frac{\bar{Y}_1 - \bar{Y}_2}{\sqrt{\hat{\theta}}}.$$

If the two population means are equal, t has approximately the t-distribution with η degrees of freedom. For more details about the Behrens–Fisher test, see for example, Brownlee (1965, Section 9.8).]

7.37. Suppose that a parabola of the form $\mu = \beta_0 + \beta_1 x + \beta_2 x^2$ is fitted to a set of paired data, $(x_1, y_1), (x_2, y_2), \ldots, (x_n, y_n)$. Obtain estimates of β_0, β_1, and β_2 by minimizing $\sum_{i=1}^{n}[y_i - (\beta_0 + \beta_1 x_i + \beta_2 x_i^2)]^2$ with respect to β_0, β_1, and β_2.
[*Note:* The estimates obtained in this manner are the least-squares estimates of β_0, β_1, and β_2.]

7.38. Suppose that we have k disjoint events $A_1, A_2, \ldots, A_k$ such that the probability of A_i is p_i $(i = 1, 2, \ldots, k)$ and $\sum_{i=1}^{k} p_i = 1$. Furthermore, suppose that among n independent trials there are $X_1, X_2, \ldots, X_k$ outcomes associated with $A_1, A_2, \ldots, A_k$, respectively. The joint probability that $X_1 = x_1, X_2 = x_2, \ldots, X_k = x_k$ is given by the likelihood function

$$L(\mathbf{x}, \mathbf{p}) = \frac{n!}{x_1! x_2! \cdots x_k!} p_1^{x_1} p_2^{x_2} \cdots p_k^{x_k},$$

where $x_i = 0, 1, 2, \ldots, n$ for $i = 1, 2, \ldots, k$ such that $\sum_{i=1}^{k} x_i = n$, $\mathbf{x} = (x_1, x_2, \ldots, x_k)'$, $\mathbf{p} = (p_1, p_2, \ldots, p_k)'$. This defines a joint distribution for $X_1, X_2, \ldots, X_k$ known as the multinomial distribution.

Find the maximum likelihood estimates of $p_1, p_2, \ldots, p_k$ by maximizing $L(\mathbf{x}, \mathbf{p})$ subject to $\sum_{i=1}^{k} p_i = 1$.
[*Hint:* Maximize the natural logarithm of $p_1^{x_1} p_2^{x_2} \cdots p_k^{x_k}$ subject to $\sum_{i=1}^{k} p_i = 1$.]

7.39. Let $\phi(y)$ be a positive, even, and continuous function on $(-\infty, \infty)$ such that $\phi(y)$ is strictly decreasing on $(0, \infty)$, and $\int_{-\infty}^{\infty} \phi(y)\, dy = 1$. Consider the following bivariate density function:

$$f(x, y) = \begin{cases} 1 + x/\phi(y), & -\phi(y) \leq x < 0, \\ 1 - x/\phi(y), & 0 \leq x \leq \phi(y), \\ 0 & \text{otherwise}. \end{cases}$$

(a) Show that $f(x, y)$ is continuous for $-\infty < x, y < \infty$.

(b) Let $F(x, y)$ be the corresponding cumulative distribution function,

$$F(x, y) = \int_{-\infty}^{x} \int_{-\infty}^{y} f(s, t)\, ds\, dt.$$

Show that if $0 < \Delta x < \phi(0)$, then

$$F(\Delta x, 0) - F(0, 0) \geq \int_{0}^{\phi^{-1}(\Delta x)} \int_{0}^{\Delta x} \left[1 - \frac{s}{\phi(t)}\right] ds\, dt$$

$$\geq \tfrac{1}{2} \Delta x\, \phi^{-1}(\Delta x),$$

where ϕ^{-1} is the inverse function of $\phi(y)$ for $0 \leq y < \infty$.

(c) Use part (b) to show that

$$\lim_{\Delta x \to 0^+} \frac{F(\Delta x, 0) - F(0,0)}{\Delta x} = \infty.$$

Hence, $\partial F(x, y)/\partial x$ does not exist at $(0,0)$.

(d) Deduce from part (c) that the equality

$$f(x, y) = \frac{\partial^2 F(x, y)}{\partial x\, \partial y}$$

does not hold in this example.

[*Note:* This example was given by Wen (2001) to demonstrate that continuity of $f(x, y)$ is not sufficient for the existence of $\partial F/\partial x$, and hence for the validity of the equality in part (d).]

CHAPTER 8

Optimization in Statistics

Optimization is an essential feature in many problems in statistics. This is apparent in almost all fields of statistics. Here are few examples, some of which will be discussed in more detail in this chapter.

1. In the theory of estimation, an estimator of an unknown parameter is sought that satisfies a certain optimality criterion such as minimum variance, maximum likelihood, or minimum average risk (as in the case of a Bayes estimator). Some of these criteria were already discussed in Section 7.11. For example, in regression analysis, estimates of the parameters of a fitted model are obtained by minimizing a certain expression that measures the closeness of the fit of the model. One common example of such an expression is the sum of the squared residuals (these are deviations of the predicted response values, as specified by the model, from the corresponding observed response values). This particular expression is used in the method of ordinary least squares. A more general class of parameter estimators is the class of M-estimators. See Huber (1973, 1981). The name "M-estimator" comes from "generalized maximum likelihood." They are based on the idea of replacing the squared residuals by another symmetric function of the residuals that has a unique minimum at zero. For example, minimizing the sum of the absolute values of the residuals produces the so-called *least absolute values* (LAV) estimators.

2. Estimates of the variance components associated with random or mixed models are obtained by using several methods. In some of these methods, the estimates are given as solutions to certain optimization problems as in maximum likelihood (ML) estimation and minimum norm quadratic unbiased estimation (MINQUE). In the former method, the likelihood function is maximized under the assumption of normally distributed data [see Hartley and Rao (1967)]. A completely different approach is used in the latter method, which was proposed by Rao (1970, 1971). This method does not require the normality assumption.

For a review of methods of estimating variance components, see Khuri and Sahai (1985).

3. In statistical inference, tests are constructed so that they are optimal in a certain sense. For example, in the Neyman–Pearson lemma (see, for example, Roussas, 1973, Chapter 13), a test is obtained by minimizing the probability of Type II error while holding the probability of Type I error at a certain level.

4. In the field of response surface methodology, design settings are chosen to minimize the prediction variance inside a region of interest, or to minimize the bias that occurs from fitting the "wrong" model. Other optimality criteria can also be considered. For example, under the D-optimality criterion, the determinant of the variance–covariance matrix of the least-squares estimator of the vector of unknown parameters (of a fitted model) is minimized with respect to the design settings.

5. Another objective of response surface methodology is the determination of optimum operating conditions on the input variables that produce maximum, or minimum, response values inside a region of interest. For example, in a particular chemical reaction setting, it may be of interest to determine the reaction temperature and the reaction time that maximize the percentage yield of a product. Optimum seeking methods in response surface methodology will be discussed in detail in Section 8.3.

6. Several response variables may be observed in an experiment for each setting of a group of input variables. Such an experiment is called a multiresponse experiment. In this case, optimization involves a number of response functions and is therefore referred to as simultaneous (or multiresponse) optimization. For example, it may be of interest to maximize the yield of a certain chemical compound while reducing the production cost. Multiresponse optimization will be discussed in Section 8.7.

7. In multivariate analysis, a large number of measurements may be available as a result of some experiment. For convenience in the analysis and interpretation of such data, it would be desirable to work with fewer of the measurements, without loss of much information. This problem of data reduction is dealt with by choosing certain linear functions of the measurements in an optimal manner. Such linear functions are called *principal components*.

Optimization of a multivariable function was discussed in Chapter 7. However, there are situations in which the optimum cannot be obtained explicitly by simply following the methods described in Chapter 7. Instead, iterative procedures may be needed. In this chapter, we shall first discuss some commonly used iterative optimization methods. A number of these methods require the explicit evaluation of the partial derivatives of the function to be optimized (objective function). These are referred to as the gradient methods. Three other optimization techniques that rely solely on the values of the objective function will also be discussed. They are called direct search methods.

8.1. THE GRADIENT METHODS

Let $f(\mathbf{x})$ be a real-valued function of k variables $x_1, x_2, \ldots, x_k$, where $\mathbf{x} = (x_1, x_2, \ldots, x_k)'$. The gradient methods are based on approximating $f(\mathbf{x})$ with a low-degree polynomial, usually of degree one or two, using Taylor's expansion. The first- and second-order partial derivatives of $f(\mathbf{x})$ are therefore assumed to exist at every point $\mathbf{x}$ in the domain of f. Without loss of generality, we shall consider that f is to be minimized.

8.1.1. The Method of Steepest Descent

This method is based on a first-order approximation of $f(\mathbf{x})$ with a polynomial of degree one using Taylor's theorem (see Section 7.5). Let $\mathbf{x}_0$ be an initial point in the domain of $f(\mathbf{x})$. Let $\mathbf{x}_0 + t\mathbf{h}_0$ be a neighboring point, where $t\mathbf{h}_0$ represents a small change in the direction of a unit vector $\mathbf{h}_0$ (that is, $t > 0$). The corresponding change in $f(\mathbf{x})$ is $f(\mathbf{x}_0 + t\mathbf{h}_0) - f(\mathbf{x}_0)$. A first-order approximation of this change is given by

$$f(\mathbf{x}_0 + t\mathbf{h}_0) - f(\mathbf{x}_0) \approx t\mathbf{h}_0' \nabla f(\mathbf{x}_0), \qquad (8.1)$$

as can be seen from applying formula (7.27). If the objective is to minimize $f(\mathbf{x})$, then $\mathbf{h}_0$ must be chosen so as to obtain the largest value for $-t\mathbf{h}_0' \nabla f(\mathbf{x}_0)$. This is a constrained maximization problem, since $\mathbf{h}_0$ has unit length. For this purpose we use the method of Lagrange multipliers. Let F be the function

$$F = -t\mathbf{h}_0' \nabla f(\mathbf{x}_0) + \lambda(\mathbf{h}_0' \mathbf{h}_0 - 1).$$

By differentiating F with respect to the elements of $\mathbf{h}_0$ and equating the derivatives to zero we obtain

$$\mathbf{h}_0 = \frac{t}{2\lambda} \nabla f(\mathbf{x}_0). \qquad (8.2)$$

Using the constraint $\mathbf{h}_0' \mathbf{h}_0 = 1$, we find that λ must satisfy the equation

$$\lambda^2 = \frac{t^2}{4} \|\nabla f(\mathbf{x}_0)\|_2^2, \qquad (8.3)$$

where $\|\nabla f(\mathbf{x}_0)\|_2$ is the Euclidean norm of $\nabla f(\mathbf{x}_0)$. In order for $-t\mathbf{h}_0' \nabla f(\mathbf{x}_0)$

to have a maximum, λ must be negative. From formula (8.3) we then have

$$\lambda = -\frac{t}{2}\|\nabla f(\mathbf{x}_0)\|_2.$$

By substituting this expression in formula (8.2) we get

$$\mathbf{h}_0 = -\frac{\nabla f(\mathbf{x}_0)}{\|\nabla f(\mathbf{x}_0)\|_2}. \tag{8.4}$$

Thus for a given $t > 0$, we can achieve a maximum reduction in $f(\mathbf{x}_0)$ by moving from $\mathbf{x}_0$ in the direction specified by $\mathbf{h}_0$ in formula (8.4). The value of t is now determined by performing a linear search in the direction of $\mathbf{h}_0$. This is accomplished by increasing the value of t (starting from zero) until no further reduction in the values of f is obtained. Let such a value of t be denoted by t_0. The corresponding value of $\mathbf{x}$ is given by

$$\mathbf{x}_1 = \mathbf{x}_0 - t_0 \frac{\nabla f(\mathbf{x}_0)}{\|\nabla f(\mathbf{x}_0)\|_2}.$$

Since the direction of $\mathbf{h}_0$ is in general not toward the location $\mathbf{x}^*$ of the true minimum of f, the above process must be performed iteratively. Thus if at stage i we have an approximation $\mathbf{x}_i$ for $\mathbf{x}^*$, then at stage $i + 1$ we have the approximation

$$\mathbf{x}_{i+1} = \mathbf{x}_i + t_i \mathbf{h}_i, \qquad i = 0, 1, 2, \dots,$$

where

$$\mathbf{h}_i = -\frac{\nabla f(\mathbf{x}_i)}{\|\nabla f(\mathbf{x}_i)\|_2}, \qquad i = 0, 1, 2, \dots,$$

and t_i is determined by a linear search in the direction of $\mathbf{h}_i$, that is, t_i is the value of t that minimizes $f(\mathbf{x}_i + t\mathbf{h}_i)$. Note that if it is desired to maximize f, then for each i (≥ 0) we need to move in the direction of $-\mathbf{h}_i$. In this case, the method is called the method of steepest ascent.

Convergence of the method of steepest descent can be very slow, since frequent changes of direction may be necessary. Another reason for slow convergence is that the direction of $\mathbf{h}_i$ at the ith iteration may be nearly perpendicular to the direction toward the minimum. Furthermore, the method becomes inefficient when the first-order approximation of f is no longer adequate. In this case, a second-order approximation should be attempted. This will be described in the next section.

8.1.2. The Newton–Raphson Method

Let $\mathbf{x}_0$ be an initial point in the domain of $f(\mathbf{x})$. By a Taylor's expansion of f in a neighborhood of $\mathbf{x}_0$ (see Theorem 7.5.1), it is possible to approximate $f(\mathbf{x})$ with the quadratic function $\phi(\mathbf{x})$ given by

$$\phi(\mathbf{x}) = f(\mathbf{x}_0) + (\mathbf{x} - \mathbf{x}_0)'\nabla f(\mathbf{x}_0) + \frac{1}{2!}(\mathbf{x} - \mathbf{x}_0)'\mathbf{H}_f(\mathbf{x}_0)(\mathbf{x} - \mathbf{x}_0), \quad (8.5)$$

where $\mathbf{H}_f(\mathbf{x}_0)$ is the Hessian matrix of f evaluated at $\mathbf{x}_0$.

On the basis of formula (8.5) we can obtain a reasonable approximation to the minimum of $f(\mathbf{x})$ by using the minimum of $\phi(\mathbf{x})$. If $\phi(\mathbf{x})$ attains a local minimum at $\mathbf{x}_1$, then we must necessarily have $\nabla\phi(\mathbf{x}_1) = \mathbf{0}$ (see Section 7.7), that is,

$$\nabla f(\mathbf{x}_0) + \mathbf{H}_f(\mathbf{x}_0)(\mathbf{x}_1 - \mathbf{x}_0) = \mathbf{0}. \quad (8.6)$$

If $\mathbf{H}_f(\mathbf{x}_0)$ is nonsingular, then from equation (8.6) we obtain

$$\mathbf{x}_1 = \mathbf{x}_0 - \mathbf{H}_f^{-1}(\mathbf{x}_0)\nabla f(\mathbf{x}_0).$$

If we now approximate $f(\mathbf{x})$ with another quadratic function, by again applying Taylor's expansion in a neighborhood of $\mathbf{x}_1$, and then repeat the same process as before with $\mathbf{x}_1$ used instead of $\mathbf{x}_0$, we obtain the point

$$\mathbf{x}_2 = \mathbf{x}_1 - \mathbf{H}_f^{-1}(\mathbf{x}_1)\nabla f(\mathbf{x}_1).$$

Further repetitions of this process lead to a sequence of points, $\mathbf{x}_0, \mathbf{x}_1, \mathbf{x}_2, \ldots, \mathbf{x}_i, \ldots$, such that

$$\mathbf{x}_{i+1} = \mathbf{x}_i - \mathbf{H}_f^{-1}(\mathbf{x}_i)\nabla f(\mathbf{x}_i), \quad i = 0, 1, 2, \ldots. \quad (8.7)$$

The Newton–Raphson method requires finding the inverse of the Hessian matrix $\mathbf{H}_f$ at each iteration. This can be computationally involved, especially if the number of variables, k, is large. Furthermore, the method may fail to converge if $\mathbf{H}_f(\mathbf{x}_i)$ is not positive definite. This can occur, for example, when $\mathbf{x}_i$ is far from the location $\mathbf{x}^*$ of the true minimum. If, however, the initial point $\mathbf{x}_0$ is close to $\mathbf{x}^*$, then convergence occurs at a rapid rate.

8.1.3. The Davidon–Fletcher–Powell Method

This method is basically similar to the one in Section 8.1.1 except that at the ith iteration we have

$$\mathbf{x}_{i+1} = \mathbf{x}_i - \theta_i \mathbf{G}_i \nabla f(\mathbf{x}_i), \quad i = 0, 1, 2, \ldots,$$

where $\mathbf{G}_i$ is a positive definite matrix that serves as the ith approximation to the inverse of the Hessian matrix $\mathbf{H}_f(\mathbf{x}_i)$, and θ_i is a scalar determined by a linear search from $\mathbf{x}_i$ in the direction of $-\mathbf{G}_i\nabla f(\mathbf{x}_i)$, similar to the one for the steepest descent method. The initial choice $\mathbf{G}_0$ of the matrix $\mathbf{G}$ can be any positive definite matrix, but is usually taken to be the identity matrix. At the $(i + 1)$st iteration, $\mathbf{G}_i$ is updated by using the formula

$$\mathbf{G}_{i+1} = \mathbf{G}_i + \mathbf{L}_i + \mathbf{M}_i, \qquad i = 0, 1, 2, \ldots,$$

where

$$\mathbf{L}_i = -\frac{\mathbf{G}_i[\nabla f(\mathbf{x}_{i+1}) - \nabla f(\mathbf{x}_i)][\nabla f(\mathbf{x}_{i+1}) - \nabla f(\mathbf{x}_i)]'\mathbf{G}_i}{[\nabla f(\mathbf{x}_{i+1}) - \nabla f(\mathbf{x}_i)]'\mathbf{G}_i[\nabla f(\mathbf{x}_{i+1}) - \nabla f(\mathbf{x}_i)]},$$

$$\mathbf{M}_i = -\frac{\theta_i[\mathbf{G}_i\nabla f(\mathbf{x}_i)][\mathbf{G}_i\nabla f(\mathbf{x}_i)]'}{[\mathbf{G}_i\nabla f(\mathbf{x}_i)]'[\nabla f(\mathbf{x}_{i+1}) - \nabla f(\mathbf{x}_i)]}.$$

The justification for this method is given in Fletcher and Powell (1963). See also Bunday (1984, Section 4.3). Note that if $\mathbf{G}_i$ is initially chosen as the identity, then the first increment is in the steepest descent direction $-\nabla f(\mathbf{x}_0)$.

This is a powerful optimization method and is considered to be very efficient for most functions.

8.2. THE DIRECT SEARCH METHODS

The direct search methods do not require the evaluation of any partial derivatives of the objective function. For this reason they are suited for situations in which it is analytically difficult to provide expressions for the partial derivatives, such as the minimization of the maximum absolute deviation. Three such methods will be discussed here, namely, the Nelder–Mead simplex method, Price's controlled random search procedure, and generalized simulated annealing.

8.2.1. The Nelder–Mead Simplex Method

Let $f(\mathbf{x})$, where $\mathbf{x} = (x_1, x_2, \ldots, x_k)'$, be the function to be minimized. The simplex method is based on a comparison of the values of f at the $k + 1$ vertices of a general simplex followed by a move away from the vertex with the highest function value. By definition, a general simplex is a geometric figure formed by a set of $k + 1$ points called vertices in a k-dimensional space. Originally, the simplex method was proposed by Spendley, Hext, and Himsworth (1962), who considered a regular simplex, that is, a simplex with mutually equidistant points such as an equilateral triangle in a two-dimensional space ($k = 2$). Nelder and Mead (1965) modified this method by allowing the simplex to be nonregular. This modified version of the simplex method will be described here.

The simplex method follows a sequential search procedure. As was mentioned earlier, it begins by evaluating f at the $k+1$ points that form a general simplex. Let these points be denoted by $x_1, x_2, \ldots, x_{k+1}$. Let f_h and f_l denote, respectively, the largest and the smallest of the values $f(x_1), f(x_2), \ldots, f(x_{k+1})$. Let us also denote the points where f_h and f_l are attained by x_h and x_l, respectively.

Obviously, if we are interested in minimizing f, then a move away from x_h will be in order. Let us therefore define x_c as the centroid of all the points with the exclusion of x_h. Thus

$$x_c = \frac{1}{k} \sum_{i \neq h} x_i.$$

In order to move away from x_h, we reflect x_h with respect to x_c to obtain the point x_h^*. More specifically, the latter point is defined by the relation

$$x_h^* - x_c = r(x_c - x_h),$$

or equivalently,

$$x_h^* = (1+r)x_c - rx_h,$$

where r is a positive constant called the reflection coefficient and is given by

$$r = \frac{\|x_h^* - x_c\|_2}{\|x_c - x_h\|_2}.$$

The points x_h, x_c, and x_h^* are depicted in Figure 8.1. Let us consider the

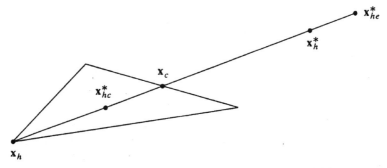

Figure 8.1. A two-dimensional simplex with the reflection (x_h^*), expansion (x_{he}^*), and contraction (x_{hc}^*) points.

following cases:

a. If $f_l < f(x_h^*) < f_h$, replace x_h by x_h^* and start the process again with the new simplex (that is, evaluate f at the vertices of the simplex which has the same points as the original simplex, but with x_h^* substituted for x_h).

b. If $f(x_h^*) < f_l$, then the move from x_c to x_h^* is in the right direction and should therefore be expanded. In this case, x_h^* is expanded to x_{he}^* defined by the relation

$$x_{he}^* - x_c = \gamma(x_h^* - x_c),$$

that is,

$$x_{he}^* = \gamma x_h^* + (1 - \gamma)x_c,$$

where γ (> 1) is an expansion coefficient given by

$$\gamma = \frac{\|x_{he}^* - x_c\|_2}{\|x_h^* - x_c\|_2}$$

(see Figure 8.1). This operation is called expansion. If $f(x_{he}^*) < f_l$, replace x_h by x_{he}^* and restart the process. However, if $f(x_{he}^*) > f_l$, then expansion is counterproductive. In this case, x_{he}^* is dropped, x_h is replaced by x_h^*, and the process is restarted.

c. If upon reflecting x_h to x_h^* we discover that $f(x_h^*) > f(x_i)$ for all $i \neq h$, then replacing x_h by x_h^* would leave $f(x_h^*)$ as the maximum in the new simplex. In this case, a new x_h is defined to be either the old x_h or x_h^*, whichever has the lower value. A point x_{hc}^* is then found such that

$$x_{hc}^* - x_c = \beta(x_h - x_c),$$

that is,

$$x_{hc}^* = \beta x_h + (1 - \beta)x_c,$$

where β ($0 < \beta < 1$) is a contraction coefficient given by

$$\beta = \frac{\|x_{hc}^* - x_c\|_2}{\|x_h - x_c\|_2}.$$

Next, x_{hc}^* is substituted for x_h and the process is restarted unless $f(x_{hc}^*) > \min[f_h, f(x_h^*)]$, that is, the contracted point is worse than the better of f_h and $f(x_h^*)$. When such a contraction fails, the size of the simplex is reduced by halving the distance of each point of the simplex from x_l, where, if we recall, x_l is the point generating the lowest function value. Thus x_i is replaced by $x_l + \frac{1}{2}(x_i - x_l)$, that is, by $\frac{1}{2}(x_i + x_l)$. The process is then restarted with the new reduced simplex.

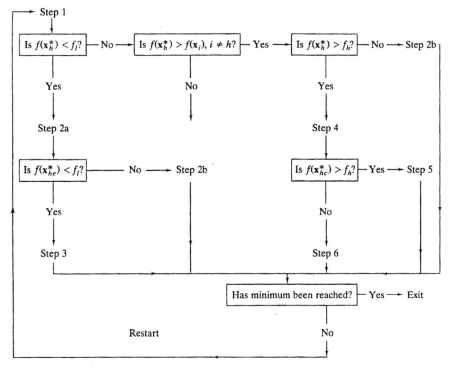

Step 1. Select initial points, $x_1, x_2, \ldots, x_{k+1}$, and calculate $f(x_i)$, $i = 1, 2, \ldots, k + 1$.
   Determine x_l, x_h and calculate $x_c = \sum_{i \neq h} x_i / k$. Select $r > 0$, say $r = \frac{1}{2}, \frac{2}{3}$, or 1, find
   $x_h^* = (1 + r)x_c - rx_h$, and calculate $f(x_h^*)$.
Step 2. (a) Calculate $x_{he}^* = \gamma x_h^* + (1 - \gamma)x_c$ by choosing $\gamma > 1$, say $\gamma = 1.5$, then calculate
   $f(x_{he}^*)$.
   (b) Replace x_h with x_h^*.
Step 3. Replace x_h with x_{he}^*.
Step 4. Calculate $x_{hc}^* = \beta x_h + (1 - \beta)x_c$ by choosing $0 < \beta < 1$, say $\beta = 0.5$, then calculate
   $f(x_{hc}^*)$.
Step 5. Replace all the x_i's with $(x_i + x_l)/2$.
Step 6. Replace x_h with x_{hc}^*.

Figure 8.2. Flow diagram for the Nelder–Mead simplex method. *Source:* Nelder and Mead (1965). Reproduced with permission of Oxford University Press.

Thus at each stage in the minimization process, x_h, the point at which f has the highest value, is replaced by a new point according to one of three operations, namely, reflection, contraction, and expansion. As an aid to illustrating this step-by-step procedure, a flow diagram is shown in Figure 8.2. This flow diagram is similar to one given by Nelder and Mead (1965, page 309). Figure 8.2 lists the explanations of steps 1 through 6.

The criterion used to stop the search procedure is based on the variation in the function values over the simplex. At each step, the standard error of these values in the form

$$s = \left[\frac{\sum_{i=1}^{k+1} \left(f_i - \bar{f} \right)^2}{k} \right]^{1/2}$$

is calculated and compared with some preselected value d, where $f_1, f_2, \ldots, f_{k+1}$ denote the function values at the vertices of the simplex at hand and $\bar{f} = \sum_{i=1}^{k+1} f_i / (k+1)$. The search is halted when $s < d$. The reasoning behind this criterion is that when $s < d$, all function values are very close together. This hopefully indicates that the points of the simplex are near the minimum.

Bunday (1984) provided the listing of a computer program which can be used to implement the steps described in the flow diagram.

Olsson and Nelson (1975) demonstrated the usefulness of this method by using it to solve six minimization problems in statistics. The robustness of the method itself and its advantages relative to other minimization techniques were reported in Nelson (1973).

8.2.2. Price's Controlled Random Search Procedure

The controlled random search procedure was introduced by Price (1977). It is capable of finding the absolute (or global) minimum of a function within a constrained region R. It is therefore well suited for a multimodal function, that is, a function that has several local minima within the region R.

The essential features of Price's algorithm are outlined in the flow diagram of Figure 8.3. A predetermined number, N, of trial points are randomly chosen inside the region R. The value of N must be greater than k, the number of variables. The corresponding function values are obtained and stored in an array $\mathbf{A}$ along with the coordinates of the N chosen points. At each iteration, $k+1$ distinct points, $\mathbf{x}_1, \mathbf{x}_2, \ldots, \mathbf{x}_{k+1}$, are chosen at random from the N points in storage. These $k+1$ points form a simplex in a k-dimensional space. The point $\mathbf{x}_{k+1}$ is arbitrarily taken as the pole (designated vertex) of the simplex, and the next trial point $\mathbf{x}_t$ is obtained as the image (reflection) point of the pole with respect to the centroid $\mathbf{x}_c$ of the remaining k points. Thus

$$\mathbf{x}_t = 2\mathbf{x}_c - \mathbf{x}_{k+1}.$$

The point $\mathbf{x}_t$ must satisfy the constraints of the region R. The value of the function f at $\mathbf{x}_t$ is then compared with $f_{\max}$, the largest function value in storage. Let $\mathbf{x}_{\max}$ denote the point at which $f_{\max}$ is achieved. If $f(\mathbf{x}_t) < f_{\max}$, then $\mathbf{x}_{\max}$ is replaced in the array $\mathbf{A}$ by $\mathbf{x}_t$. If $\mathbf{x}_t$ fails to satisfy the constraints of the region R, or if $f(\mathbf{x}_t) > f_{\max}$, then $\mathbf{x}_t$ is discarded and a new point is

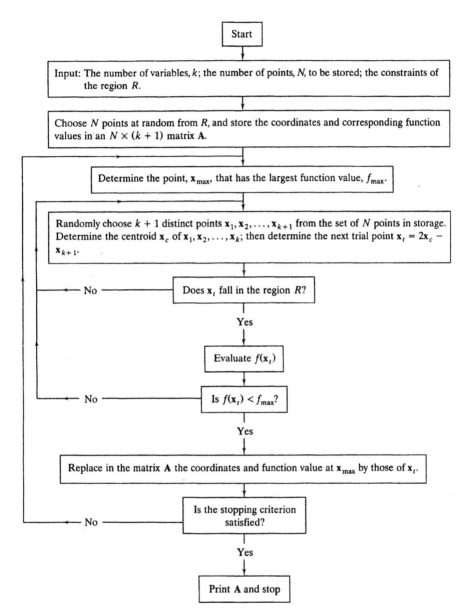

Figure 8.3. A flow diagram for Price's procedure. *Source:* Price (1977). Reproduced with permission of Oxford University Press.

chosen by following the same procedure as the one used to obtain $\mathbf{x}_t$. As the algorithm proceeds, the N points in storage tend to cluster around points at which the function values are lower than the current value of f_{max}. Price did not specify a particular stopping rule. He left it to the user to do so. A possible stopping criterion is to terminate the search when the N points in storage cluster in a small region of the k-dimensional space, that is, when f_{max} and f_{min} are close together, where f_{min} is the smallest function value in storage. Another possibility is to stop after a specified number of function evaluations have been made. In any case, the rate of convergence of the procedure depends on the value of N, the complexity of the function f, the nature of the constraints, and the way in which the set of trial points is chosen.

Price's procedure is simple and does not necessarily require a large value of N. It is sufficient that N should increase linearly with k. Price chose, for example, the value $N = 50$ for $k = 2$. The value $N = 10k$ has proved useful for many functions. Furthermore, the region constraints can be quite complex. A FORTRAN program for the implementation of Price's algorithm was written by Conlon (1991).

8.2.3. The Generalized Simulated Annealing Method

This method derives its name from the annealing of metals, in which many final crystalline configurations (corresponding to different energy states) are possible, depending on the rate of cooling (see Kirkpartrick, Gelatt, and Vechhi, 1983). The method can be applied to find the absolute (or global) optimum of a multimodal function f within a constrained region R in a k-dimensional space.

Bohachevsky, Johnson, and Stein (1986) presented a generalization of the method of simulated annealing for function optimization. The following is a description of their algorithm for function minimization (a similar one can be used for function maximization): Let f_m be some tentative estimate of the minimum of f over the region R. The method proceeds according to the following steps (reproduced with permission of the American Statistical Association):

1. Select an initial point $\mathbf{x}_0$ in R. This point can be chosen at random or specified depending on available information.
2. Calculate $f_0 = f(\mathbf{x}_0)$. If $|f_0 - f_m| < \epsilon$, where ϵ is a specified small constant, then stop.
3. Choose a random direction of search by generating k independent standard normal variates $z_1, z_2, \ldots, z_k$; then compute the elements of the random vector $\mathbf{u} = (u_1, u_2, \ldots, u_k)'$, where

$$u_i = \frac{z_i}{\left(z_1^2 + z_2^2 + \cdots + z_k^2\right)^{1/2}}, \qquad i = 1, 2, \ldots, k,$$

and k is the number of variables in the function.

4. Set $\mathbf{x}_1 = \mathbf{x}_0 + \Delta r\ \mathbf{u}$, where Δr is the size of a step to be taken in the direction of $\mathbf{u}$. The magnitude of Δr depends on the properties of the objective function and on the desired accuracy.

5. If $\mathbf{x}_1$ does not belong to R, return to step 3. Otherwise, compute $f_1 = f(\mathbf{x}_1)$ and $\Delta f = f_1 - f_0$.

6. if $f_1 \leq f_0$, set $\mathbf{x}_0 = \mathbf{x}_1$ and $f_0 = f_1$. If $|f_0 - f_m| < \epsilon$, stop. Otherwise, go to step 3.

7. If $f_1 > f_0$, set a probability value given by $p = \exp(-\beta f_0^g \Delta f)$, where β is a positive number such that $0.50 < \exp(-\beta \Delta f) < 0.90$, and g is an arbitrary negative number. Then, generate a random number v from the uniform distribution $U(0, 1)$. If $v \geq p$, go to step 3. Otherwise, if $v < p$, set $\mathbf{x}_0 = \mathbf{x}_1$, $f_0 = f_1$, and go to step 3.

From steps 6 and 7 we note that beneficial steps (that is, $f_1 \leq f_0$) are accepted unconditionally, but detrimental steps ($f_1 > f_0$) are accepted according to a probability value p described in step 7. If $v < p$, then the step leading to $\mathbf{x}_1$ is accepted; otherwise, it is rejected and a step in a new random direction is attempted. Thus the probability of accepting an increment of f depends on the size of the increment: the larger the increment, the smaller the probability of its acceptance.

Several possible values of the tentative estimate f_m can be attempted. For a given f_m we proceed with the search until $f - f_m$ becomes negative. Then, we decrease f_m, continue the search, and repeat the process when necessary. Bohachevsky, Johnson, and Stein gave an example in optimal design theory to illustrate the application of their algorithm.

Price's (1977) controlled random search algorithm produces results comparable to those of simulated annealing, but with fewer tuning parameters. It is also better suited for problems with constrained regions.

8.3. OPTIMIZATION TECHNIQUES IN RESPONSE SURFACE METHODOLOGY

Response surface methodology (RSM) is an area in the design and analysis of experiments. It consists of a collection of techniques that encompasses:

1. Conducting a series of experiments based on properly chosen settings of a set of input variables, denoted by $x_1, x_2, \ldots, x_k$, that influence a response of interest y. The choice of these settings is governed by certain criteria whose purpose is to produce adequate and reliable information about the response. The collection of all such settings constitutes a matrix $\mathbf{D}$ of order $n \times k$, where n is the number of experimental runs. The matrix $\mathbf{D}$ is referred to as a response surface design.

2. Determining a mathematical model that best fits the data collected under the design chosen in (1). Regression techniques can be used to evaluate the adequacy of fit of the model and to conduct appropriate tests concerning the model's parameters.

3. Determining optimal operating conditions on the input variables that produce maximum (or minimum) response value within a region of interest R.

This last aspect of RSM can help the experimenter in determining the best combinations of the input variables that lead to desirable response values. For example, in drug manufacturing, two drugs are tested with regard to reducing blood pressure in humans. A series of clinical trials involving a certain number of high blood pressure patients is set up, and each patient is given some predetermined combination of the two drugs. After a period of time the patient's blood pressure is checked. This information can be used to find the specific combination of the drugs that results in the greatest reduction in the patient's blood pressure within some specified time interval.

In this section we shall describe two well-known optimum-seeking procedures in RSM. These include the method of steepest ascent (or descent) and ridge analysis.

8.3.1. The Method of Steepest Ascent

This is an adaptation of the method described in Section 8.1.1 to a response surface environment; here the objective is to increase the value of a certain response function.

The method of steepest ascent requires performing a sequence of sets of trials. Each set is obtained as a result of proceeding sequentially along a path of maximum increase in the values of a given response y, which can be observed in an experiment. This method was first introduced by Box and Wilson (1951) for the general area of RSM.

The procedure of steepest ascent depends on approximating a response surface with a hyperplane in some restricted region. The hyperplane is represented by a first-order model which can be fitted to a data set obtained as a result of running experimental trials using a first-order design such as a complete 2^k factorial design, where k is the number of input variables in the model. A fraction of this design can also be used if k is large [see, for example, Section 3.3.2 in Khuri and Cornell (1996)]. The fitted first-order model is then used to determine a path along which one may initially observe increasing response values. However, due to curvature in the response surface, the initial increase in the response will likely be followed by a leveling off, and then a decrease. At this stage, a new series of experiments is performed (using again a first-order design) and the resulting data are used to fit another first-order model. A new path is determined along which increasing response values may be observed. This process continues until it becomes evident that little or no additional increase in the response can be gained.

Let us now consider more specific details of this sequential procedure. Let $y(\mathbf{x})$ be a response function that depends on k input variables, $x_1, x_2, \ldots, x_k$, which form the elements of a vector $\mathbf{x}$. Suppose that in some restricted region $y(\mathbf{x})$ is adequately represented by a first-order model of the form

$$y(\mathbf{x}) = \beta_0 + \sum_{i=1}^{k} \beta_i x_i + \epsilon, \tag{8.8}$$

where $\beta_0, \beta_1, \ldots, \beta_k$ are unknown parameters and ϵ is a random error. This model is fitted using data collected under a first-order design (for example, a 2^k factorial design or a fraction thereof). The data are utilized to calculate the least-squares estimates $\hat{\beta}_0, \hat{\beta}_1, \ldots, \hat{\beta}_k$ of the model's parameters. These are elements of $\hat{\boldsymbol{\beta}} = (\mathbf{X}'\mathbf{X})^{-1}\mathbf{X}'\mathbf{y}$, where $\mathbf{X} = [\mathbf{1}_n : \mathbf{D}]$ with $\mathbf{1}_n$ being a vector of ones of order $n \times 1$, $\mathbf{D}$ is the design matrix of order $n \times k$, and $\mathbf{y}$ is the corresponding vector of response values. It is assumed that the random errors associated with the n response values are independently distributed with means equal to zero and a common variance σ^2. The predicted response $\hat{y}(\mathbf{x})$ is then given by

$$\hat{y}(\mathbf{x}) = \hat{\beta}_0 + \sum_{i=1}^{k} \hat{\beta}_i x_i. \tag{8.9}$$

The input variables are coded so that the design center coincides with the origin of the coordinates system.

The next step is to move a distance of r units away from the design center (or the origin) such that a maximum increase in $\hat{y}$ can be obtained. To determine the direction to be followed to achieve such an increase, we need to maximize $\hat{y}(\mathbf{x})$ subject to the constraint $\sum_{i=1}^{k} x_i^2 = r^2$ using the method of Lagrange multipliers. Consider therefore the function

$$Q(\mathbf{x}) = \hat{\beta}_0 + \sum_{i=1}^{k} \hat{\beta}_i x_i - \lambda \left(\sum_{i=1}^{k} x_i^2 - r^2 \right), \tag{8.10}$$

where λ is a Lagrange multiplier. Setting the partial derivatives of Q equal to zero produces the equations

$$x_i = \frac{1}{2\lambda} \hat{\beta}_i, \qquad i = 1, 2, \ldots, k.$$

For a maximum, λ must be positive. Using the equality constraint, we conclude that

$$\lambda = \frac{1}{2r} \left(\sum_{i=1}^{k} \hat{\beta}_i^2 \right)^{1/2}.$$

A local maximum is then achieved at the point whose coordinates are given by

$$x_i = \frac{r\hat{\beta}_i}{\left(\sum_{i=1}^{k} \hat{\beta}_i^2\right)^{1/2}}, \qquad i = 1, 2, \ldots, k,$$

which can be written as

$$x_i = re_i, \qquad i = 1, 2, \ldots, k, \tag{8.11}$$

where $e_i = \hat{\beta}_i(\sum_{i=1}^{k} \hat{\beta}_i^2)^{-1/2}$, $i = 1, 2, \ldots, k$. Thus $\mathbf{e} = (e_1, e_2, \ldots, e_k)'$ is a unit vector in the direction of $(\hat{\beta}_1, \hat{\beta}_2, \ldots, \hat{\beta}_k)'$. Equations (8.11) indicate that at a distance of r units away from the origin, a maximum increase in $\hat{y}$ occurs along a path in the direction of $\mathbf{e}$. Since this is the only local maximum on the hypersphere of radius r, it must be the absolute maximum.

If the actual response value (that is, the value of y) at the point $\mathbf{x} = r\mathbf{e}$ exceeds its value at the origin, then a move along the path determined by $\mathbf{e}$ is in order. A series of experiments is then conducted to obtain response values at several points along the path until no additional increase in the response is evident. At this stage, a new first-order model is fitted using data collected under a first-order design centered at a point in the vicinity of the point at which that first drop in the response was observed along the path. This model leads to a new direction similar to the one given by formula (8.11). As before, a series of experiments are conducted along the new path until no further increase in the value of y can be observed. The process of moving along different paths continues until it becomes evident that little or no additional increase in y can be gained. This usually occurs when the first-order model becomes inadequate as the method progresses, due to curvature in the response surface. It is therefore necessary to test each fitted first-order model for lack of fit at every stage of the process. This can be accomplished by taking repeated observations at the center of each first-order design and at possibly some other design points in order to obtain an independent estimate of the error variance that is needed for the lack of fit test [see, for example, Sections 2.6 and 3.4 in Khuri and Cornell (1996)]. If the lack of fit test is significant, indicating an inadequate model, then the process is stopped and a more elaborate experiment must be conducted to fit a higher-order model, as will be seen in the next section.

Examples that illustrate the application of the method of steepest ascent can be found in Box and Wilson (1951), Bayne and Rubin (1986, Section 5.2), Khuri and Cornell (1996, Chapter 5), and Myers and Khuri (1979). In the last reference, the authors present a stopping rule along a path that takes into account random error variation in the observed response. We recall that a search along a path is discontinued as soon as a drop in the response is first observed. Since response values are subject to random error, the decision to stop can be premature due to a false drop in the observed response. The stopping rule by Myers and Khuri (1979) protects against taking too many observations along a path when in fact the true mean response (that is, the mean of y) is decreasing. It also protects against stopping prematurely when the true mean response is increasing.

It should be noted that the procedure of steepest ascent is not invariant with respect to the scales of the input variables $x_1, x_2, \ldots, x_k$. This is evident from the fact that a path taken by the procedure is determined by the least-squares estimates $\hat{\beta}_1, \hat{\beta}_2, \ldots, \hat{\beta}_k$ [see equations (8.11)], which depend on the scales of the x_i's.

There are situations in which it is of interest to determine conditions that lead to a decrease in the response, instead of an increase. For example, in a chemical investigation it may be desired to decrease the level of impurity or the unit cost. In this case, a path of steepest descent will be needed. This can be accomplished by changing the sign of the response y, followed by an application of the method of steepest ascent. Thus any steepest descent problem can be handled by the method of steepest ascent.

8.3.2. The Method of Ridge Analysis

The method of steepest ascent is most often used as a maximum-region-seeking procedure. By this we mean that it is used as a preliminary tool to get quickly to the region where the maximum of the mean response is located. Since the first-order approximation of the mean response will eventually break down, a better estimate of the maximum can be obtained by fitting a second-order model in the region of the maximum. The method of ridge analysis, which was introduced by Hoerl (1959) and formalized by Draper (1963), is used for this purpose.

Let us suppose that inside a region of interest R, the true mean response is adequately represented by the second-order model

$$y(\mathbf{x}) = \beta_0 + \sum_{i=1}^{k} \beta_i x_i + \sum_{\substack{i=1 \\ i<j}}^{k-1} \sum_{j=2}^{k} \beta_{ij} x_i x_j + \sum_{i=1}^{k} \beta_{ii} x_i^2 + \epsilon, \qquad (8.12)$$

where the β's are unknown parameters and ϵ is a random error with mean zero and variance σ^2. Model (8.12) can be written as

$$y(\mathbf{x}) = \beta_0 + \mathbf{x}'\boldsymbol{\beta} + \mathbf{x}'\mathbf{B}\mathbf{x} + \epsilon, \qquad (8.13)$$

where $\boldsymbol{\beta} = (\beta_1, \beta_2, \ldots, \beta_k)'$ and $\mathbf{B}$ is a symmetric $k \times k$ matrix of the form

$$
\mathbf{B} = \begin{bmatrix}
\beta_{11} & \frac{1}{2}\beta_{12} & \frac{1}{2}\beta_{13} & \cdots & \frac{1}{2}\beta_{1k} \\
 & \beta_{22} & \frac{1}{2}\beta_{23} & \cdots & \frac{1}{2}\beta_{2k} \\
 & & \ddots & \ddots & \vdots \\
 & & & \ddots & \frac{1}{2}\beta_{k-1,k} \\
\text{symmetric} & & & & \beta_{kk}
\end{bmatrix}.
$$

Least-squares estimates of the parameters in model (8.13) can be obtained by using data collected according to a second-order design. A description of potential second-order designs can be found in Khuri and Cornell (1996, Chapter 4).

Let $\hat{\beta}_0$, $\hat{\boldsymbol{\beta}}$, and $\hat{\mathbf{B}}$ denote the least-squares estimates of β_0, $\boldsymbol{\beta}$, and $\mathbf{B}$, respectively. The predicted response $\hat{y}(\mathbf{x})$ inside the region R is then given by

$$
\hat{y}(\mathbf{x}) = \hat{\beta}_0 + \mathbf{x}'\hat{\boldsymbol{\beta}} + \mathbf{x}'\hat{\mathbf{B}}\mathbf{x}. \tag{8.14}
$$

The input variables are coded so that the design center coincides with the origin of the coordinates system.

The method of ridge analysis is used to find the optimum (maximum or minimum) of $\hat{y}(\mathbf{x})$ on concentric hyperspheres of varying radii inside the region R. It is particularly useful in situations in which the unconstrained optimum of $\hat{y}(\mathbf{x})$ falls outside the region R, or if a saddle point occurs inside R.

Let us now proceed to optimize $\hat{y}(\mathbf{x})$ subject to the constraint

$$
\sum_{i=1}^{k} x_i^2 = r^2, \tag{8.15}
$$

where r is the radius of a hypersphere centered at the origin and is contained inside the region R. Using the method of Lagrange multipliers, let us consider the function

$$
F = \hat{y}(\mathbf{x}) - \lambda\left(\sum_{i=1}^{k} x_i^2 - r^2\right), \tag{8.16}
$$

where λ is a Lagrange multiplier. Differentiating F with respect to x_i

$(i = 1, 2, \ldots, k)$ and equating the partial derivatives to zero, we obtain

$$\frac{\partial F}{\partial x_1} = 2\left(\hat{\beta}_{11} - \lambda \right)x_1 + \hat{\beta}_{12}x_2 + \cdots + \hat{\beta}_{1k}x_k + \hat{\beta}_1 = 0,$$

$$\frac{\partial F}{\partial x_2} = \hat{\beta}_{12}x_1 + 2\left(\hat{\beta}_{22} - \lambda \right)x_2 + \cdots + \hat{\beta}_{2k}x_k + \hat{\beta}_2 = 0,$$

$$\vdots$$

$$\frac{\partial F}{\partial x_k} = \hat{\beta}_{1k}x_1 + \hat{\beta}_{2k}x_2 + \cdots + 2\left(\hat{\beta}_{kk} - \lambda \right)x_k + \hat{\beta}_k = 0.$$

These equations can be expressed as

$$\left(\hat{\mathbf{B}} - \lambda \mathbf{I}_k\right)\mathbf{x} = -\tfrac{1}{2}\hat{\boldsymbol{\beta}}. \tag{8.17}$$

Equations (8.15) and (8.17) need to be solved for $x_1, x_2, \ldots, x_k$ and λ. This traditional approach, however, requires calculations that are somewhat involved. Draper (1963) proposed the following simpler, yet equivalent procedure:

   i. Regard r as a variable, but fix λ instead.
   ii. Insert the selected value of λ in equation (8.17) and solve for $\mathbf{x}$. The solution is used in steps iii and iv.
   iii. Compute $r = (\mathbf{x}'\mathbf{x})^{1/2}$.
   iv. Evaluate $\hat{y}(\mathbf{x})$.

Several values of λ can give rise to several stationary points which lie on the same hypersphere of radius r. This can be seen from the fact that if λ is chosen to be different from any eigenvalue of $\hat{\mathbf{B}}$, then equation (8.17) has a unique solution given by

$$\mathbf{x} = -\tfrac{1}{2}\left(\hat{\mathbf{B}} - \lambda \mathbf{I}_k\right)^{-1}\hat{\boldsymbol{\beta}}. \tag{8.18}$$

By substituting $\mathbf{x}$ in equation (8.15) we obtain

$$\hat{\boldsymbol{\beta}}'\left(\hat{\mathbf{B}} - \lambda \mathbf{I}_k\right)^{-2}\hat{\boldsymbol{\beta}} = 4r^2. \tag{8.19}$$

Hence, each value of r gives rise to at most $2k$ corresponding values of λ.

The choice of λ has an effect on the nature of the stationary point. Some values of λ produce points at each of which $\hat{y}$ has a maximum. Other values of λ cause $\hat{y}$ to have minimum values. More specifically, suppose that λ_1 and

λ_2 are two values substituted for λ in equation (8.18). Let x_1, x_2 and r_1, r_2 be the corresponding values of x and r, respectively. The following results, which were established in Draper (1963), can be helpful in selecting the value of λ that produces a particular type of stationary point:

RESULT 1. If $r_1 = r_2$ and $\lambda_1 > \lambda_2$, then $\hat{y}_1 > \hat{y}_2$, where $\hat{y}_1$ and $\hat{y}_2$ are the values of $\hat{y}(x)$ at x_1 and x_2, respectively.

This result means that for two stationary points that have the same distance from the origin, $\hat{y}$ will be larger at the stationary point with the larger value of λ.

RESULT 2. Let M be the matrix of second-order partial derivatives of F in formula (8.16), that is,

$$M = 2(\hat{B} - \lambda I_k). \tag{8.20}$$

If $r_1 = r_2$, and if M is positive definite for x_1 and is indefinite (that is, neither positive definite nor negative definite) for x_2, then $\hat{y}_1 < \hat{y}_2$.

RESULT 3. If λ_1 is larger than the largest eigenvalue of $\hat{B}$, then the corresponding solution x_1 in formula (8.18) is a point of absolute maximum for $\hat{y}$ on a hypersphere of radius $r_1 = (x_1'x_1)^{1/2}$. If, on the other hand, λ_1 is smaller than the smallest eigenvalue of $\hat{B}$, then x_1 is a point of absolute minimum for $\hat{y}$ on the same hypersphere.

On the basis of Result 3 we can select several values of λ that exceed the largest eigenvalue of $\hat{B}$. The resulting values of the k elements of x and $\hat{y}$ can be plotted against the corresponding values of r. This produces $k + 1$ plots called ridge plots (see Myers, 1976, Section 5.3). They are useful in that an experimenter can determine, for a particular r, the maximum of $\hat{y}$ within a region R and the operating conditions (that is, the elements of x) that give rise to the maximum. Similar plots can be obtained for the minimum of $\hat{y}$ (here, values of λ that are smaller than the smallest eigenvalue of $\hat{B}$ must be chosen). Obviously, the portions of the ridge plots that fall outside R should not be considered.

EXAMPLE 8.3.1. An experiment was conducted to investigate the effects of three fertilizer ingredients on the yield of snap beans under field conditions. The fertilizer ingredients and actual amounts applied were nitrogen (N), from 0.94 to 6.29 lb/plot; phosphoric acid (P_2O_5), from 0.59 to 2.97 lb/plot; and potash (K_2O), from 0.60 to 4.22 lb/plot. The response of interest is the average yield in pounds per plot of snap beans.

Five levels of each fertilizer were used. The levels are coded using the following linear transformations:

$$x_1 = \frac{X_1 - 3.62}{1.59}, \qquad x_2 = \frac{X_2 - 1.78}{0.71}, \qquad x_3 = \frac{X_3 - 2.42}{1.07}.$$

Here, X_1, X_2, and X_3 denote the actual levels of nitrogen, phosphoric acid, and potash, respectively, used in the experiment, and x_1, x_2, x_3 the corresponding coded values. In this particular coding scheme, 3.62, 1.78, and 2.42 are the averages of the experimental levels of X_1, X_2, and X_3, respectively, that is, they represent the centers of the values of nitrogen, phosphoric acid, and potash, respectively. The denominators of x_1, x_2, and x_3 were chosen so that the second and fourth levels of each X_i correspond to the values -1 and 1, respectively, for x_i $(i = 1, 2, 3)$. One advantage of such a coding scheme is to make the levels of the three fertilizers scale free (this is necessary in general, since the input variables can have different units of measurement). The measured and coded levels for the three fertilizers are shown below:

| | Levels of x_i $(i = 1, 2, 3)$ | | | | |
Fertilizer	-1.682	-1.000	0.000	1.000	1.682
N	0.94	2.03	3.62	5.21	6.29
P_2O_5	0.59	1.07	1.78	2.49	2.97
K_2O	0.60	1.35	2.42	3.49	4.22

Combinations of the levels of the three fertilizers were applied according to the experimental design shown in Table 8.1, in which the design settings are given in terms of the coded levels. Six center-point replications were run in order to obtain an estimate of the experimental error variance. This particular design is called a central composite design [for a description of this design and its properties, see Khuri and Cornell (1996, Section 4.5.3)], which has the rotatability property. By this we mean that the prediction variance, that is, Var[$\hat{y}(\mathbf{x})$], is constant at all points that are equidistant from the design center [see Khuri and Cornell (1996, Section 2.8.3) for more detailed information concerning rotatability]. The corresponding response (yield) values are given in Table 8.1.

A second-order model of the form given by formula (8.12) was fitted to the data set in Table 8.1. Thus in terms of the coded variables we have the model

$$y(\mathbf{x}) = \beta_0 + \sum_{i=1}^{3} \beta_i x_i + \beta_{12} x_1 x_2 + \beta_{13} x_1 x_3 + \beta_{23} x_2 x_3 + \sum_{i=1}^{3} \beta_{ii} x_i^2 + \epsilon. \quad (8.21)$$

Table 8.1. The Coded and Actual Settings of the Three Fertilizers and the Corresponding Response Values

x_1	x_2	x_3	N	P_2O_5	K_2O	Yield y
-1	-1	-1	2.03	1.07	1.35	11.28
1	-1	-1	5.21	1.07	1.35	8.44
-1	1	-1	2.03	2.49	1.35	13.19
1	1	-1	5.21	2.49	1.35	7.71
-1	-1	1	2.03	1.07	3.49	8.94
1	-1	1	5.21	1.07	3.49	10.90
-1	1	1	2.03	2.49	3.49	11.85
1	1	1	5.21	2.49	3.49	11.03
-1.682	0	0	0.94	1.78	2.42	8.26
1.682	0	0	6.29	1.78	2.42	7.87
0	-1.682	0	3.62	0.59	2.42	12.08
0	1.682	0	3.62	2.97	2.42	11.06
0	0	-1.682	3.62	1.78	0.60	7.98
0	0	1.682	3.62	1.78	4.22	10.43
0	0	0	3.62	1.78	2.42	10.14
0	0	0	3.62	1.78	2.42	10.22
0	0	0	3.62	1.78	2.42	10.53
0	0	0	3.62	1.78	2.42	9.50
0	0	0	3.62	1.78	2.42	11.53
0	0	0	3.62	1.78	2.42	11.02

Source: A. I. Khuri and J. A. Cornell (1996). Reproduced with permission of Marcel Dekker, Inc.

The resulting prediction equation is given by

$$\hat{y}(\mathbf{x}) = 10.462 - 0.574x_1 + 0.183x_2 + 0.456x_3 - 0.678x_1x_2 + 1.183x_1x_3$$

$$+ 0.233x_2x_3 - 0.676x_1^2 + 0.563x_2^2 - 0.273x_3^2. \qquad (8.22)$$

Here, $\hat{y}(\mathbf{x})$ is the predicted yield at the point $\mathbf{x} = (x_1, x_2, x_3)'$. Equation (8.22) can be expressed in matrix form as in equation (8.14), where $\hat{\boldsymbol{\beta}} = (-0.574, 0.183, 0.456)'$ and $\hat{\mathbf{B}}$ is the matrix

$$\hat{\mathbf{B}} = \begin{bmatrix} -0.676 & -0.339 & 0.592 \\ -0.339 & 0.563 & 0.117 \\ 0.592 & 0.117 & -0.273 \end{bmatrix}. \qquad (8.23)$$

The coordinates of the stationary point x_0 of $\hat{y}(x)$ satisfy the equation

$$\frac{\partial \hat{y}}{\partial x_1} = -0.574 + 2(-0.676x_1 - 0.339x_2 + 0.592x_3) = 0,$$

$$\frac{\partial \hat{y}}{\partial x_2} = 0.183 + 2(-0.339x_1 + 0.563x_2 + 0.117x_3) = 0,$$

$$\frac{\partial \hat{y}}{\partial x_3} = 0.456 + 2(0.592x_1 + 0.117x_2 - 0.273x_3) = 0,$$

which can be expressed as

$$\hat{\beta} + 2\hat{B}x_0 = 0. \tag{8.24}$$

Hence,

$$x_0 = -\tfrac{1}{2}\hat{B}^{-1}\hat{\beta} = (-0.394, -0.364, -0.175)'.$$

The eigenvalues of $\hat{B}$ are $\tau_1 = 0.6508$, $\tau_2 = 0.1298$, $\tau_3 = -1.1678$. The matrix $\hat{B}$ is therefore neither positive definite nor negative definite, that is, x_0 is a saddle point (see Corollary 7.7.1). This point falls inside the experimental region R, which, in the space of the coded variables x_1, x_2, x_3, is a sphere centered at the origin of radius $\sqrt{3}$.

Let us now apply the method of ridge analysis to maximize $\hat{y}$ inside the region R. For this purpose we choose values of λ [the Lagrange multiplier in equation (8.16)] larger than $\tau_1 = 0.6508$, the largest eigenvalue of $\hat{B}$. For each such value of λ, equation (8.17) has a solution for x that represents a point of absolute maximum of $\hat{y}(x)$ on a sphere of radius $r = (x'x)^{1/2}$ inside R. The results are displayed in Table 8.2. We note that at the point $(-0.558, 1.640, 0.087)$, which is located near the periphery of the region R, the maximum value of $\hat{y}$ is 13.021. By expressing the coordinates of this point in terms of the actual values of the three fertilizers we obtain $X_1 = 2.733$ lb/plot, $X_2 = 2.944$ lb/plot, and $X_3 = 2.513$ lb/plot. We conclude that a combination of nitrogen, phosphoric acid, and potash fertilizers at the rates

Table 8.2. Ridge Analysis Values

λ	1.906	1.166	0.979	0.889	0.840	0.808	0.784	0.770	0.754	0.745	0.740
x_1	−0.106	−0.170	−0.221	−0.269	−0.316	−0.362	−0.408	−0.453	−0.499	−0.544	−0.558
x_2	0.102	0.269	0.438	0.605	0.771	0.935	1.099	1.263	1.426	1.589	1.640
x_3	0.081	0.110	0.118	0.120	0.117	0.113	0.108	0.102	0.096	0.089	0.087
r	0.168	0.337	0.505	0.673	0.841	1.009	1.177	1.346	1.514	1.682	1.734
$\hat{y}$	10.575	10.693	10.841	11.024	11.243	11.499	11.790	12.119	12.484	12.886	13.021

of 2.733, 2.944, and 2.513 lb/plot, respectively, results in an estimated maximum yield of snap beans of 13.021 lb/plot.

8.3.3. Modified Ridge Analysis

Optimization of $\hat{y}(\mathbf{x})$ on a hypersphere S by the method of ridge analysis is justified provided that the prediction variance on S is relatively small. Furthermore, it is desirable that this variance remain constant on S. If not, then it is possible to obtain poor estimates of the optimum response, especially when the dispersion in the prediction variances on S is large. Thus the reliability of ridge analysis as an optimum-seeking procedure depends very much on controlling the size and variability of the prediction variance. If the design is rotatable, then the prediction variance, $\mathrm{Var}[\hat{y}(\mathbf{x})]$, is constant on S. It is then easy to attain small prediction variances by restricting the procedure to hyperspheres of small radii. However, if the design is not rotatable, then $\mathrm{Var}[\hat{y}(\mathbf{x})]$ may vary widely on S, which, as was mentioned earlier, can adversely affect the quality of estimation of the optimum response. This suggests that the prediction variance should be given serious consideration in the strategy of ridge analysis if the design used is not rotatable.

Khuri and Myers (1979) proposed a certain modification to the method of ridge analysis: one that optimizes $\hat{y}(\mathbf{x})$ subject to a particular constraint on the prediction variance. The following is a description of their proposed modification:

Consider model (8.12), which can be written as

$$y(\mathbf{x}) = \mathbf{f}'(\mathbf{x})\boldsymbol{\gamma} + \epsilon, \qquad (8.25)$$

where

$$\mathbf{f}(\mathbf{x}) = \left(1, x_1, x_2, \ldots, x_k, x_1 x_2, x_1 x_3, \ldots, x_{k-1} x_k, x_1^2, x_2^2, \ldots, x_k^2\right)',$$

$$\boldsymbol{\gamma} = \left(\beta_0, \beta_1, \beta_2, \ldots, \beta_k, \beta_{12}, \beta_{13}, \ldots, \beta_{k-1,k}, \beta_{11}, \beta_{22}, \ldots, \beta_{kk}\right)'.$$

The predicted response is given by

$$\hat{y}(\mathbf{x}) = \mathbf{f}'(\mathbf{x})\hat{\boldsymbol{\gamma}}, \qquad (8.26)$$

where $\hat{\boldsymbol{\gamma}}$ is the least-squares estimator of $\boldsymbol{\gamma}$, namely,

$$\hat{\boldsymbol{\gamma}} = (\mathbf{X}'\mathbf{X})^{-1}\mathbf{X}'\mathbf{y}, \qquad (8.27)$$

where $\mathbf{X} = [\mathbf{f}(\mathbf{x}_1), \mathbf{f}(\mathbf{x}_2), \ldots, \mathbf{f}(\mathbf{x}_n)]'$ with $\mathbf{x}_i$ being the vector of design settings at the ith experimental run ($i = 1, 2, \ldots, n$, where n is the number of runs used in the experiment), and $\mathbf{y}$ is the corresponding vector of n observations. Since

$$\mathrm{Var}(\hat{\boldsymbol{\gamma}}) = (\mathbf{X}'\mathbf{X})^{-1}\sigma^2, \qquad (8.28)$$

where σ^2 is the error variance, then from equation (8.26), the prediction variance is of the form

$$\text{Var}[\hat{y}(\mathbf{x})] = \sigma^2 \mathbf{f}'(\mathbf{x})(\mathbf{X}'\mathbf{X})^{-1}\mathbf{f}(\mathbf{x}). \tag{8.29}$$

The number of unknown parameters in model (8.25) is $p = (k + 1)(k + 2)/2$, where k is the number of input variables. Let $\nu_1, \nu_2, \ldots, \nu_p$ denote the eigenvalues of $\mathbf{X}'\mathbf{X}$. Then from equation (8.29) and Theorem 2.3.16 we have

$$\frac{\sigma^2 \mathbf{f}'(\mathbf{x})\mathbf{f}(\mathbf{x})}{\nu_{\max}} \leq \text{Var}[\hat{y}(\mathbf{x})] \leq \frac{\sigma^2 \mathbf{f}'(\mathbf{x})\mathbf{f}(\mathbf{x})}{\nu_{\min}},$$

where $\nu_{\min}$ and $\nu_{\max}$ are, respectively, the smallest and largest of the ν_i's. This double inequality shows that the prediction variance can be inflated if $\mathbf{X}'\mathbf{X}$ has small eigenvalues. This occurs when the columns of $\mathbf{X}$ are multi-collinear (see, for example, Myers, 1990, pages 125–126 and Chapter 8). Now, by the spectral decomposition theorem (Theorem 2.3.10), $\mathbf{X}'\mathbf{X} = \mathbf{V}\Lambda\mathbf{V}'$, where $\mathbf{V}$ is an orthogonal matrix of orthonormal eigenvectors of $\mathbf{X}'\mathbf{X}$ and $\Lambda = \text{Diag}(\nu_1, \nu_2, \ldots, \nu_p)$ is a diagonal matrix of eigenvalues of $\mathbf{X}'\mathbf{X}$. Equation (8.29) can then be written as

$$\text{Var}[\hat{y}(\mathbf{x})] = \sigma^2 \sum_{j=1}^{p} \frac{[\mathbf{f}'(\mathbf{x})\mathbf{v}_j]^2}{\nu_j}, \tag{8.30}$$

where $\mathbf{v}_j$ is the jth column of $\mathbf{V}$ ($j = 1, 2, \ldots, p$). If we denote the elements of $\mathbf{v}_j$ by $v_{0j}, v_{1j}, \ldots, v_{kj}, v_{12j}, v_{13j}, \ldots, v_{k-1\,kj}, v_{11j}, v_{22j}, \ldots, v_{kkj}$, then $\mathbf{f}'(\mathbf{x})\mathbf{v}_j$ can be expressed as

$$\mathbf{f}'(\mathbf{x})\mathbf{v}_j = v_{0j} + \mathbf{x}'\boldsymbol{\tau}_j + \mathbf{x}'\mathbf{T}_j\mathbf{x}, \qquad j = 1, 2, \ldots, p, \tag{8.31}$$

where $\boldsymbol{\tau}_j = (v_{1j}, v_{2j}, \ldots, v_{kj})'$ and

$$\mathbf{T}_j = \begin{bmatrix} v_{11j} & \frac{1}{2}v_{12j} & \frac{1}{2}v_{13j} & \cdots & \frac{1}{2}v_{1kj} \\ & v_{22j} & \frac{1}{2}v_{23j} & \cdots & \frac{1}{2}v_{2kj} \\ & & \ddots & \ddots & \vdots \\ & & & \ddots & \frac{1}{2}v_{k-1\,kj} \\ \text{symmetric} & & & & v_{kkj} \end{bmatrix}, \qquad j = 1, 2, \ldots, p.$$

We note that the form of $\mathbf{f}'(\mathbf{x})\mathbf{v}_j$, as given by formula (8.31), is identical to that of a second-order model. Formula (8.30) can then be written as

$$\text{Var}[\hat{y}(\mathbf{x})] = \sigma^2 \sum_{j=1}^{p} \frac{(v_{0j} + \mathbf{x}'\boldsymbol{\tau}_j + \mathbf{x}'\mathbf{T}_j\mathbf{x})^2}{\nu_j}. \tag{8.32}$$

As was noted earlier, small values of ν_j $(j = 1, 2, \ldots, p)$ cause $\hat{y}(\mathbf{x})$ to have large variances.

To reduce the size of the prediction variance within the region explored by ridge analysis, we can consider putting constraints on the portion of $\text{Var}[\hat{y}(\mathbf{x})]$ that corresponds to $\nu_{\min}$. It makes sense to optimize $\hat{y}(\mathbf{x})$ subject to the constraints

$$\mathbf{x}'\mathbf{x} = r^2, \tag{8.33}$$

$$|v_{0m} + \mathbf{x}'\boldsymbol{\tau}_m + \mathbf{x}'\mathbf{T}_m\mathbf{x}| \leq q, \tag{8.34}$$

where v_{0m}, $\boldsymbol{\tau}_m$, and $\mathbf{T}_m$ are the values of v_0, $\boldsymbol{\tau}$, and $\mathbf{T}$ that correspond to $\nu_{\min}$. Here, q is a positive constant chosen small enough to offset the small value of $\nu_{\min}$. Khuri and Myers (1979) suggested that q be equal to the largest value taken by $|v_{0m} + \mathbf{x}'\boldsymbol{\tau}_m + \mathbf{x}'\mathbf{T}_m\mathbf{x}|$ at the n design points. The rationale behind this rule of thumb is that the prediction variance is smaller at the design points than at other points in the experimental region.

The modification suggested by Khuri and Myers (1979) amounts to adding the constraint (8.34) to the usual procedure of ridge analysis. In this way, some control can be maintained on the size of prediction variance during the optimization process. The mathematical algorithm needed for this constrained optimization is based on a technique introduced by Myers and Carter (1973) for a dual response system in which a *primary* second-order response function is optimized subject to the condition that a *constrained* second-order response function takes on some specified or desirable values. Here, the primary response is $\hat{y}(\mathbf{x})$ and the constrained response is $v_{0m} + \mathbf{x}'\boldsymbol{\tau}_m + \mathbf{x}'\mathbf{T}_m\mathbf{x}$.

Myers and Carter's (1973) procedure is based on the method of Lagrange multipliers, which uses the function

$$L = \hat{\beta}_0 + \mathbf{x}'\hat{\boldsymbol{\beta}} + \mathbf{x}'\hat{\mathbf{B}}\mathbf{x} - \mu(v_{0m} + \mathbf{x}'\boldsymbol{\tau}_m + \mathbf{x}'\mathbf{T}_m\mathbf{x} - \omega) - \lambda(\mathbf{x}'\mathbf{x} - r^2),$$

where $\hat{\beta}_0$, $\hat{\boldsymbol{\beta}}$, and $\hat{\mathbf{B}}$ are the same as in model (8.14), μ and λ are Lagrange multipliers, ω is such that $|\omega| \leq q$ [see inequality (8.34)], and r is the radius of a hypersphere centered at the origin and contained inside a region of interest R. By differentiating L with respect to $x_1, x_2, \ldots, x_k$ and equating the derivatives to zero, we obtain

$$(\hat{\mathbf{B}} - \mu\mathbf{T}_m - \lambda\mathbf{I}_k)\mathbf{x} = \tfrac{1}{2}(\mu\boldsymbol{\tau}_m - \hat{\boldsymbol{\beta}}). \tag{8.35}$$

As in the method of ridge analysis, to solve equation (8.35), values of μ and λ are chosen directly in such a way that the solution represents a point of maximum (or minimum) for $\hat{y}(\mathbf{x})$. Thus for a given value of μ, the matrix of second-order partial derivatives of L, namely $2(\hat{\mathbf{B}} - \mu\mathbf{T}_m - \lambda\mathbf{I}_k)$, is made negative definite [and hence a maximum of $\hat{y}(\mathbf{x})$ is achieved] by selecting λ

larger than the largest eigenvalue of $\hat{\mathbf{B}} - \mu\mathbf{T}_m$. Values of λ smaller than the smallest eigenvalue of $\hat{\mathbf{B}} - \mu\mathbf{T}_m$ should be considered in order for $\hat{y}(\mathbf{x})$ to attain a minimum. It follows that for such an assignment of values for μ and λ, the corresponding solution of equation (8.35) produces an optimum for $\hat{y}$ subject to a fixed $r = (\mathbf{x}'\mathbf{x})^{1/2}$ and a fixed value of $v_{0m} + \mathbf{x}'\boldsymbol{\tau}_m + \mathbf{x}'\mathbf{T}_m\mathbf{x}$.

EXAMPLE 8.3.2. An attempt was made to design an experiment from which one could find conditions on concentration of three basic substances that maximize a certain mechanical modular property of a solid propellant. The initial intent was to construct and use a central composite design (see Khuri and Cornell, 1996, Section 4.5.3) for the three components in the system. However, certain experimental difficulties prohibited the use of the design as planned, and the design used led to problems with multicollinearity as far as the fitting of the second-order model is concerned. The design settings and corresponding response values are given in Table 8.3.

In this example, the smallest eigenvalue of $\mathbf{X}'\mathbf{X}$ is $v_{\min} = 0.0321$. Correspondingly, the values of v_{0m}, $\boldsymbol{\tau}_m$, and $\mathbf{T}_m$ in inequality (8.34) are $v_{0m} = -0.2935$, $\boldsymbol{\tau}_m = (0.0469, 0.4081, 0.4071)'$, and

$$\mathbf{T}_m = \begin{bmatrix} 0.1129 & 0.0095 & 0.2709 \\ 0.0095 & -0.1382 & -0.0148 \\ 0.2709 & -0.0148 & 0.6453 \end{bmatrix}.$$

As for q in inequality (8.34), values of $|v_{0m} + \mathbf{x}'\boldsymbol{\tau}_m + \mathbf{x}'\mathbf{T}_m\mathbf{x}|$ were computed at each of the 15 design points in Table 8.3. The largest value was found to

Table 8.3. Design Settings and Response Values for Example 8.3.2

x_1	x_2	x_3	y
-1.020	-1.402	-0.998	13.5977
0.900	0.478	-0.818	12.7838
0.870	-1.282	0.882	16.2780
-0.950	0.458	0.972	14.1678
-0.930	-1.242	-0.868	9.2461
0.750	0.498	-0.618	17.0167
0.830	-1.092	0.732	13.4253
-0.950	0.378	0.832	16.0967
1.950	-0.462	0.002	14.5438
-2.150	-0.402	-0.038	20.9534
-0.550	0.058	-0.518	11.0411
-0.450	1.378	0.182	21.2088
0.150	1.208	0.082	25.5514
0.100	1.768	-0.008	33.3793
1.450	-0.342	0.182	15.4341

Source: Khuri and Myers (1979). Reproduced with permission of the American Statistical Association.

Table 8.4. Results of Modified Ridge Analysis

r	0.848	1.162	1.530	1.623	1.795	1.850	1.904	1.935	2.000
$|\omega|$	0.006	0.074	0.136	1.139	0.048	0.086	0.126	0.146	0.165
$\text{Var}[\hat{y}(\mathbf{x})]/\sigma^2$	1.170	1.635	2.922	3.147	1.305	2.330	3.510	4.177	5.336
x_1	0.410	0.563	0.773	0.785	0.405	0.601	0.750	0.820	0.965
x_2	0.737	1.015	1.320	1.422	1.752	1.751	1.750	1.752	1.752
x_3	0.097	0.063	0.019	0.011	0.000	0.000	0.012	0.015	−0.030
$\hat{y}(\mathbf{x})$	22.420	27.780	35.242	37.190	37.042	40.222	42.830	44.110	46.260

Source: Khuri and Myers (1979). Reproduced with permission of the American Statistical Association.

Table 8.5. Results of Standard Ridge Analysis

r	0.140	0.379	0.698	0.938	1.146	1.394	1.484	1.744	1.944	1.975	2.000
$|\omega|$	0.241	0.124	0.104	0.337	0.587	0.942	1.085	1.553	1.958	2.025	2.080
$\omega^2/\nu_{\min}$	1.804	0.477	0.337	3.543	10.718	27.631	36.641	75.163	119.371	127.815	134.735
$\text{Var}[\hat{y}(\mathbf{x})]/\sigma^2$	2.592	1.554	2.104	6.138	14.305	32.787	42.475	83.38	129.834	138.668	145.907
x_1	0.037	0.152	0.352	0.515	0.660	0.835	0.899	1.085	1.227	1.249	1.265
x_2	0.103	0.255	0.422	0.531	0.618	0.716	0.749	0.845	0.916	0.927	0.936
x_3	0.087	0.235	0.431	0.577	0.705	0.858	0.912	1.074	1.197	1.217	1.232
$\hat{y}(\mathbf{x})$	12.796	16.021	21.365	26.229	31.086	37.640	40.197	48.332	55.147	56.272	57.176

Source: Khuri and Myers (1979). Reproduced with permission of the American Statistical Association.

be 0.087. Hence, the value of $|v_{0m} + \mathbf{x}'\boldsymbol{\tau}_m + \mathbf{x}'\mathbf{T}_m\mathbf{x}|$ should not grow much larger than 0.09 in the experimental region. Furthermore, r in equation (8.33) must not exceed the value 2, since most of the design points are contained inside a sphere of radius 2. The results of maximizing $\hat{y}(\mathbf{x})$ subject to this dual constraint are given in Table 8.4. For the sake of comparison, the results of applying the standard procedure of ridge analysis (that is, without the additional constraint concerning $v_{0m} + \mathbf{x}'\boldsymbol{\tau}_m + \mathbf{x}'\mathbf{T}_m\mathbf{x}$) are displayed in Table 8.5.

It is clear from Tables 8.4 and 8.5 that the extra constraint concerning $v_{0m} + \mathbf{x}'\boldsymbol{\tau}_m + \mathbf{x}'\mathbf{T}_m\mathbf{x}$ has profoundly improved the precision of $\hat{y}$ at the estimated maxima. At a specified radius, the value of $\hat{y}$ obtained under standard ridge analysis is higher than the one obtained under modified ridge analysis. However, the prediction variance values under the latter procedure are much smaller, as can be seen from comparing Tables 8.4 and 8.5. While the tradeoff that exists between a high response value and a small prediction variance is a bit difficult to cope with from a decision making standpoint, there is a clear superiority of the results displayed in Table 8.4. For example, one would hardly choose any operating conditions in Table 8.5 that indicate $\hat{y} \geq 50$, due to the accompanying large prediction variances. On the other hand, Table 8.4 reveals that at radius $r = 2.000$, $\hat{y} = 46.26$ with $\text{Var}[\hat{y}(\mathbf{x})]/\sigma^2$

= 5.336, while a rival set of coordinates at $r = 1.744$ for standard ridge analysis gives $\hat{y} = 48.332$ with $\mathrm{Var}[\hat{y}(\mathbf{x})]/\sigma^2 = 83.38$.

Row 3 of Table 8.5 gives values of $\omega^2/\nu_{\min}$, which should be compared with the corresponding values in row 4 of the same table. One can easily see that in this example, $\omega^2/\nu_{\min}$ accounts for a large portion of $\mathrm{Var}[\hat{y}(\mathbf{x})]/\sigma^2$.

8.4. RESPONSE SURFACE DESIGNS

We recall from Section 8.3 that one of the objectives of response surface methodology is the selection of a response surface design according to a certain optimality criterion. The design selection entails the specification of the settings of a group of input variables that can be used as experimental runs in a given experiment.

The proper choice of a response surface design can have a profound effect on the success of a response surface exploration. To see this, let us suppose that the fitted model is linear of the form

$$\mathbf{y} = \mathbf{X}\boldsymbol{\beta} + \boldsymbol{\epsilon}, \tag{8.36}$$

where $\mathbf{y}$ is an $n \times 1$ vector of observations, $\mathbf{X}$ is an $n \times p$ known matrix that depends on the design settings, $\boldsymbol{\beta}$ is a vector of p unknown parameters, and $\boldsymbol{\epsilon}$ is a vector of random errors in the elements of $\mathbf{y}$. Typically, $\boldsymbol{\epsilon}$ is assumed to have the normal distribution $N(\mathbf{0}, \sigma^2 \mathbf{I}_n)$, where σ^2 is unknown. In this case, the vector $\boldsymbol{\beta}$ is estimated by the least-squares estimator $\hat{\boldsymbol{\beta}}$, which is given by

$$\hat{\boldsymbol{\beta}} = (\mathbf{X}'\mathbf{X})^{-1}\mathbf{X}'\mathbf{y}. \tag{8.37}$$

If $x_1, x_2, \ldots, x_k$ are the input variables for the model under consideration, then the predicted response at a point $\mathbf{x} = (x_1, x_2, \ldots, x_k)'$ in a region of interest R is written as

$$\hat{y}(\mathbf{x}) = \mathbf{f}'(\mathbf{x})\hat{\boldsymbol{\beta}}, \tag{8.38}$$

where $\mathbf{f}(\mathbf{x})$ is a $p \times 1$ vector whose first element is equal to one and whose remaining $p - 1$ elements are functions of $x_1, x_2, \ldots, x_k$. These functions are in the form of powers and cross products of powers of the x_i's up to degree d. In this case, the model is said to be of order d. At the uth experimental run, $\mathbf{x}_u = (x_{u1}, x_{u1}, \ldots, x_{uk})'$ and the corresponding response value is $y_u(u = 1, 2, \ldots, n)$. Then $n \times k$ matrix $\mathbf{D} = [\mathbf{x}_1 : \mathbf{x}_2 : \cdots : \mathbf{x}_n]'$ is the design matrix. Thus by a choice of design we mean the specification of the elements of $\mathbf{D}$.

If model (8.36) is correct, then $\hat{\boldsymbol{\beta}}$ is an unbiased estimator of $\boldsymbol{\beta}$ and its variance–covariance matrix is given by

$$\mathrm{Var}(\hat{\boldsymbol{\beta}}) = (\mathbf{X}'\mathbf{X})^{-1}\sigma^2. \tag{8.39}$$

Hence, from formula (8.38), the prediction variance can be written as

$$\text{Var}[\hat{y}(\mathbf{x})] = \sigma^2 \mathbf{f}'(\mathbf{x})(\mathbf{X}'\mathbf{X})^{-1}\mathbf{f}(\mathbf{x}). \tag{8.40}$$

The design $\mathbf{D}$ is rotatable if $\text{Var}[\hat{y}(\mathbf{x})]$ remains constant at all points that are equidistant from the design center, as we may recall from Section 8.3. The input variables are coded so that the center of the design coincides with the origin of the coordinates system (see Khuri and Cornell, 1996, Section 2.8).

8.4.1. First-Order Designs

If model (8.36) is of the first order (that is, $d = 1$), then the matrix $\mathbf{X}$ is of the form $\mathbf{X} = [\mathbf{1}_n : \mathbf{D}]$, where $\mathbf{1}_n$ is a vector of ones of order $n \times 1$ [see model (8.8)]. The input variables can be coded in such a way that the sum of the elements in each column of $\mathbf{D}$ is equal to zero. Consequently, the prediction variance in formula (8.40) can be written as

$$\text{Var}[\hat{y}(\mathbf{x})] = \sigma^2 \left[\frac{1}{n} + \mathbf{x}'(\mathbf{D}'\mathbf{D})^{-1}\mathbf{x} \right]. \tag{8.41}$$

Formula (8.41) clearly shows the dependence of the prediction variance on the design matrix.

A reasonable criterion for the choice of $\mathbf{D}$ is the minimization of $\text{Var}[\hat{y}(\mathbf{x})]$, or equivalently, the minimization of $\mathbf{x}'(\mathbf{D}'\mathbf{D})^{-1}\mathbf{x}$ within the region R. To accomplish this we first note that for any $\mathbf{x}$ in the region R,

$$\mathbf{x}'(\mathbf{D}'\mathbf{D})^{-1}\mathbf{x} \leq \|\mathbf{x}\|_2^2 \|(\mathbf{D}'\mathbf{D})^{-1}\|_2, \tag{8.42}$$

where $\|\mathbf{x}\|_2 = (\mathbf{x}'\mathbf{x})^{1/2}$ and $\|(\mathbf{D}'\mathbf{D})^{-1}\|_2 = [\sum_{i=1}^{k}\sum_{j=1}^{k}(d^{ij})^2]^{1/2}$ is the Euclidean norm of $(\mathbf{D}'\mathbf{D})^{-1}$ with d^{ij} being its (i, j)th element $(i, j = 1, 2, \ldots, k)$. Inequality (8.42) follows from applying Theorems 2.3.16 and 2.3.20. Thus by choosing the design $\mathbf{D}$ so that it minimizes $\|(\mathbf{D}'\mathbf{D})^{-1}\|_2$, the quantity $\mathbf{x}'(\mathbf{D}'\mathbf{D})^{-1}\mathbf{x}$, and hence the prediction variance, can be reduced throughout the region R.

Theorem 8.4.1. For a given number n of experimental runs, $\|(\mathbf{D}'\mathbf{D})^{-1}\|_2$ attains its minimum if the columns $\mathbf{d}_1, \mathbf{d}_2, \ldots, \mathbf{d}_k$ of $\mathbf{D}$ are such that $\mathbf{d}_i'\mathbf{d}_j = 0$, $i \neq j$, and $\mathbf{d}_i'\mathbf{d}_i$ is as large as possible inside the region R.

Proof. We have that $\mathbf{D} = [\mathbf{d}_1 : \mathbf{d}_2 : \cdots : \mathbf{d}_k]$. The elements of $\mathbf{d}_i$ are the n design settings of the input variable x_i $(i = 1, 2, \ldots, k)$. Suppose that the region R places the following restrictions on these settings:

$$\mathbf{d}_i'\mathbf{d}_i \leq c_i^2, \qquad i = 1, 2, \ldots, k, \tag{8.43}$$

where c_i is some fixed constant. This means that the spread of the design in the direction of the ith coordinate axis is bounded by c_i^2 $(i = 1, 2, \ldots, k)$.

□

Now, if d_{ii} denotes the ith diagonal element of $\mathbf{D'D}$, then $d_{ii} = \mathbf{d}_i'\mathbf{d}_i$ $(i = 1, 2, \ldots, k)$. Furthermore,

$$d^{ii} \geq \frac{1}{d_{ii}}, \qquad i = 1, 2, \ldots, k, \tag{8.44}$$

where d^{ii} is the ith diagonal element of $(\mathbf{D'D})^{-1}$.

To prove inequality (8.44), let $\mathbf{D}_i$ be a matrix of order $n \times (k - 1)$ obtained from $\mathbf{D}$ by removing its ith column $\mathbf{d}_i$ $(i = 1, 2, \ldots, k)$. The cofactor of d_{ii} in $\mathbf{D'D}$ is then $\det(\mathbf{D}_i'\mathbf{D}_i)$. Hence, from Section 2.3.3,

$$d^{ii} = \frac{\det(\mathbf{D}_i'\mathbf{D}_i)}{\det(\mathbf{D'D})}, \qquad i = 1, 2, \ldots, k. \tag{8.45}$$

There exists an orthogonal matrix $\mathbf{E}_i$ of order $k \times k$ (whose determinant has an absolute value of one) such that the first column of $\mathbf{DE}_i$ is $\mathbf{d}_i$ and the remaining columns are the same as those of $\mathbf{D}_i$, that is,

$$\mathbf{DE}_i = [\mathbf{d}_i : \mathbf{D}_i], \qquad i = 1, 2, \ldots, k.$$

It follows that [see property 7 in Section 2.3.3]

$$\det(\mathbf{D'D}) = \det(\mathbf{E}_i'\mathbf{D'DE}_i)$$

$$= \det(\mathbf{D}_i'\mathbf{D}_i)\left[\mathbf{d}_i'\mathbf{d}_i - \mathbf{d}_i'\mathbf{D}_i(\mathbf{D}_i'\mathbf{D}_i)^{-1}\mathbf{D}_i'\mathbf{d}_i\right].$$

Hence, from (8.45) we obtain

$$d^{ii} = \left[d_{ii} - \mathbf{d}_i'\mathbf{D}_i(\mathbf{D}_i'\mathbf{D}_i)^{-1}\mathbf{D}_i'\mathbf{d}_i\right]^{-1}, \qquad i = 1, 2, \ldots, k. \tag{8.46}$$

Inequality (8.44) now follows from formula (8.46), since $\mathbf{d}_i'\mathbf{D}_i(\mathbf{D}_i'\mathbf{D}_i)^{-1}\mathbf{D}_i'\mathbf{d}_i \geq 0$. We can therefore write

$$\left\|(\mathbf{D'D})^{-1}\right\|_2 \geq \left[\sum_{i=1}^k (d^{ii})^2\right]^{1/2} \geq \left[\sum_{i=1}^k \frac{1}{d_{ii}^2}\right]^{1/2}.$$

Using the restrictions (8.43) we then have

$$\left\| (\mathbf{D}'\mathbf{D})^{-1} \right\|_2 \geq \left[\sum_{i=1}^{k} \frac{1}{c_i^4} \right]^{1/2} .$$

Equality is achieved if the columns of $\mathbf{D}$ are orthogonal to one another and $d_{ii} = c_i^2$ $(i = 1, 2, \ldots, k)$. This follows from the fact that $d^{ii} = 1/d_{ii}$ if and only if $\mathbf{d}_i' \mathbf{D}_i = 0$ $(i = 1, 2, \ldots, k)$, as can be seen from formula (8.46).

Definition 8.4.1. A design for fitting a fist-order model is said to be orthogonal if its columns are orthogonal to one another. □

Corollary 8.4.1. For a given number n of experimental runs, $\mathrm{Var}(\hat{\beta}_i)$ attains a minimum if and only if the design is orthogonal, where $\hat{\beta}_i$ is the least-squares estimator of β_i in model (8.8), $i = 1, 2, \ldots, k$.

Proof. This follows directly from Theorem 8.4.1 and the fact that $\mathrm{Var}(\hat{\beta}_i) = \sigma^2 d^{ii}$ $(i = 1, 2, \ldots, k)$, as can be seen from formula (8.39). □

From Theorem 8.4.1 and Corollary 8.4.1 we conclude that an orthogonal design for fitting a first-order model has optimal variance properties. Another advantage of orthogonal first-order designs is that the effects of the k input variables in model (8.8), as measured by the values of the β_i's $(i = 1, 2, \ldots, k)$, can be estimated independently. This is because the off-diagonal elements of the variance–covariance matrix of $\hat{\boldsymbol{\beta}}$ in formula (8.39) are zero. This means that the elements of $\hat{\boldsymbol{\beta}}$ are uncorrelated and hence statistically independent under the assumption of normality of the random error vector $\boldsymbol{\epsilon}$ in model (8.36).

Examples of first-order orthogonal designs are given in Khuri and Cornell (1996, Chapter 3). Prominent among these designs are the 2^k factorial design (each input variable has two levels, and the number of all possible combinations of these levels is 2^k) and the Plackett–Burman design, which was introduced in Plackett and Burman (1946). In the latter design, the number of design points is equal to $k + 1$, which must be a multiple of 4.

8.4.2. Second-Order Designs

These designs are used to fit second-order models of the form given by (8.12). Since the number of parameters in this model is $p = (k + 1)(k + 2)/2$, the number of experimental runs (or design points) in a second-order design must at least be equal to p. The most frequently used second-order designs include the 3^k design (each input variable has three levels, and the number of all possible combinations of these levels is 3^k), the central composite design (CCD), and the Box–Behnken design.

The CCD was introduced by Box and Wilson (1951). It is made up of a factorial portion consisting of a 2^k factorial design, an axial portion of k pairs of points with the ith pair consisting of two symmetric points on the ith coordinate axis $(i = 1, 2, \ldots, k)$ at a distance of α (> 0) from the design center (which coincides with the center of the coordinates system by the coding scheme), and n_0 (≥ 1) center-point runs. The values of α and n_0 can be chosen so that the CCD acquires certain desirable features (see, for example, Khuri and Cornell, 1996, Section 4.5.3). In particular, if $\alpha = F^{1/4}$, where F denotes the number of points in the factorial portion, then the CCD is rotatable. The choice of n_0 can affect the stability of the prediction variance.

The Box–Behnken design, introduced in Box and Behnken (1960), is a subset of a 3^k factorial design and, in general, requires many fewer points. It also compares favorably with the CCD. A thorough description of this design is given in Box and Draper (1987, Section 15.4).

Other examples of second-order designs are given in Khuri and Cornell (1996, Chapter 4).

8.4.3. Variance and Bias Design Criteria

We have seen that the minimization of the prediction variance represents an important criterion for the selection of a response surface design. This criterion, however, presumes that the fitted model is correct. There are many situations in which bias in the predicted response can occur due to fitting the wrong model. We refer to this as model bias.

Box and Draper (1959, 1963) presented convincing arguments in favor of recognizing bias as an important design criterion—in certain cases, even more important than the variance criterion.

Consider again model (8.36). The response value at a point $\mathbf{x} = (x_1, x_2, \ldots, x_k)'$ in a region R is represented as

$$y(\mathbf{x}) = \mathbf{f}'(\mathbf{x})\boldsymbol{\beta} + \boldsymbol{\epsilon}, \tag{8.47}$$

where $\mathbf{f}'(\mathbf{x})$ is the same as in model (8.38). While it is hoped that model (8.47) is correct, there is always a fear that the true model is different. Let us therefore suppose that in reality the true mean response at $\mathbf{x}$, denoted by $\eta(\mathbf{x})$, is given by

$$\eta(\mathbf{x}) = \mathbf{f}'(\mathbf{x})\boldsymbol{\beta} + \mathbf{g}'(\mathbf{x})\boldsymbol{\delta} \tag{8.48}$$

where the elements of $\mathbf{g}'(\mathbf{x})$ depend on $\mathbf{x}$ and consist of powers and cross products of powers of $x_1, x_2, \ldots, x_k$ of degree $d' > d$, with d being the order of model (8.47), and $\boldsymbol{\delta}$ is a vector of q unknown parameters. For a given design $\mathbf{D}$ of n experimental runs, we then have the model

$$\boldsymbol{\eta} = \mathbf{X}\boldsymbol{\beta} + \mathbf{Z}\boldsymbol{\delta},$$

where $\boldsymbol{\eta}$ is the vector of true means (or expected values) of the elements of $\mathbf{y}$ at the n design points, $\mathbf{X}$ is the same as in model (8.36), and $\mathbf{Z}$ is a matrix of order $n \times q$ whose uth row is equal to $\mathbf{g}'(\mathbf{x}_u)$. Here, $\mathbf{x}'_u$ denotes the uth row of $\mathbf{D}$ ($u = 1, 2, \ldots, n$).

At each point $\mathbf{x}$ in R, the mean squared error (MSE) of $\hat{y}(\mathbf{x})$, where $\hat{y}(\mathbf{x})$ is the predicted response as given by formula (8.38), is defined as

$$\text{MSE}[\hat{y}(\mathbf{x})] = E[\hat{y}(\mathbf{x}) - \eta(\mathbf{x})]^2.$$

This can be expressed as

$$\text{MSE}[\hat{y}(\mathbf{x})] = \text{Var}[\hat{y}(\mathbf{x})] + \text{Bias}^2[\hat{y}(\mathbf{x})], \qquad (8.49)$$

where $\text{Bias}[\hat{y}(\mathbf{x})] = E[\hat{y}(\mathbf{x})] - \eta(\mathbf{x})$. The fundamental philosophy of Box and Draper (1959, 1963) is centered around the consideration of the integrated mean squared error (IMSE) of $\hat{y}(\mathbf{x})$. This is denoted by J and is defined in terms of a k-tuple Riemann integral over the region R, namely,

$$J = \frac{n\Omega}{\sigma^2} \int_R \text{MSE}[\hat{y}(\mathbf{x})] \, d\mathbf{x}, \qquad (8.50)$$

where $\Omega^{-1} = \int_R d\mathbf{x}$ and σ^2 is the error variance. The partitioning of $\text{MSE}[\hat{y}(\mathbf{x})]$ as in formula (8.49) enables us to separate J into two parts:

$$J = \frac{n\Omega}{\sigma^2} \int_R \text{Var}[\hat{y}(\mathbf{x})] \, d\mathbf{x} + \frac{n\Omega}{\sigma^2} \int_R \text{Bias}^2[\hat{y}(\mathbf{x})] \, d\mathbf{x} = V + B. \qquad (8.51)$$

The quantities V and B are called the average variance and average squared bias of $\hat{y}(\mathbf{x})$, respectively. Both V and B depend on the design $\mathbf{D}$. Thus a reasonable choice of design is one that minimizes (1) V alone, (2) B alone, or (3) $J = V + B$.

Now, using formula (8.40), V can be written as

$$V = n\Omega \int_R \mathbf{f}'(\mathbf{x})(\mathbf{X}'\mathbf{X})^{-1}\mathbf{f}(\mathbf{x}) \, d\mathbf{x}$$

$$= \text{tr}\left\{ n(\mathbf{X}'\mathbf{X})^{-1}\left[\Omega \int_R \mathbf{f}(\mathbf{x})\mathbf{f}'(\mathbf{x}) \, d\mathbf{x} \right] \right\}$$

$$= \text{tr}\left[n(\mathbf{X}'\mathbf{X})^{-1}\boldsymbol{\Gamma}_{11} \right], \qquad (8.52)$$

where

$$\mathbf{\Gamma}_{11} = \Omega \int_R \mathbf{f}(\mathbf{x})\mathbf{f}'(\mathbf{x}) \, d\mathbf{x}.$$

As for B, we note from formula (8.37) that

$$E(\hat{\boldsymbol{\beta}}) = (\mathbf{X}'\mathbf{X})^{-1}\mathbf{X}'\boldsymbol{\eta}$$
$$= \boldsymbol{\beta} + \mathbf{A}\boldsymbol{\delta},$$

where $\mathbf{A} = (\mathbf{X}'\mathbf{X})^{-1}\mathbf{X}'\mathbf{Z}$. Thus from formula (8.38) we have

$$E[\hat{y}(\mathbf{x})] = \mathbf{f}'(\mathbf{x})(\boldsymbol{\beta} + \mathbf{A}\boldsymbol{\delta}).$$

Using the expression for $\eta(\mathbf{x})$ in formula (8.48), B can be written as

$$B = \frac{n\Omega}{\sigma^2} \int_R [\mathbf{f}'(\mathbf{x})\boldsymbol{\beta} + \mathbf{f}'(\mathbf{x})\mathbf{A}\boldsymbol{\delta} - \mathbf{f}'(\mathbf{x})\boldsymbol{\beta} - \mathbf{g}'(\mathbf{x})\boldsymbol{\delta}]^2 \, d\mathbf{x}$$

$$= \frac{n\Omega}{\sigma^2} \int_R [\mathbf{f}'(\mathbf{x})\mathbf{A}\boldsymbol{\delta} - \mathbf{g}'(\mathbf{x})\boldsymbol{\delta}]^2 \, d\mathbf{x}$$

$$= \frac{n\Omega}{\sigma^2} \int_R \boldsymbol{\delta}'[\mathbf{A}'\mathbf{f}(\mathbf{x}) - \mathbf{g}(\mathbf{x})][\mathbf{f}'(\mathbf{x})\mathbf{A} - \mathbf{g}'(\mathbf{x})]\boldsymbol{\delta} \, d\mathbf{x}$$

$$= \frac{n\Omega}{\sigma^2} \int_R \boldsymbol{\delta}'[\mathbf{A}'\mathbf{f}(\mathbf{x})\mathbf{f}'(\mathbf{x})\mathbf{A} - \mathbf{g}(\mathbf{x})\mathbf{f}'(\mathbf{x})\mathbf{A} - \mathbf{A}'\mathbf{f}(\mathbf{x})\mathbf{g}'(\mathbf{x}) + \mathbf{g}(\mathbf{x})\mathbf{g}'(\mathbf{x})]\boldsymbol{\delta} \, d\mathbf{x}$$

$$= \frac{n}{\sigma^2}\boldsymbol{\delta}'\boldsymbol{\Delta}\boldsymbol{\delta}, \tag{8.53}$$

where

$$\boldsymbol{\Delta} = \mathbf{A}'\mathbf{\Gamma}_{11}\mathbf{A} - \mathbf{\Gamma}_{12}'\mathbf{A} - \mathbf{A}'\mathbf{\Gamma}_{12} + \mathbf{\Gamma}_{22},$$

$$\mathbf{\Gamma}_{12} = \Omega \int_R \mathbf{f}(\mathbf{x})\mathbf{g}'(\mathbf{x}) \, d\mathbf{x},$$

$$\mathbf{\Gamma}_{22} = \Omega \int_R \mathbf{g}(\mathbf{x})\mathbf{g}'(\mathbf{x}) \, d\mathbf{x}.$$

The matrices $\mathbf{\Gamma}_{11}$, $\mathbf{\Gamma}_{12}$, and $\mathbf{\Gamma}_{22}$ are called region moments. By adding and subtracting the matrix $\mathbf{\Gamma}_{12}'\mathbf{\Gamma}_{11}^{-1}\mathbf{\Gamma}_{12}$ from $\boldsymbol{\Delta}$ in formula (8.53), B can be expressed as

$$B = \frac{n}{\sigma^2}\boldsymbol{\delta}'\left[\left(\mathbf{\Gamma}_{22} - \mathbf{\Gamma}_{12}'\mathbf{\Gamma}_{11}^{-1}\mathbf{\Gamma}_{12}\right) + \left(\mathbf{A} - \mathbf{\Gamma}_{11}^{-1}\mathbf{\Gamma}_{12}\right)'\mathbf{\Gamma}_{11}\left(\mathbf{A} - \mathbf{\Gamma}_{11}^{-1}\mathbf{\Gamma}_{12}\right)\right]\boldsymbol{\delta}. \tag{8.54}$$

We note that the design $\mathbf{D}$ affects only the second expression inside brackets on the right-hand side of formula (8.54). Thus to minimize B, the design $\mathbf{D}$ should be chosen such that

$$\mathbf{A} - \boldsymbol{\Gamma}_{11}^{-1}\boldsymbol{\Gamma}_{12} = \mathbf{0}. \tag{8.55}$$

Since $\mathbf{A} = (\mathbf{X}'\mathbf{X})^{-1}\mathbf{X}'\mathbf{Z}$, a sufficient (but not necessary) condition for the minimization of B is

$$\mathbf{M}_{11} = \boldsymbol{\Gamma}_{11}, \qquad \mathbf{M}_{12} = \boldsymbol{\Gamma}_{12}, \tag{8.56}$$

where $\mathbf{M}_{11} = (1/n)\mathbf{X}'\mathbf{X}$, $\mathbf{M}_{12} = (1/n)\mathbf{X}'\mathbf{Z}$ are the so-called design moments. Thus a sufficient condition for the minimization of B is the equality of the design moments, $\mathbf{M}_{11}$ and $\mathbf{M}_{12}$, to the corresponding region moments, $\boldsymbol{\Gamma}_{11}$ and $\boldsymbol{\Gamma}_{12}$.

The minimization of $J = V + B$ is not possible without the specification of δ/σ. Box and Draper (1959, 1963) showed that unless V is considerably larger than B, the optimal design that minimizes J has characteristics similar to those of a design that minimizes just B.

Examples of designs that minimize V alone or B alone can be found in Box and Draper (1987, Chapter 13), Khuri and Cornell (1996, Chapter 6), and Myers (1976, Chapter 9).

8.5. ALPHABETIC OPTIMALITY OF DESIGNS

Let us again consider model (8.47), which we now assume to be correct, that is, the true mean response, $\eta(\mathbf{x})$, is equal to $\mathbf{f}'(\mathbf{x})\boldsymbol{\beta}$. In this case, the matrix $\mathbf{X}'\mathbf{X}$ plays an important role in the determination of an optimal design, since the elements of $(\mathbf{X}'\mathbf{X})^{-1}$ are proportional to the variances and covariances of the least-squares estimators of the model's parameters [see formula (8.39)].

The mathematical theory of optimal designs, which was developed by Kiefer (1958, 1959, 1960, 1961, 1962a, b), is concerned with the choice of designs that minimize certain functions of the elements of $(\mathbf{X}'\mathbf{X})^{-1}$. The kernel of Kiefer's approach is based on the concept of design measure, which represents a generalization of the traditional design concept. So far, each of the designs that we have considered for fitting a response surface model has consisted of a set of n points in a k-dimensional space ($k \geq 1$). Suppose that $\mathbf{x}_1, \mathbf{x}_2, \ldots, \mathbf{x}_m$ are distinct points of an n-point design ($m \leq n$) with the lth ($l = 1, 2, \ldots, m$) point being replicated n_l (≥ 1) times (that is, n_l repeated observations are taken at this point). The design can therefore be regarded as a collection of points in a region of interest R with the lth point being assigned the weight n_l/n ($l = 1, 2, \ldots, m$), where $n = \sum_{l=1}^{m} n_l$. Kiefer generalized this setup using the so-called continuous design measure, which is basically a probability measure $\xi(\mathbf{x})$ defined on R and satisfies the conditions

$$\xi(\mathbf{x}) \geq 0 \quad \text{for all } \mathbf{x} \in R \quad \text{and} \quad \int_R d\xi(\mathbf{x}) = 1.$$

In particular, the measure induced by a traditional design $\mathbf{D}$ with n points is called a discrete design measure and is denoted by ξ_n. It should be noted that while a discrete design measure is realizable in practice, the same is not true of a general continuous design measure. For this reason, the former design is called *exact* and the latter design is called *approximate*.

By definition, the moment matrix of a design measure ξ is a symmetric matrix of the form $\mathbf{M}(\xi) = [m_{ij}(\xi)]$, where

$$m_{ij}(\xi) = \int_R f_i(\mathbf{x}) f_j(\mathbf{x}) \, d\xi(\mathbf{x}). \tag{8.57}$$

Here, $f_i(\mathbf{x})$ is the ith element of $\mathbf{f}(\mathbf{x})$ in formula (8.47), $i = 1, 2, \ldots, p$. For a discrete design measure ξ_n, the (i, j)th element of the moment matrix is

$$m_{ij}(\xi_n) = \frac{1}{n} \sum_{l=1}^{m} n_l f_i(\mathbf{x}_l) f_j(\mathbf{x}_l), \tag{8.58}$$

where m is the number of distinct design points and n_l is the number of replications at the lth point ($l = 1, 2, \ldots, m$). In this special case, the matrix $\mathbf{M}(\xi)$ reduces to the usual moment matrix $(1/n)\mathbf{X}'\mathbf{X}$, where $\mathbf{X}$ is the same matrix as in formula (8.36).

For a general design measure ξ, the standardized prediction variance, denoted by $d(\mathbf{x}, \xi)$, is defined as

$$d(\mathbf{x}, \xi) = \mathbf{f}'(\mathbf{x})[\mathbf{M}(\xi)]^{-1}\mathbf{f}(\mathbf{x}), \tag{8.59}$$

where $\mathbf{M}(\xi)$ is assumed to be nonsingular. In particular, for a discrete design measure ξ_n, the prediction variance in formula (8.40) is equal to $(\sigma^2/n) d(\mathbf{x}, \xi_n)$.

Let H denote the class of all design measures defined on the region R. A prominent design criterion that has received a great deal of attention is that of D-optimality, in which the determinant of $\mathbf{M}(\xi)$ is maximized. Thus a design measure ξ_d is D-optimal if

$$\det[\mathbf{M}(\xi_d)] = \sup_{\xi \in H} \det[\mathbf{M}(\xi)]. \tag{8.60}$$

The rationale behind this criterion has to do with the minimization of the generalized variance of the least-squares estimator $\hat{\boldsymbol{\beta}}$ of the parameter vector $\boldsymbol{\beta}$. By definition, the generalized variance of $\hat{\boldsymbol{\beta}}$ is the same as the determinant of the variance–covariance matrix of $\hat{\boldsymbol{\beta}}$. This is based on the fact that under the normality assumption, the content (volume) of a fixed-level confidence region on $\boldsymbol{\beta}$ is proportional to $[\det(\mathbf{X}'\mathbf{X})]^{-1/2}$. The review articles by St. John and Draper (1975), Ash and Hedayat (1978), and Atkinson (1982, 1988) contain many references on D-optimality.

Another design criterion that is closely related to D-optimality is G-optimality, which is concerned with the prediction variance. By definition, a design measure ξ_g is G-optimal if it minimizes over H the maximum standardized prediction variance over the region R, that is,

$$\sup_{\mathbf{x} \in R} d(\mathbf{x}, \xi_g) = \inf_{\zeta \in H} \left\{ \sup_{\mathbf{x} \in R} d(\mathbf{x}, \zeta) \right\}. \tag{8.61}$$

Kiefer and Wolfowitz (1960) showed that D-optimality and G-optimality, as defined by formulas (8.60) and (8.61), are equivalent. Furthermore, a design measure ξ^* is G-optimal (or D-optimal) if and only if

$$\sup_{\mathbf{x} \in R} d(\mathbf{x}, \xi^*) = p, \tag{8.62}$$

where p is the number of parameters in the model. Formula (8.62) can be conveniently used to determine if a given design measure is D-optimal, since in general $\sup_{\mathbf{x} \in R} d(\mathbf{x}, \xi) \geq p$ for any design measure $\xi \in H$. If equality can be achieved by a design measure, then it must be G-optimal, and hence D-optimal.

EXAMPLE 8.5.1.   Consider fitting a second-order model in one input variable x over the region $R = [-1, 1]$. In this case, model (8.47) takes the form $y(x) = \beta_0 + \beta_1 x + \beta_{11} x^2 + \epsilon$, that is, $\mathbf{f}'(x) = (1, x, x^2)$. Suppose that the design measure used is defined as

$$\xi(x) = \begin{cases} \frac{1}{3}, & x = -1, 0, 1, \\ 0 & \text{otherwise}. \end{cases} \tag{8.63}$$

Thus ξ is a discrete design measure that assigns one-third of the experimental runs to each of the points -1, 0, and 1. This design measure is D-optimal. To verify this claim, we first need to determine the values of the elements of the moment matrix $\mathbf{M}(\xi)$. Using formula (8.58) with $n_l/n = \frac{1}{3}$ for $l = 1, 2, 3$, we find that $m_{11} = 1$, $m_{12} = 0$, $m_{13} = \frac{2}{3}$, $m_{22} = \frac{2}{3}$, $m_{23} = 0$, and $m_{33} = \frac{2}{3}$. Hence,

$$\mathbf{M}(\xi) = \begin{bmatrix} 1 & 0 & \frac{2}{3} \\ 0 & \frac{2}{3} & 0 \\ \frac{2}{3} & 0 & \frac{2}{3} \end{bmatrix},$$

$$\mathbf{M}^{-1}(\xi) = \begin{bmatrix} 3 & 0 & -3 \\ 0 & \frac{3}{2} & 0 \\ -3 & 0 & \frac{9}{2} \end{bmatrix}.$$

By applying formula (8.59) we find that $d(x, \xi) = 3 - \frac{9}{2}x^2 + \frac{9}{2}x^4, -1 \leq x \leq 1$. We note that $d(x, \xi) \leq 3$ for all x in $[-1, 1]$ with $d(0, \xi) = 3$. Thus $\sup_{x \in R} d(x, \xi) = 3$. Since 3 is the number of parameters in the model, then by condition (8.62) we conclude that the design measure defined by formula (8.63) is D-optimal.

In addition to the D- and G-optimality criteria, other variance-related design criteria have also been investigated. These include A- and E-optimality. By definition, a design measure is A-optimal if it maximizes the trace of $\mathbf{M}(\xi)$. This is equivalent to minimizing the sum of the variances of the least-squares estimators of the fitted model's parameters. In E-optimality, the smallest eigenvalue of $\mathbf{M}(\xi)$ is maximized. The rationale behind this criterion is based on the fact that

$$d(\mathbf{x}, \xi) \leq \frac{\mathbf{f}'(\mathbf{x})\mathbf{f}(\mathbf{x})}{\lambda_{\min}},$$

as can be seen from formula (8.59), where $\lambda_{\min}$ is the smallest eigenvalue of $\mathbf{M}(\xi)$. Hence, $d(\mathbf{x}, \xi)$ can be reduced by maximizing $\lambda_{\min}$.

The efficiency of a design measure $\zeta \in H$ with respect to a D-optimal design is defined as

$$D\text{-efficiency} = \left\{ \frac{\det[\mathbf{M}(\zeta)]}{\sup_{\xi \in H} \det[\mathbf{M}(\xi)]} \right\}^{1/p},$$

where p is the number of parameters in the model. Similarly, the G-efficiency of ζ is defined as

$$G\text{-efficiency} = \frac{p}{\sup_{\mathbf{x} \in R} d(\mathbf{x}, \zeta)}.$$

Both D- and G-efficiency values fall within the interval $[0, 1]$. The closer these values are to 1, the more efficient their corresponding designs are. Lucas (1976) compared several second-order designs (such as central composite and Box–Behnken designs) on the basis of their D- and G-efficiency values.

The equivalence theorem of Kiefer and Wolfowitz (1960) can be applied to construct a D-optimal design using a sequential procedure. This procedure is described in Wynn (1970, 1972) and goes as follows: Let $\mathbf{D}_{n_0}$ denote an initial response surface design with n_0 points, $\mathbf{x}_1, \mathbf{x}_2, \ldots, \mathbf{x}_{n_0}$, for which the matrix $\mathbf{X}'\mathbf{X}$ is nonsingular. A point $\mathbf{x}_{n_0+1}$ is found in the region R such that

$$d(\mathbf{x}_{n_0+1}, \xi_{n_0}) = \sup_{\mathbf{x} \in R} d(\mathbf{x}, \xi_{n_0}),$$

where ξ_{n_0} is the discrete design measure that represents $\mathbf{D}_{n_0}$. By augmenting $\mathbf{D}_{n_0}$ with $\mathbf{x}_{n_0+1}$ we obtain the design $\mathbf{D}_{n_0+1}$. Then, another point $\mathbf{x}_{n_0+2}$ is chosen such that

$$d(\mathbf{x}_{n_0+2}, \xi_{n_0+1}) = \sup_{\mathbf{x} \in R} d(\mathbf{x}, \xi_{n_0+1}),$$

where ξ_{n_0+1} is the discrete design measure that represents $\mathbf{D}_{n_0+1}$. The point $\mathbf{x}_{n_0+2}$ is then added to $\mathbf{D}_{n_0+1}$ to obtain the design $\mathbf{D}_{n_0+2}$. By continuing this process we obtain a sequence of discrete design measures, namely, $\xi_{n_0}, \xi_{n_0+1}, \xi_{n_0+2}, \ldots$. Wynn (1970) showed that this sequence converges to the D-optimal design ξ_d, that is,

$$\det[\mathbf{M}(\xi_{n_0+n})] \to \det[\mathbf{M}(\xi_d)]$$

as $n \to \infty$. An example is given in Wynn (1970, Section 5) to illustrate this sequential procedure.

The four design criteria, A-, D-, E-, and G-optimality, are referred to as alphabetic optimality. More detailed information about these criteria can be found in Atkinson (1982, 1988), Fedorov (1972), Pazman (1986), and Silvey (1980). Recall that to perform an actual experiment, one must use a discrete design. It is possible to find a discrete design measure ξ_n that approximates an optimal design measure. The approximation is good whenever n is large with respect to p (the number of parameters in the model).

Note that the equivalence theorem of Kiefer and Wolfowitz (1960) applies to general design measures and not necessarily to discrete design measures, that is, D- and G-optimality criteria are not equivalent for the class of discrete design measures. Optimal n-point discrete designs, however, can still be found on the basis of maximizing the determinant of $\mathbf{X}'\mathbf{X}$, for example. In this case, finding an optimal n-point design requires a search involving nk variables, where k is the number of input variables. Several algorithms have been introduced for this purpose. For example, the DETMAX algorithm by Mitchell (1974) is used to maximize $\det(\mathbf{X}'\mathbf{X})$. A review of algorithms for constructing optimal discrete designs can be found in Cook and Nachtsheim (1980) (see also Johnson and Nachtsheim, 1983).

One important criticism of the alphabetic optimality approach is that it is set within a rigid framework governed by a set of assumptions. For example, a specific model for the response function must be assumed as the "true" model. Optimal design measures can be quite sensitive to this assumption. Box (1982) presented a critique to this approach. He argued that in a response surface situation, it may not be realistic to assume that a model such as (8.47) represents the true response function exactly. Some protection against bias in the model should therefore be considered when choosing a response surface design. On the other hand, Kiefer (1975) criticized certain aspects of the preoccupation with bias, pointing out examples in which the variance criterion is compromised for the sake of the bias criterion. It follows that design selection should be guided by more than one single criterion (see Kiefer, 1975, page 286; Box, 1982, Section 7). A reasonable approach is to select compromise designs that are sufficiently good (but not necessarily optimal) from the viewpoint of several criteria that are important to the user.

8.6. DESIGNS FOR NONLINEAR MODELS

The models we have considered so far in the area of response surface methodology were linear in the parameters; hence the term linear models. There are, however, many experimental situations in which linear models do not adequately represent the true mean response. For example, the growth of an organism is more appropriately depicted by a nonlinear model. By definition, a nonlinear model is one of the form

$$y(\mathbf{x}) = h(\mathbf{x}, \boldsymbol{\theta}) + \epsilon, \tag{8.64}$$

where $\mathbf{x} = (x_1, x_2, \ldots, x_k)'$ is a vector of k input variables, $\boldsymbol{\theta} = (\theta_1, \theta_2, \ldots, \theta_p)'$ is a vector of p unknown parameters, ϵ is a random error, and $h(\mathbf{x}, \boldsymbol{\theta})$ is a known function, nonlinear in at least one element of $\boldsymbol{\theta}$. An example of a nonlinear model is

$$h(x, \boldsymbol{\theta}) = \frac{\theta_1 x}{\theta_2 + x}.$$

Here, $\boldsymbol{\theta} = (\theta_1, \theta_2)'$ and θ_2 is a nonlinear parameter. This particular model is known as the Michaelis–Menten model for enzyme kinetics. It relates the initial velocity of an enzymatic reaction to the substrate concentration x.

In contrast to linear models, nonlinear models have not received a great deal of attention in response surface methodology, especially in the design area. The main design criterion for nonlinear models is the D-optimality criterion, which actually applies to a linearized form of the nonlinear model. More specifically, this criterion depends on the assumption that in some neighborhood of a specified value $\boldsymbol{\theta}_0$ of $\boldsymbol{\theta}$, the function $h(\mathbf{x}, \boldsymbol{\theta})$ is approximately linear in $\boldsymbol{\theta}$. In this case, a first-order Taylor's expansion of $h(\mathbf{x}, \boldsymbol{\theta})$ yields the following approximation of $h(\mathbf{x}, \boldsymbol{\theta})$:

$$h(\mathbf{x}, \boldsymbol{\theta}) \approx h(\mathbf{x}, \boldsymbol{\theta}_0) + \sum_{i=1}^{p} (\theta_i - \theta_{i0}) \frac{\partial h(\mathbf{x}, \boldsymbol{\theta}_0)}{\partial \theta_i}.$$

Thus if $\boldsymbol{\theta}$ is close enough to $\boldsymbol{\theta}_0$, then we have approximately the linear model

$$z(\mathbf{x}) = \sum_{i=1}^{p} \psi_i \frac{\partial h(\mathbf{x}, \boldsymbol{\theta}_0)}{\partial \theta_i} + \epsilon, \tag{8.65}$$

where $z(\mathbf{x}) = y(\mathbf{x}) - h(\mathbf{x}, \boldsymbol{\theta}_0)$, and ψ_i is the ith element of $\boldsymbol{\psi} = \boldsymbol{\theta} - \boldsymbol{\theta}_0$ ($i = 1, 2, \ldots, p$).

For a given design consisting of n experimental runs, model (8.65) can be written in vector form as

$$\mathbf{z} = \mathbf{H}(\mathbf{\theta}_0)\mathbf{\psi} + \mathbf{\epsilon},$$ (8.66)

where $\mathbf{H}(\mathbf{\theta}_0)$ is an $n \times p$ matrix whose (u, i)th element is $\partial h(\mathbf{x}_u, \mathbf{\theta}_0)/\partial\theta_i$ with $\mathbf{x}_u$ being the vector of design settings for the k input variables at the uth experimental run $(i = 1, 2, \ldots, p; u = 1, 2, \ldots, n)$. Using the linearized form given by model (8.66), a design is chosen to maximize the determinant $\det[\mathbf{H}'(\mathbf{\theta}_0)\mathbf{H}(\mathbf{\theta}_0)]$. This is known as the Box–Lucas criterion (see Box and Lucas, 1959).

It can be easily seen that a nonlinear design obtained on the basis of the Box–Lucas criterion depends on the value of $\mathbf{\theta}_0$. This is an undesirable characteristic of nonlinear models, since a design is supposed to be used for estimating the unknown parameter vector $\mathbf{\theta}$. By contrast, designs for linear models are not dependent on the fitted model's parameters. Several procedures have been proposed for dealing with the problem of design dependence on the parameters of a nonlinear model. These procedures are mentioned in the review article by Myers, Khuri, and Carter (1989). See also Khuri and Cornell (1996, Section 10.5).

EXAMPLE 8.6.1. Let us again consider the Michaelis–Menten model mentioned earlier. The partial derivatives of $h(x, \mathbf{\theta})$ with respect to θ_1 and θ_2 are

$$\frac{\partial h(x, \mathbf{\theta})}{\partial\theta_1} = \frac{x}{\theta_2 + x},$$

$$\frac{\partial h(x, \mathbf{\theta})}{\partial\theta_2} = \frac{-\theta_1 x}{(\theta_2 + x)^2}.$$

Suppose that it is desired to find a two-point design that consists of the settings x_1 and x_2 using the Box–Lucas criterion. In this case,

$$\mathbf{H}(\mathbf{\theta}_0) = \begin{bmatrix} \dfrac{\partial h(x_1, \mathbf{\theta}_0)}{\partial\theta_1} & \dfrac{\partial h(x_1, \mathbf{\theta}_0)}{\partial\theta_2} \\[3mm] \dfrac{\partial h(x_2, \mathbf{\theta}_0)}{\partial\theta_1} & \dfrac{\partial h(x_2, \mathbf{\theta}_0)}{\partial\theta_2} \end{bmatrix}$$

$$= \begin{bmatrix} \dfrac{x_1}{\theta_{20} + x_1} & \dfrac{-\theta_{10}x_1}{(\theta_{20} + x_1)^2} \\[3mm] \dfrac{x_2}{\theta_{20} + x_2} & \dfrac{-\theta_{10}x_2}{(\theta_{20} + x_2)^2} \end{bmatrix},$$

where θ_{10} and θ_{20} are the elements of $\mathbf{\theta}_0$. In this example, $\mathbf{H}(\mathbf{\theta}_0)$ is a square matrix. Hence,

$$
\begin{aligned}
\det\left[\mathbf{H}'(\mathbf{\theta}_0)\mathbf{H}(\mathbf{\theta}_0)\right] &= \left\{\det\left[\mathbf{H}(\mathbf{\theta}_0)\right]\right\}^2 \\
&= \frac{\theta_{10}^2 x_1^2 x_2^2 (x_2 - x_1)^2}{(\theta_{20} + x_1)^4 (\theta_{20} + x_2)^4}.
\end{aligned}
\tag{8.67}
$$

To determine the maximum of this determinant, let us first equate its partial derivatives with respect to x_1 and x_2 to zero. It can be verified that the solution of the resulting equations (that is, the stationary point) falls outside the region of feasible values for x_1 and x_2 (both x_1 and x_2 must be nonnegative). Let us therefore restrict our search for the maximum within the region $R = \{(x_1, x_2) \mid 0 \le x_1 \le x_{max}, 0 \le x_2 \le x_{max}\}$, where x_{max} is the maximum allowable substrate concentration. Since the partial derivatives of the determinant in formula (8.67) do not vanish in R, then its maximum must be attained on the boundary of R. On $x_1 = 0$, or $x_2 = 0$, the value of the determinant is zero. If $x_1 = x_{max}$, then

$$
\det\left[\mathbf{H}'(\mathbf{\theta}_0)\mathbf{H}(\mathbf{\theta}_0)\right] = \frac{\theta_{10}^2 x_{max}^2 x_2^2 (x_2 - x_{max})^2}{(\theta_{20} + x_{max})^4 (\theta_{20} + x_2)^4}.
$$

It can be verified that this function of x_2 has a maximum at the point $x_2 = \theta_{20} x_{max}/(2\theta_{20} + x_{max})$ with a value given by

$$
\max_{x_1 = x_{max}} \left\{\det\left[\mathbf{H}'(\mathbf{\theta}_0)\mathbf{H}(\mathbf{\theta}_0)\right]\right\} = \frac{\theta_{10}^2 x_{max}^6}{16\theta_{20}^2 (\theta_{20} + x_{max})^6}.
\tag{8.68}
$$

Similarly, if $x_2 = x_{max}$, then

$$
\det\left[\mathbf{H}'(\mathbf{\theta}_0)\mathbf{H}(\mathbf{\theta}_0)\right] = \frac{\theta_{10}^2 x_{max}^2 x_1^2 (x_{max} - x_1)^2}{(\theta_{20} + x_1)^4 (\theta_{20} + x_{max})^4},
$$

which attains the same maximum value as in formula (8.68) at the point $x_1 = \theta_{20} x_{max}/(2\theta_{20} + x_{max})$. We conclude that the maximum of $\det[\mathbf{H}'(\mathbf{\theta}_0)\mathbf{H}(\mathbf{\theta}_0)]$ over the region R is achieved when $x_1 = x_{max}$ and $x_2 = \theta_{20} x_{max}/(2\theta_{20} + x_{max})$, or when $x_1 = \theta_{20} x_{max}/(2\theta_{20} + x_{max})$ and $x_2 = x_{max}$.

We can clearly see in this example the dependence of the design settings on θ_2, but not on θ_1. This is attributed to the fact that θ_1 appears linearly in the model, but θ_2 does not. In this case, the model is said to be partially nonlinear. Its D-optimal design depends only on those parameters that do not appear linearly. More details concerning partially nonlinear models can be found in Khuri and Cornell (1996, Section 10.5.3).

8.7. MULTIRESPONSE OPTIMIZATION

By definition, a multiresponse experiment is one in which a number of responses can be measured for each setting of a group of input variables. For example, in a skim milk extrusion process, the responses, y_1 = percent residual lactose and y_2 = percent ash, are known to depend on the input variables, x_1 = pH level, x_2 = temperature, x_3 = concentration, and x_4 = time (see Fichtali, Van De Voort, and Khuri, 1990).

As in single-response experiments, one of the objectives of a multiresponse experiment is the determination of conditions on the input variables that optimize the predicted responses. The definition of an optimum in a multiresponse situation, however, is more complex than in the single-response case. The reason for this is that when two or more response variables are considered simultaneously, the meaning of an optimum becomes unclear, since there is no unique way to order the values of a multiresponse function. To overcome this difficulty, Khuri and Conlon (1981) introduced a multiresponse optimization technique called the generalized distance approach. The following is an outline of this approach:

Let r be the number of responses, and n be the number of experimental runs for all the responses. Suppose that these responses can be represented by the linear models

$$\mathbf{y}_i = \mathbf{X}\boldsymbol{\beta}_i + \boldsymbol{\epsilon}_i, \qquad i = 1, 2, \ldots, r,$$

where $\mathbf{y}_i$ is a vector of observations on the ith response, $\mathbf{X}$ is a known matrix of order $n \times p$ and rank p, $\boldsymbol{\beta}_i$ is a vector of p unknown parameters, and $\boldsymbol{\epsilon}_i$ is a random error vector associated with the ith response $(i = 1, 2, \ldots, r)$. It is assumed that the rows of the error matrix $[\boldsymbol{\epsilon}_1 : \boldsymbol{\epsilon}_2 : \cdots : \boldsymbol{\epsilon}_r]$ are statistically independent with each having a zero mean vector and a common variance–covariance matrix $\boldsymbol{\Sigma}$. Note that the matrix $\mathbf{X}$ is assumed to be the same for all the responses.

Let $x_1, x_2, \ldots, x_k$ be input variables that influence the r responses. The predicted response value at a point $\mathbf{x} = (x_1, x_2, \ldots, x_k)'$ in a region R for the ith response is given by $\hat{y}_i(\mathbf{x}) = \mathbf{f}'(\mathbf{x})\hat{\boldsymbol{\beta}}_i$, where $\hat{\boldsymbol{\beta}}_i = (\mathbf{X}'\mathbf{X})^{-1}\mathbf{X}'\mathbf{y}_i$ is the least-squares estimator of $\boldsymbol{\beta}_i$ $(i = 1, 2, \ldots, r)$. Here, $\mathbf{f}'(\mathbf{x})$ is of the same form as a row of $\mathbf{X}$, except that it is evaluated at the point $\mathbf{x}$. It follows that

$$\mathrm{Var}[\hat{y}_i(\mathbf{x})] = \sigma_{ii}\mathbf{f}'(\mathbf{x})(\mathbf{X}'\mathbf{X})^{-1}\mathbf{f}(\mathbf{x}), \qquad i = 1, 2, \ldots, r,$$

$$\mathrm{Cov}[\hat{y}_i(\mathbf{x}), \hat{y}_j(\mathbf{x})] = \sigma_{ij}\mathbf{f}'(\mathbf{x})(\mathbf{X}'\mathbf{X})^{-1}\mathbf{f}(\mathbf{x}), \qquad i \neq j = 1, 2, \ldots, r,$$

where σ_{ij} is the (i, j)th element of $\boldsymbol{\Sigma}$. The variance–covariance matrix of $\hat{\mathbf{y}}(\mathbf{x}) = [\hat{y}_1(\mathbf{x}), \hat{y}_2(\mathbf{x}), \ldots, \hat{y}_r(\mathbf{x})]'$ is then of the form

$$\mathrm{Var}[\hat{\mathbf{y}}(\mathbf{x})] = \mathbf{f}'(\mathbf{x})(\mathbf{X}'\mathbf{X})^{-1}\mathbf{f}(\mathbf{x})\boldsymbol{\Sigma}.$$

Since $\boldsymbol{\Sigma}$ is in general unknown, an unbiased estimator, $\hat{\boldsymbol{\Sigma}}$, of $\boldsymbol{\Sigma}$ can be used instead, where

$$\hat{\boldsymbol{\Sigma}} = \frac{1}{n-p} \mathbf{Y}' \left[\mathbf{I}_n - \mathbf{X}(\mathbf{X}'\mathbf{X})^{-1}\mathbf{X}' \right] \mathbf{Y},$$

and $\mathbf{Y} = [\mathbf{y}_1 : \mathbf{y}_2 : \cdots : \mathbf{y}_r]$. The matrix $\hat{\boldsymbol{\Sigma}}$ is nonsingular provided that $\mathbf{Y}$ is of rank $r \le n - p$. An estimate of $\mathrm{Var}[\hat{\mathbf{y}}(\mathbf{x})]$ is then given by

$$\widehat{\mathrm{Var}}[\hat{\mathbf{y}}(\mathbf{x})] = \mathbf{f}'(\mathbf{x})(\mathbf{X}'\mathbf{X})^{-1}\mathbf{f}(\mathbf{x})\hat{\boldsymbol{\Sigma}}. \tag{8.69}$$

Let ϕ_i denote the optimum value of $\hat{y}_i(\mathbf{x})$ optimized individually over the region R ($i = 1, 2, \ldots, r$). Let $\boldsymbol{\phi} = (\phi_1, \phi_2, \ldots, \phi_r)'$. These individual optima do not in general occur at the same location in R. To achieve a *compromise* optimum, we need to find $\mathbf{x}$ that minimizes $\rho[\hat{\mathbf{y}}(\mathbf{x}), \boldsymbol{\phi}]$, where ρ is some metric that measures the distance of $\hat{\mathbf{y}}(\mathbf{x})$ from $\boldsymbol{\phi}$. One possible choice for ρ is the metric

$$\rho[\hat{\mathbf{y}}(\mathbf{x}), \boldsymbol{\phi}] = \left[[\hat{\mathbf{y}}(\mathbf{x}) - \boldsymbol{\phi}]' \{\widehat{\mathrm{Var}}[\hat{\mathbf{y}}(\mathbf{x})]\}^{-1} [\hat{\mathbf{y}}(\mathbf{x}) - \boldsymbol{\phi}] \right]^{1/2},$$

which, by formula (8.69), can be written as

$$\rho[\hat{\mathbf{y}}(\mathbf{x}), \boldsymbol{\phi}] = \left[\frac{[\hat{\mathbf{y}}(\mathbf{x}) - \boldsymbol{\phi}]' \hat{\boldsymbol{\Sigma}}^{-1} [\hat{\mathbf{y}}(\mathbf{x}) - \boldsymbol{\phi}]}{\mathbf{f}'(\mathbf{x})(\mathbf{X}'\mathbf{X})^{-1}\mathbf{f}(\mathbf{x})} \right]^{1/2}. \tag{8.70}$$

We note that $\rho = 0$ if and only if $\hat{\mathbf{y}}(\mathbf{x}) = \boldsymbol{\phi}$, that is, when all the responses attain their individual optima at the same point; otherwise, $\rho > 0$. Such a point (if it exists) is called a point of *ideal* optimum. In general, an ideal optimum rarely exists.

In order to have conditions that are as close as possible to an ideal optimum, we need to minimize ρ over the region R. Let us suppose that the minimum occurs at the point $\mathbf{x}_0 \in R$. Then, at $\mathbf{x}_0$ the experimental conditions can be described as being near optimal for each of the r response functions. We therefore refer to $\mathbf{x}_0$ as a point of compromise optimum.

Note that the elements of $\boldsymbol{\phi}$ in formula (8.70) are random variables since they are the individual optima of $\hat{y}_1(\mathbf{x}), \hat{y}_2(\mathbf{x}), \ldots, \hat{y}_r(\mathbf{x})$. If the variation associated with $\boldsymbol{\phi}$ is large, then the metric ρ may not accurately measure the deviation of $\hat{\mathbf{y}}(\mathbf{x})$ from the true ideal optimum. In this case, some account should be taken of the randomness of $\boldsymbol{\phi}$ in the development of the metric. To do so, let $\boldsymbol{\zeta} = (\zeta_1, \zeta_2, \ldots, \zeta_r)'$, where ζ_i is the optimum value of the true mean of the ith response optimized individually over the region R ($i = 1, 2, \ldots, r$). Let D_ζ be a confidence region for $\boldsymbol{\zeta}$. For a fixed $\mathbf{x} \in R$ and whenever $\boldsymbol{\zeta} \in D_\zeta$, we obviously have

$$\rho[\hat{\mathbf{y}}(\mathbf{x}), \boldsymbol{\zeta}] \le \max_{\boldsymbol{\eta} \in D_\zeta} \rho[\hat{\mathbf{y}}(\mathbf{x}), \boldsymbol{\eta}]. \tag{8.71}$$

The right-hand side of this inequality serves as an upper bound on $\rho[\hat{y}(x), \zeta]$, which represents the distance of $\hat{y}(x)$ from the true ideal optimum. It follows that

$$\min_{x \in R} \rho[\hat{y}(x), \zeta] \leq \min_{x \in R} \left\{ \max_{\eta \in D_\zeta} \rho[\hat{y}(x), \eta] \right\}. \tag{8.72}$$

The right-hand side of this inequality provides a conservative measure of distance between the compromise and ideal optima.

The confidence region D_ζ can be determined in a variety of ways. Khuri and Conlon (1981) considered a rectangular confidence region of the form

$$\gamma_{1i} \leq \zeta_i \leq \gamma_{2i}, \qquad i = 1, 2, \ldots, r,$$

where

$$\begin{aligned} \gamma_{1i} &= \phi_i - g_i(\xi_i) \, MS_i^{1/2} \, t_{\alpha/2, n-p}, \\ \gamma_{2i} &= \phi_i + g_i(\xi_i) \, MS_i^{1/2} \, t_{\alpha/2, n-p}, \end{aligned} \tag{8.73}$$

where MS_i is the error mean square for the ith response, ξ_i is the point at which $\hat{y}_i(x)$ attains the individual optimum ϕ_i, $t_{\alpha/2, n-p}$ is the upper $(\alpha/2) \times 100$th percentile of the t-distribution with $n - p$ degrees of freedom, and $g_i(\xi_i)$ is given by

$$g_i(\xi_i) = \left[f'(\xi_i)(X'X)^{-1} f(\xi_i) \right]^{1/2}, \qquad i = 1, 2, \ldots, r.$$

Khuri and Conlon (1981) showed that such a rectangular confidence region has approximately a confidence coefficient of at least $1 - \alpha^*$, where $\alpha^* = 1 - (1 - \alpha)^r$.

It should be noted that the evaluation of the right-hand side of inequality (8.72) requires that $\rho[\hat{y}(x), \eta]$ be maximized first with respect to η over D_ζ for a given $x \in R$. The maximum value thus obtained, being a function of x, is then minimized over the region R. A computer program for the implementation of this min−max procedure is described in Conlon and Khuri (1992). A complete electronic copy of the code, along with examples, can be downloaded from the Internet at ftp://ftp.stat.ufl.edu/pub/mr.tar.Z.

Numerical examples that illustrate the application of the generalized distance approach for multiresponse optimization can be found in Khuri and Conlon (1981) and Khuri and Cornell (1996, Chapter 7).

8.8. MAXIMUM LIKELIHOOD ESTIMATION AND THE EM ALGORITHM

We recall from Section 7.11.2 that the maximum likelihood (ML) estimates of a set of parameters, $\theta_1, \theta_2, \ldots, \theta_p$, for a given distribution maximize the likelihood function of a sample, $X_1, X_2, \ldots, X_n$, of size n from the distribution. The ML estimates of the θ_i's denoted by $\hat{\theta}_1, \hat{\theta}_2, \ldots, \hat{\theta}_p$, can be found by solving the likelihood equations (the likelihood function must be differentiable and unimodal)

$$\frac{\partial \log[L(\mathbf{x}, \hat{\boldsymbol{\theta}})]}{\partial \theta_i} = 0, \qquad i = 1, 2, \ldots, p, \tag{8.74}$$

where $\hat{\boldsymbol{\theta}} = (\hat{\theta}_1, \hat{\theta}_2, \ldots, \hat{\theta}_p)'$, $\mathbf{x} = (x_1, x_2, \ldots, x_n)'$, and $L(\mathbf{x}, \boldsymbol{\theta}) = f(\mathbf{x}, \boldsymbol{\theta})$ with $f(\mathbf{x}, \boldsymbol{\theta})$ being the density function (or probability mass function) of $\mathbf{X} = (X_1, X_2, \ldots, X_n)'$. Note that $f(\mathbf{x}, \boldsymbol{\theta})$ can be written as $\prod_{i=1}^{n} g(x_i, \boldsymbol{\theta})$, where $g(x, \boldsymbol{\theta})$ is the density function (or probability mass function) associated with the distribution.

Equations (8.74) may not have a closed-form solution. For example, consider the so-called truncated Poisson distribution whose probability mass function is of the form (see Everitt, 1987, page 29)

$$g(x, \theta) = \frac{e^{-\theta}\theta^x}{(1 - e^{-\theta})x!}, \qquad x = 1, 2, \ldots . \tag{8.75}$$

In this case,

$$\log L(\mathbf{x}, \theta) = \log\left[\prod_{i=1}^{n} g(x_i, \theta)\right]$$

$$= -n\theta + (\log \theta) \sum_{i=1}^{n} x_i - \sum_{i=1}^{n} \log x_i! - n \log(1 - e^{-\theta}).$$

Hence,

$$\frac{\partial L^*(\mathbf{x}, \theta)}{\partial \theta} = -n + \frac{1}{\theta} \sum_{i=1}^{n} x_i - \frac{n e^{-\theta}}{1 - e^{-\theta}}, \tag{8.76}$$

where $L^*(\mathbf{x}, \theta) = \log L(\mathbf{x}, \theta)$ is the log-likelihood function. The likelihood equation, which results from equating the right-hand side of formula (8.76) to zero, has no closed-form solution for θ.

In general, if equations (8.74) do not have a closed-form solution, then, as was seen in Section 8.1, iterative methods can be applied to maximize $L(\mathbf{x}, \boldsymbol{\theta})$ [or $L^*(\mathbf{x}, \boldsymbol{\theta})$]. Using, for example, the Newton–Raphson method (see Section 8.1.2), if $\hat{\boldsymbol{\theta}}_0$ is an initial estimate of $\boldsymbol{\theta}$ and $\hat{\boldsymbol{\theta}}_i$ is the estimate at the ith iteration, then by applying formula (8.7) we have

$$\hat{\boldsymbol{\theta}}_{i+1} = \hat{\boldsymbol{\theta}}_i - \mathbf{H}_{L^*}^{-1}(\mathbf{x}, \hat{\boldsymbol{\theta}}_i)\nabla L^*(\mathbf{x}, \hat{\boldsymbol{\theta}}_i), \qquad i = 0, 1, 2, \ldots,$$

where $\mathbf{H}_{L^*}(\mathbf{x}, \boldsymbol{\theta})$ and $\nabla L^*(\mathbf{x}, \boldsymbol{\theta})$ are, respectively, the Hessian matrix and gradient vector of the log-likelihood function. Several iterations can be made until a certain convergence criterion is satisfied. A modification of this procedure is the so-called Fisher's method of scoring, where $\mathbf{H}_{L^*}$ is replaced by its expected value, that is,

$$\hat{\boldsymbol{\theta}}_{i+1} = \hat{\boldsymbol{\theta}}_i - \left\{E\left[\mathbf{H}_{L^*}\left(\mathbf{x}, \hat{\boldsymbol{\theta}}_i\right)\right]\right\}^{-1} \nabla L^*\left(\mathbf{x}, \hat{\boldsymbol{\theta}}_i\right), \qquad i = 0, 1, 2, \ldots . \quad (8.77)$$

Here, the expected value is taken with respect to the given distribution.

EXAMPLE 8.8.1. (Everitt, 1987, pages 30–31). Consider the truncated Poisson distribution described in formula (8.75). In this case, since we only have one parameter θ, the gradient takes the form $\nabla L^*(\mathbf{x}, \theta) = \partial L^*(\mathbf{x}, \theta)/\partial \theta$, which is given by formula (8.76). Hence, the Hessian matrix is

$$\mathbf{H}_{L^*}(\mathbf{x}, \theta) = \frac{\partial^2 L^*(\mathbf{x}, \theta)}{\partial \theta^2}$$

$$= -\frac{1}{\theta^2} \sum_{i=1}^n x_i + \frac{ne^{-\theta}}{(1 - e^{-\theta})^2}.$$

Furthermore, if X denotes the truncated Poisson random variable, then

$$E(X) = \sum_{x=1}^{\infty} \frac{xe^{-\theta}\theta^x}{(1 - e^{-\theta})x!}$$

$$= \frac{\theta}{1 - e^{-\theta}}.$$

Thus

$$E\left[\mathbf{H}_{L^*}(\mathbf{x}, \theta)\right] = E\left[\frac{\partial^2 L^*(\mathbf{x}, \theta)}{\partial \theta^2}\right]$$

$$= -\frac{1}{\theta^2}\frac{n\theta}{1 - e^{-\theta}} + \frac{ne^{-\theta}}{(1 - e^{-\theta})^2}$$

$$= \frac{ne^{-\theta}(1 + \theta) - n}{\theta(1 - e^{-\theta})^2}.$$

Suppose now we have the sample $1, 2, 3, 4, 5, 6$ from this distribution. Let $\hat{\theta}_0 = 1.5118$ be an initial estimate of θ. Several iterations are made by applying formula (8.77), and the results are shown in Table 8.6. The final

Table 8.6. Fisher's Method of Scoring for the Truncated Poisson Distribution

Iteration	$\dfrac{\partial L^*}{\partial \theta}$	$\dfrac{\partial^2 L^*}{\partial \theta^2}$	θ	L^*
1	− 685.5137	− 1176.7632	1.5118	− 1545.5549
2	− 62.0889	− 1696.2834	0.9293	− 1303.3340
3	− 0.2822	− 1750.5906	0.8927	− 1302.1790
4	0.0012	− 1750.8389	0.8925	− 1302.1792

Source: Everitt (1987, page 31). Reproduced with permission of Chapman and Hall, London.

estimate of θ is 0.8925, which is considered to be the maximum likelihood estimate of θ for the given sample. The convergence criterion used here is $|\hat{\theta}_{i+1} - \hat{\theta}_i| < 0.001$.

8.8.1. The EM Algorithm

The EM algorithm is a general iterative procedure for maximum likelihood estimation in incomplete data problems. This encompasses situations involving missing data, or when the actual data are viewed as forming a subset of a larger system of quantities.

The term EM was introduced by Dempster, Laird, and Rubin (1977). The reason for this terminology is that each iteration in this algorithm consists of two steps called the expectation step (E-step) and the maximization step (M-step). In the E-step, the conditional expectations of the missing data are found given the observed data and the current estimates of the parameters. These expected values are then substituted for the missing data and used to complete the data. In the M-step, maximum likelihood estimation of the parameters is performed in the usual manner using the completed data. More generally, missing sufficient statistics can be estimated rather than the individual missing data. The estimated parameters are then used to reestimate the missing data (or missing sufficient statistics), which in turn lead to new parameter estimates. This defines an iterative procedure, which can be carried out until convergence is achieved.

More details concerning the theory of the EM algorithm can be found in Dempster, Laird, and Rubin (1977), and in Little and Rubin (1987, Chapter 7). The following two examples, given in the latter reference, illustrate the application of this algorithm:

EXAMPLE 8.8.2. (Little and Rubin, 1987, pages 130–131). Consider a sample of size n from a normal distribution with a mean μ and a variance σ^2. Suppose that $x_1, x_2, \ldots, x_m$ are observed data and that $X_{m+1}, X_{m+2}, \ldots, X_n$ are missing data. Let $\mathbf{x}_{\text{obs}} = (x_1, x_2, \ldots, x_m)'$. For $i = m + 1, m + 2, \ldots, n$, the expected value of X_i given $\mathbf{x}_{\text{obs}}$ and $\boldsymbol{\theta} = (\mu, \sigma^2)'$ is μ. Now, from Example 7.11.3, the log-likelihood function for the complete data set is

$$L^*(\mathbf{x}, \boldsymbol{\theta}) = -\frac{n}{2}\log(2\pi\sigma^2) - \frac{1}{2\sigma^2}\left(\sum_{i=1}^{n} x_i^2 - 2\mu \sum_{i=1}^{n} x_i + n\mu^2\right), \quad (8.78)$$

where $\mathbf{x} = (x_1, x_2, \ldots, x_n)'$. We note that $\sum_{i=1}^{n} X_i^2$ and $\sum_{i=1}^{n} X_i$ are sufficient statistics. Therefore, to apply the E-step of the algorithm, we only have to find the conditional expectations of these statistics given $\mathbf{x}_{\text{obs}}$ and the current estimate of $\boldsymbol{\theta}$. We thus have

$$E\left(\sum_{i=1}^{n} X_i | \hat{\boldsymbol{\theta}}_j, \mathbf{x}_{\text{obs}}\right) = \sum_{i=1}^{m} x_i + (n - m)\hat{\mu}_j, \qquad j = 0, 1, 2, \ldots, \quad (8.79)$$

$$E\left(\sum_{i=1}^{n} X_i^2 | \hat{\boldsymbol{\theta}}_j, \mathbf{x}_{\text{obs}}\right) = \sum_{i=1}^{m} x_i^2 + (n - m)\left(\hat{\mu}_j^2 + \hat{\sigma}_j^2\right), \qquad j = 0, 1, 2, \ldots, \quad (8.80)$$

where $\hat{\boldsymbol{\theta}}_j = (\hat{\mu}_j, \hat{\sigma}_j^2)'$ is the estimate of $\boldsymbol{\theta}$ at the jth iteration with $\hat{\boldsymbol{\theta}}_0$ being an initial estimate.

From Section 7.11 we recall that the maximum likelihood estimates of μ and σ^2 based on the complete data set are $(1/n)\sum_{i=1}^{n} x_i$ and $(1/n)\sum_{i=1}^{n} x_i^2 - [(1/n)\sum_{i=1}^{n} x_i]^2$. Thus in the M-step, these same expressions are used, except that the current expectations of the sufficient statistics in formulas (8.79) and (8.80) are substituted for the missing data portion of the sufficient statistics. In other words, the estimates of μ and σ^2 at the $(j + 1)$th iteration are given by

$$\hat{\mu}_{j+1} = \frac{1}{n}\left[\sum_{i=1}^{m} x_i + (n - m)\hat{\mu}_j\right], \qquad j = 0, 1, 2, \ldots, \quad (8.81)$$

$$\hat{\sigma}_{j+1}^2 = \frac{1}{n}\left[\sum_{i=1}^{m} x_i^2 + (n - m)\left(\hat{\mu}_j^2 + \hat{\sigma}_j^2\right)\right] - \hat{\mu}_{j+1}^2, \qquad j = 0, 1, 2, \ldots. \quad (8.82)$$

By setting $\hat{\mu}_j = \hat{\mu}_{j+1} = \hat{\mu}$ and $\hat{\sigma}_j = \hat{\sigma}_{j+1} = \hat{\sigma}$ in equations (8.81) and (8.82), we find that the iterations converge to

$$\hat{\mu} = \frac{1}{m}\sum_{i=1}^{m} x_i,$$

$$\hat{\sigma}^2 = \frac{1}{m}\sum_{i=1}^{m} x_i^2 - \hat{\mu}^2,$$

which are the maximum likelihood estimates of μ and σ^2 from $\mathbf{x}_{\text{obs}}$.

The EM algorithm is unnecessary in this example, since the maximum likelihood estimates of μ and σ^2 can be obtained explicitly.

EXAMPLE 8.8.3.   (Little and Rubin, 1987, pages 131–132). This example was originally given in Dempster, Laird, and Rubin (1977). It involves a multinomial $\mathbf{x} = (x_1, x_2, x_3, x_4)'$ with cell probabilities $(\frac{1}{2} - \frac{1}{2}\theta, \frac{1}{4}\theta, \frac{1}{4}\theta, \frac{1}{2})'$, where $0 \leq \theta \leq 1$. Suppose that the observed data consist of $\mathbf{x}_{\text{obs}} = (38, 34, 125)'$ such that $x_1 = 38, x_2 = 34, x_3 + x_4 = 125$. The likelihood function for the complete data is

$$L(\mathbf{x}, \theta) = \frac{(x_1 + x_2 + x_3 + x_4)!}{x_1! x_2! x_3! x_4!} \left(\tfrac{1}{2} - \tfrac{1}{2}\theta\right)^{x_1} \left(\tfrac{1}{4}\theta\right)^{x_2} \left(\tfrac{1}{4}\theta\right)^{x_3} \left(\tfrac{1}{2}\right)^{x_4}.$$

The log-likelihood function is of the form

$$L^*(\mathbf{x}, \theta) = \log\left[\frac{(x_1 + x_2 + x_3 + x_4)!}{x_1! x_2! x_3! x_4!}\right] + x_1 \log\left(\tfrac{1}{2} - \tfrac{1}{2}\theta\right)$$
$$+ x_2 \log\left(\tfrac{1}{4}\theta\right) + x_3 \log\left(\tfrac{1}{4}\theta\right) + x_4 \log\left(\tfrac{1}{2}\right).$$

By differentiating $L^*(\mathbf{x}, \theta)$ with respect to θ and equating the derivative to zero we obtain

$$-\frac{x_1}{1 - \theta} + \frac{x_2}{\theta} + \frac{x_3}{\theta} = 0.$$

Hence, the maximum likelihood estimate of θ for the complete data set is

$$\hat{\theta} = \frac{x_2 + x_3}{x_1 + x_2 + x_3}. \tag{8.83}$$

Let us now find the conditional expectations of X_1, X_2, X_3, X_4 given the observed data and the current estimate of θ:

$$E\left(X_1 | \hat{\theta}_i, \mathbf{x}_{\text{obs}}\right) = 38,$$

$$E\left(X_2 | \hat{\theta}_i, \mathbf{x}_{\text{obs}}\right) = 34,$$

$$E\left(X_3 | \hat{\theta}_i, \mathbf{x}_{\text{obs}}\right) = \frac{125\left(\tfrac{1}{4}\hat{\theta}_i\right)}{\tfrac{1}{2} + \tfrac{1}{4}\hat{\theta}_i},$$

$$E\left(X_4 | \hat{\theta}_i, \mathbf{x}_{\text{obs}}\right) = \frac{125\left(\tfrac{1}{2}\right)}{\tfrac{1}{2} + \tfrac{1}{4}\hat{\theta}_i}.$$

Table 8.7. The EM Algorithm for Example 8.8.3

Iteration	$\hat{\theta}$
0	0.500000000
1	0.608247423
2	0.624321051
3	0.626488879
4	0.626777323
5	0.626815632
6	0.626820719
7	0.626821395
8	0.626821484

Source: Little and Rubin (1987, page 132). Reproduced
with permission of John Wiley & Sons, Inc.

Thus at the $(i+1)$st iteration we have

$$\hat{\theta}_{i+1} = \frac{34 + (125)\left(\frac{1}{4}\hat{\theta}_i\right)/\left(\frac{1}{2} + \frac{1}{4}\hat{\theta}_i\right)}{38 + 34 + (125)\left(\frac{1}{4}\hat{\theta}_i\right)/\left(\frac{1}{2} + \frac{1}{4}\hat{\theta}_i\right)}, \tag{8.84}$$

as can be seen from applying formula (8.83) using the conditional expectation
of X_3 instead of x_3. Formula (8.84) can be used iteratively to obtain the
maximum likelihood estimate of θ on the basis of the observed data. Using
an initial estimate $\hat{\theta}_0 = \frac{1}{2}$, the results of this iterative procedure are given in
Table 8.7. Note that if we set $\hat{\theta}_{i+1} = \hat{\theta}_i = \hat{\theta}$ in formula (8.84) we obtain the
quadratic equation,

$$197\hat{\theta}^2 - 15\hat{\theta} - 68 = 0$$

whose only positive root is $\hat{\theta} = 0.626821498$, which is very close to the value
obtained in the last iteration in Table 8.7.

8.9. MINIMUM NORM QUADRATIC UNBAISED ESTIMATION OF VARIANCE COMPONENTS

Consider the linear model

$$\mathbf{y} = \mathbf{X}\boldsymbol{\alpha} + \sum_{i=1}^{c} \mathbf{U}_i\boldsymbol{\beta}_i, \tag{8.85}$$

where $\mathbf{y}$ is a vector of n observations; $\boldsymbol{\alpha}$ is a vector of fixed effects;
$\boldsymbol{\beta}_1, \boldsymbol{\beta}_2, \ldots, \boldsymbol{\beta}_c$ are vectors of random effects; $\mathbf{X}, \mathbf{U}_1, \mathbf{U}_2, \ldots, \mathbf{U}_c$ are known
matrices of constants with $\boldsymbol{\beta}_c = \boldsymbol{\epsilon}$, the vector of random errors; and $\mathbf{U}_c = \mathbf{I}_n$.
We assume that the $\boldsymbol{\beta}_i$'s are uncorrelated with zero mean vectors and
variance–covariance matrices $\sigma_i^2\mathbf{I}_{m_i}$, where m_i is the number of columns of

$\mathbf{U}_i$ $(i = 1, 2, \ldots, c)$. The variances $\sigma_1^2, \sigma_2^2, \ldots, \sigma_c^2$ are referred to as variance components. Model (8.85) can be written as

$$\mathbf{y} = \mathbf{X}\boldsymbol{\alpha} + \mathbf{U}\boldsymbol{\beta}, \tag{8.86}$$

where $\mathbf{U} = [\mathbf{U}_1 : \mathbf{U}_2 : \cdots : \mathbf{U}_c]$, $\boldsymbol{\beta} = (\boldsymbol{\beta}_1', \boldsymbol{\beta}_2', \ldots, \boldsymbol{\beta}_c')'$. From model (8.86) we have

$$E(\mathbf{y}) = \mathbf{X}\boldsymbol{\alpha},$$

$$\text{Var}(\mathbf{y}) = \sum_{i=1}^{c} \sigma_i^2 \mathbf{V}_i, \tag{8.87}$$

with $\mathbf{V}_i = \mathbf{U}_i \mathbf{U}_i'$.

Let us consider the estimation of a linear function of the variance components, namely, $\sum_{i=1}^{c} a_i \sigma_i^2$, where the a_i's are known constants, by a quadratic estimator of the form $\mathbf{y}'\mathbf{A}\mathbf{y}$. Here, $\mathbf{A}$ is a symmetric matrix to be determined so that $\mathbf{y}'\mathbf{A}\mathbf{y}$ satisfies certain criteria, which are the following:

1. *Translation Invariance.* If instead of $\boldsymbol{\alpha}$ we consider $\boldsymbol{\gamma} = \boldsymbol{\alpha} - \boldsymbol{\alpha}_0$, then from model (8.86) we have

$$\mathbf{y} - \mathbf{X}\boldsymbol{\alpha}_0 = \mathbf{X}\boldsymbol{\gamma} + \mathbf{U}\boldsymbol{\beta}.$$

In this case, $\sum_{i=1}^{c} a_i \sigma_i^2$ is estimated by $(\mathbf{y} - \mathbf{X}\boldsymbol{\alpha}_0)'\mathbf{A}(\mathbf{y} - \mathbf{X}\boldsymbol{\alpha}_0)$. The estimator $\mathbf{y}'\mathbf{A}\mathbf{y}$ is said to be translation invariant if

$$\mathbf{y}'\mathbf{A}\mathbf{y} = (\mathbf{y} - \mathbf{X}\boldsymbol{\alpha}_0)'\mathbf{A}(\mathbf{y} - \mathbf{X}\boldsymbol{\alpha}_0).$$

In order for this to be true we must have

$$\mathbf{A}\mathbf{X} = \mathbf{0}. \tag{8.88}$$

2. *Unbiasedness.* $E(\mathbf{y}'\mathbf{A}\mathbf{y}) = \sum_{i=1}^{c} a_i \sigma_i^2$. Using a result in Searle (1971, Theorem 1, page 55), the expected value of the quadratic form $\mathbf{y}'\mathbf{A}\mathbf{y}$ is given by

$$E(\mathbf{y}'\mathbf{A}\mathbf{y}) = \boldsymbol{\alpha}'\mathbf{X}'\mathbf{A}\mathbf{X}\boldsymbol{\alpha} + \text{tr}[\mathbf{A}\,\text{Var}(\mathbf{y})], \tag{8.89}$$

since $E(\mathbf{y}) = \mathbf{X}\boldsymbol{\alpha}$. From formulas (8.87), (8.88), and (8.89) we then have

$$E(\mathbf{y}'\mathbf{A}\mathbf{y}) = \sum_{i=1}^{c} \sigma_i^2 \, \text{tr}(\mathbf{A}\mathbf{V}_i). \tag{8.90}$$

By comparison with $\sum_{i=1}^{c} a_i \sigma_i^2$, the condition for unbiasedness is

$$a_i = \text{tr}(\mathbf{A}\mathbf{V}_i), \qquad i = 1, 2, \ldots, c. \tag{8.91}$$

3. *Minimum Norm.* If $\boldsymbol{\beta}_1, \boldsymbol{\beta}_2, \ldots, \boldsymbol{\beta}_c$ in model (8.85) were observable, then a natural unbiased estimator of $\sum_{i=1}^{c} a_i \sigma_i^2$ would be $\sum_{i=1}^{c} a_i \boldsymbol{\beta}_i' \boldsymbol{\beta}_i / m_i$, since $E(\boldsymbol{\beta}_i' \boldsymbol{\beta}_i) = \mathrm{tr}(\mathbf{I}_{m_i} \sigma_i^2) = m_i \sigma_i^2$, $i = 1, 2, \ldots, c$. This estimator can be written as $\boldsymbol{\beta}' \boldsymbol{\Delta} \boldsymbol{\beta}$, where $\boldsymbol{\Delta}$ is the block-diagonal matrix

$$\boldsymbol{\Delta} = \mathrm{Diag}\left(\frac{a_1}{m_1} \mathbf{I}_{m_1}, \frac{a_2}{m_2} \mathbf{I}_{m_2}, \ldots, \frac{a_c}{m_c} \mathbf{I}_{m_c} \right).$$

The difference between this estimator and $\mathbf{y}' \mathbf{A} \mathbf{y}$ is

$$\mathbf{y}' \mathbf{A} \mathbf{y} - \boldsymbol{\beta}' \boldsymbol{\Delta} \boldsymbol{\beta} = \boldsymbol{\beta}'(\mathbf{U}' \mathbf{A} \mathbf{U} - \boldsymbol{\Delta}) \boldsymbol{\beta},$$

since $\mathbf{A}\mathbf{X} = \mathbf{0}$. This difference can be made small by minimizing the Euclidean norm $\| \mathbf{U}' \mathbf{A} \mathbf{U} - \boldsymbol{\Delta} \|_2$.

The quadratic estimator $\mathbf{y}' \mathbf{A} \mathbf{y}$ is said to be a minimum norm quadratic unbiased estimator (MINQUE) of $\sum_{i=1}^{c} a_i \sigma_i^2$ if the matrix $\mathbf{A}$ is determined so that $\| \mathbf{U}' \mathbf{A} \mathbf{U} - \boldsymbol{\Delta} \|_2$ attains a minimum subject to the conditions given in formulas (8.88) and (8.91). Such an estimator was introduced by Rao (1971, 1972).

The minimization of $\| \mathbf{U}' \mathbf{A} \mathbf{U} - \boldsymbol{\Delta} \|_2$ is equivalent to that of $\mathrm{tr}(\mathbf{A}\mathbf{V}\mathbf{A}\mathbf{V})$, where $\mathbf{V} = \sum_{i=1}^{c} \mathbf{V}_i$. The reason for this is the following:

$$\| \mathbf{U}' \mathbf{A} \mathbf{U} - \boldsymbol{\Delta} \|_2^2 = \mathrm{tr}\left[(\mathbf{U}' \mathbf{A} \mathbf{U} - \boldsymbol{\Delta})(\mathbf{U}' \mathbf{A} \mathbf{U} - \boldsymbol{\Delta}) \right]$$

$$= \mathrm{tr}(\mathbf{U}' \mathbf{A} \mathbf{U} \mathbf{U}' \mathbf{A} \mathbf{U}) - 2\,\mathrm{tr}(\mathbf{U}' \mathbf{A} \mathbf{U} \boldsymbol{\Delta}) + \mathrm{tr}(\boldsymbol{\Delta}^2). \quad (8.92)$$

Now,

$$\mathrm{tr}(\mathbf{U}' \mathbf{A} \mathbf{U} \boldsymbol{\Delta}) = \mathrm{tr}(\mathbf{A} \mathbf{U} \boldsymbol{\Delta} \mathbf{U}')$$

$$= \mathrm{tr}\left(\mathbf{A} \sum_{i=1}^{c} \mathbf{U}_i \frac{a_i}{m_i} \mathbf{I}_{m_i} \mathbf{U}_i' \right)$$

$$= \mathrm{tr}\left(\sum_{i=1}^{c} \frac{a_i}{m_i} \mathbf{A} \mathbf{U}_i \mathbf{U}_i' \right)$$

$$= \mathrm{tr}\left(\sum_{i=1}^{c} \frac{a_i}{m_i} \mathbf{A} \mathbf{V}_i \right)$$

$$= \sum_{i=1}^{c} \frac{a_i}{m_i} \mathrm{tr}(\mathbf{A} \mathbf{V}_i)$$

$$= \sum_{i=1}^{c} \frac{a_i^2}{m_i}, \qquad \text{by (8.91)}$$

$$= \mathrm{tr}(\boldsymbol{\Delta}^2).$$

Formula (8.92) can then be written as

$$\|\mathbf{U}'\mathbf{A}\mathbf{U} - \boldsymbol{\Delta}\|_2^2 = \text{tr}(\mathbf{U}'\mathbf{A}\mathbf{U}\mathbf{U}'\mathbf{A}\mathbf{U}) - \text{tr}(\boldsymbol{\Delta}^2)$$
$$= \text{tr}(\mathbf{A}\mathbf{V}\mathbf{A}\mathbf{V}) - \text{tr}(\boldsymbol{\Delta}^2),$$

since $\mathbf{V} = \sum_{i=1}^{c}\mathbf{V}_i = \sum_{i=1}^{c}\mathbf{U}_i\mathbf{U}_i' = \mathbf{U}\mathbf{U}'$. The trace of $\boldsymbol{\Delta}^2$ does not involve $\mathbf{A}$; hence the problem of MINQUE reduces to finding $\mathbf{A}$ that minimizes $\text{tr}(\mathbf{A}\mathbf{V}\mathbf{A}\mathbf{V})$ subject to conditions (8.88) and (8.91). Rao (1971) showed that the solution to this optimization problem is of the form

$$\mathbf{A} = \sum_{i=1}^{c} \lambda_i \mathbf{R}\mathbf{V}_i\mathbf{R}, \tag{8.93}$$

where

$$\mathbf{R} = \mathbf{V}^{-1} - \mathbf{V}^{-1}\mathbf{X}(\mathbf{X}'\mathbf{V}^{-1}\mathbf{X})^{-}\mathbf{X}'\mathbf{V}^{-1}$$

with $(\mathbf{X}'\mathbf{V}^{-1}\mathbf{X})^{-}$ being a generalized inverse of $\mathbf{X}'\mathbf{V}^{-1}\mathbf{X}$, and the λ_i's are obtained from solving the equations

$$\sum_{i=1}^{c} \lambda_i \text{tr}(\mathbf{R}\mathbf{V}_i\mathbf{R}\mathbf{V}_j) = a_j, \qquad j = 1, 2, \ldots, c,$$

which can be expressed as

$$\boldsymbol{\lambda}'\mathbf{S} = \mathbf{a}', \tag{8.94}$$

where $\boldsymbol{\lambda} = (\lambda_1, \lambda_2, \ldots, \lambda_c)'$, $\mathbf{S}$ is the $c \times c$ matrix (s_{ij}) with $s_{ij} = \text{tr}(\mathbf{R}\mathbf{V}_i\mathbf{R}\mathbf{V}_j)$, and $\mathbf{a} = (a_1, a_2, \ldots, a_c)'$. The MINQUE of $\sum_{i=1}^{c}a_i\sigma_i^2$ can then be written as

$$\mathbf{y}'\left(\sum_{i=1}^{c} \lambda_i \mathbf{R}\mathbf{V}_i\mathbf{R}\right)\mathbf{y} = \sum_{i=1}^{c} \lambda_i \mathbf{y}'\mathbf{R}\mathbf{V}_i\mathbf{R}\mathbf{y}$$
$$= \boldsymbol{\lambda}'\mathbf{q},$$

where $\mathbf{q} = (q_1, q_2, \ldots, q_c)'$ with $q_i = \mathbf{y}'\mathbf{R}\mathbf{V}_i\mathbf{R}\mathbf{y}$ ($i = 1, 2, \ldots, c$). But, from formula (8.94), $\boldsymbol{\lambda}' = \mathbf{a}'\mathbf{S}^{-}$, where $\mathbf{S}^{-}$ is a generalized inverse of $\mathbf{S}$. Hence, $\boldsymbol{\lambda}'\mathbf{q} = \mathbf{a}'\mathbf{S}^{-}\mathbf{q} = \mathbf{a}'\hat{\boldsymbol{\sigma}}$, where $\hat{\boldsymbol{\sigma}} = (\hat{\sigma}_1^2, \hat{\sigma}_2^2, \ldots, \hat{\sigma}_c^2)'$ is a solution of the equation

$$\mathbf{S}\hat{\boldsymbol{\sigma}} = \mathbf{q}. \tag{8.95}$$

This equation has a unique solution if and only if the individual variance components are unbiasedly estimable (see Rao, 1972, page

114). Thus the MINQUEs of the σ_i^2's are obtained from solving equation (8.95).

If the random effects in model (8.85) are assumed to be normally distributed, then the MINQUEs of the variance components reduce to the so-called minimum variance quadratic unbiased estimators (MIVQUEs). An example that shows how to compute these estimators in the case of a random one-way classification model is given in Swallow and Searle (1978). See also Milliken and Johnson (1984, Chapter 19).

8.10. SCHEFFÉ'S CONFIDENCE INTERVALS

Consider the linear model

$$\mathbf{y} = \mathbf{X}\boldsymbol{\beta} + \boldsymbol{\epsilon}, \tag{8.96}$$

where $\mathbf{y}$ is a vector of n observations, $\mathbf{X}$ is a known matrix of order $n \times p$ and rand r ($\leq p$), $\boldsymbol{\beta}$ is a vector of unknown parameters, and $\boldsymbol{\epsilon}$ is a random error vector. It is assumed that $\boldsymbol{\epsilon}$ has the normal distribution with a mean $\mathbf{0}$ and a variance–covariance matrix $\sigma^2 \mathbf{I}_n$. Let $\psi = \mathbf{a}'\boldsymbol{\beta}$ be an estimable linear function of the elements of $\boldsymbol{\beta}$. By this we mean that there exists a linear function $\mathbf{t}'\mathbf{y}$ of $\mathbf{y}$ such that $E(\mathbf{t}'\mathbf{y}) = \psi$, where $\mathbf{t}$ is some constant vector. A necessary and sufficient condition for ψ to be estimable is that $\mathbf{a}'$ belongs to the row space of $\mathbf{X}$, that is, $\mathbf{a}'$ is a linear combination of the rows of $\mathbf{X}$ (see, for example, Searle, 1971, page 181). Since the rank of $\mathbf{X}$ is r, the row space of $\mathbf{X}$, denoted by $\rho(\mathbf{X})$, is an r-dimensional subspace of the p-dimensional Euclidean space R^p. Thus $\mathbf{a}'\boldsymbol{\beta}$ estimable if and only if $\mathbf{a}' \in \rho(\mathbf{X})$.

Suppose that $\mathbf{a}'$ is an arbitrary vector in a q-dimensional subspace $\mathscr{L}$ of $\rho(\mathbf{X})$, where $q \leq r$. Then $\mathbf{a}'\hat{\boldsymbol{\beta}} = \mathbf{a}'(\mathbf{X}'\mathbf{X})^-\mathbf{X}'\mathbf{y}$ is the best linear unbiased estimator of $\mathbf{a}'\boldsymbol{\beta}$, and its variance is given by

$$\mathrm{Var}\left(\mathbf{a}'\hat{\boldsymbol{\beta}}\right) = \sigma^2 \mathbf{a}'(\mathbf{X}'\mathbf{X})^- \mathbf{a},$$

where $(\mathbf{X}'\mathbf{X})^-$ is a generalized inverse of $\mathbf{X}'\mathbf{X}$ (see, for example, Searle, 1971, pages 181–182). Both $\mathbf{a}'\hat{\boldsymbol{\beta}}$ and $\mathbf{a}'(\mathbf{X}'\mathbf{X})^-\mathbf{a}$ are invariant to the choice of $(\mathbf{X}'\mathbf{X})^-$, since $\mathbf{a}'\boldsymbol{\beta}$ is estimable (see, for example, Searle, 1971, page 181). In particular, if $r = p$, then $\mathbf{X}'\mathbf{X}$ is of full rank and $(\mathbf{X}'\mathbf{X})^- = (\mathbf{X}'\mathbf{X})^{-1}$.

Theorem 8.10.1. Simultaneous $(1 - \alpha)100\%$ confidence intervals on $\mathbf{a}'\boldsymbol{\beta}$ for all $\mathbf{a}' \in \mathscr{L}$, where $\mathscr{L}$ is a q-dimensional subspace of $\rho(\mathbf{X})$, are of the form

$$\mathbf{a}'\hat{\boldsymbol{\beta}} \mp \left(q\,MS_E\,F_{\alpha,q,n-r}\right)^{1/2}\left[\mathbf{a}'(\mathbf{X}'\mathbf{X})^-\mathbf{a}\right]^{1/2}, \tag{8.97}$$

where $F_{\alpha,q,n-r}$ is the upper $\alpha100$th percentile of the F-distribution with q and $n-r$ degrees of freedom, and MS_E is the error mean square given by

$$MS_E = \frac{1}{n-r}\mathbf{y}'\left[\mathbf{I}_n - \mathbf{X}(\mathbf{X}'\mathbf{X})^-\mathbf{X}'\right]\mathbf{y}. \qquad (8.98)$$

In Theorem 8.10.1, the word "simultaneous" means that with probability $1-\alpha$, the values of $\mathbf{a}'\boldsymbol{\beta}$ for all $\mathbf{a}' \in \mathscr{L}$ satisfy the double inequality

$$\mathbf{a}'\hat{\boldsymbol{\beta}} - \left(qMS_E\, F_{\alpha,q,n-r}\right)^{1/2}\left[\mathbf{a}'(\mathbf{X}'\mathbf{X})^-\mathbf{a}\right]^{1/2}$$

$$\leq \mathbf{a}'\boldsymbol{\beta} \leq \mathbf{a}'\hat{\boldsymbol{\beta}} + \left(qMS_E\, F_{\alpha,q,n-r}\right)^{1/2}\left[\mathbf{a}'(\mathbf{X}'\mathbf{X})^-\mathbf{a}\right]^{1/2}. \qquad (8.99)$$

A proof of this theorem is given in Scheffé (1959, Section 3.5). Another proof is presented here using the method of Lagrange multipliers. This proof is based on the following lemma:

Lemma 8.10.1.   Let C be the set $\{\mathbf{x} \in R^q | \mathbf{x}'\mathbf{A}\mathbf{x} \leq 1\}$, where $\mathbf{A}$ is a positive definite matrix of order $q \times q$. Then $\mathbf{x} \in C$ if and only if $|\mathbf{l}'\mathbf{x}| \leq (\mathbf{l}'\mathbf{A}^{-1}\mathbf{l})^{1/2}$ for all $\mathbf{l} \in R^q$.

Proof. Suppose that $\mathbf{x} \in C$. Since $\mathbf{A}$ is positive definite, the boundary of C is an ellipsoid in a q-dimensional space. For any $\mathbf{l} \in R^q$, let $\mathbf{e}$ be a unit vector in its direction. The projection of $\mathbf{x}$ on an axis in the direction of $\mathbf{l}$ is given by $\mathbf{e}'\mathbf{x}$. Consider optimizing $\mathbf{e}'\mathbf{x}$ with respect to $\mathbf{x}$ over the set C. The minimum and maximum values of $\mathbf{e}'\mathbf{x}$ are obviously determined by the end points of the projection of C on the $\mathbf{l}$-axis. This is equivalent to optimizing $\mathbf{e}'\mathbf{x}$ subject to the constraint $\mathbf{x}'\mathbf{A}\mathbf{x} = 1$, since the projection of C on the $\mathbf{l}$-axis is the same as the projection of its boundary, the ellipsoid $\mathbf{x}'\mathbf{A}\mathbf{x} = 1$. This constrained optimization problem can be solved by using the method of Lagrange multipliers.

Let $G = \mathbf{e}'\mathbf{x} + \lambda(\mathbf{x}'\mathbf{A}\mathbf{x} - 1)$, where λ is a Lagrange multiplier. By differentiating G with respect to $x_1, x_2, \ldots, x_q$, where x_i is the ith element of $\mathbf{x}$ $(i = 1, 2, \ldots, q)$, and equating the derivatives to zero, we obtain the equation $\mathbf{e} + 2\lambda\mathbf{A}\mathbf{x} = \mathbf{0}$, whose solution is $\mathbf{x} = -(1/2\lambda)\mathbf{A}^{-1}\mathbf{e}$. If we substitute this value of $\mathbf{x}$ into the equation $\mathbf{x}'\mathbf{A}\mathbf{x} = 1$ and then solve for λ, we obtain the two solutions $\lambda_1 = -\frac{1}{2}(\mathbf{e}'\mathbf{A}^{-1}\mathbf{e})^{1/2}$, $\lambda_2 = \frac{1}{2}(\mathbf{e}'\mathbf{A}^{-1}\mathbf{e})^{1/2}$. But, $\mathbf{e}'\mathbf{x} = -2\lambda$, since $\mathbf{x}'\mathbf{A}\mathbf{x} = 1$. It follows that the minimum and maximum values of $\mathbf{e}'\mathbf{x}$ under the constraint $\mathbf{x}'\mathbf{A}\mathbf{x} = 1$ are $-(\mathbf{e}'\mathbf{A}^{-1}\mathbf{e})^{1/2}$ and $(\mathbf{e}'\mathbf{A}^{-1}\mathbf{e})^{1/2}$, respectively. Hence,

$$|\mathbf{e}'\mathbf{x}| \leq (\mathbf{e}'\mathbf{A}^{-1}\mathbf{e})^{1/2}. \qquad (8.100)$$

Since $\mathbf{l} = \|\mathbf{l}\|_2\mathbf{e}$, where $\|\mathbf{l}\|_2$ is the Euclidean norm of $\mathbf{l}$, multiplying the two sides of inequality (8.100) by $\|\mathbf{l}\|_2$ yields

$$|\mathbf{l}'\mathbf{x}| \leq (\mathbf{l}'\mathbf{A}^{-1}\mathbf{l})^{1/2}. \tag{8.101}$$

Vice versa, if inequality (8.101) is true for all $\mathbf{l} \in R^q$, then by choosing $\mathbf{l}' = \mathbf{x}'\mathbf{A}$ we obtain

$$|\mathbf{x}'\mathbf{A}\mathbf{x}| \leq (\mathbf{x}'\mathbf{A}\mathbf{A}^{-1}\mathbf{A}\mathbf{x})^{1/2},$$

which is equivalent to $\mathbf{x}'\mathbf{A}\mathbf{x} \leq 1$, that is, $\mathbf{x} \in C$.   □

Proof of Theorem 8.10.1. Let $\mathbf{L}$ be a $q \times p$ matrix of rank q whose rows form a basis for the q-dimensional subspace $\mathscr{L}$ of $\rho(\mathbf{X})$. Since $\mathbf{y}$ in model (8.96) is distributed as $N(\mathbf{X}\boldsymbol{\beta}, \sigma^2 \mathbf{I}_n)$, $\mathbf{L}\hat{\boldsymbol{\beta}} = \mathbf{L}(\mathbf{X}'\mathbf{X})^-\mathbf{X}'\mathbf{y}$ is distributed as $N[\mathbf{L}\boldsymbol{\beta}, \sigma^2\mathbf{L}(\mathbf{X}'\mathbf{X})^-\mathbf{L}']$. Thus the random variable

$$F = \frac{\left[\mathbf{L}(\hat{\boldsymbol{\beta}} - \boldsymbol{\beta})\right]'\left[\mathbf{L}(\mathbf{X}'\mathbf{X})^-\mathbf{L}'\right]^{-1}\left[\mathbf{L}(\hat{\boldsymbol{\beta}} - \boldsymbol{\beta})\right]}{q\,MS_E}$$

has the F-distribution with q and $n - r$ degrees of freedom (see, for example, Searle, 1971, page 190). It follows that

$$P\left(F \leq F_{\alpha, q, n-r}\right) = 1 - \alpha. \tag{8.102}$$

By applying Lemma 8.10.1 to formula (8.102) with $\mathbf{x} = \mathbf{L}(\hat{\boldsymbol{\beta}} - \boldsymbol{\beta})$ and $\mathbf{A} = [\mathbf{L}(\mathbf{X}'\mathbf{X})^-\mathbf{L}']^{-1}/(q\,MS_E\,F_{\alpha, q, n-r})$ we obtain the equivalent probability statement

$$P\left\{\left|\mathbf{l}'\mathbf{L}(\hat{\boldsymbol{\beta}} - \boldsymbol{\beta})\right| \leq (q\,MS_E\,F_{\alpha, q, n-r})^{1/2}\left[\mathbf{l}'\mathbf{L}(\mathbf{X}'\mathbf{X})^-\mathbf{L}'\mathbf{l}\right]^{1/2} \; \forall \mathbf{l} \in R^q\right\} = 1 - \alpha.$$

Let $\mathbf{a}' = \mathbf{l}'\mathbf{L}$. We then have

$$P\left\{\left|\mathbf{a}'(\hat{\boldsymbol{\beta}} - \boldsymbol{\beta})\right| \leq (q\,MS_E\,F_{\alpha, q, n-r})^{1/2}\left[\mathbf{a}'(\mathbf{X}'\mathbf{X})^-\mathbf{a}\right]^{1/2} \; \forall \mathbf{a}' \in \mathscr{L}\right\} = 1 - \alpha.$$

We conclude that the values of $\mathbf{a}'\boldsymbol{\beta}$ satisfy the double inequality (8.99) for all $\mathbf{a}' \in \mathscr{L}$ with probability $1 - \alpha$. Simultaneous $(1 - \alpha)100\%$ confidence intervals on $\mathbf{a}'\boldsymbol{\beta}$ are therefore given by formula (8.97). We refer to these intervals as Scheffé's confidence intervals.

Theorem 8.10.1 can be used to obtain simultaneous confidence intervals on all contrasts among the elements of $\boldsymbol{\beta}$. By definition, the linear function $\mathbf{a}'\boldsymbol{\beta}$ is a contrast among the elements of $\boldsymbol{\beta}$ if $\sum_{i=1}^p a_i = 0$, where a_i is the ith element of $\mathbf{a}$ $(i = 1, 2, \ldots, p)$. If $\mathbf{a}'$ is in the row space of $\mathbf{X}$, then it must belong to a q-dimensional subspace of $\rho(\mathbf{X})$, where $q = r - 1$. Hence, simultaneous $(1 - \alpha)100\%$ confidence intervals on all such contrasts can be obtained from formula (8.97) by replacing q with $r - 1$. □

8.10.1. The Relation of Scheffé's Confidence Intervals to the F-Test

There is a relationship between the confidence intervals (8.97) and the F-test used to test the hypothesis H_0: $\mathbf{L}\boldsymbol{\beta} = \mathbf{0}$ versus H_a: $\mathbf{L}\boldsymbol{\beta} \neq \mathbf{0}$, where $\mathbf{L}$ is the matrix whose rows form a basis for the q-dimensional subspace $\mathscr{L}$ of $\rho(\mathbf{X})$. The test statistic for testing H_0 is given by (see Searle, 1971, Section 5.5)

$$F = \frac{\hat{\boldsymbol{\beta}}'\mathbf{L}'\left[\mathbf{L}(\mathbf{X}'\mathbf{X})^{-}\mathbf{L}'\right]^{-1}\mathbf{L}\hat{\boldsymbol{\beta}}}{q\,MS_E},$$

which under H_0 has the F-distribution with q and $n - r$ degrees of freedom. The hypothesis H_0 can be rejected at the α-level of significance if $F > F_{\alpha, q, n-r}$. In this case, by Lemma 8.10.1, there exits at least one $\mathbf{l} \in R^q$ such that

$$|\mathbf{l}'\mathbf{L}\hat{\boldsymbol{\beta}}| > \left(q\,MS_E\,F_{\alpha, q, n-r}\right)^{1/2}\left[\mathbf{l}'\mathbf{L}(\mathbf{X}'\mathbf{X})^{-}\mathbf{L}'\mathbf{l}\right]^{1/2}. \tag{8.103}$$

It follows that the F-test rejects H_0 if and only if there exists a linear combination $\mathbf{a}'\hat{\boldsymbol{\beta}}$, where $\mathbf{a}' = \mathbf{l}'\mathbf{L}$ for some $\mathbf{l} \in R^q$, for which the confidence interval in formula (8.97) does not contain the value zero. In this case, $\mathbf{a}'\hat{\boldsymbol{\beta}}$ is said to be significantly different from zero.

It is easy to see that inequality (8.103) holds for some $\mathbf{l} \in R^q$ if and only if

$$\sup_{\mathbf{l} \in R^q} \frac{|\mathbf{l}'\mathbf{L}\hat{\boldsymbol{\beta}}|^2}{\mathbf{l}'\mathbf{L}(\mathbf{X}'\mathbf{X})^{-}\mathbf{L}'\mathbf{l}} > q\,MS_E\,F_{\alpha, q, n-r},$$

or equivalently,

$$\sup_{\mathbf{l} \in R^q} \frac{\mathbf{l}'\mathbf{G}_1\mathbf{l}}{\mathbf{l}'\mathbf{G}_2\mathbf{l}} > q\,MS_E\,F_{\alpha, q, n-r}, \tag{8.104}$$

where

$$\mathbf{G}_1 = \mathbf{L}\hat{\boldsymbol{\beta}}\hat{\boldsymbol{\beta}}'\mathbf{L}', \tag{8.105}$$

$$\mathbf{G}_2 = \mathbf{L}(\mathbf{X}'\mathbf{X})^{-}\mathbf{L}'. \tag{8.106}$$

However, by Theorem 2.3.17,

$$\sup_{\mathbf{l} \in R^q} \frac{\mathbf{l}'\mathbf{G}_1\mathbf{l}}{\mathbf{l}'\mathbf{G}_2\mathbf{l}} = e_{\max}\left(\mathbf{G}_2^{-1}\mathbf{G}_1\right)$$

$$= \hat{\boldsymbol{\beta}}'\mathbf{L}'\left[\mathbf{L}(\mathbf{X}'\mathbf{X})^{-}\mathbf{L}'\right]^{-1}\mathbf{L}\hat{\boldsymbol{\beta}}, \tag{8.107}$$

where $e_{max}(\mathbf{G}_2^{-1}\mathbf{G}_1)$ is the largest eigenvalue of $\mathbf{G}_2^{-1}\mathbf{G}_1$. The second equality in (8.107) is true because the nonzero eigenvalues of $[\mathbf{L}(\mathbf{X}'\mathbf{X})^-\mathbf{L}']^{-1}\mathbf{L}\hat{\boldsymbol{\beta}}\hat{\boldsymbol{\beta}}'\mathbf{L}'$ are the same as those of $\hat{\boldsymbol{\beta}}'\mathbf{L}'[\mathbf{L}(\mathbf{X}'\mathbf{X})^-\mathbf{L}']^{-1}\mathbf{L}\hat{\boldsymbol{\beta}}$ by Theorem 2.3.9. Note that the latter expression is the numerator sum of squares of the F-test statistic for H_0.

The eigenvector of $\mathbf{G}_2^{-1}\mathbf{G}_1$ corresponding to $e_{max}(\mathbf{G}_2^{-1}\mathbf{G}_1)$ is of special interest. Let $\mathbf{l}^*$ be such an eigenvector. Then

$$\frac{\mathbf{l}^{*\prime}\mathbf{G}_1\mathbf{l}^*}{\mathbf{l}^{*\prime}\mathbf{G}_2\mathbf{l}^*} = e_{max}(\mathbf{G}_2^{-1}\mathbf{G}_1). \tag{8.108}$$

This follows from the fact that $\mathbf{l}^*$ satisfies the equation

$$(\mathbf{G}_1 - e_{max}\mathbf{G}_2)\mathbf{l}^* = \mathbf{0},$$

where e_{max} is an abbreviation for $e_{max}(\mathbf{G}_2^{-1}\mathbf{G}_1)$. It is easy to see that $\mathbf{l}^*$ can be chosen to be the vector $\mathbf{G}_2^{-1}\mathbf{L}\hat{\boldsymbol{\beta}}$, since

$$\mathbf{G}_2^{-1}\mathbf{G}_1\left(\mathbf{G}_2^{-1}\mathbf{L}\hat{\boldsymbol{\beta}}\right) = \mathbf{G}_2^{-1}\mathbf{L}\hat{\boldsymbol{\beta}}\hat{\boldsymbol{\beta}}'\mathbf{L}'\left(\mathbf{G}_2^{-1}\mathbf{L}\hat{\boldsymbol{\beta}}\right)$$

$$= \left(\hat{\boldsymbol{\beta}}'\mathbf{L}'\mathbf{G}_2^{-1}\mathbf{L}\hat{\boldsymbol{\beta}}\right)\mathbf{G}_2^{-1}\mathbf{L}\hat{\boldsymbol{\beta}}$$

$$= e_{max}\mathbf{G}_2^{-1}\mathbf{L}\hat{\boldsymbol{\beta}}.$$

This shows that $\mathbf{G}_2^{-1}\mathbf{L}\hat{\boldsymbol{\beta}}$ is an eigenvector of $\mathbf{G}_2^{-1}\mathbf{G}_1$ for the eigenvalue e_{max}.

From inequality (8.104) and formula (8.108) we conclude that if the F-test rejects H_0 at the α-level, then

$$|\mathbf{l}^{*\prime}\mathbf{L}\hat{\boldsymbol{\beta}}| > \left(q\,MS_E\,F_{\alpha,q,n-r}\right)^{1/2}\left[\mathbf{l}^{*\prime}\mathbf{L}(\mathbf{X}'\mathbf{X})^-\mathbf{L}'\mathbf{l}^*\right]^{1/2}.$$

This means that the linear combination $\mathbf{a}^{*\prime}\hat{\boldsymbol{\beta}}$, where $\mathbf{a}^{*\prime} = \mathbf{l}^{*\prime}\mathbf{L}$, is significantly different from zero. Let us express $\mathbf{a}^{*\prime}\hat{\boldsymbol{\beta}}$ as

$$\mathbf{l}^{*\prime}\mathbf{L}\hat{\boldsymbol{\beta}} = \sum_{i=1}^{q} l_i^*\hat{\gamma}_i, \tag{8.109}$$

where l_i^* and $\hat{\gamma}_i$ are the ith elements of $\mathbf{l}^*$ and $\hat{\boldsymbol{\gamma}} = \mathbf{L}\hat{\boldsymbol{\beta}}$, respectively $(i = 1, 2, \ldots, q)$. If we divide $\hat{\gamma}_i$ by its estimated standard error $\hat{\kappa}_i$ [which is equal to the square root of the ith diagonal element of the variance–covariance matrix of $\mathbf{L}\hat{\boldsymbol{\beta}}$, namely, $\sigma^2\mathbf{L}(\mathbf{X}'\mathbf{X})^-\mathbf{L}'$ with σ^2 replaced by the error mean square MS_E in formula (8.98)], then formula (8.109) can be written as

$$\mathbf{l}^{*\prime}\mathbf{L}\hat{\boldsymbol{\beta}} = \sum_{i=1}^{q} l_i^*\hat{\kappa}_i\hat{\tau}_i, \tag{8.110}$$

where $\hat{\tau}_i = \hat{\gamma}_i/\hat{\kappa}_i$, $i = 1, 2, \ldots, q$. Consequently, large values of $|l_i^*|\hat{\kappa}_i$ identify those elements of $\hat{\boldsymbol{\gamma}}$ that are influential contributors to the significance of the F-test concerning H_0. Note that the elements of $\gamma = \mathbf{L}\boldsymbol{\beta}$ form a set of linearly independent estimable linear functions of $\boldsymbol{\beta}$.

We conclude from the previous arguments that the eigenvector $\mathbf{l}^*$, which corresponds to the largest eigenvalue of $\mathbf{G}_2^{-1}\mathbf{G}_1$, can be conveniently used to identify an estimable linear function of $\boldsymbol{\beta}$ that is significantly different from zero whenever the F-test rejects H_0.

It should be noted that if model (8.96) is a response surface model (in this case, the matrix $\mathbf{X}$ in the model is of full column rank, that is, $r = p$) whose input variables, $x_1, x_2, \ldots, x_k$, have different units of measurement, then these variables must be made scale free. This is accomplished as follows: If x_{ui} denotes the uth measurement on x_i, then we may consider the transformation

$$z_{ui} = \frac{x_{ui} - \bar{x}_i}{s_i}, \qquad i = 1, 2, \ldots, k; u = 1, 2, \ldots, n,$$

where $\bar{x}_i = (1/n)\sum_{u=1}^{n} x_{ui}$, $s_i = [\sum_{u=1}^{n}(x_{ui} - \bar{x}_i)^2]^{1/2}$, and n is the total number of observations. One advantage of this scaling convention, besides making the input variables scale free, is that it can greatly improve the conditioning of the matrix $\mathbf{X}$ with regard to multicollinearity (see, for example, Belsley, Kuh, and Welsch, 1980, pages 183–185).

EXAMPLE 8.10.1.   Let us consider the one-way classification model

$$y_{ij} = \mu + \alpha_i + \epsilon_{ij}, \qquad i = 1, 2, \ldots, m; j = 1, 2, \ldots, n_i, \qquad (8.111)$$

where μ and α_i are unknown parameters with the latter representing the effect of the ith level of a certain factor at m levels; n_i observations are obtained at the ith level. The ϵ_{ij}'s are random errors assumed to be independent and normally distributed with zero means and a common variance σ^2.

Model (8.111) can be represented in vector form as model (8.96). Here, $\mathbf{y} = (y_{11}, y_{12}, \ldots, y_{1n_1}, y_{21}, y_{22}, \ldots, y_{2n_2}, \ldots, y_{m1}, y_{m2}, \ldots, y_{mn_m})'$, $\boldsymbol{\beta} = (\mu, \alpha_1, \alpha_2, \ldots, \alpha_m)'$, and $\mathbf{X}$ is of order $n \times (m + 1)$ of the form $\mathbf{X} = [\mathbf{1}_n : \mathbf{T}]$, where $\mathbf{1}_n$ is a vector of ones of order $n \times 1$, $n = \sum_{i=1}^{m} n_i$, and $\mathbf{T} = \text{Diag}(\mathbf{1}_{n_1}, \mathbf{1}_{n_2}, \ldots, \mathbf{1}_{n_m})$. The rank of $\mathbf{X}$ is $r = m$. For such a model, the hypothesis of interest is

$$H_0: \alpha_1 = \alpha_2 = \cdots = \alpha_m,$$

which can be expressed as $H_0: \mathbf{L}\boldsymbol{\beta} = \mathbf{0}$, where $\mathbf{L}$ is a matrix of order

$(m-1) \times (m+1)$ and rank $m-1$ of the form

$$
\mathbf{L} = \begin{bmatrix}
0 & 1 & -1 & 0 & \cdots & 0 \\
0 & 1 & 0 & -1 & \cdots & 0 \\
\vdots & \vdots & \vdots & \vdots & & \vdots \\
0 & 1 & 0 & 0 & \cdots & -1
\end{bmatrix}.
$$

This hypothesis states that the factor under consideration has no effect on the response. Note that each row of $\mathbf{L}$ is a linear combination of the rows of $\mathbf{X}$. For example, the ith row of $\mathbf{L}$—whose elements are equal to zero except for the second and the $(i+2)$th elements, which are equal to 1 and -1, respectively—is the difference between rows 1 and $v_i + 1$ of $\mathbf{X}$, where $v_i = \Sigma_{j=1}^{i} n_j$, $i = 1, 2, \ldots, m-1$. Thus the rows of $\mathbf{L}$ form a basis for a q-dimensional subspace $\mathscr{L}$ of $\rho(\mathbf{X})$, the row space of $\mathbf{X}$, where $q = m-1$.

Let $\mu_i = \mu + \alpha_i$. Then μ_i is the mean of the ith level of the factor $(i = 1, 2, \ldots, m)$. Consider the contrast $\psi = \Sigma_{i=1}^{m} c_i \mu_i$, that is, $\Sigma_{i=1}^{m} c_i = 0$. We can write $\psi = \mathbf{a}'\boldsymbol{\beta}$, where $\mathbf{a}' = (0, c_1, c_2, \ldots, c_m)$ belongs to a q-dimensional subspace of R^{m+1}. This subspace is the same as $\mathscr{L}$, since each row of $\mathbf{L}$ is of the form $(0, c_1, c_2, \ldots, c_m)$ with $\Sigma_{i=1}^{m} c_i = 0$. Vice versa, if $\psi = \mathbf{a}'\boldsymbol{\beta}$ is such that $\mathbf{a}' = (0, c_1, c_2, \ldots, c_m)$ with $\Sigma_{i=1}^{m} c_i = 0$, then $\mathbf{a}'$ can be expressed as

$$
\mathbf{a}' = (-c_2, -c_3, \ldots, -c_m)\mathbf{L},
$$

since $c_1 = -\Sigma_{i=2}^{m} c_i$. Hence, $\mathbf{a}' \in \mathscr{L}$. It follows that $\mathscr{L}$ is a subspace associated with all contrasts among the means $\mu_1, \mu_2, \ldots, \mu_m$ of the m levels of the factor.

Simultaneous $(1-\alpha)100\%$ confidence intervals on all contrasts of the form $\psi = \Sigma_{i=1}^{m} c_i \mu_i$ can be obtained by applying formula (8.97). Here, $q = m-1$, $r = m$, and a generalized inverse of $\mathbf{X}'\mathbf{X}$ is of the form

$$
(\mathbf{X}'\mathbf{X})^- = \begin{bmatrix} 0 & \mathbf{0}' \\ \mathbf{0} & \mathbf{D} \end{bmatrix},
$$

where $\mathbf{D} = \mathrm{Diag}(n_1^{-1}, n_2^{-1}, \ldots, n_m^{-1})$ and $\mathbf{0}$ is a zero vector of order $m \times 1$. Hence,

$$
\mathbf{a}'\hat{\boldsymbol{\beta}} = (0, c_1, c_2, \ldots, c_m)(\mathbf{X}'\mathbf{X})^- \mathbf{X}'\mathbf{y}
$$

$$
= \sum_{i=1}^{m} c_i \bar{y}_{i.},
$$

where $\bar{y}_{i.} = (1/n_i)\sum_{j=1}^{n_i} y_{ij}$, $i = 1, 2, \ldots, m$. Furthermore,

$$\mathbf{a}'(\mathbf{X}'\mathbf{X})^{-}\mathbf{a} = \sum_{i=1}^{m} \frac{c_i^2}{n_i}.$$

By making the substitution in formula (8.97) we obtain

$$\sum_{i=1}^{m} c_i \bar{y}_{i.} \mp \left[(m-1) MS_E F_{\alpha, m-1, n-m}\right]^{1/2} \left(\sum_{i=1}^{m} \frac{c_i^2}{n_i}\right)^{1/2}. \quad (8.112)$$

Now, if the F-test rejects H_0 at the α-level, then there exists a contrast $\sum_{i=1}^{m} c_i^* \bar{y}_{i.} = \mathbf{a}^{*'}\hat{\boldsymbol{\beta}}$, which is significantly different from zero, that is, the interval (8.112) for $c_i = c_i^*$ ($i = 1, 2, \ldots, m$) does not contain the value zero. Here, $\mathbf{a}^{*'} = \mathbf{l}^{*'}\mathbf{L}$, where $\mathbf{l}^* = \mathbf{G}_2^{-1}\mathbf{L}\hat{\boldsymbol{\beta}}$ is an eigenvector of $\mathbf{G}_2^{-1}\mathbf{G}_1$ corresponding to $e_{\text{mam}}(\mathbf{G}_2^{-1}\mathbf{G}_1)$. We have that

$$\mathbf{G}_2^{-1} = \left[\mathbf{L}(\mathbf{X}'\mathbf{X})^{-}\mathbf{L}'\right]^{-1} = \left(\frac{1}{n_1}\mathbf{J}_{m-1} + \boldsymbol{\Lambda}\right)^{-1},$$

where $\mathbf{J}_{m-1}$ is a matrix of ones of order $(m-1) \times (m-1)$, and $\boldsymbol{\Lambda} = \text{Diag}(n_2^{-1}, n_3^{-1}, \ldots, n_m^{-1})$. By applying the Sherman–Morrison–Woodbury formula (see Exercise 2.15), we obtain

$$\mathbf{G}_2^{-1} = n_1 \left(n_1 \boldsymbol{\Lambda} + \mathbf{1}_{m-1}\mathbf{1}'_{m-1}\right)^{-1}$$

$$= \boldsymbol{\Lambda}^{-1} - \frac{\boldsymbol{\Lambda}^{-1}\mathbf{1}_{m-1}\mathbf{1}'_{m-1}\boldsymbol{\Lambda}^{-1}}{n_1 + \mathbf{1}'_{m-1}\boldsymbol{\Lambda}^{-1}\mathbf{1}_{m-1}}$$

$$= \text{Diag}(n_2, n_2, \ldots, n_m) - \frac{1}{n}\begin{bmatrix} n_2 \\ n_3 \\ \vdots \\ n_m \end{bmatrix}[n_2, n_3, \ldots, n_m].$$

Also,

$$\hat{\boldsymbol{\gamma}} = \mathbf{L}\hat{\boldsymbol{\beta}} = \mathbf{L}(\mathbf{X}'\mathbf{X})^{-}\mathbf{X}'\mathbf{y} = \begin{bmatrix} \bar{y}_{1.} - \bar{y}_{2.} \\ \bar{y}_{1.} - \bar{y}_{3.} \\ \vdots \\ \bar{y}_{1.} - \bar{y}_{m.} \end{bmatrix}.$$

It can be verified that the ith element of $\mathbf{l}^* = \mathbf{G}_2^{-1}\mathbf{L}\hat{\boldsymbol{\beta}}$ is given by

$$l_i^* = n_{i+1}(\bar{y}_{1.} - \bar{y}_{i+1.}) - \frac{n_{i+1}}{n}\sum_{j=2}^{m} n_j(\bar{y}_{1.} - \bar{y}_{j.}), \qquad i = 1, 2, \ldots, m-1. \quad (8.113)$$

The estimated standard error, $\hat{\kappa}_i$, of the ith element of $\hat{\gamma}$ ($i = 1, 2, \ldots, m-1$) is the square root of the ith diagonal element of $MS_E \, \mathbf{L}(\mathbf{X}'\mathbf{X})^- \mathbf{L}' = [(1/n_1)\mathbf{J}_{m-1} + \mathbf{\Lambda}]MS_E$, that is,

$$\hat{\kappa}_i = \left[\left(\frac{1}{n_1} + \frac{1}{n_{i+1}} \right) MS_E \right]^{1/2}, \qquad i = 1, 2, \ldots, m-1.$$

Thus by formula (8.110), large values of $|l_i^*|\hat{\kappa}_i$ identify those elements of $\hat{\gamma} = \mathbf{L}\hat{\beta}$ that are influential contributors to the significance of the F-test. In particular, if the data set used to analyze model (8.111) is balanced, that is, $n_i = n/m$ for $i = 1, 2, \ldots, m$, then

$$|l_i^*|\hat{\kappa}_i = |\bar{y}_{i+1.} - \bar{y}_{..}| \left(\frac{2n}{m} MS_E \right)^{1/2}, \qquad i = 1, 2, \ldots, m-1,$$

where $\bar{y}_{..} = (1/m)\sum_{i=1}^m \bar{y}_{i.}$.

Alternatively, the contrast $\mathbf{a}^{*\prime}\boldsymbol{\beta}$ can be expressed as

$$\mathbf{a}^{*\prime}\hat{\boldsymbol{\beta}} = \mathbf{l}^{*\prime}\mathbf{L}(\mathbf{X}'\mathbf{X})^- \mathbf{X}'\mathbf{y}$$

$$= \left(0, \sum_{i=1}^{m-1} l_i^*, -l_1^*, -l_2^*, \ldots, -l_{m-1}^* \right) (0, \bar{y}_{1.}, \bar{y}_{2.}, \ldots, \bar{y}_{m.})'$$

$$= \sum_{i=1}^m c_i^* \bar{y}_{i.},$$

where l_i^* is given in formula (8.113) and

$$c_i^* = \begin{cases} \displaystyle\sum_{j=1}^{m-1} l_j^*, & i = 1, \\ -l_{i-1}^*, & i = 2, 3, \ldots, m. \end{cases} \tag{8.114}$$

Since the estimated standard error of $\bar{y}_{i.}$ is $(MS_E/n_i)^{1/2}$, $i = 1, 2, \ldots, m$, by dividing $\bar{y}_{i.}$ by this value we obtain

$$\mathbf{a}^{*\prime}\hat{\boldsymbol{\beta}} = \sum_{i=1}^m \left(\frac{1}{n_i} MS_E \right)^{1/2} c_i^* w_i,$$

where $w_i = \bar{y}_{i.}/(MS_E/n_i)^{1/2}$ is a scaled value of $\bar{y}_{i.}$ ($i = 1, 2, \ldots, m$). Hence, large values of $(MS_E/n_i)^{1/2}|c_i^*|$ identify those $\bar{y}_{i.}$'s that contribute significantly to the rejection of H_0. In particular, for a balanced data set,

$$l_i^* = \frac{n}{m}(\bar{y}_{..} - \bar{y}_{i+1.}), \qquad i = 1, 2, \ldots, m-1.$$

Thus from formula (8.114) we get

$$\left(\frac{MS_E}{n_i}\right)^{1/2} |c_i^*| = \left(\frac{n}{m}MS_E\right)^{1/2} |\bar{y}_{i.} - \bar{y}_{..}|, \qquad i = 1, 2, \ldots, m.$$

We conclude that large values of $|\bar{y}_{i.} - \bar{y}_{..}|$ are responsible for the rejection of H_0 by the F-test. This is consistent with the fact that the numerator sum of squares of the F-test statistic for H_0 is proportional to $\sum_{i=1}^{m}(\bar{y}_{i.} - \bar{y}_{..})^2$ when the data set is balanced.

FURTHER READING AND ANNOTATED BIBLIOGRAPHY

Adby, P. R., and M. A. H. Dempster (1974). *Introduction of Optimization Methods.* Chapman and Hall, London. (This book is an introduction to nonlinear methods of optimization. It covers basic optimization techniques such as steepest descent and the Newton–Raphson method.)

Ash, A., and A. Hedayat (1978). "An introduction to design optimality with an overview of the literature." *Comm. Statist. Theory Methods*, **7**, 1295–1325.

Atkinson, A. C. (1982). "Developments in the design of experiments." *Internat. Statist. Rev.*, **50**, 161–177.

Atkinson, A. C. (1988). "Recent developments in the methods of optimum and related experimental designs." *Internat. Statist. Rev.*, **56**, 99–115.

Bates, D. M., and D. G. Watts (1988). *Nonlinear Regression Analysis and its Applications.* Wiley, New York. (Estimation of parameters in a nonlinear model is addressed in Chaps. 2 and 3. Design aspects for nonlinear models are briefly discussed in Section 3.14.)

Bayne, C. K., and I. B. Rubin (1986). *Practical Experimental Designs and Optimization Methods for Chemists.* VCH Publishers, Deerfield Beach, Florida. (Steepest ascent and the simplex method are discussed in Chap. 5. A bibliography of optimization and response surface methods, as actually applied in 17 major fields of chemistry, is provided in Chap. 7.)

Belsley, D. A., E. Kuh, and R. E. Welsch (1980). *Regression Diagnostics.* Wiley, New York. (Chap. 3 is devoted to the diagnosis of multicollinearity among the columns of the matrix in a regression model. Multicollinearity renders the model's least-squares parameter estimates less precise and less useful than would otherwise be the case.)

Biles, W. E., and J. J. Swain (1980). *Optimization and Industrial Experimentation.* Wiley-Interscience, New York. (Chaps. 4 and 5 discuss optimization techniques that are directly applicable in response surface methodology.)

Bohachevsky, I. O., M. E. Johnson, and M. L. Stein (1986). "Generalized simulated annealing for function optimization." *Technometrics*, **28**, 209–217.

Box, G. E. P. (1982). "Choice of response surface design and alphabetic optimality." *Utilitas Math.*, **21B**, 11–55.

Box, G. E. P., and D. W. Behnken (1960). "Some new three level designs for the study of quantitative variables." *Technometrics*, **2**, 455–475.

Box, G. E. P., and N. R. Draper (1959). "A basis for the selection of a response surface design." *J. Amer. Statist. Assoc.*, **55**, 622–654.

Box, G. E. P., and N. R. Draper (1963). "The choice of a second order rotatable design." *Biometrika*, **50**, 335–352.

Box, G. E. P., and N. R. Draper (1965). "The Bayesian estimation of common parameters from several responses." *Biometrika*, **52**, 355–365.

Box, G. E. P., and N. R. Draper (1987). *Empirical Model-Building and Response Surfaces*. Wiley, New York. (Chap. 9 introduces the exploration of maxima with second-order models; the alphabetic optimality approach is critically considered in Chap. 14. Many examples are given throughout the book.)

Box, G. E. P., and H. L. Lucas (1959). "Design of experiments in nonlinear situations." *Biometrika*, **46**, 77–90.

Box, G. E. P., and K. B. Wilson (1951). "On the experimental attainment of optimum conditions." *J. Roy. Statist. Soc. Ser. B*, **13**, 1–45.

Bunday, B. D. (1984). *Basic Optimization Methods*. Edward Arnold Ltd., Victoria, Australia. (Chaps. 3 and 4 discuss basic optimization techniques such as the Nelder–Mead simplex method and the Davidon–Fletcher–Powell method.)

Conlon, M. (1991). "The controlled random search procedure for function optimization." Personal communication. (This is a FORTRAN file for implementing Price's controlled random search procedure.)

Conlon, M., and A. I. Khuri (1992). "Multiple response optimization." Technical Report, Department of Statistics, University of Florida, Gainesville, Florida.

Cook, R. D., and C. J. Nachtsheim (1980). "A comparison of algorithms for constructing exact *D*-optimal designs." *Technometrics*, **22**, 315–324.

Dempster, A. P., N. M. Laird, and D. B. Rubin (1977). "Maximum likelihood from incomplete data via the EM algorithm." *J. Roy. Statist. Soc. Ser. B*, **39**, 1–38.

Draper, N. R. (1963). "Ridge analysis of response surfaces." *Technometrics*, **5**, 469–479.

Everitt, B. S. (1987). *Introduction to Optimization Methods and Their Application in Statistics*. Chapman and Hall, London. (This book gives a brief introduction to optimization methods and their use in several areas of statistics. These include maximum likelihood estimation, nonlinear regression estimation, and applied multivariate analysis.)

Fedorov, V. V. (1972). *Theory of Optimal Experiments*. Academic Press, New York. (This book is a translation of a monograph in Russian. It presents the mathematical apparatus of experimental design for a regression model.)

Fichtali, J., F. R. Van De Voort, and A. I. Khuri (1990). "Multiresponse optimization of acid casein production." *J. Food Process Eng.*, **12**, 247–258.

Fletcher, R. (1987). *Practical Methods of Optimization*, 2nd ed. Wiley, New York. (This book gives a detailed study of several unconstrained and constrained optimization techniques.)

Fletcher, R., and M. J. D. Powell (1963). "A rapidly convergent descent method for minimization." *Comput. J.*, **6**, 163–168.

Hartley, H. O., and J. N. K. Rao (1967). "Maximum likelihood estimation for the mixed analysis of variance model." *Biometrika*, **54**, 93–108.

Hoerl, A. E. (1959). "Optimum solution of many variables equations." *Chem. Eng. Prog.*, **55**, 69–78.

Huber, P. J. (1973). "Robust regression: Asymptotics, conjectures and Monte Carlo." *Ann. Statist.*, **1**, 799–821.

Huber, P. J. (1981). *Robust Statistics*. Wiley, New York. (This book gives a solid foundation in robustness in statistics. Chap. 3 introduces and discusses *M*-estimation; Chap. 7 addresses *M*-estimation for a regression model.)

Johnson, M. E., and C. J. Nachtsheim (1983). "Some guidelines for constructing exact *D*-optimal designs on convex design spaces." *Technometrics*, **25**, 271–277.

Jones, E. R., and T. J. Mitchell (1978). "Design criteria for detecting model inadequacy." *Biometrika*, **65**, 541–551.

Karson, M. J., A. R. Manson, and R. J. Hader (1969). "Minimum bias estimation and experimental design for response surfaces." *Technometrics*, **11**, 461–475.

Khuri, A. I., and M. Conlon (1981). "Simultaneous optimization of multiple responses represented by polynomial regression functions." *Technometric*, **23**, 363–375.

Khuri, A. I., and J. A. Cornell (1996). *Response Surfaces*, 2nd ed. Marcel Dekker, New York. (Optimization techniques in response surface methodology are discussed in Chap. 5.)

Khuri, A. I., and R. H. Myers (1979). "Modified ridge analysis." *Technometrics*, **21**, 467–473.

Khuri, A. I., and H. Sahai (1985). "Variance components analysis: A selective literature survey." *Internat. Statist. Rev.*, **53**, 279–300.

Kiefer, J. (1958). "On the nonrandomized optimality and the randomized nonoptimality of symmetrical designs." *Ann. Math. Statist.*, **29**, 675–699.

Kiefer, J. (1959). "Optimum experimental designs" (with discussion). *J. Roy. Statist. Soc. Ser. B*, **21**, 272–319.

Kiefer, J. (1960). "Optimum experimental designs *V*, with applications to systematic and rotatable designs." In *Proceedings of the Fourth Berkeley Symposium on Mathematical Statistics and Probability*, Vol. 1. University of California Press, Berkeley, pp. 381–405.

Kiefer, J. (1961). "Optimum designs in regression problems II." *Ann. Math. Statist.*, **32**, 298–325.

Kiefer, J. (1962a). "Two more criteria equivalent to *D*-optimality of designs." *Ann. Math. Statist.*, **33**, 792–796.

Kiefer, J. (1962b). "An extremum result." *Canad. J. Math.*, **14**, 597–601.

Kiefer, J. (1975). "Optimal design: Variation in structure and performance under change of criterion." *Biometrika*, **62**, 277–288.

Kiefer, J., and J. Wolfowitz (1960). "The equivalence of two extremum problems." *Canad. J. Math.*, **12**, 363–366.

Kirkpatrick, S., C. D. Gelatt, and M. P. Vechhi (1983). "Optimization by simulated annealing." *Science*, **220**, 671–680.

Little, R. J. A., and D. B. Rubin (1987). *Statistical Analysis with Missing Data*. Wiley, New York. (The theory of the EM algorithm is introduced in Chap. 7. The book presents a systematic approach to the analysis of data with missing values, where inferences are based on likelihoods derived from formal statistical models for the data.)

Lucas, J. M. (1976). "Which response surface design is best." *Technometrics*, **18**, 411–417.

Miller, R. G., Jr. (1981). *Simultaneous Statistical Inference*, 2nd ed. Springer-Verlag, New York. (Scheffé's simultaneous confidence intervals are derived in Chap. 2.)

Milliken, G. A., and D. E. Johnson (1984). *Analysis of Messy Data*. Lifetime Learning Publications, Belmont, California. (This book presents several techniques and methods for analyzing unbalanced data.)

Mitchell, T. J. (1974). "An algorithm for the construction of *D*-optimal experimental designs." *Technometrics*, **16**, 203–210.

Myers, R. H. (1976). *Response Surface Methodology*. Author, Blacksburg, Virginia. (Chap. 5 discusses the determination of optimum operating conditions in response surface methodology; designs for fitting first-order and second-order models are discussed in Chaps. 6 and 7, respectively; Chap. 9 presents the *J*-criterion for choosing a response surface design.)

Myers, R. H. (1990). *Classical and Modern Regression with Applications*, 2nd ed., PWS-Kent, Boston. (Chap. 3 discusses the effects and hazards of multicollinearity in a regression model. Methods for detecting and combating multicollinearity are given in Chap. 8.)

Myers, R. H., and W. H. Carter, Jr. (1973). "Response surface techniques for dual response systems." *Technometrics*, **15**, 301–317.

Myers, R. H., and A. I. Khuri (1979). "A new procedure for steepest ascent." *Comm. Statist. Theory Methods*, **8**, 1359–1376.

Myers, R. H., A. I. Khuri, and W. H. Carter, Jr. (1989). "Response surface methodology: 1966–1988." *Technometrics*, **31**, 137–157.

Nelder, J. A., and R. Mead (1965). "A simplex method for function minimization." *Comput. J.*, **7**, 308–313.

Nelson, L. S. (1973). "A sequential simplex procedure for non-linear least-squares estimation and other function minimization problems." In *27th Annual Technical Conference Transaction*, American Society for Quality Control, pp. 107–117.

Olsson, D. M., and L. S. Nelson (1975). "The Nelder–Mead simplex procedure for function minimization." *Technometrics*, **17**, 45–51.

Pazman, A. (1986). *Foundations of Optimum Experimental Design*. D. Reidel, Dordrecht, Holland.

Plackett, R. L., and J. P. Burman (1946). "The design of optimum multifactorial experiments." *Biometrika*, **33**, 305–325.

Price, W. L. (1977). "A controlled random search procedure for global optimization." *Comput. J.*, **20**, 367–370.

Rao, C. R. (1970). "Estimation of heteroscedastic variances in linear models." *J. Amer. Statist. Assoc.*, **65**, 161–172.

Rao, C. R. (1971). "Estimation of variance and covariance components—MINQUE theory." *J. Multivariate Anal.*, **1**, 257–275.

Rao, C. R. (1972). "Estimation of variance and covariance components in linear models." *J. Amer. Statist. Assoc.*, **67**, 112–115.

Roussas, G. G. (1973). *A First Course in Mathematical Statistics*. Addison-Wesley, Reading, Massachusetts.

Rustagi, J. S., ed. (1979). *Optimizing Methods in Statistics*. Academic Press, New York.

Scheffé, H. (1959). *The Analysis of Variance*. Wiley, New York. (This classic book presents the basic theory of analysis of variance, mainly in the balanced case.)

Searle, S. R. (1971). *Linear Models*. Wiley, New York. (This book describes general procedures of estimation and hypothesis testing for linear models. Estimable linear functions for models that are not of full rank are discussed in Chap. 5.)

Seber, G. A. F. (1984). *Multivariate Observations,* Wiley, New York. (This book gives a comprehensive survey of the subject of multivariate analysis and provides many useful references.)

Silvey, S. D. (1980). *Optimal Designs*. Chapman and Hall, London.

Spendley, W., G. R. Hext, and F. R. Himsworth (1962). "Sequential application of simplex designs in optimization and evolutionary operation." *Technometrics*, **4**, 441–461.

St. John, R. C., and N. R. Draper (1975). "*D*-Optimality for regression designs: A review." *Technometrics*, **17**, 15–23.

Swallow, W. H., and S. R. Searle (1978). "Minimum variance quadratic unbiased estimation (MIVQUE) of variance components." *Technometrics*, **20**, 265–272.

Watson, G. S. (1964). "A note on maximum likelihood." *Sankhyā Ser. A*, **26**, 303–304.

Wynn, H. P. (1970). "The sequential generation of *D*-optimum experimental designs." *Ann. Math. Statist.*, **41**, 1655–1664.

Wynn, H. P. (1972). "Results in the theory and construction of *D*-optimum experimental designs." *J. Roy. Statist. Soc. Ser. B*, **34**, 133–147.

Zanakis, S. H., and J. S. Rustagi, eds. (1982). *Optimization in Statistics*. North-Holland, Amsterdam, Holland. (This is Volume 19 in *Studies in the Management Sciences*. It contains 21 articles that address applications of optimization in three areas of statistics, namely, regression and correlation; multivariate data analysis and design of experiments; and statistical estimation, reliability, and quality control.)

EXERCISES

8.1. Consider the function

$$f(x_1, x_2) = 8x_1^2 - 4x_1 x_2 + 5x_2^2.$$

Minimize $f(x_1, x_2)$ using the method of steepest descent with $\mathbf{x}_0 = (5, 2)'$ as an initial point.

8.2. Conduct a simulated steepest ascent exercise as follows: Use the function

$$\eta(x_1, x_2) = 47.9 + 3x_1 - x_2 + 4x_1^2 + 4x_1 x_2 + 3x_2^2$$

as the true mean response, which depends on two input variables x_1 and x_2. Generate response values by using the model

$$y(\mathbf{x}) = \eta(\mathbf{x}) + \epsilon,$$

where ϵ has the normal distribution with mean 0 and variance 2.25, and $\mathbf{x} = (x_1, x_2)'$. Fit a first-order model in x_1 and x_2 in a neighborhood of the origin using a 2^2 factorial design along with the corresponding simulated response values. Make sure that replications are taken at the origin in order to test for lack of fit of the fitted model. Determine the path of steepest ascent, then proceed along it using simulated response values. Conduct additional experiments as described in Section 8.3.1.

8.3. Two types of fertilizers were applied to experimental plots to assess their effects on the yield of a certain variety of potato. The design settings used in the experiment along with the corresponding yield values are given in the following table:

Original Settings		Coded Settings		Yield y
Fertilizer 1	Fertilizer 2	x_1	x_2	(lb/plot)
50.0	15.0	-1	-1	24.30
120.0	15.0	1	-1	35.82
50.0	25.0	-1	1	40.50
120.0	25.0	1	1	50.94
35.5	20.0	$-2^{1/2}$	0	30.60
134.5	20.0	$2^{1/2}$	0	42.90
85.0	12.9	0	$-2^{1/2}$	22.50
85.0	27.1	0	$2^{1/2}$	50.40
85.0	20.0	0	0	45.69

(a) Fit a second-order model in the coded variables

$$x_1 = \frac{F_1 - 85}{35}, \qquad x_2 = \frac{F_2 - 20}{5}$$

to the yield data, where F_1 and F_2 are the original settings of fertilizers 1 and 2, respectively, used in the experiment.

(b) Apply the method of ridge analysis to determine the settings of the two fertilizers that are needed to maximize the predicted yield (in the space of the coded input variables, the region R is the interior and boundary of a circle centered at the origin with a radius equal to $2^{1/2}$).

8.4. Suppose that λ_1 and λ_2 are two values of the Lagrange multiplier λ used in the method of ridge analysis. Let $\hat{y}_1$ and $\hat{y}_2$ be the corresponding values of $\hat{y}$ on the two spheres $\mathbf{x}'\mathbf{x} = r_1^2$ and $\mathbf{x}'\mathbf{x} = r_2^2$, respectively. Show that if $r_1 = r_2$ and $\lambda_1 > \lambda_2$, then $\hat{y}_1 > \hat{y}_2$.

8.5. Consider again Exercise 8.4. Let $\mathbf{x}_1$ and $\mathbf{x}_2$ be the stationary points corresponding to λ_1 and λ_2, respectively. Consider also the matrix

$$\mathbf{M}(\mathbf{x}_i) = 2(\hat{\mathbf{B}} - \lambda_i \mathbf{I}), \qquad i = 1, 2.$$

Show that if $r_1 = r_2$, $\mathbf{M}(\mathbf{x}_1)$ is positive definite, and $\mathbf{M}(\mathbf{x}_2)$ is indefinite, then $\hat{y}_1 < \hat{y}_2$.

8.6. Consider once more the method of ridge analysis. Let $\mathbf{x}$ be a stationary point that corresponds to the radius r.

(a) Show that

$$r^3 \frac{\partial^2 r}{\partial \lambda^2} = 2r^2 \sum_{i=1}^{k} \left(\frac{\partial x_i}{\partial \lambda} \right)^2 + \left[r^2 \sum_{i=1}^{k} \left(\frac{\partial x_i}{\partial \lambda} \right)^2 - \left(\sum_{i=1}^{k} x_i \frac{\partial x_i}{\partial \lambda} \right)^2 \right],$$

where x_i is the ith element of $\mathbf{x}$ $(i = 1, 2, \ldots, k)$.

(b) Make use of part (a) to show that

$$\frac{\partial^2 r}{\partial \lambda^2} > 0 \qquad \text{if } r \neq 0.$$

8.7. Suppose that the "true" mean response $\eta(\mathbf{x})$ is represented by a model of order d_2 in k input variables $x_1, x_2, \ldots, x_k$ of the form

$$\eta(\mathbf{x}) = \mathbf{f}'(\mathbf{x})\boldsymbol{\beta} + \mathbf{g}'(\mathbf{x})\boldsymbol{\delta},$$

where $\mathbf{x} = (x_1, x_2, \ldots, x_k)'$. The fitted model is of order d_1 $(< d_2)$ of the form

$$\hat{y}(\mathbf{x}) = \mathbf{f}'(\mathbf{x})\hat{\boldsymbol{\lambda}},$$

where $\hat{\boldsymbol{\lambda}}$ is an estimator of $\boldsymbol{\beta}$, not necessarily obtained by the method of least squares. Let $\boldsymbol{\gamma} = E(\hat{\boldsymbol{\lambda}})$.

(a) Give an expression for B, the average squared bias of $\hat{y}(\mathbf{x})$, in terms of $\boldsymbol{\beta}$, $\boldsymbol{\delta}$, and $\boldsymbol{\gamma}$.

(b) Show that B achieves its minimum value if and only if $\boldsymbol{\gamma}$ is of the form $\boldsymbol{\gamma} = \mathbf{C}\boldsymbol{\tau}$, where $\boldsymbol{\tau} = (\boldsymbol{\beta}', \boldsymbol{\delta}')'$ and $\mathbf{C} = [\mathbf{I} : \boldsymbol{\Gamma}_{11}^{-1}\boldsymbol{\Gamma}_{12}]$. The matrices $\boldsymbol{\Gamma}_{11}$ and $\boldsymbol{\Gamma}_{12}$ are the region moments used in formula (8.54).

(c) Deduce from part (b) that B achieves its minimum value if and only if $\mathbf{C}\boldsymbol{\tau}$ is an estimable linear function (see Searle, 1971, Section 5.4).

(d) Use part (c) to show that B achieves its minimum value if and only if there exists a matrix $\mathbf{L}$ such that $\mathbf{C} = \mathbf{L}[\mathbf{X}: \mathbf{Z}]$, where $\mathbf{X}$ and $\mathbf{Z}$ are matrices consisting of the values taken by $\mathbf{f}'(\mathbf{x})$ and $\mathbf{g}'(\mathbf{x})$, respectively, at n experimental runs.

(e) Deduce from part (d) that B achieves its minimum for any design for which the row space of $[\mathbf{X}: \mathbf{Z}]$ contains the rows of $\mathbf{C}$.

(f) Show that if $\hat{\boldsymbol{\lambda}}$ is the least-squares estimator of $\boldsymbol{\beta}$, that is, $\hat{\boldsymbol{\lambda}} = (\mathbf{X}'\mathbf{X})^{-1}\mathbf{X}'\mathbf{y}$, where $\mathbf{y}$ is the vector of response values at the n experimental runs, then the design property stated in part (e) holds for any design that satisfies the conditions described in equations (8.56).

[*Note:* This problem is based on an article by Karson, Manson, and Hader (1969), who introduced the so-called minimum bias estimation to minimize the average squared bias B.]

8.8. Consider again Exercise 8.7. Suppose that $\mathbf{f}'(\mathbf{x})\boldsymbol{\beta} = \beta_0 + \sum_{i=1}^{3} \beta_i x_i$ is a first-order model in three input variables fitted to a data set obtained by using the design

$$
\mathbf{D} = \begin{bmatrix} -g & -g & -g \\ g & g & -g \\ g & -g & g \\ -g & g & g \end{bmatrix},
$$

where g is a scale factor. The region of interest is a sphere of radius 1. Suppose that the "true" model is of the form

$$
\eta(\mathbf{x}) = \beta_0 + \sum_{i=1}^{3} \beta_i x_i + \beta_{12} x_1 x_2 + \beta_{13} x_1 x_3 + \beta_{23} x_2 x_3.
$$

(a) Can g be chosen so that $\mathbf{D}$ satisfies the conditions described in equations (8.56)?

(b) Can g be chosen so that $\mathbf{D}$ satisfies the minimum bias property described in part (e) of Exercise 8.7?

8.9. Consider the function

$$
h(\boldsymbol{\delta}, \mathbf{D}) = \boldsymbol{\delta}'\boldsymbol{\Delta}\boldsymbol{\delta},
$$

where $\boldsymbol{\delta}$ is a vector of unknown parameters as in model (8.48), $\boldsymbol{\Delta}$ is the matrix in formula (8.53), namely $\boldsymbol{\Delta} = \mathbf{A}'\boldsymbol{\Gamma}_{11}\mathbf{A} - \boldsymbol{\Gamma}_{12}'\mathbf{A} - \mathbf{A}'\boldsymbol{\Gamma}_{12} + \boldsymbol{\Gamma}_{22}$, and $\mathbf{D}$ is the design matrix.

(a) Show that for a given $\mathbf{D}$, the maximum of $h(\boldsymbol{\delta}, \mathbf{D})$ over the region $\psi = \{\boldsymbol{\delta} | \boldsymbol{\delta}' \boldsymbol{\delta} \le r^2\}$ is equal to $r^2 e_{\max}(\boldsymbol{\Delta})$, where $e_{\max}(\boldsymbol{\Delta})$ is the largest eigenvalue of $\boldsymbol{\Delta}$.

(b) Deduce from part (a) a design criterion for choosing $\mathbf{D}$.

8.10. Consider fitting the model

$$y(\mathbf{x}) = \mathbf{f}'(\mathbf{x})\boldsymbol{\beta} + \boldsymbol{\epsilon},$$

where $\boldsymbol{\epsilon}$ is a random error with a zero mean and a variance σ^2. Suppose that the "true" mean response is given by

$$\eta(\mathbf{x}) = \mathbf{f}'(\mathbf{x})\boldsymbol{\beta} + \mathbf{g}'(\mathbf{x})\boldsymbol{\delta}.$$

Let $\mathbf{X}$ and $\mathbf{Z}$ be the same matrices defined in part (d) of Exercise 8.7. Consider the function $\lambda(\boldsymbol{\delta}, \mathbf{D}) = \boldsymbol{\delta}' \mathbf{S} \boldsymbol{\delta}$, where

$$\mathbf{S} = \mathbf{Z}' \left[\mathbf{I} - \mathbf{X}(\mathbf{X}'\mathbf{X})^{-1}\mathbf{X}' \right] \mathbf{Z},$$

and $\mathbf{D}$ is the design matrix. The quantity $\lambda(\boldsymbol{\delta}, \mathbf{D})/\sigma^2$ is the noncentrality parameter associated with the lack of fit F-test for the fitted model (see Khuri and Cornell, 1996, Section 2.6). Large values of λ/σ^2 increase the power of the lack of fit test. By formula (8.54), the minimum value of B is given by

$$B_{\min} = \frac{n}{\sigma^2} \boldsymbol{\delta}' \mathbf{T} \boldsymbol{\delta},$$

where $\mathbf{T} = \boldsymbol{\Gamma}_{22} - \boldsymbol{\Gamma}_{12}' \boldsymbol{\Gamma}_{11}^{-1} \boldsymbol{\Gamma}_{12}$. The fitted model is considered to be inadequate if there exists some constant $\kappa > 0$ such that $\boldsymbol{\delta}' \mathbf{T} \boldsymbol{\delta} \ge \kappa$. Show that

$$\inf_{\boldsymbol{\delta} \in \Phi} \boldsymbol{\delta}' \mathbf{S} \boldsymbol{\delta} = \kappa \, e_{\min}(\mathbf{T}^{-1}\mathbf{S}),$$

where $e_{\min}(\mathbf{T}^{-1}\mathbf{S})$ is the smallest eigenvalue of $\mathbf{T}^{-1}\mathbf{S}$ and Φ is the region $\{\boldsymbol{\delta} | \boldsymbol{\delta}' \mathbf{T} \boldsymbol{\delta} \ge \kappa\}$.

[*Note:* On the basis of this problem, we can define a new design criterion, that which maximizes $e_{\min}(\mathbf{T}^{-1}\mathbf{S})$ with respect to $\mathbf{D}$. A design chosen according to this criterion is called Λ_1-optimal (see Jones and Mitchell, 1978).]

8.11. A second-order model of the form

$$y(\mathbf{x}) = \beta_0 + \beta_1 x_1 + \beta_2 x_2 + \beta_{11} x_1^2 + \beta_{22} x_2^2 + \beta_{12} x_1 x_2 + \boldsymbol{\epsilon}$$

is fitted using a rotatable central composite design **D**, which consists of a factorial 2^2 portion, an axial portion with an axial parameter $\alpha = 2^{1/2}$, and n_0 center-point replications. The settings of the 2^2 factorial portion are ± 1. The region of interest R consists of the interior and boundary of a circle of radius $2^{1/2}$ centered at the origin.

(a) Express V, the average variance of the predicted response given by formula (8.52), as a function of n_0.

(b) Can n_0 be chosen so that it minimizes V?

8.12. Suppose that we have r response functions represented by the models

$$\mathbf{y}_i = \mathbf{X}\boldsymbol{\beta}_i + \boldsymbol{\epsilon}_i, \qquad i = 1, 2, \ldots, r,$$

where $\mathbf{X}$ is a known matrix of order $n \times p$ and rank p. The random error vectors have the same variance–covariance structure as in Section 8.7. Let $\mathbf{F} = E(\mathbf{Y}) = \mathbf{XB}$, where $\mathbf{Y} = [\mathbf{y}_1 : \mathbf{y}_2 : \cdots : \mathbf{y}_r]$ and $\mathbf{B} = [\boldsymbol{\beta}_1 : \boldsymbol{\beta}_2 : \cdots : \boldsymbol{\beta}_r]$.

Show that the determinant of $(\mathbf{Y} - \mathbf{F})'(\mathbf{Y} - \mathbf{F})$ attains a minimum value when $\mathbf{B} = \hat{\mathbf{B}}$, where $\hat{\mathbf{B}}$ is obtained by replacing each $\boldsymbol{\beta}_i$ in $\mathbf{B}$ with $\hat{\boldsymbol{\beta}}_i = (\mathbf{X}'\mathbf{X})^{-1}\mathbf{X}'\mathbf{y}_i$ $(i = 1, 2, \ldots, r)$.

[*Note:* The minimization of the determinant of $(\mathbf{Y} - \mathbf{F})'(\mathbf{Y} - \mathbf{F})$ with respect to $\mathbf{B}$ represents a general multiresponse estimation criterion known as the Box–Draper determinant criterion (see Box and Draper, 1965).]

8.13. Let $\mathbf{A}$ be a $p \times p$ matrix with nonnegative eigenvalues. Show that

$$\det(\mathbf{A}) \leq \exp\left[\operatorname{tr}(\mathbf{A} - \mathbf{I}_p)\right].$$

[*Note:* This inequality is proved in an article by Watson (1964). It is based on the simple inequality $a \leq \exp(a - 1)$, which can be easily proved for any real number a.]

8.14. Let $\mathbf{x}_1, \mathbf{x}_2, \ldots, \mathbf{x}_n$ be a sample of n independently distributed random vectors from a p-variate normal distribution $N(\boldsymbol{\mu}, \mathbf{V})$. The corresponding likelihood function is

$$L = \frac{1}{(2\pi)^{np/2}[\det(\mathbf{V})]^{n/2}} \exp\left[-\frac{1}{2}\sum_{i=1}^{n}(\mathbf{x}_i - \boldsymbol{\mu})'\mathbf{V}^{-1}(\mathbf{x}_i - \boldsymbol{\mu})\right].$$

It is known that the maximum likelihood estimate of $\boldsymbol{\mu}$ is $\bar{\mathbf{x}}$, where $\bar{\mathbf{x}} = (1/n)\sum_{i=1}^{n}\mathbf{x}_i$ (see, for example, Seber, 1984, pages 59–61). Let $\mathbf{S}$ be the matrix

$$\mathbf{S} = \frac{1}{n}\sum_{i=1}^{n}(\mathbf{x}_i - \bar{\mathbf{x}})(\mathbf{x}_i - \bar{\mathbf{x}})'.$$

Show that S is the maximum likelihood estimate of V by proving that

$$\frac{1}{[\det(V)]^{n/2}} \exp\left[-\frac{1}{2}\sum_{i=1}^{n}(x_i - \bar{x})'V^{-1}(x_i - \bar{x})\right]$$

$$\leq \frac{1}{[\det(S)]^{n/2}} \exp\left[-\frac{1}{2}\sum_{i=1}^{n}(x_i - \bar{x})'S^{-1}(x_i - \bar{x})\right],$$

or equivalently,

$$[\det(SV^{-1})]^{n/2} \exp\left[-\frac{n}{2}\operatorname{tr}(SV^{-1})\right] \leq \exp\left[-\frac{n}{2}\operatorname{tr}(I_p)\right].$$

[*Hint:* Use the inequality given in Exercise 8.13.]

8.15. Consider the random one-way classification model

$$y_{ij} = \mu + \alpha_i + \epsilon_{ij}, \qquad i = 1, 2, \ldots, a; \, j = 1, 2, \ldots, n_i,$$

where the α_i's and ϵ_{ij}'s are independently distributed as $N(0, \sigma_\alpha^2)$ and $N(0, \sigma_\epsilon^2)$. Determine the matrix S and the vector q in equation (8.95) that can be used to obtain the MINQUEs of σ_α^2 and σ_ϵ^2.

8.16. Consider the linear model

$$y = X\beta + \epsilon,$$

where X is a known matrix of order $n \times p$ and rank p, and ϵ is normally distributed with a zero mean vector and a variance–covariance matrix $\sigma^2 I_n$. Let $\hat{y}(x)$ denote the predicted response at a point x in a region of interest R.

Use Scheffé's confidence intervals given by formula (8.97) to obtain simultaneous confidence intervals on the mean response values at the points $x_1, x_2, \ldots, x_m$ ($m \leq p$) in R. What is the joint confidence coefficient for these intervals?

8.17. Consider the fixed-effects two-way classification model

$$y_{ijk} = \mu + \alpha_i + \beta_j + (\alpha\beta)_{ij} + \epsilon_{ijk},$$

$$i = 1, 2, \ldots, a; \, j = 1, 2, \ldots, b; \, k = 1, 2, \ldots, m,$$

where α_i and β_j are unknown parameters, $(\alpha\beta)_{ij}$ is the interaction effect, and ϵ_{ijk} is a random error that has the normal distribution with a zero mean and a variance σ^2.

(a) Use Scheffé's confidence intervals to obtain simultaneous confidence intervals on all contrasts among the μ_i's, where $\mu_i = E(\bar{y}_{i..})$ and $\bar{y}_{i..} = (1/bm)\sum_{j=1}^{b}\sum_{k=1}^{m} y_{ijk}$.

(b) Identify those $\bar{y}_{i..}$'s that are influential contributors to the significance of the F-test concerning the hypothesis H_0: $\mu_1 = \mu_2 = \cdots = \mu_a$.

CHAPTER 9

Approximation of Functions

The class of polynomials is undoubtedly the simplest class of functions. In this chapter we shall discuss how to use polynomials to approximate continuous functions. Piecewise polynomial functions (splines) will also be discussed. Attention will be primarily confined to real-valued functions of a single variable x.

9.1. WEIERSTRASS APPROXIMATION

We may recall from Section 4.3 that if a function $f(x)$ has derivatives of all orders in some neighborhood of the origin, then it can be represented by a power series of the form $\sum_{n=0}^{\infty} a_n x^n$. If ρ is the radius of convergence of this series, then the series converges uniformly for $|x| \leq r$, where $r < \rho$ (see Theorem 5.4.4). It follows that for a given $\epsilon > 0$ we can take sufficiently many terms of this power series and obtain a polynomial $p_n(x) = \sum_{k=0}^{n} a_k x^k$ of degree n for which $|f(x) - p_n(x)| < \epsilon$ for $|x| \leq r$. But a function that is not differentiable of all orders does not have a power series representation. However, if the function is continuous on the closed interval $[a, b]$, then it can be approximated uniformly by a polynomial. This is guaranteed by the following theorem:

Theorem 9.1.1 (Weierstrass Approximation Theorem). Let $f: [a, b] \to R$ be a continuous function. Then, for any $\epsilon > 0$, there exists a polynomial $p(x)$ such that

$$|f(x) - p(x)| < \epsilon \qquad \text{for all } x \in [a, b].$$

Proof. Without loss of generality we can consider $[a, b]$ to be the interval $[0, 1]$. This can always be achieved by making a change of variable of the form

$$t = \frac{x - a}{b - a}.$$

As x varies from a to b, t varies from 0 to 1. Thus, if necessary, we consider that such a linear transformation has been made and that t has been renamed as x.

For each n, let $b_n(x)$ be defined as a polynomial of degree n of the form

$$b_n(x) = \sum_{k=0}^{n} \binom{n}{k} x^k (1-x)^{n-k} f\left(\frac{k}{n}\right), \qquad (9.1)$$

where

$$\binom{n}{k} = \frac{n!}{k!(n-k)!}.$$

We have that

$$\sum_{k=0}^{n} \binom{n}{k} x^k (1-x)^{n-k} = 1, \qquad (9.2)$$

$$\sum_{k=0}^{n} \frac{k}{n} \binom{n}{k} x^k (1-x)^{n-k} = x, \qquad (9.3)$$

$$\sum_{k=0}^{n} \frac{k^2}{n^2} \binom{n}{k} x^k (1-x)^{n-k} = \left(1 - \frac{1}{n}\right) x^2 + \frac{x}{n}. \qquad (9.4)$$

These identities can be shown as follows: Let Y_n be a binomial random variable $B(n, x)$. Thus Y_n represents the number of successes in a sequence of n independent Bernoulli trials with x the probability of success on a single trial. Hence, $E(Y_n) = nx$ and $\text{Var}(Y_n) = nx(1-x)$ (see, for example, Harris, 1966, page 104; see also Exercise 5.30). It follows that

$$\sum_{k=0}^{n} \binom{n}{k} x^k (1-x)^{n-k} = P(0 \le Y_n \le n) = 1. \qquad (9.5)$$

Furthermore,

$$\sum_{k=0}^{n} k \binom{n}{k} x^k (1-x)^{n-k} = E(Y_n) = nx, \qquad (9.6)$$

$$\sum_{k=0}^{n} k^2 \binom{n}{k} x^k (1-x)^{n-k} = E(Y_n^2) = \text{Var}(Y_n) + \left[E(Y_n)\right]^2$$

$$= nx(1-x) + n^2 x^2 = n^2 \left[\left(1 - \frac{1}{n}\right) x^2 + \frac{x}{n}\right]. \qquad (9.7)$$

Identities (9.2)–(9.4) follow directly from (9.5)–(9.7).

Let us now consider the difference $f(x) - b_n(x)$, which with the help of identity (9.2) can be written as

$$f(x) - b_n(x) = \sum_{k=0}^{n} \left[f(x) - f\left(\frac{k}{n}\right) \right] \binom{n}{k} x^k (1-x)^{n-k}. \qquad (9.8)$$

Since $f(x)$ is continuous on $[0, 1]$, then it must be bounded and uniformly continuous there (see Theorems 3.4.5 and 3.4.6). Hence, for the given $\epsilon > 0$, there exist numbers δ and m such that

$$|f(x_1) - f(x_2)| \le \frac{\epsilon}{2} \qquad \text{if } |x_1 - x_2| < \delta$$

and

$$|f(x)| < m \qquad \text{for all } x \in [0, 1].$$

From formula (9.8) we then have

$$|f(x) - b_n(x)| \le \sum_{k=0}^{n} \left| f(x) - f\left(\frac{k}{n}\right) \right| \binom{n}{k} x^k (1-x)^{n-k}.$$

If $|x - k/n| < \delta$, then $|f(x) - f(k/n)| < \epsilon/2$; otherwise, we have $|f(x) - f(k/n)| < 2m$ for $0 \le x \le 1$. Consequently, by using identities (9.2)–(9.4) we obtain

$$|f(x) - b_n(x)| \le \sum_{|x - k/n| < \delta} \frac{\epsilon}{2} \binom{n}{k} x^k (1-x)^{n-k}$$

$$+ \sum_{|x - k/n| \ge \delta} 2m \binom{n}{k} x^k (1-x)^{n-k}$$

$$\le \frac{\epsilon}{2} + 2m \sum_{|x - k/n| \ge \delta} \frac{(k/n - x)^2}{(k/n - x)^2} \binom{n}{k} x^k (1-x)^{n-k}$$

$$\le \frac{\epsilon}{2} + \frac{2m}{\delta^2} \sum_{|x - k/n| \ge \delta} \left(\frac{k}{n} - x\right)^2 \binom{n}{k} x^k (1-x)^{n-k}$$

$$\le \frac{\epsilon}{2} + \frac{2m}{\delta^2} \sum_{k=0}^{n} \left(\frac{k^2}{n^2} - \frac{2kx}{n} + x^2\right) \binom{n}{k} x^k (1-x)^{n-k}$$

$$\le \frac{\epsilon}{2} + \frac{2m}{\delta^2} \left[\left(1 - \frac{1}{n}\right) x^2 + \frac{x}{n} - 2x^2 + x^2 \right].$$

Hence,

$$|f(x) - b_n(x)| \le \frac{\epsilon}{2} + \frac{2m}{\delta^2} \frac{x(1-x)}{n}$$

$$\le \frac{\epsilon}{2} + \frac{m}{2n\delta^2}, \tag{9.9}$$

since $x(1-x) \le \frac{1}{4}$ for $0 \le x \le 1$. By choosing n large enough that

$$\frac{m}{2n\delta^2} < \frac{\epsilon}{2},$$

we conclude that

$$|f(x) - b_n(x)| < \epsilon$$

for all $x \in [0, 1]$. The proof of the theorem follows by taking $p(x) = b_n(x)$.

$\square$

Definition 9.1.1. Let $f(x)$ be defined on $[0, 1]$. The polynomial $b_n(x)$ defined by formula (9.1) is called the Bernstein polynomial of degree n for $f(x)$. $\square$

By the proof of Theorem 9.1.1 we conclude that the sequence $\{b_n(x)\}_{n=1}^{\infty}$ of Bernstein polynomials converges uniformly to $f(x)$ on $[0, 1]$. These polynomials are useful in that they not only prove the existence of an approximating polynomial for $f(x)$, but also provide a simple explicit representation for it. Another advantage of Bernstein polynomials is that if $f(x)$ is continuously differentiable on $[0, 1]$, then the derivative of $b_n(x)$ converges also uniformly to $f'(x)$. A more general statement is given by the next theorem, whose proof can be found in Davis (1975, Theorem 6.3.2, page 113).

Theorem 9.1.2. Let $f(x)$ be p times differentiable on $[0, 1]$. If the pth derivative is continuous there, then

$$\lim_{n \to \infty} \frac{d^p b_n(x)}{dx^p} = \frac{d^p f(x)}{dx^p}$$

uniformly on $[0, 1]$.

Obviously, the knowledge that the sequence $\{b_n(x)\}_{n=1}^{\infty}$ converges uniformly to $f(x)$ on $[0, 1]$ is not complete without knowing something about the rate of convergence. For this purpose we need to define the so-called *modulus of continuity* of $f(x)$ on $[a, b]$.

Definition 9.1.2.   If $f(x)$ is continuous on $[a, b]$, then, for any $\delta > 0$, the modulus of continuity of $f(x)$ on $[a, b]$ is

$$\omega(\delta) = \sup_{|x_1 - x_2| \leq \delta} |f(x_1) - f(x_2)|,$$

where x_1 and x_2 are points in $[a, b]$.   □

On the basis of Definition 9.1.2 we have the following properties concerning the modulus of continuity:

Lemma 9.1.1.   If $0 < \delta_1 \leq \delta_2$, then $\omega(\delta_1) \leq \omega(\delta_2)$.

Lemma 9.1.2.   For a function $f(x)$ to be uniformly continuous on $[a, b]$ it is necessary and sufficient that $\lim_{\delta \to 0} \omega(\delta) = 0$.

The proofs of Lemmas 9.1.1 and 9.1.2 are left to the reader.

Lemma 9.1.3.   For any $\lambda > 0$, $\omega(\lambda\delta) \leq (\lambda + 1)\omega(\delta)$.

Proof. Suppose that $\lambda > 0$ is given. We can find an integer n such that $n \leq \lambda < n + 1$. By Lemma 9.1.1, $\omega(\lambda\delta) \leq \omega[(n + 1)\delta]$. Let x_1 and x_2 be two points in $[a, b]$ such that $x_1 < x_2$ and $|x_1 - x_2| \leq (n + 1)\delta$. Let us also divide the interval $[x_1, x_2]$ into $n + 1$ equal parts, each of length $(x_2 - x_1)/(n + 1)$, by means of the partition points

$$y_i = x_1 + i\frac{(x_2 - x_1)}{n + 1}, \qquad i = 0, 1, \ldots, n + 1.$$

Then

$$|f(x_1) - f(x_2)| = |f(x_2) - f(x_1)|$$

$$= \left| \sum_{i=0}^{n} [f(y_{i+1}) - f(y_i)] \right|$$

$$\leq \sum_{i=0}^{n} |f(y_{i+1}) - f(y_i)|$$

$$\leq (n + 1)\omega(\delta),$$

since $|y_{i+1} - y_i| = [1/(n + 1)]|x_2 - x_1| \leq \delta$ for $i = 0, 1, \ldots, n$. It follows that

$$\omega[(n + 1)\delta] = \sup_{|x_1 - x_2| \leq (n+1)\delta} |f(x_1) - f(x_2)| \leq (n + 1)\omega(\delta).$$

Consequently,

$$\omega(\lambda\delta) \le \omega[(n+1)\delta] \le (n+1)\omega(\delta)$$

$$\le (\lambda+1)\omega(\delta). \qquad \square$$

Theorem 9.1.3. Let $f(x)$ be continuous on $[0,1]$, and let $b_n(x)$ be the Bernstein polynomial defined by formula (9.1). Then

$$|f(x) - b_n(x)| \le \frac{3}{2}\omega\left(\frac{1}{\sqrt{n}}\right)$$

for all $x \in [0,1]$, where $\omega(\delta)$ is the modulus of continuity of $f(x)$ on $[0,1]$.

Proof. Using formula (9.1) and identity (9.2), we have that

$$|f(x) - b_n(x)| = \left| \sum_{k=0}^{n} \left[f(x) - f\left(\frac{k}{n}\right) \right] \binom{n}{k} x^k (1-x)^{n-k} \right|$$

$$\le \sum_{k=0}^{n} \left| f(x) - f\left(\frac{k}{n}\right) \right| \binom{n}{k} x^k (1-x)^{n-k}$$

$$\le \sum_{k=0}^{n} \omega\left(\left| x - \frac{k}{n} \right|\right) \binom{n}{k} x^k (1-x)^{n-k}.$$

Now, by applying Lemma 9.1.3 we can write

$$\omega\left(\left| x - \frac{k}{n} \right|\right) = \omega\left(n^{1/2} \left| x - \frac{k}{n} \right| n^{-1/2} \right)$$

$$\le \left(1 + n^{1/2} \left| x - \frac{k}{n} \right| \right) \omega(n^{-1/2}).$$

Thus

$$|f(x) - b_n(x)| \le \sum_{k=0}^{n} \left(1 + n^{1/2} \left| x - \frac{k}{n} \right| \right) \omega(n^{-1/2}) \binom{n}{k} x^k (1-x)^{n-k}$$

$$\le \omega(n^{-1/2}) \left[1 + n^{1/2} \sum_{k=0}^{n} \left| x - \frac{k}{n} \right| \binom{n}{k} x^k (1-x)^{n-k} \right].$$

But, by the Cauchy–Schwarz inequality (see part 1 of Theorem 2.1.2), we have

$$\sum_{k=0}^{n}\left|x-\frac{k}{n}\right|\binom{n}{k}x^k(1-x)^{n-k}$$

$$=\sum_{k=0}^{n}\left|x-\frac{k}{n}\right|\left[\binom{n}{k}x^k(1-x)^{n-k}\right]^{1/2}\left[\binom{n}{k}x^k(1-x)^{n-k}\right]^{1/2}$$

$$\leq\left[\sum_{k=0}^{n}\left(x-\frac{k}{n}\right)^2\binom{n}{k}x^k(1-x)^{n-k}\right]^{1/2}\left[\sum_{k=0}^{n}\binom{n}{k}x^k(1-x)^{n-k}\right]^{1/2}$$

$$=\left[\sum_{k=0}^{n}\left(x-\frac{k}{n}\right)^2\binom{n}{k}x^k(1-x)^{n-k}\right]^{1/2}, \qquad \text{by identity (9.2)}$$

$$=\left[x^2-2x^2+\left(1-\frac{1}{n}\right)x^2+\frac{x}{n}\right]^{1/2}, \qquad \text{by identities (9.3) and (9.4)}$$

$$=\left[\frac{x(1-x)}{n}\right]^{1/2}$$

$$\leq\left[\frac{1}{4n}\right]^{1/2}, \qquad \text{since } x(1-x)\leq\frac{1}{4}.$$

It follows that

$$|f(x)-b_n(x)|\leq\omega(n^{-1/2})\left[1+n^{1/2}\left(\frac{1}{4n}\right)^{1/2}\right],$$

that is,

$$|f(x)-b_n(x)|\leq\tfrac{3}{2}\omega(n^{-1/2})$$

for all $x\in[0,1]$. $\square$

We note that Theorem 9.1.3 can be used to prove Theorem 9.1.1 as follows: If $f(x)$ is continuous on $[0,1]$, then $f(x)$ is uniformly continuous on $[0,1]$. Hence, by Lemma 9.1.2, $\omega(n^{-1/2})\to 0$ as $n\to\infty$.

Corollary 9.1.1. If $f(x)$ is a Lipschitz continuous function Lip(K,α) on $[0,1]$, then

$$|f(x)-b_n(x)|\leq\tfrac{3}{2}Kn^{-\alpha/2} \qquad (9.10)$$

for all $x\in[0,1]$.

Proof. By Definition 3.4.6,

$$|f(x_1) - f(x_2)| \leq K|x_1 - x_2|^\alpha$$

for all x_1, x_2 in $[0, 1]$. Thus

$$\omega(\delta) \leq K\delta^\alpha.$$

By Theorem 9.1.3 we then have

$$|f(x) - b_n(x)| \leq \tfrac{3}{2}Kn^{-\alpha/2}$$

for all $x \in [0, 1]$.   □

Theorem 9.1.4 (Voronovsky's Theorem).   If $f(x)$ is bounded on $[0, 1]$ and has a second-order derivative at a point x_0 in $[0, 1]$, then

$$\lim_{n \to \infty} n[b_n(x_0) - f(x_0)] = \tfrac{1}{2}x_0(1 - x_0)f''(x_0).$$

Proof. See Davis (1975, Theorem 6.3.6, page 117).   □

We note from Corollary 9.1.1 and Voronovsky's theorem that the convergence of Bernstein polynomials can be very slow. For example, if $f(x)$ satisfies the conditions of Voronovsky's theorem, then at every point $x \in [0, 1]$ where $f''(x) \neq 0$, $b_n(x)$ converges to $f(x)$ just like c/n, where c is a constant.

EXAMPLE 9.1.1.   We recall from Section 3.4.2 that $f(x) = \sqrt{x}$ is $\mathrm{Lip}(1, \tfrac{1}{2})$ for $x \geq 0$. Then, by Corollary 9.1.1,

$$|\sqrt{x} - b_n(x)| \leq \frac{3}{2n^{1/4}},$$

for $0 \leq x \leq 1$, where

$$b_n(x) = \sum_{k=0}^{n} \binom{n}{k} x^k (1-x)^{n-k} \left(\frac{k}{n}\right)^{1/2}.$$

9.2. APPROXIMATION BY POLYNOMIAL INTERPOLATION

One possible method to approximate a function $f(x)$ with a polynomial $p(x)$ is to select such a polynomial so that both $f(x)$ and $p(x)$ have the same values at a certain number of points in the domain of $f(x)$. This procedure is called interpolation. The rationale behind it is that if $f(x)$ agrees with $p(x)$ at some known points, then the two functions should be close to one another at intermediate points.

Let us first consider the following result given by the next theorem.

Theorem 9.2.1. Let $a_0, a_1, \ldots, a_n$ be $n + 1$ distinct points in R, the set of real numbers. Let $b_0, b_1, \ldots, b_n$ be any given set of $n + 1$ real numbers. Then, there exists a unique polynomial $p(x)$ of degree $\leq n$ such that $p(a_i) = b_i$, $i = 0, 1, \ldots, n$.

Proof. Since $p(x)$ is a polynomial of degree $\leq n$, it can be represented as $p(x) = \sum_{j=0}^{n} c_j x^j$. We must then have

$$\sum_{j=0}^{n} c_j a_i^j = b_i, \qquad i = 0, 1, \ldots, n.$$

These equations can be written in vector form as

$$
\begin{bmatrix}
1 & a_0 & a_0^2 & \cdots & a_0^n \\
1 & a_1 & a_1^2 & \cdots & a_1^n \\
\vdots & \vdots & \vdots & & \vdots \\
1 & a_n & a_n^2 & \cdots & a_n^n
\end{bmatrix}
\begin{bmatrix}
c_0 \\ c_1 \\ \vdots \\ c_n
\end{bmatrix}
=
\begin{bmatrix}
b_0 \\ b_1 \\ \vdots \\ b_n
\end{bmatrix}. \tag{9.11}
$$

The determinant of the $(n + 1) \times (n + 1)$ matrix on the left side of equation (9.11) is known as Vandermonde's determinant and is equal to $\prod_{i > j}^{n} (a_i - a_j)$. The proof of this last assertion can be found in, for example, Graybill (1983, Theorem 8.12.2, page 266). Since the a_i's are distinct, this determinant is different from zero. It follows that this matrix is nonsingular. Hence, equation (9.11) provides a unique solution for $c_0, c_1, \ldots, c_n$. $\square$

Corollary 9.2.1. The polynomial $p(x)$ in Theorem 9.2.1 can be represented as

$$p(x) = \sum_{i=0}^{n} b_i l_i(x), \tag{9.12}$$

where

$$l_i(x) = \prod_{\substack{j=0 \\ j \neq i}}^{n} \frac{x - a_j}{a_i - a_j}, \qquad i = 0, 1, \ldots, n. \tag{9.13}$$

Proof. We have that $l_i(x)$ is a polynomial of degree n ($i = 0, 1, \ldots, n$). Furthermore, $l_i(a_j) = 0$ if $i \neq j$, and $l_i(a_i) = 1$ ($i = 0, 1, \ldots, n$). It follows that $\sum_{i=0}^{n} b_i l_i(x)$ is a polynomial of degree $\leq n$ and assumes the values $b_0, b_1, \ldots, b_n$ at $a_0, a_1, \ldots, a_n$, respectively. This polynomial is unique by Theorem 9.2.1. $\square$

Definition 9.2.1. The polynomial defined by formula (9.12) is called a *Lagrange interpolating polynomial.* The points $a_0, a_1, \ldots, a_n$ are called points of interpolation (or nodes), and $l_i(x)$ in formula (9.13) is called the ith Lagrange polynomial associated with the a_i's.   □

The values $b_0, b_1, \ldots, b_n$ in formula (9.12) are frequently the values of some function $f(x)$ at the points $a_0, a_1, \ldots, a_n$. Thus $f(x)$ and the polynomial $p(x)$ in formula (9.12) attain the same values at these points. The polynomial $p(x)$, which can be written as

$$p(x) = \sum_{i=0}^{n} f(a_i) l_i(x), \tag{9.14}$$

provides therefore an approximation for $f(x)$ over $[a_0, a_n]$.

EXAMPLE 9.2.1.   Consider the function $f(x) = x^{1/2}$. Let $a_0 = 60$, $a_1 = 70$, $a_2 = 85$, $a_3 = 105$ be interpolation points. Then

$$p(x) = 7.7460 l_0(x) + 8.3666 l_1(x) + 9.2195 l_2(x) + 10.2470 l_3(x),$$

where

$$l_0(x) = \frac{(x - 70)(x - 85)(x - 105)}{(60 - 70)(60 - 85)(60 - 105)},$$

$$l_1(x) = \frac{(x - 60)(x - 85)(x - 105)}{(70 - 60)(70 - 85)(70 - 105)},$$

$$l_2(x) = \frac{(x - 60)(x - 70)(x - 105)}{(85 - 60)(85 - 70)(85 - 105)},$$

$$l_3(x) = \frac{(x - 60)(x - 70)(x - 85)}{(105 - 60)(105 - 70)(105 - 85)}.$$

Table 9.1. Approximation of $f(x) = x^{1/2}$ by the Lagrange Interpolating Polynomial $p(x)$

x	$f(x)$	$p(x)$
60	7.74597	7.74597
64	8.00000	7.99978
68	8.24621	8.24611
70	8.36660	8.36660
74	8.60233	8.60251
78	8.83176	8.83201
82	9.05539	9.05555
85	9.21954	9.21954
90	9.48683	9.48646
94	9.69536	9.69472
98	9.89949	9.89875
102	10.09950	10.09899
105	10.24695	10.24695

Using $p(x)$ as an approximation of $f(x)$ over the interval $[60, 105]$, tabulated values of $f(x)$ and $p(x)$ were obtained at several points inside this interval. The results are given in Table 9.1.

9.2.1. The Accuracy of Lagrange Interpolation

Let us now address the question of evaluating the accuracy of Lagrange interpolation. The answer to this question is given in the next theorem.

Theorem 9.2.2. Suppose that $f(x)$ has n continuous derivatives on the interval $[a, b]$, and its $(n + 1)$st derivative exists on (a, b). Let $a = a_0 < a_1 < \cdots < a_n = b$ be $n + 1$ points in $[a, b]$. If $p(x)$ is the Lagrange interpolating polynomial defined by formula (9.14), then there exists a point $c \in (a, b)$ such that for any $x \in [a, b]$, $x \neq a_i$ $(i = 0, 1, \ldots, n)$,

$$f(x) - p(x) = \frac{1}{(n + 1)!} f^{(n+1)}(c) g_{n+1}(x), \tag{9.15}$$

where

$$g_{n+1}(x) = \prod_{i=0}^{n} (x - a_i).$$

Proof. Define the function $h(t)$ as

$$h(t) = f(t) - p(t) - [f(x) - p(x)] \frac{g_{n+1}(t)}{g_{n+1}(x)}.$$

If $t = x$, then $h(x) = 0$. For $t = a_i$ $(i = 0, 1, \ldots, n)$,

$$h(a_i) = f(a_i) - p(a_i) - [f(x) - p(x)] \frac{g_{n+1}(a_i)}{g_{n+1}(x)} = 0.$$

The function $h(t)$ has n continuous derivatives on $[a, b]$, and its $(n + 1)$st derivative exists on (a, b). Furthermore, $h(t)$ vanishes at x and at all $n + 1$ interpolation points, that is, it has at least $n + 2$ different zeros in $[a, b]$. By Rolle's theorem (Theorem 4.2.1), $h'(t)$ vanishes at least once between any two zeros of $h(t)$ and thus has at least $n + 1$ different zeros in (a, b). Also by Rolle's theorem, $h''(t)$ has at least n different zeros in (a, b). By continuing this argument, we see that $h^{(n+1)}(t)$ has at least one zero in (a, b), say at the point c. But,

$$h^{(n+1)}(t) = f^{(n+1)}(t) - p^{(n+1)}(t) - \frac{f(x) - p(x)}{g_{n+1}(x)} g_{n+1}^{(n+1)}(t)$$

$$= f^{(n+1)}(t) - \frac{f(x) - p(x)}{g_{n+1}(x)} (n + 1)!,$$

since $p(t)$ is a polynomial of degree $\leq n$ and $g_{n+1}(t)$ is a polynomial of the form $t^{n+1} + \lambda_1 t^n + \lambda_2 t^{n-1} + \cdots + \lambda_{n+1}$ for suitable constants $\lambda_1, \lambda_2, \ldots, \lambda_{n+1}$. We thus have

$$f^{(n+1)}(c) - \frac{f(x) - p(x)}{g_{n+1}(x)}(n+1)! = 0,$$

from which we can conclude formula (9.15). $\quad\square$

Corollary 9.2.2. Suppose that $f^{(n+1)}(x)$ is continuous on $[a, b]$. Let $\tau_{n+1} = \sup_{a \leq x \leq b}|f^{(n+1)}(x)|$, $\kappa_{n+1} = \sup_{a \leq x \leq b}|g_{n+1}(x)|$. Then

$$\sup_{a \leq x \leq b} |f(x) - p(x)| \leq \frac{\tau_{n+1} \kappa_{n+1}}{(n+1)!}.$$

Proof. This follows directly from formula (9.15) and the fact that $f(x) - p(x) = 0$ for $x = a_0, a_1, \ldots, a_n$. $\quad\square$

We note that κ_{n+1}, being the supremum of $|g_{n+1}(x)| = |\prod_{i=0}^{n}(x - a_i)|$ over $[a, b]$, is a function of the location of the a_i's. From Corollary 9.2.2 we can then write

$$\sup_{a \leq x \leq b} |f(x) - p(x)| \leq \frac{\phi(a_0, a_1, \ldots, a_n)}{(n+1)!} \sup_{a \leq x \leq b} |f^{(n+1)}(x)|, \quad (9.16)$$

where $\phi(a_0, a_1, \ldots, a_n) = \sup_{a \leq x \leq b}|g_{n+1}(x)|$. This inequality provides us with an upper bound on the error of approximating $f(x)$ with $p(x)$ over the interpolation region. We refer to this error as interpolation error. The upper bound clearly shows that the interpolation error depends on the location of the interpolation points.

Corollary 9.2.3. If, in Corollary 9.2.2, $n = 2$, and if $a_1 - a_0 = a_2 - a_1 = \delta$, then

$$\sup_{a \leq x \leq b} |f(x) - p(x)| \leq \frac{\sqrt{3}}{27} \tau_3 \delta^3.$$

Proof. Consider $g_3(x) = (x - a_0)(x - a_1)(x - a_2)$, which can be written as $g_3(x) = z(z^2 - \delta^2)$, where $z = x - a_1$. This function is symmetric with respect to $x = a_1$. It is easy to see that $|g_3(x)|$ attains an absolute maximum over $a_0 \leq x \leq a_2$, or equivalently, $-\delta \leq z \leq \delta$, when $z = \pm\delta/\sqrt{3}$. Hence,

$$\kappa_3 = \sup_{a_0 \leq x \leq a_2} |g_3(x)| = \max_{-\delta \leq z \leq \delta} |z(z^2 - \delta^2)|$$

$$= \left| \frac{\delta}{\sqrt{3}} \left(\frac{\delta^2}{3} - \delta^2 \right) \right| = \frac{2}{3\sqrt{3}} \delta^3.$$

By applying Corollary 9.2.2 we obtain

$$\sup_{a_0 \le x \le a_2} |f(x) - p(x)| \le \frac{\sqrt{3}}{27} \tau_3 \delta^3.$$

We have previously noted that the interpolation error depends on the choice of the interpolation points. This leads us to the following important question: How can the interpolation points be chosen so as to minimize the interpolation error? The answer to this question lies in inequality (9.16). One reasonable criterion for the choice of interpolation points is the minimization of $\phi(a_0, a_1, \ldots, a_n)$ with respect to $a_0, a_1, \ldots, a_n$. It turns out that the optimal locations of $a_0, a_1, \ldots, a_n$ are given by the zeros of the Chebyshev polynomial (of the first kind) of degree $n + 1$ (see Section 10.4.1). $\square$

Definition 9.2.2. The Chebyshev polynomial of degree n is defined by

$$T_n(x) = \cos(n \, \text{Arccos} \, x)$$

$$= x^n + \binom{n}{2} x^{n-2}(x^2 - 1) + \binom{n}{4} x^{n-4}(x^2 - 1)^2 + \cdots,$$

$$n = 0, 1, \ldots .(9.17)$$

Obviously, by the definition of $T_n(x)$, $-1 \le x \le 1$. One of the properties of $T_n(x)$ is that it has simple zeros at the following n points:

$$\zeta_i = \cos\left[\frac{(2i-1)}{2n}\pi\right], \qquad i = 1, 2, \ldots, n. \qquad \square$$

The proof of this property is given in Davis (1975, pages 61–62).

We can consider Chebyshev polynomials defined on the interval $[a, b]$ by making the transformation

$$x = \frac{a + b}{2} + \frac{b - a}{2}t,$$

which transforms the interval $-1 \le t \le 1$ into the interval $a \le x \le b$. In this case, the zeros of the Chebyshev polynomial of degree n over the interval $[a, b]$ are given by

$$z_i = \frac{a + b}{2} + \frac{b - a}{2}\cos\left[\left(\frac{2i-1}{2n}\right)\pi\right], \qquad i = 1, 2, \ldots, n.$$

We refer to the z_i's as Chebyshev points. These points can be obtained geometrically by subdividing the semicircle over the interval $[a, b]$ into n

equal arcs and then projecting the midpoint of each arc onto the interval (see De Boor, 1978, page 26).

Chebyshev points have a very interesting property that pertains to the minimization of $\phi(a_0, a_1, \ldots, a_n)$ in inequality (9.16). This property is described in Theorem 9.2.3, whose proof can be found in Davis (1975, Section 3.3); see also De Boor (1978, page 30).

Theorem 9.2.3. The function

$$\phi(a_0, a_1, \ldots, a_n) = \sup_{a \le x \le b} \left| \prod_{i=0}^{n} (x - a_i) \right|,$$

where the a_i's belong to the interval $[a, b]$, achieves its minimum at the zeros of the Chebyshev polynomial of degree $n + 1$, that is, at

$$z_i = \frac{a+b}{2} + \frac{b-a}{2} \cos\left[\left(\frac{2i+1}{2n+2}\right)\pi\right], \qquad i = 0, 1, \ldots, n, \qquad (9.18)$$

and

$$\min_{a_0, a_1, \ldots, a_n} \phi(a_0, a_1, \ldots, a_n) = \frac{2(b-a)^{n+1}}{4^{n+1}}.$$

From Theorem 9.2.3 and inequality (9.16) we conclude that the choice of the Chebyshev points given in formula (9.18) is optimal in the sense of reducing the interpolation error. In other words, among all sets of interpolation points of size $n + 1$ each, Chebyshev points produce a Lagrange polynomial approximation for $f(x)$ over the interval $[a, b]$ with a minimum upper bound on the error of approximation. Using inequality (9.16), we obtain the following interesting result:

$$\sup_{a \le x \le b} |f(x) - p(x)| \le \frac{2}{(n+1)!} \left(\frac{b-a}{4}\right)^{n+1} \sup_{a \le x \le b} |f^{(n+1)}(x)|. \quad (9.19)$$

The use of Chebyshev points in the construction of Lagrange interpolating polynomial $p(x)$ for the function $f(x)$ over $[a, b]$ produces an approximation which, for all practical purposes, differs very little from the best possible approximation of $f(x)$ by a polynomial of the same degree. This was shown by Powell (1967). More explicitly, let $p^*(x)$ be the best approximating polynomial of $f(x)$ of the same degree as $p(x)$ over $[a, b]$. Then, obviously,

$$\sup_{a \le x \le b} |f(x) - p^*(x)| \le \sup_{a \le x \le b} |f(x) - p(x)|.$$

De Boor (1978, page 31) pointed out that for $n \le 20$,

$$\sup_{a \le x \le b} |f(x) - p(x)| \le 4 \sup_{a \le x \le b} |f(x) - p^*(x)|.$$

This indicates that the error of interpolation which results from the use of Lagrange polynomials in combination with Chebyshev points does not exceed the minimum approximation error by more than a factor of 4 for $n \le 20$. This is a very useful result, since the derivation of the best approximating polynomial can be tedious and complicated, whereas a polynomial approximation obtained by Lagrange interpolation that uses Chebyshev points as interpolation points is simple and straightforward.

9.2.2. A Combination of Interpolation and Approximation

In Section 9.1 we learned how to approximate a continuous function f: $[a, b] \to R$ with a polynomial by applying the Weierstrass theorem. In this section we have seen how to interpolate values of f on $[a, b]$ by using Lagrange polynomials. We now show that these two processes can be combined. More specifically, suppose that we are given $n + 1$ distinct points in $[a, b]$, which we denote by $a_0, a_1, \ldots, a_n$ with $a_0 = a$ and $a_n = b$. Let $\epsilon > 0$ be given. We need to find a polynomial $q(x)$ such that $|f(x) - q(x)| < \epsilon$ for all x in $[a, b]$, and $f(a_i) = q(a_i)$, $i = 0, 1, \ldots, n$.

By Theorem 9.1.1 there exists a polynomial $p(x)$ such that

$$|f(x) - p(x)| < \epsilon' \qquad \text{for all } x \in [a, b],$$

where $\epsilon' < \epsilon/(1 + M)$, and M is a nonnegative number to be described later. Furthermore, by Theorem 9.2.1 there exists a unique polynomial $u(x)$ such that

$$u(a_i) = f(a_i) - p(a_i), \qquad i = 0, 1, \ldots, n.$$

This polynomial is given by

$$u(x) = \sum_{i=0}^{n} [f(a_i) - p(a_i)] l_i(x), \qquad (9.20)$$

where $l_i(x)$ is the ith Lagrange polynomial defined in formula (9.13). Using formula (9.20) we obtain

$$\max_{a \le x \le b} |u(x)| \le \sum_{i=0}^{n} |f(a_i) - p(a_i)| \max_{a \le x \le b} |l_i(x)|$$

$$\le \epsilon' M,$$

where $M = \sum_{i=0}^{n} \max_{a \le x \le b} |l_i(x)|$, which is some finite nonnegative number. Note that M depends only on $[a, b]$ and $a_0, a_1, \ldots, a_n$. Now, define $q(x)$ as $q(x) = p(x) + u(x)$. Then

$$q(a_i) = p(a_i) + u(a_i) = f(a_i), \qquad i = 0, 1, \ldots, n.$$

Furthermore,

$$|f(x) - q(x)| \leq |f(x) - p(x)| + |u(x)|$$
$$< \epsilon' + \epsilon'M$$
$$< \epsilon \qquad \text{for all } x \in [a, b].$$

9.3. APPROXIMATION BY SPLINE FUNCTIONS

Approximation of a continuous function $f(x)$ with a single polynomial $p(x)$ may not be quite adequate in situations in which $f(x)$ represents a real physical relationship. The behavior of such a function in one region may be unrelated to its behavior in another region. This type of behavior may not be satisfactorily matched by any polynomial. This is attributed to the fact that the behavior of a polynomial everywhere is governed by its behavior in any small region. In such situations, it would be more appropriate to partition the domain of $f(x)$ into several intervals and then use a different approximating polynomial, usually of low degree, in each subinterval. These polynomial segments can be joined in a smooth way, which leads to what is called a piecewise polynomial function.

By definition, a spline function is a piecewise polynomial of degree n. The various polynomial segments (all of degree n) are joined together at points called knots in such a way that the entire spline function is continuous and its first $n - 1$ derivatives are also continuous. Spline functions were first introduced by Schoenberg (1946).

9.3.1. Properties of Spline Functions

Let $[a, b]$ can be interval, and let $a = \tau_0 < \tau_1 < \cdots < \tau_m < \tau_{m+1} = b$ be partition points in $[a, b]$. A spline function $s(x)$ of degree n with knots at the points $\tau_1, \tau_2, \ldots, \tau_m$ has the following properties:

i. $s(x)$ is a polynomial of degree not exceeding n on each subinterval $[\tau_{i-1}, \tau_i]$, $1 \leq i \leq m + 1$.

ii. $s(x)$ has continuous derivatives up to order $n - 1$ on $[a, b]$.

In particular, if $n = 1$, then the spline function is called a linear spline and can be represented as

$$s(x) = \sum_{i=1}^{m} a_i |x - \tau_i|,$$

where $a_1, a_2, \ldots, a_m$ are fixed numbers. We note that between any two knots, $|x - \tau_i|$, $i = 1, 2, \ldots, m$, represents a straight-line segment. Thus the graph of $s(x)$ is made up of straight-line segments joined at the knots.

We can obtain a linear spline that resembles Lagrange interpolation: Let $\theta_0, \theta_1, \ldots, \theta_{m+1}$ be given real numbers. For $1 \le i \le m$, consider the functions

$$l_0(x) = \begin{cases} \dfrac{x - \tau_1}{\tau_0 - \tau_1}, & \tau_0 \le x \le \tau_1, \\ 0, & \tau_1 \le x \le \tau_{m+1}, \end{cases}$$

$$l_i(x) = \begin{cases} 0, & x \notin [\tau_{i-1}, \tau_{i+1}], \\ \dfrac{x - \tau_{i-1}}{\tau_i - \tau_{i-1}}, & \tau_{i-1} \le x \le \tau_i, \\ \dfrac{\tau_{i+1} - x}{\tau_{i+1} - \tau_i}, & \tau_i \le x \le \tau_{i+1}, \end{cases}$$

$$l_{m+1}(x) = \begin{cases} 0, & \tau_0 \le x \le \tau_m, \\ \dfrac{x - \tau_m}{\tau_{m+1} - \tau_m}, & \tau_m \le x \le \tau_{m+1}. \end{cases}$$

Then the linear spline

$$s(x) = \sum_{i=0}^{m+1} \theta_i l_i(x) \tag{9.21}$$

has the property that $s(\tau_i) = \theta_i$, $0 \le i \le m + 1$. It can be shown that the linear spline having this property is unique.

Another special case is the cubic spline for $n = 3$. This is a widely used spline function in many applications. It can be represented as $s(x) = s_i(x) = a_i + b_i x + c_i x^2 + d_i x^3$, $\tau_{i-1} \le x \le \tau_i$, $i = 1, 2, \ldots, m + 1$, such that for $i = 1, 2, \ldots, m$,

$$s_i(\tau_i) = s_{i+1}(\tau_i),$$

$$s_i'(\tau_i) = s_{i+1}'(\tau_i),$$

$$s_i''(\tau_i) = s_{i+1}''(\tau_i).$$

In general, a spline of degree n with knots at $\tau_1, \tau_2, \ldots, \tau_m$ is represented as

$$s(x) = \sum_{i=1}^{m} e_i (x - \tau_i)_+^n + p(x), \tag{9.22}$$

where $e_1, e_2, \ldots, e_m$ are constants, $p(x)$ is a polynomial of degree n, and

$$(x - \tau_i)_+^n = \begin{cases} (x - \tau_i)^n, & x \geq \tau_i, \\ 0, & x \leq \tau_i. \end{cases}$$

For an illustration, consider the cubic spline

$$s(x) = \begin{cases} a_1 + b_1 x + c_1 x^2 + d_1 x^3, & a \leq x \leq \tau, \\ a_2 + b_2 x + c_2 x^2 + d_2 x^3, & \tau \leq x \leq b. \end{cases}$$

Here, $s(x)$ along with its first and second derivatives must be continuous at $x = \tau$. Therefore, we must have

$$a_1 + b_1\tau + c_1\tau^2 + d_1\tau^3 = a_2 + b_2\tau + c_2\tau^2 + d_2\tau^3, \tag{9.23}$$

$$b_1 + 2c_1\tau + 3d_1\tau^2 = b_2 + 2c_2\tau + 3d_2\tau^2, \tag{9.24}$$

$$2c_1 + 6d_1\tau = 2c_2 + 6d_2\tau. \tag{9.25}$$

Equation (9.25) can be written as

$$c_1 - c_2 = 3(d_2 - d_1)\tau. \tag{9.26}$$

From equations (9.24) and (9.26) we get

$$b_1 - b_2 + 3(d_2 - d_1)\tau^2 = 0. \tag{9.27}$$

Using now equations (9.26) and (9.27) in equation (9.23), we obtain $a_1 - a_2 + 3(d_1 - d_2)\tau^3 + 3(d_2 - d_1)\tau^3 + (d_1 - d_2)\tau^3 = 0$, or equivalently,

$$a_1 - a_2 + (d_1 - d_2)\tau^3 = 0. \tag{9.28}$$

We conclude that

$$d_2 - d_1 = \frac{1}{\tau^3}(a_1 - a_2) = \frac{1}{3\tau^2}(b_2 - b_1) = \frac{1}{3\tau}(c_1 - c_2). \tag{9.29}$$

Let us now express $s(x)$ in the form given by equation (9.22), that is,

$$s(x) = e_1(x - \tau)_+^3 + \alpha_0 + \alpha_1 x + \alpha_2 x^2 + \alpha_3 x^3.$$

In this case,

$$\alpha_0 = a_1, \qquad \alpha_1 = b_1, \qquad \alpha_2 = c_1, \qquad \alpha_3 = d_1,$$

and

$$-e_1\tau^3 + \alpha_0 = a_2, \tag{9.30}$$

$$3e_1\tau^2 + \alpha_1 = b_2, \tag{9.31}$$

$$-3e_1\tau + \alpha_2 = c_2, \tag{9.32}$$

$$e_1 + \alpha_3 = d_2. \tag{9.33}$$

In light of equation (9.29), equations (9.30)–(9.33) have a common solution for e_1 given by $e_1 = d_2 - d_1 = (1/3\tau)(c_1 - c_2) = (1/3\tau^2)(b_2 - b_1) = (1/\tau^3)(a_1 - a_2)$.

9.3.2. Error Bounds for Spline Approximation

Let $a = \tau_0 < \tau_1 < \cdots < \tau_m < \tau_{m+1} = b$ be a partition of $[a, b]$. We recall that the linear spline $s(x)$ given by formula (9.21) has the property that $s(\tau_i) = \theta_i$, $0 \le i \le m + 1$, where $\theta_0, \theta_1, \ldots, \theta_{m+1}$ are any given real numbers. In particular, if θ_i is the value at τ_i of a function $f(x)$ defined on the interval $[a, b]$, then $s(x)$ provides a spline approximation of $f(x)$ over $[a, b]$ which agrees with $f(x)$ at $\tau_0, \tau_1, \ldots, \tau_{m+1}$. If $f(x)$ has continuous derivatives up to order 2 over $[a, b]$, then an upper bound on the error of approximation is given by (see De Boor, 1978, page 40)

$$\max_{a \le x \le b} |f(x) - s(x)| \le \tfrac{1}{8}\left(\max_i \Delta\tau_i\right)^2 \max_{a \le x \le b} |f^{(2)}(x)|,$$

where $\Delta\tau_i = \tau_{i+1} - \tau_i$, $i = 0, 1, \ldots, m$. This error bound can be made small by reducing the value of $\max_i \Delta\tau_i$.

A more efficient and smoother spline approximation than the one provided by the linear spline is the commonly used cubic spline approximation. We recall that a cubic spline defined on $[a, b]$ is a piecewise cubic polynomial that is twice continuously differentiable. Let $f(x)$ be defined on $[a, b]$. There exists a unique cubic spline $s(x)$ that satisfies the following interpolatory constraints:

$$s(\tau_i) = f(\tau_i), \qquad i = 0, 1, \ldots, m + 1,$$

$$s'(\tau_0) = f'(\tau_0),$$

$$s'(\tau_{m+1}) = f'(\tau_{m+1}), \tag{9.34}$$

(see Prenter, 1975, Section 4.2).

If $f(x)$ has continuous derivatives up to order 4 on $[a, b]$, then information on the error of approximation, which results from using a cubic spline, can be obtained from the following theorem, whose proof is given in Hall (1968):

Theorem 9.3.1. Let $a = \tau_0 < \tau_1 < \cdots < \tau_m < \tau_{m+1} = b$ be a partition of $[a, b]$. Let $s(x)$ be a cubic spline associated with $f(x)$ and satisfies the constraints described in (9.34). If $f(x)$ has continuous derivatives up to order 4 on $[a, b]$, then

$$\max_{a \le x \le b} |f(x) - s(x)| \le \tfrac{5}{384} \left(\max_i \Delta \tau_i \right)^4 \max_{a \le x \le b} |f^{(4)}(x)|,$$

where $\Delta \tau_i = \tau_{i+1} - \tau_i$, $i = 0, 1, \ldots, m$.

Another advantage of cubic spline approximation is the fact that it can be used to approximate the first-order and second-order derivatives of $f(x)$. Hall and Meyer (1976) proved that if $f(x)$ satisfies the conditions of Theorem 9.3.1, then

$$\max_{a \le x \le b} |f'(x) - s'(x)| \le \tfrac{1}{24} \left(\max_i \Delta \tau_i \right)^3 \max_{a \le x \le b} |f^{(4)}(x)|,$$

$$\max_{a \le x \le b} |f''(x) - s''(x)| \le \tfrac{3}{8} \left(\max_i \Delta \tau_i \right)^2 \max_{a \le x \le b} |f^{(4)}(x)|.$$

Furthermore, the bounds concerning $|f(x) - s(x)|$ and $|f'(x) - s'(x)|$ are best possible.

9.4. APPLICATIONS IN STATISTICS

There is a wide variety of applications of polynomial approximation in statistics. In this section, we discuss the use of Lagrange interpolation in optimal design theory and the role of spline approximation in regression analysis. Other applications will be seen later in Chapter 10 (Section 10.9).

9.4.1. Approximate Linearization of Nonlinear Models by Lagrange Interpolation

We recall from Section 8.6 that a nonlinear model is one of the form

$$y(\mathbf{x}) = h(\mathbf{x}, \boldsymbol{\theta}) + \epsilon, \tag{9.35}$$

where $\mathbf{x} = (x_1, x_2, \ldots, x_k)'$ is a vector of k input variables, $\boldsymbol{\theta} = (\theta_1, \theta_2, \ldots, \theta_p)'$ is a vector of p unknown parameters, ϵ is a random error, and $h(\mathbf{x}, \boldsymbol{\theta})$ is a known function which is nonlinear in at least one element of $\boldsymbol{\theta}$.

We also recall that the choice of design for model (9.35), on the basis of the Box–Lucas criterion, depends on the values of the elements of $\boldsymbol{\theta}$ that appear nonlinearly in the model. To overcome this undesirable design dependence problem, one possible approach is to construct an approximation to the mean response function $h(\mathbf{x}, \boldsymbol{\theta})$ with a Lagrange interpolating polynomial. This approximation can then be utilized to obtain a design for parameter estimation which does not depend on the parameter vector $\boldsymbol{\theta}$. We shall restrict our consideration of model (9.35) to the case of a single input variable x.

Let us suppose that the region of interest, R, is the interval $[a, b]$, and that $\boldsymbol{\theta}$ belongs to a parameter space Ω. We assume that:

a. $h(x, \boldsymbol{\theta})$ has continuous partial derivatives up to order $r + 1$ with respect to x over $[a, b]$ for all $\boldsymbol{\theta} \in \Omega$, where r is such that $r + 1 \geq p$ with p being the number of parameters in model (9.35), and is large enough so that

$$\frac{2}{(r+1)!} \left(\frac{b-a}{4} \right)^{r+1} \sup_{a \leq x \leq b} \left| \frac{\partial^{r+1} h(x, \boldsymbol{\theta})}{\partial x^{r+1}} \right| < \delta \qquad (9.36)$$

for all $\boldsymbol{\theta} \in \Omega$, where δ is a small positive constant chosen appropriately so that the Lagrange interpolation of $h(x, \boldsymbol{\theta})$ achieves a certain accuracy.

b. $h(x, \boldsymbol{\theta})$ has continuous first-order partial derivatives with respect to the elements of $\boldsymbol{\theta}$.

c. For any set of distinct points, $x_0, x_1, \ldots, x_r$, such that $a \leq x_0 < x_1 < \cdots < x_r \leq b$, where r is the integer defined in (a), the $p \times (r + 1)$ matrix

$$\mathbf{U}(\boldsymbol{\theta}) = \left[\nabla h(x_0, \boldsymbol{\theta}) : \nabla h(x_1, \boldsymbol{\theta}) : \cdots : \nabla h(x_r, \boldsymbol{\theta}) \right]$$

is of rank p, where $\nabla h(x_i, \boldsymbol{\theta})$ is the vector of partial derivatives of $h(x_i, \boldsymbol{\theta})$ with respect to the elements of $\boldsymbol{\theta}$ $(i = 0, 1, \ldots, r)$.

Let us now consider the points $z_0, z_1, \ldots, z_r$, where z_i is the ith Chebyshev point defined by formula (9.18). Let $p_r(x, \boldsymbol{\theta})$ denote the corresponding Lagrange interpolating polynomial for $h(x, \boldsymbol{\theta})$ over $[a, b]$, which utilizes the z_i's as interpolation points. Then, by formula (9.14) we have

$$p_r(x, \boldsymbol{\theta}) = \sum_{i=0}^{r} h(z_i, \boldsymbol{\theta}) l_i(x), \qquad (9.37)$$

where $l_i(x)$ is a polynomial of degree r which can be obtained from formula (9.13) by substituting z_i for a_i $(i = 0, 1, \ldots, r)$. By inequality (9.19), an upper bound on the error of approximating $h(x, \boldsymbol{\theta})$ with $p_r(x, \boldsymbol{\theta})$ is given by

$$\sup_{a \leq x \leq b} |h(x, \boldsymbol{\theta}) - p_r(x, \boldsymbol{\theta})| \leq \frac{2}{(r+1)!} \left(\frac{b-a}{4} \right)^{r+1} \sup_{a \leq x \leq b} \left| \frac{\partial^{r+1} h(x, \boldsymbol{\theta})}{\partial x^{r+1}} \right|.$$

However, by inequality (9.36), this upper bound is less than δ. We then have

$$\sup_{a \leq x \leq b} |h(x,\theta) - p_r(x,\theta)| < \delta \tag{9.38}$$

for all $\theta \in \Omega$. This provides the desired accuracy of approximation.

On the basis of the above arguments, an approximate representation of model (9.35) is given by

$$y(x) = p_r(x,\theta) + \epsilon. \tag{9.39}$$

Model (9.39) will now be utilized in place of $h(x,\theta)$ to construct an optimal design for estimating θ.

Let us now apply the Box–Lucas criterion described in Section 8.6 to approximate the mean response in model (9.39). In this case, the matrix $H(\theta)$ [see model (8.66)] is an $n \times p$ matrix whose (u, i)th element is $\partial p_r(x_u, \theta)/\partial\theta_i$, where x_u is the design setting at the uth experimental run ($u = 1, 2, \ldots, n$) and n is the number of experimental runs. From formula (9.37) we than have

$$\frac{\partial p_r(x_u, \theta)}{\partial\theta_i} = \sum_{j=0}^{r} \frac{\partial h(z_j, \theta)}{\partial\theta_i} l_j(x_u), \qquad i = 1, 2, \ldots, p.$$

These equations can be written as

$$\nabla p_r(x_u, \theta) = U(\theta)\lambda(x_u),$$

where $\lambda(x_u) = [l_0(x_u), l_1(x_u), \ldots, l_r(x_u)]'$ and $U(\theta)$ is the $p \times (r+1)$ matrix

$$U(\theta) = \left[\nabla h(z_0, \theta) : \nabla h(z_1, \theta) : \cdots : \nabla h(z_r, \theta) \right].$$

By assumption (c), $U(\theta)$ is of rank p. The matrix $H(\theta)$ is therefore of the form

$$H(\theta) = \Lambda U'(\theta),$$

where

$$\Lambda' = \left[\lambda(x_1) : \lambda(x_2) : \cdots : \lambda(x_n) \right].$$

Thus

$$H'(\theta)H(\theta) = U(\theta) \Lambda'\Lambda U'(\theta). \tag{9.40}$$

If $n \geq r+1$ and at least $r+1$ of the design points (that is, $x_1, x_2, \ldots, x_n$) are distinct, then $\Lambda'\Lambda$ is a nonsingular matrix. To show this, it is sufficient to prove that Λ is of full column rank $r+1$. If not, then there must exist constants $\alpha_0, \alpha_1, \ldots, \alpha_r$, not all equal to zero, such that

$$\sum_{i=0}^{r} \alpha_i l_i(x_u) = 0, \qquad u = 1, 2, \ldots, n.$$

This indicates that the rth degree polynomial $\sum_{i=0}^{r} \alpha_i l_i(x)$ has n roots, namely, $x_1, x_2, \ldots, x_n$. This is not possible, because $n \geq r + 1$ and at least $r + 1$ of the x_u's $(u = 1, 2, \ldots, n)$ are distinct (a polynomial of degree r has at most r distinct roots). This contradiction implies that Λ is of full column rank and $\Lambda'\Lambda$ is therefore nonsingular.

Applying the Box–Lucas design criterion to the approximating model (9.37) amounts to finding the design settings that maximize $\det[\mathbf{H}'(\boldsymbol{\theta})\mathbf{H}(\boldsymbol{\theta})]$. From formula (9.40) we have

$$\det[\mathbf{H}'(\boldsymbol{\theta})\mathbf{H}(\boldsymbol{\theta})] = \det[\mathbf{U}(\boldsymbol{\theta})\,\Lambda'\Lambda\mathbf{U}'(\boldsymbol{\theta})]. \tag{9.41}$$

We note that the matrix $\Lambda'\Lambda = \sum_{u=1}^{n} \boldsymbol{\lambda}(x_u)\boldsymbol{\lambda}'(x_u)$ depends only on the design settings. Let $\nu_{\min}(x_1, x_2, \ldots, x_n)$ and $\nu_{\max}(x_1, x_2, \ldots, x_n)$ denote, respectively, the smallest and the largest eigenvalue of $\Lambda'\Lambda$. These eigenvalues are positive, since $\Lambda'\Lambda$ is positive definite by the fact that $\Lambda'\Lambda$ is nonsingular, as was shown earlier. From formula (9.41) we conclude that

$$\det[\mathbf{U}(\boldsymbol{\theta})\mathbf{U}'(\boldsymbol{\theta})]\,\nu_{\min}^{p}(x_1, x_2, \ldots, x_n) \leq \det[\mathbf{H}'(\boldsymbol{\theta})\mathbf{H}(\boldsymbol{\theta})]$$
$$\leq \det[\mathbf{U}(\boldsymbol{\theta})\mathbf{U}'(\boldsymbol{\theta})]\,\nu_{\max}^{p}(x_1, x_2, \ldots, x_n). \tag{9.42}$$

This double inequality follows from the fact that the matrices $\nu_{\max}(x_1, x_2, \ldots, x_n)\mathbf{U}(\boldsymbol{\theta})\mathbf{U}'(\boldsymbol{\theta}) - \mathbf{H}'(\boldsymbol{\theta})\mathbf{H}(\boldsymbol{\theta})$ and $\mathbf{H}'(\boldsymbol{\theta})\mathbf{H}(\boldsymbol{\theta}) - \nu_{\min}(x_1, x_2, \ldots, x_n)\mathbf{U}(\boldsymbol{\theta})\mathbf{U}'(\boldsymbol{\theta})$ are positive semidefinite. An application of Theorem 2.3.19(1) to these matrices results in the double inequality (9.42) (why?). Note that the determinant of $\mathbf{U}(\boldsymbol{\theta})\mathbf{U}'(\boldsymbol{\theta})$ is not zero, since $\mathbf{U}(\boldsymbol{\theta})\mathbf{U}'(\boldsymbol{\theta})$, which is of order $p \times p$, is of rank p by assumption (c).

Now, from the double inequality (9.42) we deduce that there exists a number γ, $0 \leq \gamma \leq 1$, such that

$$\det[\mathbf{H}'(\boldsymbol{\theta})\mathbf{H}(\boldsymbol{\theta})] = \left[\gamma\nu_{\min}^{p}(x_1, x_2, \ldots, x_n) + (1 - \gamma)\nu_{\max}^{p}(x_1, x_2, \ldots, x_n)\right]$$
$$\times \det[\mathbf{U}(\boldsymbol{\theta})\mathbf{U}'(\boldsymbol{\theta})].$$

If γ is integrated out, we obtain

$$\int_0^1 \det[\mathbf{H}'(\boldsymbol{\theta})\mathbf{H}(\boldsymbol{\theta})]\, d\gamma = \tfrac{1}{2}\left[\nu_{\min}^{p}(x_1, x_2, \ldots, x_n) + \nu_{\max}^{p}(x_1, x_2, \ldots, x_n)\right]$$
$$\times \det[\mathbf{U}(\boldsymbol{\theta})\mathbf{U}'(\boldsymbol{\theta})].$$

Consequently, to construct an optimal design we can consider finding $x_1, x_2, \ldots, x_n$ that maximize the function

$$\psi(x_1, x_2, \ldots, x_n) = \tfrac{1}{2}\left[\nu_{\min}^{p}(x_1, x_2, \ldots, x_n) + \nu_{\max}^{p}(x_1, x_2, \ldots, x_n)\right]. \tag{9.43}$$

This is a modified version of the Box–Lucas criterion. Its advantage is that the optimal design is free of $\boldsymbol{\theta}$. We therefore call such a design a parameter-free design. The maximization of $\psi(x_1, x_2, \ldots, x_n)$ can be conveniently carried out by using a FORTRAN program written by Conlon (1991), which is based on Price's (1977) controlled random search procedure.

EXAMPLE 9.4.1.   Let us consider the nonlinear model used by Box and Lucas (1959) of a consecutive first-order chemical reaction in which a raw material A reacts to form a product B, which in turn decomposes to form substance C. After time x has elapsed, the mean yield of the intermediate product B is given by

$$ h(x, \boldsymbol{\theta}) = \frac{\theta_1}{\theta_1 - \theta_2} (e^{-\theta_2 x} - e^{-\theta_1 x}), $$

where θ_1 and θ_2 are the rate constants for the reactions $A \to B$ and $B \to C$, respectively.

Suppose that the region of interest R is the interval $[0, 10]$. Let the parameter space Ω be such that $0 \leq \theta_1 \leq 1, 0 \leq \theta_2 \leq 1$. It can be verified that

$$ \frac{\partial^{r+1} h(x, \boldsymbol{\theta})}{\partial x^{r+1}} = (-1)^{r+1} \frac{\theta_1}{\theta_1 - \theta_2} (\theta_2^{r+1} e^{-\theta_2 x} - \theta_1^{r+1} e^{-\theta_1 x}). $$

Let us consider the function $\omega(x, \phi) = \phi^{r+1} e^{-\phi x}$. By the mean value theorem (Theorem 4.2.2),

$$ \theta_2^{r+1} e^{-\theta_2 x} - \theta_1^{r+1} e^{-\theta_1 x} = (\theta_2 - \theta_1) \frac{\partial w(x, \theta_*)}{\partial \phi}, $$

where $\partial w(x, \theta_*) / \partial \phi$ is the partial derivative of $w(x, \phi)$ with respect to ϕ evaluated at θ_*, and where θ_* is between θ_1 and θ_2. Thus

$$ \theta_2^{r+1} e^{-\theta_2 x} - \theta_1^{r+1} e^{-\theta_1 x} = (\theta_2 - \theta_1)[(r+1)\theta_*^r e^{-\theta_* x} - \theta_*^{r+1} x e^{-\theta_* x}]. $$

Hence,

$$ \sup_{0 \leq x \leq 10} \left| \frac{\partial^{r+1} h(x, \boldsymbol{\theta})}{\partial x^{r+1}} \right| \leq \theta_1 \theta_*^r \sup_{0 \leq x \leq 10} [e^{-\theta_* x} |r + 1 - x\theta_*|] $$

$$ \leq \sup_{0 \leq x \leq 10} |r + 1 - x\theta_*|. $$

However,

$$|r + 1 - x\theta_*| = \begin{cases} r + 1 - x\theta_* & \text{if } r + 1 \geq x\theta_*, \\ -r - 1 + x\theta_* & \text{if } r + 1 < x\theta_*. \end{cases}$$

Since $0 \leq x\theta_* \leq 10$, then

$$\sup_{0 \leq x \leq 10} |r + 1 - x\theta_*| \leq \max(r + 1, 9 - r).$$

We then have

$$\sup_{0 \leq x \leq 10} \left| \frac{\partial^{r+1} h(x, \boldsymbol{\theta})}{\partial x^{r+1}} \right| \leq \max(r + 1, 9 - r).$$

By inequality (9.36), the integer r is determined such that

$$\frac{2}{(r + 1)!} \left(\frac{10}{4} \right)^{r+1} \max(r + 1, 9 - r) < \delta. \tag{9.44}$$

If we choose $\delta = 0.053$, for example, then it can be verified that the smallest positive integer that satisfies inequality (9.44) is $r = 9$. The Chebyshev points in formula (9.18) that correspond to this value of r are given in Table 9.2. On choosing n, the number of design points, to be equal to $r + 1 = 10$, where all ten design points are distinct, the matrix $\boldsymbol{\Lambda}$ in formula (9.40) will be nonsingular of order 10×10. Using Conlon's (1991) FORTRAN program for the maximization of the function ψ in formula (9.43) with $p = 2$, it can be shown that the maximum value of ψ is 17.457. The corresponding optimal values of $x_1, x_2, \ldots, x_{10}$ are given in Table 9.2.

Table 9.2. Chebyshev Points and Optimal Design Points for Example 9.4.1

Chebyshev Points	Optimal Design Points
9.938	9.989
9.455	9.984
8.536	9.983
7.270	9.966
5.782	9.542
4.218	7.044
2.730	6.078
1.464	4.038
0.545	1.381
0.062	0.692

9.4.2. Splines in Statistics

There is a broad variety of work on splines in statistics. Spline functions are quite suited in practical applications involving data that arise from the physical world rather than the mathematical world. It is therefore only natural that splines have many useful applications in statistics. Some of these applications will be discussed in this section.

9.4.2.1. The Use of Cubic Splines in Regression

Let us consider fitting the model

$$y = g(x) + \epsilon, \tag{9.45}$$

where $g(x)$ is the mean response at x and ϵ is a random error. Suppose that the domain of x is divided into a set of $m + 1$ intervals by the points $\tau_0 < \tau_1 < \cdots < \tau_m < \tau_{m+1}$ such that on the ith interval $(i = 1, 2, \ldots, m + 1)$, $g(x)$ is represented by the cubic spline

$$s_i(x) = a_i + b_i x + c_i x^2 + d_i x^3, \qquad \tau_{i-1} \leq x \leq \tau_i. \tag{9.46}$$

As was seen earlier in Section 9.3.1, the parameters a_i, b_i, c_i, d_i $(i = 1, 2, \ldots, m + 1)$ are subject to the following continuity restrictions:

$$a_i + b_i \tau_i + c_i \tau_i^2 + d_i \tau_i^3 = a_{i+1} + b_{i+1} \tau_i + c_{i+1} \tau_i^2 + d_{i+1} \tau_i^3, \tag{9.47}$$

that is, $s_i(\tau_i) = s_{i+1}(\tau_i)$, $i = 1, 2, \ldots, m$;

$$b_i + 2 c_i \tau_i + 3 d_i \tau_i^2 = b_{i+1} + 2 c_{i+1} \tau_i + 3 d_{i+1} \tau_i^2, \tag{9.48}$$

that is, $s_i'(\tau_i) = s_{i+1}'(\tau_i)$, $i = 1, 2, \ldots, m$; and

$$2 c_i + 6 d_i \tau_i = 2 c_{i+1} + 6 d_{i+1} \tau_i, \tag{9.49}$$

that is, $s_i''(\tau_i) = s_{i+1}''(\tau_i)$, $i = 1, 2, \ldots, m$. The number of unknown parameters in model (9.45) is therefore equal to $4(m + 1)$. The continuity restrictions (9.47)–(9.49) reduce the dimensionality of the parameter space to $m + 4$. However, only $m + 2$ parameters can be estimated. This is because the spline method does not estimate the parameters of the s_i's directly, but estimates the ordinates of the s_i's at the points $\tau_0, \tau_1, \ldots, \tau_{m+1}$, that is, $s_1(\tau_0)$ and $s_i(\tau_i)$, $i = 1, 2, \ldots, m + 1$. Two additional restrictions are therefore needed. These are chosen to be of the form (see Poirier, 1973, page 516; Buse and Lim, 1977, page 64):

$$s_1''(\tau_0) = \pi_0 s_1''(\tau_1),$$

or

$$2 c_1 + 6 d_1 \tau_0 = \pi_0 (2 c_1 + 6 d_1 \tau_1), \tag{9.50}$$

and

$$s''_{m+1}(\tau_{m+1}) = \pi_{m+1} s''_{m+1}(\tau_m),$$

or

$$2c_{m+1} + 6d_{m+1}\tau_{m+1} = \pi_{m+1}(2c_{m+1} + 6d_{m+1}\tau_m), \tag{9.51}$$

where π_0 and π_{m+1} are known.

Let $y_1, y_2, \ldots, y_n$ be n observations on the response y, where $n > m + 2$, such that n_i observations are taken in the ith interval $[\tau_{i-1}, \tau_i]$, $i = 1, 2, \ldots, m + 1$. Thus $n = \sum_{i=1}^{m+1} n_i$. If $y_{i1}, y_{i2}, \ldots, y_{in_i}$ are the observations in the ith interval ($i = 1, 2, \ldots, m + 1$), then from model (9.45) we have

$$y_{ij} = g(x_{ij}) + \epsilon_{ij}, \qquad i = 1, 2, \ldots, m + 1; \ j = 1, 2, \ldots, n_i, \tag{9.52}$$

where x_{ij} is the setting of x for which $y = y_{ij}$, and the ϵ_{ij}'s are distributed independently with means equal to zero and a common variance σ^2. The estimation of the parameters of model (9.45) is then reduced to a restricted least-squares problem with formulas (9.47)–(9.51) representing linear restrictions on the $4(m + 1)$ parameters of the model [see, for example, Searle (1971, Section 3.6), for a discussion concerning least-squares estimation under linear restrictions on the fitted model's parameters]. Using matrix notation, model (9.52) and the linear restrictions (9.47)–(9.51) can be expressed as

$$y = X\beta + \epsilon, \tag{9.53}$$

$$C\beta = \delta, \tag{9.54}$$

where $y = (y_1': y_2': \cdots : y_{m+1}')'$ with $y_i = (y_{i1}, y_{i2}, \ldots, y_{in_i})'$, $i = 1, 2, \ldots, m + 1$, $X = \text{Diga}(X_1, X_2, \ldots, X_{m+1})$ is a block-diagonal matrix of order $n \times [4(m + 1)]$ with X_i being a matrix of order $n_i \times 4$ whose jth row is of the form $(1, x_{ij}, x_{ij}^2, x_{ij}^3)$, $j = 1, 2, \ldots, n_i$; $i = 1, 2, \ldots, m + 1$; $\beta = (\beta_1': \beta_2': \cdots : \beta_{m+1}')'$ with $\beta_i = (a_i, b_i, c_i, d_i)'$, $i = 1, 2, \ldots, m + 1$; and $\epsilon = (\epsilon_1': \epsilon_2': \cdots : \epsilon_{m+1}')'$, where ϵ_j is the vector of random errors associated with the observations in the ith interval, $i = 1, 2, \ldots, m + 1$. Furthermore, $C = [C_0': C_1': C_2': C_3']'$, where, for $l = 0, 1, 2$,

$$C_l = \begin{bmatrix} -e_{1l}' & e_{1l}' & 0' & \cdots & 0' & 0' \\ 0' & -e_{2l}' & e_{2l}' & \cdots & 0' & 0' \\ \vdots & \vdots & \vdots & & \vdots & \vdots \\ 0' & 0' & 0' & \cdots & -e_{ml}' & e_{ml}' \end{bmatrix}$$

is a matrix of order $m \times [4(m+1)]$ such that $\mathbf{e}'_{i0} = (1, \tau_i, \tau_i^2, \tau_i^3)$, $\mathbf{e}'_{i1} = (0, 1, 2\tau_i, 3\tau_i^2)$, $\mathbf{e}'_{i2} = (0, 0, 2, 6\tau_i)$, $i = 1, 2, \ldots, m$, and

$$
\mathbf{C}_3 = \begin{bmatrix} 0 & 0 & 2(\pi_0 - 1) & 6(\pi_0\tau_1 - \tau_0) & \cdots & 0 & 0 & 0 & 0 \\ 0 & 0 & 0 & 0 & \cdots & 0 & 0 & 2(\pi_{m+1} - 1) & 6(\pi_{m+1}\tau_m - \tau_{m+1}) \end{bmatrix}
$$

is a $2 \times [4(m+1)]$ matrix. Finally, $\boldsymbol{\delta} = (\boldsymbol{\delta}'_0 : \boldsymbol{\delta}'_1 : \boldsymbol{\delta}'_2 : \boldsymbol{\delta}'_3)' = \mathbf{0}$, where the partitioning of $\boldsymbol{\delta}$ into $\boldsymbol{\delta}_0$, $\boldsymbol{\delta}_1$, $\boldsymbol{\delta}_2$, and $\boldsymbol{\delta}_3$ conforms to that of $\mathbf{C}$. Consequently, and on the basis of formula (103) in Searle (1971, page 113), the least-squares estimator of $\boldsymbol{\beta}$ for model (9.53) under the restriction described by formula (9.54) is given by

$$
\begin{aligned}
\hat{\boldsymbol{\beta}}_r &= \hat{\boldsymbol{\beta}} - (\mathbf{X'X})^{-1}\mathbf{C'}\left[\mathbf{C}(\mathbf{X'X})^{-1}\mathbf{C'}\right]^{-1}(\mathbf{C}\hat{\boldsymbol{\beta}} - \boldsymbol{\delta}) \\
&= \hat{\boldsymbol{\beta}} - (\mathbf{X'X})^{-1}\mathbf{C'}\left[\mathbf{C}(\mathbf{X'X})^{-1}\mathbf{C'}\right]^{-1}\mathbf{C}\hat{\boldsymbol{\beta}},
\end{aligned}
$$

where $\hat{\boldsymbol{\beta}} = (\mathbf{X'X})^{-1}\mathbf{X'y}$ is the ordinary least-squares estimator of $\boldsymbol{\beta}$.

This estimation procedure, which was developed by Buse and Lim (1977), demonstrates that the fitting of a cubic spline regression model can be reduced to a restricted least-squares problem. Buse and Lim presented a numerical example based on Indianapolis 500 race data over the period (1911–1971) to illustrate the implementation of their procedure.

Other papers of interest in the area of regression splines include those of Poirier (1973) and Gallant and Fuller (1973). The paper by Poirier discusses the basic theory of cubic regression splines from an economic point of view. In the paper by Gallant and Fuller, the knots are treated as unknown parameters rather than being fixed. Thus in their procedure, the knots must be estimated, which causes the estimation process to become nonlinear.

9.4.2.2. Designs for Fitting Spline Models

A number of papers have addressed the problem of finding a design to estimate the parameters of model (9.45), where $g(x)$ is represented by a spline function. We shall make a brief reference to some of these papers.

Agarwal and Studden (1978) considered a representation of $g(x)$ over $0 \leq x \leq 1$ by a linear spline $s(x)$, which has the form given by (9.21). Here, $g''(x)$ is assumed to be continuous. If we recall, the θ_i coefficients in formula (9.21) are the values of s at $\tau_0, \tau_1, \ldots, \tau_{m+1}$.

Let $x_1, x_2, \ldots, x_r$ be r design points in $[0, 1]$. Let $\bar{y}_i$ denote the average of n_i observations taken at x_i ($i = 1, 2, \ldots, r$). The vector $\boldsymbol{\theta} = (\theta_0, \theta_1, \ldots, \theta_{m+1})'$ can therefore be estimated by

$$
\hat{\boldsymbol{\theta}} = \mathbf{A}\bar{\mathbf{y}}, \tag{9.55}
$$

where $\bar{\mathbf{y}} = (\bar{y}_1, \bar{y}_2, \ldots, \bar{y}_r)'$ and $\mathbf{A}$ is an $(m+2) \times r$ matrix. Hence, an estimate of $g(x)$ is given by

$$
\hat{g}(x) = \mathbf{l'}(x)\hat{\boldsymbol{\theta}} = \mathbf{l'}(x)\mathbf{A}\bar{\mathbf{y}}, \tag{9.56}
$$

where $\mathbf{l}(x) = [l_0(x), l_1(x), \ldots, l_{m+1}(x)]'$.

Now, $E(\hat{\boldsymbol{\theta}}) = \mathbf{Ag}_r$, where $\mathbf{g}_r = [g(x_1), g(x_2), \ldots, g(x_r)]'$. Thus $E[\hat{g}(x)] = \mathbf{l}'(x)\mathbf{Ag}_r$, and the variance of $\hat{g}(x)$ is

$$\mathrm{Var}[\hat{g}(x)] = E\left[\mathbf{l}'(x)\hat{\boldsymbol{\theta}} - \mathbf{l}'(x)\mathbf{Ag}_r\right]^2$$

$$= \frac{\sigma^2}{n}\mathbf{l}'(x)\mathbf{AD}^{-1}\mathbf{A}'\mathbf{l}(x),$$

where $\mathbf{D}$ is an $r \times r$ diagonal matrix with diagonal elements $n_1/n, n_2/n, \ldots, n_r/n$. The mean squared error of $\hat{g}(x)$ is the variance plus the squared bias of $\hat{g}(x)$. It follows that the integrated mean squared error (IMSE) of $\hat{g}(x)$ (see Section 8.4.3) is

$$J = \frac{n\omega}{\sigma^2}\int_0^1 \mathrm{Var}[\hat{g}(x)]\,dx + \frac{n\omega}{\sigma^2}\int_0^1 \mathrm{Bias}^2[\hat{g}(x)]\,dx$$

$$= V + B,$$

where $\omega = (\int_0^1 dx)^{-1} = \frac{1}{2}$, and

$$\mathrm{Bias}^2[\hat{g}(x)] = [g(x) - \mathbf{l}'(x)\mathbf{Ag}_r]^2.$$

Thus

$$J = \frac{1}{2}\int_0^1 \mathbf{l}'(x)\mathbf{AD}^{-1}\mathbf{A}'\mathbf{l}(x)\,dx + \frac{n}{2\sigma^2}\int_0^1 [g(x) - \mathbf{l}'(x)\mathbf{Ag}_r]^2\,dx$$

$$= \frac{1}{2}\mathrm{tr}(\mathbf{AD}^{-1}\mathbf{A}'\mathbf{M}) + \frac{n}{2\sigma^2}\int_0^1 [g(x) - \mathbf{l}'(x)\mathbf{Ag}_r]^2\,dx,$$

where $\mathbf{M} = \int_0^1 \mathbf{l}(x)\mathbf{l}'(x)\,dx$.

Agarwal and Studden (1978) proposed to minimize J with respect to (i) the design (that is, $x_1, x_2, \ldots, x_r$ as well as $n_1, n_2, \ldots, n_r$), (ii) the matrix $\mathbf{A}$, and (iii) the knots $\tau_1, \tau_2, \ldots, \tau_m$, assuming that g is known.

Park (1978) adopted the D-optimality criterion (see Section 8.5) for the choice of design when $g(x)$ is represented by a spline of the form given by formula (9.22) with only one intermediate knot.

Draper, Guttman, and Lipow (1977) extended the design criterion based on the minimization of the average squared bias B (see Section 8.4.3) to situations involving spline models. In particular, they considered fitting first-order or second-order models when the true mean response is of the second order or the third order, respectively.

9.4.2.3. Other Applications of Splines in Statistics

Spline functions have many other useful applications in both theoretical and applied statistical research. For example, splines are used in nonparametric regression and data smoothing, nonparametric density estimation, and time series analysis. They are also utilized in the analysis of response curves in agriculture and economics. The review articles by Wegman and Wright (1983) and Ramsay (1988) contain many references on the various uses of splines in statistics (see also the article by Smith, 1979). An overview of the role of splines in regression analysis is given in Eubank (1984).

FURTHER READING AND ANNOTATED BIBLIOGRAPHY

Agarwal, G. G., and W. J. Studden (1978), "Asymptotic design and estimation using linear splines." *Comm. Statist. Simulation Comput.*, **7**, 309–319.

Box, G. E. P., and H. L. Lucas (1959). "Design of experiments in nonlinear situations." *Biometrika*, **46**, 77–90.

Buse, A., and L. Lim (1977). "Cubic splines as a special case of restricted least squares." *J. Amer. Statist. Assoc.*, **72**, 64–68.

Cheney, E. W. (1982). *Introduction to Approximation Theory*, 2nd ed. Chelsea, New York. (The Weierstrass approximation theorem and Lagrange interpolation are covered in Chap. 3; least-squares approximation is discussed in Chap. 4.)

Conlon, M. (1991). "The controlled random search procedure for function optimization." Personal communication. (This is a FORTRAN file for implementing Price's controlled random search procedure.)

Cornish, E. A., and R. A. Fisher (1937). "Moments and cumulants in the specification of distribution." *Rev. Internat. Statist. Inst.*, **5**, 307–320.

Cramér, H. (1946). *Mathematical Methods of Statistics*. Princeton University Press, Princeton. (This classic book provides the mathematical foundation of statistics. Chap. 17 is a good source for approximation of density functions.)

Davis, P. J. (1975). *Interpolation and Approximation*. Dover, New York. (Chaps. 2, 3, 6, 8, and 10 are relevant to the material on Lagrange interpolation, least-squares approximation, and orthogonal polynomials.)

De Boor, C. (1978). *A Practical Guide to Splines*. Springer-Verlag, New York. (Chaps. 1 and 2 provide a good coverage of Lagrange interpolation, particularly with regard to the use of Chebyshev points. Chap. 4 discusses cubic spline approximation.)

Draper, N. R., I. Guttman, and P. Lipow (1977). "All-bias designs for spline functions joined at the axes." *J. Amer. Statist. Assoc.*, **72**, 424–429.

Eubank, R. L. (1984). "Approximate regression models and splines." *Comm. Statist. Theory Methods*, **13**, 433–484.

Gallant, A. R., and W. A. Fuller (1973). "Fitting segmented polynomial regression models whose join points have to be estimated." *J. Amer. Statist. Assoc.*, **68**, 144–147.

Graybill, F. A. (1983). *Matrices with Applications in Statistics*, 2nd ed. Wadsworth, Belmont, California.

Hall, C. A. (1968). "On error bounds for spline interpolation." *J. Approx. Theory*, **1**, 209–218.

Hall, C. A., and W. W. Meyer (1976). "Optimal error bounds for cubic spline interpolation." *J. Approx. Theory*, **16**, 105–122.

Harris, B. (1966). *Theory of Probability*. Addison-Wesley, Reading, Massachusetts.

Johnson, N. L., and S. Kotz (1970). *Continuous Univariate Distributions—1*. Houghton Mifflin, Boston. (Chap. 12 contains a good discussion concerning the Cornish–Fisher expansion of percentage points.)

Kendall, M. G., and A. Stuart (1977). *The Advanced Theory of Statistics*, Vol. 1, 4th ed. Macmillan, New York. (This classic book provides a good source for learning about the Gram–Charlier series of type A and the Cornish–Fisher expansion.)

Lancaster, P., and K. Salkauskas (1986). *Curve and Surface Fitting*. Academic Press, London. (This book covers the foundations and major features of several basic methods for curve and surface fitting that are currently in use.)

Park, S. H. (1978). "Experimental designs for fitting segmented polynomial regression models." *Technometrics*, **20**, 151–154.

Poirier, D. J. (1973). "Piecewise regression using cubic splines." *J. Amer. Statist. Assoc.*, **68**, 515–524.

Powell, M. J. D. (1967). "On the maximum errors of polynomial approximation defined by interpolation and by least squares criteria." *Comput. J.*, **9**, 404–407.

Prenter, P. M. (1975). *Splines and Variational Methods*. Wiley, New York. (Lagrange interpolation is covered in Chap. 2; cubic splines are discussed in Chap. 4. An interesting feature of this book is its coverage of polynomial approximation of a function of several variables.)

Price, W. L. (1977). "A controlled random search procedure for global optimization." *Comput. J.*, **20**, 367–370.

Ramsay, J. O. (1988). "Monotone regression splines in action." *Statist. Sci.*, **3**, 425–461.

Rice, J. R. (1969). *The Approximation of Functions*, Vol. 2. Addison-Wesley, Reading, Massachusetts. (Approximation by spline functions is presented in Chap. 10.)

Rivlin, T. J. (1969). *An Introduction to the Approximation of Functions*. Dover, New York. (This book provides an introduction to some of the most significant methods of approximation of functions by polynomials. Spline approximation is also discussed.)

Schoenberg, I. J. (1946). "Contributions to the problem of approximation of equidistant data by analytic functions." *Quart. Appl. Math.*, **4**, Part A, 45–99; Part B, 112–141.

Searle, S. R. (1971). *Linear Models*. Wiley, New York.

Smith, P. L. (1979). "Splines as a useful and convenient statistical tool." *Amer. Statist.*, **33**, 57–62.

Szidarovszky, F., and S. Yakowitz (1978). *Principles and Procedures of Numerical Analysis*. Plenum Press, New York. (Chap. 2 provides a brief introduction to approximation and interpolation of functions.)

Wegman, E. J., and I. W. Wright (1983). "Splines in statistics." *J. Amer. Statist. Assoc.*, **78**, 351–365.

Wold, S. (1974). "Spline functions in data analysis." *Technometrics*, **16**, 1–11.

EXERCISES

In Mathematics

9.1. Let $f(x)$ be a function with a continuous derivative on $[0, 1]$, and let $b_n(x)$ be the nth degree Bernstein approximating polynomial of f. Then, for some constant c and for all n,

$$\sup_{0 \le x \le 1} |f(x) - b_n(x)| \le \frac{c}{n^{1/2}}.$$

9.2. Prove Lemma 9.1.1.

9.3. Prove Lemma 9.1.2.

9.4. Show that for every interval $[-a, a]$ there is a sequence of polynomials $p_n(x)$ such that $p_n(0) = 0$ and $\lim_{n \to \infty} p_n(x) = |x|$ uniformly on $[-a, a]$.

9.5. Suppose that $f(x)$ is continuous on $[0, 1]$ and that

$$\int_0^1 f(x) x^n \, dx = 0, \qquad n = 0, 1, 2, \dots .$$

Show that $f(x) = 0$ on $[0, 1]$.
[*Hint:* $\int_0^1 f(x) p_n(x) \, dx = 0$, where $p_n(x)$ is any polynomial of degree n.]

9.6. Suppose that the function $f(x)$ has $n + 1$ continuous derivatives on $[a, b]$. Let $a = a_0 < a_1 < \cdots < a_n = b$ be $n + 1$ points in $[a, b]$. Then

$$\sup_{a \le x \le b} |f(x) - p(x)| \le \frac{T_{n+1} h^{n+1}}{4(n+1)},$$

where $p(x)$ is the Lagrange polynomial defined by formula (9.14), $T_{n+1} = \sup_{a \le x \le b} |f^{(n+1)}(x)|$, and $h = \max(a_{i+1} - a_i)$, $i = 0, 1, \dots, n - 1$.
[*Hint:* Show that $|\prod_{i=0}^{n}(x - a_i)| \le n!(h^{n+1}/4)$.]

9.7. Apply Lagrange interpolation to approximate the function $f(x) = \log x$ over the interval $[3.50, 3.80]$ using $a_0 = 3.50$, $a_1 = 3.60$, $a_2 = 3.70$, and $a_3 = 3.80$ as interpolation points. Compute an upper bound on the error of approximation.

9.8. Let $a = \tau_0 < \tau_1 < \cdots < \tau_n = b$ be a partition of $[a, b]$. Suppose that $f(x)$ has continuous derivatives up to order 2 over $[a, b]$. Consider a cubic spline $s(x)$ that satisfies

$$s(\tau_i) = f(\tau_i), \qquad i = 0, 1, \ldots, n,$$
$$s'(a) = f'(a),$$
$$s'(b) = f'(b).$$

Show that

$$\int_a^b [f''(x)]^2 \, dx \geq \int_a^b [s''(x)]^2 \, dx.$$

9.9. Determine the cubic spline approximation of the function $f(x) = \cos(2\pi x)$ over the interval $[0, \pi]$ using five evenly spaced knots. Give an upper bound on the error approximation.

In Statistics

9.10. Consider the nonlinear model

$$y(x) = h(x, \boldsymbol{\theta}) + \epsilon,$$

where

$$h(x, \boldsymbol{\theta}) = \theta_1 \left(1 - \theta_2 \, e^{-\theta_3 x}\right),$$

such that $0 \leq \theta_1 \leq 50$, $0 \leq \theta_2 \leq 1$, $0 \leq \theta_3 \leq 1$. Obtain a Lagrange interpolating polynomial that approximates the mean response function $h(x, \boldsymbol{\theta})$ over the region $[0, 8]$ with an error not exceeding $\delta = 0.05$.

9.11. Consider the nonlinear model

$$y = \alpha + (0.49 - \alpha)\exp[-\beta(x - 8)] + \epsilon,$$

where ϵ is a random error with a zero mean and a variance σ^2. Suppose that the region of interest is the interval $[10, 40]$, and that the parameter space Ω is such that $0.36 \leq \alpha \leq 0.41$, $0.06 \leq \beta \leq 0.16$. Let $s(x)$ be the cubic spline that approximates the mean response, that is, $\eta(x, \alpha, \beta) = \alpha + (0.49 - \alpha)\exp[-\beta(x - 8)]$, over $[10, 40]$. Determine the number of knots needed so that

$$\max_{10 \leq x \leq 40} |\eta(x, \alpha, \beta) - s(x)| < 0.001$$

for all $(\alpha, \beta) \in \Omega$.

9.12. Consider fitting the spline model

$$y = \beta_0 + \beta_1 x + \beta_2 (x - \alpha)_+^2 + \epsilon$$

over the interval $[-1, 1]$, where α is a known constant, $-1 \le \alpha \le 1$. A three-point design consisting of x_1, x_2, x_3 with $-1 \le x_1 < \alpha \le x_2 < x_3 \le 1$ is used to fit the model. Using matrix notation, the model is written as

$$y = X\beta + \epsilon.$$

where X is the matrix

$$X = \begin{bmatrix} 1 & x_1 & 0 \\ 1 & x_2 & (x_2 - \alpha)^2 \\ 1 & x_3 & (x_3 - \alpha)^2 \end{bmatrix},$$

and $\beta = (\beta_0, \beta_1, \beta_2)'$. Determine x_1, x_2, x_3 so that the design is D-optimal, that is, it maximizes the determinant of $X'X$.
[*Note:* See Park (1978).]

CHAPTER 10

Orthogonal Polynomials

The subject of orthogonal polynomials can be traced back to the work of the French mathematician Adrien-Marie Legendre (1752–1833) on planetary motion. These polynomials have important applications in physics, quantum mechanics, mathematical statistics, and other areas in mathematics.

This chapter provides an exposition of the properties of orthogonal polynomials. Emphasis will be placed on Legendre, Chebyshev, Jacobi, Laguerre, and Hermite polynomials. In addition, applications of these polynomials in statistics will be discussed in Section 10.9.

10.1. INTRODUCTION

Suppose that $f(x)$ and $g(x)$ are two continuous functions on $[a, b]$. Let $w(x)$ be a positive function that is Riemann integrable on $[a, b]$. The dot product of $f(x)$ and $g(x)$ with respect to $w(x)$, which is denoted by $(f \cdot g)_\omega$, is defined as

$$(f \cdot g)_\omega = \int_a^b f(x) g(x) w(x) \, dx.$$

The norm of $f(x)$ with respect to $w(x)$, denoted by $\|f\|_\omega$, is defined as $\|f\|_\omega = [\int_a^b f^2(x) w(x) \, dx]^{1/2}$. The functions $f(x)$ and $g(x)$ are said to be orthogonal [with respect to $w(x)$] if $(f \cdot g)_\omega = 0$. Furthermore, a sequence $\{f_n(x)\}_{n=0}^\infty$ of continuous functions defined on $[a, b]$ are said to be orthogonal with respect to $w(x)$ if $(f_m \cdot f_n)_\omega = 0$ for $m \neq n$. If, in addition, $\|f_n\|_\omega = 1$ for all n, then the functions $f_n(x)$, $n = 0, 1, 2, \ldots$, are called orthonormal. In particular, if $S = \{p_n(x)\}_{n=0}^\infty$ is a sequence of polynomials such that $(p_n \cdot p_m)_\omega = 0$ for all $m \neq n$, then S forms a sequence of polynomials orthogonal with respect to $w(x)$.

A sequence of orthogonal polynomials can be constructed on the basis of the following theorem:

Theorem 10.1.1. The polynomials $\{p_n(x)\}_{n=0}^{\infty}$ which are defined according to the following recurrence relation are orthogonal:

$$p_0(x) = 1,$$

$$p_1(x) = x - \frac{(xp_0 \cdot p_0)_{\omega}}{\|p_0\|_{\omega}^2} = x - \frac{(x \cdot 1)_{\omega}}{\|1\|_{\omega}^2}, \tag{10.1}$$

$$\vdots$$

$$p_n(x) = (x - a_n)p_{n-1}(x) - b_n p_{n-2}(x), \qquad n = 2, 3, \ldots,$$

where

$$a_n = \frac{(xp_{n-1} \cdot p_{n-1})_{\omega}}{\|p_{n-1}\|_{\omega}^2} \tag{10.2}$$

$$b_n = \frac{(xp_{n-1} \cdot p_{n-2})_{\omega}}{\|p_{n-2}\|_{\omega}^2} \tag{10.3}$$

Proof. We show by mathematical induction on n that $(p_n \cdot p_i)_{\omega} = 0$ for $i < n$. For $n = 1$,

$$(p_1 \cdot p_0)_{\omega} = \int_a^b \left[x - \frac{(x \cdot 1)_{\omega}}{\|1\|_{\omega}^2} \right] w(x)\, dx$$

$$= (x \cdot 1)_{\omega} - (x \cdot 1)_{\omega} \frac{\|1\|_{\omega}^2}{\|1\|_{\omega}^2}$$

$$= 0.$$

Now, suppose that the assertion is true for $n - 1$ ($n \geq 2$). To show that it is true for n. We have that

$$(p_n \cdot p_i)_{\omega} = \int_a^b [(x - a_n)p_{n-1}(x) - b_n p_{n-2}(x)] p_i(x) w(x)\, dx$$

$$= (xp_{n-1} \cdot p_i)_{\omega} - a_n(p_{n-1} \cdot p_i)_{\omega} - b_n(p_{n-2} \cdot p_i)_{\omega}.$$

Thus, for $i = n - 1$,

$$(p_n \cdot p_i)_{\omega} = (xp_{n-1} \cdot p_{n-1})_{\omega} - a_n \|p_{n-1}\|_{\omega}^2 - b_n(p_{n-2} \cdot p_{n-1})_{\omega}$$

$$= 0,$$

by the definition of a_n in (10.2) and the fact that $(p_{n-2} \cdot p_{n-1})_\omega = (p_{n-1} \cdot p_{n-2})_\omega = 0$. Similarly, for $i = n - 2$,

$$(p_n \cdot p_i)_\omega = (xp_{n-1} \cdot p_{n-2})_\omega - a_n(p_{n-1} \cdot p_{n-2})_\omega - b_n(p_{n-2} \cdot p_{n-2})_\omega$$

$$= (xp_{n-1} \cdot p_{n-2})_\omega - b_n \| p_{n-2} \|_\omega^2$$

$$= 0, \qquad \text{by (10.3)}.$$

Finally, for $i < n - 2$, we have

$$(p_n \cdot p_i)_\omega = (xp_{n-1} \cdot p_i)_\omega - a_n(p_{n-1} \cdot p_i)_\omega - b_n(p_{n-2} \cdot p_i)_\omega$$

$$= \int_a^b xp_{n-1}(x)p_i(x)w(x)\,dx. \tag{10.4}$$

But, from the recurrence relation,

$$p_{i+1}(x) = (x - a_{i+1})p_i(x) - b_{i+1}p_{i-1}(x),$$

that is,

$$xp_i(x) = p_{i+1}(x) + a_{i+1}p_i(x) + b_{i+1}p_{i-1}(x).$$

It follows that

$$\int_a^b xp_{n-1}(x)p_i(x)w(x)\,dx$$

$$= \int_a^b p_{n-1}(x)[p_{i+1}(x) + a_{i+1}p_i(x) + b_{i+1}p_{i-1}(x)]w(x)\,dx$$

$$= (p_{n-1} \cdot p_{i+1})_\omega + a_{i+1}(p_{n-1} \cdot p_i)_\omega + b_{i+1}(p_{n-1} \cdot p_{i-1})_\omega$$

$$= 0.$$

Hence, by (10.4), $(p_n \cdot p_i)_\omega = 0$. $\qquad\square$

It is easy to see from the recurrence relation (10.1) that $p_n(x)$ is of degree n, and the coefficient of x^n is equal to one. Furthermore, we have the following corollaries:

Corollary 10.1.1. An arbitrary polynomial of degree $\le n$ is uniquely expressible as a linear combination of $p_0(x), p_1(x), \ldots, p_n(x)$.

Corollary 10.1.2. The coefficient of x^{n-1} in $p_n(x)$ is $-\sum_{i=1}^n a_i$ $(n \ge 1)$.

Proof. If d_n denotes the coefficient of x^{n-1} in $p_n(x)$ $(n \ge 2)$, then by comparing the coefficients of x^{n-1} on both sides of the recurrence relation

(10.1), we obtain

$$d_n = d_{n-1} - a_n, \qquad n = 2, 3, \ldots. \tag{10.5}$$

The result follows from (10.5) and by noting that $d_1 = -a_1$ □

Another property of orthogonal polynomials is given by the following theorem:

Theorem 10.1.2. If $\{p_n(x)\}_{n=0}^{\infty}$ is a sequence of orthogonal polynomials with respect to $w(x)$ on $[a, b]$, then the zeros of $p_n(x)$ $(n \geq 1)$ are all real, distinct, and located in the interior of $[a, b]$.

Proof. Since $(p_n \cdot p_0)_\omega = 0$ for $n \geq 1$, then $\int_a^b p_n(x)w(x)\,dx = 0$. This indicates that $p_n(x)$ must change sign at least once in (a, b) [recall that $w(x)$ is positive]. Suppose that $p_n(x)$ changes sign between a and b at just k points, denoted by $x_1, x_2, \ldots, x_k$. Let $g(x) = (x - x_1)(x - x_2) \cdots (x - x_k)$. Then, $p_n(x)g(x)$ is a polynomial with no zeros of odd multiplicity in (a, b). Hence, $\int_a^b p_n(x)g(x)w(x)\,dx \neq 0$, that is, $(p_n \cdot g)_\omega \neq 0$. If $k < n$, then we have a contradiction by the fact that p_n is orthogonal to $g(x)$ [$g(x)$, being a polynomial of degree k, can be expressed as a linear combination of $p_0(x), p_1(x), \ldots, p_k(x)$ by Corollary 10.1.1]. Consequently, $k = n$, and $p_n(x)$ has n distinct zeros in the interior of $[a, b]$.   □

Particular orthogonal polynomials can be derived depending on the choice of the interval $[a, b]$, and the weight function $w(x)$. For example, the well-known orthogonal polynomials listed below are obtained by the following selections of $[a, b]$ and $w(x)$:

Orthogonal Polynomial	a	b	$w(x)$
Legendre	-1	1	1
Jacobi	-1	1	$(1 - x)^\alpha (1 + x)^\beta$, $\alpha, \beta > -1$
Chebyshev of the first kind	-1	1	$(1 - x^2)^{-1/2}$
Chebyshev of the second kind	-1	1	$(1 - x^2)^{1/2}$
Hermite	$-\infty$	∞	$e^{-x^2/2}$
Laguerre	0	∞	$e^{-x} x^\alpha$, $\alpha > -1$

These polynomials are called classical orthogonal polynomials. We shall study their properties and methods of derivation.

10.2. LEGENDRE POLYNOMIALS

These polynomials are derived by applying the so-called *Rodrigues formula*

$$p_n(x) = \frac{1}{2^n n!} \frac{d^n (x^2 - 1)^n}{dx^n}, \qquad n = 0, 1, 2, \ldots.$$

Thus, for $n = 0, 1, 2, 3, 4$, for example, we have

$$p_0(x) = 1,$$
$$p_1(x) = x,$$
$$p_2(x) = \tfrac{3}{2}x^2 - \tfrac{1}{2},$$
$$p_3(x) = \tfrac{5}{2}x^3 - \tfrac{3}{2}x,$$
$$p_4(x) = \tfrac{35}{8}x^4 - \tfrac{30}{8}x^2 + \tfrac{3}{8}.$$

From the Rodrigues formula it follows that $p_n(x)$ is a polynomial of degree n and the coefficient of x^n is $\binom{2n}{n}/2^n$. We can multiply $p_n(x)$ by $2^n/\binom{2n}{n}$ to make the coefficient of x^n equal to one ($n = 1, 2, \dots$).

Another definition of Legendre polynomials is obtained by means of the generating function,

$$g(x, r) = \frac{1}{(1 - 2rx + r^2)^{1/2}},$$

by expanding it as a power series in r for sufficiently small values of r. The coefficient of r^n in this expansion is $p_n(x)$, $n = 0, 1, \dots$, that is,

$$g(x, r) = \sum_{n=0}^{\infty} p_n(x) r^n.$$

To demonstrate this, let us consider expanding $(1 - z)^{-1/2}$ in a neighborhood of zero, where $z = 2xr - r^2$:

$$\frac{1}{(1 - z)^{1/2}} = 1 + \tfrac{1}{2}z + \tfrac{3}{8}z^2 + \tfrac{5}{16}z^3 + \cdots, \qquad |z| < 1,$$

$$= 1 + \tfrac{1}{2}(2xr - r^2) + \tfrac{3}{8}(2xr - r^2)^2 + \tfrac{5}{16}(2xr - r^2)^3 + \cdots$$

$$= 1 + xr + \left(\tfrac{3}{2}x^2 - \tfrac{1}{2}\right)r^2 + \left(\tfrac{5}{2}x^3 - \tfrac{3}{2}x\right)r^3 + \cdots .$$

We note that the coefficients of $1, r, r^2$, and r^3 are the same as $p_0(x)$, $p_1(x)$, $p_2(x)$, $p_3(x)$, as was seen earlier. In general, it is easy to see that the coefficient of r^n is $p_n(x)$ ($n = 0, 1, 2, \dots$).

By differentiating $g(x, r)$ with respect to r, it can be seen that

$$(1 - 2rx + r^2)\frac{\partial g(x, r)}{\partial r} - (x - r)g(x, r) = 0.$$

By substituting $g(x, r) = \sum_{n=0}^{\infty} p_n(x) r^n$ in this equation, we obtain

$$(1 - 2rx + r^2) \sum_{n=1}^{\infty} np_n(x) r^{n-1} - (x - r) \sum_{n=0}^{\infty} p_n(x) r^n = 0.$$

The coefficient of r^n must be zero for each n and for all values of x ($n = 1, 2, \ldots$). We thus have the following identity:

$$(n + 1) p_{n+1}(x) - (2n + 1) xp_n(x) + np_{n-1}(x) = 0, \qquad n = 1, 2, \ldots \quad (10.6)$$

This is a recurrence relation that connects any three successive Legendre polynomials. For example, for $p_2(x) = \frac{3}{2}x^2 - \frac{1}{2}$, $p_3(x) = \frac{5}{2}x^3 - \frac{3}{2}x$, we find from (10.6) that

$$p_4(x) = \frac{1}{4}[7xp_3(x) - 3p_2(x)]$$

$$= \frac{35}{8}x^4 - \frac{30}{8}x^2 + \frac{3}{8}.$$

10.2.1. Expansion of a Function Using Legendre Polynomials

Suppose that $f(x)$ is a function defined on $[-1, 1]$ such that $\int_{-1}^{1} f(x) p_n(x) \, dx$ exists for $n = 0, 1, 2, \ldots$. Consider the series expansion

$$f(x) = \sum_{i=0}^{\infty} a_i p_i(x). \tag{10.7}$$

Multiplying both sides of (10.7) by $p_n(x)$ and then integrating from -1 to 1, we obtain, by the orthogonality of Legendre polynomials,

$$a_n = \left[\int_{-1}^{1} p_n^2(x) \, dx \right]^{-1} \int_{-1}^{1} f(x) p_n(x) \, dx, \qquad n = 0, 1, \ldots .$$

It can be shown that (see Jackson, 1941, page 52)

$$\int_{-1}^{1} p_n^2(x) \, dx = \frac{2}{2n + 1}, \qquad n = 0, 1, 2, \ldots .$$

Hence, the coefficient of $p_n(x)$ in (10.7) is given by

$$a_n = \frac{2n + 1}{2} \int_{-1}^{1} f(x) p_n(x) \, dx. \tag{10.8}$$

If $s_n(x)$ denotes the partial sum $\sum_{i=0}^{n} a_i p_i(x)$ of the series in (10.7), then

$$s_n(x) = \frac{n+1}{2} \int_{-1}^{1} \frac{f(t)}{t-x} [p_{n+1}(t)p_n(x) - p_n(t)p_{n+1}(x)] dt,$$

$$n = 0, 1, 2, \ldots . \quad (10.9)$$

This is known as *Christoffel's identity* (see Jackson, 1941, page 55). If $f(x)$ is continuous on $[-1, 1]$ and has a derivative at $x = x_0$, then $\lim_{n \to \infty} s_n(x_0) = f(x_0)$, and hence the series in (10.7) converges at x_0 to the value $f(x_0)$ (see Jackson, 1941, pages 64–65).

10.3. JACOBI POLYNOMIALS

Jacobi polynomials, named after the German mathematician Karl Gustav Jacobi(1804–1851), are orthogonal on $[-1, 1]$ with respect to the weight function $w(x) = (1-x)^{\alpha}(1+x)^{\beta}$, $\alpha > -1$, $\beta > -1$. The restrictions on α and β are needed to guarantee integrability of $w(x)$ over the interval $[-1, 1]$. These polynomials, which we denote by $p_n^{(\alpha, \beta)}(x)$, can be derived by applying the Rodrigues formula:

$$p_n^{(\alpha, \beta)}(x) = \frac{(-1)^n}{2^n n!} (1-x)^{-\alpha}(1+x)^{-\beta} \frac{d^n\left[(1-x)^{\alpha+n}(1+x)^{\beta+n}\right]}{dx^n},$$

$$n = 0, 1, 2, \ldots . \quad (10.10)$$

This formula reduces to the one for Legendre polynomials when $\alpha = \beta = 0$. Thus, Legendre polynomials represent a special class of Jacobi polynomials.

Applying the so-called Leibniz formula (see Exercise 4.2 in Chapter 4) concerning the nth derivative of a product of two functions, namely, $f_n(x) = (1-x)^{\alpha+n}$ and $g_n(x) = (1+x)^{\beta+n}$ in (10.10), we obtain

$$\frac{d^n\left[(1-x)^{\alpha+n}(1+x)^{\beta+n}\right]}{dx^n} = \sum_{i=0}^{n} \binom{n}{i} f_n^{(i)}(x) g_n^{(n-i)}(x), \quad n = 0, 1, \ldots,$$

$$(10.11)$$

where for $i = 0, 1, \ldots, n$, $f_n^{(i)}(x)$ is a constant multiple of $(1-x)^{\alpha+n-i} = (1-x)^{\alpha}(1-x)^{n-i}$, and $g_n^{(n-i)}(x)$ is a constant multiple of $(1+x)^{\beta+i} = (1+x)^{\beta}(1+x)^{i}$. Thus, the nth derivative in (10.11) has $(1-x)^{\alpha}(1+x)^{\beta}$ as a factor. Using formula (10.10), it can be shown that $p_n^{(\alpha, \beta)}(x)$ is a polynomial of degree n with the leading coefficient (that is, the coefficient of x^n) equal to $(1/2^n n!)\Gamma(2n + \alpha + \beta + 1)/\Gamma(n + \alpha + \beta + 1)$.

10.4. CHEBYSHEV POLYNOMIALS

These polynomials were named after the Russian mathematician Pafnuty Lvovich Chebyshev (1821–1894). In this section, two kinds of Chebyshev polynomials will be studied, called, Chebyshev polynomials of the first kind and of the second kind.

10.4.1. Chebyshev Polynomials of the First Kind

These polynomials are denoted by $T_n(x)$ and defined as

$$T_n(x) = \cos(n \operatorname{Arccos} x), \qquad n = 0, 1, \ldots, \qquad (10.12)$$

where $0 \le \operatorname{Arccos} x \le \pi$. Note that $T_n(x)$ can be expressed as

$$T_n(x) = x^n + \binom{n}{2} x^{n-2}(x^2 - 1) + \binom{n}{4} x^{n-4}(x^2 - 1)^2 + \cdots, \qquad n = 0, 1, \ldots,$$

$$(10.13)$$

where $-1 \le x \le 1$. Historically, the polynomials defined by (10.13) were originally called Chebyshev polynomials without any qualifying expression. Using (10.13), it is easy to obtain the first few of these polynomials:

$$T_0(x) = 1,$$

$$T_1(x) = x,$$

$$T_2(x) = 2x^2 - 1,$$

$$T_3(x) = 4x^3 - 3x,$$

$$T_4(x) = 8x^4 - 8x^2 + 1,$$

$$T_5(x) = 16x^5 - 20x^3 + 5x,$$

$$\vdots$$

The following are some properties of $T_n(x)$:

1. $-1 \le T_n(x) \le 1$ for $-1 \le x \le 1$.
2. $T_n(-x) = (-1)^n T_n(x)$.
3. $T_n(x)$ has simple zeros at the following n points:

$$\xi_i = \cos\left[\frac{(2i-1)\pi}{2n}\right], \qquad i = 1, 2, \ldots, n.$$

We may recall that these zeros, also referred to as Chebyshev points, were instrumental in minimizing the error of Lagrange interpolation in Chapter 9 (see Theorem 9.2.3).

4. The weight function for $T_n(x)$ is $w(x) = (1 - x^2)^{-1/2}$. To show this, we have that for two nonnegative integers, m, n,

$$\int_0^\pi \cos m\theta \cos n\theta \, d\theta = 0, \qquad m \neq n, \tag{10.14}$$

and

$$\int_0^\pi \cos^2 n\theta \, d\theta = \begin{cases} \pi/2, & n \neq 0, \\ \pi, & n = 0 \end{cases}. \tag{10.15}$$

Making the change of variables $x = \cos \theta$ in (10.14) and (10.15), we obtain

$$\int_{-1}^1 \frac{T_m(x) T_n(x)}{(1 - x^2)^{1/2}} \, dx = 0, \qquad m \neq n,$$

$$\int_{-1}^1 \frac{T_n^2(x)}{(1 - x^2)^{1/2}} \, dx = \begin{cases} \pi/2, & n \neq 0, \\ \pi, & n = 0. \end{cases}$$

This shows that $\{T_n(x)\}_{n=0}^\infty$ forms a sequence of orthogonal polynomials on $[-1, 1]$ with respect to $w(x) = (1 - x^2)^{-1/2}$.

5. We have

$$T_{n+1}(x) = 2x T_n(x) - T_{n-1}(x), \qquad n = 1, 2, \ldots. \tag{10.16}$$

To show this recurrence relation, we use the following trigonometric identities:

$$\cos[(n + 1)\theta] = \cos n\theta \cos \theta - \sin n\theta \sin \theta,$$

$$\cos[(n - 1)\theta] = \cos n\theta \cos \theta + \sin n\theta \sin \theta.$$

Adding these identities, we obtain

$$\cos[(n + 1)\theta] = 2 \cos n\theta \cos \theta - \cos[(n - 1)\theta]. \tag{10.17}$$

If we set $x = \cos \theta$ and $\cos n\theta = T_n(x)$ in (10.17), we obtain (10.16). Recall that $T_0(x) = 1$ and $T_1(x) = x$.

10.4.2. Chebyshev Polynomials of the Second Kind

These polynomials are defined in terms of Chebyshev polynomials of the first kind as follows: Differentiating $T_n(x) = \cos n\theta$ with respect to $x = \cos \theta$, we obtain,

$$\frac{dT_n(x)}{dx} = -n \sin n\theta \frac{d\theta}{dx}$$

$$= n \frac{\sin n\theta}{\sin \theta}.$$

Let $U_n(x)$ be defined as

$$U_n(x) = \frac{1}{n+1} \frac{dT_{n+1}(x)}{dx}$$

$$= \frac{\sin[(n+1)\theta]}{\sin \theta}, \qquad n = 0, 1, \ldots \qquad (10.18)$$

This polynomial, which is of degree n, is called a Chebyshev polynomial of the second kind. Note that

$$U_n(x) = \frac{\sin n\theta \cos \theta + \cos n\theta \sin \theta}{\sin \theta}$$

$$= xU_{n-1}(x) + T_n(x), \qquad n = 1, 2, \ldots, \qquad (10.19)$$

where $U_0(x) = 1$. Formula (10.19) provides a recurrence relation for $U_n(x)$. Another recurrence relation that is free of $T_n(x)$ can be obtained from the following identity:

$$\sin[(n+1)\theta] = 2 \sin n\theta \cos \theta - \sin[(n-1)\theta].$$

Hence,

$$U_n(x) = 2xU_{n-1}(x) - U_{n-2}(x), \qquad n = 2, 3, \ldots. \qquad (10.20)$$

Using the fact that $U_0(x) = 1$, $U_1(x) = 2x$, formula (10.20) can be used to derive expressions for $U_n(x)$, $n = 2, 3, \ldots$. It is easy to see that the leading coefficient of x^n in $U_n(x)$ is 2^n.

We now show that $\{U_n(x)\}_{n=0}^\infty$ forms a sequence of orthogonal polynomials with respect to the weight function, $w(x) = (1 - x^2)^{1/2}$, over $[-1, 1]$. From the formula

$$\int_0^\pi \sin[(m+1)\theta]\sin[(n+1)\theta]\, d\theta = 0, \qquad m \neq n,$$

we get, after making the change of variables $x = \cos\theta$,

$$\int_{-1}^{1} U_m(x)U_n(x)(1-x^2)^{1/2}\,dx = 0, \qquad m \neq n.$$

This shows that $w(x) = (1-x^2)^{1/2}$ is a weight function for the sequence $\{U_n(x)\}_{n=0}^{\infty}$. Note that $\int_{-1}^{1} U_n^2(x)(1-x^2)^{1/2}\,dx = \pi/2$.

10.5. HERMITE POLYNOMIALS

Hermite polynomials, denoted by $\{H_n(x)\}_{n=0}^{\infty}$, were named after the French mathematician Charles Hermite (1822–1901). They are defined by the Rodrigues formula,

$$H_n(x) = (-1)^n e^{x^2/2} \frac{d^n(e^{-x^2/2})}{dx^n}, \qquad n = 0, 1, 2, \ldots . \qquad (10.21)$$

From (10.21), we have

$$\frac{d^n(e^{-x^2/2})}{dx^n} = (-1)^n e^{-x^2/2} H_n(x). \qquad (10.22)$$

By differentiating the two sides in (10.22), we obtain

$$\frac{d^{n+1}(e^{-x^2/2})}{dx^{n+1}} = (-1)^n \left[-xe^{-x^2/2} H_n(x) + e^{-x^2/2} \frac{dH_n(x)}{dx} \right]. \qquad (10.23)$$

But, from (10.21),

$$\frac{d^{n+1}(e^{-x^2/2})}{dx^{n+1}} = (-1)^{n+1} e^{-x^2/2} H_{n+1}(x). \qquad (10.24)$$

From (10.23) and (10.24) we then have

$$H_{n+1}(x) = xH_n(x) - \frac{dH_n(x)}{dx}, \qquad n = 0, 1, 2, \ldots,$$

which defines a recurrence relation for the sequence $\{H_n(x)\}_{n=0}^{\infty}$. Since $H_0(x) = 1$, it follows by induction, using this relation, that $H_n(x)$ is a polynomial of degree n. Its leading coefficient is equal to one.

Note that if $w(x) = e^{-x^2/2}$, then

$$w(x-t) = \exp\left(-\frac{x^2}{2} + tx - \frac{t^2}{2}\right)$$

$$= w(x)\exp\left(tx - \frac{t^2}{2}\right).$$

Applying Taylor's expansion to $w(x-t)$, we obtain

$$w(x-t) = \sum_{n=0}^{\infty} \frac{(-1)^n}{n!} t^n \frac{d^n[w(x)]}{dx^n}$$

$$= \sum_{n=0}^{\infty} \frac{t^n}{n!} H_n(x)w(x).$$

Consequently, $H_n(x)$ is the coefficient of $t^n/n!$ in the expansion of $\exp(tx - t^2/2)$. It follows that

$$H_n(x) = x^n - \frac{n^{[2]}}{2.1!}x^{n-2} + \frac{n^{[4]}}{2^2 \cdot 2!}x^{n-4} - \frac{n^{[6]}}{2^3 \cdot 3!}x^{n-6} + \cdots,$$

where $n^{[r]} = n(n-1)(n-2)\cdots(n-r+1)$. This particular representation of $H_n(x)$ is given in Kendall and Stuart (1977, page 167). For example, the first seven Hermite polynomials are

$$H_0(x) = 1,$$
$$H_1(x) = x,$$
$$H_2(x) = x^2 - 1,$$
$$H_3(x) = x^3 - 3x,$$
$$H_4(x) = x^4 - 6x^2 + 3,$$
$$H_5(x) = x^5 - 10x^3 + 15x,$$
$$H_6(x) = x^6 - 15x^4 + 45x^2 - 15,$$
$$\vdots$$

Another recurrence relation that does not use the derivative of $H_n(x)$ is given by

$$H_{n+1}(x) = xH_n(x) - nH_{n-1}(x), \qquad n = 1, 2, \ldots, \qquad (10.25)$$

with $H_0(x) = 1$ and $H_1(x) = x$. To show this, we use (10.21) in (10.25):

$$(-1)^{n+1} e^{x^2/2} \frac{d^{n+1}(e^{-x^2/2})}{dx^{n+1}}$$

$$= x(-1)^n e^{x^2/2} \frac{d^n(e^{-x^2/2})}{dx^n} - n(-1)^{n-1} e^{x^2/2} \frac{d^{n-1}(e^{-x^2/2})}{dx^{n-1}}$$

or equivalently,

$$-\frac{d^{n+1}(e^{-x^2/2})}{dx^{n+1}} = x \frac{d^n(e^{-x^2/2})}{dx^n} + n \frac{d^{n-1}(e^{-x^2/2})}{dx^{n-1}}.$$

This is true given the fact that

$$\frac{d(e^{-x^2/2})}{dx} = -xe^{-x^2/2}.$$

Hence,

$$-\frac{d^{n+1}(e^{-x^2/2})}{dx^{n+1}} = \frac{d^n(xe^{-x^2/2})}{dx^n}$$

$$= n \frac{d^{n-1}(e^{-x^2/2})}{dx^{n-1}} + x \frac{d^n(e^{-x^2/2})}{dx^n}, \qquad (10.26)$$

which results from applying Leibniz's formula to the right-hand side of (10.26).

We now show that $\{H_n(x)\}_{n=0}^\infty$ forms a sequence of orthogonal polynomials with respect to the weight function $w(x) = e^{-x^2/2}$ over $(-\infty, \infty)$. For this purpose, let m, n be nonnegative integers, and let c be defined as

$$c = \int_{-\infty}^{\infty} e^{-x^2/2} H_m(x) H_n(x) \, dx. \qquad (10.27)$$

Then, from (10.21) and (10.27), we have

$$c = (-1)^n \int_{-\infty}^{\infty} H_m(x) \frac{d^n(e^{-x^2/2})}{dx^n} \, dx.$$

Integrating by parts gives

$$c = (-1)^n \left\{ \left[H_m(x) \frac{d^{n-1}(e^{-x^2/2})}{dx^{n-1}} \right]_{-\infty}^{\infty} - \int_{-\infty}^{\infty} \frac{dH_m(x)}{dx} \frac{d^{n-1}(e^{-x^2/2})}{dx^{n-1}} \, dx \right\}$$

$$= (-1)^{n+1} \int_{-\infty}^{\infty} \frac{dH_m(x)}{dx} \frac{d^{n-1}(e^{-x^2/2})}{dx^{n-1}} \, dx. \qquad (10.28)$$

Formula (10.28) is true because $H_m(x) d^{n-1}(e^{-x^2/2})/dx^{n-1}$, which is a poly-nomial multiplied by $e^{-x^2/2}$, has a limit equal to zero as $x \to \mp\infty$. By repeating the process of integration by parts $m - 1$ more times, we obtain for $n > m$

$$c = (-1)^{m+n} \int_{-\infty}^{\infty} \frac{d^m[H_m(x)]}{dx^m} \frac{d^{n-m}(e^{-x^2/2})}{dx^{n-m}} \, dx. \qquad (10.29)$$

Note that since $H_m(x)$ is a polynomial of degree m with a leading coefficient equal to one, $d^m[H_m(x)]/dx^m$ is a constant equal to $m!$. Furthermore, since $n > m$, then

$$\int_{-\infty}^{\infty} \frac{d^{n-m}(e^{-x^2/2})}{dx^{n-m}} \, dx = 0.$$

It follows that $c = 0$. We can also arrive at the same conclusion if $n < m$. Hence,

$$\int_{-\infty}^{\infty} e^{-x^2/2} H_m(x) H_n(x) \, dx = 0, \qquad m \neq n.$$

This shows that $\{H_n(x)\}_{n=0}^{\infty}$ is a sequence of orthogonal polynomials with respect to $w(x) = e^{-x^2/2}$ over $(-\infty, \infty)$.
    Note that if $m = n$ in (10.29), then

$$c = \int_{-\infty}^{\infty} \frac{d^n[H_n(x)]}{dx^n} e^{-x^2/2} \, dx$$

$$= n! \int_{-\infty}^{\infty} e^{-x^2/2} \, dx$$

$$= 2n! \int_{0}^{\infty} e^{-x^2/2} \, dx$$

$$= n!\sqrt{2\pi}$$

By comparison with (10.27), we conclude that

$$\int_{-\infty}^{\infty} e^{-x^2/2} H_n^2(x) \, dx = n!\sqrt{2\pi}. \qquad (10.30)$$

    Hermite polynomials can be used to provide the following series expansion of a function $f(x)$:

$$f(x) = \sum_{n=0}^{\infty} c_n H_n(x), \qquad (10.31)$$

where

$$c_n = \frac{1}{n!\sqrt{2\pi}} \int_{-\infty}^{\infty} e^{-x^2/2} f(x) H_n(x)\, dx, \qquad n = 0, 1, \ldots. \qquad (10.32)$$

Formula (10.32) follows from multiplying both sides of (10.31) with $e^{-x^2/2} H_n(x)$, integrating over $(-\infty, \infty)$, and noting formula (10.30) and the orthogonality of the sequence $\{H_n(x)\}_{n=0}^{\infty}$.

10.6. LAGUERRE POLYNOMIALS

Laguerre polynomials were named after the French mathematician Edmond Laguerre (1834–1886). They are denoted by $L_n^{(\alpha)}(x)$ and are defined over the interval $(0, \infty)$, $n = 0, 1, 2, \ldots$, where $\alpha > -1$.

The development of these polynomials is based on an application of Leibniz formula to finding the nth derivative of the function

$$\phi_n(x) = x^{\alpha+n} e^{-x}.$$

More specifically, for $\alpha > -1$, $L_n^{(\alpha)}(x)$ is defined by a Rodrigues-type formula, namely,

$$L_n^{(\alpha)}(x) = (-1)^n e^x x^{-\alpha} \frac{d^n(x^{n+\alpha} e^{-x})}{dx^n}, \qquad n = 0, 1, 2, \ldots.$$

We shall henceforth use $L_n(x)$ instead of $L_n^{(\alpha)}(x)$.

From this definition, we conclude that $L_n(x)$ is a polynomial of degree n with a leading coefficient equal to one. It can also be shown that Laguerre polynomials are orthogonal with respect to the weight function $w(x) = e^{-x} x^{\alpha}$ over $(0, \infty)$, that is,

$$\int_0^{\infty} e^{-x} x^{\alpha} L_m(x) L_n(x)\, dx = 0, \qquad m \neq n$$

(see Jackson, 1941, page 185). Furthermore, if $m = n$, then

$$\int_0^{\infty} e^{-x} x^{\alpha} [L_n(x)]^2\, dx = n!\Gamma(\alpha + n + 1), \qquad n = 0, 1, \ldots.$$

A function $f(x)$ can be expressed as an infinite series of Laguerre polynomials of the form

$$f(x) = \sum_{n=0}^{\infty} c_n L_n(x),$$

where

$$c_n = \frac{1}{n!\,\Gamma(\alpha+n+1)} \int_0^{\infty} e^{-x} x^{\alpha} L_n(x) f(x)\, dx, \qquad n = 0, 1, 2, \ldots .$$

A recurrence relation for $L_n(x)$ is developed as follows: From the definition of $L_n(x)$, we have

$$(-1)^n x^{\alpha} e^{-x} L_n(x) = \frac{d^n (x^{n+\alpha} e^{-x})}{dx^n}, \qquad n = 0, 1, 2, \ldots . \quad (10.33)$$

Replacing n by $n+1$ in (10.33) gives

$$(-1)^{n+1} x^{\alpha} e^{-x} L_{n+1}(x) = \frac{d^{n+1}(x^{n+\alpha+1} e^{-x})}{dx^{n+1}}. \quad (10.34)$$

Now,

$$x^{n+\alpha+1} e^{-x} = x(x^{n+\alpha} e^{-x}). \quad (10.35)$$

Applying the Leibniz formula for the $(n+1)$st derivative of the product on the right-hand side of (10.35) and noting that the nth derivative of x is zero for $n = 2, 3, 4, \ldots$, we obtain

$$\frac{d^{n+1}(x^{n+\alpha+1} e^{-x})}{dx^{n+1}} = x\frac{d^{n+1}(x^{n+\alpha} e^{-x})}{dx^{n+1}} + (n+1)\frac{d^n(x^{n+\alpha} e^{-x})}{dx^n}$$

$$= x\frac{d}{dx}\left[\frac{d^n(x^{n+\alpha} e^{-x})}{dx^n}\right] + (n+1)\frac{d^n(x^{n+\alpha} e^{-x})}{dx^n}, \quad (10.36)$$

Using (10.33) and (10.34) in (10.36) gives

$$(-1)^{n+1} x^{\alpha} e^{-x} L_{n+1}(x)$$

$$= x\frac{d}{dx}\left[(-1)^n x^{\alpha} e^{-x} L_n(x)\right] + (-1)^n (n+1) x^{\alpha} e^{-x} L_n(x)$$

$$= (-1)^n x^{\alpha} e^{-x}\left[(\alpha+n+1-x) L_n(x) + x\frac{dL_n(x)}{dx}\right]. \quad (10.37)$$

Multiplying the two sides of (10.37) by $(-1)^{n+1} e^{x} x^{-\alpha}$, we obtain

$$L_{n+1}(x) = (x-\alpha-n-1) L_n(x) - x\frac{dL_n(x)}{dx}, \qquad n = 0, 1, 2, \ldots .$$

This recurrence relation gives $L_{n+1}(x)$ in terms of $L_n(x)$ and its derivative. Note that $L_0(x) = 1$.

Another recurrence relation that does not require using the derivative of $L_n(x)$ is given by (see Jackson, 1941, page 186)

$$L_{n+1}(x) = (x - \alpha - 2n - 1)L_n(x) - n(\alpha + n)L_{n-1}(x), \qquad n = 1, 2, \dots .$$

10.7. LEAST-SQUARES APPROXIMATION WITH ORTHOGONAL POLYNOMIALS

In this section, we consider an approximation problem concerning a continuous function. Suppose that we have a set of polynomials, $\{p_i(x)\}_{i=0}^n$, orthogonal with respect to a weight function $w(x)$ over the interval $[a, b]$. Let $f(x)$ be a continuous function on $[a, b]$. We wish to approximate $f(x)$ by the sum $\sum_{i=0}^n c_i p_i(x)$, where $c_0, c_1, \dots, c_n$ are constants to be determined by minimizing the function

$$\gamma(c_0, c_1, \dots, c_n) = \int_a^b \left[\sum_{i=0}^n c_i p_i(x) - f(x) \right]^2 w(x) \, dx,$$

that is, γ is the square of the norm, $\|\sum_{i=0}^n c_i p_i - f\|_\omega$. If we differentiate γ with respect to $c_0, c_1, \dots, c_n$ and equate the partial derivatives to zero, we obtain

$$\frac{\partial \gamma}{\partial c_i} = 2 \int_a^b \left[\sum_{j=0}^n c_j p_j(x) - f(x) \right] p_i(x) w(x) \, dx = 0, \qquad i = 0, 1, \dots, n.$$

Hence,

$$\sum_{j=0}^n \left[\int_a^b p_i(x) p_j(x) w(x) \, dx \right] c_j = \int_a^b f(x) p_i(x) w(x) \, dx, \qquad i = 0, 1, \dots, n.$$

$$(10.38)$$

Equations (10.38) can be written in vector form as

$$\mathbf{Sc} = \mathbf{u}, \qquad (10.39)$$

where $\mathbf{c} = (c_0, c_1, \dots, c_n)'$, $\mathbf{u} = (u_0, u_1, \dots, u_n)'$ with $u_i = \int_a^b f(x) p_i(x) w(x) \, dx$, and $\mathbf{S}$ is an $(n+1) \times (n+1)$ matrix whose (i, j)th element, s_{ij}, is given by

$$s_{ij} = \int_a^b p_i(x) p_j(x) w(x) \, dx, \qquad i, j = 0, 1, \dots, n.$$

Since $p_0(x), p_1(x), \dots, p_n(x)$ are orthogonal, then $\mathbf{S}$ must be a diagonal matrix with diagonal elements given by

$$s_{ii} = \int_a^b p_i^2(x) w(x) \, dx = \| p_i \|_\omega^2, \qquad i = 0, 1, \dots, n.$$

From equation (10.39) we get the solution $\mathbf{c} = \mathbf{S}^{-1}\mathbf{u}$. The ith element of $\mathbf{c}$ is therefore of the form

$$
c_i = \frac{u_i}{s_{ii}}
$$

$$
= \frac{(f \cdot p_i)_\omega}{\|p_i\|_\omega^2}, \qquad i = 0, 1, \ldots, n.
$$

For such a value of $\mathbf{c}$, γ has an absolute minimum, since $\mathbf{S}$ is positive definite. It follows that the linear combination

$$
p_n^*(x) = \sum_{i=0}^{n} c_i p_i(x)
$$

$$
= \sum_{i=0}^{n} \frac{(f \cdot p_i)_\omega}{\|p_i\|_\omega^2} p_i(x) \qquad (10.40)
$$

minimizes γ. We refer to $p_n^*(x)$ as the least-squares polynomial approximation of $f(x)$ with respect to $p_0(x), p_1(x), \ldots, p_n(x)$.

If $\{p_n(x)\}_{n=0}^\infty$ is a sequence of orthogonal polynomials, then $p_n^*(x)$ in (10.40) represents a partial sum of the infinite series $\sum_{n=0}^\infty [(f \cdot p_n)_\omega / \|p_n\|_\omega^2] p_n(x)$. This series may fail to converge point by point to $f(x)$. It converges, however, to $f(x)$ in the norm $\|\cdot\|_\omega$. This is shown in the next theorem.

Theorem 10.7.1. If $f:[a, b] \to R$ is continuous, then

$$
\int_a^b [f(x) - p_n^*(x)]^2 w(x) \, dx \to 0
$$

as $n \to \infty$, where $p_n^*(x)$ is defined by formula (10.40).

Proof. By the Weierstrass theorem (Theorem 9.1.1), there exists a polynomial $b_n(x)$ of degree n that converges uniformly to $f(x)$ on $[a, b]$, that is,

$$
\sup_{a \le x \le b} |f(x) - b_n(x)| \to 0 \qquad \text{as } n \to \infty.
$$

Hence,

$$
\int_a^b |f(x) - b_n(x)|^2 w(x) \, dx \to 0
$$

as $n \to \infty$, since

$$
\int_a^b |f(x) - b_n(x)|^2 w(x) \, dx \le \sup_{a \le x \le b} |f(x) - b_n(x)|^2 \int_a^b w(x) \, dx.
$$

Furthermore,

$$\int_a^b |f(x) - p_n^*(x)|^2 w(x)\, dx \le \int_a^b |f(x) - b_n(x)|^2 w(x)\, dx, \quad (10.41)$$

since, by definition, $p_n^*(x)$ is the least-squares polynomial approximation of $f(x)$. From inequality (10.41) we conclude that $\|f - p_n^*\|_\omega \to 0$ as $n \to \infty$. $\quad\square$

10.8. ORTHOGONAL POLYNOMIALS DEFINED ON A FINITE SET

In this section, we consider polynomials, $p_0(x), p_1(x), \ldots, p_n(x)$, defined on a finite set $D = \{x_0, x_2, \ldots, x_n\}$ such that $a \le x_i \le b$, $i = 0, 1, \ldots, n$. These polynomials are orthogonal with respect to a weight function $w^*(x)$ over D if

$$\sum_{i=0}^n w^*(x_i) p_m(x_i) p_\nu(x_i) = 0, \qquad m \ne \nu; \quad m, \nu = 0, 1, \ldots, n.$$

Such polynomials are said to be orthogonal of the discrete type. For example, the set of discrete Chebyshev polynomials, $\{t_i(j, n)\}_{i=0}^n$, which are defined over the set of integers $j = 0, 1, \ldots, n$, are orthogonal with respect to $w^*(j) = 1$, $j = 0, 1, 2, \ldots, n$, and are given by the following formula (see Abramowitz and Stegun, 1964, page 791):

$$t_i(j, n) = \sum_{k=0}^i (-1)^k \binom{i}{k} \binom{i+k}{k} \frac{j!(n-k)!}{(j-k)!n!},$$

$$i = 0, 1, \ldots, n; \quad j = 0, 1, \ldots, n. \qquad (10.42)$$

For example, for $i = 0, 1, 2$, we have

$$t_0(j, n) = 1, \qquad\qquad j = 0, 1, 2, \ldots, n,$$

$$t_1(j, n) = 1 - \frac{2}{n} j, \qquad j = 0, 1, \ldots, n,$$

$$t_2(j, n) = 1 - \frac{6}{n}\left[j - \frac{j(j-1)}{n-1} \right]$$

$$= 1 - \frac{6j}{n}\left(1 - \frac{j-1}{n-1} \right)$$

$$= 1 - \frac{6j}{n}\left(\frac{n-j}{n-1} \right), \qquad j = 0, 1, 2, \ldots, n.$$

A recurrence relation for the discrete Chebyshev polynomials is of the form

$$(i+1)(n-i)t_{i+1}(j,n) = (2i+1)(n-2j)t_i(j,n) - i(n+i+1)t_{i-1}(j,n),$$
$$i = 1,2,\ldots,n. \qquad (10.43)$$

10.9. APPLICATIONS IN STATISTICS

Orthogonal polynomials play an important role in approximating distribution functions of certain random variables. In this section, we consider only univariate distributions.

10.9.1. Applications of Hermite Polynomials

Hermite polynomials provide a convenient tool for approximating density functions and quantiles of distributions using convergent series. They are associated with the normal distribution, and it is therefore not surprising that they come up in various investigations in statistics and probability theory. Here are some examples.

10.9.1.1. Approximation of Density Functions and Quantiles of Distributions

Let $\phi(x)$ denote the density function of the standard normal distribution, that is,

$$\phi(x) = \frac{1}{\sqrt{2\pi}}e^{-x^2/2}, \qquad -\infty < x < \infty. \qquad (10.44)$$

We recall from Section 10.5 that the sequence $\{H_n(x)\}_{n=0}^{\infty}$ of Hermite polynomials is orthogonal with respect to $w(x) = e^{-x^2/2}$, and that

$$H_n(x) = x^n - \frac{n^{[2]}}{2\cdot 1!}x^{n-2} + \frac{n^{[4]}}{2^2\cdot 2!}x^{n-4} - \frac{n^{[6]}}{2^3\cdot 3!}x^{n-6} + \cdots, \quad (10.45)$$

where $n^{[r]} = n(n-1)(n-2)\cdots(n-r+1)$. Suppose now that $g(x)$ is a density function for some continuous distribution. We can represent $g(x)$ as a series of the form

$$g(x) = \sum_{n=0}^{\infty} b_n H_n(x)\phi(x), \qquad (10.46)$$

where, as in formula (10.32),

$$b_n = \frac{1}{n!}\int_{-\infty}^{\infty} g(x)H_n(x)\,dx. \qquad (10.47)$$

By substituting $H_n(x)$, as given by formula (10.45), in formula (10.47), we obtain an expression for b_n in terms of the central moments, $\mu_0, \mu_1, \ldots,$ $\mu_n, \ldots,$ of the distribution whose density function is $g(x)$. These moments are defined as

$$\mu_n = \int_{-\infty}^{\infty} (x - \mu)^n g(x)\, dx, \qquad n = 0, 1, 2, \ldots,$$

where μ is the mean of the distribution. Note that $\mu_0 = 1$, $\mu_1 = 0$, and $\mu_2 = \sigma^2$, the variance of the distribution. In particular, if $\mu = 0$, then

$$b_0 = 1,$$

$$b_1 = 0,$$

$$b_2 = \tfrac{1}{2}(\mu_2 - 1),$$

$$b_3 = \tfrac{1}{6}\mu_3,$$

$$b_4 = \tfrac{1}{24}(\mu_4 - 6\mu_2 + 3),$$

$$b_5 = \tfrac{1}{120}(\mu_5 - 10\mu_3),$$

$$b_6 = \tfrac{1}{720}(\mu_6 - 15\mu_4 + 45\mu_2 - 15),$$

$$\vdots$$

The expression for $g(x)$ in formula (10.46) can then be written as

$$g(x) = \phi(x)\left[1 + \tfrac{1}{2}(\mu_2 - 1)H_2(x) + \tfrac{1}{6}\mu_3 H_3(x)\right.$$
$$\left. + \tfrac{1}{24}(\mu_4 - 6\mu_2 + 3)H_4(x) + \cdots\right]. \qquad (10.48)$$

This expression is known as the *Gram–Charlier series of type A.*

Thus the Gram–Charlier series provides an expansion of $g(x)$ in terms of its central moments, the standard normal density, and Hermite polynomials. Using formulas (10.21) and (10.46), we note that $g(x)$ can be expressed as a series of derivatives of $\phi(x)$ of the form

$$g(x) = \sum_{n=0}^{\infty} \frac{c_n}{n!} \left(\frac{d}{dx}\right)^n \phi(x), \qquad (10.49)$$

where

$$c_n = (-1)^n \int_{-\infty}^{\infty} g(x) H_n(x)\, dx, \qquad n = 0, 1, \ldots\, .$$

Cramér (1946, page 223) gave conditions for the convergence of the series on the right-hand side of formula (10.49), namely, if $g(x)$ is continuous and of bounded variation on $(-\infty, \infty)$, and if the integral $\int_{-\infty}^{\infty} g(x) \exp(x^2/4)\, dx$ is convergent, then the series in formula (10.49) will converge for every x to $g(x)$.

We can utilize Gram–Charlier series to find the upper α-quantile, x_α, of the distribution with the density function $g(x)$. This point is defined as

$$\int_{-\infty}^{x_\alpha} g(x)\, dx = 1 - \alpha.$$

From (10.46) we have that

$$g(x) = \phi(x) + \sum_{n=2}^{\infty} b_n H_n(x) \phi(x).$$

Then

$$\int_{-\infty}^{x_\alpha} g(x)\, dx = \int_{-\infty}^{x_\alpha} \phi(x)\, dx + \sum_{n=2}^{\infty} b_n \int_{-\infty}^{x_\alpha} H_n(x) \phi(x)\, dx. \qquad (10.50)$$

However,

$$\int_{-\infty}^{x_\alpha} H_n(x) \phi(x)\, dx = -H_{n-1}(x_\alpha) \phi(x_\alpha).$$

To prove this equality we note that by formula (10.21)

$$\int_{-\infty}^{x_\alpha} H_n(x) \phi(x)\, dx = (-1)^n \int_{-\infty}^{x_\alpha} \left(\frac{d}{dx}\right)^n \phi(x)\, dx$$

$$= (-1)^n \left(\frac{d}{dx}\right)^{n-1} \phi(x_\alpha),$$

where $(d/dx)^{n-1}\phi(x_\alpha)$ denotes the value of the $(n-1)$st derivative of $\phi(x)$ at x_α. By applying formula (10.21) again we obtain

$$\int_{-\infty}^{x_\alpha} H_n(x) \phi(x)\, dx = (-1)^n (-1)^{n-1} H_{n-1}(x_\alpha) \phi(x_\alpha)$$

$$= -H_{n-1}(x_\alpha) \phi(x_\alpha).$$

By making the substitution in formula (10.50), we get

$$\int_{-\infty}^{x_\alpha} g(x)\, dx = \int_{-\infty}^{x_\alpha} \phi(x)\, dx - \sum_{n=2}^{\infty} b_n H_{n-1}(x_\alpha) \phi(x_\alpha). \qquad (10.51)$$

Now, suppose that z_α is the upper α-quantile of the standard normal distribution. Then

$$\int_{-\infty}^{x_\alpha} g(x)\, dx = 1 - \alpha = \int_{-\infty}^{z_\alpha} \phi(x)\, dx.$$

Using the expansion (10.51), we obtain

$$\int_{-\infty}^{x_\alpha} \phi(x)\, dx - \sum_{n=2}^{\infty} b_n H_{n-1}(x_\alpha)\phi(x_\alpha) = \int_{-\infty}^{z_\alpha} \phi(x)\, dx. \qquad (10.52)$$

If we expand the right-hand side of equation (10.52) using Taylor's series in a neighborhood of x_α, we get

$$\int_{-\infty}^{z_\alpha} \phi(x)\, dx = \int_{-\infty}^{x_\alpha} \phi(x)\, dx + \sum_{j=1}^{\infty} \frac{(z_\alpha - x_\alpha)^j}{j!} \left(\frac{d}{dx}\right)^{j-1} \phi(x_\alpha)$$

$$= \int_{-\infty}^{x_\alpha} \phi(x)\, dx + \sum_{j=1}^{\infty} \frac{(z_\alpha - x_\alpha)^j}{j!} (-1)^{j-1} H_{j-1}(x_\alpha)\phi(x_\alpha),$$

using formula (10.21)

$$= \int_{-\infty}^{x_\alpha} \phi(x)\, dx - \sum_{j=1}^{\infty} \frac{(x_\alpha - z_\alpha)^j}{j!} H_{j-1}(x_\alpha)\phi(x_\alpha). \qquad (10.53)$$

From formulas (10.52) and (10.53) we conclude that

$$\sum_{n=2}^{\infty} b_n H_{n-1}(x_\alpha)\phi(x_\alpha) = \sum_{j=1}^{\infty} \frac{(x_\alpha - z_\alpha)^j}{j!} H_{j-1}(x_\alpha)\phi(x_\alpha).$$

By dividing both sides by $\phi(x_\alpha)$ we obtain

$$\sum_{n=2}^{\infty} b_n H_{n-1}(x_\alpha) = \sum_{j=1}^{\infty} \frac{(x_\alpha - z_\alpha)^j}{j!} H_{j-1}(x_\alpha). \qquad (10.54)$$

This provides a relationship between x_α, the α-quantile of the distribution with the density function $g(x)$, and z_α, the corresponding quantile for the standard normal. Since the b_n's are functions of the moments associated with $g(x)$, then it is possible to use (10.54) to express x_α in terms of z_α and the moments of $g(x)$. This was carried out by Cornish and Fisher (1937). They provided an expansion for x_α in terms of z_α and the cumulants (instead of the moments) associated with $g(x)$. (See Section 5.6.2 for a definition of cumulants. Note that there is a one-to-one correspondence between moments and cumulants.) Such an expansion became known as the

Cornish–Fisher expansion. It is reported in Johnson and Kotz (1970, page 34) (see Exercise 10.11). See also Kendall and Stuart (1977, pages 175–178).

10.9.1.2. Approximation of a Normal Integral

A convergent series representing the integral

$$\psi(x) = \frac{1}{\sqrt{2\pi}} \int_0^x e^{-t^2/2} \, dt$$

was derived by Kerridge and Cook (1976). Their method is based on the fact that

$$\int_0^x f(t) \, dt = 2 \sum_{n=0}^{\infty} \frac{(x/2)^{2n+1}}{(2n+1)!} f^{(2n)}\left(\frac{x}{2}\right) \tag{10.55}$$

for any function $f(t)$ with a suitably convergent Taylor's expansion in a neighborhood of $x/2$, namely,

$$f(t) = \sum_{n=0}^{\infty} \frac{1}{n!} \left(t - \frac{x}{2}\right)^n f^{(n)}\left(\frac{x}{2}\right). \tag{10.56}$$

Formula (10.55) results from integrating (10.56) with respect to t from 0 to x and noting that the even terms vanish. Taking $f(t) = e^{-t^2/2}$, we obtain

$$\int_0^x e^{-t^2/2} \, dt = 2 \sum_{n=0}^{\infty} \frac{(x/2)^{2n+1}}{(2n+1)!} \frac{d^{2n}(e^{-t^2/2})}{dt^{2n}} \bigg|_{t=x/2}. \tag{10.57}$$

Using the Rodrigues formula for Hermite polynomials [formula (10.21)], we get

$$\frac{d^n(e^{-x^2/2})}{dx^n} = (-1)^n e^{-x^2/2} H_n(x), \qquad n = 0, 1, \ldots .$$

By making the substitution in (10.57), we find

$$\int_0^x e^{-t^2/2} \, dt = 2 \sum_{n=0}^{\infty} \frac{(x/2)^{2n+1}}{(2n+1)!} e^{-x^2/2} H_{2n}\left(\frac{x}{2}\right)$$

$$= 2e^{-x^2/8} \sum_{n=0}^{\infty} \frac{(x/2)^{2n+1}}{(2n+1)!} H_{2n}\left(\frac{x}{2}\right). \tag{10.58}$$

This expression can be simplified by letting $\Theta_n(x) = x^n H_n(x)/n!$ in (10.58), which gives

$$\int_0^x e^{-t^2/2} \, dt = x e^{-x^2/8} \sum_{n=0}^{\infty} \frac{\Theta_{2n}(x/2)}{2n+1}.$$

Hence,

$$\psi(x) = \frac{1}{\sqrt{2\pi}} x e^{-x^2/8} \sum_{n=0}^{\infty} \frac{\Theta_{2n}(x/2)}{2n+1}. \tag{10.59}$$

Note that on the basis of formula (10.25), the recurrence relation for $\Theta_n(x)$ is given by

$$\Theta_{n+1} = \frac{x^2(\Theta_n - \Theta_{n-1})}{n+1}, \qquad n = 1, 2, \ldots .$$

The $\Theta_n(x)$'s are easier to handle numerically than the Hermite polynomials, as they remain relatively small, even for large n. Kerridge and Cook (1976) report that the series in (10.59) is accurate over a wide range of x. Divgi (1979), however, states that the convergence of the series becomes slower as x increases.

10.9.1.3. Estimation of Unknown Densities

Let $X_1, X_2, \ldots, X_n$ represent a sequence of independent random variables with a common, but unknown, density function $f(x)$ assumed to be square integrable. From (10.31) we have the representation

$$f(x) = \sum_{j=0}^{\infty} c_j H_j(x),$$

or equivalently,

$$f(x) = \sum_{j=0}^{\infty} a_j h_j(x), \tag{10.60}$$

where $h_j(x)$ is the so-called normalized Hermite polynomial of degree j, namely,

$$h_j(x) = \frac{1}{(\sqrt{2\pi}\, j!)^{1/2}} e^{-x^2/4} H_j(x), \qquad j = 0, 1, \ldots,$$

and $a_j = \int_{-\infty}^{\infty} f(x) h_j(x)\, dx$, since $\int_{-\infty}^{\infty} h_j^2(x)\, dx = 1$ by virtue of (10.30).

Schwartz (1967) considered an estimate of $f(x)$ of the form

$$\hat{f}_n(x) = \sum_{j=0}^{q(n)} \hat{a}_{jn} h_j(x),$$

where

$$\hat{a}_{jn} = \frac{1}{n} \sum_{k=1}^{n} h_j(X_k),$$

and $q(n)$ is a suitably chosen integer dependent on n such that $q(n) = o(n)$ as $n \to \infty$. Under these conditions, Schwartz (1967, Theorem 1) showed that $\hat{f}_n(x)$ is a consistent estimator of $f(x)$ in the mean integrated squared error sense, that is,

$$\lim_{n \to \infty} E \int \left[\hat{f}_n(x) - f(x) \right]^2 dx = 0.$$

Under additional conditions on $f(x)$, $\hat{f}_n(x)$ is also consistent in the mean squared error sense, that is,

$$\lim_{n \to \infty} E \left[f(x) - \hat{f}_n(x) \right]^2 = 0$$

uniformly in x.

10.9.2. Applications of Jacobi and Laguerre Polynomials

Dasgupta (1968) presented an approximation to the distribution function of $X = \frac{1}{2}(r + 1)$, where r is the sample correlation coefficient, in terms of a beta density and Jacobi polynomials. Similar methods were used by Durbin and Watson (1951) in deriving an approximation of the distribution of a statistic used for testing serial correlation in least-squares regression.

Quadratic forms in random variables, which can often be regarded as having joint multivariate normal distributions, play an important role in analysis of variance and in estimation of variance components for a random or a mixed model. Approximation of the distributions of such quadratic forms can be carried out using Laguerre polynomials (see, for example, Gurland, 1953, and Johnson and Kotz, 1968). Tiku (1964a) developed Laguerre series expansions of the distribution functions of the nonnormal variance ratios used for testing the homogeneity of treatment means in the case of one-way classification for analysis of variance with nonidentical group-to-group error distributions that are not assumed to be normal. Tiku (1964b) also used Laguerre polynomials to obtain an approximation to the first negative moment of a Poisson random variable, that is, the value of $E(X^{-1})$, where X has the Poisson distribution.

More recently, Schöne and Schmid (2000) made use of Laguerre polynomials to develop a series representation of the joint density and the joint distribution of a quadratic form and a linear form in normal variables. Such a representation can be used to calculate, for example, the joint density and the joint distribution function of the sample mean and sample variance. Note that for autocorrelated variables, the sample mean and sample variance are, in general, not independent.

10.9.3. Calculation of Hypergeometric Probabilities Using Discrete Chebyshev Polynomials

The hypergeometric distribution is a discrete distribution, somewhat related to the binomial distribution. Suppose, for example, we have a lot of M items,

r of which are defective and $M - r$ of which are nondefective. Suppose that we choose at random m items without replacement from the lot $(m \leq M)$. Let X be the number of defectives found. Then, the probability that $X = k$ is given by

$$P(X = k) = \frac{\binom{r}{k}\binom{M-r}{m-k}}{\binom{M}{m}}, \tag{10.61}$$

where $\max(0, m - M + r) \leq k \leq \min(m, r)$. A random variable with the probability mass function (10.61) is said to have a hypergeometric distribution. We denote such a probability function by $h(k; m, r, M)$.

There are tables for computing the probability value in (10.61) (see, for example, the tables given by Lieberman and Owen, 1961). There are also several algorithms for computing this probability. Recently, Alvo and Cabilio (2000) proposed to represent the hypergeometric distribution in terms of discrete Chebyshev polynomials, as was seen in Section 10.8. The following is a summary of this work: Consider the sequence $\{t_n(k, m)\}_{n=0}^m$ of discrete Chebyshev polynomials defined over the set of integers $k = 0, 1, 2, \ldots, m$ [see formula (10.42)], which is given by

$$t_n(k, m) = \sum_{i=0}^{n} (-1)^i \binom{n}{i}\binom{n+i}{i}\frac{k!(m-i)!}{(k-i)!m!},$$

$$n = 0, 1, \ldots, m, \quad k = 0, 1, \ldots, m. \tag{10.62}$$

Let X have the hypergeometric distribution as in (10.61). Then according to Theorem 1 in Alvo and Cabilio (2000),

$$\sum_{k=0}^{m} t_n(k, m) h(k; m, r, M) = t_n(r, M) \tag{10.63}$$

for all $n = 0, 1, \ldots, m$ and $r = 0, 1, \ldots, M$. Let $\mathbf{t}_n = [t_n(0, m), t_n(1, m), \ldots, t_n(m, m)]'$, $n = 0, 1, \ldots, m$, be the base vectors in an $(m + 1)$-dimensional Euclidean space determined from the Chebyshev polynomials. Let $g(k)$ be any function defined over the set of integers, $k = 0, 1, \ldots, m$. Then $g(k)$ can be expressed as

$$g(k) = \sum_{n=0}^{m} g_n t_n(k, m), \quad k = 0, 1, \ldots, m, \tag{10.64}$$

where $g_n = \mathbf{g} \cdot \mathbf{t}_n / \|\mathbf{t}_n\|^2$, and $\mathbf{g} = [g(0), g(1), \ldots, g(m)]'$. Now, using the result in (10.63), the expected value of $g(X)$ is given by

$$
\begin{aligned}
E[g(X)] &= \sum_{k=0}^{m} g(k)h(k;m,r,M) \\
&= \sum_{k=0}^{m}\sum_{n=0}^{m} g_n t_n(k,m)h(k;m,r,M) \\
&= \sum_{n=0}^{m} g_n \sum_{k=0}^{m} t_n(k,m)h(k;m,r,M) \\
&= \sum_{n=0}^{m} g_n t_n(r,M).
\end{aligned}
\tag{10.65}
$$

This shows that the expected value of $g(X)$ can be computed from knowledge of the coefficients g_n and the discrete Chebyshev polynomials up to order m, evaluated at r and M.

In particular, if $g(x)$ is an indicator function taking the value one at $x = k$ and the value zero elsewhere, then

$$
\begin{aligned}
E[g(X)] &= P(X = k) \\
&= h(k;m,r,M).
\end{aligned}
$$

Applying the result in (10.65), we then obtain

$$
h(k;m,r,M) = \sum_{n=0}^{m} \frac{t_n(k,m)}{\|\mathbf{t}_n\|^2} t_n(r,M).
\tag{10.66}
$$

Because of the recurrence relation (10.43) for discrete Chebyshev polynomials, calculating the hypergeometric probability using (10.66) can be done simply on a computer.

FURTHER READING AND ANNOTATED BIBLIOGRAPHY

Abramowitz, M., and I. A. Stegun (1964). *Handbook of Mathematical Functions with Formulas, Graphs, and Mathematical Tables*. Wiley, New York. (This useful volume was prepared by the National Bureau of Standards. It was edited by Milton Abramowitz and Irene A. Stegun.)

Alvo, M., and P. Cabilio (2000). "Calculation of hypergeometric probabilities using Chebyshev polynomials." *Amer. Statist.*, **54**, 141–144.

Cheney, E. W. (1982). *Introduction to Approximation Theory*, 2nd ed. Chelsea, New York. (Least-squares polynomial approximation is discussed in Chap. 4.)

Chihara, T. S. (1978). *An Introduction to Orthogonal Polynomials*. Gordon and Breach, New York. (This text deals with the general theory of orthogonal polynomials, including recurrence relations, and some particular systems of orthogonal polynomials.)

Cornish, E. A., and R. A. Fisher (1937). "Moments and cumulants in the specification of distributions." *Rev. Internat. Statist. Inst.*, **5**, 307–320.

Cramér, H. (1946). *Mathematical Methods of Statistics*. Princeton University Press, Princeton. (This classic book provides the mathematical foundation of statistics. Chap. 17 is a good source for approximation of density functions.)

Dasgupta, P. (1968). "An approximation to the distribution of sample correlation coefficient, when the population is non-normal." *Sankhyā, Ser. B.*, **30**, 425–428.

Davis, P. J. (1975). *Interpolation and Approximation*. Dover, New York. (Chaps. 8 and 10 discuss least-squares approximation and orthogonal polynomials.)

Divgi, D. R. (1979). "Calculation of univariate and bivariate normal probability functions." *Ann. Statist.*, **7**, 903–910.

Durbin, J., and G. S. Watson (1951). "Testing for serial correlation in least-squares regression II." *Biometrika*, **38**, 159–178.

Freud, G. (1971). *Orthogonal Polynomials*. Pergamon Press, Oxford. (This book deals with fundamental properties of orthogonal polynomials, including Legendre, Chebyshev, and Jacobi polynomials. Convergence theory of series of orthogonal polynomials is discussed in Chap. 4.)

Gurland, J. (1953). "Distributions of quadratic forms and ratios of quadratic forms." *Ann. Math. Statist.*, **24**, 416–427.

Jackson, D. (1941). *Fourier Series and Orthogonal Polynomials*. Mathematical Association of America. (This classic monograph provides a good coverage of orthogonal polynomials, including Legendre, Jacobi, Hermite, and Laguerre polynomials. The presentation is informative and easy to follow.)

Johnson, N. L., and S. Kotz (1968). "Tables of distributions of positive definite quadratic forms in central normal variables." *Sankhyā, Ser. B*, **30**, 303–314.

Johnson, N. L., and S. Kotz (1970). *Continuous Univariate Distributions*—1. Houghton Mifflin, Boston. (Chap. 12 contains a good discussion concerning the Cornish–Fisher expansion of quantiles.)

Kendall, M. G., and A. Stuart (1977). *The Advanced Theory of Statistics*, Vol. 1, 4th ed. Macmillan, New York. (This classic book provides a good source for learning about the Gram–Charlier series of Type A and the Cornish–Fisher expansion.)

Kerridge, D. F., and G. W. Cook (1976). "Yet another series for the normal integral." *Biometrika*, **63**, 401–403.

Lieberman, G. J., and Owen, D. B. (1961). *Tables of the Hypergeometric Probability Distribution*. Stanford University Press, Palo Alto, California.

Ralston, A., and P. Rabinowitz (1978). *A First Course in Numerical Analysis*. McGraw-Hill, New York. (Chap. 7 discusses Chebyshev polynomials of the first kind).

Rivlin, T. (1990). *Chebyshev Polynomials*, 2nd ed. Wiley, New York. (This book gives a survey of the most important properties of Chebyshev polynomials.)

Schöne, A., and W. Schmid (2000). "On the joint distribution of a quadratic and a linear form in normal variables." *J. Mult. Analysis*, **72**, 163–182.

Schwartz, S. C. (1967). "Estimation of probability density by an orthogonal series," *Ann. Math. Statist.*, **38**, 1261–1265.

Subrahmaniam, K. (1966). "Some contributions to the theory of non-normality—I (univariate case)." *Sankhyā, Ser. A*, **28**, 389–406.

Szegö, G. (1975). *Orthogonal Polynomials*, 4th ed. Amer. Math. Soc., Providence, Rhode Island. (This much-referenced book provides a thorough coverage of orthogonal polynomials.)

Tiku, M. L. (1964a). "Approximating the general non-normal variance ratio sampling distributions." *Biometrika*, **51**, 83–95.

Tiku, M. L. (1964b). "A note on the negative moments of a truncated Poisson variate." *J. Amer. Statist. Assoc.*, **59**, 1220–1224.

Viskov, O. V. (1992). "Some remarks on Hermite polynomials." *Theory of Probability and its Applications*, **36**, 633–637.

EXERCISES

In Mathematics

10.1. Show that the sequence $\{1/\sqrt{2}, \cos x, \sin x, \cos 2x, \sin 2x, \ldots, \cos nx, \sin nx, \ldots\}$ is orthonormal with respect to $w(x) = 1/\pi$ over $[-\pi, \pi]$.

10.2. Let $\{p_n(x)\}_{n=0}^{\infty}$ be a sequence of Legendre polynomials
   (a) Use the Rodrigues formula to show that
      (i) $\int_{-1}^{1} x^m p_n(x)\, dx = 0$ for $m = 0, 1, \ldots, n-1$,
      (ii) $\int_{-1}^{1} x^n p_n(x)\, dx = \dfrac{2^{n+1}}{2n+1}\dbinom{2n}{n}^{-1}$, $n = 0, 1, 2, \ldots$
   (b) Deduce from (a) that $\int_{-1}^{1} p_n(x)\pi_{n-1}(x)\, dx = 0$, where $\pi_{n-1}(x)$ denotes an arbitrary polynomial of degree at most equal to $n-1$.
   (c) Make use of (a) and (b) to show that $\int_{-1}^{1} p_n^2(x)\, dx = 2/(2n+1)$, $n = 0, 1, \ldots$.

10.3. Let $\{T_n(x)\}_{n=0}^{\infty}$ be a sequence of Chebyshev polynomials of the first kind. Let $\zeta_i = \cos[(2i-1)\pi/2n]$, $i = 1, 2, \ldots, n$.
   (a) Verify that $\zeta_1, \zeta_2, \ldots, \zeta_n$ are zeros of $T_n(x)$, that is, $T_n(\zeta_i) = 0$, $i = 1, 2, \ldots, n$.
   (b) Show that $\zeta_1, \zeta_2, \ldots, \zeta_n$ are simple zeros of $T_n(x)$.
   [*Hint*: show that $T_n'(\zeta_i) \neq 0$ for $i = 1, 2, \ldots, n$.]

10.4. Let $\{H_n(x)\}_{n=0}^{\infty}$ be a sequence of Hermite polynomials. Show that
   (a) $\dfrac{dH_n(x)}{dx} = nH_{n-1}(x)$,
   (b) $\dfrac{d^2 H_n(x)}{dx^2} - x\,\dfrac{dH_n(x)}{dx} + nH_n(x) = 0$.

10.5. Let $\{T_n(x)\}_{n=0}^{\infty}$ and $\{U_n(x)\}_{n=0}^{\infty}$ be sequences of Chebyshev polynomials of the first and second kinds, respectively,

(a) Show that $|U_n(x)| \leq n + 1$ for $-1 \leq x \leq 1$.
[*Hint*: Use the representation (10.18) and mathematical induction on n.]

(b) Show that $|dT_n(x)/dx| \leq n^2$ for all $-1 \leq x \leq 1$, with equality holding only if $x = \mp 1$ $(n \geq 2)$.

(c) Show that

$$\int_{-1}^{x} \frac{T_n(t)}{\sqrt{1-t^2}}\, dt = -\frac{U_{n-1}(x)}{n}\sqrt{1-x^2}$$

for $-1 \leq x \leq 1$ and $n \neq 0$.

10.6. Show that the Laguerre polynomial $L_n(x)$ of degree n satisfies the differential equation

$$x\frac{d^2 L_n(x)}{dx^2} + (\alpha + 1 - x)\frac{dL_n(x)}{dx} + nL_n(x) = 0.$$

10.7. Consider the function

$$H(x,t) = \frac{1}{(1-t)^{\alpha+1}} e^{-xt/(1-t)}, \qquad \alpha > -1.$$

Expand $H(x,t)$ as a power series in t and let the coefficient of t^n be denoted by $(-1)^n g_n(x)/n!$ so that

$$H(x,t) = \sum_{n=0}^{\infty} \frac{(-1)^n}{n!} g_n(x)t^n.$$

Show that $g_n(x)$ is identical to $L_n(x)$ for all n, where $L_n(x)$ is the Laguerre polynomial of degree n.

10.8. Find the least-squares polynomial approximation of the function $f(x) = e^x$ over the interval $[-1, 1]$ by using a Legendre polynomial of degree not exceeding 4.

10.9. A function $f(x)$ defined on $-1 \leq x \leq 1$ can be represented using an infinite series of Chebyshev polynomials of the first kind, namely,

$$f(x) = \frac{c_0}{2} + \sum_{n=1}^{\infty} c_n T_n(x),$$

where

$$c_n = \frac{2}{\pi} \int_{-1}^{1} \frac{f(x)T_n(x)}{\sqrt{1-x^2}} \, dx, \qquad n = 0, 1, \ldots .$$

This series converges uniformly whenever $f(x)$ is continuous and of bounded variation on $[-1, 1]$. Approximate the function $f(x) = e^x$ using the first five terms of the above series.

In Statistics

10.10. Suppose that from a certain distribution with a mean equal to zero we have knowledge of the following central moments: $\mu_2 = 1.0$, $\mu_3 = -0.91$, $\mu_4 = 4.86$, $\mu_5 = -12.57$, $\mu_6 = 53.22$. Obtain an approximation for the density function of the distribution using Gram–Charlier series of type A.

10.11. The Cornish–Fisher expansion for x_α, the upper α-quantile of a certain distribution, standardized so that its mean and variance are equal to zero and one, respectively, is of the following form (see Johnson and Kotz, 1970, page 34):

$$
\begin{aligned}
x_\alpha = z_\alpha &+ \tfrac{1}{6}\left(z_\alpha^2 - 1\right)\kappa_3 + \tfrac{1}{24}\left(z_\alpha^3 - 3z_\alpha\right)\kappa_4 \\
&- \tfrac{1}{36}\left(2z_\alpha^3 - 5z_\alpha\right)\kappa_3^2 + \tfrac{1}{120}\left(z_\alpha^4 - 6z_\alpha^2 + 3\right)\kappa_5 \\
&- \tfrac{1}{24}\left(z_\alpha^4 - 5z_\alpha^2 + 2\right)\kappa_3\kappa_4 + \tfrac{1}{324}\left(12z_\alpha^4 - 53z_\alpha^2 + 17\right)\kappa_3^3 \\
&+ \tfrac{1}{720}\left(z_\alpha^5 - 10z_\alpha^3 + 15z_\alpha\right)\kappa_6 \\
&- \tfrac{1}{180}\left(2z_\alpha^5 - 17z_\alpha^3 + 21z_\alpha\right)\kappa_3\kappa_5 \\
&- \tfrac{1}{384}\left(3z_\alpha^5 - 24z_\alpha^3 + 29z_\alpha\right)\kappa_4^2 \\
&+ \tfrac{1}{288}\left(14z_\alpha^5 - 103z_\alpha^3 + 107z_\alpha\right)\kappa_3^2\kappa_4 \\
&- \tfrac{1}{7776}\left(252z_\alpha^5 - 1688z_\alpha^3 + 1511z_\alpha\right)\kappa_3^4 + \cdots,
\end{aligned}
$$

where z_α is the upper α-quantile of the standard normal distribution, and κ_r is the the rth cumulant of the distribution ($r = 3, 4, \ldots$). Apply this expansion to finding the upper 0.05-quantile of the central chi-squared distribution with $n = 5$ degrees of freedom.
[*Note*: The mean and variance of a central chi-squared distribution with n degrees of freedom are n and $2n$, respectively. Its rth cumu-

lant, denoted by κ'_r, is

$$\kappa'_r = n(r-1)! \, 2^{r-1}, \qquad r = 1,2,\dots .$$

Hence, the rth cumulant, κ_r, of the standardized chi-squared distribution is $\kappa_r = (2n)^{-r/2}\kappa'_r \ (r = 2,3,\dots).]$

10.12. The normal integral $\int_0^x e^{-t^2/2}\, dt$ can be calculated from the series

$$\int_0^x e^{-t^2/2}\, dt = x - \frac{x^3}{2\cdot 3\cdot 1!} + \frac{x^5}{2^2\cdot 5\cdot 2!} - \frac{x^7}{2^3\cdot 7\cdot 3!} + \cdots .$$

   (a) Use this series to obtain an approximate value for $\int_0^1 e^{-t^2/2}\, dt$.
   (b) Redo part (a) using the series given by formula (10.59), that is,

$$\int_0^x e^{-t^2/2}\, dt = xe^{-x^2/8} \sum_{n=0}^{\infty} \frac{\Theta_{2n}(x/2)}{2n+1}.$$

   (c) Compare the results from (a) and (b) with regard to the number of terms in each series needed to achieve an answer correct to five decimal places.

10.13. Show that the expansion given by formula (10.46) is equivalent to representing the density function $g(x)$ as a series of the form

$$g(x) = \sum_{n=0}^{\infty} \frac{c_n}{n!} \frac{d^n\phi(x)}{dx^n},$$

where $\phi(x)$ is the standard normal density function, and the c_n's are constant coefficients .

10.14. Consider the random variable

$$W = \sum_{i=1}^{n} X_i^2,$$

where $X_1, X_2, \dots, X_n$ are independent random variables from a distribution with the density function

$$f(x) = \phi(x) - \frac{\lambda_3}{6} \frac{d^3\phi(x)}{dx^3} + \frac{\lambda_4}{24} \frac{d^4\phi(x)}{dx^4} + \frac{\lambda_3^2}{72} \frac{d^6\phi(x)}{dx^6},$$

where $\phi(x)$ is the standard normal density function and the quantities λ_3 and λ_4 are, respectively, the standard measures of skewness and kurtosis for the distribution. Obtain the moment generating function of W, and compare it with the moment generating function of a chi-squared distribution with n degrees of freedom. (See Example 6.9.8 in Section 6.9.3.)

[*Hint*: Use Hermite polynomials.]

10.15. A lot of $M = 10$ articles contains $r = 3$ defectives and 7 good articles. Suppose that a sample of $m = 4$ articles is drawn from the lot without replacement. Let X denote the number of defective articles in the sample. Find the expected value of $g(X) = X^3$ using formula (10.65).

CHAPTER 11

Fourier Series

Fourier series were first formalized by the French mathematician Jean-Baptiste Joseph Fourier (1768–1830) as a result of his work on solving a particular partial differential equation known as the heat conduction equation. However, the actual introduction of the so-called Fourier theory was motivated by a problem in musical acoustics concerning vibrating strings. Daniel Bernoulli (1700–1782) is credited as being the first to model the motion of a vibrating string as a series of trigonometric functions in 1748, twenty years before the birth of Fourier. The actual development of Fourier theory took place in 1807 upon Fourier's return from Egypt, where he was a participant in the Egyptian campaign of 1798 under Napoleon Bonaparte.

11.1. INTRODUCTION

A series of the form

$$\frac{a_0}{2} + \sum_{n=1}^{\infty} [a_n \cos nx + b_n \sin nx] \tag{11.1}$$

is called a trigonometric series. Let $f(x)$ be a function defined and Riemann integrable on the interval $[-\pi, \pi]$. By definition, the Fourier series associated with $f(x)$ is a trigonometric series of the form (11.1), where a_n and b_n are given by

$$a_n = \frac{1}{\pi} \int_{-\pi}^{\pi} f(x)\cos nx\, dx, \qquad n = 0, 1, 2, \ldots, \tag{11.2}$$

$$b_n = \frac{1}{\pi} \int_{-\pi}^{\pi} f(x)\sin nx\, dx, \qquad n = 1, 2, \ldots. \tag{11.3}$$

In this case, we write

$$f(x) \sim \frac{a_0}{2} + \sum_{n=1}^{\infty} [a_n \cos nx + b_n \sin nx]. \tag{11.4}$$

The numbers a_n and b_n are called the Fourier coefficients of $f(x)$. The symbol $\sim$ is used here instead of equality because at this stage, nothing is known about the convergence of the series in (11.4) for all x in $[-\pi, \pi]$. Even if the series converges, it may not converge to $f(x)$.

We can also consider the following reverse approach: if the trigonometric series in (11.4) is uniformly convergent to $f(x)$ on $[-\pi, \pi]$, that is,

$$f(x) = \frac{a_0}{2} + \sum_{n=1}^{\infty} [a_n \cos nx + b_n \sin nx], \tag{11.5}$$

then a_n and b_n are given by formulas (11.2) and (11.3). In this case, the derivation of a_n and b_n is obtained by multiplying both sides of (11.5) by $\cos nx$ and $\sin nx$, respectively, followed by integration over $[-\pi, \pi]$. More specifically, to show formula (11.2), we multiply both sides of (11.5) by $\cos nx$. For $n \neq 0$, we then have

$$f(x)\cos nx = \frac{a_0}{2} \cos nx + \sum_{k=1}^{\infty} [a_k \cos kx \cos nx + b_k \sin kx \cos nx] \, dx. \tag{11.6}$$

Since the series on the right-hand side converges uniformly, it can be integrated term by term (this can be easily proved by applying Theorem 6.6.1 to the sequence whose nth term is the nth partial sum of the series). We then have

$$\int_{-\pi}^{\pi} f(x)\cos nx \, dx = \frac{a_0}{2} \int_{-\pi}^{\pi} \cos nx \, dx$$
$$+ \sum_{k=1}^{\infty} \left[\int_{-\pi}^{\pi} a_k \cos kx \cos nx \, dx + \int_{-\pi}^{\pi} b_k \sin kx \cos nx \, dx \right].$$

$$\tag{11.7}$$

But

$$\int_{-\pi}^{\pi} \cos nx \, dx = 0, \qquad n = 1, 2, \ldots, \tag{11.8}$$

$$\int_{-\pi}^{\pi} \sin kx \cos nx \, dx = 0, \tag{11.9}$$

$$\int_{-\pi}^{\pi} \cos kx \cos nx \, dx = \begin{cases} 0, & k \neq n, \\ \pi, & k = n \geq 1. \end{cases} \tag{11.10}$$

From (11.7)–(11.10) we conclude (11.2). Note that formulas (11.9) and (11.10) can be shown to be true by recalling the following trigonometric identities:

$$\sin kx \cos nx = \tfrac{1}{2}\{\sin[(k+n)x] + \sin[(k-n)x]\},$$

$$\cos kx \cos nx = \tfrac{1}{2}\{\cos[(k+n)x] + \cos[(k-n)x]\}.$$

For $n = 0$ we obtain from (11.7),

$$\int_{-\pi}^{\pi} f(x)\,dx = a_0\pi.$$

Formula (11.3) for b_n can be proved similarly. We can therefore state the following conclusion: a uniformly convergent trigonometric series is the Fourier series of its sum.

Note. If the series in (11.1) converges or diverges at a point x_0, then it converges or diverges at $x_0 + 2n\pi$ $(n = 1, 2, \ldots)$ due to the periodic nature of the sine and cosine functions. Thus, if the series (11.1) represents a function $f(x)$ on $[-\pi, \pi]$, then the series also represents the so-called periodic extension of $f(x)$ for all values of x. Geometrically speaking, the periodic extension of $f(x)$ is obtained by shifting the graph of $f(x)$ on $[-\pi, \pi]$ by $2\pi, 4\pi, \ldots$ to the right and to the left. For example, for $-3\pi < x \le -\pi$, $f(x)$ is defined by $f(x + 2\pi)$, and for $\pi < x \le 3\pi$, $f(x)$ is defined by $f(x - 2\pi)$, etc. This defines $f(x)$ for all x as a periodic function with period 2π.

EXAMPLE 11.1.1.   Let $f(x)$ be defined on $[-\pi, \pi]$ by the formula

$$f(x) = \begin{cases} 0, & -\pi \le x < 0, \\ \dfrac{x}{\pi}, & 0 \le x \le \pi. \end{cases}$$

Then, from (11.2) we have

$$a_n = \frac{1}{\pi}\int_{-\pi}^{\pi} f(x)\cos nx\,dx$$

$$= \frac{1}{\pi^2}\int_{0}^{\pi} x \cos nx\,dx$$

$$= \frac{1}{\pi^2}\left[\left.\frac{x \sin nx}{n}\right|_0^{\pi} - \frac{1}{n}\int_0^{\pi}\sin nx\,dx\right]$$

$$= \frac{1}{\pi^2 n^2}\left[(-1)^n - 1\right].$$

Thus, for $n = 1, 2, \ldots,$

$$a_n = \begin{cases} 0, & n \text{ even}, \\ -2/(\pi^2 n^2), & n \text{ odd}. \end{cases}$$

Also, from (11.3), we get

$$b_n = \frac{1}{\pi} \int_{-\pi}^{\pi} f(x) \sin nx \, dx$$

$$= \frac{1}{\pi^2} \int_{0}^{\pi} x \sin nx \, dx$$

$$= \frac{1}{\pi^2} \left[-\frac{x \cos nx}{n} \bigg|_0^{\pi} + \frac{1}{n} \int_0^{\pi} \cos nx \, dx \right]$$

$$= \frac{1}{\pi^2} \left[-\frac{\pi}{n}(-1)^n + \frac{\sin nx}{n^2} \bigg|_0^{\pi} \right]$$

$$= \frac{(-1)^{n+1}}{\pi n}.$$

For $n = 0$, a_0 is given by

$$a_0 = \frac{1}{\pi} \int_{-\pi}^{\pi} f(x) \, dx$$

$$= \frac{1}{\pi^2} \int_0^{\pi} x \, dx$$

$$= \tfrac{1}{2}.$$

The Fourier series of $f(x)$ is then of the form

$$f(x) \sim \frac{1}{4} - \frac{2}{\pi^2} \sum_{n=1}^{\infty} \frac{1}{(2n-1)^2} \cos[(2n-1)x] - \frac{1}{\pi} \sum_{n=1}^{\infty} \frac{(-1)^n}{n} \sin nx.$$

EXAMPLE 11.1.2. Let $f(x) = x^2(-\pi \le x \le \pi)$. Then

$$a_0 = \frac{1}{\pi} \int_{-\pi}^{\pi} x^2 \, dx$$

$$= \frac{2\pi^2}{3},$$

$$a_n = \frac{1}{\pi} \int_{-\pi}^{\pi} x^2 \cos nx\, dx$$

$$= \frac{x^2 \sin nx}{\pi n} \Big|_{-\pi}^{\pi} - \frac{2}{\pi n} \int_{-\pi}^{\pi} x \sin nx\, dx$$

$$= -\frac{2}{\pi n} \int_{-\pi}^{\pi} x \sin nx\, dx$$

$$= \frac{2x \cos nx}{\pi n^2} \Big|_{-\pi}^{\pi} - \frac{2}{\pi n^2} \int_{-\pi}^{\pi} \cos nx\, dx$$

$$= \frac{4 \cos n\pi}{n^2}$$

$$= \frac{4(-1)^n}{n^2}, \qquad n = 1, 2, \ldots,$$

$$b_n = \frac{1}{\pi} \int_{-\pi}^{\pi} x^2 \sin nx\, dx$$

$$= 0,$$

since $x^2 \sin nx$ is an odd function. Thus, the Fourier expansion of $f(x)$ is

$$x^2 \sim \frac{\pi^2}{3} - 4\left(\cos x - \frac{\cos 2x}{2^2} + \frac{\cos 3x}{3^2} - \cdots\right).$$

11.2. CONVERGENCE OF FOURIER SERIES

In this section, we consider the conditions under which the Fourier series of $f(x)$ converges to $f(x)$. We shall assume that $f(x)$ is Riemann integrable on $[-\pi, \pi]$. Hence, $f^2(x)$ is also Riemann integrable on $[-\pi, \pi]$ by Corollary 6.4.1. This condition will be satisfied if, for example, $f(x)$ is continuous on $[-\pi, \pi]$, or if it has a finite number of discontinuities of the first kind (see Definition 3.4.2) in this interval.

In order to study the convergence of Fourier series, the following lemmas are needed:

Lemma 11.2.1. If $f(x)$ is Riemann integrable on $[-\pi, \pi]$, then

$$\lim_{n \to \infty} \int_{-\pi}^{\pi} f(x) \cos nx\, dx = 0, \qquad (11.11)$$

$$\lim_{n \to \infty} \int_{-\pi}^{\pi} f(x) \sin nx\, dx = 0. \qquad (11.12)$$

Proof. Let $s_n(x)$ denote the following partial sum of Fourier series of $f(x)$:

$$s_n(x) = \frac{a_0}{2} + \sum_{k=1}^{n} [a_k \cos kx + b_k \sin kx], \tag{11.13}$$

where a_k $(k = 0, 1, \ldots, n)$ and b_k $(k = 1, 2, \ldots, n)$ are given by (11.2) and (11.3), respectively. Then

$$\int_{-\pi}^{\pi} f(x)s_n(x)\, dx = \frac{a_0}{2} \int_{-\pi}^{\pi} f(x)\, dx$$

$$+ \sum_{k=1}^{n} \left[a_k \int_{-\pi}^{\pi} f(x)\cos kx\, dx + b_k \int_{-\pi}^{\pi} f(x)\sin kx\, dx \right]$$

$$= \frac{\pi a_0^2}{2} + \pi \sum_{k=1}^{n} \left(a_k^2 + b_k^2 \right).$$

It can also be verified that

$$\int_{-\pi}^{\pi} s_n^2(x)\, dx = \frac{\pi a_0^2}{2} + \pi \sum_{k=1}^{n} \left(a_k^2 + b_k^2 \right).$$

Consequently,

$$\int_{-\pi}^{\pi} [f(x) - s_n(x)]^2\, dx = \int_{-\pi}^{\pi} f^2(x)\, dx - 2\int_{-\pi}^{\pi} f(x)s_n(x)\, dx + \int_{-\pi}^{\pi} s_n^2(x)\, dx$$

$$= \int_{-\pi}^{\pi} f^2(x)\, dx - \left[\frac{\pi a_0^2}{2} + \pi \sum_{k=1}^{n} \left(a_k^2 + b_k^2 \right) \right].$$

It follows that

$$\frac{a_0^2}{2} + \sum_{k=1}^{n} \left(a_k^2 + b_k^2 \right) \le \frac{1}{\pi} \int_{-\pi}^{\pi} f^2(x)\, dx. \tag{11.14}$$

Since the right-hand side of (11.14) exists and is independent of n, the series $\sum_{k=1}^{\infty}(a_k^2 + b_k^2)$ must be convergent. This follows from applying Theorem 5.1.2 and the fact that the sequence

$$s_n^* = \sum_{k=1}^{n} \left(a_k^2 + b_k^2 \right)$$

is bounded and monotone increasing. But, the convergence of $\sum_{k=1}^{\infty}(a_k^2 + b_k^2)$ implies that $\lim_{k \to \infty}(a_k^2 + b_k^2) = 0$ (see Result 5.2.1 in Chapter 5). Hence, $\lim_{k \to \infty} a_k = 0$, $\lim_{k \to \infty} b_k = 0$.   □

Corollary 11.2.1. If $\phi(x)$ is Riemann integrable on $[-\pi, \pi]$, then

$$\lim_{n \to \infty} \int_{-\pi}^{\pi} \phi(x)\sin\left[(n + \tfrac{1}{2})x\right] dx = 0. \tag{11.15}$$

Proof. We have that

$$\phi(x)\sin\left[(n + \tfrac{1}{2})x\right] = \left[\phi(x)\cos\frac{x}{2}\right]\sin nx + \left[\phi(x)\sin\frac{x}{2}\right]\cos nx.$$

Let $\phi_1(x) = \phi(x)\sin(x/2)$, $\phi_2(x) = \phi(x)\cos(x/2)$. Both $\phi_1(x)$ and $\phi_2(x)$ are Riemann integrable by Corollary 6.4.2. By applying Lemma 11.2.1 to both $\phi_1(x)$ and $\phi_2(x)$, we obtain

$$\lim_{n \to \infty} \int_{-\pi}^{\pi} \phi_1(x)\cos nx \, dx = 0, \tag{11.16}$$

$$\lim_{n \to \infty} \int_{-\pi}^{\pi} \phi_2(x)\sin nx \, dx = 0. \tag{11.17}$$

Formula (11.15) follows from the addition of (11.16) and (11.17). □

Corollary 11.2.2. If $\phi(x)$ is Riemann integrable on $[-\pi, \pi]$, then

$$\lim_{n \to \infty} \int_{-\pi}^{0} \phi(x)\sin\left[(n + \tfrac{1}{2})x\right] dx = 0,$$

$$\lim_{n \to \infty} \int_{0}^{\pi} \phi(x)\sin\left[(n + \tfrac{1}{2})x\right] dx = 0.$$

Proof. Define the functions $h_1(x)$ and $h_2(x)$ as

$$h_1(x) = \begin{cases} 0, & 0 \leq x \leq \pi, \\ \phi(x), & -\pi \leq x < 0, \end{cases}$$

$$h_2(x) = \begin{cases} \phi(x), & 0 \leq x \leq \pi, \\ 0, & -\pi \leq x < 0. \end{cases}$$

Both $h_1(x)$ and $h_2(x)$ are Riemann integrable on $[-\pi, \pi]$. Hence, by Corollary 11.2.1,

$$\lim_{n \to \infty} \int_{-\pi}^{0} \phi(x)\sin\left[(n + \tfrac{1}{2})x\right] dx = \lim_{n \to \infty} \int_{-\pi}^{\pi} h_1(x)\sin\left[(n + \tfrac{1}{2})x\right] dx$$

$$= 0$$

$$\lim_{n \to \infty} \int_{0}^{\pi} \phi(x)\sin\left[(n + \tfrac{1}{2})x\right] dx = \lim_{n \to \infty} \int_{-\pi}^{\pi} h_2(x)\sin\left[(n + \tfrac{1}{2})x\right] dx$$

$$= 0. \qquad \square$$

Lemma 11.2.2.

$$\frac{1}{2} + \sum_{k=1}^{n} \cos ku = \frac{\sin\left[\left(n + \frac{1}{2}\right)u\right]}{2\sin(u/2)}. \tag{11.18}$$

Proof. Let $G_n(u)$ be defined as

$$G_n(u) = \frac{1}{2} + \sum_{k=1}^{n} \cos ku.$$

Multiplying both sides by $2\sin(u/2)$ and using the identity

$$2\sin\frac{u}{2}\cos ku = \sin\left[\left(k + \frac{1}{2}\right)u\right] - \sin\left[\left(k - \frac{1}{2}\right)u\right],$$

we obtain

$$2\sin\frac{u}{2}\,G_n(u) = \sin\frac{u}{2} + \sum_{k=1}^{n}\left\{\sin\left[\left(k + \frac{1}{2}\right)u\right] - \sin\left[\left(k - \frac{1}{2}\right)u\right]\right\}$$

$$= \sin\left[\left(n + \frac{1}{2}\right)u\right].$$

Hence, if $\sin(u/2) \neq 0$, then

$$G_n(u) = \frac{\sin\left[\left(n + \frac{1}{2}\right)u\right]}{2\sin(u/2)}. \qquad \square$$

Theorem 11.2.1. Let $f(x)$ be Riemann integrable on $[-\pi, \pi]$, and let it be extended periodically outside this interval. Suppose that at a point x, $f(x)$ satisfies the following two conditions:

i. Both $f(x^-)$ and $f(x^+)$ exist, where $f(x^-)$ and $f(x^+)$ are the left-sided and right-sided limits of $f(x)$, and

$$f(x) = \frac{1}{2}[f(x^-) + f(x^+)]. \tag{11.19}$$

ii. Both one-sided derivatives,

$$f'(x^+) = \lim_{h \to 0^+} \frac{(x+h) - f(x^+)}{h},$$

$$f'(x^-) = \lim_{h \to 0^-} \frac{f(x+h) - f(x^-)}{h},$$

exist.

Then the Fourier series of $f(x)$ converges to $f(x)$ at x, that is,

$$s_n(x) \to \begin{cases} f(x) & \text{if } x \text{ is a point of continuity,} \\ \frac{1}{2}[f(x^+) + f(x^-)] & \text{if } x \text{ is a point of discontinuity of the first kind.} \end{cases}$$

Before proving this theorem, it should be noted that if $f(x)$ is continuous at x, then condition (11.19) is satisfied. If, however, x is a point of discontinuity of $f(x)$ of the first kind, then $f(x)$ is defined to be equal to the right-hand side of (11.19). Such a definition of $f(x)$ does not affect the values of the Fourier coefficients, a_n and b_n, in (11.2) and (11.3). Hence, the Fourier series of $f(x)$ remains unchanged.

Proof of Theorem 11.2.1. From (11.2), (11.3), and (11.13), we have

$$s_n(x) = \frac{1}{2\pi} \int_{-\pi}^{\pi} f(t)\, dt + \sum_{k=1}^{n} \left[\frac{1}{\pi} \left(\int_{-\pi}^{\pi} f(t)\cos kt\, dt \right) \cos kx \right.$$

$$\left. + \frac{1}{\pi} \left(\int_{-\pi}^{\pi} f(t)\sin kt\, dt \right) \sin kx \right]$$

$$= \frac{1}{\pi} \int_{-\pi}^{\pi} f(t) \left[\frac{1}{2} + \sum_{k=1}^{n} (\cos kt \cos kx + \sin kt \sin kx) \right] dt$$

$$= \frac{1}{\pi} \int_{-\pi}^{\pi} f(t) \left[\frac{1}{2} + \sum_{k=1}^{n} \cos k(t-x) \right] dt.$$

Using Lemma 11.2.2, $s_n(x)$ can be written as

$$s_n(x) = \frac{1}{\pi} \int_{-\pi}^{\pi} f(t) \left\{ \frac{\sin[(n + \frac{1}{2})(t-x)]}{2\sin[(t-x)/2]} \right\} dt. \tag{11.20}$$

If we make the change of variable $t - x = u$ in (11.20), we obtain

$$s_n(x) = \frac{1}{\pi} \int_{-\pi-x}^{\pi-x} f(x+u) \frac{\sin[(n + \frac{1}{2})u]}{2\sin(u/2)} du.$$

Since both $f(x+u)$ and $\sin[(n + \frac{1}{2})u]/[2\sin(u/2)]$ have period 2π with respect to u, the integral from $-\pi - x$ to $\pi - x$ has the same value as the one from $-\pi$ to π. Thus,

$$s_n(x) = \frac{1}{\pi} \int_{-\pi}^{\pi} f(x+u) \frac{\sin[(n + \frac{1}{2})u]}{2\sin(u/2)} du. \tag{11.21}$$

We now need to show that for each x,

$$\lim_{n \to \infty} s_n(x) = \tfrac{1}{2}[f(x^-) + f(x^+)].$$

Formula (11.21) can be written as

$$s_n(x) = \frac{1}{\pi} \int_{-\pi}^{0} f(x+u) \frac{\sin[(n+\tfrac{1}{2})u]}{2\sin(u/2)} du$$

$$+ \frac{1}{\pi} \int_{0}^{\pi} f(x+u) \frac{\sin[(n+\tfrac{1}{2})u]}{2\sin(u/2)} du$$

$$= \frac{1}{\pi} \int_{-\pi}^{0} [f(x+u) - f(x^-)] \frac{\sin[(n+\tfrac{1}{2})u]}{2\sin(u/2)} du$$

$$+ f(x^-) \frac{1}{\pi} \int_{-\pi}^{0} \frac{\sin[(n+\tfrac{1}{2})u]}{2\sin(u/2)} du$$

$$+ \frac{1}{\pi} \int_{0}^{\pi} [f(x+u) - f(x^+)] \frac{\sin[(n+\tfrac{1}{2})u]}{2\sin(u/2)} du$$

$$+ f(x^+) \frac{1}{\pi} \int_{0}^{\pi} \frac{\sin[(n+\tfrac{1}{2})u]}{2\sin(u/2)} du. \tag{11.22}$$

The first integral in (11.22) can be expressed as

$$\frac{1}{\pi} \int_{-\pi}^{0} [f(x+u) - f(x^-)] \frac{\sin[(n+\tfrac{1}{2})u]}{2\sin(u/2)} du$$

$$= \frac{1}{\pi} \int_{-\pi}^{0} \frac{f(x+u) - f(x^-)}{u} \cdot \frac{u}{2\sin(u/2)} \sin[(n+\tfrac{1}{2})u] du.$$

We note that the function

$$\frac{f(x+u) - f(x^-)}{u} \cdot \frac{u}{2\sin(u/2)}$$

is Riemann integrable on $[-\pi, 0]$, and at $u = 0$ it has a discontinuity of the first kind, since

$$\lim_{u \to 0^-} \frac{f(x+u) - f(x^-)}{u} = f'(x^-), \quad \text{and}$$

$$\lim_{u \to 0^-} \frac{u}{2\sin(u/2)} = 1,$$

that is, both limits are finite. Consequently, by applying Corollary 11.2.2 to the function $\{[f(x+u)-f(x^-)]/u\} \cdot u/[2\sin(u/2)]$, we get

$$\lim_{n\to\infty} \frac{1}{\pi} \int_{-\pi}^{0} [f(x+u)-f(x^-)] \frac{\sin\left[\left(n+\frac{1}{2}\right)u\right]}{2\sin(u/2)} \, du = 0. \qquad (11.23)$$

We can similarly show that the third integral in (11.22) has a limit equal to zero as $n\to\infty$, that is,

$$\lim_{n\to\infty} \frac{1}{\pi} \int_{0}^{\pi} [f(x+u)-f(x^+)] \frac{\sin\left[\left(n+\frac{1}{2}\right)u\right]}{2\sin(u/2)} \, du = 0. \qquad (11.24)$$

Furthermore, from Lemma 11.2.2, we have

$$\int_{-\pi}^{0} \frac{\sin\left[\left(n+\frac{1}{2}\right)u\right]}{2\sin(u/2)} \, du = \int_{-\pi}^{0} \left[\frac{1}{2} + \sum_{k=1}^{n} \cos ku\right] du$$

$$= \frac{\pi}{2}, \qquad (11.25)$$

$$\int_{0}^{\pi} \frac{\sin\left[\left(n+\frac{1}{2}\right)u\right]}{2\sin(u/2)} \, du = \int_{0}^{\pi} \left[\frac{1}{2} + \sum_{k=1}^{n} \cos ku\right] du$$

$$- \frac{\pi}{2}. \qquad (11.26)$$

From (11.22)–(11.26), we conclude that

$$\lim_{n\to\infty} s_n(x) = \frac{1}{2}[f(x^-) + f(x^+)]. \qquad \square$$

Definition 11.2.1.   A function $f(x)$ is said to be piecewise continuous on $[a, b]$ if it is continuous on $[a, b]$ except for a finite number of discontinuities of the first kind in $[a, b]$, and, in addition, both $f(a^+)$ and $f(b^-)$ exist. $\square$

Corollary 11.2.3.   Suppose that $f(x)$ is piecewise continuous on $[-\pi, \pi]$, and that it can be extended periodically outside this interval. In addition, if, at each interior point of $[-\pi, \pi]$, $f'(x^+)$ and $f'(x^-)$ exist and $f'(-\pi^+)$ and $f'(\pi^-)$ exist, then at a point x, the Fourier series of $f(x)$ converges to $\frac{1}{2}[f(x^-) + f(x^+)]$.

Proof. This follows directly from applying Theorem 11.2.1 and the fact that a piecewise continuous function on $[a, b]$ is Riemann integrable there.

$\square$

EXAMPLE 11.2.1.   Let $f(x) = x$ $(-\pi \le x \le \pi)$. The periodic extension of $f(x)$ (outside the interval $[-\pi, \pi]$) is defined everywhere. In this case,

$$a_n = \frac{1}{\pi} \int_{-\pi}^{\pi} x \cos nx \, dx = 0, \qquad n = 0, 1, 2, \dots,$$

$$b_n = \frac{1}{\pi} \int_{-\pi}^{\pi} x \sin nx \, dx$$

$$= \frac{2(-1)^{n+1}}{n}, \qquad n = 1, 2, \dots.$$

Hence, the Fourier series of $f(x)$ is

$$x \sim \sum_{n=1}^{\infty} \frac{2(-1)^{n+1}}{n} \sin nx.$$

This series converges to x at each x in $(-\pi, \pi)$. At $x = -\pi, \pi$ we have discontinuities of the first kind for the periodic extension of $f(x)$. Hence, at $x = \pi$, the Fourier series converges to

$$\tfrac{1}{2}[f(\pi^-) + f(\pi^+)] = \tfrac{1}{2}[\pi + (-\pi)]$$

$$= 0.$$

Similarly, at $x = -\pi$, the series converges to

$$\tfrac{1}{2}[f(-\pi^-) + f(-\pi^+)] = \tfrac{1}{2}[\pi + (-\pi)]$$

$$= 0.$$

For other values of x, the series converges to the value of the periodic extension of $f(x)$.

EXAMPLE 11.2.2.   Consider the function $f(x) = x^2$ defined in Example 11.1.2. Its Fourier series is

$$x^2 \sim \frac{\pi^2}{3} + 4 \sum_{n=1}^{\infty} \frac{(-1)^n}{n^2} \cos nx.$$

The periodic extension of $f(x)$ is continuous everywhere. We can therefore write

$$x^2 = \frac{\pi^2}{3} + 4 \sum_{n=1}^{\infty} \frac{(-1)^n}{n^2} \cos nx.$$

In particular, for $x = \mp \pi$, we have

$$\pi^2 = \frac{\pi^2}{3} + 4 \sum_{n=1}^{\infty} \frac{(-1)^{2n}}{n^2},$$

$$\frac{\pi^2}{6} = \sum_{n=1}^{\infty} \frac{1}{n^2}.$$

11.3. DIFFERENTIATION AND INTEGRATION OF FOURIER SERIES

In Section 11.2 conditions were given under which a function $f(x)$ defined on $[-\pi, \pi]$ is represented as a Fourier series. In this section, we discuss the conditions under which the series can be differentiated or integrated term by term.

Theorem 11.3.1. Let $a_0/2 + \sum_{n=1}^{\infty} [a_n \cos nx + b_n \sin nx]$ be the Fourier series of $f(x)$. If $f(x)$ is continuous on $[-\pi, \pi]$, $f(-\pi) = f(\pi)$, and $f'(x)$ is piecewise continuous on $[-\pi, \pi]$, then

a. at each point where $f''(x)$ exists, $f'(x)$ can be represented by the derivative of the Fourier series of $f(x)$, where differentiation is done term by term, that is,

$$f'(x) = \sum_{n=1}^{\infty} [nb_n \cos nx - na_n \sin nx];$$

b. the Fourier series of $f(x)$ converges uniformly and absolutely to $f(x)$ on $[-\pi, \pi]$.

Proof.

a. The Fourier series of $f(x)$ converges to $f(x)$ by Corollary 11.2.3. Thus,

$$f(x) = \frac{a_0}{2} + \sum_{n=1}^{\infty} [a_n \cos nx + b_n \sin nx].$$

The periodic extension of $f(x)$ is continuous, since $f(\pi) = f(-\pi)$. Furthermore, the derivative, $f'(x)$, of $f(x)$ satisfies the conditions of Corollary 11.2.3. Hence, the Fourier series of $f'(x)$ converges to $f'(x)$, that is,

$$f'(x) = \frac{\alpha_0}{2} + \sum_{n=1}^{\infty} [\alpha_n \cos nx + \beta_n \sin nx], \qquad (11.27)$$

where

$$\alpha_0 = \frac{1}{\pi} \int_{-\pi}^{\pi} f'(x)\, dx$$

$$= \frac{1}{\pi}[f(\pi) - f(-\pi)]$$

$$= 0,$$

$$\alpha_n = \frac{1}{\pi} \int_{-\pi}^{\pi} f'(x)\cos nx\, dx$$

$$= \frac{1}{\pi} f(x)\cos nx \Big|_{-\pi}^{\pi} + \frac{n}{\pi} \int_{-\pi}^{\pi} f(x)\sin nx\, dx$$

$$= \frac{(-1)^n}{n}[f(\pi) - f(-\pi)] + nb_n$$

$$= nb_n,$$

$$\beta_n = \frac{1}{\pi} \int_{-\pi}^{\pi} f'(x)\sin nx\, dx$$

$$= \frac{1}{\pi} f(x)\sin nx \Big|_{-\pi}^{\pi} - \frac{n}{\pi} \int_{-\pi}^{\pi} f(x)\cos nx\, dx$$

$$= -na_n.$$

By substituting α_0, α_n, and β_n in (11.27), we obtain

$$f'(x) = \sum_{n=1}^{\infty} [nb_n \cos nx - na_n \sin nx].$$

b. Consider the Fourier series of $f'(x)$ in (11.27), where $\alpha_0 = 0$, $\alpha_n = nb_n$, $\beta_n = -na_n$. Then, using inequality (11.14), we obtain

$$\sum_{n=1}^{\infty} (\alpha_n^2 + \beta_n^2) \le \frac{1}{\pi} \int_{-\pi}^{\pi} [f'(x)]^2\, dx. \qquad (11.28)$$

Inequality (11.28) indicates that $\sum_{n=1}^{\infty}(\alpha_n^2 + \beta_n^2)$ is a convergent series. Now, let $s_n(x) = a_0/2 + \sum_{k=1}^{n} [a_k \cos kx + b_k \sin kx]$. Then, for $n \ge m + 1$,

$$|s_n(x) - s_m(x)| = \left| \sum_{k=m+1}^{n} [a_k \cos kx + b_k \sin kx] \right|$$

$$\le \sum_{k=m+1}^{n} |a_k \cos kx + b_k \sin kx|.$$

Note that

$$|a_k \cos kx + b_k \sin kx| \le \left(a_k^2 + b_k^2\right)^{1/2}. \tag{11.29}$$

Inequality (11.29) follows from the fact that $a_k \cos kx + b_k \sin kx$ is the dot product, $\mathbf{u} \cdot \mathbf{v}$, of the vectors $\mathbf{u} = (a_k, b_k)'$, $\mathbf{v} = (\cos kx, \sin kx)'$, and $|\mathbf{u} \cdot \mathbf{v}| \le \|\mathbf{u}\|_2 \|\mathbf{v}\|_2$ (see Theorem 2.1.2). Hence,

$$\begin{aligned}
|s_n(x) - s_m(x)| &\le \sum_{k=m+1}^{n} \left(a_k^2 + b_k^2\right)^{1/2} \\
&= \sum_{k=m+1}^{n} \frac{1}{k}\left(\alpha_k^2 + \beta_k^2\right)^{1/2}. \tag{11.30}
\end{aligned}$$

But, by the Cauchy–Schwarz inequality,

$$\sum_{k=m+1}^{n} \frac{1}{k}\left(\alpha_k^2 + \beta_k^2\right)^{1/2} \le \left[\sum_{k=m+1}^{n} \frac{1}{k^2}\right]^{1/2} \left[\sum_{k=m+1}^{n} \left(\alpha_k^2 + \beta_k^2\right)\right]^{1/2},$$

and by (11.28),

$$\sum_{k=m+1}^{n} \left(\alpha_k^2 + \beta_k^2\right) \le \frac{1}{\pi}\int_{-\pi}^{\pi} [f'(x)]^2 \, dx.$$

In addition, $\sum_{k=1}^{\infty} 1/k^2$ is a convergent series. Hence, by the Cauchy criterion, for a given $\epsilon > 0$, there exists a positive integer N such that $\sum_{k=m+1}^{n} 1/k^2 < \epsilon^2$ if $n > m > N$. Hence,

$$\begin{aligned}
|s_n(x) - s_m(x)| &\le \sum_{k=m+1}^{n} |a_k \cos kx + b_k \sin kx| \\
&\le c\epsilon, \qquad \text{if } m > n > N, \tag{11.31}
\end{aligned}$$

where $c = \{(1/\pi)\int_{-\pi}^{\pi} [f'(x)]^2 \, dx\}^{1/2}$. The double inequality (11.31) shows that the Fourier series of $f(x)$ converges absolutely and uniformly to $f(x)$ on $[-\pi, \pi]$ by the Cauchy criterion. $\qquad\square$

Note that from (11.30) we can also conclude that $\sum_{k=1}^{\infty}(a_k^2 + b_k^2)^{1/2}$ satisfies the Cauchy criterion. This series is therefore convergent. Furthermore, it is easy to see that

$$\sum_{k=1}^{\infty} (|a_k| + |b_k|) \le \sum_{k=1}^{\infty} \left[2\left(a_k^2 + b_k^2\right)\right]^{1/2}.$$

This indicates that the series $\sum_{k=1}^{\infty}(|a_k| + |b_k|)$ is convergent by the comparison test.

Note that, in general, we should not expect that a term-by-term differentiation of a Fourier series of $f(x)$ will result in a Fourier series of $f'(x)$. For example, for the function $f(x) = x$, $-\pi \leq x \leq \pi$, the Fourier series is (see Example 11.2.1)

$$x \sim \sum_{n=1}^{\infty} \frac{2(-1)^{n+1}}{n} \sin nx.$$

Differentiating this series term by term, we obtain $\sum_{n=1}^{\infty} 2(-1)^{n+1} \cos nx$. This, however, is not the Fourier series of $f'(x) = 1$, since the Fourier series of $f'(x) = 1$ is just 1. Note that in this case, $f(\pi) \neq f(-\pi)$, which violates one of the conditions in Theorem 11.3.1.

Theorem 11.3.2. If $f(x)$ is piecewise continuous on $[-\pi, \pi]$ and has the Fourier series

$$f(x) \sim \frac{a_0}{2} + \sum_{n=1}^{\infty} [a_n \cos nx + b_n \sin nx], \qquad (11.32)$$

then a term-by-term integration of this series gives the Fourier series of $\int_{-\pi}^{x} f(t)\, dt$ for $x \in [-\pi, \pi]$, that is,

$$\int_{-\pi}^{x} f(t)\, dt = \frac{a_0(\pi + x)}{2} + \sum_{n=1}^{\infty} \left[\frac{a_n}{n} \sin nx - \frac{b_n}{n} (\cos nx - \cos n\pi) \right],$$

$$-\pi \leq x \leq \pi.$$

Furthermore, the integrated series converges uniformly to $\int_{-\pi}^{x} f(t)\, dt$.

Proof. Define the function $g(x)$ as

$$g(x) = \int_{-\pi}^{x} f(t)\, dt - \frac{a_0}{2} x. \qquad (11.33)$$

If $f(x)$ is piecewise continuous on $[-\pi, \pi]$, then it is Riemann integrable there, and by Theorem 6.4.7, $g(x)$ is continuous on $[-\pi, \pi]$. Furthermore, by Theorem 6.4.8, at each point where $f(x)$ is continuous, $g(x)$ is differentiable and

$$g'(x) = f(x) - \frac{a_0}{2}. \qquad (11.34)$$

This implies that $g'(x)$ is piecewise continuous on $[-\pi, \pi]$. In addition, $g(-\pi) = g(\pi)$. To show this, we have from (11.33)

$$g(-\pi) = \int_{-\pi}^{-\pi} f(t)\, dt + \frac{a_0}{2}\pi$$

$$= \frac{a_0}{2}\pi.$$

$$g(\pi) = \int_{-\pi}^{\pi} f(t)\, dt - \frac{a_0}{2}\pi$$

$$= a_0\pi - \frac{a_0}{2}\pi$$

$$= \frac{a_0}{2}\pi,$$

by the definition of a_0. Thus, the function $g(x)$ satisfies the conditions of Theorem 11.3.1. It follows that the Fourier series of $g(x)$ converges uniformly to $g(x)$ on $[-\pi, \pi]$. We therefore have

$$g(x) = \frac{A_0}{2} + \sum_{n=1}^{\infty} [A_n \cos nx + B_n \sin nx]. \tag{11.35}$$

Moreover, by part (a) of Theorem 11.3.1, we have

$$g'(x) = \sum_{n=1}^{\infty} [nB_n \cos nx - nA_n \sin nx]. \tag{11.36}$$

Then, from (11.32), (11.34), and (11.36), we obtain

$$a_n = nB_n, \qquad n = 1, 2, \ldots,$$

$$b_n = -nA_n, \qquad n = 1, 2, \ldots.$$

Substituting in (11.35), we get

$$g(x) = \frac{A_0}{2} + \sum_{n=1}^{\infty} \left[-\frac{b_n}{n} \cos nx + \frac{a_n}{n} \sin nx \right].$$

From (11.33) we then have

$$\int_{-\pi}^{x} f(t)\, dt = \frac{a_0 x}{2} + \frac{A_0}{2} + \sum_{n=1}^{\infty} \left[-\frac{b_n}{n} \cos nx + \frac{a_n}{n} \sin nx \right]. \tag{11.37}$$

To find the value of A_0, we set $x = -\pi$ in (11.37), which gives

$$0 = -\frac{a_0 \pi}{2} + \frac{A_0}{2} + \sum_{n=1}^{\infty} \left(-\frac{b_n}{n} \cos n\pi \right).$$

Hence,

$$\frac{A_0}{2} = \frac{a_0 \pi}{2} + \sum_{n=1}^{\infty} \frac{b_n}{n} \cos n\pi.$$

Substituting $A_0/2$ in (11.37), we finally obtain

$$\int_{-\pi}^{x} f(t) \, dt = \frac{a_0(\pi + x)}{2} + \sum_{n=1}^{\infty} \left[\frac{a_n}{n} \sin nx - \frac{b_n}{n} (\cos nx - \cos n\pi) \right]. \qquad \square$$

11.4. THE FOURIER INTEGRAL

We have so far considered Fourier series corresponding to a function defined on the interval $[-\pi, \pi]$. As was seen earlier in this chapter, if a function is initially defined on $[-\pi, \pi]$, we can extend its definition outside $[-\pi, \pi]$ by considering its periodic extension. For example, if $f(-\pi) = f(\pi)$, then we can define $f(x)$ everywhere in $(-\infty, \infty)$ by requiring that $f(x + 2\pi) = f(x)$ for all x. The choice of the interval $[-\pi, \pi]$ was made mainly for convenience. More generally, we can now consider a function $f(x)$ defined on the interval $[-c, c]$. For such a function, the corresponding Fourier series is given by

$$\frac{a_0}{2} + \sum_{n=1}^{\infty} \left[a_n \cos \left(\frac{n\pi x}{c} \right) + b_n \sin \left(\frac{n\pi x}{c} \right) \right], \qquad (11.38)$$

where

$$a_n = \frac{1}{c} \int_{-c}^{c} f(x) \cos \left(\frac{n\pi x}{c} \right) dx, \qquad n = 0, 1, 2, \ldots, \qquad (11.39)$$

$$b_n = \frac{1}{c} \int_{-c}^{c} f(x) \sin \left(\frac{n\pi x}{c} \right) dx, \qquad n = 1, 2, \ldots. \qquad (11.40)$$

Now, a question arises as to what to do when we have a function $f(x)$ that is already defined everywhere on $(-\infty, \infty)$, but is not periodic. We shall show that, under certain conditions, such a function can be represented by an infinite integral rather than by an infinite series. This integral is called a Fourier integral. We now show the development of such an integral.

Substituting the expressions for a_n and b_n given by (11.39), (11.40) into (11.38), we obtain the Fourier series

$$\frac{1}{2c} \int_{-c}^{c} f(t) \, dt + \frac{1}{c} \sum_{n=1}^{\infty} \int_{-c}^{c} f(t) \cos \left[\frac{n\pi}{c} (t - x) \right] dt. \qquad (11.41)$$

If c is finite and $f(x)$ satisfies the conditions of Corollary 11.2.3 on $[-c, c]$, then the Fourier series (11.41) converges to $\frac{1}{2}[f(x^-) + f(x^+)]$. However, this series representation of $f(x)$ is not valid outside the interval $[-c, c]$ unless $f(x)$ is periodic with the period $2c$.

In order to provide a representation that is valid for all values of x when $f(x)$ is not periodic, we need to consider extending the series in (11.41) by letting c go to infinity, assuming that $f(x)$ is absolutely integrable over the whole real line. We now show how this can be done:

As $c \to \infty$, the first term in (11.41) goes to zero provided that $\int_{-\infty}^{\infty} f(t) \, dt$ exists. To investigate the limit of the series in (11.41) as $c \to \infty$, we set $\lambda_1 = \pi/c$, $\lambda_2 = 2\pi/c$, ..., $\lambda_n = n\pi/c$, ..., $\Delta\lambda_n = \lambda_{n+1} - \lambda_n = \pi/c$, $n = 1, 2, \ldots$. We can then write

$$\frac{1}{c} \sum_{n=1}^{\infty} \int_{-c}^{c} f(t) \cos\left[\frac{n\pi}{c}(t-x)\right] dt = \frac{1}{\pi} \sum_{n=1}^{\infty} \Delta\lambda_n \int_{-c}^{c} f(t) \cos[\lambda_n(t-x)] \, dt.$$

$$(11.42)$$

When c is large, $\Delta\lambda_n$ is small, and the right-hand side of (11.42) will be an approximation of the integral

$$\frac{1}{\pi} \int_0^{\infty} \left\{ \int_{-\infty}^{\infty} f(t) \cos[\lambda(t-x)] \, dt \right\} d\lambda. \qquad (11.43)$$

This is the Fourier integral of $f(x)$. Note that (11.43) can be written as

$$\int_0^{\infty} [a(\lambda) \cos \lambda x + b(\lambda) \sin \lambda x] \, d\lambda, \qquad (11.44)$$

where

$$a(\lambda) = \frac{1}{\pi} \int_{-\infty}^{\infty} f(t) \cos \lambda t \, dt,$$

$$b(\lambda) = \frac{1}{\pi} \int_{-\infty}^{\infty} f(t) \sin \lambda t \, dt.$$

The expression in (11.44) resembles a Fourier series where the sum has been replaced by an integral and the parameter λ is used in place of the integer n. Moreover, $a(\lambda)$ and $b(\lambda)$ act like Fourier coefficients.

We now show that the Fourier integral in (11.43) provides a representation for $f(x)$ provided that $f(x)$ satisfies the conditions of the next theorem.

Theorem 11.4.1. Let $f(x)$ be piecewise continuous on every finite interval $[a, b]$. If $\int_{-\infty}^{\infty} |f(x)| \, dx$ exists, then at every point $x(-\infty < x < \infty)$ where

$f'(x^+)$ and $f'(x^-)$ exist, the Fourier integral of $f(x)$ converges to $\frac{1}{2}[f(x^-)+f(x^+)]$, that is,

$$\frac{1}{\pi}\int_0^\infty \left\{\int_{-\infty}^\infty f(t)\cos[\lambda(t-x)]\,dt\right\}d\lambda = \frac{1}{2}[f(x^-)+f(x^+)].$$

The proof of this theorem depends on the following lemmas:

Lemma 11.4.1. If $f(x)$ is piecewise continuous on $[a, b]$, then

$$\lim_{n\to\infty}\int_a^b f(x)\sin nx\,dx = 0, \qquad\qquad (11.45)$$

$$\lim_{n\to\infty}\int_a^b f(x)\cos nx\,dx = 0. \qquad\qquad (11.46)$$

Proof. Let the interval $[a, b]$ be divided into a finite number of subintervals on each of which $f(x)$ is continuous. Let any one of these subintervals be denoted by $[p, q]$. To prove formula (11.45) we only need to show that

$$\lim_{n\to\infty}\int_p^q f(x)\sin nx\,dx = 0. \qquad\qquad (11.47)$$

For this purpose, we divide the interval $[p, q]$ into k equal subintervals using the partition points $x_0 = p, x_1, x_2, \ldots, x_k = q$. We can then write the integral in (11.47) as

$$\sum_{i=0}^{k-1}\int_{x_i}^{x_{i+1}} f(x)\sin nx\,dx,$$

or equivalently as

$$\sum_{i=0}^{k-1}\left\{f(x_i)\int_{x_i}^{x_{i+1}}\sin nx\,dx + \int_{x_i}^{x_{i+1}}[f(x)-f(x_i)]\sin nx\,dx\right\}.$$

It follows that

$$\left|\int_p^q f(x)\sin nx\,dx\right| \le \sum_{i=0}^{k-1}\left|f(x_i)\frac{\cos nx_i - \cos nx_{i+1}}{n}\right|$$

$$+ \sum_{i=0}^{k-1}\int_{x_i}^{x_{i+1}}|f(x)-f(x_i)|\,dx.$$

Let M denote the maximum value of $|f(x)|$ on $[p, q]$. Then

$$\left| \int_p^q f(x)\sin nx\, dx \right| \le \frac{2Mk}{n} + \sum_{i=0}^{k-1} \int_{x_i}^{x_{i+1}} |f(x) - f(x_i)|\, dx. \quad (11.48)$$

Furthermore, since $f(x)$ is continuous on $[p, q]$, it is uniformly continuous there [if necessary, $f(x)$ can be made continuous at p, q by simply using $f(p^+)$ and $f(q^-)$ as the values of $f(x)$ at p, q, respectively]. Hence, for a given $\epsilon > 0$, there exists a $\delta > 0$ such that

$$|f(x_1) - f(x_2)| < \frac{\epsilon}{2(q-p)} \quad (11.49)$$

if $|x_1 - x_2| < \delta$, where x_1 and x_2 are points in $[p, q]$. If k is chosen large enough so that $|x_{i+1} - x_i| < \delta$, and hence $|x - x_i| < \delta$ if $x_i \le x \le x_{i+1}$, then from (11.48) we obtain

$$\left| \int_p^q f(x)\sin nx\, dx \right| \le \frac{2Mk}{n} + \frac{\epsilon}{2(q-p)} \sum_{i=0}^{k-1} \int_{x_i}^{x_{i+1}} dx,$$

or

$$\left| \int_p^q f(x)\sin nx\, dx \right| \le \frac{2Mk}{n} + \frac{\epsilon}{2},$$

since

$$\sum_{i=0}^{k-1} \int_{x_i}^{x_{i+1}} dx = \sum_{i=0}^{k-1} (x_{i+1} - x_i)$$

$$= q - p.$$

Choosing n large enough so that $2Mk/n < \epsilon/2$, we finally get

$$\left| \int_p^q f(x)\sin nx\, dx \right| < \epsilon. \quad (11.50)$$

Formula (11.45) follows from (11.50), since $\epsilon > 0$ is arbitrary. Formula (11.46) can be proved in a similar fashion. $\square$

Lemma 11.4.2. If $f(x)$ is piecewise continuous on $[0, b]$ and $f'(0^+)$ exists, then

$$\lim_{n \to \infty} \int_0^b f(x) \frac{\sin nx}{x}\, dx = \frac{\pi}{2} f(0^+).$$

Proof. We have that

$$\int_0^b f(x) \frac{\sin nx}{x} dx = f(0^+) \int_0^b \frac{\sin nx}{x} dx$$

$$+ \int_0^b \frac{f(x) - f(0^+)}{x} \sin nx\, dx. \qquad (11.51)$$

But

$$\lim_{n \to \infty} \int_0^b \frac{\sin nx}{x} dx = \lim_{n \to \infty} \int_0^{bn} \frac{\sin x}{x} dx$$

$$= \int_0^\infty \frac{\sin x}{x} dx = \frac{\pi}{2}$$

(see Gillespie, 1959, page 89). Furthermore, the function $(1/x)[f(x) - f(0^+)]$ is piecewise continuous on $[0, b]$, since $f(x)$ is, and

$$\lim_{x \to 0^+} \frac{f(x) - f(0^+)}{x} = f'(0^+),$$

which exists. Hence, by Lemma 11.4.1,

$$\lim_{n \to \infty} \int_0^b \frac{f(x) - f(0^+)}{x} \sin nx\, dx = 0.$$

From (11.51) we then have

$$\lim_{n \to \infty} \int_0^b f(x) \frac{\sin nx}{x} dx = \frac{\pi}{2} f(0^+). \qquad \square$$

Lemma 11.4.3. If $f(x)$ is piecewise continuous on $[a, b]$, and $f'(x_0^-)$, $f'(x_0^+)$ exist at x_0, $a < x_0 < b$, then

$$\lim_{n \to \infty} \int_a^b f(x) \frac{\sin[n(x - x_0)]}{x - x_0} dx = \frac{\pi}{2} [f(x_0^-) + f(x_0^+)].$$

Proof. We have that

$$\int_a^b f(x) \frac{\sin[n(x - x_0)]}{x - x_0} dx = \int_a^{x_0} f(x) \frac{\sin[n(x - x_0)]}{x - x_0} dx$$

$$+ \int_{x_0}^b f(x) \frac{\sin[n(x - x_0)]}{x - x_0} dx$$

$$= \int_0^{x_0 - a} f(x_0 - x) \frac{\sin nx}{x} dx$$

$$+ \int_0^{b - x_0} f(x_0 + x) \frac{\sin nx}{x} dx.$$

Lemma 11.4.2 applies to each of the above integrals, since the right-hand derivatives of $f(x_0 - x)$ and $f(x_0 + x)$ at $x = 0$ are $-f'(x_0^-)$ and $f'(x_0^+)$, respectively, and both derivatives exist. Furthermore,

$$\lim_{x \to 0^+} f(x_0 - x) = f(x_0^-)$$

and

$$\lim_{x \to 0^+} f(x_0 + x) = f(x_0^+).$$

It follows that

$$\lim_{n \to \infty} \int_a^b f(x) \frac{\sin[n(x - x_0)]}{x - x_0} \, dx = \frac{\pi}{2} [f(x_0^-) + f(x_0^+)]. \qquad \square$$

Proof of Theorem 11.4.1. The function $f(x)$ satisfies the conditions of Lemma 11.4.3 on the interval $[a, b]$. Hence, at any point x, $a < x < b$, where $f'(x_0^-)$ and $f'(x_0^+)$ exist,

$$\lim_{\lambda \to \infty} \int_a^b f(t) \frac{\sin[\lambda(t - x)]}{t - x} \, dt = \frac{\pi}{2} [f(x^-) + f(x^+)]. \qquad (11.52)$$

Let us now partition the integral

$$I = \int_{-\infty}^{\infty} f(t) \frac{\sin[\lambda(t - x)]}{t - x} \, dt$$

as

$$I = \int_{-\infty}^a f(t) \frac{\sin[\lambda(t - x)]}{t - x} \, dt + \int_a^b f(t) \frac{\sin[\lambda(t - x)]}{t - x} \, dt$$

$$+ \int_b^{\infty} f(t) \frac{\sin[\lambda(t - x)]}{t - x} \, dt. \qquad (11.53)$$

From the first integral in (11.53) we have

$$\left| \int_{-\infty}^a f(t) \frac{\sin[\lambda(t - x)]}{t - x} \, dt \right| \le \int_{-\infty}^a \frac{|f(t)|}{|t - x|} \, dt.$$

Since $t \le a$ and $a < x$, then $|t - x| \ge x - a$. Hence,

$$\int_{-\infty}^a \frac{|f(t)|}{|t - x|} \, dt \le \frac{1}{x - a} \int_{-\infty}^a |f(t)| \, dt. \qquad (11.54)$$

The integral on the right-hand side of (11.54) exists because $\int_{-\infty}^{\infty}|f(t)|\,dt$ does. Similarly, from the third integral in (11.53) we have, if $x < b$,

$$\left|\int_b^\infty f(t)\,\frac{\sin[\lambda(t-x)]}{t-x}\,dt\right| \le \int_b^\infty \frac{|f(t)|}{|t-x|}\,dt$$

$$\le \frac{1}{b-x}\int_b^\infty |f(t)|\,dt$$

$$\le \frac{1}{b-x}\int_{-\infty}^\infty |f(t)|\,dt.$$

Hence, the first and third integrals in (11.53) are convergent. It follows that for any $\epsilon > 0$, there exists a positive number N such that if $a < -N$ and $b > N$, then these integrals will each be less than $\epsilon/3$ in absolute value. Furthermore, by (11.52), the absolute value of the difference between the second integral in (11.53) and the value $(\pi/2)[f(x^-)+f(x^+)]$ can be made less then $\epsilon/3$, if λ is chosen large enough. Consequently, the absolute value of the difference between the value of the integral I and $(\pi/2)[f(x^-)+f(x^+)]$ will be less than ϵ, if λ is chosen large enough. Thus,

$$\lim_{\lambda\to\infty}\int_{-\infty}^\infty f(t)\,\frac{\sin[\lambda(t-x)]}{t-x}\,dt = \frac{\pi}{2}[f(x^-)+f(x^+)]. \qquad (11.55)$$

The expression $\sin[\lambda(t-x)]/(t-x)$ in (11.55) can be written as

$$\frac{\sin[\lambda(t-x)]}{t-x} = \int_0^\lambda \cos[\alpha(t-x)]\,d\alpha.$$

Formula (11.55) can then be expressed as

$$\tfrac{1}{2}[f(x^-)+f(x^+)] = \frac{1}{\pi}\lim_{\lambda\to\infty}\int_{-\infty}^\infty f(t)\,dt\int_0^\lambda \cos[\alpha(t-x)]\,d\alpha$$

$$= \frac{1}{\pi}\lim_{\lambda\to\infty}\int_0^\lambda d\alpha\int_{-\infty}^\infty f(t)\cos[\alpha(t-x)]\,dt. \qquad (11.56)$$

The change of the order of integration in (11.56) is valid because the integrand in (11.56) does not exceed $|f(t)|$ in absolute value, so that the integral $\int_{-\infty}^\infty f(t)\cos[\alpha(t-x)]\,dt$ converges uniformly for all α (see Carslaw, 1930, page 199; Pinkus and Zafrany, 1997, page 187). From (11.56) we finally obtain

$$\tfrac{1}{2}[f(x^-)+f(x^+)] = \frac{1}{\pi}\int_0^\infty \left\{\int_{-\infty}^\infty f(t)\cos[\alpha(t-x)]\,dt\right\}d\alpha. \qquad \square$$

11.5. APPROXIMATION OF FUNCTIONS BY TRIGONOMETRIC POLYNOMIALS

By a trigonometric polynomial of the nth order it is meant an expression of the form

$$t_n(x) = \frac{\alpha_0}{2} + \sum_{k=1}^{n} [\alpha_k \cos kx + \beta_k \sin kx]. \tag{11.57}$$

A theorem of Weierstrass states that any continuous function of period 2π can be uniformly approximated by a trigonometric polynomial of some order (see, for example, Tolstov, 1962, Chapter 5). Thus, for a given $\epsilon > 0$, there exists a trigonometric polynomial of the form (11.57) such that

$$|f(x) - t_n(x)| < \epsilon$$

for all values of x. In case the Fourier series for $f(x)$ is uniformly convergent, then $t_n(x)$ can be chosen to be equal to $s_n(x)$, the nth partial sum of the Fourier series. However, it should be noted that $t_n(x)$ is not merely a partial sum of the Fourier series for $f(x)$, since a continuous function may have a divergent Fourier series (see Jackson, 1941, page 26). We now show that $s_n(x)$ has a certain optimal property among all trigonometric polynomials of the same order. To demonstrate this fact, let $f(x)$ be Riemann integrable on $[-\pi, \pi]$, and let $s_n(x)$ be the partial sum of order n of its Fourier series, that is, $s_n(x) = a_0/2 + \sum_{k=1}^{n} [a_k \cos kx + b_k \sin kx]$. Let $r_n(x) = f(x) - s_n(x)$. Then, from (11.2),

$$\int_{-\pi}^{\pi} f(x) \cos kx \, dx = \int_{-\pi}^{\pi} s_n(x) \cos kx \, dx$$

$$= \pi a_k, \qquad k = 0, 1, \ldots, n.$$

Hence,

$$\int_{-\pi}^{\pi} r_n(x) \cos kx \, dx = 0 \qquad \text{for } k \leq n. \tag{11.58}$$

We can similarly show that

$$\int_{-\pi}^{\pi} r_n(x) \sin kx \, dx = 0 \qquad \text{for } k \leq n. \tag{11.59}$$

Now, let $u_n(x) = t_n(x) - s_n(x)$, where $t_n(x)$ is given by (11.57). Then

$$\int_{-\pi}^{\pi} [f(x) - t_n(x)]^2 \, dx = \int_{-\pi}^{\pi} [r_n(x) - u_n(x)]^2 \, dx$$

$$= \int_{-\pi}^{\pi} r_n^2(x) \, dx - 2\int_{-\pi}^{\pi} r_n(x)u_n(x) \, dx + \int_{-\pi}^{\pi} u_n^2(x) \, dx$$

$$= \int_{-\pi}^{\pi} [f(x) - s_n(x)]^2 \, dx + \int_{-\pi}^{\pi} u_n^2(x) \, dx, \qquad (11.60)$$

since, by (11.58) and (11.59),

$$\int_{-\pi}^{\pi} r_n(x)u_n(x) \, dx = 0.$$

From (11.60) it follows that

$$\int_{-\pi}^{\pi} [f(x) - t_n(x)]^2 \, dx \geq \int_{-\pi}^{\pi} [f(x) - s_n(x)]^2 \, dx. \qquad (11.61)$$

This shows that for all trigonometric polynomials of order n, $\int_{-\pi}^{\pi}[f(x) - t_n(x)]^2 \, dx$ is minimized when $t_n(x) = s_n(x)$.

11.5.1. Parseval's Theorem

Suppose that we have the Fourier series (11.5) for the function $f(x)$, which is assumed to be continuous of period 2π. Let $s_n(x)$ be the nth partial sum of the series. We recall from the proof of Lemma 11.2.1 that

$$\int_{-\pi}^{\pi} [f(x) - s_n(x)]^2 \, dx = \int_{-\pi}^{\pi} f^2(x) \, dx - \left[\frac{\pi a_0^2}{2} + \pi \sum_{k=1}^{n} (a_k^2 + b_k^2) \right]. \qquad (11.62)$$

We also recall that for a given $\epsilon > 0$, there exists a trigonometric polynomial $t_n(x)$ of order n such that

$$|f(x) - t_n(x)| < \epsilon.$$

Hence,

$$\int_{-\pi}^{\pi} [f(x) - t_n(x)]^2 \, dx < 2\pi\epsilon^2.$$

Applying (11.61), we obtain

$$\int_{-\pi}^{\pi} [f(x) - s_n(x)]^2 \, dx \leq \int_{-\pi}^{\pi} [f(x) - t_n(x)]^2 \, dx$$

$$< 2\pi\epsilon^2. \tag{11.63}$$

Since $\epsilon > 0$ is arbitrary, we may conclude from (11.62) and (11.63) that the limit of the right-hand side of (11.62) is zero as $n \to \infty$, that is,

$$\frac{1}{\pi} \int_{-\pi}^{\pi} f^2(x) \, dx = \frac{a_0^2}{2} + \sum_{k=1}^{\infty} \left(a_k^2 + b_k^2 \right).$$

This result is known as *Parseval's theorem* after Marc Antoine Parseval (1755–1836).

11.6. THE FOURIER TRANSFORM

In the previous sections we discussed Fourier series for functions defined on a finite interval (or periodic functions defined on R, the set of all real numbers). In this section, we study a particular transformation of functions defined on R which are not periodic.

Let $f(x)$ be defined on $R = (-\infty, \infty)$. The Fourier transform of $f(x)$ is a function defined on R as

$$F(w) = \frac{1}{2\pi} \int_{-\infty}^{\infty} f(x) e^{-iwx} \, dx, \tag{11.64}$$

where i the complex number $\sqrt{-1}$, and

$$e^{-iwx} = \cos wx - i \sin wx.$$

A proper understanding of such a transformation requires some knowledge of complex analysis, which is beyond the scope of this book. However, due to the importance and prevalence of the use of this transformation in various fields of science and engineering, some coverage of its properties is necessary. For this reason, we merely state some basic results and properties concerning this transformation. For more details, the reader is referred to standard books on Fourier series, for example, Pinkus and Zafrany (1997, Chapter 3), Kufner and Kadlec (1971, Chapter 8), and Weaver (1989, Chapter 6).

Theorem 11.6.1. If $f(x)$ is absolutely integrable on R, then its Fourier transform $F(w)$ exists.

Theorem 11.6.2. If $f(x)$ is piecewise continuous and absolutely integrable on R, then its Fourier transform $F(w)$ has the following properties:

a. $F(w)$ is a continuous function on R.
b. $\lim_{w \to \mp \infty} F(w) = 0$.

Note that $f(x)$ is piecewise continuous on R if it is piecewise continuous on each finite interval $[a, b]$.

EXAMPLE 11.6.1.　Let $f(x) = e^{-|x|}$. This function is absolutely integrable on R, since

$$\int_{-\infty}^{\infty} e^{-|x|}\, dx = 2 \int_{0}^{\infty} e^{-x}\, dx$$

$$= 2.$$

Its Fourier transform is given by

$$F(w) = \frac{1}{2\pi} \int_{-\infty}^{\infty} e^{-|x|} e^{-iwx}\, dx$$

$$= \frac{1}{2\pi} \int_{-\infty}^{\infty} e^{-|x|}(\cos wx - i \sin wx)\, dx$$

$$= \frac{1}{2\pi} \int_{-\infty}^{\infty} e^{-|x|} \cos wx\, dx$$

$$= \frac{1}{\pi} \int_{0}^{\infty} e^{-x} \cos wx\, dx.$$

Integrating by parts twice, it can be shown that

$$F(w) = \frac{1}{\pi(1 + w^2)}.$$

EXAMPLE 11.6.2.　Consider the function

$$f(x) = \begin{cases} 1 & |x| \le a, \\ 0 & \text{otherwise,} \end{cases}$$

where a is a finite positive number. This function is absolutely integrable on R, since

$$\int_{-\infty}^{\infty} |f(x)|\, dx = \int_{-a}^{a} dx$$

$$= 2a.$$

Its Fourier transform is given by

$$F(w) = \frac{1}{2\pi} \int_{-a}^{a} e^{-iwx}\, dx$$

$$= \frac{1}{2\pi iw} (e^{iwa} - e^{-iwa})$$

$$= \frac{\sin wa}{\pi w}.$$

The next theorem gives the condition that makes it possible to express the function $f(x)$ in terms of its Fourier transform using the so-called *inverse Fourier transform*.

Theorem 11.6.3. Let $f(x)$ be piecewise continuous and absolutely integrable on R. Then for every point $x \in R$ where $f'(x^-)$ and $f'(x^+)$ exist, we have

$$\tfrac{1}{2}[f(x^-) + f(x^+)] = \int_{-\infty}^{\infty} F(w)e^{iwx}\, dw.$$

In particular, if $f(x)$ is continuous on R, then

$$f(x) = \int_{-\infty}^{\infty} F(w)e^{iwx}\, dw. \tag{11.65}$$

By applying Theorem 11.6.3 to the function in Example 11.6.1, we obtain

$$e^{-|x|} = \int_{-\infty}^{\infty} \frac{e^{iwx}}{\pi(1+w^2)}\, dw$$

$$= \int_{-\infty}^{\infty} \frac{1}{\pi(1+w^2)} (\cos wx + i \sin wx)\, dw$$

$$= \int_{-\infty}^{\infty} \frac{\cos wx}{\pi(1+w^2)}\, dw$$

$$= \frac{2}{\pi} \int_{0}^{\infty} \frac{\cos wx}{1+w^2}\, dw.$$

11.6.1. Fourier Transform of a Convolution

Let $f(x)$ and $g(x)$ be absolutely integrable functions on R. By definition, the function

$$h(x) = \int_{-\infty}^{\infty} f(x-y)g(y)\, dy \tag{11.66}$$

is called the convolution of $f(x)$ and $g(x)$ and is denoted by $(f * g)(x)$.

Theorem 11.6.4. Let $f(x)$ and $g(x)$ be absolutely integrable on R. Let $F(w)$ and $G(w)$ be their respective Fourier transforms. Then, the Fourier transform of the convolution $(f * g)(x)$ is given by $2\pi F(w)G(w)$.

11.7. APPLICATIONS IN STATISTICS

Fourier series have been used in a wide variety of areas in statistics, such as time series, stochastic processes, approximation of probability distribution functions, and the modeling of a periodic response variable, to name just a few. In addition, the methods and results of Fourier analysis have been effectively utilized in the analytic theory of probability (see, for example, Kawata, 1972).

11.7.1. Applications in Time Series

A time series is a collection of observations made sequentially in time. Examples of time series can be found in a variety of fields ranging from economics to engineering. Many types of time series occur in the physical sciences, particularly in meteorology, such as the study of rainfall on successive days, as well as in marine science and geophysics.

The stimulus for the use of Fourier methods in time series analysis is the recognition that when observing data over time, some aspects of an observed physical phenomenon tend to exhibit cycles or periodicities. Therefore, when considering a model to represent such data, it is natural to use models that contain sines and cosines, that is, trigonometric models, to describe the behavior. Let $y_1, y_2, \ldots, y_n$ denote a time series consisting of n observations obtained over time. These observations can be represented by the trigonometric polynomial model

$$y_t = \frac{a_0}{2} + \sum_{k=1}^{m} [a_k \cos \omega_k t + b_k \sin \omega_k t], \qquad t = 1, 2, \ldots, n,$$

where

$$\omega_k = \frac{2\pi k}{n}, \qquad\qquad k = 0, 1, 2, \ldots, m,$$

$$a_k = \frac{2}{n} \sum_{t=1}^{n} y_t \cos \omega_k t, \qquad k = 0, 1, \ldots, m,$$

$$b_k = \frac{2}{n} \sum_{t=1}^{n} y_t \sin \omega_k t, \qquad k = 1, 2, \ldots, m.$$

The values $\omega_1, \omega_2, \ldots, \omega_m$ are called harmonic frequencies. This model provides a decomposition of the time series into a set of cycles based on the harmonic frequencies. Here, n is assumed to be odd and equal to $2m + 1$, so that the harmonic frequencies lie in the range 0 to π. The expressions for a_k $(k = 0, 1, \ldots, m)$ and b_k $(k = 1, 2, \ldots, m)$ were obtained by treating the model as a linear regression model with $2m + 1$ parameters and then fitting it to the $2m + 1$ observations by the method of least squares. See, for example, Fuller (1976, Chapter 7).

The quantity

$$I_n(\omega_k) = \frac{n}{2}(a_k^2 + b_k^2), \qquad k = 1, 2, \ldots, m, \qquad (11.67)$$

represents the sum of squares associated with the frequency ω_k. For $k = 1, 2, \ldots, m$, the quantities in (11.67) define the so-called *periodogram*.

If $y_1, y_2, \ldots, y_n$ are independently distributed as normal variates with zero means and variances σ^2, then the a_k's and b_k's, being linear combinations of the y_t's, will be normally distributed. They are also independent, since the sine and cosine functions are orthogonal. It follows that $[n/(2\sigma^2)](a_k^2 + b_k^2)$, for $k = 1, 2, \ldots, m$, are distributed as independent chi-squared variates with two degrees of freedom each. The periodogram can be used to search for cycles or periodicities in the data.

Much of time series data analysis is based on the Fourier transform and its efficient computation. For more details concerning Fourier analysis of time series, the reader is referred to Bloomfield (1976) and Otnes and Enochson (1978).

11.7.2. Representation of Probability Distributions

One of the interesting applications of Fourier series in statistics is in providing a representation that can be used to evaluate the distribution function of a random variable with a finite range. Woods and Posten (1977) introduced two such representations by combining the concepts of Fourier series and Chebyshev polynomials of the first kind (see Section 10.4.1). These representations are given by the following two theorems:

Theorem 11.7.1. Let X be a random variable with a cumulative distribution function $F(x)$ defined on $[0, 1]$. Then, $F(x)$ can be represented as a Fourier series of the form

$$F(x) = \begin{cases} 0, & x < 0, \\ 1 - \theta/\pi - \sum_{n=1}^{\infty} b_n \sin n\theta, & 0 \le x \le 1, \\ 1, & x > 1, \end{cases}$$

where $\theta = \text{Arccos}(2x - 1)$, $b_n = [2/(n\pi)]E[T_n^*(X)]$, and $E[T_n^*(X)]$ is the expected value of the random variable

$$T_n^*(X) = \cos[n\,\text{Arccos}(2X - 1)], \qquad 0 \le X \le 1. \qquad (11.68)$$

Note that $T_n^*(x)$ is basically a Chebyshev polynomial of the first kind and of the nth degree defined on $[0, 1]$.

Proof. See Theorem 1 in Woods and Posten (1977).   □

The second representation theorem is similar to Theorem 11.7.1, except that X is now assumed to be a random variable over $[-1, 1]$.

Theorem 11.7.2.   Let X be a random variable with a cumulative distribution function $F(x)$ defined on $[-1, 1]$. Then

$$F(x) = \begin{cases} 0, & x < -1, \\ 1 - \theta/\pi - \sum_{n=1}^{\infty} b_n \sin n\theta, & -1 \le x \le 1, \\ 1, & x > 1, \end{cases}$$

where $\theta = \text{Arccos}\,x$, $b_n = [2/(n\pi)]E[T_n(X)]$, $E[T_n(X)]$ is the expected value of the random variable

$$T_n(X) = \cos[n\,\text{Arccos}\,X], \qquad -1 \le X \le 1,$$

and $T_n(x)$ is Chebyshev polynomial of the first kind and the nth degree [see formula (10.12)].

Proof. See Theorem 2 in Woods and Posten (1977).   □

To evaluate the Fourier series representation of $F(x)$, we must first compute the coefficients b_n. For example, in Theorem 11.7.2, $b_n = [2/(n\pi)]E[T_n(X)]$. Since the Chebyshev polynomial $T_n(x)$ can be written in the form

$$T_n(x) = \sum_{k=0}^{n} \alpha_{nk} x^k, \qquad n = 1, 2, \dots,$$

the computation of b_n is equivalent to evaluating

$$b_n = \frac{2}{n\pi} \sum_{k=0}^{n} \alpha_{nk}\,\mu_k', \qquad n = 1, 2, \dots,$$

where $\mu_k' = E(X^k)$ is the kth noncentral moment of X. The coefficients α_{nk} can be obtained by using the recurrence relation (10.16), that is,

$$T_{n+1}(x) = 2xT_n(x) - T_{n-1}(x), \qquad n = 1, 2, \dots,$$

with $T_0(x) = 1$, $T_1(x) = x$. This allows us to evaluate the α_{nk}'s recursively. The series

$$F(x) = 1 - \frac{\theta}{\pi} - \sum_{n=1}^{\infty} b_n \sin n\theta$$

is then truncated at $n = N$. Thus

$$F(x) \approx 1 - \frac{\theta}{\pi} - \sum_{k=1}^{N} b_k \sin k\theta.$$

Several values of N can be tried to determine the sensitivity of the approximation. We note that this series expansion provides an approximation of $F(x)$ in terms of the noncentral moments of X. Good estimates of these moments should therefore be available.

It is also possible to extend the applications of Theorems 11.7.1 and 11.7.2 to a random variable X with an infinite range provided that there exists a transformation which transforms X to a random variable Y over $[0, 1]$, or over $[-1, 1]$, such that the moments of Y are known from the moments of X.

In another application, Fettis (1976) developed a Fourier series expansion for Pearson Type IV distributions. These are density functions, $f(x)$, that satisfy the differential equation

$$\frac{df(x)}{dx} = \frac{-(x+a)}{c_0 + c_1 x + c_2 x^2} f(x),$$

where a, c_0, c_1, and c_2 are constants determined from the central moments μ_1, μ_2, μ_3, and μ_4, which can be estimated from the raw data. The data are standardized so that $\mu_1 = 0$, $\mu_2 = 1$. This results in the following expressions for a, c_0, c_1, and c_2:

$$c_0 = \frac{2\alpha - 1}{2(\alpha + 1)},$$

$$c_1 = a$$

$$= \frac{\mu_3(\alpha - 1)}{\alpha + 1},$$

$$c_2 = \frac{1}{2(\alpha + 1)},$$

where

$$\alpha = \frac{3\left(\mu_4 - \mu_3^2 - 1\right)}{2\mu_4 - 3\mu_3^2 - 6}.$$

Fettis (1976) provided additional details that explain how to approximate the cumulative distribution function,

$$F(x) = \int_{-\infty}^{x} f(t)\, dt,$$

using Fourier series.

11.7.3. Regression Modeling

In regression analysis and response surface methodology, it is quite common to use polynomial models to approximate the mean η of a response variable. There are, however, situations in which polynomial models are not adequate representatives of the mean response, as when η is known to be a periodic function. In this case, it is more appropriate to use an approximating function which is itself periodic.

Kupper (1972) proposed using partial sums of Fourier series as possible models to approximate the mean response. Consider the following trigonometric polynomial of order d,

$$\eta = \alpha_0 + \sum_{u=1}^{d} \left[\alpha_n \cos n\phi + \beta_n \sin n\phi \right], \tag{11.69}$$

where $0 \le \phi \le 2\pi$ represents either a variable taking values on the real line between 0 and 2π, or the angle associated with the polar coordinates of a point on the unit circle. Let $\mathbf{u} = (u_1, u_2)'$, where $u_1 = \cos \phi$, $u_2 = \sin \phi$. Then, when $d = 2$, the model in (11.69) can be written as

$$\eta = \alpha_0 + \alpha_1 u_1 + \beta_1 u_2 + \alpha_2 u_1^2 - \alpha_2 u_2^2 + 2\beta_2 u_1 u_2, \tag{11.70}$$

since $\sin 2\phi = 2\sin\phi\cos\phi = 2u_1 u_2$, and $\cos 2\phi = \cos^2\phi - \sin^2\phi = u_1^2 - u_2^2$.

One of the objectives of response surface methodology is the determination of optimum settings of the model's control variables that result in a maximum (or minimum) predicted response. The predicted response $\hat{y}$ at a point provides an estimate of η in (11.69) and is obtained by replacing α_0, α_n, β_n in (11.69) by their least-squares estimates $\hat{\alpha}_0$, $\hat{\alpha}_n$, and $\hat{\beta}_n$, respectively, $n = 1, 2, \ldots, d$. For example, if $d = 2$, we have

$$\hat{y} = \hat{\alpha}_0 + \sum_{n=1}^{2} \left[\hat{\alpha}_n \cos n\phi + \hat{\beta}_n \sin n\phi \right], \tag{11.71}$$

which can be expressed using (11.70) as

$$\hat{y} = \hat{\alpha}_0 + \mathbf{u}'\hat{\mathbf{b}} + \mathbf{u}'\hat{\mathbf{B}}\mathbf{u},$$

where $\hat{\mathbf{b}} = (\hat{\alpha}_1, \hat{\beta}_1)'$ and

$$\hat{\mathbf{B}} = \begin{bmatrix} \hat{\alpha}_2 & \hat{\beta}_2 \\ \hat{\beta}_2 & -\hat{\alpha}_2 \end{bmatrix}$$

with $\mathbf{u}'\mathbf{u} = 1$. The method of Lagrange multipliers can then be used to determine the stationary points of $\hat{y}$ subject to the constraint $\mathbf{u}'\mathbf{u} = 1$. Details of this procedure are given in Kupper (1972, Section 3). In a follow-up paper, Kupper (1973) presented some results on the construction of optimal designs for model (11.69).

More recently, Anderson-Cook (2000) used model (11.71) in experimental situations involving cylindrical data. For such data, it is of interest to model the relationship between two correlated components, one a standard linear measurement y, and the other an angular measure ϕ. Examples of such data arise, for example, in biology (plant or animal migration patterns), and geology (direction and magnitude of magnetic fields). The fitting of model (11.71) is done by using the method of ordinary least squares with the assumption that y is normally distributed and has a constant variance. Anderson-Cook used an example, originally presented in Mardia and Sutton (1978), of a cylindrical data set in which y is temperature (measured in degrees Fahrenheit) and ϕ is wind direction (measured in radians). Based on this example, the fitted model is

$$\hat{y} = 41.33 - 2.43\cos\phi - 2.60\sin\phi + 3.05\cos 2\phi + 2.98\sin 2\phi.$$

The corresponding standard errors of $\hat{\alpha}_0$, $\hat{\alpha}_1$, $\hat{\beta}_1$, $\hat{\alpha}_2$, $\hat{\beta}_2$ are 1.1896, 1.6608, 1.7057, 1.4029, 1.7172, respectively. Both α_0 and α_2 are significant parameters at the 5% level, and β_2 is significant at the 10% level.

11.7.4. The Characteristic Function

We have seen that the moment generating function $\phi(t)$ for a random variable X is used to obtain the moments of X (see Section 5.6.2 and Example 6.9.8). It may be recalled, however, that $\phi(t)$ may not be defined for all values of t. To generate all the moments of X, it is sufficient for $\phi(t)$ to be defined in a neighborhood of $t = 0$ (see Section 5.6.2). Some well-known distributions do not have moment generating functions, such as the Cauchy distribution (see Example 6.9.1).

Another function that generates the moments of a random variable in a manner similar to $\phi(t)$, but is defined for all values of t and for all random variables, is the *characteristic function*. By definition, the characteristic function of a random variable X, denoted by $\phi_c(t)$, is

$$\phi_c(t) = E[e^{itX}]$$

$$= \int_{-\infty}^{\infty} e^{itx}\, dF(x), \tag{11.72}$$

where $F(x)$ is the cumulative distribution function of X, and i is the complex number $\sqrt{-1}$. If X is discrete and has the values $c_1, c_2, \ldots, c_n, \ldots$, then (11.72) takes the form

$$\phi_c(t) = \sum_{j=1}^{\infty} p(c_j)e^{itc_j}, \tag{11.73}$$

where $p(c_j) = P[X = c_j]$, $j = 1, 2, \ldots$. If X is continuous with the density function $f(x)$, then

$$\phi_c(t) = \int_{-\infty}^{\infty} e^{itx} f(x)\, dx. \tag{11.74}$$

The function $\phi_c(t)$ is complex-valued in general, but is defined for all values of t, since $e^{itx} = \cos tx + i \sin tx$, and both $\int_{-\infty}^{\infty} \cos tx\, dF(x)$ and $\int_{-\infty}^{\infty} \sin tx\, dF(x)$ exist by the fact that

$$\int_{-\infty}^{\infty} |\cos tx|\, dF(x) \le \int_{-\infty}^{\infty} dF(x) = 1,$$

$$\int_{-\infty}^{\infty} |\sin tx|\, dF(x) \le \int_{-\infty}^{\infty} dF(x) = 1.$$

The characteristic function and the moment generating function, when the latter exists, are related according to the formula

$$\phi_c(t) = \phi(it).$$

Furthermore, it can be shown that if X has finite moments, then they can be obtained by repeatedly differentiating $\phi_c(t)$ and evaluating the derivatives at zero, that is,

$$E(X^n) = \frac{1}{i^n} \frac{d^n \phi_c(t)}{dt^n}\Bigg|_{t=0}, \qquad n = 1, 2, \ldots .$$

Although $\phi_c(t)$ generates moments, it is mainly used as a tool to derive distributions. For example, from (11.74) we note that when X is continuous, the characteristic function is a Fourier-type transformation of the density function $f(x)$. This follows from (11.64), the Fourier transform of $f(x)$, which is given by $(1/2\pi)\int_{-\infty}^{\infty} f(x)e^{-itx}\,dx$. If we denote this transform by $G(t)$, then the relationship between $\phi_c(t)$ and $G(t)$ is given by

$$\phi_c(t) = 2\pi G(-t).$$

By Theorem 11.6.3, if $f(x)$ is continuous and absolutely integrable on R, then $f(x)$ can be derived from $\phi_c(t)$ by using formula (11.65), which can be written as

$$
\begin{aligned}
f(x) &= \int_{-\infty}^{\infty} G(t)e^{itx}\,dt \\
&= \frac{1}{2\pi}\int_{-\infty}^{\infty} \phi_c(-t)e^{itx}\,dt \\
&= \frac{1}{2\pi}\int_{-\infty}^{\infty} \phi_c(t)e^{-itx}\,dt.
\end{aligned}
\tag{11.75}
$$

This is known as the *inversion formula* for characteristic functions. Thus the distribution of X can be uniquely determined by its characteristic function. There is therefore a one-to-one correspondence between distribution functions and their corresponding characteristic functions. This provides a useful tool for deriving distributions of random variables that cannot be easily calculated, but whose characteristic functions are straightforward. Waller, Turnbull, and Hardin (1995) reviewed and discussed several algorithms for inverting characteristic functions, and gave several examples from various areas in statistics. Waller (1995) demonstrated that characteristic functions provide information beyond what is given by moment generating functions. He pointed out that moment generating functions may be of more mathematical than numerical use in characterizing distributions. He used an example to illustrate that numerical techniques using characteristic functions can differentiate between two distributions, even though their moment generating functions are very similar (see also McCullagh, 1994).

Luceño (1997) provided further and more general arguments to show that characteristic functions are superior to moment generating and probability generating functions (see Section 5.6.2) in their numerical behavior.

One of the principal uses of characteristic functions is in deriving limiting distributions. This is based on the following theorem (see, for example, Pfeiffer, 1990, page 426):

Theorem 11.7.3. Consider the sequence $\{F_n(x)\}_{n=1}^{\infty}$ of cumulative distribution functions. Let $\{\phi_{nc}(t)\}_{n=1}^{\infty}$ be the corresponding sequence of characteristic functions.

a. If $F_n(x)$ converges to a distribution function $F(x)$ at every point of continuity for $F(x)$, then $\phi_{nc}(t)$ converges to $\phi_c(t)$ for all t, where $\phi_c(t)$ is the characteristic function for $F(x)$.

b. If $\phi_{nc}(t)$ converges to $\phi_c(t)$ for all t and $\phi_c(t)$ is continuous at $t = 0$, then $\phi_c(t)$ is the characteristic function for a distribution function $F(x)$ such that $F_n(x)$ converges to $F(x)$ at each point of continuity of $F(x)$.

It should be noted that in Theorem 11.7.3 the condition that the limiting function $\phi_c(t)$ is continuous at $t = 0$ is essential for the validity of the theorem. The following example shows that if this condition is violated, then the theorem is no longer true:

Consider the cumulative distribution function

$$F_n(x) = \begin{cases} 0, & x \le -n, \\ \dfrac{x+n}{2n}, & -n < x < n, \\ 1, & x \ge n. \end{cases}$$

The corresponding characteristic function is

$$\phi_{nc}(t) = \frac{1}{2n} \int_{-n}^{n} e^{itx}\, dx$$

$$= \frac{\sin nt}{nt}.$$

As $n \to \infty$, $\phi_{nc}(t)$ converges for every t to $\phi_c(t)$ defined by

$$\phi_c(t) = \begin{cases} 1, & t = 0, \\ 0, & t \ne 0. \end{cases}$$

Thus, $\phi_c(t)$ is not continuous for $t = 0$. We note, however, that $F_n(x) \to \frac{1}{2}$ for every fixed x. Hence, the limit of $F_n(x)$ is not a cumulative distribution function.

EXAMPLE 11.7.1.   Consider the distribution defined by the density function $f(x) = e^{-x}$ for $x > 0$. Its characteristic function is given by

$$\phi_c(t) = \int_0^\infty e^{itx} e^{-x}\, dx$$

$$= \int_0^\infty e^{-x(1-it)}\, dx$$

$$= \frac{1}{1-it}.$$

EXAMPLE 11.7.2.   Consider the Cauchy density function

$$f(x) = \frac{1}{\pi(1+x^2)}, \qquad -\infty < x < \infty,$$

given in Example 6.9.1. The characteristic function is

$$\phi_c(t) = \frac{1}{\pi} \int_{-\infty}^{\infty} \frac{e^{itx}}{1+x^2} \, dx$$

$$= \frac{1}{\pi} \int_{-\infty}^{\infty} \frac{\cos tx}{1+x^2} \, dx + \frac{i}{\pi} \int_{-\infty}^{\infty} \frac{\sin tx}{1+x^2} \, dx$$

$$= \frac{1}{\pi} \int_{-\infty}^{\infty} \frac{\cos tx}{1+x^2} \, dx$$

$$= e^{-|t|}.$$

Note that this function is not differentiable at $t = 0$. We may recall from Example 6.9.1 that all moments of the Cauchy distribution do not exist.

EXAMPLE 11.7.3.   In Example 6.9.8 we saw that the moment generating function for the gamma distribution $G(\alpha, \beta)$ with the density function

$$f(x) = \frac{x^{\alpha-1} e^{-x/\beta}}{\Gamma(\alpha)\beta^\alpha}, \qquad \alpha > 0, \quad \beta > 0, \quad 0 < x < \infty,$$

is $\phi(t) = (1 - \beta t)^{-\alpha}$. Hence, its characteristic function is $\phi_c(t) = (1 - i\beta t)^{-\alpha}$.

EXAMPLE 11.7.4.   The characteristic function of the standard normal distribution with the density function

$$f(x) = \frac{1}{\sqrt{2\pi}} e^{-x^2/2}, \qquad -\infty < x < \infty,$$

is

$$\phi_c(t) = \frac{1}{\sqrt{2\pi}} \int_{-\infty}^{\infty} e^{-x^2/2} e^{itx} \, dx$$

$$= \frac{1}{\sqrt{2\pi}} \int_{-\infty}^{\infty} e^{-\frac{1}{2}(x^2 - 2itx)} \, dx$$

$$= \frac{e^{-t^2/2}}{\sqrt{2\pi}} \int_{-\infty}^{\infty} e^{-\frac{1}{2}(x - it)^2} \, dx$$

$$= e^{-t^2/2}.$$

Vice versa, the density function can be retrieved from $\phi_c(t)$ by using the inversion formula (11.75):

$$f(x) = \frac{1}{2\pi} \int_{-\infty}^{\infty} e^{-t^2/2} e^{-itx} \, dt$$

$$= \frac{1}{2\pi} \int_{-\infty}^{\infty} e^{-\frac{1}{2}(t^2 + 2itx)} \, dt$$

$$= \frac{1}{2\pi} \int_{-\infty}^{\infty} e^{-\frac{1}{2}[t^2 + 2t(ix) + (ix)^2]} e^{(ix)^2/2} \, dt$$

$$= \frac{e^{-x^2/2}}{\sqrt{2\pi}} \int_{-\infty}^{\infty} \frac{1}{\sqrt{2\pi}} e^{-\frac{1}{2}(t + ix)^2} \, dt$$

$$= \frac{e^{-x^2/2}}{\sqrt{2\pi}} \int_{-\infty}^{\infty} \frac{1}{\sqrt{2\pi}} e^{-u^2/2} \, du$$

$$= \frac{e^{-x^2/2}}{\sqrt{2\pi}}.$$

11.7.4.1. Some Properties of Characteristic Functions

The book by Lukacs (1970) provides a detailed study of characteristic functions and their properties. Proofs of the following theorems can be found in Chapter 2 of that book.

Theorem 11.7.4. Every characteristic function is uniformly continuous on the whole real line.

Theorem 11.7.5. Suppose that $\phi_{1c}(t), \phi_{2c}(t), \ldots, \phi_{nc}(t)$ are characteristic functions. Let $a_1, a_2, \ldots, a_n$ be nonnegative numbers such that $\sum_{i=1}^{n} a_i = 1$. Then $\sum_{i=1}^{n} a_i \phi_{ic}(t)$ is also a characteristic function.

Theorem 11.7.6. The characteristic function of the convolution of two distribution functions is the product of their characteristic functions.

Theorem 11.7.7. The product of two characteristic functions is a characteristic function.

FURTHER READING AND ANNOTATED BIBLIOGRAPHY

Anderson-Cook, C. M. (2000). "A second order model for cylindrical data." *J. Statist. Comput. Simul.*, **66**, 51–65.

Bloomfield, P. (1976). *Fourier Analysis of Time Series: An Introduction*. Wiley, New York. (This is an introductory text on Fourier methods written at an applied level for users of time series.)

Carslaw, H. S. (1930). *Introduction to the Theory of Fourier Series and Integrals*, 3rd ed. Dover, New York.

Churchill, R. V. (1963). *Fourier Series and Boundary Value Problems*, 2nd ed. McGraw-Hill, New York. (This text provides an introductory treatment of Fourier series and their applications to boundary value problems in partial differential equations of engineering and physics. Fourier integral representations and expansions in series of Bessel functions and Legendre polynomials are also treated.)

Davis, H. F. (1963). *Fourier Series and Orthogonal Functions*. Allyn & Bacon, Boston.

Fettis, H. E. (1976). "Fourier series expansions for Pearson Type IV distributions and probabilities." *SIAM J. Applied Math.*, **31**, 511–518.

Fuller, W. A. (1976). *Introduction to Statistical Time Series*. Wiley, New York.

Gillespie, R. P. (1959). *Integration*. Oliver and Boyd, London.

Jackson, D. (1941). *Fourier Series and Orthogonal Polynomials*. Mathematical Association of America, Washington.

Kawata, T. (1972). *Fourier Analysis in Probability Theory*. Academic Press, New York. (This text presents useful results from the theories of Fourier series, Fourier transforms, Laplace transforms, and other related topics that are pertinent to the study of probability theory.)

Kufner, A., and J. Kadlec (1971). *Fourier Series*. Iliffe Books—The Butterworth Group, London. (This is an English translation edited by G. A. Toombs.)

Kupper, L. L. (1972). "Fourier series and spherical harmonics regression." *Appl. Statist.*, **21**, 121–130.

Kupper, L. L. (1973). "Minimax designs for Fourier series and spherical harmonics regressions: A characterization of rotatable arrangements." *J. Roy. Statist. Soc., Ser. B*, **35**, 493–500.

Luceño, A. (1997). "Further evidence supporting the numerical usefulness of characteristic functions." *Amer. Statist.*, **51**, 233–234.

Lukacs, E. (1970). *Characteristic Functions*, 2nd ed. Hafner, New York. (This is a classic book covering many interesting details concerning characteristic functions.)

Mardia, K. V., and T. W. Sutton (1978). "Model for cylindrical variables with applications." *J. Roy. Statist. Soc., Ser. B*, **40**, 229–233.

McCullagh, P. (1994). "Does the moment-generating function characterize a distribution?" *Amer. Statist.*, **48**, 208.

Otnes, R. K., and L. Enochson (1978). *Applied Time Series Analysis*. Wiley, New York.

Pfeiffer, P. E. (1990). *Probability for Applications*. Springer-Verlag, New York.

Pinkus, A., and S. Zafrany (1997). *Fourier Series and Integral Transforms*. Cambridge University Press, Cambridge, England.

Tolstov, G. P. (1962). *Fourier Series*. Dover, New York. (Translated from the Russian by Richard A. Silverman.)

Waller, L. A. (1995). "Does the characteristic function numerically distinguish distributions?" *Amer. Statist.*, **49**, 150–152.

Waller, L. A., B. W. Turnbull, and J. M. Hardin (1995). "Obtaining distribution functions by numerical inversion of characteristic functions with applications." *Amer. Statist.*, **49**, 346–350.

Weaver, H. J. (1989). *Theory of Discrete and Continuous Fourier Analysis.* Wiley, New York.

Woods, J. D., and H. O. Posten (1977). "The use of Fourier series in the evaluation of probability distribution functions." *Commun. Statist.—Simul. Comput.*, **6**, 201–219.

EXERCISES

In Mathematics

11.1. Expand the following functions using Fourier series:

(a) $f(x) = |x|$, $-\pi \leq x \leq \pi$.

(b) $f(x) = |\sin x|$.

(c) $f(x) = x + x^2$, $-\pi \leq x \leq \pi$.

11.2. Show that

$$\sum_{n=1}^{\infty} \frac{1}{(2n-1)^2} = \frac{\pi^2}{8}.$$

[*Hint*: Use the Fourier series for x^2.]

11.3. Let a_n and b_n be the Fourier coefficients for a continuous function $f(x)$ defined on $[-\pi, \pi]$ such that $f(-\pi) = f(\pi)$, and $f'(x)$ is piecewise continuous on $[-\pi, \pi]$. Show that

(a) $\lim_{n \to \infty}(na_n) = 0$,

(b) $\lim_{n \to \infty}(nb_n) = 0$.

11.4. If $f(x)$ is continuous on $[-\pi, \pi]$, $f(-\pi) = f(\pi)$, and $f'(x)$ is piecewise continuous on $[-\pi, \pi]$, then show that

$$|f(x) - s_n(x)| \leq \frac{c}{\sqrt{n}},$$

where

$$s_n(x) = \frac{a_0}{2} + \sum_{k=1}^{n} [a_k \cos kx + b_k \sin kx],$$

and

$$c^2 = \frac{1}{\pi} \int_{-\pi}^{\pi} f'^2(x)\, dx.$$

11.5. Suppose that $f(x)$ is piecewise continuous on $[-\pi, \pi]$ and has the Fourier series given in (11.32).

(a) Show that $\sum_{n=1}^{\infty}(-1)^n b_n/n$ is a convergent series.

(b) Show that $\sum_{n=1}^{\infty} b_n/n$ is convergent.
[*Hint*: Use Theorem 11.3.2.]

11.6. Show that the trigonometric series, $\sum_{n=2}^{\infty}(\sin nx)/\log n$, is not a Fourier series of any integrable function.
[*Hint*: If it were a Fourier series of a function $f(x)$, then $b_n = 1/\log n$ would be the Fourier coefficient of an odd function. Apply now part (b) of Exercise 11.5 and show that this assumption leads to a contradiction.]

11.7. Consider the Fourier series of $f(x)=x$ given in Example 11.2.1.

(a) Show that

$$\frac{x^2}{4} = \frac{\pi^2}{12} - \sum_{n=1}^{\infty} \frac{(-1)^{n+1} \cos nx}{n^2} \qquad \text{for } -\pi < x < \pi.$$

[*Hint*: Consider the Fourier series of $\int_0^x f(t)\,dt$.]

(b) Deduce that

$$\sum_{n=1}^{\infty} \frac{(-1)^{n+1}}{n^2} = \frac{\pi^2}{12}.$$

11.8. Make use of the result in Exercise 11.7 to find the sum of the series $\sum_{n=1}^{\infty}[(-1)^{n+1} \sin nx]/n^3$.

11.9. Show that the Fourier transform of $f(x) = e^{-x^2}$, $-\infty < x < \infty$, is given by

$$F(w) = \frac{1}{2\sqrt{\pi}} e^{-w^2/4}.$$

[*Hint*: Show that $F'(w) + \frac{1}{2}wF(w) = 0$.]

11.10. Prove Theorem 11.6.4 using Fubini's theorem.

11.11. Use the Fourier transform to solve the integral equation

$$\int_{-\infty}^{\infty} f(x-y)f(y)\,dy = e^{-x^2/2}$$

for the function $f(x)$.

514

11.12. Consider the function $f(x)=x$, $-\pi<x<\pi$, with $f(-\pi)=f(\pi)=0$, and $f(x)$ 2π-periodic defined on $(-\infty,\infty)$. The Fourier series of $f(x)$ is

$$\sum_{n=1}^{\infty} \frac{2(-1)^{n+1}\sin nx}{n}.$$

Let $s_n(x)$ be the nth partial sum of this series, that is,

$$s_n(x) = \sum_{k=1}^{n} \frac{2(-1)^{k+1}}{k}\sin kx.$$

Let $x_n = \pi - \pi/n$.
(a) Show that

$$s_n(x_n) = \sum_{k=1}^{n} \frac{2\sin(k\pi/n)}{k}.$$

(b) Show that

$$\lim_{n\to\infty} s_n(x_n) = 2\int_0^{\pi} \frac{\sin x}{x}\,dx$$

$$\approx 1.18\pi.$$

Note: As $n\to\infty$, $x_n\to\pi^-$. Hence, for n sufficiently large,

$$s_n(x_n) - f(x_n) \approx 1.18\pi - \pi = 0.18\pi.$$

Thus, near $x=\pi$ [a point of discontinuity for $f(x)$], the partial sums of the Fourier series exceed the value of this function by approximately the amount $0.18\pi = 0.565$. This illustrates the so-called *Gibbs phenomenon* according to which the Fourier series of $f(x)$ "overshoots" the value of $f(x)$ in a small neighborhood to the left of the point of discontinuity of $f(x)$. It can also be shown that in a small neighborhood to the right of $x=-\pi$, the Fourier series of $f(x)$ "undershoots" the value of $f(x)$.

In Statistics

11.13. In the following table, two observations of the resistance in ohms are recorded at each of six equally spaced locations on the perimeter of a new type of solid circular coil (see Kupper, 1972, Table 1):

ϕ (radians)	Resistance (ohms)
0	13.62, 14.40
$\pi/3$	10.552, 10.602
$2\pi/3$	2.196, 3.696
π	6.39, 7.25
$4\pi/3$	8.854, 10.684
$5\pi/3$	5.408, 8.488

(a) Use the method of least squares to estimate the parameters in the following trigonometric polynomial of order 2:

$$\eta = \alpha_0 + \sum_{n=1}^{2} [\alpha_n \cos n\phi + \beta_n \sin n\phi],$$

where $0 \le \phi \le 2\pi$, and η denotes the average resistance at location ϕ.

(b) Use the prediction equation obtained in part (a) to determine the points of minimum and maximum resistance on the perimeter of the circular coil.

11.14. Consider the following circular data set in which ϕ is wind direction and y is temperature (see Anderson–Cook, 2000, Table 1).

ϕ (radians)	y (°F)	ϕ (radians)	y (°F)
4.36	52	4.54	38
3.67	41	2.62	40
4.36	41	2.97	49
1.57	31	4.01	48
3.67	53	4.19	37
3.67	47	5.59	37
6.11	43	5.59	33
5.93	43	3.32	47
0.52	41	3.67	51
3.67	46	1.22	42
3.67	48	4.54	53
3.32	52	4.19	46
4.89	43	3.49	51
3.14	46	4.71	39

Fit a second − order trigonometric polynomial to this data set, and verify that the prediction equation is given by

$$\hat{y} = 41.33 - 2.43 \cos \phi - 2.60 \sin \phi + 3.05 \cos 2\phi + 2.98 \sin 2\phi.$$

11.15. Let $\{Y_n\}_{n=1}^{\infty}$ be a sequence of independent, indentically distributed random variables with mean μ and variance σ^2. Let

$$s_n^* = \frac{\bar{Y}_n - \mu}{\sigma/\sqrt{n}},$$

where $\bar{Y}_n = (1/n)\sum_{i=1}^{n}Y_i$.

(a) Find the characteristic function of s_n^*.

(b) Use Theorem 11.7.3 and part (a) to show that the limiting distribution of s_n^* as $n \to \infty$ is the standard normal distribution.

Note: Part (b) represents the statement of the well-known *central limit theorem*, which asserts that for large n, the arithmetic mean $\bar{Y}_n$ of a sample of independent, identically distributed random variables is approximately normally distributed with mean μ and standard deviation $\sigma/\sqrt{n}$.

C H A P T E R   1 2

Approximation of Integrals

Integration plays an important role in many fields of science and engineering. For applications, numerical values of integrals are often required. However, in many cases, the evaluation of integrals, or *quadrature*, by elementary functions may not be feasible. Hence, approximating the value of an integral in a reliable fashion is a problem of utmost importance. Numerical quadrature is in fact one of the oldest branches of mathematics: the determination, approximately or exactly, of the areas of regions bounded by lines or curves, a subject which was studied by the ancient Babylonians (see Haber, 1970). The word "quadrature" indicates the process of measuring an area inside a curve by finding a square having the same area. Probably no other problem has exercised a greater or a longer attraction than that of constructing a square equal in area to a given circle. Thousands of people have worked on this problem, including the ancient Egyptians as far back as 1800 B.C.

In this chapter, we provide an exposition of methods for approximating integrals, including those that are multidimensional.

12.1. THE TRAPEZOIDAL METHOD

This is the simplest method of approximating an integral of the form $\int_a^b f(x)\,dx$, which represents the area bounded by the curve of the function $y = f(x)$ and the two lines $x = a$, $x = b$. The method is based on approximating the curve by a series of straight line segments. As a result, the area is approximated with a series of trapezoids. For this purpose, the interval from a to b is divided into n equal parts by the partition points $a = x_0, x_1, x_2, \ldots, x_n = b$. For the ith trapezoid, which lies between x_{i-1} and x_i, its width is $h = (1/n)(b - a)$ and its area is given by

$$A_i = \frac{h}{2}[f(x_{i-1}) + f(x_i)], \qquad i = 1, 2, \ldots, n. \qquad (12.1)$$

517

The sum, S_n, of $A_1, A_2, \ldots, A_n$ provides an approximation to the integral $\int_a^b f(x)\, dx$.

$$S_n = \sum_{i=1}^{n} A_i$$

$$= \frac{h}{2}\{[f(x_0) + f(x_1)] + [f(x_1) + f(x_2)] + \cdots + [f(x_{n-1}) + f(x_n)]\}$$

$$= \frac{h}{2}\left[f(x_0) + f(x_n) + 2 \sum_{i=1}^{n-1} f(x_i)\right]. \tag{12.2}$$

12.1.1. Accuracy of the Approximation

The accuracy in the trapezoidal method depends on the number n of trapezoids we take. The next theorem provides information concerning the error or approximation.

Theorem 12.1.1. Suppose that $f(x)$ has a continuous second derivative on $[a, b]$, and $|f''(x)| \le M_2$ for all x in $[a, b]$. Then

$$\left|\int_a^b f(x)\, dx - S_n\right| \le \frac{(b-a)^3 M_2}{12 n^2},$$

where S_n is given by formula (12.2).

Proof. Consider the partition points $a = x_0, x_1, x_2, \ldots, x_n = b$ such that $h = x_i - x_{i-1} = (1/n)(b-a)$, $i = 1, 2, \ldots, n$. The integral of $f(x)$ from x_{i-1} to x_i is

$$I_i = \int_{x_{i-1}}^{x_i} f(x)\, dx. \tag{12.3}$$

Now, in the trapezoidal method, $f(x)$ is approximated in the interval $[x_{i-1}, x_i]$ by the right-hand side of the straight-line equation,

$$p_i(x) = f(x_{i-1}) + \frac{1}{h}[f(x_i) - f(x_{i-1})](x - x_{i-1})$$

$$= \frac{x_i - x}{h} f(x_{i-1}) + \frac{x - x_{i-1}}{h} f(x_i)$$

$$= \frac{x - x_i}{x_{i-1} - x_i} f(x_{i-1}) + \frac{x - x_{i-1}}{x_i - x_{i-1}} f(x_i), \qquad i = 1, 2, \ldots, n.$$

Note that $p_i(x)$ is a linear Lagrange interpolating polynomial (of degree $n = 1$) with x_{i-1} and x_i as its points of interpolation [see formula (9.14)].

Using Theorem 9.2.2, the error of interpolation resulting from approximating $f(x)$ with $p_i(x)$ over $[x_{i-1}, x_i]$ is given by

$$f(x) - p_i(x) = \frac{1}{2!}f''(\xi_i)(x - x_{i-1})(x - x_i), \qquad i = 1, 2, \ldots, n, \quad (12.4)$$

where $x_{i-1} < \xi_i < x_i$. Formula (12.4) results from applying formula (9.15). Hence, the error of approximating I_i with A_i in (12.1) is

$$I_i - A_i = \int_{x_{i-1}}^{x_i} [f(x) - p_i(x)]\, dx$$

$$= \frac{1}{2!}f''(\xi_i) \int_{x_{i-1}}^{x_i} (x - x_{i-1})(x - x_i)\, dx$$

$$= \frac{1}{2!}f''(\xi_i)\left[\tfrac{1}{3}(x_i^3 - x_{i-1}^3) - \tfrac{1}{2}(x_{i-1} + x_i)(x_i^2 - x_{i-1}^2) + x_{i-1}x_i(x_i - x_{i-1})\right]$$

$$= \frac{h}{2!}f''(\xi_i)\left[\tfrac{1}{3}(x_i^2 + x_{i-1}x_i + x_{i-1}^2) - \tfrac{1}{2}(x_{i-1} + x_i)^2 + x_{i-1}x_i\right]$$

$$= \frac{h}{2!}f''(\xi_i)\left[\tfrac{1}{6}(2x_{i-1}x_i - x_{i-1}^2 - x_i^2)\right]$$

$$= -\frac{h^3}{12}f''(\xi_i), \qquad i = 1, 2, \ldots, n. \qquad (12.5)$$

The total error of approximating $\int_a^b f(x)\, dx$ with S_n is then given by

$$\int_a^b f(x)\, dx - S_n = -\frac{h^3}{12}\sum_{i=1}^{n} f''(\xi_i).$$

It follows that

$$\left|\int_a^b f(x)\, dx - S_n\right| \le \frac{nh^3 M_2}{12}$$

$$= \frac{(b-a)^3 M_2}{12n^2}. \qquad \square \qquad (12.6)$$

An alternative procedure to approximating the integral $\int_a^b f(x)\, dx$ by a sum of trapezoids is to approximate $\int_{x_{i-1}}^{x_i} f(x)\, dx$ by a trapezoid bounded from above by the tangent to the curve of $y = f(x)$ at the point $x_{i-1} + h/2$, which is the midpoint of the interval $[x_{i-1}, x_i]$. In this case, the area of the ith trapezoid is

$$A_i^* = hf\left(x_{i-1} + \frac{h}{2}\right), \qquad i = 1, 2, \ldots, n.$$

Hence,

$$\int_a^b f(x)\, dx \approx h \sum_{i=1}^{n} f\left(x_{i-1} + \frac{h}{2}\right), \tag{12.7}$$

and $f(x)$ is approximated in the interval $[x_{i-1}, x_i]$ by

$$p_i^*(x) = f\left(x_{i-1} + \frac{h}{2}\right) + \left(x - x_{i-1} - \frac{h}{2}\right) f'\left(x_{i-1} + \frac{h}{2}\right).$$

By applying Taylor's theorem (Theorem 4.3.1) to $f(x)$ in a neighborhood of $x_{i-1} + h/2$, we obtain

$$f(x) = f\left(x_{i-1} + \frac{h}{2}\right) + \left(x - x_{i-1} - \frac{h}{2}\right) f'\left(x_{i-1} + \frac{h}{2}\right)$$

$$+ \frac{1}{2!}\left(x - x_{i-1} - \frac{h}{2}\right)^2 f''(\eta_i)$$

$$= p_i^*(x) + \frac{1}{2!}\left(x - x_{i-1} - \frac{h}{2}\right)^2 f''(\eta_i),$$

where η_i lies between $x_{i-1} + h/2$ and x. The error of approximating $\int_{x_{i-1}}^{x_i} f(x)\, dx$ with A_i^* is then given by

$$\int_{x_{i-1}}^{x_i} [f(x) - p_i^*(x)]\, dx = \frac{1}{2!} \int_{x_{i-1}}^{x_i} \left(x - x_{i-1} - \frac{h}{2}\right)^2 f''(\eta_i)\, dx$$

$$\leq \frac{M_2}{2!} \int_{x_{i-1}}^{x_i} \left(x - x_{i-1} - \frac{h}{2}\right)^2 dx$$

$$= \frac{h^3 M_2}{24}.$$

Consequently, the absolute value of the total error in this case has an upper bound of the form

$$\left| \sum_{i=1}^{n} \int_{x_{i-1}}^{x_i} [f(x) - p_i^*(x)]\, dx \right| \leq \frac{n h^3 M_2}{24}$$

$$= \frac{(b-a)^3 M_2}{24 n^2}. \tag{12.8}$$

We note that the upper bound in (12.8) is half as large as the one in (12.6). This alternative procedure is therefore slightly more precise than the original one. Both procedures produce an approximating error that is $O(1/n^2)$. It should be noted that this error does not include the roundoff errors in the computation of the areas of the approximating trapezoids.

EXAMPLE 12.1.1. Consider approximating the integral $\int_1^2 dx/x$, which has an exact value equal to $\log 2 \approx 0.693147$. Let us divide the interval $[1, 2]$ into $n = 10$ subintervals of length $h = \frac{1}{10}$. Hence, $x_0 = 1$, $x_1 = 1.1, \ldots, x_{10} = 2.0$. Using the first trapezoidal method (formula 12.2), we obtain

$$\int_1^2 \frac{dx}{x} \approx \frac{1}{20} \left[1 + \frac{1}{2} + 2 \sum_{i=1}^{9} \frac{1}{x_i} \right]$$

$$= 0.69377.$$

Using now the second trapezoidal method [formula (12.7)], we get

$$\int_1^2 \frac{dx}{x} \approx \frac{1}{10} \sum_{i=1}^{10} \frac{1}{x_{i-1} + 0.05}$$

$$= 0.69284.$$

12.2. SIMPSON'S METHOD

Let us again consider the integral $\int_a^b f(x) \, dx$. Let $a = x_0 < x_1 < \cdots < x_{2n-1} < x_{2n} = b$ be a sequence of equally spaced points that partition the interval $[a, b]$ such that $x_i - x_{i-1} = h$, $i = 1, 2, \ldots, 2n$. Simpson's method is based on approximating the graph of the function $f(x)$ over the interval $[x_{i-1}, x_{i+1}]$ by a parabola which agrees with $f(x)$ at the points x_{i-1}, x_i, and x_{i+1}. Thus, over $[x_{i-1}, x_{i+1}]$, $f(x)$ is approximated with a Lagrange interpolating polynomial of degree 2 of the form [see formula (9.14)]

$$q_i(x) = f(x_{i-1}) \frac{(x - x_i)(x - x_{i+1})}{(x_{i-1} - x_i)(x_{i-1} - x_{i+1})}$$

$$+ f(x_i) \frac{(x - x_{i-1})(x - x_{i+1})}{(x_i - x_{i-1})(x_i - x_{i+1})} + f(x_{i+1}) \frac{(x - x_{i-1})(x - x_i)}{(x_{i+1} - x_{i-1})(x_{i+1} - x_i)}$$

$$= f(x_{i-1}) \frac{(x - x_i)(x - x_{i+1})}{2h^2} - f(x_i) \frac{(x - x_{i-1})(x - x_{i+1})}{h^2}$$

$$+ f(x_{i+1}) \frac{(x - x_{i-1})(x - x_i)}{2h^2}.$$

It follows that

$$\int_{x_{i-1}}^{x_{i+1}} f(x)\,dx \approx \int_{x_{i-1}}^{x_{i+1}} q_i(x)\,dx$$

$$= \frac{f(x_{i-1})}{2h^2}\int_{x_{i-1}}^{x_{i+1}}(x-x_i)(x-x_{i+1})\,dx$$

$$- \frac{f(x_i)}{h^2}\int_{x_{i-1}}^{x_{i+1}}(x-x_{i-1})(x-x_{i+1})\,dx$$

$$+ \frac{f(x_{i+1})}{2h^2}\int_{x_{i-1}}^{x_{i+1}}(x-x_{i-1})(x-x_i)\,dx$$

$$= \frac{f(x_{i-1})}{2h^2}\left(\frac{2h^3}{3}\right) - \frac{f(x_i)}{h^2}\left(\frac{-4h^3}{3}\right) + \frac{f(x_{i+1})}{2h^2}\left(\frac{2h^3}{3}\right)$$

$$= \frac{h}{3}\left[f(x_{i-1}) + 4f(x_i) + f(x_{i+1})\right], \qquad i = 1,3,\ldots,2n-1.$$

$$(12.9)$$

By adding up all the approximations in (12.9) for $i = 1,3,\ldots,2n-1$, we obtain

$$\int_a^b f(x)\,dx \approx \frac{h}{3}\left[f(x_0) + 4\sum_{i=1}^{n} f(x_{2i-1}) + 2\sum_{i=1}^{n-1} f(x_{2i}) + f(x_{2n})\right]. \quad (12.10)$$

As before, in the case of the trapezoidal method, the accuracy of the approximation in (12.10) can be figured out by using formula (9.15). Courant and John (1965, page 487), however, stated that the error of approximation can be improved by one order of magnitude by using a cubic interpolating polynomial which agrees with $f(x)$ at x_{i-1}, x_i, x_{i+1}, and whose derivative at x_i is equal to $f'(x_i)$. Such a polynomial gives a better approximation to $f(x)$ over $[x_{i-1}, x_{i+1}]$ than the quadratic one, and still provides the same approximation formula (12.l0) for the integral. If $q_i(x)$ is chosen as such, then the error of interpolation resulting from approximating $f(x)$ with $q_i(x)$ over $[x_{i-1}, x_{i+1}]$ is given by $f(x) - q_i(x) = (1/4!)f^{(4)}(\xi_i)(x-x_{i-1})(x-x_i)^2(x-x_{i+1})$, where $x_{i-1} < \xi_i < x_{i+1}$, provided that $f^{(4)}(x)$ exists and is continuous on $[a, b]$. This is equivalent to using formula (9.15) with $n = 3$ and with two of the interpolation points coincident at x_i. We then have

$$\left|\int_{x_{i-1}}^{x_{i+1}}[f(x) - q_i(x)]\,dx\right| \le \frac{M_4}{4!}\int_{x_{i-1}}^{x_{i+1}}(x-x_{i-1})(x-x_i)^2(x-x_{i+1})\,dx,$$

$$i = 1,3,\ldots,2n-1, \quad (12.11)$$

where M_4 is an upper bound on $|f^{(4)}(x)|$ for $a \leq x \leq b$. By computing the integral in (12.11) we obtain

$$\left| \int_{x_{i-1}}^{x_{i+1}} [f(x) - q_i(x)] \, dx \right| \leq \frac{M_4 h^5}{90}, \qquad i = 1, 3, \ldots, 2n - 1.$$

Consequently, the total error of approximation in (12.10) is less than or equal to

$$\frac{n M_4 h^5}{90} = \frac{M_4 (b - a)^5}{2880 n^4}, \tag{12.12}$$

since $h = (b - a)/2n$. Thus the error of approximation with Simpson's method is $O(1/n^4)$, where n is half the number of subintervals into which $[a, b]$ is divided. Hence, Simpson's method yields a much more accurate approximation than the trapezoidal method.

As an example, let us apply Simpson's method to the calculation of the integral in Example 12.1.1 using the same division of $[1, 2]$ into 10 subintervals, each of length $h = \frac{1}{10}$. By applying formula (12.10) we obtain

$$\int_1^2 \frac{dx}{x} \approx \frac{0.10}{3} \left[1 + 4 \left(\frac{1}{1.1} + \frac{1}{1.3} + \frac{1}{1.5} + \frac{1}{1.7} + \frac{1}{1.9} \right) \right.$$

$$\left. + 2 \left(\frac{1}{1.2} + \frac{1}{1.4} + \frac{1}{1.6} + \frac{1}{1.8} \right) + \frac{1}{2} \right]$$

$$= 0.69315.$$

12.3. NEWTON–COTES METHODS

The trapezoidal and Simpson's methods are two special cases of a general series of approximate integration methods of the so-called Newton–Cotes type. In the trapezoidal method, straight line segments were used to approximate the graph of the function $f(x)$ between a and b. In Simpson's method, the approximation was carried out using a series of parabolas. We can refine this approximation even further by considering a series of cubic curves, quartic curves, and so on. For cubic approximations, four equally spaced points are used to subdivide each subinterval of $[a, b]$ (instead of two points for the trapezoidal method and three points for Simpson's method), whereas five points are needed for quartic approximation, and so on. All such approximations are of the Newton–Cotes type.

12.4. GAUSSIAN QUADRATURE

All Newton–Cotes methods require the use of equally spaced points, as was seen in the cases of the trapezoidal method and Simpson's method. If this requirement is waived, then it is possible to select the points in a manner that reduces the approximation error.

Let $x_0 < x_1 < x_2 < \cdots < x_n$ be $n+1$ distinct points in $[a, b]$. Consider the approximation

$$\int_a^b f(x)\, dx \approx \sum_{i=0}^n \omega_i f(x_i), \tag{12.13}$$

where the coefficients, $\omega_0, \omega_1, \ldots, \omega_n$, are to be determined along with the points $x_0, x_1, \ldots, x_n$. The total number of unknown quantities in (12.13) is $2n + 2$. Hence, $2n + 2$ conditions must be specified. According to the so-called Gaussian integration rule, the ω_i's and x_i's are chosen such that the approximation in (12.13) will be exact for all polynomials of degrees not exceeding $2n + 1$. This is equivalent to requiring that the approximation (12.13) be exact for $f(x) = x^j$, $j = 0, 1, 2, \ldots, 2n + 1$, that is,

$$\int_a^b x^j\, dx = \sum_{i=0}^n \omega_i x_i^j, \qquad j = 0, 1, 2, \ldots, 2n + 1. \tag{12.14}$$

This process produces $2n + 2$ equations to be solved for the ω_i's and x_i's. In particular, if the limits of integration are $a = -1$, $b = 1$, then it can be shown (see Phillips and Taylor, 1973, page 140) that the x_i-values will be the $n + 1$ zeros of the Legendre polynomial $p_{n+1}(x)$ of degree $n + 1$ (see Section 10.2). The ω_i-values can be easily found by solving the system of equations (12.14), which is linear in the ω_i's.

For example, for $n = 1$, the zeros of $p_2(x) = \frac{1}{2}(3x^2 - 1)$ are $x_0 = -1/\sqrt{3}$, $x_1 = 1/\sqrt{3}$. Applying (12.14), we obtain

$$\int_{-1}^1 dx = \omega_0 + \omega_1 \quad \Rightarrow \quad \omega_0 + \omega_1 = 2,$$

$$\int_{-1}^1 x\, dx = \omega_0 x_0 + \omega_1 x_1 \quad \Rightarrow \quad \frac{1}{\sqrt{3}}(-\omega_0 + \omega_1) = 0,$$

$$\int_{-1}^1 x^2\, dx = \omega_0 x_0^2 + \omega_1 x_1^2 \quad \Rightarrow \quad \frac{1}{3}(\omega_0 + \omega_1) = \frac{2}{3},$$

$$\int_{-1}^1 x^3\, dx = \omega_0 x_0^3 + \omega_1 x_1^3 \quad \Rightarrow \quad \frac{1}{3\sqrt{3}}(-\omega_0 + \omega_1) = 0.$$

We note that the last two equations are identical to the first two. Solving the latter for ω_0 and ω_1, we get $\omega_0 = \omega_1 = 1$. Hence, we have the approximation

$$\int_{-1}^{1} f(x)\,dx \approx f\left(-\frac{1}{\sqrt{3}}\right) + f\left(\frac{1}{\sqrt{3}}\right),$$

which is exact if $f(x)$ is a polynomial of degree not exceeding $2n + 1 = 3$.

If the limits of integration are not equal to $-1, 1$, we can easily convert the integral $\int_a^b f(x)\,dx$ to one with the limits $-1, 1$ by making the change of variable

$$z = \frac{2x - (a+b)}{b-a}.$$

This converts the general integral $\int_a^b f(x)\,dx$ to the integral $[(b - a)/2]\int_{-1}^{1} g(z)\,dz$, where

$$g(z) = f\left[\frac{(b-a)z + b + a}{2}\right].$$

We therefore have the approximation

$$\int_a^b f(x)\,dx \approx \frac{b-a}{2} \sum_{i=0}^{n} \omega_i g(z_i), \tag{12.15}$$

where the z_i's are the zeros of the Legendre polynomial $p_{n+1}(z)$.

It can be shown that (see Davis and Rabinowitz, 1975, page 75) that when $a = -1$, $b = 1$, the error of approximation in (12.13) is given by

$$\int_a^b f(x)\,dx - \sum_{i=0}^{n} \omega_i f(x_i) = \frac{(b-a)^{2n+3}[(n+1)!]^4}{(2n+3)[(2n+2)!]^3} f^{(2n+2)}(\xi), \quad a < \xi < b,$$

$$\tag{12.16}$$

provided that $f^{(2n+2)}(x)$ is continuous on $[a, b]$. This error decreases rapidly as n increases. Thus this Gaussian quadrature provides a very good approximation with a formula of the type given in (12.13).

There are several extensions of the approximation in (12.13). These extensions are of the form

$$\int_a^b \lambda(x)f(x)\,dx \approx \sum_{i=0}^{n} \omega_i f(x_i), \tag{12.17}$$

where $\lambda(x)$ is a particular positive weight function. As before, the coefficients $\omega_0, \omega_1, \ldots, \omega_n$ and the points $x_0, x_1, \ldots, x_n$, which belong to $[a, b]$, are chosen so that (12.17) is exact for all polynomials of degrees not exceeding $2n + 1$. The choice of the x_i's depends on the form of $\lambda(x)$. It can be shown that the values of x_i are the zeros of a polynomial of degree $n + 1$ belonging to the sequence of polynomials that are orthogonal on $[a, b]$ with respect to $\lambda(x)$ (see Davis and Rabinowitz, 1975, page 74; Phillips and Taylor, 1973, page 142). For example, if $a = -1$, $b = 1$, $\lambda(x) = (1 - x)^\alpha (1 + x)^\beta$, $\alpha > -1$, $\beta > -1$, then the x_i's are the zeros of the Jacobi polynomial $p_{n+1}^{(\alpha, \beta)}(x)$ (see Section 10.3). Also, if $a = -1$, $b = 1$, $\lambda(x) = (1 - x^2)^{-1/2}$, then the x_i's are the zeros of the Chebyshev polynomial of the first kind, $T_{n+1}(x)$ (see Section 10.4), and so on. For those two cases, formula (12.17) is called the *Gauss–Jacobi quadrature* formula and the *Gauss–Chebyshev quadrature* formula, respectively. The choice $\lambda(x) = 1$ results in the original formula (12.13), which is now referred to as the *Gauss–Legendre quadrature* formula.

EXAMPLE 12.4.1.   Consider the integral $\int_0^1 dx/(1 + x)$, which has the exact value $\log 2 = 0.69314718$. Applying formula (12.15), we get

$$\int_0^1 \frac{dx}{1 + x} \approx \frac{1}{2} \sum_{i=0}^{n} \frac{\omega_i}{1 + \frac{1}{2}(z_i + 1)}$$

$$= \sum_{i=0}^{n} \frac{\omega_i}{3 + z_i}, \qquad -1 \le z_i \le 1, \qquad (12.18)$$

where the z_i's are the zeros of the Legendre polynomial $p_{n+1}(z)$, $z = 2x - 1$.

Let $n = 1$; then $p_2(z) = \frac{1}{2}(3z^2 - 1)$ with zeros equal to $z_0 = -1/\sqrt{3}$, $z_1 = 1/\sqrt{3}$. We have seen earlier that $\omega_0 = 1$, $\omega_1 = 1$; hence, from (12.18), we obtain

$$\int_0^1 \frac{dx}{1 + x} \approx \sum_{i=0}^{1} \frac{\omega_i}{3 + z_i}$$

$$= \frac{\sqrt{3}}{3\sqrt{3} - 1} + \frac{\sqrt{3}}{3\sqrt{3} + 1}$$

$$= 0.692307691.$$

Let us now use $n = 2$ in (12.18). Then $p_3(z) = \frac{1}{2}(5z^3 - 3z)$ (see Section 10.2). Its zeros are $z_0 = -(\frac{3}{5})^{1/2}$, $z_1 = 0$, $z_2 = (\frac{3}{5})^{1/2}$. To find the ω_i's, we apply formula (12.14) using $a = -1$, $b = 1$, and z in place of x. For

$j = 0, 1, 2, 3, 4, 5$, we have

$$\int_{-1}^{1} dz = \omega_0 + \omega_1 + \omega_2,$$

$$\int_{-1}^{1} z \, dz = \omega_0 z_0 + \omega_1 z_1 + \omega_2 z_2,$$

$$\int_{-1}^{1} z^2 \, dz = \omega_0 z_0^2 + \omega_1 z_1^2 + \omega_2 z_2^2,$$

$$\int_{-1}^{1} z^3 \, dz = \omega_0 z_0^3 + \omega_1 z_1^3 + \omega_2 z_2^3,$$

$$\int_{-1}^{1} z^4 \, dz = \omega_0 z_0^4 + \omega_1 z_1^4 + \omega_2 z_2^4,$$

$$\int_{-1}^{1} z^5 \, dz = \omega_0 z_0^5 + \omega_1 z_1^5 + \omega_2 z_2^5.$$

These equations can be written as

$$\omega_0 + \omega_1 + \omega_2 = 2,$$

$$\left(\tfrac{3}{5}\right)^{1/2}(-\omega_0 + \omega_2) = 0,$$

$$\tfrac{3}{5}(\omega_0 + \omega_2) = \tfrac{2}{3},$$

$$\left(\tfrac{3}{5}\right)^{3/2}(-\omega_0 + \omega_2) = 0,$$

$$\tfrac{9}{25}(\omega_0 + \omega_2) = \tfrac{2}{5},$$

$$\tfrac{9}{25}\left(\tfrac{3}{5}\right)^{1/2}(-\omega_0 + \omega_2) = 0.$$

The above six equations can be reduced to only three that are linearly independent, namely,

$$\omega_0 + \omega_1 + \omega_2 = 2,$$

$$-\omega_0 + \omega_2 = 0,$$

$$\omega_0 + \omega_2 = \tfrac{10}{9},$$

the solution of which is $\omega_0 = \tfrac{5}{9}$, $\omega_1 = \tfrac{8}{9}$, $\omega_2 = \tfrac{5}{9}$. Substituting the ω_i's and z_i's in (12.18), we obtain

$$\int_0^1 \frac{dx}{1+x} \approx \frac{\omega_0}{3+z_0} + \frac{\omega_1}{3+z_1} + \frac{\omega_2}{3+z_2}$$

$$= \frac{\frac{5}{9}}{3-\left(\frac{3}{5}\right)^{1/2}} + \frac{\frac{8}{9}}{3} + \frac{\frac{5}{9}}{3+\left(\frac{3}{5}\right)^{1/2}}$$

$$= 0.693121685.$$

For higher values of n, the zeros of Legendre polynomials and the corresponding values of ω_i can be found, for example, in Shoup (1984, Table 7.5) and Krylov (1962, Appendix A).

12.5. APPROXIMATION OVER AN INFINITE INTERVAL

Consider an integral of the form $\int_a^\infty f(x)\,dx$, which is improper of the first kind (see Section 6.5). It can be approximated by using the integral $\int_a^b f(x)\,dx$ for a sufficiently large value of b, provided, of course, that $\int_a^\infty f(x)\,dx$ is convergent. The methods discussed earlier in Sections 12.1–12.4 can then be applied to $\int_a^b f(x)\,dx$.

For improper integrals of the first kind, of the form $\int_0^\infty \lambda(x)f(x)\,dx$, $\int_{-\infty}^\infty \lambda(x)f(x)\,dx$, we have the following Gaussian approximations:

$$\int_0^\infty \lambda(x)f(x)\,dx \approx \sum_{i=0}^n \omega_i f(x_i), \qquad (12.19)$$

$$\int_{-\infty}^\infty \lambda(x)f(x)\,dx \approx \sum_{i=0}^n \omega_i f(x_i), \qquad (12.20)$$

where, as before, the x_i's and ω_i's are chosen so that (12.19) and (12.20) are exact for all polynomials of degrees not exceeding $2n+1$. For the weight function $\lambda(x)$ in (12.19), the choice $\lambda(x) = e^{-x}$ gives the *Gauss–Laguerre quadrature*, for which the x_i's are the zeros of the Laguerre polynomial $L_{n+1}(x)$ of degree $n+1$ and $\alpha=0$ (see Section 10.6). The associated error of approximation is given by (see Davis and Rabinowitz, 1975, page 173)

$$\int_0^\infty e^{-x}f(x)\,dx - \sum_{i=0}^n \omega_i f(x_i) = \frac{[(n+1)!]^2}{(2n+2)!} f^{(2n+2)}(\xi), \qquad 0 < \xi < \infty.$$

Choosing $\lambda(x) = e^{-x^2}$ in (12.20) gives the *Gauss–Hermite quadrature*, and the corresponding x_i's are the zeros of the Hermite polynomial $H_{n+1}(x)$ of degree $n+1$ (see Section 10.5). The associated error of approximation is of the form (see Davis and Rabinowitz, 1975, page 174)

$$\int_{-\infty}^\infty e^{-x^2}f(x)\,dx - \sum_{i=0}^n \omega_i f(x_i) = \frac{(n+1)!\sqrt{\pi}}{2^{n+1}(2n+2)!} f^{(2n+2)}(\xi), \qquad -\infty < \xi < \infty.$$

We can also use the Gauss-Laguerre and the Gauss–Hermite quadrature formulas to approximate convergent integrals of the form $\int_0^\infty f(x)\,dx$, $\int_{-\infty}^\infty f(x)\,dx$:

$$\int_0^\infty f(x)\,dx = \int_0^\infty e^{-x}e^x f(x)\,dx$$

$$\approx \sum_{i=0}^n \omega_i e^{x_i} f(x_i),$$

$$\int_{-\infty}^\infty f(x)\,dx = \int_{-\infty}^\infty e^{-x^2}e^{x^2} f(x)\,dx$$

$$\approx \sum_{i=0}^n \omega_i e^{x_i^2} f(x_i).$$

EXAMPLE 12.5.1. Consider the integral

$$I = \int_0^\infty \frac{e^{-x}x}{1-e^{-2x}}\,dx$$

$$\approx \sum_{i=0}^n \omega_i f(x_i), \tag{12.21}$$

where $f(x) = x(1-e^{-2x})^{-1}$ and the x_i's are the zeros of the Laguerre polynomial $L_{n+1}(x)$. To find expressions for $L_n(x)$, $n = 0, 1, 2, \ldots$, we can use the recurrence relation (10.37) with $\alpha = 0$, which can be written as

$$L_{n+1}(x) = (x-n-1)L_n(x) - x\frac{dL_n(x)}{dx}, \qquad n = 0, 1, 2, \ldots. \tag{12.22}$$

Recall that $L_0(x) = 1$. Choosing $n = 1$ in (12.21), we get

$$I \approx \omega_0 f(x_0) + \omega_1 f(x_1). \tag{12.23}$$

From (12.22) we have

$$L_1(x) = (x-1)L_0(x)$$
$$= x - 1,$$

$$L_2(x) = (x-2)L_1(x) - x\frac{d(x-1)}{dx}$$

$$= (x-2)(x-1) - x$$

$$= x^2 - 4x + 2.$$

The zeros of $L_2(x)$ are $x_0 = 2 - \sqrt{2}$, $x_1 = 2 + \sqrt{2}$. To find ω_0 and ω_1, formula (12.19) must be exact for all polynomials of degrees not exceeding

$2n + 1 = 3$. This is equivalent to requiring that

$$\int_0^\infty e^{-x}\, dx = \omega_0 + \omega_1,$$

$$\int_0^\infty e^{-x} x\, dx = \omega_0 x_0 + \omega_1 x_1,$$

$$\int_0^\infty e^{-x} x^2\, dx = \omega_0 x_0^2 + \omega_1 x_1^2,$$

$$\int_0^\infty e^{-x} x^3\, dx = \omega_0 x_0^3 + \omega_1 x_1^3.$$

Only two equations are linearly independent, the solution of which is $\omega_0 = 0.853553$, $\omega_1 = 0.146447$. From (12.23) we then have

$$I \approx \frac{\omega_0 x_0}{1 - e^{-2x_0}} + \frac{\omega_1 x_1}{1 - e^{-2x_1}}$$

$$= 1.225054.$$

Let us now calculate (12.21) using $n = 2, 3, 4$. The zeros of Laguerre polynomials $L_3(x)$, $L_4(x)$, $L_5(x)$, and the values of ω_i for each n are shown in Table 12.1. These values are given in Ralston and Rabinowitz (1978, page 106) and also in Krylov (1962, Appendix C). The corresponding approximate values of I are given in Table 12.1. It can be shown that the exact value of I is $\pi^2/8 \approx 1.2337$.

Table 12.1. Zeros of Laguerre Polynomials (x_i), Values of ω_i, and the Corresponding Approximate Values[a] of I

n	x_i	ω_i	I
1	0.585786	0.853553	
	3.414214	0.146447	1.225054
2	0.415775	0.711093	
	2.294280	0.278518	
	6.289945	0.010389	1.234538
3	0.322548	0.603154	
	1.745761	0.357419	
	4.536620	0.038888	
	9.395071	0.000539	1.234309
4	0.263560	0.521756	
	1.413403	0.398667	
	3.596426	0.075942	
	7.085810	0.003612	
	12.640801	0.000023	1.233793

[a]See (12.21).

12.6. THE METHOD OF LAPLACE

This method is used to approximate integrals of the form

$$I(\lambda) = \int_a^b \varphi(x) e^{\lambda h(x)} \, dx, \qquad (12.24)$$

where λ is a large positive constant, $\varphi(x)$ is continuous on $[a, b]$, and the first and second derivatives of $h(x)$ are continuous on $[a, b]$. The limits a and b may be finite or infinite. This integral was used by Laplace in his original development of the central limit theorem (see Section 4.5.1). More specifically, if $X_1, X_2, \ldots, X_n, \ldots$ is a sequence of independent and identically distributed random variables with a common density function, then the density function of the sum $S_n = \sum_{i=1}^{n} X_i$ can be represented in the form (12.24) (see Wong, 1989, Chapter 2).

Suppose that $h(x)$ has a single maximum in the interval $[a, b]$ at $x = t$, $a \leq t < b$, where $h'(t) = 0$ and $h''(t) < 0$. Hence, $e^{\lambda h(x)}$ is maximized at t for any $\lambda > 0$. Suppose further that $e^{\lambda h(x)}$ becomes very strongly peaked at $x = t$ and decreases rapidly away from $x = t$ on $[a, b]$ as $\lambda \to \infty$. In this case, the major portion of $I(\lambda)$ comes from integrating the function $\varphi(x) e^{\lambda h(x)}$ over a small neighborhood around $x = t$. Under these conditions, it can be shown that if $a < t < b$, and as $\lambda \to \infty$,

$$I(\lambda) \sim \varphi(t) e^{\lambda h(t)} \left[\frac{-2\pi}{\lambda h''(t)} \right]^{1/2}, \qquad (12.25)$$

where $\sim$ denotes asymptotic equality (see Section 3.3). Formula (12.25) is known as *Laplace's approximation*.

A heuristic derivation of (12.25) can be arrived at by replacing $\varphi(x)$ and $h(x)$ by the leading terms in their Taylor's series expansions around $x = t$. The integration limits are then extended to $-\infty$ and ∞, that is,

$$\int_a^b \varphi(x) e^{\lambda h(x)} \, dx \approx \int_a^b \varphi(t) \exp\left[\lambda h(t) + \frac{\lambda}{2} (x - t)^2 h''(t) \right] dx$$

$$\approx \int_{-\infty}^{\infty} \varphi(t) \exp\left[\lambda h(t) + \frac{\lambda}{2} (x - t)^2 h''(t) \right] dx$$

$$= \varphi(t) e^{\lambda h(t)} \int_{-\infty}^{\infty} \exp\left[\frac{\lambda}{2} (x - t)^2 h''(t) \right] dx \qquad (12.26)$$

$$= \varphi(t) e^{\lambda h(t)} \left[\frac{-2\pi}{\lambda h''(t)} \right]^{1/2}. \qquad (12.27)$$

Formula (12.27) follows from (12.26) by making use of the fact that $\int_0^\infty e^{-x^2} \, dx = \frac{1}{2} \Gamma(\frac{1}{2}) = \sqrt{\pi}/2$, where $\Gamma(\cdot)$ is the gamma function (see Example 6.9.6), or by simply evaluating the integral of a normal density function.

If $t = a$, then it can be shown that as $\lambda \to \infty$,

$$I(\lambda) \sim \varphi(a) e^{\lambda h(a)} \left[\frac{-\pi}{2\lambda h''(a)} \right]^{1/2}. \tag{12.28}$$

Rigorous proofs of (12.27) and (12.28) can be found in Wong (1989, Chapter 2), Copson (1965, Chapter 5), Fulks (1978, Chapter 18), and Lange (1999, Chapter 4).

EXAMPLE 12.6.1.   Consider the gamma function

$$\Gamma(n + 1) = \int_0^\infty e^{-x} x^n \, dx, \qquad n > -1. \tag{12.29}$$

Let us find an approximation for $\Gamma(n + 1)$ when n is large an positive, but not necessarily an integer. Let $x = nz$; then (12.29) can be written as

$$\Gamma(n + 1) = n \int_0^\infty e^{-nz} \exp[n \log(nz)] \, dz$$

$$= n \int_0^\infty e^{-nz} \exp[n \log n + n \log z] \, dz$$

$$= n^{n+1} \int_0^\infty \exp[n(-z + \log z)] \, dz. \tag{12.30}$$

Let $h(z) = -z + \log z$. Then $h(z)$ has a unique maximum at $z = 1$ with $h'(1) = 0$ and $h''(1) = -1$. Applying formula (12.27) to (12.30), we obtain

$$\Gamma(n + 1) \sim n^{n+1} e^{-n} \left[\frac{-2\pi}{n(-1)} \right]^{1/2}$$

$$= e^{-n} n^n \sqrt{2\pi n}, \tag{12.31}$$

as $n \to \infty$. Formula (12.31) is known as *Stirling's formula*.

EXAMPLE 12.6.2.   Consider the integral,

$$I_n(\lambda) = \frac{1}{\pi} \int_0^\pi \exp(\lambda \cos x) \cos nx \, dx,$$

as $\lambda \to \infty$. This integral looks like (12.24) with $h(x) = \cos x$, which has a single maximum at $x = 0$ in $[0, \pi]$. Since $h''(0) = -1$, then by applying (12.28) we obtain, as $\lambda \to \infty$,

$$I_n(\lambda) \sim \frac{1}{\pi} e^\lambda \left[\frac{-\pi}{2\lambda(-1)} \right]^{1/2}$$

$$= \frac{e^\lambda}{\sqrt{2\pi\lambda}}.$$

EXAMPLE 12.6.3.   Consider the integral

$$I(\lambda) = \int_0^\pi \exp[\lambda \sin x] \, dx$$

as $\lambda \to \infty$. Here, $h(x) = \sin x$, which has a single maximum at $x = \pi/2$ in $[0, \pi]$, and $h''(\pi/2) = -1$. From (12.27), we get

$$I(\lambda) \sim e^\lambda \left[\frac{-2\pi}{\lambda(-1)} \right]^{1/2}$$

$$= e^\lambda \sqrt{\frac{2\pi}{\lambda}}$$

as $\lambda \to \infty$.

12.7. MULTIPLE INTEGRALS

We recall that integration of a multivariable function was discussed in Section 7.9. In the present section, we consider approximate integration formulas for an n-tuple Riemann integral over a region D in an n-dimensional Euclidean space R^n.

For example, let us consider the double integral $I = \iint_D f(x_1, x_2) \, dx_1 \, dx_2$, where $D \subset R^2$ is the region

$$D = \{(x_1, x_2) | a \le x_1 \le b, \psi(x_1) \le x_2 \le \phi(x_1)\}.$$

Then

$$I = \int_a^b \left[\int_{\psi(x_1)}^{\phi(x_1)} f(x_1, x_2) \, dx_2 \right] dx_1$$

$$= \int_a^b g(x_1) \, dx_1, \tag{12.32}$$

where

$$g(x_1) = \int_{\psi(x_1)}^{\phi(x_1)} f(x_1, x_2) \, dx_2. \tag{12.33}$$

Let us now apply a Gaussian integration rule to (12.32) using the points $x_1 = z_0, z_1, \ldots, z_m$ with the matching coefficients $\omega_0, \omega_1, \ldots, \omega_m$,

$$\int_a^b g(x_1)\, dx_1 \approx \sum_{i=0}^{m} \omega_i g(z_i).$$

Thus

$$I \approx \sum_{i=0}^{m} \omega_i \int_{\psi(z_i)}^{\phi(z_i)} f(z_i, x_2)\, dx_2. \tag{12.34}$$

For the ith of the $m+1$ integrals in (12.34) we can apply a Gaussian integration rule using the points $y_{i0}, y_{i1}, \ldots, y_{in}$ with the corresponding coefficients $v_{i0}, v_{i1}, \ldots, v_{in}$. We then have

$$\int_{\psi(z_i)}^{\phi(z_i)} f(z_i, x_2)\, dx_2 \approx \sum_{j=0}^{n} v_{ij} f(z_i, y_{ij}).$$

Hence,

$$I = \sum_{i=0}^{m} \sum_{j=0}^{n} \omega_i v_{ij} f(z_i, y_{ij}).$$

This procedure can obviously be generalized to higher-order multiple integrals. More details can be found in Stroud (1971) and Davis and Rabinowitz (1975, Chapter 5).

The method of Laplace in Section 12.6 can be extended to an n-dimensional integral of the form

$$I(\lambda) = \int_D \varphi(\mathbf{x}) e^{\lambda h(\mathbf{x})}\, d\mathbf{x},$$

which is a multidimensional version of the integral in (12.24). Here, D is a region in R^n, which may be bounded or unbounded, λ is a large positive constant, and $\mathbf{x} = (x_1, x_2, \ldots, x_n)'$. As before, it is assumed that:

a. $\varphi(\mathbf{x})$ is continuous in D.

b. $h(\mathbf{x})$ has continuous first-order and second-order partial derivatives with respect to $x_1, x_2, \ldots, x_n$ in D.

c. $h(\mathbf{x})$ has a single maximum in D at $\mathbf{x} = \mathbf{t}$.

If $\mathbf{t}$ is an interior point of D, then it is also a stationary point of $h(\mathbf{x})$, that is, $\partial h/\partial x_i|_{\mathbf{x}=\mathbf{t}} = 0$, $i = 1, 2, \ldots, n$, since $\mathbf{t}$ is a point of maximum and the partial derivatives of $h(\mathbf{x})$ exist. Furthermore, the Hessian matrix,

$$\mathbf{H}_h(\mathbf{t}) = \left(\frac{\partial^2 h(\mathbf{x})}{\partial x_i \partial x_j} \right)_{\mathbf{x}=\mathbf{t}},$$

is negative definite. Then, for large λ, $I(\lambda)$ is approximately equal to

$$I(\lambda) \sim \left(\frac{2\pi}{\lambda} \right)^{n/2} \varphi(\mathbf{t})\{-\det[\mathbf{H}_h(\mathbf{t})]\}^{-1/2} e^{\lambda h(\mathbf{t})}.$$

A proof of this approximation can be found in Wong (1989, Section 9.5). We note that this expression is a generalization of Laplace's formula (12.25) to an n-tuple Riemann integral.

If $\mathbf{t}$ happens to be on the boundary of D and still satisfies the conditions that $\partial h/\partial x_i|_{\mathbf{x}=\mathbf{t}} = 0$ for $i = 1, 2, \ldots, n$, and $\mathbf{H}_h(\mathbf{t})$ is negative definite, then it can be shown that for large λ,

$$I(\lambda) \sim \frac{1}{2} \left(\frac{2\pi}{\lambda} \right)^{n/2} \varphi(\mathbf{t})\{-\det[\mathbf{H}_h(\mathbf{t})]\}^{-1/2} e^{\lambda h(\mathbf{t})},$$

which is one-half of the previous approximation for $I(\lambda)$ (see Wong,1989, page 498).

12.8. THE MONTE CARLO METHOD

A new approach to approximate integration arose in the 1940s as part of the Monte Carlo method of S. Ulam and J. von Neumann (Haber, 1970). The basic idea of the Monte Carlo method for integrals is described as follows: suppose that we need to compute the integral

$$I = \int_a^b f(x) \, dx. \tag{12.35}$$

We consider I as the expected value of a certain stochastic process. An estimate of I can be obtained by random sampling from this process, and the estimate is then used as an approximation to I. For example, let X be a continuous random variable that has the uniform distribution $U(a, b)$ over the interval $[a, b]$. The expected value of $f(X)$ is

$$E[f(X)] = \frac{1}{b-a} \int_a^b f(x) \, dx$$

$$= \frac{I}{b-a}.$$

Let $x_1, x_2, \ldots, x_n$ be a random sample from $U(a, b)$. An estimate of $E[f(X)]$ is given by $(1/n)\sum_{i=1}^{n} f(x_i)$. Hence, an approximate value of I, denoted by $\hat{I}_n$, can be obtained as

$$\hat{I}_n = \frac{b-a}{n} \sum_{i=1}^{n} f(x_i). \tag{12.36}$$

The justification for using $\hat{I}_n$ as an approximation to I is that $\hat{I}_n$ is a consistent estimator of I, that is, for a given $\epsilon > 0$,

$$\lim_{n \to \infty} P\left(\left|\hat{I}_n - I\right| \geq \epsilon\right) = 0.$$

This is true because $(1/n)\sum_{i=1}^{n} f(x_i)$ converges in probability to $E[f(X)]$ as $n \to \infty$, according to the law of large numbers (see Sections 3.7 and 5.6.3). In other words, the probability that $\hat{I}_n$ will be different from I can be made arbitrarily close to zero if n is chosen large enough. In fact, we even have the stronger result that $\hat{I}_n$ converges strongly, or almost surely, to I, as $n \to \infty$, by the strong law of large numbers (see Section 5.6.3).

The closeness of $\hat{I}_n$ to I depends on the variance of $\hat{I}_n$, which is equal to

$$\mathrm{Var}\left(\hat{I}_n\right) = (b-a)^2 \mathrm{Var}\left[\frac{1}{n} \sum_{i=1}^{n} f(x_i)\right]$$

$$= \frac{(b-a)^2 \sigma_f^2}{n}, \tag{12.37}$$

where σ_f^2 is the variance of the random variable $f(X)$, that is,

$$\sigma_f^2 = \mathrm{Var}[f(X)]$$

$$= E[f^2(X)] - \{E[f(X)]\}^2$$

$$= \frac{1}{b-a} \int_a^b f^2(x)\, dx - \left(\frac{I}{b-a}\right)^2. \tag{12.38}$$

By the central limit theorem (see Section 4.5.1), if n is large enough, then $\hat{I}_n$ is approximately normally distributed with mean I and variance $(1/n)(b-a)^2\sigma_f^2$. Thus,

$$\frac{\hat{I}_n - I}{(b-a)\sigma_f/\sqrt{n}} \xrightarrow{d} Z,$$

where Z has the standard normal distribution, and the symbol $\xrightarrow{d}$ denotes convergence in distribution (see Section 4.5.1). It follows that for a given $\tau > 0$,

$$P\left[\left|\hat{I}_n - I\right| \leq \frac{\tau}{\sqrt{n}}(b-a)\sigma_f\right] \approx \frac{1}{\sqrt{2\pi}} \int_{-\tau}^{\tau} e^{-x^2/2} \, dx. \qquad (12.39)$$

The right-hand side of (12.39) is the probability that a standard normal distribution attains values between $-\tau$ and τ. Let us denote this probability by $1 - \alpha$. Then $\tau = z_{\alpha/2}$, which is the upper $(\alpha/2)100$th percentile of the standard normal distribution. If we denote the error of approximation, $\hat{I}_n - I$, by E_n, then formula (12.39) indicates that for large n,

$$|E_n| \leq \frac{1}{\sqrt{n}}(b-a)\sigma_f z_{\alpha/2} \qquad (12.40)$$

with an approximate probability equal to $1 - \alpha$, which is called the confidence coefficient. Thus, for a fixed α, the error bound in (12.40) is proportional to σ_f and is inversely proportional to $\sqrt{n}$. For example, if $1 - \alpha = 0.90$, then $z_{\alpha/2} = 1.645$, and

$$|E_n| \leq \frac{1.645}{\sqrt{n}}(b-a)\sigma_f$$

with an approximate confidence coefficient equal to 0.90. Also, if $1 - \alpha = 0.95$, then $z_{\alpha/2} = 1.96$, and

$$|E_n| \leq \frac{1.96}{\sqrt{n}}(b-a)\sigma_f$$

with an approximate confidence coefficient equal to 0.95.

In order to compute the error bound in (12.40), an estimate of σ_f^2 is needed. Using (12.38) and the random sample $x_1, x_2, \ldots, x_n$, an estimate of σ_f^2 is given by

$$\hat{\sigma}_{fn}^2 = \frac{1}{n}\sum_{i=1}^{n} f^2(x_i) - \left[\frac{1}{n}\sum_{i=1}^{n} f(x_i)\right]^2. \qquad (12.41)$$

12.8.1. Variance Reduction

In order to increase the accuracy of the Monte Carlo approximation of I, the error bound in (12.40) should be reduced for a fixed value of α. We can achieve this by increasing n. Alternatively, we can reduce σ_f by considering a distribution other than the uniform distribution. This can be accomplished by using the so-called method of *importance sampling*, a description of which follows.

Let $g(x)$ be a density function that is positive over the interval $[a, b]$. Thus, $g(x) > 0$, $a \le x \le b$, and $\int_a^b g(x)\, dx = 1$. The integral in (12.35) can be written as

$$I = \int_a^b \frac{f(x)}{g(x)} g(x)\, dx. \tag{12.42}$$

In this case, I is the expected value of $f(X)/g(X)$, where X is a continuous random variable with the density function $g(x)$. Using now a random sample, $x_1, x_2, \ldots, x_n$, from this distribution, an estimate of I can be obtained as

$$\hat{I}_n^* = \frac{1}{n} \sum_{i=1}^n \frac{f(x_i)}{g(x_i)}. \tag{12.43}$$

The variance of $\hat{I}_n^*$ is then given by

$$\mathrm{Var}\left(\hat{I}_n^*\right) = \frac{\sigma_{fg}^2}{n},$$

where

$$\sigma_{fg}^2 = \mathrm{Var}\left[\frac{f(X)}{g(X)}\right]$$

$$= E\left\{\left[\frac{f(X)}{g(X)}\right]^2\right\} - \left\{E\left[\frac{f(X)}{g(X)}\right]\right\}^2$$

$$= \int_a^b \frac{f^2(x)}{g^2(x)} g(x)\, dx - I^2. \tag{12.44}$$

As before, the error bound can be derived on the basis of the central limit theorem, using the fact that $\hat{I}_n^*$ is approximately normally distributed with mean I and variance $(1/n)\sigma_{fg}^2$ for large n. Hence, as in (12.40),

$$|E_n^*| \le \frac{1}{\sqrt{n}} \sigma_{fg} z_{\alpha/2}$$

with an approximate probability equal to $1 - \alpha$, where $E_n^* = \hat{I}_n^* - I$. The density $g(x)$ should therefore be chosen so that an error bound smaller than

Table 12.2. Approximate Values of $I = \int_1^2 x^2\, dx$ Using Formula (12.36)

n	$\hat{I}_n$
50	2.3669
100	2.5087
150	2.2227
200	2.3067
250	2.3718
300	2.3115
350	2.3366

the one in (12.40) can be achieved. For example, if $f(x) > 0$, and if

$$g(x) = \frac{f(x)}{\int_a^b f(x)\, dx}, \qquad (12.45)$$

then $\sigma_{fg}^2 = 0$ as can be seen from formula (12.44).

Unfortunately, since the exact value of $\int_a^b f(x)\, dx$ is the one we seek to compute, formula (12.45) cannot be used. However, by choosing $g(x)$ to behave approximately as $f(x)$ [assuming $f(x)$ is positive], we should expect a reduction in the variance. Note that the generation of random variables from the $g(x)$ distribution is more involved than just using a random sample from the uniform distribution $U(a, b)$.

EXAMPLE 12.8.1. Consider the integral $I = \int_1^2 x^2\, dx = \frac{7}{3} \approx 2.3333$. Suppose that we use a sample of 50 points from the uniform distribution $U(1, 2)$. Applying formula (12.36), we obtain $\hat{I}_{50} = 2.3669$. Repeating this process several times using higher values of n, we obtain the values in Table 12.2.

EXAMPLE 12.8.2. Let us now apply the method of importance sampling to the integral $I = \int_0^1 e^x\, dx \approx 1.7183$. Consider the density function $g(x) = \frac{2}{3}(1 + x)$ over the interval $[0, 1]$. Using the method described in Section 3.7, a random sample, $x_1, x_2, \ldots, x_n$, can be generated from this distribution as follows: the cumulative distribution function for $g(x)$ is $y = G(x) = P[X \leq x]$, that is,

$$y = G(x) = \int_0^x g(t)\, dt$$

$$= \frac{2}{3} \int_0^x (1 + t)\, dt$$

$$= \frac{2}{3}\left(x + \frac{x^2}{2}\right), \qquad 0 \leq x \leq 1.$$

Table 12.3. Approximate Values of $I = \int_0^1 e^x \, dx$

n	$\hat{I}_n^*$ [Formula (12.46)]	$\hat{I}_n$ [Formula (12.36)]
50	1.7176	1.7156
100	1.7137	1.7025
150	1.7063	1.6854
200	1.7297	1.7516
250	1.7026	1.6713
300	1.7189	1.7201
350	1.7093	1.6908
400	1.7188	1.7192

The only solution of $y = G(x)$ in $[0, 1]$ is

$$x = -1 + (1 + 3y)^{1/2}, \qquad 0 \le y \le 1.$$

Hence, the inverse function of $G(x)$ is

$$G^{-1}(y) = -1 + (1 + 3y)^{1/2}, \qquad 0 \le y \le 1.$$

If $y_1, y_2, \ldots, y_n$ form a random sample of n values from the uniform distribution $U(0, 1)$, then $x_1 = G^{-1}(y_1)$, $x_2 = G^{-1}(y_2), \ldots, x_n = G^{-1}(y_n)$ will form a sample from the distribution with the density function $g(x)$. Formula (12.43) can then be applied to approximate the value of I using the estimate

$$\hat{I}_n^* = \frac{3}{2n} \sum_{i=1}^{n} \frac{e^{x_i}}{1 + x_i}. \tag{12.46}$$

Table 12.3 gives $\hat{I}_n^*$ for several values of n. For the sake of comparison, values of $\hat{I}_n$ from formula (12.36) [with $f(x) = e^x$, $a = 0$, $b = 1$] were also computed using a sample from the uniform distribution $U(0, 1)$. The results are shown in Table 12.3. We note that the values of $\hat{I}_n^*$ are more stable and closer to the true value of I than those of $\hat{I}_n$.

12.8.2. Integrals in Higher Dimensions

The Monte Carlo method can be extended to multidimensional integrals. Consider computing the integral $I = \int_D f(\mathbf{x}) \, d\mathbf{x}$, where D is a bounded region in R^n, the n-dimensional Euclidean space, and $\mathbf{x} = (x_1, x_2, \ldots, x_n)'$. As before, we consider I as the expected value of a stochastic process having a certain distribution over D. For example, we may take $\mathbf{X}$ to be a continuous random vector uniformly distributed over D. By this we mean that the probability of $\mathbf{X}$ being in D is $1/v(D)$, where $v(D)$ denotes the volume of D, and the probability of $\mathbf{X}$ being outside D is zero. Hence, the expected value of $f(\mathbf{X})$ is

$$E[f(\mathbf{X})] = \frac{1}{v(D)} \int_D f(\mathbf{x})\, d\mathbf{x}$$

$$= \frac{I}{v(D)}.$$

The variance of $f(\mathbf{X})$, denoted by σ_f^2, is

$$\sigma_f^2 = E[f^2(\mathbf{X})] - \{E[f(\mathbf{X})]\}^2$$

$$= \frac{1}{v(D)} \int_D f^2(\mathbf{x})\, d\mathbf{x} - \left[\frac{I}{v(D)}\right]^2.$$

Let us now take a sample of N independent observations on $\mathbf{X}$, namely, $\mathbf{X}_1, \mathbf{X}_2, \ldots, \mathbf{X}_N$. Then a consistent estimator of $E[f(\mathbf{X})]$ is $(1/N)\sum_{i=1}^N f(\mathbf{X}_i)$, and hence, an estimate of I is given by

$$\hat{I}_N = \frac{v(D)}{N} \sum_{i=1}^N f(\mathbf{X}_i),$$

which can be used to approximate I. The variance of $\hat{I}_N$ is

$$\operatorname{Var}(\hat{I}_N) = \frac{\sigma_f^2}{N} v^2(D). \tag{12.47}$$

If N is large enough, then by the central limit theorem, $\hat{I}_N$ is approximately normally distributed with mean I and variance as in (12.47). It follows that

$$|E_N| \le \frac{v(D)}{\sqrt{N}} \sigma_f z_{\alpha/2}$$

with an approximate probability equal to $1 - \alpha$, where $E_N = I - \hat{I}_N$ is the error of approximation. This formula is analogous to formula (12.40).

The method of importance sampling can also be applied here to reduce the error of approximation. The application of this method is similar to the case of a single-variable integral as seen earlier.

12.9. APPLICATIONS IN STATISTICS

Approximation of integrals is a problem of substantial concern for statisticians. The statistical literature in this area has grown significantly in the last 20 years, particularly in connection with integrals that arise in Bayesian statistics. Evans and Swartz (1995) presented a survey of the major techniques and approaches available for the numerical approximation of integrals in statistics. The proceedings edited by Flournoy and Tsutakawa (1991) includes several interesting articles on statistical multiple integration, including a detailed description of available software to compute multidimensional integrals (see the article by Kahaner, 1991, page 9).

12.9.1. The Gauss-Hermite Quadrature

The Gauss–Hermite quadrature mentioned earlier in Section 12.5 is often used for numerical integration in statistics because of its relation to Gaussian (normal) densities. We recall that this quadrature is defined in terms of integrals of the form $\int_{-\infty}^{\infty} e^{-x^2} f(x)\, dx$. Using formula (12.20), we have approximately

$$\int_{-\infty}^{\infty} e^{-x^2} f(x)\, dx \approx \sum_{i=0}^{n} \omega_i f(x_i),$$

where the x_i's are the zeros of the Hermite polynomial $H_{n+1}(x)$ of degree $n+1$, and the ω_i's are suitably corresponding weights. Tables of x_i and ω_i values are given by Abramowitz and Stegun (1972, page 924) and by Krylov (1962, Appendix B).

Liu and Pierce (1994) applied the Gauss–Hermite quadrature to integrals of the form $\int_{-\infty}^{\infty} g(t)\, dt$, which can be expressed in the form

$$\int_{-\infty}^{\infty} g(t)\, dt = \int_{-\infty}^{\infty} f(t)\phi(t, \mu, \sigma)\, dt,$$

where $\phi(t, \mu, \sigma)$ is the normal density

$$\phi(t, \mu, \sigma) = \frac{1}{\sqrt{2\pi\sigma^2}} \exp\left[-\frac{1}{2\sigma^2}(t - \mu)^2 \right],$$

and $f(t) = g(t)/\phi(t, \mu, \sigma)$. Thus,

$$\int_{-\infty}^{\infty} g(t)\, dt = \int_{-\infty}^{\infty} \frac{1}{\sqrt{2\pi\sigma^2}} f(t)\exp\left[-\frac{1}{2\sigma^2}(t - \mu)^2 \right] dt$$

$$= \frac{1}{\sqrt{\pi}} \int_{-\infty}^{\infty} f(\mu + \sqrt{2}\,\sigma x)e^{-x^2}\, dx$$

$$\approx \frac{1}{\sqrt{\pi}} \sum_{i=0}^{n} \omega_i f(\mu + \sqrt{2}\,\sigma x_i), \tag{12.48}$$

where the x_i's are the zeros of the Hermite polynomial $H_{n+1}(x)$ of degree $n + 1$. We may recall that the x_i's and ω_i's are chosen so that this approximation will be exact for all polynomials of degrees not exceeding $2n + 1$. For this reason, Liu and Pierce (1994) recommend choosing μ and σ in (12.48) so that $f(t)$ is well approximated by a low-order polynomial in the region where the values of $\mu + \sqrt{2}\,\sigma x_i$ are taken. More specifically, the $(n + 1)$th-order Gauss–Hermite quadrature in (12.48) will be highly effective if the ratio of $g(t)$ to the normal density $\phi(t, \mu, \sigma^2)$ can be well approximated by a polynomial of degree not exceeding $2n + 1$ in the region where $g(t)$ is substantial. This arises frequently, for example, when $g(t)$ is a likelihood function [if $g(t) > 0$], or the product of a likelihood function and a normal density, as was pointed out by Liu and Pierce (1994), who gave several examples to demonstrate the usefulness of the approximation in (12.48).

12.9.2. Minimum Mean Squared Error Quadrature

Correlated observations may arise in some experimental work (Piegorsch and Bailer, 1993). Consider, for example, the model

$$y_{qr} = f(t_q) + \epsilon_{qr},$$

where y_{qr} represents the observed value from experimental unit r at time t_q $(q = 0, 1, \ldots, m; r = 1, 2, \ldots, n)$, $f(t_q)$ is the underlying response function, and ϵ_{qr} is a random error term. It is assumed that $E(\epsilon_{qr}) = 0$, $\text{Cov}(\epsilon_{pr}, \epsilon_{qr}) = \sigma_{pq}$, and $\text{Cov}(\epsilon_{pr}, \epsilon_{qs}) = 0$ $(r \neq s)$ for all p, q. The area under the response curve over the interval $t_0 < t < t_m$ is

$$A = \int_{t_0}^{t_m} f(t)\, dt. \tag{12.49}$$

This is an important measure for a variety of experimental situations, including the assessment of chemical bioavailability in drug disposition studies (Gibaldi and Perrier, 1982, Chapters 2, 7) and other clinical settings. If the functional form of $f(\cdot)$ is unknown, the integral in (12.49) is estimated by numerical methods. This is accomplished using a quadrature approximation of the integral in (12.49). By definition, a quadrature estimator of this integral is

$$\hat{A} = \sum_{q=0}^{m} \phi_q \hat{f}_q, \tag{12.50}$$

where $\hat{f}_q$ is some unbiased estimator of $f_q = f(t_q)$ with $\text{Cov}(\hat{f}_p, \hat{f}_q) = \sigma_{pq}/n$, and the ϕ_q's form a set of quadrature coefficients.

The expected value of $\hat{A}$,

$$E(\hat{A}) = \sum_{q=0}^{m} \phi_q f_q,$$

may not necessarily be equal to A due to the quadrature approximation employed in calculating $\hat{A}$. The bias in estimating A is

$$\text{bias} = E(\hat{A}) - A,$$

and the mean squared error (MSE) of $\hat{A}$ is given by

$$\text{MSE}(\hat{A}) = \text{Var}\,\hat{A} + \left[E(\hat{A}) - A\right]^2$$

$$= \sum_{p=0}^{m} \sum_{q=0}^{m} \phi_p \phi_q \frac{\sigma_{pq}}{n} + \left(\sum_{q=0}^{m} \phi_q f_q - A\right)^2,$$

which can be written as

$$\text{MSE}(\hat{A}) = \frac{1}{n}\boldsymbol{\phi}'\mathbf{V}\boldsymbol{\phi} + (\mathbf{f}'\boldsymbol{\phi} - A)^2,$$

where $\mathbf{V} = (\sigma_{pq})$, $\mathbf{f} = (f_0, f_1, \ldots, f_m)'$, and $\boldsymbol{\phi}' = (\phi_0, \phi_1, \ldots, \phi_m)$. Hence,

$$\text{MSE}(\hat{A}) = \boldsymbol{\phi}'\left[\frac{1}{n}\mathbf{V} + \mathbf{f}\mathbf{f}'\right]\boldsymbol{\phi} - 2A\mathbf{f}'\boldsymbol{\phi} + A^2. \qquad (12.51)$$

Let us now seek an optimum value of $\boldsymbol{\phi}$ that minimizes $\text{MSE}(\hat{A})$ in (12.51). For this purpose, we equate the gradient of $\text{MSE}(\hat{A})$ with respect to $\boldsymbol{\phi}$, namely $\boldsymbol{\nabla}_\phi \text{MSE}(\hat{A})$, to zero. We obtain

$$\boldsymbol{\nabla}_\phi \text{MSE}(\hat{A}) = 2\left[\left(\frac{1}{n}\mathbf{V} + \mathbf{f}\mathbf{f}'\right)\boldsymbol{\phi} - A\mathbf{f}\right] = \mathbf{0}. \qquad (12.52)$$

In order for this equation to have a solution, the vector $A\mathbf{f}$ must belong to the column space of the matrix $(1/n)\mathbf{V} + \mathbf{f}\mathbf{f}'$. Note that this matrix is positive semidefinite. If its inverse exists, then it will be positive definite, and the only solution, $\boldsymbol{\phi}^*$, to (12.52), namely,

$$\boldsymbol{\phi}^* = A\left(\frac{1}{n}\mathbf{V} + \mathbf{f}\mathbf{f}'\right)^{-1}\mathbf{f}, \qquad (12.53)$$

yields a unique minimum of $\text{MSE}(\hat{A})$ (see Section 7.7). Using $\boldsymbol{\phi}^*$ in (12.50), we obtain the following estimate of A:

$$\hat{A}^* = \hat{\mathbf{f}}'\boldsymbol{\phi}^*$$

$$= A\hat{\mathbf{f}}'\left[\frac{1}{n}\mathbf{V} + \mathbf{f}\mathbf{f}'\right]^{-1}\mathbf{f}. \tag{12.54}$$

Using the Sherman−Morrison formula (see Exercise 2.14), (12.54) can be written as

$$\hat{A}^* = nA\left[\hat{\mathbf{f}}'\mathbf{V}^{-1}\mathbf{f} - \frac{n\hat{\mathbf{f}}'\mathbf{V}^{-1}\mathbf{f}\mathbf{f}'\mathbf{V}^{-1}\mathbf{f}}{1 + n\mathbf{f}'\mathbf{V}^{-1}\mathbf{f}}\right], \tag{12.55}$$

where $\hat{\mathbf{f}} = (\hat{f}_0, \hat{f}_1, \ldots, \hat{f}_m)'$. We refer to $\hat{A}^*$ as an MSE-optimal estimator of A. Replacing $\mathbf{f}$ with its unbiased estimator $\hat{\mathbf{f}}$, and A with some initial estimate, say $A_0 = \hat{\mathbf{f}}'\boldsymbol{\phi}_0$, where $\boldsymbol{\phi}_0$ is an initial value of $\boldsymbol{\phi}$, yields the approximate MSE-optimal estimator

$$\hat{A}^{**} = \frac{c}{1 + c}A_0, \tag{12.56}$$

where c is the quadratic form,

$$c = n\hat{\mathbf{f}}'\mathbf{V}^{-1}\hat{\mathbf{f}}.$$

This estimator has the form $\hat{\mathbf{f}}'\tilde{\boldsymbol{\phi}}$, where $\tilde{\boldsymbol{\phi}} = [c/(1 + c)]\boldsymbol{\phi}_0$. Since $c \geq 0$, $\hat{A}^{**}$ provides a shrinkage of the initial estimate in A_0 toward zero. This creates a biased estimator of A with a smaller variance, which results in an overall reduction in MSE. A similar approach is used in ridge regression to estimate the parameters of a linear model when the columns of the corresponding matrix $\mathbf{X}$ are multicollinear (see Section 5.6.6).

We note that this procedure requires knowledge of $\mathbf{V}$. In many applications, it may not be possible to specify $\mathbf{V}$. Piegorsch and Bailer (1993) used an estimate of $\mathbf{V}$ when $\mathbf{V}$ was assumed to have certain particular patterns, such as $\mathbf{V} = \sigma^2\mathbf{I}$ or $\mathbf{V} = \sigma^2[(1 - \rho)\mathbf{I} + \rho\mathbf{J}]$, where $\mathbf{I}$ and $\mathbf{J}$ are, respectively, the identity matrix and the square matrix of ones, and σ^2 and ρ are unknown constants. For example, under the equal variance assumption $\mathbf{V} = \sigma^2\mathbf{I}$, the ratio $c/(1 + c)$ is equal to

$$\frac{c}{1 + c} = \left(1 + \frac{\sigma^2}{n\hat{\mathbf{f}}'\hat{\mathbf{f}}}\right)^{-1}.$$

If $\hat{\mathbf{f}}$ is normally distributed, $\hat{\mathbf{f}} \sim N[\mathbf{f}, (1/n)\mathbf{V}]$—as is the case when $\hat{\mathbf{f}}$ is a vector of means, $\hat{\mathbf{f}} = (\bar{y}_0, \bar{y}_1, \ldots, \bar{y}_m)'$ with $\bar{y}_q = (1/n)\sum_{r=1}^{n} y_{qr}$—then an unbiased estimate of σ^2 is given by

$$s^2 = \frac{1}{(m+1)(n-1)} \sum_{q=0}^{m} \sum_{r=1}^{n} \left(y_{qr} - \bar{y}_q \right)^2.$$

Substituting s^2 in place of σ^2 in (12.56), we obtain the area estimator

$$\hat{A}^{**} = A_0 \left(1 + \frac{s^2}{n \hat{\mathbf{f}}' \hat{\mathbf{f}}} \right)^{-1}.$$

A similar quadrature approximation of A in (12.49) was considered earlier by Katz and D'Argenio (1983) using a trapezoidal approximation to A. The quadrature points were selected so that they minimize the expectation of the square of the difference between the exact integral and the quadrature approximation. This approach was applied to simulated pharmacokinetic problems.

12.9.3. Moments of a Ratio of Quadratic Forms

Consider the ratio of quadratic forms,

$$Q = \frac{\mathbf{y}'\mathbf{A}\mathbf{y}}{\mathbf{y}'\mathbf{B}\mathbf{y}}, \tag{12.57}$$

where $\mathbf{A}$ and $\mathbf{B}$ are symmetric matrices, $\mathbf{B}$ is positive definite, and $\mathbf{y}$ is an $n \times 1$ random vector. Ratios such as Q are frequently encountered in statistics and econometrics. In general, their exact distributions are mathematically intractable, especially when the quadratic forms are not independent. For this reason, the derivation of the moments of such ratios, for the purpose of approximating their distributions, is of interest. Sutradhar and Bartlett (1989) obtained approximate expressions for the first four moments of the ratio Q for a normally distributed $\mathbf{y}$. The moments were utilized to approximate the distribution of Q. This approximation was then applied to calculate the percentile points of a modified F-test statistic for testing treatment effects in a one-way model under correlated observations. Morin(1992) derived exact, but complicated, expressions for the first four moments of Q for a normally distributed $\mathbf{y}$. The moments are expressed in terms of confluent hypergeometric functions of many variables.

If $\mathbf{y}$ is not normally distributed, then no tractable formulas exist for the moments of Q in (12.57). Hence, manageable and computable approximations for these moments would be helpful. Lieberman (1994) used the method of Laplace to provide general approximations for the moments of Q without making the normality assumption on $\mathbf{y}$. Lieberman showed that if $E(\mathbf{y}'\mathbf{B}\mathbf{y})$ and $E[(\mathbf{y}'\mathbf{A}\mathbf{y})^k]$ exist for $k \geq 1$, then the Laplace approximation of

$E(Q^k)$, the kth noncentral moment of Q, is given by

$$E_L(Q^k) = \frac{E\left[(\mathbf{y}'\mathbf{A}\mathbf{y})^k\right]}{\left[E(\mathbf{y}'\mathbf{B}\mathbf{y})\right]^k}.$$

(12.58)

In particular, if $\mathbf{y} \sim N(\boldsymbol{\mu}, \boldsymbol{\Sigma})$, then the Laplace approximation for the mean and second noncentral moment of Q are written explicitly as

$$E_L(Q) = \frac{\text{tr}(\mathbf{A}\boldsymbol{\Sigma}) + \boldsymbol{\mu}'\mathbf{A}\boldsymbol{\mu}}{\text{tr}(\mathbf{B}\boldsymbol{\Sigma}) + \boldsymbol{\mu}'\mathbf{B}\boldsymbol{\mu}}$$

$$E_L(Q^2) = \frac{E\left[(\mathbf{y}'\mathbf{A}\mathbf{y})^2\right]}{\left[E(\mathbf{y}'\mathbf{B}\mathbf{y})\right]^2}$$

$$= \frac{\text{Var}(\mathbf{y}'\mathbf{A}\mathbf{y}) + \left[E(\mathbf{y}'\mathbf{A}\mathbf{y})\right]^2}{\left[E(\mathbf{y}'\mathbf{B}\mathbf{y})\right]^2}$$

$$= \frac{2\text{tr}\left[(\mathbf{A}\boldsymbol{\Sigma})^2\right] + 4\boldsymbol{\mu}'\mathbf{A}\boldsymbol{\Sigma}\mathbf{A}\boldsymbol{\mu} + \left[\text{tr}(\mathbf{A}\boldsymbol{\Sigma}) + \boldsymbol{\mu}'\mathbf{A}\boldsymbol{\mu}\right]^2}{\left[\text{tr}(\mathbf{B}\boldsymbol{\Sigma}) + \boldsymbol{\mu}'\mathbf{B}\boldsymbol{\mu}\right]^2}$$

(see Searle, 1971, Section 2.5 for expressions for the mean and variance of a quadratic form in normal variables).

EXAMPLE 12.9.1. Consider the linear model, $\mathbf{y} = \mathbf{X}\boldsymbol{\beta} + \boldsymbol{\epsilon}$, where $\mathbf{X}$ is $n \times p$ of rank p, $\boldsymbol{\beta}$ is a vector of unknown parameters, and $\boldsymbol{\epsilon}$ is a random error vector. Under the null hypothesis of no serial correlation in the elements of $\boldsymbol{\epsilon}$, we have $\boldsymbol{\epsilon} \sim N(\mathbf{0}, \sigma^2\mathbf{I})$. The corresponding Durbin-Watson (1950, 1951) test statistic is given by

$$d = \frac{\boldsymbol{\epsilon}'\mathbf{P}\mathbf{A}_1\mathbf{P}\boldsymbol{\epsilon}}{\boldsymbol{\epsilon}'\mathbf{P}\boldsymbol{\epsilon}},$$

where $\mathbf{P} = \mathbf{I} - \mathbf{X}(\mathbf{X}'\mathbf{X})^{-1}\mathbf{X}'$, and $\mathbf{A}_1$ is the matrix

$$\mathbf{A}_1 = \begin{bmatrix} 1 & -1 & 0 & 0 & \cdots & 0 & 0 \\ -1 & 2 & -1 & 0 & \cdots & 0 & 0 \\ 0 & -1 & 2 & -1 & \cdots & 0 & 0 \\ \vdots & \vdots & \ddots & \ddots & \ddots & \vdots & \vdots \\ 0 & 0 & \cdots & -1 & 2 & -1 & 0 \\ 0 & 0 & \cdots & 0 & -1 & 2 & -1 \\ 0 & 0 & \cdots & 0 & 0 & -1 & 1 \end{bmatrix}.$$

Then, by (12.58), the Laplace approximation of $E(d)$ is

$$E_L(d) = \frac{E(\epsilon' \mathbf{PA}_1 \mathbf{P}\epsilon)}{E(\epsilon' \mathbf{P}\epsilon)}. \qquad (12.59)$$

Durbin and Watson (1951) showed that d is distributed independently of its own denominator, so that the moments of the ratio d are the ratios of the corresponding moments of the numerator and denominator, that is,

$$E(d^k) = \frac{E\left[(\epsilon' \mathbf{PA}_1 \mathbf{P}\epsilon)^k\right]}{E\left[(\epsilon' \mathbf{P}\epsilon)^k\right]}. \qquad (12.60)$$

From (12.59) and (12.60) we note that the Laplace approximation for the mean, $E(d)$, is exact. For $k \geq 2$, Lieberman (1994) showed that

$$\frac{E_L(d^k)}{E(d^k)} = 1 + O\left(\frac{1}{n}\right).$$

Thus, the relative error of approximating higher-order moments of d is $O(1/n)$, regardless of the matrix $\mathbf{X}$.

12.9.4. Laplace's Approximation in Bayesian Statistics

Suppose (Kass, Tierney, and Kadane, 1991) that a data vector $\mathbf{y} = (y_1, y_2, \ldots, y_n)'$ has a distribution with the density function $p(\mathbf{y}|\theta)$, where θ is an unknown parameter. Let $L(\theta)$ denote the corresponding likelihood function, which is proportional to $p(\mathbf{y}|\theta)$. In Bayesian statistics, a prior density, $\pi(\theta)$, is assumed on θ, and inferences are based on the posterior density $q(\theta|\mathbf{y})$, which is proportional to $L(\theta)\pi(\theta)$, where the proportionality constant is determined by requiring that $q(\theta|\mathbf{y})$ integrate to one. For a given real-valued function $g(\theta)$, its posterior expectation is given by

$$E[g(\theta)|\mathbf{y}] = \frac{\int g(\theta) L(\theta) \pi(\theta)\, d\theta}{\int L(\theta) \pi(\theta)\, d\theta}. \qquad (12.61)$$

Tierney, Kass, and Kadane (1989) expressed the integrands in (12.61) as follows:

$$g(\theta) L(\theta) \pi(\theta) = b_N(\theta) \exp\left[-n h_N(\theta)\right],$$

$$L(\theta) \pi(\theta) = b_D(\theta) \exp\left[-n h_D(\theta)\right],$$

where $b_N(\theta)$ and $b_D(\theta)$ are smooth functions that do not depend on n and $h_N(\theta)$ and $h_D(\theta)$ are constant-order functions of n, as $n \to \infty$. Formula (12.61) can then be written as

$$E[g(\theta)|\mathbf{y}] = \frac{\int b_N(\theta) \exp[-nh_N(\theta)] \, d\theta}{\int b_D(\theta) \exp[-nh_D(\theta)] \, d\theta}. \qquad (12.62)$$

Applying Laplace's approximation in (12.25) to the integrals in (12.62), we obtain, if n is large,

$$E[g(\theta)|\mathbf{y}] \approx \left[\frac{h_D''(\hat{\theta}_D)}{h_N''(\hat{\theta}_N)}\right]^{1/2} \frac{b_N(\hat{\theta}_N) \exp[-nh_N(\hat{\theta}_N)]}{b_D(\hat{\theta}_D) \exp[-nh_D(\hat{\theta}_D)]}, \qquad (12.63)$$

where $\hat{\theta}_N$ and $\hat{\theta}_D$ are the locations of the single maxima of $-h_N(\theta)$ and $-h_D(\theta)$, respectively. In particular, if we choose $h_N(\theta) = h_D(\theta) = -(1/n)\log[L(\theta)\pi(\theta)]$, $b_N(\theta) = g(\theta)$ and $b_D(\theta) = 1$, then (12.63) reduces to

$$E[g(\theta)|\mathbf{y}] \approx g(\hat{\theta}), \qquad (12.64)$$

where $\hat{\theta}$ is the point at which $(1/n)\log[L(\theta)\pi(\theta)]$ attains its maximum.

Formula (12.64) provides a first-order approximation of $E[g(\theta)|\mathbf{y}]$. This approximation is often called the *modal approximation* because $\hat{\theta}$ is the mode of the posterior density. A more accurate second-order approximation of $E[g(\theta)|\mathbf{y}]$ was given by Kass, Tierney, and Kadane (1991).

12.9.5. Other Methods of Approximating Integrals in Statistics

There are several major techniques and approaches available for the numerical approximation of integrals in statistics that are beyond the scope of this book. These techniques, which include the *saddlepoint approximation* and *Markov chain Monte Carlo*, have received a great deal of attention in the statistical literature in recent years.

The saddlepoint method is designed to approximate integrals of the Laplace type in which both the integrand and contour of integration are allowed to be complex valued. It is a powerful tool for obtaining accurate expressions for densities and distribution functions. A good introduction to the basic principles underlying this method was given by De Bruijn (1961, Chapter 5). Daniels (1954) is credited with having introduced it in statistics in the context of approximating the density of a sample mean of independent and identically distributed random variables.

Markov chain Monte Carlo (MCMC) is a general method for the simulation of stochastic processes having probability densities known up to a constant of proportionality. It generally deals with high-dimensional statistical problems, and has come into prominence in statistical applications during the past several years. Although MCMC has potential applications in several areas of statistics, most attention to date has been focused on Bayesian applications.

For a review of these techniques, see, for example, Geyer (1992), Evans and Swartz (1995), Goutis and Casella (1999), and Strawderman (2000).

FURTHER READING AND ANNOTATED BIBLIOGRAPHY

Abramowitz, M., and I. A. Stegun, eds. (1972). *Handbook of Mathematical Functions with Formulas, Graphs, and Mathematical Tables*. Wiley, New York. (This volume is an excellent source for a wide variety of numerical tables of mathematical functions. Chap. 25 gives zeros of Legendre, Hermite, and Laguerre polynomials along with their corresponding weight factors.)

Copson, E. T. (1965). *Asymptotic Expansions*. Cambridge University Press, London. (The method of Laplace is discussed in Chap. 5.)

Courant, R., and F. John (1965). *Introduction to Calculus and Analysis*, Volume 1. Wiley, New York. (The trapezoidal and Simpson's methods are discussed in Chap. 6.)

Daniels, H. (1954). "Saddlepoint approximation in statistics." *Ann. Math. Statist.*, **25**, 631–650.

Davis, P. J., and P. Rabinowitz (1975). *Methods of Numerical Integration*. Academic Press, New York. (This book presents several useful numerical integration methods, including approximate integrations over finite or infinite intervals as well as integration in two or more dimensions.)

De Bruijn, N. G. (1961). *Asymptotic Methods in Analysis*, 2nd ed. North-Holland, Amsterdam. (Chap. 4 covers the method of Laplace, and the saddlepoint method is the topic of Chap. 5.)

Durbin, J., and G. S. Watson (1950). "Testing for serial correlation in least squares regression, I." *Biometrika*, **37**, 409–428.

Durbin, J., and G. S. Watson (1951). "Testing for serial correlation in least squares regression, II." *Biometrika*, **38**, 159–178.

Evans, M., and T. Swartz (1995). "Methods for approximating integrals in statistics with special emphasis on Bayesian integration problems." *Statist. Sci.*, **10**, 254–272.

Flournoy, N., and R. K. Tsutakawa, eds. (1991). *Statistical Multiple Integration, Contemporary Mathematics* **115**. Amer. Math. Soc., Providence, Rhode Island. (This volume contains the proceedings of an AMS–IMS–SIAM joint summer research conference on statistical multiple integration, which was held at Humboldt University, Arcata, California, June 17–23, 1989.)

Fulks, W. (1978). *Advanced Calculus*, 3rd ed. Wiley, New York. (Section 18.3 of this book contains proofs associated with the method of Laplace.)

Geyer, C. J. (1992). "Practical Markov chain Monte Carlo." *Statist. Sci.*, **7**, 473–511.

Ghazal, G. A. (1994). "Moments of the ratio of two dependent quadratic forms." *Statist. Prob. Letters*, **20**, 313–319.

Gibaldi, M., and D. Perrier (1982). *Pharmacokinetics*, 2nd ed. Dekker, New York.

Goutis, C., and G. Casella (1999). "Explaining the saddlepoint approximation." *Amer. Statist.*, **53**, 216–224.

Haber, S. (1970). "Numerical evaluation of multiple integrals." *SIAM Rev.*, **12**, 481–526.

Kahaner, D. K. (1991). "A survey of existing multidimensional quadrature routines." In *Statistical Multiple Integration*, Contemporary Mathematics **115**, N. Flournoy and R. K. Tsutakawa, eds. Amer. Math. Soc., Providence, pp. 9–22.

Kass, R. E., L. Tierney, and J. B. Kadane (1991). "Laplace's method in Bayesian analysis." In *Statistical Multiple Integration*, Contemporary Mathematics **115**, N. Flournoy and R. K. Tsutakawa, eds. Amer. Math. Soc., Providence, pp. 89–99.

Katz, D., and D. Z. D'Argenio (1983). "Experimental design for estimating integrals by numerical quadrature, with applications to pharmacokinetic studies." *Biometrics*, **39**, 621–628.

Krylov, V. I. (1962). *Approximate Calculation of Integrals*. Macmillan, New York. (This book considers only the problem of approximate integration of functions of a single variable. It was translated from the Russian by A. H. Stroud.)

Lange, K. (1999). *Numerical Analysis for Statisticians*. Springer, New York. (This book contains a wide variety of topics on numerical analysis of potential interest to statisticians, including recent topics such as bootstrap calculations and the Markov chain Monte Carlo method.)

Lieberman, O. (1994). "A Laplace approximation to the moments of a ratio of quadratic forms." *Biometrika*, **81**, 681–690.

Liu, Q., and D. A. Pierce (1994). "A note on Gauss–Hermite quadrature." *Biometrika*, **81**, 624–629.

Morin, D. (1992). "Exact moments of ratios of quadratic forms." *Metron*, **50**, 59–78.

Morland, T. (1998). "Approximations to the normal distribution function." *Math. Gazette*, **82**, 431–437.

Nonweiler, T. R. F. (1984). *Computational Mathematics*. Ellis Horwood, Chichester, England. (Numerical quadrature is covered in Chap. 5.)

Phillips, C., and B. Cornelius (1986). *Computational Numerical Methods*. Ellis Horwood, Chichester, England. (Numerical integration is the subject of Chap. 6.).

Phillips, G. M., and P. J. Taylor (1973). *Theory and Applications of Numerical Analysis*. Academic Press, New York. (Gaussian quadrature is covered in Chap. 6.).

Piegorsch, W. W., and A. J. Bailer (1993). "Minimum mean-square error quadrature." *J. Statist. Comput. Simul*, **46**, 217–234.

Ralston, A., and P. Rabinowitz (1978). *A First Course in Numerical Analysis*. McGraw-Hill, New York. (Gaussian quadrature is covered in Chap. 4.)

Reid, W. H., and S. J. Skates (1986). "On the asymptotic approximation of integrals." *SIAM J. Appl. Math.*, **46**, 351–358.

Roussas, G. G. (1973). *A First Course in Mathematical Statistics*. Addison-Wesley, Reading, Massachusetts.

Searle, S. R. (1971). *Linear Models*. Wiley, New York.

Shoup, T. E. (1984). *Applied Numerical Methods for the Microcomputer.* Prentice-Hall, Englewood Cliffs, New Jersey.

Stark, P. A. (1970). *Introduction to Numerical Methods.* Macmillan, London.

Strawderman, R. L. (2000). "Higher-order asymptotic approximation: Laplace, saddlepoint, and related methods." *J. Amer. Statist. Assoc.*, **95**, 1358–1364.

Stroud, A. H. (1971). *Approximate Calculation of Multiple Integrals.* Prentice-Hall, Englewood Cliffs, New Jersey.

Sutradhar, B. C., and R. F. Bartlett (1989). "An approximation to the distribution of the ratio of two general quadratic forms with application to time series valued designs." *Comm. Statist. Theory Methods*, **18**, 1563–1588.

Tierney, L., R. E. Kass, and J. B. Kadane (1989). "Fully exponential Laplace approximations to expectations and variances of nonpositive functions." *J. Amer. Statist. Assoc.*, **84**, 710–716.

Wong, R. (1989). *Asymptotic Approximations of Integrals.* Academic Press, New York. (This is a useful reference book on the method of Laplace and Mellin transform techniques for multiple integrals. All results are accompanied by error bounds.)

EXERCISES

In Mathematics

12.1. Consider the integral

$$I_n = \int_1^n \log x \, dx.$$

It is easy to show that

$$I_n = n \log n - n + 1.$$

(a) Approximate I_n by using the trapezoidal method and the partition points $x_0 = 1$, $x_1 = 2, \ldots, x_n = n$, and verify that

$$I_n \approx \log(n!) - \tfrac{1}{2} \log n.$$

(b) Deduce from (a) that $n!$ and $n^{n+1/2} e^{-n}$ are of the same order of magnitude, which is essentially what is stated in Stirling's formula (see Example 12.6.1)

12.2. Obtain an approximation of the integral $\int_0^1 dx/(1+x)$ by Simpson's method for the following values of n: 2, 4, 8, 16. Show that when $n = 8$, the error of approximation is less than or equal to 0.000002.

12.3. Use Gauss–Legendre quadrature with $n = 2$ to approximate the value of the integral $\int_0^{\pi/2} \sin\theta\, d\theta$. Give an upper bound on the error of approximation using formula (12.16).

12.4. (a) Show that

$$\int_m^\infty e^{-x^2}\, dx < \frac{1}{m} e^{-m^2}, \qquad m > 0.$$

[*Hint*: Use the inequality $e^{-x^2} < e^{-mx}$ for $x > m$.]

(b) Find a value of m so that the upper bound in (a) is smaller than 10^{-4}.

(c) Use part (b) to find an approximation for $\int_0^\infty e^{-x^2}\, dx$ correct to three decimal places.

12.5. Obtain an approximate value for the integral $\int_0^\infty x(e^x + e^{-x} - 1)^{-1}\, dx$ using the Gauss–Laguerre quadrature. [*Hint*: Use the tables in Appendix C of the book by Krylov (1962, page 347) giving the zeros of Laguerre polynomials and the corresponding values of ω_i.]

12.6. Consider the indefinite integral

$$I(x) = \int_0^x \frac{dt}{1 + t^2} = \text{Arctan } x.$$

(a) Make an appropriate change of variables to show that $I(x)$ can be written as

$$I(x) = 2x \int_{-1}^1 \frac{du}{4 + x^2(u + 1)^2}$$

(b) Use a five-point Gauss–Legendre quadrature to provide an approximation for $I(x)$.

12.7. Investigate the asymptotic behavior of the integral

$$I(\lambda) = \int_0^1 (\cos x)^\lambda\, dx$$

as $\lambda \to \infty$.

12.8. (a) Use the Gauss-Laguerre quadrature to approximate the integral $\int_0^\infty e^{-6x} \sin x\, dx$ using $n = 1, 2, 3, 4$, and compare the results with the true value of the integral.

(b) Use the Gauss-Hermite quadrature to approximate the integral $\int_{-\infty}^{\infty} |x| \exp(-3x^2)\, dx$ using $n = 1, 2, 3, 4$, and compare the results with the true value of the integral.

12.9. Give an approximation to the double integral

$$\int_0^1 \int_0^{(1-x_1^2)^{1/2}} \left(1 - x_1^2 - x_2^2\right)^{1/2} dx_1\, dx_2$$

by applying the Gauss–Legendre rule to formulas (12.32) and (12.33).

12.10. Show that $\int_0^\infty (1 + x^2)^{-n}\, dx$ is asymptotically equal to $\frac{1}{2}(\pi/n)^{1/2}$ as $n \to \infty$.

In Statistics

12.11. Consider a sample, $X_1, X_2, \ldots, X_n$, of independent and indentically distributed random variables from the standard normal distribution. Suppose that n is odd. The sample median is the $(m + 1)$th order statistic $X_{(m+1)}$, where $m = (n - 1)/2$. It is known that $X_{(m+1)}$ has the density function

$$f(x) = n \binom{n-1}{m} \Phi^m(x)[1 - \Phi(x)]^m \phi(x),$$

where

$$\phi(x) = \frac{1}{\sqrt{2\pi}} \exp\left(-\frac{x^2}{2}\right)$$

is the standard normal density function, and

$$\Phi(x) = \int_{-\infty}^x \phi(t)\, dt$$

(see Roussas, 1973, page 194). Since the mean of $X_{(m+1)}$ is zero, the variance of $X_{(m+1)}$ is given by

$$\text{Var}[X_{(m+1)}] = n \binom{n-1}{m} \int_{-\infty}^{\infty} x^2 \Phi^m(x)[1 - \Phi(x)]^m \phi(x)\, dx.$$

Obtain an approximation for this variance using the Gauss–Hermite quadrature for $n = 11$ and varying numbers of quadrature points.

12.12. (Morland, 1998.) Consider the density function

$$\phi(x) = \frac{1}{\sqrt{2\pi}}e^{-x^2/2}$$

for the standard normal distribution. Show that if $x \geq 0$, then the cumulative distribution function,

$$\Phi(x) = \frac{1}{\sqrt{2\pi}}\int_{-\infty}^{x}e^{-t^2/2}\,dt,$$

can be represented as the sum of the series

$$\Phi(x) = \frac{1}{2} + \frac{x}{\sqrt{2\pi}}\left[1 - \frac{x^2}{6} + \frac{x^4}{40} - \frac{x^6}{336}\right.$$

$$\left. + \frac{x^8}{3456} - \cdots + (-1)^n\frac{x^{2n}}{(2n+1)2^n n!} + \cdots\right].$$

[*Note*: By truncating this series, we can obtain a polynomial approximation of $\Phi(x)$. For example, we have the following approximation of order 11:

$$\Phi(x) \approx \frac{1}{2} + \frac{x}{\sqrt{2\pi}}\left[1 - \frac{x^2}{6} + \frac{x^4}{40} - \frac{x^6}{336} + \frac{x^8}{3456} - \frac{x^{10}}{42240}\right].]$$

12.13. Use the result in Exercise 12.12 to show that there exist constants a, b, c, d, such that

$$\Phi(x) \approx \frac{1}{2} + \frac{x}{\sqrt{2\pi}}\frac{1 + ax^2 + bx^4}{1 + cx^2 + dx^4}.$$

12.14. Apply the method of importance sampling to approximate the value of the integral $\int_0^2 e^{x^2}\,dx$ using a sample of size $n = 150$ from the distribution whose density function is $g(x) = \frac{1}{4}(1+x)$, $0 \leq x \leq 2$. Compare your answer with the one you would get from applying formula (12.36).

12.15. Consider the density function

$$f(x) = \frac{\Gamma\left(\dfrac{n+1}{2}\right)}{\sqrt{n\pi}\,\Gamma\left(\dfrac{n}{2}\right)}\left(1 + \frac{x^2}{n}\right)^{-\frac{1}{2}(n+1)}, \qquad -\infty < x < \infty,$$

for a t-distribution with n degrees of freedom (see Exercise 6.28). Let $F(x) = \int_{-\infty}^{x} f(t)\,dt$ be its cumulative distribution function. Show that for large n,

$$F(x) \approx \frac{1}{\sqrt{2\pi}} \int_{-\infty}^{x} e^{-t^2/2}\,dt,$$

which is the cumulative distribution function for the standard normal. [*Hint*: Apply Stirling's formula.]

APPENDIX

Solutions to Selected Exercises

CHAPTER 1

1.3. $A \cup C \subset B$ implies $A \subset B$. Hence, $A = A \cap B$. Thus, $A \cap B \subset \overline{C}$ implies $A \subset \overline{C}$. It follows that $A \cap C = \emptyset$.

1.4. (a) $x \in A \vartriangle B$ implies that $x \in A$ but $\notin B$, or $x \in B$ but $\notin A$; thus $x \in A \cup B - A \cap B$. Vice versa, if $x \in A \cup B - A \cap B$, then $x \in A \vartriangle B$.

(c) $x \in A \cap (B \vartriangle D)$ implies that $x \in A$ and $x \in B \vartriangle D$, so that either $x \in A \cap B$ but $\notin A \cap D$, or $x \in A \cap D$ but $\notin A \cap B$, so that $x \in (A \cap B) \vartriangle (A \cap D)$. Vice versa, if $x \in (A \cap B) \vartriangle (A \cap D)$, then either $x \in A \cap B$ but $\notin A \cap D$, or $x \in A \cap D$ but $\notin A \cap B$, so that either x is in A and B but $\notin D$, or x is in A and D, but $\notin B$; thus $x \in A \cap (B \vartriangle D)$.

1.5. It is obvious that ρ is reflexive, symmetric, and transitive. Hence, it is an equivalence relation. If (m_0, n_0) is an element in A, then its equivalence class is the set of all pairs (m, n) in A such that $m/m_0 = n/n_0$, that is, $m/n = m_0/n_0$.

1.6. The equivalence class of $(1, 2)$ consists of all pairs (m, n) in A such that $m - n = -1$.

1.7. The first elements in all four pairs are distinct.

1.8. (a) If $y \in f(\bigcup_{i=1}^{n} A_i)$, then $y = f(x)$, where $x \in \bigcup_{i=1}^{n} A_i$. Hence, if $x \in A_i$ and $f(x) \in f(A_i)$ for some i, then $f(x) \in \bigcup_{i=1}^{n} f(A_i)$; thus $f(\bigcup_{i=1}^{n} A_i) \subset \bigcup_{i=1}^{n} f(A_i)$. Vice versa, it is easy to show that $\bigcup_{i=1}^{n} f(A_i) \subset f(\bigcup_{i=1}^{n} A_i)$.

(b) If $y \in f(\bigcap_{i=1}^{n} A_i)$, then $y = f(x)$, where $x \in A_i$ for all i; then $f(x) \in f(A_i)$ for all i; then $f(x) \in \bigcap_{i=1}^{n} f(A_i)$. Equality holds if f is one-to-one.

1.11. Define $f: J^+ \to A$ such that $f(n) = 2n^2 + 1$. Then f is one-to-one and onto, so A is countable.

1.13. $a + \sqrt{b} = c + \sqrt{d} \Rightarrow a - c = \sqrt{d} - \sqrt{b}$. If $a = c$, then $b = d$. If $a \neq c$, then $\sqrt{d} - \sqrt{b}$ is a nonzero rational number and $\sqrt{d} + \sqrt{b} = (d - b)/(\sqrt{d} - \sqrt{b})$. It follows that both $\sqrt{d} - \sqrt{b}$ and $\sqrt{d} + \sqrt{b}$ are rational numbers, and therefore $\sqrt{b}$ and $\sqrt{d}$ must be rational numbers.

1.14. Let $g = \inf(A)$. Then $g \leq x$ for all x in A, so $-g \geq -x$, hence, $-g$ is an upper bound of $-A$ and is the least upper bound: if $-g'$ is any other upper bound of $-A$, then $-g' \geq -x$, so $x \geq g'$, hence, g' is a lower bound of A, so $g' \leq g$, that is, $-g' \geq -g$, so $-g = \sup(-A)$.

1.15. Suppose that $b \notin A$. Since b is the least upper bound of A, it must be a limit point of A: for any $\epsilon > 0$, there exists an element $a \in A$ such that $a > b - \epsilon$. Furthermore, $a < b$, since $b \notin A$. Hence, b is a limit point of A. But A is closed; hence, by Theorem 1.6.4, $b \in A$.

1.17. Suppose that G is a basis for $\mathcal{F}$, and let $p \in B$. Then $B = \bigcup_\alpha U_\alpha$, where U_α belongs to G. Hence, there is at least one U_α such that $p \in U_\alpha \subset B$. Vice versa, if for each $B \in \mathcal{F}$ and each $p \in B$ there is a $U \in G$ such that $p \in U \subset B$, then G must be a basis for $\mathcal{F}$: for each $p \in B$ we can find a set $U_p \in G$ such that $p \in U_p \subset B$; then $B = \bigcup \{U_p \mid p \in B\}$, so G is a basis.

1.18. Let p be a limit point of $A \cup B$. Then p is a limit point of A or of B. In either case, $p \in A \cup B$. Hence, by Theorem 1.6.4, $A \cup B$ is a closed set.

1.19. Let $\{C_\alpha\}$ be an open covering of B. Then $\bar{B}$ and $\{C_\alpha\}$ form an open covering of A. Since A is compact, a finite subcollection of the latter covering covers A. Furthermore, since $\bar{B}$ does not cover B, the members of this finite covering that contain B are all in $\{C_\alpha\}$. Hence, B is compact.

1.20. No. Let $(A, \mathcal{F})$ be a topological space such that A consists of two points a and b and $\mathcal{F}$ consists of A and the empty set $\varnothing$. Then A is compact, and the point a is a compact subset; but it is not closed, since the complement of a, namely b, is not a member of $\mathcal{F}$, and is therefore not open.

1.21. (a) Let $w \in A$. Then $X(w) \leq x < x + 3^{-n}$ for all n, so $w \in \bigcap_{n=1}^\infty A_n$. Vice versa, if $w \in \bigcap_{n=1}^\infty A_n$, then $X(w) < x + 3^{-n}$ for all n. To show that $X(w) \leq x$: if $X(w) > x$, then $X(w) > x + 3^{-n}$ for some n, a contradiction. Therefore, $X(w) \leq x$, and $w \in A$.

(b) Let $w \in B$. Then $X(w) < x$, so $X(w) < x - 3^{-n}$ for some n, hence, $w \in \bigcup_{n=1}^{\infty} B_n$. Vice versa, if $w \in \bigcup_{n=1}^{\infty} B_n$, then $X(w) \le x - 3^{-n}$ for some n, so $X(w) < x$: if $X(w) \ge x$, then $X(w) > x - 3^{-n}$ for all n, a contradiction. Therefore, $w \in B$.

1.22. **(a)** $P(X \ge 2) \le \lambda/2$, since $\mu = \lambda$.

(b)

$$P(X \ge 2) = 1 - p(X < 2)$$
$$= 1 - p(0) - p(1)$$
$$= 1 - e^{-\lambda}(\lambda + 1).$$

To show that

$$1 - e^{-\lambda}(\lambda + 1) < \frac{\lambda}{2}.$$

Let $\phi(\lambda) = \lambda/2 + e^{-\lambda}(\lambda + 1) - 1$. Then
i. $\phi(0) = 0$, and
ii. $\phi'(\lambda) = \frac{1}{2} - \lambda e^{-\lambda} > 0$ for all λ.
From (i) and (ii) we conclude that $\phi(\lambda) > 0$ for all $\lambda > 0$.

1.23. **(a)**

$$p(|X - \mu| \ge c) = P\left[(X - \mu)^2 \ge c^2\right]$$

$$\le \frac{\sigma^2}{c^2},$$

by Markov's inequality and the fact that $E(X - \mu)^2 = \sigma^2$.
(b) Use $c = k\sigma$ in (a).
(c)

$$P(|X - \mu| < k\sigma) = 1 - P(|X - \mu| \ge k\sigma)$$

$$\ge 1 - \frac{1}{k^2}.$$

1.24. $\mu = E(X) = \int_{-1}^{1} x(1 - |x|)\, dx = 0$, $\sigma^2 = \int_{-1}^{1} x^2(1 - |x|)\, dx = \frac{1}{6}$.
(a)
 i. $P(|X| \ge \frac{1}{2}) \le \sigma^2/\frac{1}{4} = \frac{2}{3}$.
 ii. $P(|X| > \frac{1}{3}) = P(|X| \ge \frac{1}{3}) \le \sigma^2/\frac{1}{9} = \frac{3}{2}$.
(b)

$$P(|X| \ge \frac{1}{2}) = P(X \ge \frac{1}{2}) + P(X \le -\frac{1}{2})$$

$$= \frac{1}{4} < \frac{2}{3}.$$

1.25. (a) True, since $X_{(1)} \geq x$ if and only if $X_i \geq x$ for all i.
   (b) True, since $X_{(n)} \leq x$ if and only if $X_i \leq x$ for all i.
   (c)

$$P(X_{(1)} \leq x) = 1 - P(X_{(1)} > x)$$
$$= 1 - [1 - F(x)]^n.$$

   (d) $P(X_{(n)} \leq x) = [F(x)]^n.$

1.26. $P(2 \leq X_{(1)} \leq 3) = P(X_{(1)} \leq 3) - P(X_{(1)} \leq 2)$. Hence,

$$P(2 \leq X_{(1)} \leq 3) = 1 - [1 - F(3)]^5 - 1 + [1 - F(2)]^5$$
$$= [1 - F(2)]^5 - [1 - F(3)]^5.$$

But $F(x) = \int_0^x 2e^{-2t}\, dt = 1 - e^{-2x}$. Hence,

$$P(2 \leq X_{(1)} \leq 3) = (e^{-4})^5 - (e^{-6})^5$$
$$= e^{-20} - e^{-30}.$$

CHAPTER 2

2.1. If $m > n$, then the rank of the $n \times m$ matrix $\mathbf{U} = [\mathbf{u}_1 : \mathbf{u}_2 : \cdots : \mathbf{u}_m]$ is less than or equal to n. Hence, the number of linearly independent columns of $\mathbf{U}$ is less than or equal to n, so the m columns of $\mathbf{U}$ must be linearly dependent.

2.2. If $\mathbf{u}_1, \mathbf{u}_2, \ldots, \mathbf{u}_n$ and $\mathbf{v}$ are linearly dependent, then $\mathbf{v}$ must belong to W, a contradiction.

2.5. Suppose that $\mathbf{u}_1, \mathbf{u}_2, \ldots, \mathbf{u}_n$ are linearly independent in U, and that $\sum_{i=1}^n \alpha_i T(\mathbf{u}_i) = \mathbf{0}$ for some constants $\alpha_1, \alpha_2, \ldots, \alpha_n$. Then $T(\sum_{i=1}^n \alpha_i \mathbf{u}_i)$ $= \mathbf{0}$. If T is one-to-one, then $\sum_{i=1}^n \alpha_i \mathbf{u}_i = \mathbf{0}$, hence, $\alpha_i = 0$ for all i, since the $\mathbf{u}_i$'s are linearly independent. It follows that $T(\mathbf{u}_1)$, $T(\mathbf{u}_2), \ldots, T(\mathbf{u}_n)$ must also be linearly independent. Vice versa, let $\mathbf{u} \in U$ such that $T(\mathbf{u}) = \mathbf{0}$. Let $\mathbf{e}_1, \mathbf{e}_2, \ldots, \mathbf{e}_n$ be a basis for U. Then $\mathbf{u} = \sum_{i=1}^n \tau_i \mathbf{e}_i$ for some constants $\tau_1, \tau_2, \ldots, \tau_n$. $T(\mathbf{u}) = \mathbf{0} \Rightarrow \sum_{i=1}^n \tau_i T(\mathbf{e}_i)$ $= \mathbf{0} \Rightarrow \tau_i = 0$ for all i, since $T(\mathbf{e}_1), T(\mathbf{e}_2), \ldots, T(\mathbf{e}_n)$ are linearly independent. It follows that $\mathbf{u} = \mathbf{0}$ and T is one-to-one.

2.7. If $\mathbf{A} = (a_{ij})$, then $\operatorname{tr}(\mathbf{A'A}) = \sum_i \sum_j a_{ij}^2$. Hence, $\mathbf{A} = \mathbf{0}$ if and only if $\operatorname{tr}(\mathbf{A'A}) = 0$.

2.8. It is sufficient to show that $\mathbf{Av} = \mathbf{0}$ if $\mathbf{v}'\mathbf{Av} = 0$. If $\mathbf{v}'\mathbf{Av} = 0$, then $\mathbf{A}^{1/2}\mathbf{v} = \mathbf{0}$, and hence $\mathbf{Av} = \mathbf{0}$. [*Note*: $\mathbf{A}^{1/2}$ is defined as follows: since $\mathbf{A}$ is symmetric, it can be written as $\mathbf{A} = \mathbf{P}\mathbf{\Lambda}\mathbf{P}'$ by Theorem 2.3.10, where $\mathbf{\Lambda}$ is a diagonal matrix whose diagonal elements are the eigenvalues of $\mathbf{A}$, and $\mathbf{P}$ is an orthogonal matrix. Furthermore, since $\mathbf{A}$ is positive semidefinite, its eigenvalues are greater than or equal to zero by Theorem 2.3.13. The matrix $\mathbf{A}^{1/2}$ is defined as $\mathbf{P}\mathbf{\Lambda}^{1/2}\mathbf{P}'$, where the diagonal elements of $\mathbf{\Lambda}^{1/2}$ are the square roots of the corresponding diagonal elements of $\mathbf{\Lambda}$.]

2.9. By Theorem 2.3.15, there exists an orthogonal matrix $\mathbf{P}$ and diagonal matrices $\mathbf{\Lambda}_1, \mathbf{\Lambda}_2$ such that $\mathbf{A} = \mathbf{P}\mathbf{\Lambda}_1\mathbf{P}'$, $\mathbf{B} = \mathbf{P}\mathbf{\Lambda}_2\mathbf{P}'$. The diagonal elements of $\mathbf{\Lambda}_1$ and $\mathbf{\Lambda}_2$ are nonnegative. Hence,

$$\mathbf{AB} = \mathbf{P}\mathbf{\Lambda}_1\mathbf{\Lambda}_2\mathbf{P}'$$

is positive semidefinite, since the diagonal elements of $\mathbf{\Lambda}_1\mathbf{\Lambda}_2$ are nonnegative.

2.10. Let $\mathbf{C} = \mathbf{AB}$. Then $\mathbf{C}' = -\mathbf{BA}$. Since $\text{tr}(\mathbf{AB}) = \text{tr}(\mathbf{BA})$, we have $\text{tr}(\mathbf{C}) = \text{tr}(-\mathbf{C}') = -\text{tr}(\mathbf{C})$ and thus $\text{tr}(\mathbf{C}) = 0$.

2.11. Let $\mathbf{B} = \mathbf{A} - \mathbf{A}'$. Then

$$\begin{aligned}
\text{tr}(\mathbf{B}'\mathbf{B}) &= \text{tr}[(\mathbf{A}' - \mathbf{A})(\mathbf{A} - \mathbf{A}')] \\
&= \text{tr}[(\mathbf{A}' - \mathbf{A})\mathbf{A}] - \text{tr}[(\mathbf{A}' - \mathbf{A})\mathbf{A}'] \\
&= \text{tr}[(\mathbf{A}' - \mathbf{A})\mathbf{A}] - \text{tr}[\mathbf{A}(\mathbf{A} - \mathbf{A}')] \\
&= \text{tr}[(\mathbf{A}' - \mathbf{A})\mathbf{A}] - \text{tr}[(\mathbf{A} - \mathbf{A}')\mathbf{A}] \\
&= 2\,\text{tr}[(\mathbf{A}' - \mathbf{A})\mathbf{A}] = 0,
\end{aligned}$$

since $\mathbf{A}' - \mathbf{A}$ is skew-symmetric. Hence, by Exercise 2.7, $\mathbf{B} = \mathbf{0}$ and thus $\mathbf{A} = \mathbf{A}'$.

2.12. This follows easily from Theorem 2.3.9.

2.13. By Theorem 2.3.10, we can write

$$\mathbf{A} - \lambda\mathbf{I}_n = \mathbf{P}(\mathbf{\Lambda} - \lambda\mathbf{I}_n)\mathbf{P}',$$

where $\mathbf{P}$ is an orthogonal matrix and $\mathbf{\Lambda}$ is diagonal with diagonal elements equal to the eigenvalues of $\mathbf{A}$. The diagonal elements of $\mathbf{\Lambda} - \lambda\mathbf{I}_n$ are the eigenvalues of $\mathbf{A} - \lambda\mathbf{I}_n$. Now, k diagonal elements of $\mathbf{\Lambda} - \lambda\mathbf{I}_n$ are equal to zero, and the remaining $n - k$ elements must be different from zero. Hence, $\mathbf{A} - \lambda\mathbf{I}_n$ has rank $n - k$.

2.17. If $AB = BA = B$, then $(A - B)^2 = A^2 - AB - BA + B^2 = A - 2B + B = A - B$. Vice versa, if $A - B$ is idempotent, then $A - B = (A - B)^2 = A^2 - AB - BA + B^2 = A - AB - BA + B$. Thus,

$$AB + BA = 2B,$$

so $AB + ABA = 2AB$ and thus $AB = ABA$. We also have

$$ABA + BA = 2BA, \qquad \text{hence}, \quad BA = ABA.$$

It follows that $AB = BA$. We finally conclude that

$$B = AB = BA.$$

2.18. If A is an $n \times n$ orthogonal matrix with determinant 1, then its eigenvalues are of the form $e^{\pm i\phi_1}, e^{\pm i\phi_2}, \ldots, e^{\pm i\phi_q}, 1$, where 1 is of multiplicity $n - 2q$ and none of the real numbers ϕ_j is a multiple of 2π $(j = 1, 2, \ldots, q)$ (those that are odd multiples of π give an even number of eigenvalues equal to -1).

2.19. Using Theorem 2.3.10, it is easy to show that

$$e_{\min}(A)I_n \le A \le e_{\max}(A)I_n,$$

where the inequality on the right means that $e_{\max}(A)I_n - A$ is positive semidefinite, and the one on the left means that $A - e_{\min}(A)I_n$ is positive semidefinite. It follows that

$$e_{\min}(A)L'L \le L'AL \le e_{\max}(A)L'L$$

Hence, by Theorem 2.3.19(1),

$$e_{\min}(A)\,\mathrm{tr}(L'L) \le \mathrm{tr}(L'AL) \le e_{\max}(A)\,\mathrm{tr}(L'L).$$

2.20. (a) We have that $A \ge e_{\min}(A)I_n$ by Theorem 2.3.10. Hence, $L'AL \ge e_{\min}(A)L'L$, and therefore, $e_{\min}(L'AL) \ge e_{\min}(A)e_{\min}(L'L)$ by Theorem 2.3.18 and the fact that $e_{\min}(A) \ge 0$ and $L'L$ is nonnegative definite.
(b) This is similar to (a).

2.21. We have that

$$e_{\min}(B)A \le A^{1/2}BA^{1/2} \le e_{\max}(B)A,$$

since A is nonnegative definite. Hence,

$$e_{\min}(B)\,\mathrm{tr}(A) \le \mathrm{tr}(A^{1/2}BA^{1/2}) \le e_{\max}(B)\,\mathrm{tr}(A).$$

The result follows by noting that

$$\text{tr}(\mathbf{AB}) = \text{tr}(\mathbf{A}^{1/2}\,\mathbf{BA}^{1/2}).$$

2.23. (a) $\mathbf{A} = \mathbf{I}_n - (1/n)\mathbf{J}_n$, where $\mathbf{J}_n$ is the matrix of ones of order $n \times n$. $\mathbf{A}^2 = \mathbf{A}$, since $\mathbf{J}_n^2 = n\mathbf{J}_n$. The rank of $\mathbf{A}$ is the same as its trace, which is equal to $n - 1$.

(b) $(n-1)s^2/\sigma^2 = (1/\sigma^2)\mathbf{y}'\mathbf{A}\mathbf{y} \sim \chi^2_{n-1}$, since $(1/\sigma^2)\mathbf{A}(\sigma^2\mathbf{I}_n)$ is idempotent of rank $n - 1$, and the noncentrality parameter is zero.

(c)

$$\text{Cov}(\bar{y}, \mathbf{A}\mathbf{y}) = \text{Cov}\left(\frac{1}{n}\mathbf{1}'_n\mathbf{y}, \mathbf{A}\mathbf{y}\right)$$

$$= \frac{1}{n}\mathbf{1}'_n(\sigma^2\mathbf{I}_n)\mathbf{A}$$

$$= \frac{\sigma^2}{n}\mathbf{1}'_n\mathbf{A}$$

$$= \mathbf{0}'.$$

Since $\mathbf{y}$ is normally distributed, both $\bar{y}$ and $\mathbf{A}\mathbf{y}$, being linear transformations of $\mathbf{y}$, are also normally distributed. Hence, they must be independent, since they are uncorrelated. We can similarly show that $\bar{y}$ and $\mathbf{A}^{1/2}\mathbf{y}$ are uncorrelated, and hence independent, by the fact that $\mathbf{1}'_n\mathbf{A} = \mathbf{0}'$ and thus $\mathbf{1}'_n\mathbf{A}\mathbf{1}_n = 0$, which means $\mathbf{1}'_n\mathbf{A}^{1/2} = \mathbf{0}'$. It follows that $\bar{y}$ and $\mathbf{y}'\mathbf{A}^{1/2}\mathbf{A}^{1/2}\mathbf{y} = \mathbf{y}'\mathbf{A}\mathbf{y}$ are independent.

2.24. (a) The model is written as

$$\mathbf{y} = \mathbf{X}\mathbf{g} + \boldsymbol{\epsilon},$$

where $\mathbf{X} = [\mathbf{1}_N : \oplus_{i=1}^{a}\mathbf{1}_{n_i}]$, $\oplus$ denotes the direct sum of matrices, $\mathbf{g} = (\mu, \alpha_1, \alpha_2, \ldots, \alpha_a)'$, and $N = \Sigma_{i=1}^{a}n_i$. Now, $\mu + \alpha_i$ is estimable, since it can be written as $\mathbf{a}'_i\mathbf{g}$, where $\mathbf{a}'_i$ is a row of $\mathbf{X}$, $i = 1, 2, \ldots, a$. It follows that $\alpha_i - \alpha_{i'} = (\mathbf{a}'_i - \mathbf{a}'_{i'})\mathbf{g}$ is estimable, since the vector $\mathbf{a}'_i - \mathbf{a}'_{i'}$ belongs to the row space of $\mathbf{X}$.

(b) Suppose that μ is estimable; then it can be written as

$$\mu = \sum_{i=1}^{a} \tau_i(\mu + \alpha_i),$$

where $\tau_1, \tau_2, \ldots, \tau_a$ are constants, since $\mu + \alpha_1, \mu + \alpha_2, \ldots, \mu + \alpha_a$ form a basic set of estimable linear functions. We must therefore have $\Sigma_{i=1}^{a}\tau_i = 1$, and $\tau_i = 0$ for all i, a contradiction. Therefore, μ is nonestimable.

2.25. **(a)** $X(X'X)^-X'X(X'X)^-X' = X(X'X)^-X'$ by Theorem 2.3.3(2).

(b) $E(\boldsymbol{\ell}'y) = \boldsymbol{\lambda}'\boldsymbol{\beta} \Rightarrow \boldsymbol{\ell}'X\boldsymbol{\beta} = \boldsymbol{\lambda}'\boldsymbol{\beta}$. Hence, $\boldsymbol{\lambda}' = \boldsymbol{\ell}'X$. Now,

$$
\begin{aligned}
\text{Var}(\boldsymbol{\lambda}'\hat{\boldsymbol{\beta}}) &= \boldsymbol{\lambda}'(X'X)^-X'X(X'X)^-\boldsymbol{\lambda}\sigma^2 \\
&= \boldsymbol{\ell}'X(X'X)^-X'X(X'X)^-X'\boldsymbol{\ell}\sigma^2 \\
&= \boldsymbol{\ell}'X(X'X)^-X'\boldsymbol{\ell}\sigma^2, \qquad \text{by Theorem 2.3.3(2)} \\
&\leq \boldsymbol{\ell}'\boldsymbol{\ell}\sigma^2,
\end{aligned}
$$

since $I_n - X(X'X)^-X'$ is positive semidefinite. The result follows from the last inequality, since $\text{Var}(\boldsymbol{\ell}'y) = \boldsymbol{\ell}'\boldsymbol{\ell}\sigma^2$,

2.26. $\boldsymbol{\beta}^* = P(\Lambda + kI_p)^{-1}P'X'y$. But $\hat{\boldsymbol{\beta}} = P\Lambda^{-1}P'X'y$, so $\Lambda P'\hat{\boldsymbol{\beta}} = P'X'y$. Hence,

$$
\begin{aligned}
\boldsymbol{\beta}^* &= P(\Lambda + kI_p)^{-1}\Lambda P'\hat{\boldsymbol{\beta}} \\
&= PDP'\hat{\boldsymbol{\beta}}.
\end{aligned}
$$

2.27.

$$
\begin{aligned}
\sup_{x,y} \rho^2 &= \sup_{\nu,\tau} \left(\nu'C_1^{-1}AC_2^{-1}\tau\right)^2 \\
&= \sup_{\nu}\left\{\sup_{\tau}\left(\nu'C_1^{-1}AC_2^{-1}\tau\right)^2\right\} \\
&= \sup_{\nu}\left\{\sup_{\tau}(b'\tau)^2\right\} \qquad (b' = \nu'C_1^{-1}AC_2^{-1}) \\
&= \sup_{\nu}\left\{\sup_{\tau}(\tau'bb'\tau)\right\} \\
&= \sup_{\nu}\{e_{\max}(bb')\}, \qquad \text{by applying (2.9)} \\
&= \sup_{\nu}\{e_{\max}(b'b)\} \\
&= \sup_{\nu}\{\nu'C_1^{-1}AC_2^{-2}A'C_1^{-1}\nu\} \\
&= e_{\max}(C_1^{-1}AC_2^{-2}A'C_1^{-1}), \qquad \text{by (2.9)} \\
&= e_{\max}(C_1^{-2}AC_2^{-2}A') \\
&= e_{\max}(B_1^{-1}AB_2^{-1}A').
\end{aligned}
$$

CHAPTER 3

3.1. (a)

$$\lim_{x \to 1} \frac{x^5 - 1}{x - 1} = \lim_{x \to 1} (1 + x + x^2 + x^3 + x^4)$$

$$= 5.$$

(b) $|x\sin(1/x)| \le |x| \Rightarrow \lim_{x \to 0} x\sin(1/x) = 0.$

(c) The limit does not exist, since the function is equal to 1 except when $\sin(1/x) = 0$, that is, when $x = \mp 1/\pi, \mp 1/2\pi, \ldots, \mp 1/n\pi, \ldots$. For such values, the function takes the form $0/0$, and is therefore undefined. To have a limit at $x = 0$, the function must be defined at all points in a deleted neighborhood of $x = 0$.

(d) $\lim_{x \to 0^-} f(x) = \frac{1}{2}$ and $\lim_{x \to 0^+} f(x) = 1$. Therefore, the function $f(x)$ does not have a limit as $x \to 0$.

3.2. (a) To show that $(\tan x^3)/x^2 \to 0$ as $x \to 0$: $(\tan x^3)/x^2 = (1/\cos x^3)(\sin x^3)/x^2$. But $|\sin x^3| \le |x^3|$. Hence, $|(\tan x^3)/x^2| \le |x|/|\cos x^3| \to 0$ as $x \to 0$.

(b) $x/\sqrt{x} \to 0$ as $x \to 0$.

(c) $O(1)$ is bounded as $x \to \infty$. Therefore, $O(1)/x \to 0$ as $x \to \infty$, so $O(1) = o(x)$ as $x \to \infty$.

(d)

$$f(x)g(x) = \left[x + o(x^2)\right]\left[\frac{1}{x^2} + o\left(\frac{1}{x}\right)\right]$$

$$= \frac{1}{x} + x\,O\left(\frac{1}{x}\right) + \frac{1}{x^2}\,o(x^2) + o(x^2)O\left(\frac{1}{x}\right).$$

Now, by definition, $|xO(1/x)|$ is bounded by a positive constant as $x \to 0$, and $o(x^2)/x^2 \to 0$ as $x \to 0$. Hence, $o(x^2)/x^2$ is bounded, that is, $O(1)$. Furthermore, $o(x^2)O(1/x) = x[o(x^2)/x^2]xO(1/x)$, which goes to zero as $x \to 0$, is also bounded. It follows that

$$f(x)g(x) = \frac{1}{x} + O(1).$$

3.3. (a) $f(0) = 0$ and $\lim_{x \to 0} f(x) = 0$. Hence, $f(x)$ is continuous at $x = 0$. It is also continuous at all other values of x.

(b) The function $f(x)$ is defined on $[1, 2]$, but $\lim_{x \to 2^-} f(x) = \infty$, so $f(x)$ is not continuous at $x = 2$.

(c) If n is odd, then $f(x)$ is continuous everywhere, except at $x = 0$. If n is even, $f(x)$ will be continuous only when $x > 0$. (m and n must be expressed in their lowest terms so that the only common divisor is 1).

(d) $f(x)$ is continuous for $x \neq 1$.

3.6.

$$f(x) = \begin{cases} -3, & x \neq 0, \\ 0, & x = 0. \end{cases}$$

Thus $f(x)$ is continuous everywhere except at $x = 0$.

3.7. $\lim_{x \to 0^-} f(x) = 2$ and $\lim_{x \to 0^+} f(x) = 0$. Therefore, $f(x)$ cannot be made continuous at $x = 0$.

3.8. Letting $a = b = 0$ in $f(a + b) = f(a) + f(b)$, we conclude that $f(0) = 0$. Now, for any $x_1, x_2 \in R$,

$$f(x_1) = f(x_2) + f(x_1 - x_2).$$

Let $z = x_1 - x_2$. Then

$$|f(x_1) - f(x_2)| = |f(z)|$$
$$= |f(z) - f(0)|.$$

Since $f(x)$ is continuous at $x = 0$, for a given $\epsilon > 0$ there exists a $\delta > 0$ such that $|f(z) - f(0)| < \epsilon$ if $|z| < \delta$. Hence, $|f(x_1) - f(x_2)| < \epsilon$ for all x_1, x_2 such that $|x_1 - x_2| < \delta$. Thus $f(x)$ is uniformly continuous everywhere in R.

3.9. $\lim_{x \to 1^-} f(x) = \lim_{x \to 1^+} f(x) = f(1) = 1$, so $f(x)$ is continuous at $x = 1$. It is also continuous everywhere else on $[0, 2]$, which is closed and bounded. Therefore, it is uniformly continuous by Theorem 3.4.6.

3.10.

$$|\cos x_1 - \cos x_2| = \left| 2\sin\left(\frac{x_2 - x_1}{2}\right) \sin\left(\frac{x_2 + x_1}{2}\right) \right|$$
$$\leq 2\left| \sin\left(\frac{x_2 - x_1}{2}\right) \right|$$
$$\leq |x_1 - x_2| \quad \text{for all } x_1, x_2 \text{ in } R.$$

Thus, for a given $\epsilon > 0$, there exists a $\delta > 0$ (namely, $\delta < \epsilon$) such that $|\cos x_1 - \cos x_2| < \epsilon$ whenever $|x_1 - x_2| < \delta$.

3.13. $f(x) = 0$ if x is a rational number in $[a, b]$. If x is an irrational number, then it is a limit of a sequence $\{y_n\}_{n=1}^{\infty}$ of rational numbers (any neighborhood of x contains infinitely many rationals). Hence,

$$f(x) = f\left(\lim_{n \to \infty} y_n\right) = \lim_{n \to \infty} f(y_n) = 0,$$

since $f(x)$ is a continuous function. Thus $f(x) = 0$ for every x in $[a, b]$.

3.14. $f(x)$ can be written as

$$f(x) = \begin{cases} 3 - 2x, & x \le -1, \\ 5, & -1 < x < 1, \\ 3 + 2x, & x \ge 1. \end{cases}$$

Hence, $f(x)$ has a unique inverse for $x \le -1$ and for $x \ge 1$.

3.15. The inverse function is

$$x = f^{-1}(y) = \begin{cases} y, & y \le 1, \\ \frac{1}{2}(y+1), & y > 1. \end{cases}$$

3.16. (a) The inverse of $f(x)$ is $f^{-1}(y) = 2 - \sqrt{y/2}$.
 (b) The inverse of $f(x)$ is $f^{-1}(y) = 2 + \sqrt{y/2}$.

3.17. By Theorem 3.6.1,

$$\sum_{i=1}^{n} |f(a_i) - f(b_i)| \le K \sum_{i=1}^{n} |a_i - b_i|$$

$$< K\delta.$$

Choosing δ such that $\delta < \epsilon/K$, we get

$$\sum_{i=1}^{n} |f(a_i) - f(b_i)| < \epsilon,$$

for any given $\epsilon > 0$.

3.18. This inequality can be proved by using mathematical induction: it is obviously true for $n = 1$. Suppose that it is true for $n = m$. To show that it is true for $n = m + 1$. For $n = m$, we have

$$f\left(\frac{\sum_{i=1}^{m} a_i x_i}{A_m}\right) \le \frac{\sum_{i=1}^{m} a_i f(x_i)}{A_m},$$

where $A_m = \sum_{i=1}^{m} a_i$. Let

$$b_m = \frac{\sum_{i=1}^{m} a_i x_i}{A_m}.$$

Then

$$f\left(\frac{\sum_{i=1}^{m+1} a_i x_i}{A_{m+1}}\right) = f\left(\frac{A_m b_m + a_{m+1} x_{m+1}}{A_{m+1}}\right)$$

$$\leq \frac{A_m}{A_{m+1}} f(b_m) + \frac{a_{m+1}}{A_{m+1}} f(x_{m+1}).$$

But $f(b_m) \leq (1/A_m)\sum_{i=1}^{m} a_i f(x_i)$. Hence,

$$f\left(\frac{\sum_{i=1}^{m+1} a_i x_i}{A_{m+1}}\right) \leq \frac{\sum_{i=1}^{m} a_i f(x_i) + a_{m+1} f(x_{m+1})}{A_{m+1}}$$

$$= \frac{\sum_{i=1}^{m+1} a_i f(x_i)}{A_{m+1}}.$$

3.19. Let a be a limit point of S. There exists a sequence $\{a_n\}_{n=1}^{\infty}$ in S such that $\lim_{n \to \infty} a_n = a$ (if S is finite, then S is closed already). Hence, $f(a) = f(\lim_{n \to \infty} a_n) = \lim_{n \to \infty} f(a_n) = 0$. It follows that $a \in S$, and S is therefore a closed set.

3.20. Let $g(x) = \exp[f(x)]$. We have that

$$f[\lambda x_1 + (1 - \lambda)x_2] \leq \lambda f(x_1) + (1 - \lambda)f(x_2)$$

for all x_1, x_2 in D, $0 \leq \lambda \leq 1$. Hence,

$$g[\lambda x_1 + (1 - \lambda)x_2] \leq \exp[\lambda f(x_1) + (1 - \lambda)f(x_2)]$$

$$\leq \lambda g(x_1) + (1 - \lambda)g(x_2),$$

since the function e^x is convex. Hence, $g(x)$ is convex on D.

3.22. (a) We have that $E(|X|) \geq |E(X)|$. Let $X \sim N(\mu, \sigma^2)$. Then

$$E(X) = \frac{1}{\sqrt{2\pi\sigma^2}} \int_{-\infty}^{\infty} \exp\left[-\frac{1}{2\sigma^2}(x - \mu)^2\right] x\, dx.$$

Therefore,

$$|E(X)| \le \frac{1}{\sqrt{2\pi\sigma^2}} \int_{-\infty}^{\infty} \exp\left[-\frac{1}{2\sigma^2}(x-\mu)^2\right] |x| \, dx$$

$$= E(|X|).$$

(b) We have that $E(e^{-X}) \ge e^{-\mu}$, where $\mu = E(X)$, since e^{-x} is a convex function. The density function of the exponential distribution with mean μ is

$$g(x) = \frac{1}{\mu} e^{-x/\mu}, \qquad 0 < x < \infty.$$

Hence,

$$E(e^{-X}) = \frac{1}{\mu} \int_0^{\infty} e^{-x/\mu} e^{-x} \, dx$$

$$= \frac{1}{\mu} \frac{1}{1 + 1/\mu} = \frac{1}{\mu + 1}.$$

But

$$e^{\mu} = 1 + \mu + \frac{1}{2!} \mu^2 + \cdots + \frac{1}{n!} \mu^n + \cdots$$

$$\ge 1 + \mu.$$

It follows that

$$e^{-\mu} \le \frac{1}{\mu + 1}$$

and thus

$$E(e^{-X}) \ge e^{-\mu}.$$

3.23. $P(|X_n^2| \ge \epsilon) = P(|X_n| \ge \epsilon^{1/2}) \to 0$ as $n \to \infty$, since X_n converges in probability to zero.

3.24.

$$\sigma^2 = E\left[(X - \mu)^2\right]$$

$$= E\left[|X - \mu|^2\right]$$

$$\ge E^2[|X - \mu|],$$

hence,

$$\sigma \ge E[|X - \mu|].$$

CHAPTER 4

4.1.

$$\lim_{h \to 0} \frac{f(h) - f(-h)}{2h} = \lim_{h \to 0} \frac{f(h) - f(0)}{2h} + \lim_{h \to 0} \frac{f(0) - f(-h)}{2h}$$

$$= \frac{f'(0)}{2} + \frac{1}{2} \lim_{h \to 0} \frac{f(-h) - f(0)}{(-h)}$$

$$= \frac{f'(0)}{2} + \frac{f'(0)}{2} = f'(0).$$

The converse is not true: let $f(x) = |x|$. Then

$$\frac{f(h) - f(-h)}{2h} = 0,$$

and its limit is zero. But $f'(0)$ does not exist.

4.3. For a given $\epsilon > 0$, there exists a $\delta > 0$ such that for any $x \in N_\delta(x_0)$, $x \neq x_0$,

$$\left| \frac{f(x) - f(x_0)}{x - x_0} - f'(x_0) \right| < \epsilon.$$

Hence,

$$\left| \frac{f(x) - f(x_0)}{x - x_0} \right| < |f'(x_0)| + \epsilon$$

and thus

$$|f(x) - f(x_0)| < A|x - x_0|,$$

where $A = |f'(x_0)| + \epsilon$.

4.4. $g(x) = f(x + 1) - f(x) = f'(\xi)$, where ξ is between x and $x + 1$. As $x \to \infty$, we have $\xi \to \infty$ and $f'(\xi) \to 0$, hence, $g(x) \to 0$ as $x \to \infty$.

4.5. We have that $f'(1) = \lim_{x \to 1^+} (x^3 - 2x + 1)/(x - 1) = \lim_{x \to 1^-} (ax^2 - bx + 1 + 1)/(x - 1)$, and thus $1 = 2a - b$. Furthermore, since $f(x)$ is continuous at $x = 1$, we must have $a - b + 1 = -1$. It follows that $a = 3$, $b = 5$, and

$$f'(x) = \begin{cases} 3x^2 - 2, & x \geq 1, \\ 6x - 5, & x < 1, \end{cases}$$

which is continuous everywhere.

4.6. Let $x > 0$.

(a) Taylor's theorem gives

$$f(x + 2h) = f(x) + 2hf'(x) + \frac{(2h)^2}{2!}f''(\xi),$$

where $x < \xi < x + 2h$, and $h > 0$. Hence,

$$f'(x) = \frac{1}{2h}[f(x + 2h) - f(x)] - hf''(\xi),$$

so that $|f'(x)| \le m_0/h + hm_2$.

(b) Since m_1 is the least upper bound of $|f'(x)|$,

$$m_1 \le \frac{m_0}{h} + hm_2.$$

Equivalently,

$$m_2 h^2 - m_1 h + m_0 \ge 0$$

This inequality is valid for all $h > 0$ provided that the discriminant

$$\Delta = m_1^2 - 4m_0 m_2$$

is less than or equal to zero, that is, $m_1^2 \le 4m_0 m_2$.

4.7. No, unless $f'(x)$ is continuous at x_0. For example, the function

$$f(x) = \begin{cases} x + 1, & x > 0, \\ 0, & x = 0, \\ x - 1, & x < 0, \end{cases}$$

does not have a derivative at $x = 0$, but $\lim_{x \to 0} f'(x) = 1$.

4.8.

$$D'(y_j) = \lim_{a \to y_j} \frac{|y_j - a| + \Sigma_{i \ne j}|y_i - a| - \Sigma_{i=1}^n |y_i - y_j|}{a - y_j}$$

$$= \lim_{a \to y_j} \frac{|y_j - a| + \Sigma_{i \ne j}|y_i - a| - \Sigma_{i \ne j}|y_i - y_j|}{a - y_j},$$

which does not exist, since $\lim_{a \to y_j}|y_j - a|/(a - y_j)$ does not exist.

4.9.

$$\frac{d}{dx}\left[\frac{f(x)}{x}\right] = \frac{xf'(x) - f(x)}{x^2}.$$

$xf'(x) - f(x) \to 0$ as $x \to 0$. Hence, by l'Hospital's rule,

$$\lim_{x \to 0} \frac{d}{dx}\left[\frac{f(x)}{x}\right] = \lim_{x \to 0} \frac{f'(x) + xf''(x) - f'(x)}{2x}$$

$$= \frac{1}{2}f''(0).$$

4.10. Let $x_1, x_2 \in R$. Then,

$$|f(x_1) - f(x_2)| = |(x_1 - x_2)f'(\xi)|,$$

here ξ is between x_1 and x_2. Since $f'(x)$ is bounded for all x, $|f'(\xi)| < M$ for some positive constant M. Hence, for a given $\epsilon > 0$, there exists a $\delta > 0$, where $M\delta < \epsilon$, such that $|f(x_1) - f(x_2)| < \epsilon$ if $|x_1 - x_2| < \delta$. Thus $f(x)$ is uniformly continuous on R.

4.11. $f'(x) = 1 + cg'(x)$, and $|cg'(x)| < cM$. Choose c such that $cM < \frac{1}{2}$. In this case,

$$|cg'(x)| < \frac{1}{2},$$

so

$$-\frac{1}{2} < cg'(x) < \frac{1}{2},$$

and,

$$\frac{1}{2} < f'(x) < \frac{3}{2}.$$

Hence, $f'(x)$ is positive and $f(x)$ is therefore strictly monotone increasing, thus $f(x)$ is one-to-one.

4.12. It is sufficient to show that $g'(x) \geq 0$ on $(0, \infty)$:

$$g'(x) = \frac{xf'(x) - f(x)}{x^2}, \qquad x > 0.$$

By the mean value theorem,

$$f(x) = f(0) + xf'(c), \qquad 0 < c < x$$

$$= xf'(c).$$

Hence,

$$g'(x) = \frac{xf'(x) - xf'(c)}{x^2} = \frac{f'(x) - f'(c)}{x}, \qquad x > 0.$$

Since $f'(x)$ is monotone increasing, we have for all $x > 0$,

$$f'(x) \geq f'(c) \quad \text{and thus} \quad g'(x) \geq 0$$

4.14. Let $y = (1 + 1/x)^x$. Then,

$$\log y = x \log\left(1 + \frac{1}{x}\right) = \frac{\log(1 + 1/x)}{1/x}.$$

Applying l'Hospital's rule, we get

$$\lim_{x \to \infty} \log y = \lim_{x \to \infty} \frac{-\dfrac{1}{x^2}\left(1 + \dfrac{1}{x}\right)^{-1}}{-\dfrac{1}{x^2}} = 1.$$

Therefore, $\lim_{x \to \infty} y = e$.

4.15. (a) $\lim_{x \to 0^+} f(x) = 1$, where $f(x) = (\sin x)^x$.
(b) $\lim_{x \to 0^+} g(x) = 0$, where $g(x) = e^{-1/x}/x$.

4.16. Let $f(x) = [1 + ax + o(x)]^{1/x}$, and let $y = ax + o(x)$. Then $y = x[a + o(1)]$, and $[1 + ax + o(x)]^{1/x} = (1 + y)^{a/y}(1 + y)^{o(1)/y}$. Now, as $x \to 0$, we have $y \to 0$, $(1 + y)^{a/y} \to e^a$, and $(1 + y)^{o(1)/y} \to 1$, since

$$\frac{o(1)}{y}\log(1 + y) \to 0 \qquad \text{as } y \to 0.$$

It follows that $f(x) \to e^a$.

4.17. No, because both $f'(x)$ and $g'(x)$ vanish at $x = 0.541$ in $(0, 1)$.

4.18. Let $g(x) = f(x) - \gamma(x - a)$. Then

$$g'(x) = f'(x) - \gamma,$$
$$g'(a) = f'(a) - \gamma < 0,$$
$$g'(b) = f'(b) - \gamma > 0.$$

The function $g(x)$ is continuous on $[a, b]$. Therefore, it must achieve its absolute minimum at some point ξ in $[a, b]$. This point cannot be a or b, since $g'(a) < 0$ and $g'(b) > 0$. Hence, $a < \xi < b$, and $g'(\xi) = 0$, so $f'(\xi) = \gamma$.

4.19. Define $g(x) = f(x) - \tau(x - a)$, where

$$\tau = \sum_{i=1}^{n} \lambda_i f'(x_i).$$

There exist c_1, c_2 such that

$$\max_i f'(x_i) = f'(c_2), \qquad a < c_2 < b,$$

$$\min_i f'(x_i) = f'(c_1), \qquad a < c_1 < b.$$

If $f'(x_1) = f'(x_2) = \cdots = f'(x_n)$, then the result is obviously true. Let us therefore assume that these n derivatives are not all equal. In this case,

$$f'(c_1) < \tau < f'(c_2).$$

Apply now the result in Exercise 4.18 to conclude that there exists a point c between c_1 and c_2 such that $f'(c) = \tau$.

4.20. We have that

$$\sum_{i=1}^{n} [f(y_i) - f(x_i)] = \sum_{i=1}^{n} f'(c_i)(y_i - x_i),$$

where c_i is between x_i and y_i $(i = 1, 2, \ldots, n)$. Using Exercise 4.19, there exists a point c in (a, b) such that

$$\frac{\sum_{i=1}^{n}(y_i - x_i)f'(c_i)}{\sum_{i=1}^{n}(y_i - x_i)} = f'(c).$$

Hence,

$$\sum_{i=1}^{n} [f(y_i) - f(x_i)] = f'(c) \sum_{i=1}^{n}(y_i - x_i).$$

4.21. $\log(1 + x) = \sum_{n=1}^{\infty} (-1)^{n-1} x^n / n, \qquad |x| < 1.$

4.23. $f(x)$ has an absolute minimum at $x = \frac{1}{3}$.

4.24. The function $f(x)$ is bounded if $x^2 + ax + b \neq 0$ for all x in $[-1, 1]$. Let $\Delta = a^2 - 4b$. If $\Delta < 0$, then $x^2 + ax + b > 0$ for all x. The denominator has an absolute minimum at $x = -a/2$. Thus, if $-1 \leq -a/2 \leq 1$, then $f(x)$ will have an absolute maximum at $x = -a/2$. Otherwise, $f(x)$ attains its absolute maximum at $x = -1$ or $x = 1$. If $\Delta = 0$, then

$$f(x) = \frac{1}{\left(x + \dfrac{a}{2}\right)^2}.$$

In this case, the point $-a/2$ must fall outside $[-1, 1]$, and the absolute maximum of $f(x)$ is attained at $x = -1$ or $x = 1$. Finally, if $\Delta > 0$, then

$$f(x) = \frac{1}{(x - x_1)(x - x_2)},$$

where $x_1 = \frac{1}{2}(-a - \sqrt{\Delta})$, $x_2 = \frac{1}{2}(-a + \sqrt{\Delta})$. Both x_1 and x_2 must fall outside $[-1, 1]$. In this case, the point $x = -a/2$ is equal to $\frac{1}{2}(x_1 + x_2)$, which falls outside $[-1, 1]$. Thus $f(x)$ attains its absolute maximum at $x = -1$ or $x = 1$.

4.25. Let $H(y)$ denote the cumulative distribution function of $G^{-1}[F(Y)]$. Then

$$
\begin{aligned}
H(y) &= P\{G^{-1}[F(Y)] \leq y\} \\
&= P[F(Y) \leq G(y)] \\
&= P\{Y \leq F^{-1}[G(y)]\} \\
&= F\{F^{-1}[G(y)]\} \\
&= G(y).
\end{aligned}
$$

4.26. Let $g(w)$ be the density function of W. Then $g(w) = 2we^{-w^2}$, $w \geq 0$.

4.27. **(a)** Let $g(w)$ be the density function of W. Then

$$g(w) = \frac{1}{3w^{2/3}\sqrt{0.08\pi}} \exp\left[-\frac{1}{0.08}(w^{1/3} - 1)^2\right], \qquad w \neq 0.$$

(b) Exact mean is $E(W) = 1.12$. Exact variance is $\text{Var}(W) = 0.42$.

(c) $E(w) \approx 1$, $\text{Var}(w) \approx 0.36$.

4.28. Let $G(y)$ be the cumulative distribution function of Y. Then

$$
\begin{aligned}
G(y) &= P(Y \leq y) \\
&= P(Z^2 \leq y) \\
&= P(|Z| \leq y^{1/2}) \\
&= P(-y^{1/2} \leq Z \leq y^{1/2}) \\
&= 2F(y^{1/2}) - 1,
\end{aligned}
$$

where $F(\cdot)$ is the cumulative distribution function of Z. Thus the density function of Y is $g(y) = 1/\sqrt{2\pi y}\, e^{-y/2}$, $y > 0$. This represents the density function of a chi-squared distribution with one degree of freedom.

4.29. (a) failure rate $= \dfrac{F(x+h) - F(x)}{1 - F(x)}$.

(b)

$$
\begin{aligned}
\text{Hazard rate} &= \frac{dF(x)/dx}{1 - F(x)} \\
&= \frac{1}{e^{-x/\sigma}} \left(\frac{1}{\sigma} e^{-x/\sigma} \right) \\
&= \frac{1}{\sigma}.
\end{aligned}
$$

(c) If

$$
\frac{dF(x)/dx}{1 - F(x)} = c,
$$

then

$$
-\log[1 - F(x)] = cx + c_1,
$$

hence,

$$
1 - F(x) = c_2 e^{-cx}.
$$

Since $F(0) = 0$, $c_2 = 1$. Therefore,

$$
F(x) = 1 - e^{-cx}.
$$

4.30.

$$P(Y_n = r) = \frac{n(n-1)\cdots(n-r+1)}{r!}\left(\frac{\lambda t}{n}\right)^r \left(1 - \frac{\lambda t}{n}\right)^{n-r}$$

$$= \frac{n(n-1)\cdots(n-r+1)}{n^r} \frac{(\lambda t)^r}{r!}\left(1 - \frac{\lambda t}{n}\right)^{-r}\left(1 - \frac{\lambda t}{n}\right)^n.$$

As $n \to \infty$, the first r factors on the right tend to 1, the next factor is fixed, the next tends to 1, and the last factor tends to $e^{-\lambda t}$. Hence,

$$\lim_{n \to \infty} P(Y_n = r) = \frac{e^{-\lambda t}(\lambda t)^r}{r!}.$$

CHAPTER 5

5.1. (a) The sequence $\{b_n\}_{n=1}^\infty$ is monotone increasing, since

$$\max\{a_1, a_2, \ldots, a_n\} \le \max\{a_1, a_2, \ldots, a_{n+1}\}.$$

It is also bounded. Therefore, it is convergent by Theorem 5.1.2. Its limit is $\sup_{n \ge 1} a_n$, since $a_n \le b_n \le \sup_{n \ge 1} a_n$ for $n \ge 1$.

(b) Let $d_i = \log a_i - \log c$, $i = 1, 2, \ldots$. Then

$$\log c_n = \frac{1}{n}\sum_{i=1}^n \log a_i = \log c + \frac{1}{n}\sum_{i=1}^n d_i.$$

To show that $(1/n)\sum_{i=1}^n d_i \to 0$ as $n \to \infty$: We have that $d_i \to 0$ as $i \to \infty$. Therefore, for a given $\epsilon > 0$, there exists a positive integer N_1 such that $|d_i| < \epsilon/2$ if $i > N_1$. Thus, for $n > N_1$,

$$\left|\frac{1}{n}\sum_{i=1}^n d_i\right| \le \frac{1}{n}\sum_{i=1}^{N_1} |d_i| + \frac{\epsilon}{2n}(n - N_1)$$

$$< \frac{1}{n}\sum_{i=1}^{N_1} |d_i| + \frac{\epsilon}{2}.$$

Furthermore, there exists a positive integer N_2 such that

$$\frac{1}{n}\sum_{i=1}^{N_1} |d_i| < \frac{\epsilon}{2},$$

if $n > N_2$. Hence, if $n > \max(N_1, N_2)$, $|(1/n)\sum_{i=1}^n d_i| < \epsilon$, which implies that $(1/n)\sum_{i=1}^n d_i \to 0$, that is, $\log c_n \to \log c$, and $c_n \to c$ as $n \to \infty$.

5.2. Let a and b be the limits of $\{a_n\}_{n=1}^{\infty}$ and $\{b_n\}_{n=1}^{\infty}$, respectively. These limits exist because the two sequences are of the Cauchy type. To show that $d_n \to d$, where $d = |a - b|$:

$$|d_n - d| = |\, |a_n - b_n| - |a - b|\, |$$
$$\leq |(a_n - b_n) - (a - b)|$$
$$\leq |a_n - a| + |b_n - b|.$$

It is now obvious that $d_n \to d$ as $a_n \to a$ and $b_n \to b$.

5.3. Suppose that $a_n \to c$. Then, for a given $\epsilon > 0$, there exists a positive integer N such that $|a_n - c| < \epsilon$ if $n > N$. Let $b_n = a_{k_n}$ be the nth term of a subsequence. Since $k_n \geq n$, we have $|b_n - c| < \epsilon$ if $n > N$. Hence, $b_n \to c$. Vice versa, if every subsequence converges to c, then $a_n \to c$, since $\{a_n\}_{n=1}^{\infty}$ is a subsequence of itself.

5.4. If E is not bounded, then there exists a subsequence $\{b_n\}_{n=1}^{\infty}$ such that $b_n \to \infty$, where $b_n = a_{k_n}$, $k_1 < k_2 < \cdots < k_n < \cdots$. This is a contradiction, since $\{a_n\}_{n=1}^{\infty}$ is bounded.

5.5. (a) For a given $\epsilon > 0$, there exists a positive integer N such that $|a_n - c| < \epsilon$ if $n > N$. Thus, there is a positive constant M such that $|a_n - c| < M$ for all n. Now,

$$\left| \frac{\sum_{i=1}^{n} \alpha_i a_i}{\sum_{i=1}^{n} \alpha_i} - c \right| = \frac{1}{\sum_{i=1}^{n} \alpha_i} \left| \sum_{i=1}^{n} \alpha_i (a_i - c) \right|$$

$$= \frac{1}{\sum_{i=1}^{n} \alpha_i} \left| \sum_{i=1}^{N} \alpha_i (a_i - c) \right| + \frac{1}{\sum_{i=1}^{n} \alpha_i} \left| \sum_{i=N+1}^{n} \alpha_i (a_i - c) \right|$$

$$\leq \frac{1}{\sum_{i=1}^{n} \alpha_i} \sum_{i=1}^{N} \alpha_i |a_i - c| + \frac{1}{\sum_{i=1}^{n} \alpha_i} \sum_{i=N+1}^{n} \alpha_i |a_i - c|$$

$$\leq \frac{M}{\sum_{i=1}^{n} \alpha_i} \sum_{i=1}^{N} \alpha_i + \frac{\epsilon}{\sum_{i=1}^{n} \alpha_i} \sum_{i=N+1}^{n} \alpha_i$$

$$< M \frac{\sum_{i=1}^{N} \alpha_i}{\sum_{i=1}^{n} \alpha_i} + \epsilon.$$

Hence, as $n \to \infty$,

$$\left| \frac{\sum_{i=1}^{n} \alpha_i a_i}{\sum_{i=1}^{n} \alpha_i} - c \right| \to 0.$$

(b) Let $a_n = (-1)^n$. Then, $\{a_n\}_{n=1}^{\infty}$ does not converge. But $(1/n)\sum_{i=1}^{n} a_i$ is equal to zero if n is even, and is equal to $-1/n$ if n is odd. Hence, $(1/n)\sum_{i=1}^{n} a_i$ goes to zero as $n \to \infty$.

5.6. For a given $\epsilon > 0$, there exists a positive integer N such that

$$\left| \frac{a_{n+1}}{a_n} - b \right| < \epsilon$$

if $n > N$. Since $b < 1$, we can choose ϵ so that $b + \epsilon < 1$. Then

$$\frac{a_{n+1}}{a_n} < b + \epsilon < 1, \qquad n > N.$$

Hence, for $n \geq N + 1$,

$$a_{N+2} < a_{N+1}(b + \epsilon),$$

$$a_{N+3} < a_{N+2}(b + \epsilon) < a_{N+1}(b + \epsilon)^2,$$

$$\vdots$$

$$a_n < a_{N+1}(b + \epsilon)^{n-N-1} = \frac{a_{N+1}}{(b + \epsilon)^{N+1}}(b + \epsilon)^n.$$

Letting $c = a_{N+1}/(b + \epsilon)^{N+1}$, $r = b + \epsilon$, we get $a_n < cr^n$, where $0 < r < 1$, for $n \geq N + 1$.

5.7. We first note that for each n, $a_n > 0$, which can be proved by induction. Let us consider two cases.

1. $b > 1$. In this case, we can show that the sequence is (i) bounded from above, and (ii) monotone increasing:
   i. The sequence is bounded from above by $\sqrt{b}$, that is, $a_n < \sqrt{b}$: $a_n^2 < b$ is true for $n = 1$ because $a_1 = 1 < b$. Suppose now that $a_n^2 < b$; to show that $a_{n+1}^2 < b$:

$$b - a_{n+1}^2 = b - \frac{a_n^2(3b + a_n^2)^2}{(3a_n^2 + b)^2}$$

$$= \frac{(b - a_n^2)^3}{(3a_n^2 + b)^2} > 0.$$

Thus $a_{n+1}^2 < b$, and the sequence is bounded from above by $\sqrt{b}$.

ii. The sequence is monotone increasing: we have that $(3b + a_n^2)/(3a_n^2 + b) > 1$, since $a_n^2 < b$, as was seen earlier. Hence, $a_{n+1} > a_n$ for all n.

By Corollary 5.1.1(1), the sequence must be convergent. Let c be its limit. We then have the equation

$$c = \frac{c(3b + c^2)}{3c^2 + b},$$

which results from taking the limit of both sides of $a_{n+1} = a_n(3b + a_n^2)/(3a_n^2 + b)$ and noting that $\lim_{n \to \infty} a_{n+1} = \lim_{n \to \infty} a_n = c$. The only solution to the above equation is $c = \sqrt{b}$.

2. $b < 1$. In this case, we can similarly show that the sequence is bounded from below and is monotone decreasing. Therefore, by Corollary 5.1.1(2) it must be convergent. Its limit is equal to $\sqrt{b}$.

5.8. The sequence is bounded from above by 3, that is, $a_n < 3$ for all n: This is true for $n = 1$. If it is true for n, then

$$a_{n+1} = (2 + a_n)^{1/2} < (2 + 3)^{1/2} < 3.$$

By induction, $a_n < 3$ for all n. Furthermore, the sequence is monotone increasing: $a_1 \le a_2$, since $a_1 = 1$, $a_2 = \sqrt{3}$. If $a_n \le a_{n+1}$, then

$$a_{n+2} = (2 + a_{n+1})^{1/2} \ge (2 + a_n)^{1/2} = a_{n+1}.$$

By induction, $a_n \le a_{n+1}$ for all n. By Corollary 5.1.1(1) the sequence must be convergent. Let c be its limit, which can be obtained by solving the equation

$$c = (2 + c)^{1/2}.$$

The only solution is $c = 2$.

5.10. Let $m = 2n$. Then $a_m - a_n = 1/(n + 1) + 1/(n + 2) + \cdots + 1/2n$. Hence,

$$a_m - a_n > \frac{n}{2n} = \frac{1}{2}.$$

Therefore, $|a_m - a_n|$ cannot be made less than $\frac{1}{2}$ no matter how large m and n are, if $m = 2n$. This violates the condition for a sequence to be Cauchy.

5.11. For $m > n$, we have that

$$
\begin{aligned}
|a_m - a_n| &= |(a_m - a_{m-1}) + (a_{m-1} - a_{m-2}) \\
&\quad + \cdots + (a_{n+2} - a_{n+1}) + (a_{n+1} - a_n)| \\
&< br^{m-1} + br^{m-2} + \cdots + br^n \\
&= br^n(1 + r + r^2 + \cdots + r^{m-n-1}) \\
&= \frac{br^n(1 - r^{m-n})}{1 - r} < \frac{br^n}{1 - r}.
\end{aligned}
$$

For a given $\epsilon > 0$, we choose n large enough such that $br^n/(1 - r) < \epsilon$, which implies that $|a_m - a_n| < \epsilon$, that is, the sequence satisfies the Cauchy criterion; hence, it is convergent.

5.12. If $\{s_n\}_{n=1}^{\infty}$ is bounded from above, then it must be convergent, since it is monotone increasing and thus $\sum_{n=1}^{\infty} a_n$ converges. Vice versa, if $\sum_{n=1}^{\infty} a_n$ is convergent, then $\{s_n\}_{n=1}^{\infty}$ is a convergent sequence; hence, it is bounded, by Theorem 5.1.1.

5.13. Let s_n be the nth partial sum of the series. Then

$$
\begin{aligned}
s_n &= \frac{1}{3} \sum_{i=1}^{n} \left(\frac{1}{3i - 1} - \frac{1}{3i + 2} \right) \\
&= \frac{1}{3} \left(\frac{1}{2} - \frac{1}{5} + \frac{1}{5} - \frac{1}{8} + \cdots + \frac{1}{3n - 1} - \frac{1}{3n + 2} \right) \\
&= \frac{1}{3} \left(\frac{1}{2} - \frac{1}{3n + 2} \right) \\
&\to \frac{1}{6} \qquad \text{as } n \to \infty.
\end{aligned}
$$

5.14. For $n > 2$, the binomial theorem gives

$$
\begin{aligned}
\left(1 + \frac{1}{n} \right)^n &= 1 + \binom{n}{1} \frac{1}{n} + \binom{n}{2} \frac{1}{n^2} + \cdots + \frac{1}{n^n} \\
&= 2 + \frac{1}{2!} \left(1 - \frac{1}{n} \right) + \frac{1}{3!} \left(1 - \frac{1}{n} \right) \left(1 - \frac{2}{n} \right) + \cdots + \frac{1}{n^n} \\
&< 2 + \sum_{i=2}^{n} \frac{1}{i!} \\
&< 2 + \sum_{i=2}^{n} \frac{1}{2^{i-1}} \\
&< 3 \\
&< n.
\end{aligned}
$$

Hence,

$$\left(1 + \frac{1}{n}\right) < n^{1/n},$$

$$(n^{1/n} - 1)^p > \frac{1}{n^p}.$$

Therefore, the series $\sum_{n=1}^{\infty}(n^{1/n} - 1)^p$ is divergent by the comparison test, since $\sum_{n=1}^{\infty} 1/n^p$ is divergent for $p \leq 1$.

5.15. (a) Suppose that $a_n < M$ for all n, where M is a positive constant. Then

$$\frac{a_n}{1 + a_n} > \frac{a_n}{1 + M}.$$

The series $\sum_{n=1}^{\infty} a_n/(1 + a_n)$ is divergent by the comparison test, since $\sum_{n=1}^{\infty} a_n$ is divergent.

(b) We have that

$$\frac{a_n}{1 + a_n} = 1 - \frac{1}{1 + a_n}.$$

If $\{a\}_{n=1}^{\infty}$ is not bounded, then

$$\lim_{n \to \infty} a_n = \infty, \qquad \text{hence,} \qquad \lim_{n \to \infty} \frac{a_n}{1 + a_n} = 1 \neq 0.$$

Therefore, $\sum_{n=1}^{\infty} a_n/(1 + a_n)$ is divergent.

5.16. The sequence $\{s_n\}_{n=1}^{\infty}$ is monotone increasing. Hence, for $n = 2, 3, \ldots,$

$$\frac{a_n}{s_n^2} \leq \frac{a_n}{s_n s_{n-1}} = \frac{s_n - s_{n-1}}{s_n s_{n-1}} = \frac{1}{s_{n-1}} - \frac{1}{s_n},$$

$$\sum_{i=1}^{n} \frac{a_i}{s_i^2} = \frac{a_1}{s_1^2} + \sum_{i=2}^{n} \frac{a_i}{s_i^2}$$

$$\leq \frac{a_1}{s_1^2} + \left(\frac{1}{s_1} - \frac{1}{s_2}\right) + \cdots + \left(\frac{1}{s_{n-1}} - \frac{1}{s_n}\right)$$

$$= \frac{a_1}{s_1^2} + \frac{1}{s_1} - \frac{1}{s_n} \to \frac{a_1}{s_1^2} + \frac{1}{s_1},$$

since $s_n \to \infty$ by the divergence of $\sum_{n=1}^{\infty} a_n$. It follows that $\sum_{n=1}^{\infty} a_n/s_n^2$ is a convergent series.

5.17. Let A denote the sum of the series $\sum_{n=1}^{\infty} a_n$. Then $r_n = A - s_{n-1}$, where $s_n = \sum_{i=1}^{n} a_i$. The sequence $\{r_n\}_{n=2}^{\infty}$ is monotone decreasing. Hence,

$$\sum_{i=m}^{n} \frac{a_i}{r_i} > \frac{a_m + a_{m+1} + \cdots + a_{n-1}}{r_m} = \frac{r_m - r_n}{r_m}$$

$$= 1 - \frac{r_n}{r_m}.$$

Since $r_n \to 0$, we have $1 - r_n/r_m \to 1$ as $n \to \infty$. Therefore, for $0 < \epsilon < 1$, there exists a positive integer k such that

$$\frac{a_m}{r_m} + \frac{a_{m+1}}{r_{m+1}} + \cdots + \frac{a_{m+k}}{r_{m+k}} > \epsilon.$$

This implies that the series $\sum_{n=1}^{\infty} a_n/r_n$ does not satisfy the Cauchy criterion. Hence, it is divergent.

5.18. $(1/n)\log(1/n) = -(\log n)/n \to 0$ as $n \to \infty$. Hence, $(1/n)^{1/n} \to 1$. Similarly, $(1/n)\log(1/n^2) = -(2\log n)/n \to 0$, which implies $(1/n^2)^{1/n} \to 1$.

5.19. **(a)** $a_n^{1/n} = n^{1/n} - 1 \to 0 < 1$ as $n \to \infty$. Therefore, $\sum_{n=1}^{\infty} a_n$ is convergent by the root test.

(b) $a_n < [\log(1+n)]/\log e^{n^2} = [\log(1+n)]/n^2$. Since

$$\int_1^{\infty} \frac{\log(1+x)}{x^2} \, dx = -\frac{1}{x}\log(1+x)\Big|_1^{\infty} + \int_1^{\infty} \frac{dx}{x(x+1)}$$

$$= \log 2 + \log\left(\frac{x}{x+1}\right)\Big|_1^{\infty}$$

$$= 2\log 2,$$

the series $\sum_{n=1}^{\infty} [\log(1+n)]/n^2$ is convergent by the integral test. Hence, $\sum_{n=1}^{\infty} a_n$ converges by the comparison test.

(c) $a_n/a_{n+1} = (2n+2)(2n+3)/(2n+1)^2 \Rightarrow \lim_{n \to \infty} n\,(a_n/a_{n+1} - 1) = \frac{3}{2} > 1$. The series $\sum_{n=1}^{\infty} a_n$ is convergent by Raabe's test.

(d)

$$a_n = \sqrt{n + 2\sqrt{n}} - \sqrt{n}$$

$$= \frac{n + 2\sqrt{n} - n}{\sqrt{n + 2\sqrt{n}} + \sqrt{n}} \to 1 \qquad \text{as } n \to \infty.$$

Therefore, $\sum_{n=1}^{\infty} a_n$ is divergent.

(e) $|a_n|^{1/n} = 4/n \to 0 < 1$. The series is absolutely covergent by the root test.

(f) $a_n = (-1)^n \sin(\pi/n)$. For large n, $\sin(\pi/n) \sim \pi/n$. Therefore, $\sum_{n=1}^{\infty} a_n$ is conditionally convergent.

5.20. (a) Applying the ratio test, we have the condition

$$|x|^2 \lim_{n \to \infty} \left| \frac{a_{n+1}}{a_n} \right| < 1,$$

which is equivalent to

$$\frac{|x|^2}{3} < 1, \qquad \text{that is,} \quad |x| < \sqrt{3}.$$

The series is divergent if $|x| = \sqrt{3}$. Hence, it is uniformly convergent on $[-r, r]$ where $r < \sqrt{3}$.

(b) Using the ratio test, we get

$$|x| \lim_{n \to \infty} \left| \frac{a_{n+1}}{a_n} \right| < 1,$$

where $a_n = 10^n/n$. This is equivalent to

$$10|x| < 1.$$

The series is divergent if $|x| = \frac{1}{10}$. Hence, it is uniformly convergent on $[-r, r]$ where $r < \frac{1}{10}$.

(c) The series is uniformly convergent on $[-r, r]$ where $r < 1$.

(d) The series is uniformly convergent everywhere by Weierstrass's M-test, since

$$\left| \frac{\cos nx}{n(n^2 + 1)} \right| \le \frac{1}{n(n^2 + 1)} \qquad \text{for all } x,$$

and the series whose nth term is $1/n(n^2 + 1)$ is convergent.

5.21. The series $\sum_{n=1}^{\infty} a_n$ is convergent by Theorem 5.2.14. Let s denote its sum:

$$s = 1 - \tfrac{1}{2} + \tfrac{1}{3} - \left(\tfrac{1}{4} - \tfrac{1}{5} \right) - \left(\tfrac{1}{6} - \tfrac{1}{7} \right) - \cdots$$

$$< 1 - \tfrac{1}{2} + \tfrac{1}{3} = \tfrac{5}{6} = \tfrac{10}{12}.$$

Now,

$$\sum_{n=1}^{\infty} b_n = \left(1 + \tfrac{1}{3} - \tfrac{1}{2}\right) + \left(\tfrac{1}{5} + \tfrac{1}{7} - \tfrac{1}{4}\right) + \cdots$$

$$+ \left(\frac{1}{4n-3} + \frac{1}{4n-1} - \frac{1}{2n}\right) + \cdots .$$

Let s'_{3n} denote the sum of the first $3n$ terms of $\sum_{n=1}^{\infty} b_n$. Then,

$$s'_{3n} = s_{2n} + u_n,$$

where s_{2n} is the sum of the first $2n$ terms of $\sum_{n=1}^{\infty} a_n$, and

$$u_n = \frac{1}{2n+1} + \frac{1}{2n+3} + \cdots + \frac{1}{4n-1}, \qquad n = 1, 2, \ldots$$

The sequence $\{u_n\}_{n=1}^{\infty}$ is monotone increasing and is bounded from above by $\tfrac{1}{2}$, since

$$u_n < \frac{n}{2n+1}, \qquad n = 1, 2, \ldots$$

(the number of terms that make up u_n is equal to n). Thus $\{s'_{3n}\}_{n=1}^{\infty}$ is convergent, which implies convergence of $\sum_{n=1}^{\infty} b_n$. Let t denote the sum of this series. Note that

$$t > \left(1 + \tfrac{1}{3} - \tfrac{1}{2}\right) + \left(\tfrac{1}{5} + \tfrac{1}{7} - \tfrac{1}{4}\right) = \tfrac{11}{12},$$

since $1/(4n-3) + 1/(4n-1) - 1/2n > 0$ for all n.

5.22. Let c_n denote the nth term of Cauchy's product. Then

$$c_n = \sum_{k=0}^{n} a_k a_{n-k}$$

$$= (-1)^n \sum_{k=0}^{n} \frac{1}{[(n-k+1)(k+1)]^{1/2}}.$$

Since $n - k + 1 \leq n + 1$ and $k + 1 \leq n + 1$,

$$|c_n| \geq \sum_{k=0}^{n} \frac{1}{n+1} = 1.$$

Hence, c_n does not go to zero as $n \to \infty$. Therefore, $\sum_{n=0}^{\infty} c_n$ is divergent.

5.23. $f_n(x) \to f(x)$, where

$$f(x) = \begin{cases} 0, & x = 0, \\ 1/x, & x > 0. \end{cases}$$

The convergence is not uniform on $[0, \infty)$, since $f_n(x)$ is continuous on $[0, \infty)$ for all n, but $f(x)$ is discontinuous at $x = 0$.

5.24. **(a)** $1/n^x \leq 1/n^{1+\delta}$. Since $\sum_{n=1}^{\infty} 1/n^{1+\delta}$ is a convergent series, $\sum_{n=1}^{\infty} 1/n^x$ is uniformly convergent on $[1 + \delta, \infty)$ by Weierstrass's M-test.

(b) $d/dx(1/n^x) = -(\log n)/n^x$. The series $\sum_{n=1}^{\infty} (\log n)/n^x$ is uniformly convergent on $[1 + \delta, \infty)$:

$$\left| \frac{\log n}{n^x} \right| \leq \frac{1}{n^{1+\delta/2}} \frac{\log n}{n^{\delta/2}}, \qquad x \geq \delta + 1.$$

Since $(\log n)/n^{\delta/2} \to 0$ as $n \to \infty$, there exists a positive integer N such that

$$\frac{\log n}{n^{\delta/2}} < 1 \qquad \text{if } n > N.$$

Therefore,

$$\left| \frac{\log n}{n^x} \right| < \frac{1}{n^{1+\delta/2}}$$

if $n > N$ and $x \geq \delta + 1$. But, $\sum_{n=1}^{\infty} 1/n^{1+\delta/2}$ is convergent. Hence, $\sum_{n=1}^{\infty} (\log n)/n^x$ is uniformly convergent on $[1 + \delta, \infty)$ by Weierstrass's M-test. If $\zeta(x)$ denotes the sum of the series $\sum_{n=1}^{\infty} 1/n^x$, then

$$\zeta'(x) = - \sum_{n=1}^{\infty} \frac{\log n}{n^x}, \qquad x \geq \delta + 1.$$

5.25. We have that

$$\sum_{n=0}^{\infty} \binom{n+k-1}{n} x^n = \frac{1}{(1-x)^k}, \qquad k = 1, 2, \ldots,$$

if $-1 < x < 1$ (see Example 5.4.4). Differentiating this series term by term, we get

$$\sum_{n=1}^{\infty} n x^{n-1} \binom{n+k-1}{n} = \frac{k}{(1-x)^{k+1}}.$$

It follows that

$$\sum_{n=0}^{\infty} n \binom{n+r-1}{r} p^r (1-p)^n = p^r \frac{r(1-p)}{p^{r+1}}$$

$$= \frac{r(1-p)}{p}.$$

Taking now second derivatives, we obtain

$$\sum_{n=2}^{\infty} n(n-1) x^{n-2} \binom{n+k-1}{n} = \frac{k(k+1)}{(1-x)^{k+2}}.$$

From this we conclude that

$$\sum_{n=1}^{\infty} n^2 \binom{n+k-1}{n} x^n = \frac{kx(1+kx)}{(1-x)^{k+2}}.$$

Hence,

$$\sum_{n=0}^{\infty} n^2 \binom{n+r-1}{n} p^r (1-p)^n = \frac{p^r (1-p) r [1 + r(1-p)]}{p^{r+2}}$$

$$= \frac{r(1-p)(1+r-rp)}{p^2}.$$

5.26.

$$\phi(t) = \sum_{n=0}^{\infty} e^{nt} \binom{n+r-1}{n} p^r q^n$$

$$= p^r \sum_{n=0}^{\infty} \binom{n+r-1}{n} (qe^t)^n$$

$$= \frac{p^r}{1 - qe^t}, \qquad qe^t < 1.$$

The series converges uniformly on $(-\infty, s]$ where $s < -\log q$. Yes, formula (5.63) can be applied, since there exists a neighborhood $N_\delta(0)$ contained inside the interval of convergence by the fact that $-\log q > 0$.

5.28. It is sufficient to show that the series $\sum_{n=1}^{\infty}(\mu'_n/n!)\tau^n$ is absolutely convergent for some $\tau > 0$ (see Theorem 5.6.1). Using the root test,

$$\rho = \lim_{n \to \infty} \sup \frac{|\mu'_n|^{1/n}\tau}{(n!)^{1/n}}$$

$$= \tau e \lim_{n \to \infty} \sup \frac{|\mu'_n|^{1/n}}{(2\pi)^{1/2n}n^{1+1/2n}}$$

$$= \tau e \lim_{n \to \infty} \sup \frac{|\mu'_n|^{1/n}}{n}.$$

Let

$$m = \lim_{n \to \infty} \sup \frac{|\mu'_n|^{1/n}}{n}.$$

If $m > 0$ and finite, then the series $\sum_{n=1}^{\infty}(\mu'_n/n!)\tau^n$ is absolutely convergent if $\rho < 1$, that is, if $\tau < 1/(em)$.

5.30. (a)

$$E(X_n) = \sum_{k=0}^{n} k\binom{n}{k}p^k(1-p)^{n-k}$$

$$= p\sum_{k=1}^{n} k\binom{n}{k}p^{k-1}(1-p)^{n-k}$$

$$= p\sum_{k=0}^{n-1}(k+1)\binom{n}{k+1}p^k(1-p)^{n-k-1}$$

$$= p\sum_{k=0}^{n-1}\frac{n!}{k!(n-k-1)!}p^k(1-p)^{n-k-1}$$

$$= np(p+1-p)^{n-1}$$

$$= np.$$

We can similarly show that

$$E(X_n^2) = np(1-p) + n^2p^2.$$

Hence,

$$\text{Var}(X_n) = E(X_n^2) - n^2 p^2$$
$$\text{Var}(X_n) = np(1-p).$$

(b) We have that $P(|Y_n| \geq r\sigma) \leq 1/r^2$. Let $r\sigma = \epsilon$, where $\sigma^2 = p(1 - p)/n$. Then

$$P(|Y_n| \geq \epsilon) \leq \frac{p(1-p)}{n\epsilon^2}.$$

(c) As $n \to \infty$, $P(|Y_n| \geq \epsilon) \to 0$, which implies that Y_n converges in probability to zero.

5.31. (a)

$$\phi_n(t) = (p_n e^t + q_n)^n, \qquad q_n = 1 - p_n$$
$$= [1 + p_n(e^t - 1)]^n.$$

Let $np_n = r_n$. Then

$$\phi_n(t) = \left[1 + \frac{r_n}{n}(e^t - 1)\right]^n$$
$$\to \exp[\mu(e^t - 1)]$$

as $n \to \infty$. The limit of $\phi_n(t)$ is the moment generating function of a Poisson distribution with mean μ. Thus S_n has a limiting Poisson distribution with mean μ.

5.32. We have that

$$Q = (I_n - H_1 + kH_2 - k^2 H_3 + k^3 H_4 - \cdots)^2$$
$$= (I_n - H_1) + k^2 H_3 - 2k^3 H_4 + \cdots.$$

Thus

$$y'Qy = y'(I_n - H_1)y + k^2 y' H_3 y - 2k^3 y' H_4 y + \cdots$$
$$= SS_E + k^2 S_3 - 2k^3 S_4 + \cdots$$
$$= SS_E + \sum_{i=3}^{\infty} (i-2)(-k)^{i-1} S_i.$$

CHAPTER 6

6.1. If $f(x)$ is Riemann integrable, then inequality (6.1) is satisfied; hence, equality (6.6) is true, as was shown to be the case in the proof of Theorem 6.2.1. Vice versa, if equality (6.6) is satisfied, then by the double inequality (6.2), $S(P, f)$ must have a limit as $\Delta_p \to 0$, which implies that $f(x)$ is Riemann integrable on $[a, b]$.

6.2. Consider the function $f(x)$ on $[0, 1]$ such that $f(x) = 0$ if $0 \le x < \frac{1}{2}$, $f(x) = 2$ if $x = 1$, and $f(x) = \sum_{i=0}^{n} 1/2^i$ if $(n+1)/(n+2) \le x < (n+2)/(n+3)$ for $n = 0, 1, 2, \dots$. This function has a countable number of discontinuities at $\frac{1}{2}, \frac{2}{3}, \frac{3}{4}, \dots$, but is Riemann integrable on $[0, 1]$, since it is monotone increasing (see Theorem 6.3.2).

6.3. Suppose that $f(x)$ has one discontinuity of the first kind at $x = c$, $a < c < b$. Let $\lim_{x \to c^-} f(x) = L_1$, $\lim_{x \to c^+} f(x) = L_2$, $L_1 \ne L_2$. In any partition P of $[a, b]$, the point c appears in at most two subintervals. The contribution to $US_P(f) - LS_P(f)$ from these intervals is less than $2(M - m)\Delta_p$, where M, m are the supremum and infimum of $f(x)$ on $[a, b]$, and this can be made as small as we please. Furthermore, $\sum_i M_i \Delta x_i - \sum_i m_i \Delta x_i$ for the remaining subintervals can also be made as small as we please. Hence, $US_P(f) - LS_P(f)$ can be made smaller than ϵ for any given $\epsilon > 0$. Therefore, $f(x)$ is Riemann integrable on $[a, b]$. A similar argument can be used if $f(x)$ has a finite number of discontinuities of the first kind in $[a, b]$.

6.4. Consider the following partition of $[0, 1]$: $P = \{0, 1/2n, 1/(2n-1), \dots, 1/3, 1/2, 1\}$. The number of partition points is $2n + 1$ including 0 and 1. In this case,

$$\sum_{i=1}^{2n} |\Delta f_i| = \left| \frac{1}{2n} \cos \pi n - 0 \right|$$

$$+ \left| \frac{1}{2n-1} \cos\left[\frac{\pi}{2}(2n-1) \right] - \frac{1}{2n} \cos \pi n \right|$$

$$+ \left| \frac{1}{2n-2} \cos[\pi(n-1)] - \frac{1}{2n-1} \cos\left[\frac{\pi}{2}(2n-1) \right] \right| + \cdots$$

$$+ \left| \frac{1}{2} \cos \pi - \frac{1}{3} \cos\left(\frac{3\pi}{2} \right) \right| + \left| \cos\left(\frac{\pi}{2} \right) - \frac{1}{2} \cos \pi \right|$$

$$= \frac{1}{n} + \frac{1}{n-1} + \frac{1}{n-2} + \cdots + 1.$$

As $n \to \infty$, $\sum_{i=1}^{2n} |\Delta f_i| \to \infty$, since $\sum_{n=1}^{\infty} 1/n$ is divergent. Hence, $f(x)$ is not of bounded variation on $[0, 1]$.

6.5. **(a)** For a given $\epsilon > 0$, there exists a constant $M > 0$ such that

$$\left| \frac{f'(x)}{g'(x)} - L \right| < \epsilon$$

if $x > M$. Since $g'(x) > 0$,

$$|f'(x) - Lg'(x)| < \epsilon g'(x).$$

Hence, if λ_1 and λ_2 are chosen larger than M, then

$$\left| \int_{\lambda_1}^{\lambda_2} [f'(x) - Lg'(x)] \, dx \right| \leq \int_{\lambda_1}^{\lambda_2} |f'(x) - Lg'(x)| \, dx$$

$$< \epsilon \int_{\lambda_1}^{\lambda_2} g'(x) \, dx.$$

(b) From (a) we get

$$|f(\lambda_2) - Lg(\lambda_2) - f(\lambda_1) + Lg(\lambda_1)| < \epsilon [g(\lambda_2) - g(\lambda_1)].$$

Divide both sides by $g(\lambda_2)$ [which is positive for large λ_2, since $g(x) \to \infty$ as $x \to \infty$], we obtain

$$\left| \frac{f(\lambda_2)}{g(\lambda_2)} - L - \frac{f(\lambda_1)}{g(\lambda_2)} + L \frac{g(\lambda_1)}{g(\lambda_2)} \right| < \epsilon \left[1 - \frac{g(\lambda_1)}{g(\lambda_2)} \right]$$

$$< \epsilon.$$

Hence,

$$\left| \frac{f(\lambda_2)}{g(\lambda_2)} - L \right| < \epsilon + \frac{|f(\lambda_1)|}{g(\lambda_2)} + |L| \frac{g(\lambda_1)}{g(\lambda_2)}.$$

(c) For sufficiently large λ_2, the second and third terms on the right-hand side of the above inequality can each be made smaller than ϵ; hence,

$$\left| \frac{f(\lambda_2)}{g(\lambda_2)} - L \right| < 3\epsilon.$$

6.6. We have that

$$mg(x) \leq f(x)g(x) \leq Mg(x),$$

where m and M are the infimum and supremum of $f(x)$ on $[a, b]$, respectively. Let ξ and η be points in $[a, b]$ such that $m = f(\xi)$, $M = f(\eta)$. We conclude that

$$f(\xi) \leq \frac{\int_a^b f(x)g(x)\,dx}{\int_a^b g(x)\,dx} \leq f(\eta).$$

By the intermediate-value theorem (Theorem 3.4.4), there exists a constant c, between ξ and η, such that

$$\frac{\int_a^b f(x)g(x)\,dx}{\int_a^b g(x)\,dx} = f(c).$$

Note that c can be equal to ξ or η in case equality is attained at the lower end or the upper end of the above double inequality.

6.9. Integration by parts gives

$$\int_a^b f(x)\,dg(x) = f(b)g(b) - f(a)g(a) + \int_a^b g(x)\,d[-f(x)].$$

Since $g(x)$ is bounded, $f(b)g(b) \to 0$ as $b \to \infty$. Let us now establish convergence of $\int_a^b g(x)\,d[-f(x)]$ as $b \to \infty$: let $M > 0$ be such that $|g(x)| \leq M$ for all $x \geq a$. In addition,

$$\int_a^b M\,d[-f(x)] = Mf(a) - Mf(b)$$

$$\to Mf(a) \qquad \text{as } b \to \infty.$$

Hence, $\int_a^\infty M\,d[-f(x)]$ is a convergent integral. Since $-f(x)$ is monotone increasing on $[a, \infty)$, then

$$\int_a^\infty |g(x)|\,d[-f(x)] \leq \int_a^\infty M\,d[-f(x)].$$

This implies absolute convergence of $\int_a^\infty g(x)\,d[-f(x)]$. It follows that the integral $\int_a^\infty f(x)\,dg(x)$ is convergent.

6.10. Let n be a positive integer. Then

$$\int_0^{n\pi} \left|\frac{\sin x}{x}\right| dx = \int_0^{\pi} \frac{\sin x}{x} dx - \int_{\pi}^{2\pi} \frac{\sin x}{x} dx$$

$$+ \cdots + (-1)^{n-1} \int_{(n-1)\pi}^{n\pi} \frac{\sin x}{x} dx$$

$$= \int_0^{\pi} \sin x \left[\frac{1}{x} + \frac{1}{x+\pi} + \cdots + \frac{1}{x+(n-1)\pi}\right] dx$$

$$> \left(\frac{1}{\pi} + \frac{1}{2\pi} + \cdots + \frac{1}{n\pi}\right) \int_0^{\pi} \sin x \, dx$$

$$= \frac{2}{\pi}\left(1 + \frac{1}{2} + \cdots + \frac{1}{n}\right) \to \infty \qquad \text{as } n \to \infty.$$

6.11. (a) This is convergent by the integral test, since $\int_1^{\infty} (\log x)/(x\sqrt{x}) \, dx = 4\int_0^{\infty} e^{-x} x \, dx = 4$ by a proper change of variable.

(b) $(n+4)/(2n^3+1) \sim 1/2n^2$. But $\sum_{n=1}^{\infty} 1/n^2$ is convergent by the integral test, since $\int_1^{\infty} dx/x^2 = 1$. Hence, this series is convergent by the comparison test.

(c) $1/(\sqrt{n+1}-1) \sim 1/\sqrt{n}$. By the integral test, $\sum_{n=1}^{\infty} 1/\sqrt{n}$ is divergent, since $\int_1^{\infty} dx/\sqrt{x} = 2\sqrt{x}\,\big|_1^{\infty} = \infty$. Hence, $\sum_{n=1}^{\infty} 1/(\sqrt{n+1}-1)$ is divergent.

6.13. (a) Near $x = 0$ we have $x^{m-1}(1-x)^{n-1} \sim x^{m-1}$, and $\int_0^1 x^{m-1} \, dx = 1/m$. Also, near $x = 1$ we have $x^{m-1}(1-x)^{n-1} \sim (1-x)^{n-1}$, and $\int_0^1 (1-x)^{n-1} \, dx = 1/n$. Hence, $B(m,n)$ converges if $m > 0$, $n > 0$.

(b) Let $\sqrt{x} = \sin\theta$. Then

$$\int_0^1 x^{m-1}(1-x)^{n-1} \, dx = 2\int_0^{\frac{\pi}{2}} \sin^{2m-1}\theta\cos^{2n-1}\theta \, d\theta.$$

(c) Let $x = 1/(1+y)$. Then

$$\int_0^1 x^{m-1}(1-x)^{n-1} \, dx = \int_0^{\infty} \frac{y^{n-1}}{(1+y)^{m+n}} \, dy.$$

Letting $z = 1-x$, we get

$$\int_0^1 x^{m-1}(1-x)^{n-1} \, dx = -\int_1^0 (1-z)^{m-1} z^{n-1} \, dz$$

$$= \int_0^1 x^{n-1}(1-x)^{m-1} \, dx.$$

Hence, $B(m, n) = B(n, m)$. It follows that

$$B(m, n) = \int_0^\infty \frac{x^{m-1}}{(1+x)^{m+n}} \, dx.$$

(d)

$$B(m, n) = \int_0^\infty \frac{x^{n-1}}{(1+x)^{m+n}} \, dx$$

$$= \int_0^1 \frac{x^{n-1}}{(1+x)^{m+n}} \, dx + \int_1^\infty \frac{x^{n-1}}{(1+x)^{m+n}} \, dx.$$

Let $y = 1/x$ in the second integral. We get

$$\int_1^\infty \frac{x^{n-1}}{(1+x)^{m+n}} \, dx = \int_0^1 \frac{y^{m-1}}{(1+y)^{m+n}} \, dy$$

Therefore,

$$B(m, n) = \int_0^1 \frac{x^{n-1}}{(1+x)^{m+n}} \, dx + \int_0^1 \frac{x^{m-1}}{(1+x)^{m+n}} \, dx$$

$$= \int_0^1 \frac{x^{n-1} + x^{m-1}}{(1+x)^{m+n}} \, dx.$$

6.14. (a)

$$\int_0^\infty \frac{dx}{\sqrt{1+x^3}} = \int_0^1 \frac{dx}{\sqrt{1+x^3}} + \int_1^\infty \frac{dx}{\sqrt{1+x^3}}.$$

The first integral exists because the integrand is continuous. The second integral is convergent because

$$\int_1^\infty \frac{dx}{\sqrt{1+x^3}} < \int_1^\infty \frac{dx}{x^{3/2}} = 2.$$

Hence, $\int_0^\infty dx/\sqrt{1+x^3}$ is convergent.

(b) Divergent, since for large x we have $1/(1+x^3)^{1/3} \sim 1/(1+x)$, and $\int_0^\infty dx/(1+x) = \infty$.

(c) Convergent, since if $f(x) = 1/(1-x^3)^{1/3}$ and $g(x) = 1/(1-x)^{1/3}$, then

$$\lim_{x \to 1^-} \frac{f(x)}{g(x)} = \left(\frac{1}{3}\right)^{1/3}.$$

But $\int_0^1 g(x) \, dx = \frac{3}{2}$. Hence, $\int_0^1 f(x) \, dx$ is convergent.

(d) Partition the integral as the sum of

$$\int_0^1 \frac{dx}{\sqrt{x}\,(1+x)} \quad \text{and} \quad \int_1^\infty \frac{dx}{\sqrt{x}\,(1+2x)}.$$

The first integral is convergent, since near $x = 0$ we have $1/[\sqrt{x}\,(1+x)] \sim 1/\sqrt{x}$, and $\int_0^1 dx/\sqrt{x} = 2$. The second integral is also convergent, since as $x \to \infty$ we have $1/[\sqrt{x}\,(1+2x)] \sim 1/2x^{3/2}$, and $\int_1^\infty dx/x^{3/2} = 2$.

6.15. The proof of this result is similar to the proof of Corollary 6.4.2.

6.16. It is is easy to show that

$$h'_n(x) = -\frac{(b-x)^{n-1}}{(n-1)!} f^{(n)}(x).$$

Hence,

$$h_n(a) = h_n(b) - \int_a^b h'_n(x)\, dx$$

$$= \frac{1}{(n-1)!} \int_a^b (b-x)^{n-1} f^{(n)}(x)\, dx,$$

since $h_n(b) = 0$. It follows that

$$f(b) = f(a) + (b-a)\, f'(a) + \cdots + \frac{(b-a)^{n-1}}{(n-1)!} f^{(n-1)}(a)$$

$$+ \frac{1}{(n-1)!} \int_a^b (b-x)^{n-1} f^{(n)}(x)\, dx.$$

6.17. Let $G(x) = \int_a^x g(t)\, dt$. Then, $G(x)$ is uniformly continuous and $G'(x) = g(x)$ by Theorem 6.4.8. Thus

$$\int_a^b f(x) g(x)\, dx = f(x) G(x)\big|_a^b - \int_a^b G(x) f'(x)\, dx$$

$$= f(b) G(b) - \int_a^b G(x) f'(x)\, dx.$$

Now, since $G(x)$ is continuous, then by Corollary 3.4.1,

$$G(\xi) \le G(x) \le G(\eta)$$

for all x in $[a, b]$, where ξ, η are points in $[a, b]$ at which $G(x)$ achieves its infimum and supremum in $[a, b]$, respectively. Furthermore, since $f(x)$ is monotone (say monotone increasing) and $f'(x)$ exists, then $f'(x) \geq 0$ by Theorem 4.2.3. It follows that

$$G(\xi) \int_a^b f'(x)\, dx \leq \int_a^b G(x) f'(x)\, dx \leq G(\eta) \int_a^b f'(x)\, dx.$$

This implies that

$$\int_a^b G(x) f'(x)\, dx = \lambda \int_a^b f'(x)\, dx,$$

where $G(\xi) \leq \lambda \leq G(\eta)$. By Theorem 3.4.4, there exists a point c between ξ and η such that $\lambda = G(c)$. Hence,

$$\int_a^b G(x) f'(x)\, dx = G(c) \int_a^b f'(x)\, dx$$

$$= [f(b) - f(a)] \int_a^c g(x)\, dx.$$

Consequently,

$$\int_a^b f(x) g(x)\, dx = f(b) G(b) - \int_a^b G(x) f'(x)\, dx$$

$$= f(b) \int_a^b g(x)\, dx - [f(b) - f(a)] \int_a^c g(x)\, dx$$

$$= f(a) \int_a^c g(x)\, dx + f(b) \int_c^b g(x)\, dx$$

6.18. This follows directly from applying the result in Exercise 6.17 and letting $f(x) = 1/x$, $g(x) = \sin x$.

$$\int_a^b \frac{\sin x}{x}\, dx = \frac{1}{a} \int_a^c \sin x\, dx + \frac{1}{b} \int_c^b \sin x\, dx$$

$$= \frac{1}{a} (\cos a - \cos c) + \frac{1}{b} (\cos c - \cos b).$$

Therefore,

$$\left| \int_a^b \frac{\sin x}{x}\, dx \right| \leq \frac{1}{a} |\cos a - \cos c| + \frac{1}{a} |\cos c - \cos b|$$

$$\leq \frac{4}{a}.$$

6.21. Let $\epsilon > 0$ be given. Choose $\eta > 0$ such that

$$\eta[g(b) - g(a)] < \epsilon.$$

Since $f(x)$ is uniformly continuous on $[a, b]$ (see Theorem 3.4.6), there exists a $\delta > 0$ such that

$$|f(x) - f(z)| < \eta$$

if $|x - z| < \delta$ for all x, z in $[a, b]$. Let us now choose a partition P of $[a, b]$ whose norm Δ_p is smaller than δ. Then,

$$US_P(f, g) - LS_P(f, g) = \sum_{i=1}^{n} (M_i - m_i)\Delta g_i$$

$$\leq \eta \sum_{i=1}^{n} \Delta g_i$$

$$= \eta[g(b) - g(a)]$$

$$< \epsilon.$$

It follows that $f(x)$ is Riemann–Stieltjes integrable on $[a, b]$.

6.22. Since $g(x)$ is continuous, for any positive integer n we can choose a partition P such that

$$\Delta g_i = \frac{g(b) - g(a)}{n}, \qquad i = 1, 2, \ldots, n.$$

Also, since $f(x)$ is monotone increasing, then $M_i = f(x_i)$, $m_i = f(x_{i-1})$, $i = 1, 2, \ldots, n$. Hence,

$$US_P(f, g) - LS_P(f, g) = \frac{g(b) - g(a)}{n} \sum_{i=1}^{n} [f(x_i) - f(x_{i-1})]$$

$$= \frac{g(b) - g(a)}{n} [f(b) - f(a)].$$

This can be made less than ϵ, for any given $\epsilon > 0$, if n is chosen large enough. It follows that $f(x)$ is Riemann–Stieltjes integrable with respect to $g(x)$.

6.23. Let $f(x) = x^k/(1 + x^2)$, $g(x) = x^{k-2}$. Then $\lim_{x \to \infty} f(x)/g(x) = 1$. The integral $\int_0^\infty x^{k-2} \, dx$ is divergent, since $\int_0^\infty x^{k-2} \, dx = x^{k-1}/(k-1)|_0^\infty = \infty$ if $k > 1$. If $k = 1$, then $\int_0^\infty x^{k-2} \, dx = \int_0^\infty dx/x = \infty$. By Theorem 6.5.3, $\int_0^\infty f(x) \, dx$ must be divergent if $k \geq 1$. A similar procedure can be used to show that $\int_{-\infty}^0 f(x) \, dx$ is divergent.

6.24. Since $E(X) = 0$, then $\text{Var}(X) = E(X^2)$, which is given by

$$E(X^2) = \int_{-\infty}^{\infty} \frac{x^2 e^x}{(1 + e^x)^2} \, dx.$$

Let $u = e^x/(1 + e^x)$. The integral becomes

$$E(X^2) = \int_0^1 \left[\log\left(\frac{u}{1 - u}\right) \right]^2 du$$

$$= \left\{ \log\left(\frac{u}{1 - u}\right) \left[u \log u + (1 - u) \log(1 - u) \right] \right\}_0^1$$

$$- \int_0^1 \frac{\left[u \log u + (1 - u) \log(1 - u) \right]}{u(1 - u)} \, du$$

$$= - \int_0^1 \frac{\log u}{1 - u} \, du - \int_0^1 \frac{\log(1 - u)}{u} \, du$$

$$= -2 \int_0^1 \frac{\log u}{1 - u} \, du.$$

But

$$\int_0^1 \frac{\log u}{1 - u} \, du = \int_0^1 (1 + u + u^2 + \cdots + u^n + \cdots) \log u \, du,$$

and

$$\int_0^1 u^n \log u \, du = \frac{1}{n + 1} u^{n+1} \log u \Big|_0^1 - \frac{1}{n + 1} \int_0^1 u^n \, du$$

$$= - \frac{1}{(n + 1)^2}.$$

Hence,

$$\int_0^1 \frac{\log u}{1 - u} \, du = - \sum_{n=0}^{\infty} \frac{1}{(n + 1)^2} = - \frac{\pi^2}{6},$$

and $E(X^2) = \pi^2/3$.

6.26. Let $g(x) = f(x)/x^\alpha$, $h(x) = 1/x^\alpha$. We have that

$$\lim_{x \to 0^+} \frac{g(x)}{h(x)} = f(0) < \infty.$$

Since $\int_0^\delta h(x)\,dx = \delta^{1-\alpha}/(1-\alpha) < \infty$ for any $\delta > 0$, by Theorem 6.5.7, $\int_0^\delta f(x)/x^\alpha\,dx$ exists. Furthermore,

$$\int_\delta^\infty \frac{f(x)}{x^\alpha}\,dx \le \frac{1}{\delta^\alpha}\int_\delta^\infty f(x)\,dx \le \frac{1}{\delta^\alpha}.$$

Hence,

$$E(X^{-\alpha}) = \int_0^\infty \frac{f(x)}{x^\alpha}\,dx$$

exists.

6.27. Let $g(x) = f(x)/x^{1+\alpha}$, $h(x) = 1/x$. Then

$$\lim_{x\to 0^+} \frac{g(x)}{h(x)} = k < \infty.$$

Since $\int_0^\delta h(x)\,dx$ is divergent for any $\delta > 0$, then so is $\int_0^\delta g(x)\,dx$ by Theorem 6.5.7, and hence

$$E[X^{-(1+\alpha)}] = \int_0^\delta g(x)\,dx + \int_\delta^\infty g(x)\,dx = \infty.$$

6.28. Applying formula (6.84), the density function of W is given by

$$g(w) = \frac{2\Gamma\left(\dfrac{n+1}{2}\right)}{\sqrt{n\pi}\,\Gamma\left(\dfrac{n}{2}\right)}\left(1 + \frac{w^2}{n}\right)^{-(n+1)/2}, \qquad w \ge 0.$$

6.29. **(a)** From Theorem 6.9.2, we have

$$P(|X - \mu| \ge u\sigma) \le \frac{1}{u^2}.$$

Letting $u\sigma = r$, we get

$$P(|X - \mu| \ge r) \le \frac{\sigma^2}{r^2}.$$

(b) This follows from (a) and the fact that

$$E(\bar{X}_n) = \mu, \qquad \mathrm{Var}(\bar{X}_n) = \frac{\sigma^2}{n}.$$

(c)

$$P(|\bar{X}_n - \mu| \geq \epsilon) \leq \frac{\sigma^2}{n\epsilon^2}.$$

Hence, as $n \to \infty$, $P(|\bar{X}_n - \mu| \geq \epsilon) \to 0$.

6.30. (a) Using the hint, for any u and $v > 0$ we have

$$u^2 v_{k-1} + 2uv v_k + v^2 v_{k+1} \geq 0.$$

Since $v_{k-1} \geq 0$ for $k \geq 1$, we must therefore have

$$v^2 v_k^2 - v^2 v_{k-1} v_{k+1} \leq 0, \qquad k \geq 1,$$

that is,

$$v_k^2 \leq v_{k-1} v_{k+1}, \qquad k = 1, 2, \ldots, n-1.$$

(b) $v_0 = 1$. For $k = 1$, $v_1^2 \leq v_2$. Hence, $v_1 \leq v_2^{1/2}$. Let us now use mathematical induction to show that $v_n^{1/n} \leq v_{n+1}^{1/(n+1)}$ for all n for which v_n and v_{n+1} exist. The statement is true for $n = 1$. Suppose that

$$v_{n-1}^{1/(n-1)} \leq v_n^{1/n}.$$

To show that $v_n^{1/n} \leq v_{n+1}^{1/(n+1)}$: We have that

$$v_n^2 \leq v_{n-1} v_{n+1}$$

$$\leq v_n^{(n-1)/n} v_{n+1}.$$

Thus,

$$v_n^{(n+1)/n} \leq v_{n+1}.$$

Hence,

$$v_n^{1/n} \leq v_{n+1}^{1/(n+1)}.$$

CHAPTER 7

7.1. (a) Along $x_1 = 0$, we have $f(0, x_2) = 0$ for all x_2, and the limit is zero as $\mathbf{x} \to \mathbf{0}$. Any line through the origin can be represented by the equation $x_1 = tx_2$. Along this line, we have

$$f(tx_2, x_2) = \frac{|tx_2|}{x_2^2} \exp\left(-\frac{|t||x_2|}{x_2^2}\right)$$

$$= \frac{|t|}{|x_2|} \exp\left(-\frac{|t|}{|x_2|}\right), \qquad x_2 \neq 0.$$

Using l'Hospital's rule (Theorem 4.2.6), $f(tx_2, x_2) \to 0$ as $x_2 \to 0$.

(b) Along the parabola $x_1 = x_2^2$, we have

$$f(x_1, x_2) = \frac{x_2^2}{x_2^2} \exp\left(-\frac{x_2^2}{x_2^2}\right)$$

$$= e^{-1}, \qquad x_2 \neq 0,$$

which does not go to zero as $x_2 \to 0$. By Definition 7.2.1, $f(x_1, x_2)$ does not have a limit as $\mathbf{x} \to \mathbf{0}$.

7.5. (a) This function has no limit as $x \to 0$, as was shown in Example 7.2.2. Hence, it is not continuous at the origin.

(b)

$$\left.\frac{\partial f(x_1, x_2)}{\partial x_1}\right|_{x=0} = \lim_{\Delta x_1 \to 0}\left\{\frac{1}{\Delta x_1}\left[\frac{0\Delta x_1}{(\Delta x_1)^2 + 0} - 0\right]\right\} = 0,$$

$$\left.\frac{\partial f(x_1, x_2)}{\partial x_2}\right|_{x=0} = \lim_{\Delta x_2 \to 0}\left\{\frac{1}{\Delta x_2}\left[\frac{0\Delta x_2}{0 + (\Delta x_2)^2} - 0\right]\right\} = 0.$$

7.6. Applying formula (7.12), we get

$$\frac{df}{dt} = \sum_{i=1}^{k} x_i \frac{\partial f(\mathbf{u})}{\partial u_i} = nt^{n-1}f(\mathbf{x}),$$

where u_i is the ith element of $\mathbf{u} = t\mathbf{x}$, $i = 1, 2, \ldots, k$. But, on one hand,

$$\frac{\partial f(\mathbf{u})}{\partial x_i} = \sum_{j=1}^{k} \frac{\partial f(\mathbf{u})}{\partial u_j} \frac{\partial u_j}{\partial x_i}$$

$$= t \frac{\partial f(\mathbf{u})}{\partial u_i},$$

and on the other hand,

$$\frac{\partial f(\mathbf{u})}{\partial x_i} = t^n \frac{\partial f(\mathbf{x})}{\partial x_i}.$$

Hence,

$$\frac{\partial f(\mathbf{u})}{\partial u_i} = t^{n-1} \frac{\partial f(\mathbf{x})}{\partial x_i}.$$

It follows that

$$\sum_{i=1}^{k} x_i \frac{\partial f(\mathbf{u})}{\partial u_i} = nt^{n-1}f(\mathbf{x})$$

can be written as

$$t^{n-1} \sum_{i=1}^{k} x_i \frac{\partial f(\mathbf{x})}{\partial x_i} = n t^{n-1} f(\mathbf{x}),$$

that is,

$$\sum_{i=1}^{k} x_i \frac{\partial f(\mathbf{x})}{\partial x_i} = n f(\mathbf{x}).$$

7.7. (a) $f(x_1, x_2)$ is not continuous at the origin, since along $x_1 = 0$ and $x_2 \neq 0$, $f(x_1, x_2) = 0$, which has a limit equal to zero as $x_2 \to 0$. But, along the parabola $x_2 = x_1^2$,

$$f(x_1, x_2) = \frac{x_1^4}{x_1^4 + x_1^4} = \frac{1}{2} \qquad \text{if } x_1 \neq 0.$$

Hence, $\lim_{x_1 \to 0} f(x_1, x_1^2) = \frac{1}{2} \neq 0$.

(b) The directional derivative in the direction of the unit vector $\mathbf{v} = (v_1, v_2)'$ at the origin is given by

$$v_1 \frac{\partial f(\mathbf{x})}{\partial x_1}\bigg|_{\mathbf{x}=\mathbf{0}} + v_2 \frac{\partial f(\mathbf{x})}{\partial x_2}\bigg|_{\mathbf{x}=\mathbf{0}}.$$

This derivative is equal to zero, since

$$\frac{\partial f(\mathbf{x})}{\partial x_1}\bigg|_{\mathbf{x}=\mathbf{0}} = \lim_{\Delta x_1 \to 0} \frac{1}{\Delta x_1} \left[\frac{(\Delta x_1)^2 0}{(\Delta x_1)^4 + 0} - 0 \right] = 0,$$

$$\frac{\partial f(\mathbf{x})}{\partial x_2}\bigg|_{\mathbf{x}=\mathbf{0}} = \lim_{\Delta x_2 \to 0} \frac{1}{\Delta x_2} \left[\frac{0 \Delta x_2}{0 + (\Delta x_2)^2} - 0 \right] = 0.$$

7.8. The directional derivative of f at a point $\mathbf{x}$ on C in the direction of $\mathbf{v}$ is $\sum_{i=1}^{k} v_i \, \partial f(\mathbf{x})/\partial x_i$. But $\mathbf{v}$ is given by

$$\mathbf{v} = \frac{d\mathbf{x}}{dt} \bigg/ \left\| \frac{d\mathbf{x}}{dt} \right\|$$

$$= \frac{d\mathbf{x}}{dt} \bigg/ \left\{ \sum_{i=1}^{k} \left[\frac{dg_i(t)}{dt} \right]^2 \right\}^{1/2}$$

$$= \frac{d\mathbf{x}}{dt} \bigg/ \left\| \frac{d\mathbf{g}}{dt} \right\|,$$

where **g** is the vector whose ith element is $g_i(t)$. Hence,

$$\sum_{i=1}^{k} v_i \frac{\partial f(\mathbf{x})}{\partial x_i} = \frac{1}{\|d\mathbf{g}/dt\|} \sum_{i=1}^{k} \frac{dg_i(t)}{dt} \frac{\partial f(\mathbf{x})}{\partial x_i}.$$

Furthermore,

$$\frac{ds}{dt} = \left\| \frac{d\mathbf{g}}{dt} \right\|.$$

It follows that

$$\sum_{i=1}^{k} v_i \frac{\partial f(\mathbf{x})}{\partial x_i} = \frac{dt}{ds} \sum_{i=1}^{k} \frac{dg_i(t)}{dt} \frac{\partial f(\mathbf{x})}{\partial x_i}$$

$$= \sum_{i=1}^{k} \frac{dx_i}{ds} \frac{\partial f(\mathbf{x})}{\partial x_i}$$

$$= \frac{df(\mathbf{x})}{ds}.$$

7.9. (a)

$$f(x_1, x_2) \approx f(0,0) + \left(x_1 \frac{\partial}{\partial x_1} + x_2 \frac{\partial}{\partial x_2} \right) f(0,0)$$

$$+ \frac{1}{2!} \left(x_1^2 \frac{\partial^2}{\partial x_1^2} + 2x_1 x_2 \frac{\partial^2}{\partial x_1 \partial x_2} + x_2^2 \frac{\partial^2}{\partial x_2^2} \right) f(0,0)$$

$$= 1 + x_1 x_2.$$

(b)

$$f(x_1, x_2, x_3) \approx f(0,0,0) + \left(\sum_{i=1}^{3} x_i \frac{\partial}{\partial x_i} \right) f(0,0,0)$$

$$+ \frac{1}{2!} \left(\sum_{i=1}^{3} x_i \frac{\partial}{\partial x_i} \right)^2 f(0,0,0)$$

$$= \sin(1) + x_1 \cos(1) + \frac{1}{2!} \{ x_1^2 [\cos(1) - \sin(1)] + 2x_2^2 \cos(1) \}.$$

(c)

$$f(x_1, x_2) \approx f(0,0) + \left(x_1 \frac{\partial}{\partial x_1} + x_2 \frac{\partial}{\partial x_2} \right) f(0,0)$$

$$+ \frac{1}{2!} \left(x_1^2 \frac{\partial^2}{\partial x_1^2} + 2x_1 x_2 \frac{\partial^2}{\partial x_1 \partial x_2} + x_2^2 \frac{\partial^2}{\partial x_2^2} \right) f(0,0)$$

$$= 1,$$

since all first-order and second-order partial derivatives vanish at $\mathbf{x} = \mathbf{0}$.

7.10. (a) If $u_1 = f(x_1, x_2)$ and $\partial f / \partial x_1 \neq 0$ at $\mathbf{x}_0$, then by the implicit function theorem (Theorem 7.6.2), there is neighborhood of $\mathbf{x}_0$ in which the equation $u_1 = f(x_1, x_2)$ can be solved uniquely for x_1 in terms of x_2 and u_1, that is, $x_1 = h(u_1, x_2)$. Thus,

$$f[h(u_1, x_2), x_2] \equiv u_1.$$

Hence, by differentiating this identity with respect to x_2, we get

$$\frac{\partial f}{\partial x_1} \frac{\partial h}{\partial x_2} + \frac{\partial f}{\partial x_2} = 0,$$

which gives

$$\frac{\partial h}{\partial x_2} = -\frac{\partial f}{\partial x_2} \bigg/ \frac{\partial f}{\partial x_1}$$

in a neighborhood of $\mathbf{x}_0$.

(b) On the basis of part (a), we can consider $g(x_1, x_2)$ as a function of x_2 and u, since $x_1 = h(u_1, x_2)$ in a neighborhood of $\mathbf{x}_0$. In such a neighborhood, the partial derivative of g with respect to x_2 is

$$\frac{\partial g}{\partial x_1} \frac{\partial h}{\partial x_2} + \frac{\partial g}{\partial x_2} = \frac{\partial(f, g)}{\partial(x_1, x_2)} \bigg/ \frac{\partial f}{\partial x_1},$$

which is equal to zero because $\partial(f, g)/\partial(x_1, x_2) = 0$. [Recall that $\partial h / \partial x_2 = -(\partial f / \partial x_2)(\partial h / \partial x_1)$].

(c) From part (b), $g[h(u_1, x_2), x_2]$ is independent of x_2 in a neighborhood of $\mathbf{x}_0$. We can then write $\phi(u_1) = g[h(u_1, x_2), x_2]$. Since $x_1 = h(u_1, x_2)$ is equivalent to $u_1 = f(x_1, x_2)$ in a neighborhood of $\mathbf{x}_0$, $\phi[f(x_1, x_2)] = g(x_1, x_2)$ in this neighborhood.

(d) Let $G(f, g) = g(x_1, x_2) - \phi[f(x_1, x_2)]$. Then, in a neighborhood of $\mathbf{x}_0$, $G(f, g) = 0$. Hence,

$$\frac{\partial G}{\partial f} \frac{\partial f}{\partial x_1} + \frac{\partial G}{\partial g} \frac{\partial g}{\partial x_1} \equiv 0,$$

$$\frac{\partial G}{\partial f} \frac{\partial f}{\partial x_2} + \frac{\partial G}{\partial g} \frac{\partial g}{\partial x_2} \equiv 0.$$

In order for these two identities to be satisfied by values of $\partial G / \partial f$, $\partial G / \partial g$ not all zero, it is necessary that $\partial(f, g)/\partial(x_1, x_2)$ be equal to zero in this neighborhood of $\mathbf{x}_0$.

7.11. $u(x_1, x_2, x_3) = u(\xi_1\xi_3, \xi_2\xi_3, \xi_3)$. Hence,

$$\frac{\partial u}{\partial \xi_3} = \frac{\partial u}{\partial x_1}\frac{\partial x_1}{\partial \xi_3} + \frac{\partial u}{\partial x_2}\frac{\partial x_2}{\partial \xi_3} + \frac{\partial u}{\partial x_3}\frac{\partial x_3}{\partial \xi_3}$$

$$= \xi_1\frac{\partial u}{\partial x_1} + \xi_2\frac{\partial u}{\partial x_2} + \frac{\partial u}{\partial x_3},$$

so that

$$\xi_3\frac{\partial u}{\partial \xi_3} = \xi_1\xi_3\frac{\partial u}{\partial x_1} + \xi_2\xi_3\frac{\partial u}{\partial x_2} + \xi_3\frac{\partial u}{\partial x_3}$$

$$= x_1\frac{\partial u}{\partial x_1} + x_2\frac{\partial u}{\partial x_2} + x_3\frac{\partial u}{\partial x_3} = nu.$$

Integrating this partial differential equation with respect to ξ_3, we get

$$\log u = n\log \xi_3 + \psi(\xi_1, \xi_2),$$

where $\psi(\xi_1, \xi_2)$ is a function of ξ_1, ξ_2. Hence,

$$u = \xi_3^n\exp[\psi(\xi_1, \xi_2)]$$
$$= x_3^n F(\xi_1, \xi_2)$$
$$= x_3^n F\left(\frac{x_1}{x_3}, \frac{x_2}{x_3}\right),$$

where $F(\xi_1, \xi_2) = \exp[\psi(\xi_1, \xi_2)]$.

7.13. (a) The Jacobian determinant is $27x_1^2x_2^2x_3^2$, which is zero in any subset of R^3 that contains points on any of the coordinate planes.

(b) The unique inverse function is given by

$$x_1 = u_1^{1/3}, \qquad x_2 = u_2^{1/3}, \qquad x_3 = u_3^{1/3}.$$

7.14. Since $\partial(g_1, g_2)/\partial(x_1, x_2) \neq 0$, we can solve for x_1, x_2 uniquely in terms of y_1 and y_2 by Theorem 7.6.2. Differentiating g_1 and g_2 with respect to y_1, we obtain

$$\frac{\partial g_1}{\partial x_1}\frac{\partial x_1}{\partial y_1} + \frac{\partial g_1}{\partial x_2}\frac{\partial x_2}{\partial y_1} + \frac{\partial g_1}{\partial y_1} = 0,$$

$$\frac{\partial g_2}{\partial x_1}\frac{\partial x_1}{\partial y_1} + \frac{\partial g_2}{\partial x_2}\frac{\partial x_2}{\partial y_1} + \frac{\partial g_2}{\partial y_1} = 0,$$

Solving for $\partial x_1/\partial y_1$ and $\partial x_2/\partial y_1$ yields the desired result.

7.15. This is similar to Exercise 7.14. *Since $\partial(f, g)/\partial(x_1, x_2) \neq 0$, we can solve for x_1, x_2 in terms of x_3, and we get

$$\frac{dx_1}{dx_3} = -\frac{\partial(f, g)}{\partial(x_3, x_2)} \bigg/ \frac{\partial(f, g)}{\partial(x_1, x_2)},$$

$$\frac{dx_2}{dx_3} = -\frac{\partial(f, g)}{\partial(x_1, x_3)} \bigg/ \frac{\partial(f, g)}{\partial(x_1, x_2)}.$$

But

$$\frac{\partial(f, g)}{\partial(x_3, x_2)} = -\frac{\partial(f, g)}{\partial(x_2, x_3)}, \quad \frac{\partial(f, g)}{\partial(x_1, x_3)} = -\frac{\partial(f, g)}{\partial(x_3, x_1)}.$$

Hence,

$$\frac{dx_1}{\partial(f, g)/\partial(x_2, x_3)} = \frac{dx_3}{\partial(f, g)/\partial(x_1, x_2)} = \frac{dx_2}{\partial(f, g)/\partial(x_3, x_1)}.$$

7.16. (a) $(-\frac{1}{3}, -\frac{1}{3})$ is a point of local minimum.
(b)

$$\frac{\partial f}{\partial x_1} = 4\alpha x_1 - x_2 + 1 = 0,$$

$$\frac{\partial f}{\partial x_2} = -x_1 + 2x_2 - 1 = 0.$$

The solution of these two equations is $x_1 = 1/(1 - 8\alpha)$, $x_2 = (4\alpha - 1)/(8\alpha - 1)$. The Hessian matrix of f is

$$A = \begin{bmatrix} 4\alpha & -1 \\ -1 & 2 \end{bmatrix}.$$

Here, $f_{11} = 4\alpha$, and $\det(A) = 8\alpha - 1$.
  i. Yes, if $\alpha > \frac{1}{8}$.
  ii. No, it is not possible.
  iii. Yes, if $\alpha < \frac{1}{8}$, $\alpha \neq 0$.
(c) The stationary points are $(-2, -2)$, $(4, 4)$. The first is a saddle point, and the second is a point of local minimum.
(d) The stationary points are $(\sqrt{2}, -\sqrt{2})$, $(-\sqrt{2}, \sqrt{2})$, and $(0, 0)$. The first two are points of local minimum. At $(0, 0)$, $f_{11}f_{22} - f_{12}^2 = 0$. In

this case, $\mathbf{h'Ah}$ has the same sign as $f_{11}|_{(0,0)} = -4$ for all values of $\mathbf{h} = (h_1, h_2)'$, except when

$$h_1 + h_2 \frac{f_{12}}{f_{11}} = 0$$

or $h_1 - h_2 = 0$, in which case $\mathbf{h'Ah} = 0$. For such values of $\mathbf{h}$, $(\mathbf{h' \nabla})^3 f(0,0) = 0$, but $(\mathbf{h' \nabla})^4 f(0,0) = 24(h_1^4 + h_2^4) = 48h_1^4$, which is nonnegative. Hence, the point $(0,0)$ is a saddlepoint.

7.19. The Lagrange equations are

$$(2 + 8\lambda)x_1 + 12x_2 = 0,$$

$$12x_1 + (4 + 2\lambda)x_2 = 0,$$

$$4x_1^2 + x_2^2 = 25.$$

We must have

$$(4\lambda + 1)(\lambda + 2) - 36 = 0.$$

The solutions are $\lambda_1 = -4.25$, $\lambda_2 = 2$. For $\lambda_1 = -4.25$, we have the points
    i. $x_1 = 1.5$, $x_2 = 4.0$,
    ii. $x_1 = -1.5$, $x_2 = -4.0$.
For $\lambda_2 = 2$, we have the points
    iii. $x_1 = 2$, $x_2 = -3$,
    iv. $x_1 = -2$, $x_2 = 3$.
The matrix $\mathbf{B}_1$ has the value

$$\mathbf{B}_1 = \begin{bmatrix} 2 + 8\lambda & 12 & 8x_1 \\ 12 & 4 + 2\lambda & 2x_2 \\ 8x_1 & 2x_2 & 0 \end{bmatrix}.$$

The determinant of $\mathbf{B}_1$ is Δ_1.
    At (i), $\Delta_1 = 5000$; the point is a local maximum.
    At (ii), $\Delta_1 = 5000$; the point is a local maximum.
    At (iii) and (iv), $\Delta_1 = -5000$; the points are local minima.

7.21. Let $F = x_1^2 x_2^2 x_3^2 + \lambda(x_1^2 + x_2^2 + x_3^2 - c^2)$. Then

$$\frac{\partial F}{\partial x_1} = 2x_1 \, x_2^2 x_3^2 + 2\lambda x_1 = 0,$$

$$\frac{\partial F}{\partial x_2} = 2x_1^2 x_2 x_3^2 + 2\lambda x_2 = 0,$$

$$\frac{\partial F}{\partial x_3} = 2x_1^2 x_2^2 x_3 + 2\lambda x_3 = 0,$$

$$x_1^2 + x_2^2 + x_3^2 = c^2.$$

Since x_1, x_2, x_3 cannot be equal to zero, we must have

$$x_2^2 x_3^2 + \lambda = 0,$$

$$x_1^2 x_3^2 + \lambda = 0,$$

$$x_1^2 x_2^2 + \lambda = 0.$$

These equations imply that $x_1^2 = x_2^2 = x_3^2 = c^2/3$ and $\lambda = -c^4/9$. For these values, it can be verified that $\Delta_1 < 0$, $\Delta_2 > 0$, where Δ_1 and Δ_2 are the determinants of $\mathbf{B}_1$ and $\mathbf{B}_2$, and

$$\mathbf{B}_1 = \begin{bmatrix} 2x_2^2 x_3^2 + 2\lambda & 4x_1 x_2 x_3^2 & 4x_1 x_2^2 x_3 & 2x_1 \\ 4x_1 x_2 x_3^2 & 2x_1^2 x_3^2 + 2\lambda & 4x_1^2 x_2 x_3 & 2x_2 \\ 4x_1 x_2^2 x_3 & 4x_1^2 x_2 x_3 & 2x_1^2 x_2^2 + 2\lambda & 2x_3 \\ 2x_1 & 2x_2 & 2x_3 & 0 \end{bmatrix},$$

$$\mathbf{B}_2 = \begin{bmatrix} 2x_1^2 x_3^2 + 2\lambda & 4x_1^2 x_2 x_3 & 2x_2 \\ 4x_1^2 x_2 x_3 & 2x_1^2 x_2^2 + 2\lambda & 2x_3 \\ 2x_2 & 2x_3 & 0 \end{bmatrix}.$$

Since $n = 3$ is odd, the function $x_1^2 x_2^2 x_3^2$ must attain a maximum value given by $(c^2/3)^3$. It follows that for all values of x_1, x_2, x_3,

$$x_1^2 x_2^2 x_3^2 \le \left(\frac{c^2}{3}\right)^3,$$

that is,

$$\left(x_1^2 x_2^2 x_3^2\right)^{1/3} \le \frac{x_1^2 + x_2^2 + x_3^2}{3}.$$

7.24. The domain of integration, D, is the region bounded from above by the parabola $x_2 = 4x_1 - x_1^2$ and from below by the parabola $x_2 = x_1^2$. The two parabolas intersect at $(0,0)$ and $(2,4)$. It is then easy to see that

$$\int_0^2 \left[\int_{x_1^2}^{4x_1 - x_1^2} f(x_1, x_2) \, dx_2 \right] dx_1 = \int_0^4 \left[\int_{2 - \sqrt{4 - x_2}}^{\sqrt{x_2}} f(x_1, x_2) \, dx_1 \right] dx_2.$$

7.25. (a) $I = \int_0^1 \left[\int_{1 - x_2}^{\sqrt{1 - x_2}} f(x_1, x_2) \, dx_1 \right] dx_2.$

(b) $dg/dx_1 = \int_{1 - x_1}^{1 - x_1^2} \dfrac{\partial f(x_1, x_2)}{\partial x_1} \, dx_2 - 2x_1 f(x_1, 1 - x_1^2) + f(x_1, 1 - x_1).$

7.26. Let $u = x_2^2/x_1$, $v = x_1^2/x_2$. Then $uv = x_1 x_2$, and

$$\iint_D x_1 x_2 \, dx_1 \, dx_2 = \int_1^2 \left[\int_1^2 uv \left| \frac{\partial(x_1, x_2)}{\partial(u, v)} \right| du \right] dv.$$

But

$$\left| \frac{\partial(x_1, x_2)}{\partial(u, v)} \right| = \left| \frac{\partial(u, v)}{\partial(x_1, x_2)} \right|^{-1} = \frac{1}{3}.$$

Hence,

$$\iint_D x_1 x_2 \, dx_1 \, dx_2 = \frac{3}{4}.$$

7.28. Let $I(a) = \int_0^{\sqrt{3}} dx/(a + x^2)$. Then

$$\frac{d^2 I(a)}{da^2} = 2 \int_0^{\sqrt{3}} \frac{dx}{(a + x^2)^3}.$$

On the other hand,

$$I(a) = \frac{1}{\sqrt{a}} \operatorname{Arctan} \sqrt{\frac{3}{a}}$$

$$\frac{d^2 I(a)}{da^2} = \frac{3}{4} a^{-5/2} \operatorname{Arctan} \sqrt{\frac{3}{a}} + \frac{\sqrt{3}}{4} \frac{1}{a^2(a + 3)}$$

$$+ \frac{\sqrt{3}}{2} \frac{(2a + 3)}{a^2(a + 3)^2}.$$

Thus

$$\int_0^{\sqrt{3}} \frac{dx}{(a+x^2)^3} = \frac{3}{8} a^{-5/2} \text{Arctan} \sqrt{\frac{3}{a}} + \frac{\sqrt{3}}{8} \frac{1}{a^2(a+3)}$$

$$+ \frac{\sqrt{3}}{4} \frac{(2a+3)}{a^2(a+3)^2}.$$

Putting $a = 1$, we obtain

$$\int_0^{\sqrt{3}} \frac{dx}{(1+x^2)^3} = \frac{3}{8} \text{Arctan} \sqrt{3} + \frac{7\sqrt{3}}{64}.$$

7.29. (a) The marginal density functions of X_1 and X_2 are

$$f_1(x_1) = \int_0^1 (x_1 + x_2) \, dx_2$$

$$= x_1 + \tfrac{1}{2}, \qquad 0 < x_1 < 1,$$

$$f_2(x_2) = \int_0^1 (x_1 + x_2) \, dx_1$$

$$= x_2 + \tfrac{1}{2}, \qquad 0 < x_2 < 1,$$

$$f(x_1, x_2) \neq f_1(x_1) f_2(x_2).$$

The random variables X_1 and X_2 are therefore not independent.

(b)

$$E(X_1 X_2) = \int_0^1 \int_0^1 x_1 x_2 (x_1 + x_2) \, dx_1 \, dx_2$$

$$= \tfrac{1}{3}.$$

7.30. The marginal densities of X_1 and X_2 are

$$f_1(x_1) = \begin{cases} 1 + x_1, & -1 < x_1 < 0, \\ 1 - x_1, & 0 < x_1 < 1, \\ 0, & \text{otherwise}, \end{cases}$$

$$f_2(x_2) = 2x_2, \qquad 0 < x_2 < 1.$$

Note that $f(x_1, x_2) \neq f_1(x_1)f_2(x_2)$. Hence, X_1 and X_2 are not independent. But

$$E(X_1 X_2) = \int_0^1 \left[\int_{-x_2}^{x_2} x_1 x_2 \, dx_1 \right] dx_2 = 0,$$

$$E(X_1) = \int_{-1}^0 x_1(1 + x_1) \, dx_1 + \int_0^1 x_1(1 - x_1) \, dx_1 = 0,$$

$$E(X_2) = \int_0^1 2x_2^2 \, dx_2 = \tfrac{2}{3}.$$

Hence,

$$E(X_1 X_2) = E(X_1)E(X_2) = 0.$$

7.31. **(a)** The joint density function of Y_1 and Y_2 is

$$g(y_1, y_2) = \frac{1}{\Gamma(\alpha)\Gamma(\beta)} y_1^{\alpha-1}(1 - y_1)^{\beta-1} y_2^{\alpha+\beta-1} e^{-y_2}.$$

(b) The marginal densities of Y_1 and Y_2 are, respectively,

$$g_1(y_1) = \frac{\Gamma(\alpha + \beta)}{\Gamma(\alpha)\Gamma(\beta)} y_1^{\alpha-1}(1 - y_1)^{\beta-1}, \qquad 0 < y_1 < 1,$$

$$g_2(y_2) = \frac{B(\alpha, \beta)}{\Gamma(\alpha)\Gamma(\beta)} y_2^{\alpha+\beta-1} e^{-y_2}, \qquad 0 < y_2 < \infty.$$

(c) Since $B(\alpha, \beta) = \Gamma(\alpha)\Gamma(\beta)/\Gamma(\alpha + \beta)$,

$$g(y_1, y_2) = g_1(y_1)g_2(y_2),$$

and Y_1 and Y_2 are therefore independent.

7.32. Let $U = X_1$, $W = X_1 X_2$. Then, $X_1 = U$, $X_2 = W/U$. The joint density function of U and W is

$$g(u, w) = \frac{10w^2}{u^2}, \qquad w < u < \sqrt{w}, \quad 0 < w < 1.$$

The marginal density function of W is

$$g_1(w) = 10w^2 \int_w^{\sqrt{w}} \frac{du}{u^2}$$

$$= 10w^2 \left(\frac{1}{w} - \frac{1}{\sqrt{w}} \right), \qquad 0 < w < 1.$$

7.34. The inverse of this transformation is

$$X_1 = Y_1,$$

$$X_2 = Y_2 - Y_1,$$

$$\vdots$$

$$X_n = Y_n - Y_{n-1},$$

and the absolute value of the Jacobian determinant of this transformation is 1. Therefore, the joint density function of the Y_i's is

$$g(y_1, y_2, \ldots, y_n) = e^{-y_1} e^{-(y_2 - y_1)} \cdots e^{-(y_n - y_{n-1})}$$

$$= e^{-y_n}, \qquad 0 < y_1 < y_2 < \cdots < y_n < \infty.$$

The marginal density function of Y_n is

$$g_n(y_n) = e^{-y_n} \int_0^{y_n} \int_0^{y_{n-1}} \cdots \int_0^{y_2} dy_1 \, dy_2 \cdots dy_{n-1}$$

$$= \frac{y_n^{n-1}}{(n-1)!} e^{-y_n}, \qquad 0 < y_n < \infty.$$

7.37. Let $\boldsymbol{\beta} = (\beta_0, \beta_1, \beta_2)'$. The least-squares estimate of $\boldsymbol{\beta}$ is given by

$$\hat{\boldsymbol{\beta}} = (\mathbf{X}'\mathbf{X})^{-1} \mathbf{X}' \mathbf{y},$$

where $\mathbf{y} = (y_1, y_2, \ldots, y_n)'$, and $\mathbf{X}$ is a matrix of order $n \times 3$ whose first column is the column of ones, and whose second and third columns are given by the values of x_i, x_i^2, $i = 1, 2, \ldots, n$.

7.38. Maximizing $L(\mathbf{x}, \mathbf{p})$ is equivalent to maximizing its natural logarithm. Let us therefore consider maximizing $\log L$ subject to $\sum_{i=1}^k p_i = 1$. Using the method of Lagrange multipliers, let

$$F = \log L + \lambda \left(\sum_{i=1}^k p_i - 1 \right).$$

Differentiating F with respect to $p_1, p_2, \ldots, p_k$, and equating the derivatives to zero, we obtain, $x_i/p_i + \lambda = 0$ for $i = 1, 2, \ldots, k$. Combining these equations with the constraint $\sum_{i=1}^{k} p_i = 1$, we obtain the maximum likelihood estimates

$$\hat{p}_i = \frac{x_i}{n}, \qquad i = 1, 2, \ldots, k.$$

CHAPTER 8

8.1. Let $\mathbf{x}_i = (x_{i1}, x_{i2})'$, $\mathbf{h}_i = (h_{i1}, h_{i2})'$, $i = 0, 1, 2, \ldots$. Then

$$f(\mathbf{x}_i + t\mathbf{h}_i) = 8(x_{i1} + th_{i1})^2 - 4(x_{i1} + th_{i1})(x_{i2} + th_{i2}) + 5(x_{i2} + th_{i2})^2$$

$$= a_i t^2 + b_i t + c_i,$$

where

$$a_i = 8h_{i1}^2 - 4h_{i1}h_{i2} + 5h_{i2}^2,$$

$$b_i = 16x_{i1}h_{i1} - 4(x_{i1}h_{i2} + x_{i2}h_{i1}) + 10x_{i2}h_{i2},$$

$$c_i = 8x_{i1}^2 - 4x_{i1}x_{i2} + 5x_{i2}^2.$$

The value of t that minimizes $f(\mathbf{x}_i + t\mathbf{h}_i)$ in the direction of $\mathbf{h}_i$ is given by the solution of $\partial f/\partial t = 0$, namely, $t_i = -b_i/(2a_i)$. Hence,

$$\mathbf{x}_{i+1} = \mathbf{x}_i - \frac{b_i}{2a_i}\mathbf{h}_i, \qquad i = 0, 1, 2, \ldots,$$

where

$$\mathbf{h}_i = -\frac{\nabla f(\mathbf{x}_i)}{\|\nabla f(\mathbf{x}_i)\|_2}, \qquad i = 0, 1, 2, \ldots,$$

and

$$\nabla f(\mathbf{x}_i) = (16x_{i1} - 4x_{i2}, -4x_{i1} + 10x_{i2})', \qquad i = 0, 1, 2, \ldots.$$

The results of the iterative minimization procedure are given in the following table:

Iteration i	$\mathbf{x}_i$	$\mathbf{h}_i$	t_i	a_i	$f(\mathbf{x}_i)$
0	$(5, 2)'$	$(-1, 0)'$	4.5	8	180
1	$(0.5, 2)'$	$(0, -1)'$	1.8	5	18
2	$(0.5, 0.2)'$	$(-1, 0)'$	0.45	8	1.8
3	$(0.05, 0.2)'$	$(0, -1)'$	0.18	5	0.18
4	$(0.05, 0.02)'$	$(-1, 0)'$	0.045	8	0.018
5	$(0.005, 0.02)'$	$(0, -1)'$	0.018	5	0.0018
6	$(0.005, 0.002)'$	$(-1, 0)'$	0.0045	8	0.00018

Note that $a_i > 0$, which confirms that $t_i = -b_i/(2a_i)$ does indeed minimize $f(\mathbf{x}_i + t\mathbf{h}_i)$ in the direction of $\mathbf{h}_i$. It is obvious that if we were to continue with this iterative procedure, the point $\mathbf{x}_i$ would converge to $\mathbf{0}$.

8.3. (a)

$$\hat{y}(\mathbf{x}) = 45.690 + 4.919x_1 + 8.847x_2$$

$$- 0.270x_1x_2 - 4.148x_1^2 - 4.298x_2^2.$$

(b) The matrix $\hat{\mathbf{B}}$ is

$$\hat{\mathbf{B}} = \begin{bmatrix} -4.148 & -0.135 \\ -0.135 & -4.298 \end{bmatrix}.$$

Its eigenvalues are $\tau_1 = -8.136$, $\tau_2 = -8.754$. This makes $\hat{\mathbf{B}}$ a negative definite matrix. The results of applying the method of ridge analysis inside the region R are given in the following table:

λ	x_1	x_2	r	$\hat{y}$
31.4126	0.0687	0.1236	0.1414	47.0340
13.5236	0.1373	0.2473	0.2829	48.2030
7.5549	0.2059	0.3709	0.4242	49.1969
4.5694	0.2745	0.4946	0.5657	50.0158
2.7797	0.3430	0.6184	0.7072	50.6595
1.5873	0.4114	0.7421	0.8485	51.1282
0.7359	0.4797	0.8659	0.9899	51.4218
0.0967	0.5480	0.9898	1.1314	51.5404
-0.4009	0.6163	1.1137	1.2729	51.4839
-0.7982	0.6844	1.2376	1.4142	51.2523

Note that the stationary point, $\mathbf{x}_0 = (0.5601, 1.0117)'$, is a point of absolute maximum since the eigenvalues of $\hat{\mathbf{B}}$ are negative. This point falls inside R. The corresponding maximum value of $\hat{y}$ is 51.5431.

8.4. We know that

$$\left(\hat{\mathbf{B}} - \lambda_1 \mathbf{I}_k\right)\mathbf{x}_1 = -\tfrac{1}{2}\hat{\boldsymbol{\beta}},$$

$$\left(\hat{\mathbf{B}} - \lambda_2 \mathbf{I}_k\right)\mathbf{x}_2 = -\tfrac{1}{2}\hat{\boldsymbol{\beta}},$$

where $\mathbf{x}_1$ and $\mathbf{x}_2$ correspond to λ_1 and λ_2, respectively, and are such that $\mathbf{x}_1'\mathbf{x}_1 = \mathbf{x}_2'\mathbf{x}_2 = r^2$, with r^2 being the common value of r_1^2 and r_2^2. The corresponding values of $\hat{y}$ are

$$\hat{y}_1 = \hat{\beta}_0 + \mathbf{x}_1'\hat{\boldsymbol{\beta}} + \mathbf{x}_1'\hat{\mathbf{B}}\mathbf{x}_1,$$

$$\hat{y}_2 = \hat{\beta}_0 + \mathbf{x}_2'\hat{\boldsymbol{\beta}} + \mathbf{x}_2'\hat{\mathbf{B}}\mathbf{x}_2.$$

We then have

$$\mathbf{x}_1'\hat{\mathbf{B}}\mathbf{x}_1 - \mathbf{x}_2'\hat{\mathbf{B}}\mathbf{x}_2 + \tfrac{1}{2}(\mathbf{x}_1' - \mathbf{x}_2')\hat{\boldsymbol{\beta}} = (\lambda_1 - \lambda_2)r^2,$$

$$\hat{y}_1 - \hat{y}_2 = \tfrac{1}{2}(\mathbf{x}_1' - \mathbf{x}_2')\hat{\boldsymbol{\beta}} + (\lambda_1 - \lambda_2)r^2.$$

Furthermore, from the equations defining $\mathbf{x}_1$ and $\mathbf{x}_2$, we have

$$(\lambda_2 - \lambda_1)\mathbf{x}_1'\mathbf{x}_2 = \tfrac{1}{2}(\mathbf{x}_1' - \mathbf{x}_2')\hat{\boldsymbol{\beta}}.$$

Hence,

$$\hat{y}_1 - \hat{y}_2 = (\lambda_1 - \lambda_2)\left(r^2 - \mathbf{x}_1'\mathbf{x}_2\right).$$

But $r^2 = \|\mathbf{x}_1\|_2 \|\mathbf{x}_2\|_2 > \mathbf{x}_1'\mathbf{x}_2$, since $\mathbf{x}_1$ and $\mathbf{x}_2$ are not parallel vectors. We conclude that $\hat{y}_1 > \hat{y}_2$ whenever $\lambda_1 > \lambda_2$.

8.5. If $\mathbf{M}(\mathbf{x}_1)$ is positive definite, then λ_1 must be smaller than all the eigenvalues of $\hat{\mathbf{B}}$ (this is based on applying the spectral decomposition theorem to $\hat{\mathbf{B}}$ and using Theorem 2.3.12). Similarly, if $\mathbf{M}(\mathbf{x}_2)$ is indefinite, then λ_2 must be larger than the smallest eigenvalue of $\hat{\mathbf{B}}$ and also smaller than its largest eigenvalue. It follows that $\lambda_1 < \lambda_2$. Hence, by Exercise 8.4, $\hat{y}_1 < \hat{y}_2$.

8.6. (a) We know that

$$\left(\hat{\mathbf{B}} - \lambda \mathbf{I}_k\right)\mathbf{x} = -\tfrac{1}{2}\hat{\boldsymbol{\beta}}.$$

Differentiating with respect to λ gives

$$\left(\hat{\mathbf{B}} - \lambda \mathbf{I}_k\right)\frac{\partial \mathbf{x}}{\partial \lambda} = \mathbf{x},$$

and since $\mathbf{x}'\mathbf{x} = r^2$,

$$\mathbf{x}'\frac{\partial \mathbf{x}}{\partial \lambda} = r\frac{\partial r}{\partial \lambda}.$$

A second differentiation with respect to λ yields

$$(\hat{\mathbf{B}} - \lambda\mathbf{I}_k)\frac{\partial^2 \mathbf{x}}{\partial \lambda^2} = 2\frac{\partial \mathbf{x}}{\partial \lambda},$$

$$\mathbf{x}'\frac{\partial^2 \mathbf{x}}{\partial \lambda^2} + \frac{\partial \mathbf{x}'}{\partial \lambda}\frac{\partial \mathbf{x}}{\partial \lambda} = r\frac{\partial^2 r}{\partial \lambda^2} + \left(\frac{\partial r}{\partial \lambda}\right)^2.$$

If we premultiply the second equation by $\partial^2 \mathbf{x}'/\partial \lambda^2$ and the fourth equation by $\partial \mathbf{x}'/\partial \lambda$, subtract, and transpose, we obtain

$$\mathbf{x}'\frac{\partial^2 \mathbf{x}}{\partial \lambda^2} - 2\frac{\partial \mathbf{x}'}{\partial \lambda}\frac{\partial \mathbf{x}}{\partial \lambda} = 0.$$

Substituting this in the fifth equation, we get

$$r\frac{\partial^2 r}{\partial \lambda^2} = 3\frac{\partial \mathbf{x}'}{\partial \lambda}\frac{\partial \mathbf{x}}{\partial \lambda} - \left(\frac{\partial r}{\partial \lambda}\right)^2.$$

Now, since

$$\frac{\partial r}{\partial \lambda} = \frac{\partial}{\partial \lambda}(\mathbf{x}'\mathbf{x})^{1/2}$$

$$= \mathbf{x}'\frac{\partial \mathbf{x}/\partial \lambda}{(\mathbf{x}'\mathbf{x})^{1/2}},$$

we conclude that

$$r^3\frac{\partial^2 r}{\partial \lambda^2} = 3r^2\frac{\partial \mathbf{x}'}{\partial \lambda}\frac{\partial \mathbf{x}}{\partial \lambda} - \left(\mathbf{x}'\frac{\partial \mathbf{x}}{\partial \lambda}\right)^2.$$

(b) The expression in (a) can be written as

$$r^3\frac{\partial^2 r}{\partial \lambda^2} = 2r^2\frac{\partial \mathbf{x}'}{\partial \lambda}\frac{\partial \mathbf{x}}{\partial \lambda} + \left[r^2\frac{\partial \mathbf{x}'}{\partial \lambda}\frac{\partial \mathbf{x}}{\partial \lambda} - \left(\mathbf{x}'\frac{\partial \mathbf{x}}{\partial \lambda}\right)^2\right].$$

The first part on the right-hand side is nonnegative and is zero only when $r = 0$ or when $\partial \mathbf{x}/\partial \lambda = \mathbf{0}$. The second part is nonnegative by the fact that

$$\left(\mathbf{x}'\frac{\partial \mathbf{x}}{\partial \lambda}\right)^2 \le r^2 \left\|\frac{\partial \mathbf{x}}{\partial \lambda}\right\|_2^2$$

$$= r^2 \frac{\partial \mathbf{x}'}{\partial \lambda}\frac{\partial \mathbf{x}}{\partial \lambda}.$$

Equality occurs only when $\mathbf{x} = \mathbf{0}$, that is, $r = 0$, or when $\partial \mathbf{x}/\partial \lambda = \mathbf{0}$. But when $\partial \mathbf{x}/\partial \lambda = \mathbf{0}$, we have $\mathbf{x} = \mathbf{0}$ if λ is different from all the eigenvalues of $\hat{\mathbf{B}}$, and thus $r = 0$. It follows that $\partial^2 r/\partial \lambda^2 > 0$ except when $r = 0$, where it takes the value zero.

8.7. (a)

$$B = \frac{n\Omega}{\sigma^2}\int_R \{E[\hat{y}(\mathbf{x})] - \eta(\mathbf{x})\}^2\, d\mathbf{x}$$

$$= \frac{n}{\sigma^2}\{(\boldsymbol{\gamma} - \boldsymbol{\beta})'\boldsymbol{\Gamma}_{11}(\boldsymbol{\gamma} - \boldsymbol{\beta}) - 2(\boldsymbol{\gamma} - \boldsymbol{\beta})'\boldsymbol{\Gamma}_{12}\boldsymbol{\delta} + \boldsymbol{\delta}'\boldsymbol{\Gamma}_{22}\boldsymbol{\delta}\},$$

where $\boldsymbol{\Gamma}_{11}$, $\boldsymbol{\Gamma}_{12}$, $\boldsymbol{\Gamma}_{22}$ are the region moments defined in Section 8.4.3.

(b) To minimize B we differentiate it with respect to $\boldsymbol{\gamma}$, equate the derivative to zero, and solve for $\boldsymbol{\gamma}$. We get

$$2\boldsymbol{\Gamma}_{11}(\boldsymbol{\gamma} - \boldsymbol{\beta}) - 2\boldsymbol{\Gamma}_{12}\boldsymbol{\delta} = \mathbf{0},$$

$$\boldsymbol{\gamma} = \boldsymbol{\beta} + \boldsymbol{\Gamma}_{11}^{-1}\boldsymbol{\Gamma}_{12}\boldsymbol{\delta}$$

$$= \mathbf{C}\boldsymbol{\tau}.$$

This solution minimizes B, since $\boldsymbol{\Gamma}_{11}$ is positive definite.

(c) B is minimized if and only if $\boldsymbol{\gamma} = \mathbf{C}\boldsymbol{\tau}$. This is equivalent to stating that $\mathbf{C}\boldsymbol{\tau}$ is estimable, since $E(\hat{\boldsymbol{\lambda}}) = \boldsymbol{\gamma}$.

(d) Writing $\hat{\boldsymbol{\lambda}}$ as a linear function of the vector $\mathbf{y}$ of observations of the form $\hat{\boldsymbol{\lambda}} = \mathbf{Ly}$, we obtain

$$E(\hat{\boldsymbol{\lambda}}) = \mathbf{L}E(\mathbf{y})$$

$$= \mathbf{L}(\mathbf{X}\boldsymbol{\beta} + \mathbf{Z}\boldsymbol{\delta})$$

$$= \mathbf{L}[\mathbf{X}:\mathbf{Z}]\boldsymbol{\tau}.$$

But $\boldsymbol{\gamma} = E(\hat{\boldsymbol{\lambda}}) = \mathbf{C}\boldsymbol{\tau}$. We conclude that $\mathbf{C} = \mathbf{L}[\mathbf{X}:\mathbf{Z}]$.

(e) It is obvious from part (d) that the rows of $\mathbf{C}$ are spanned by the rows of $[\mathbf{X}:\mathbf{Z}]$.

(f) The matrix $\mathbf{L}$ defined by $\mathbf{L} = (\mathbf{X}'\mathbf{X})^{-1}\mathbf{X}'$ satisfies the equation

$$\mathbf{L}[\mathbf{X}:\mathbf{Z}] = \mathbf{C},$$

since

$$\begin{aligned}
\mathbf{L}[\mathbf{X}:\mathbf{Z}] &= (\mathbf{X}'\mathbf{X})^{-1}\mathbf{X}'[\mathbf{X}:\mathbf{Z}] \\
&= \left[\mathbf{I}:(\mathbf{X}'\mathbf{X})^{-1}\mathbf{X}'\mathbf{Z}\right] \\
&= \left[\mathbf{I}:\mathbf{M}_{11}^{-1}\mathbf{M}_{12}\right] \\
&= \left[\mathbf{I}:\boldsymbol{\Gamma}_{11}^{-1}\boldsymbol{\Gamma}_{12}\right] \\
&= \mathbf{C}.
\end{aligned}$$

8.8. If the region R is a sphere of radius 1, then $3g^2 \le 1$. Now,

$$\boldsymbol{\Gamma}_{11} = \begin{bmatrix} 1 & 0 & 0 & 0 \\ 0 & \frac{1}{5} & 0 & 0 \\ 0 & 0 & \frac{1}{5} & 0 \\ 0 & 0 & 0 & \frac{1}{5} \end{bmatrix},$$

$$\boldsymbol{\Gamma}_{12} = \begin{bmatrix} 0 & 0 & 0 \\ 0 & 0 & 0 \\ 0 & 0 & 0 \\ 0 & 0 & 0 \end{bmatrix}.$$

Hence, $\mathbf{C} = [\mathbf{I}_4 : \mathbf{O}_{4\times 3}]$. Furthermore,

$$\mathbf{X} = [\mathbf{1}_4 : \mathbf{D}],$$

$$\mathbf{Z} = \begin{bmatrix} g^2 & g^2 & g^2 \\ g^2 & -g^2 & -g^2 \\ -g^2 & g^2 & -g^2 \\ -g^2 & -g^2 & g^2 \end{bmatrix}.$$

Hence,

$$\mathbf{M}_{11} = \tfrac{1}{4}\mathbf{X}'\mathbf{X}$$

$$= \begin{bmatrix} 1 & 0 & 0 & 0 \\ 0 & g^2 & 0 & 0 \\ 0 & 0 & g^2 & 0 \\ 0 & 0 & 0 & g^2 \end{bmatrix},$$

$$\mathbf{M}_{12} = \tfrac{1}{4}\mathbf{X}'\mathbf{Z}$$

$$= \begin{bmatrix} 0 & 0 & 0 \\ 0 & 0 & -g^3 \\ 0 & -g^3 & 0 \\ -g^3 & 0 & 0 \end{bmatrix}.$$

(a)

$$\mathbf{M}_{11} = \mathbf{\Gamma}_{11} \quad \Rightarrow \quad g^2 = \tfrac{1}{5}$$

$$\mathbf{M}_{12} = \mathbf{\Gamma}_{12} \quad \Rightarrow \quad g = 0$$

Thus, it is not possible to choose g so that $\mathbf{D}$ satisfies the conditions in (8.56).

(b) Suppose that there exists a matrix $\mathbf{L}$ of order 4×4 such that

$$\mathbf{C} = \mathbf{L}[\mathbf{X} : \mathbf{Z}].$$

Then

$$\mathbf{I}_4 = \mathbf{L}\mathbf{X},$$

$$\mathbf{0} = \mathbf{L}\mathbf{Z}.$$

The second equation implies that $\mathbf{L}$ is of rank 1, while the first equation implies that the rank of $\mathbf{L}$ is greater than or equal to the rank of $\mathbf{I}_4$, which is equal to 4. Therefore, it is not possible to find a matrix such as $\mathbf{L}$. Hence, g cannot be chosen so that D satisfies the minimum bias property described in part (e) of Exercise 8.7.

8.9. (a) Since $\mathbf{\Delta}$ is symmetric, it can be written as $\mathbf{\Delta} = \mathbf{P}\mathbf{\Lambda}\mathbf{P}'$, where $\mathbf{\Lambda}$ is a diagonal matrix of eigenvalues of $\mathbf{\Delta}$ and the columns of $\mathbf{P}$ are the corresponding orthogonal eigenvectors of $\mathbf{\Delta}$, each of length equal to 1 (see Theorem 2.3.10). It is easy to see that over the region ψ,

$$h(\mathbf{\delta}, \mathbf{D}) \le \mathbf{\delta}'\mathbf{\delta} e_{max}(\mathbf{\Delta}) \le r^2 e_{max}(\mathbf{\Delta}),$$

by the fact that $e_{max}(\mathbf{\Delta})\mathbf{I} - \mathbf{P}\mathbf{\Lambda}\mathbf{P}'$ is positive semidefinite. Without loss of generality, we consider that the diagonal elements of $\mathbf{\Lambda}$ are written in descending order. The upper bound in the above inequality is attained by $h(\mathbf{\delta}, \mathbf{D})$ for $\mathbf{\delta} = r\mathbf{P}_1$, where $\mathbf{P}_1$ is the first column of $\mathbf{P}$, which corresponds to $e_{max}(\mathbf{\Delta})$.

(b) The design D can be chosen so that $e_{max}(\mathbf{\Delta})$ is minimized over the region R.

8.10. This is similar to Exercise 8.9. Write $\mathbf{T}$ as $\mathbf{P}\mathbf{\Lambda}\mathbf{P}'$, where $\mathbf{\Lambda}$ is a diagonal matrix of eigenvalues of $\mathbf{T}$, and $\mathbf{P}$ is an orthogonal matrix of corresponding eigenvectors. Then $\mathbf{\delta}'\mathbf{T}\mathbf{\delta} = \mathbf{u}'\mathbf{u}$, where $\mathbf{u} = \mathbf{\Lambda}^{1/2}\mathbf{P}'\mathbf{\delta}$, and

$\lambda(\delta, \mathbf{D}) = \delta' \mathbf{S} \delta = \mathbf{u}' \Lambda^{-1/2} \mathbf{P}' \mathbf{S} \mathbf{P} \Lambda^{-1/2} \mathbf{u}$. Hence, over the region Φ,

$$\delta' \mathbf{S} \delta \geq \kappa e_{\min}(\Lambda^{-1/2} \mathbf{P}' \mathbf{S} \mathbf{P} \Lambda^{-1/2}).$$

But, if $\mathbf{S}$ is positive definite, then by Theorem 2.3.9,

$$e_{\min}(\Lambda^{-1/2} \mathbf{P}' \mathbf{S} \mathbf{P} \Lambda^{-1/2}) = e_{\min}(\mathbf{P} \Lambda^{-1} \mathbf{P}' \mathbf{S})$$
$$= e_{\min}(\mathbf{T}^{-1} \mathbf{S}).$$

8.11. (a) Using formula (8.52), it can be shown that (see Khuri and Cornell, 1996, page 229)

$$V = \frac{1}{\lambda_4(k+2) - \lambda_2^2 k} \left[\lambda_4(k+2) - 2k \left(\lambda_2 - \frac{\lambda_4}{\lambda_2} \right) \right.$$
$$\left. + 2k \frac{\lambda_4(k+1) - \lambda_2^2(k-1)}{\lambda_4(k+4)} \right],$$

where k is the number of input variables (that is, $k = 2$),

$$\lambda_2 = \frac{1}{n} \sum_{u=1}^{n} x_{ui}^2, \qquad i = 1, 2$$

$$= \frac{1}{n}(2^k + 2\alpha^2)$$

$$= \frac{8}{n},$$

$$\lambda_4 = \frac{1}{n} \sum_{u=1}^{n} x_{ui}^2 x_{uj}^2, \qquad i \neq j$$

$$= \frac{2^k}{n}$$

$$= \frac{4}{n},$$

and $n = 2^k + 2k + n_0 = 8 + n_0$ is the total number of observations. Here, x_{ui} denotes the design setting of variable x_i, $i = 1, 2$; $u = 1, 2, \ldots, n$.

(b) The quantity V, being a function of n_0, can be minimized with respect to n_0.

8.12. We have that

$$(\mathbf{Y} - \mathbf{XB})'(\mathbf{Y} - \mathbf{XB}) = (\mathbf{Y} - \mathbf{X\hat{B}} + \mathbf{X\hat{B}} - \mathbf{XB})'(\mathbf{Y} - \mathbf{X\hat{B}} + \mathbf{X\hat{B}} - \mathbf{XB})$$

$$= (\mathbf{Y} - \mathbf{X\hat{B}})'(\mathbf{Y} - \mathbf{X\hat{B}}) + (\mathbf{X\hat{B}} - \mathbf{XB})'(\mathbf{X\hat{B}} - \mathbf{XB}),$$

since

$$(\mathbf{Y} - \mathbf{X\hat{B}})'(\mathbf{X\hat{B}} - \mathbf{XB}) = (\mathbf{X'Y} - \mathbf{X'X\hat{B}})'(\mathbf{\hat{B}} - \mathbf{B}) = \mathbf{0}.$$

Furthermore, since $(\mathbf{X\hat{B}} - \mathbf{XB})'(\mathbf{X\hat{B}} - \mathbf{XB})$ is positive semidefinite, then by Theorem 2.3.19,

$$e_i[(\mathbf{Y} - \mathbf{XB})'(\mathbf{Y} - \mathbf{XB})] \geq e_i\big[(\mathbf{Y} - \mathbf{X\hat{B}})'(\mathbf{Y} - \mathbf{X\hat{B}})\big], \qquad i = 1, 2, \ldots, r,$$

where $e_i(\cdot)$ denotes the ith eigenvalue of a square matrix. If $(\mathbf{Y} - \mathbf{XB})'(\mathbf{Y} - \mathbf{XB})$ and $(\mathbf{Y} - \mathbf{X\hat{B}})'(\mathbf{Y} - \mathbf{X\hat{B}})$ are nonsingular, then by multiplying the eigenvalues on both sides of the inequality, we obtain

$$\det[(\mathbf{Y} - \mathbf{XB})'(\mathbf{Y} - \mathbf{XB})] \geq \det\big[(\mathbf{Y} - \mathbf{X\hat{B}})'(\mathbf{Y} - \mathbf{X\hat{B}})\big].$$

Equality holds when $\mathbf{B} = \mathbf{\hat{B}}$.

8.13. For any b, $1 + b \leq e^b$. Let $a = 1 + b$. Then $a \leq e^{a-1}$. Now, let λ_i denote the ith eigenvalue of $\mathbf{A}$. Then $\lambda_i \geq 0$, and by the previous inequality,

$$\prod_{i=1}^{p} \lambda_i \leq \exp\left(\sum_{i=1}^{p} \lambda_i - p\right).$$

Hence,

$$\det(\mathbf{A}) \leq \exp\big[\text{tr}(\mathbf{A} - \mathbf{I}_p)\big].$$

8.14. The likelihood function is proportional to

$$[\det(\mathbf{V})]^{-n/2} \exp\left[-\frac{n}{2}\text{tr}(\mathbf{SV}^{-1})\right].$$

Now, by Exercise 8.13,

$$\det(\mathbf{SV}^{-1}) \leq \exp\big[\text{tr}(\mathbf{SV}^{-1} - \mathbf{I}_p)\big],$$

since $\det(\mathbf{SV}^{-1}) = \det(\mathbf{V}^{-1/2}\,\mathbf{SV}^{-1/2})$, $\mathrm{tr}(\mathbf{SV}^{-1}) = \mathrm{tr}(\mathbf{V}^{-1/2}\,\mathbf{SV}^{-1/2})$, and $\mathbf{V}^{-1/2}\,\mathbf{SV}^{-1/2}$ is positive semidefinite. Hence,

$$\left[\det(\mathbf{SV}^{-1})\right]^{n/2}\exp\left[-\frac{n}{2}\mathrm{tr}(\mathbf{SV}^{-1})\right] \le \exp\left[-\frac{n}{2}\mathrm{tr}(\mathbf{I}_p)\right].$$

This results in the following inequality:

$$\left[\det(\mathbf{V})\right]^{-n/2}\exp\left[-\frac{n}{2}\mathrm{tr}(\mathbf{SV}^{-1})\right] \le \left[\det(\mathbf{S})\right]^{-n/2}\exp\left[-\frac{n}{2}\mathrm{tr}(\mathbf{I}_p)\right],$$

which is the desired result.

8.16. $\hat{y}(\mathbf{x}) = \mathbf{f}'(\mathbf{x})\hat{\boldsymbol{\beta}}$, where $\hat{\boldsymbol{\beta}} = (\mathbf{X}'\mathbf{X})^{-1}\mathbf{X}'\mathbf{y}$, and $\mathbf{f}'(\mathbf{x})$ is as in model (8.47). Simultaneous $(1 - \alpha) \times 100\%$ confidence intervals on $\mathbf{f}'(\mathbf{x})\boldsymbol{\beta}$ for all $\mathbf{x}$ in R are of the form

$$\mathbf{f}'(\mathbf{x})\hat{\boldsymbol{\beta}} \mp \left(p\,MS_E\,F_{\alpha,\,p,\,n-p}\right)^{1/2}\left[\mathbf{f}'(\mathbf{x})(\mathbf{X}'\mathbf{X})^{-1}\mathbf{f}(\mathbf{x})\right]^{1/2}.$$

For the points $\mathbf{x}_1, \mathbf{x}_2, \ldots, \mathbf{x}_m$, the joint confidence coefficient is at least $1 - \alpha$.

CHAPTER 9

9.1. If $f(x)$ has a continuous derivative on $[0, 1]$, then by Theorems 3.4.5 and 4.2.2 we can find a positive constant A such that

$$\left|f(x_1) - f(x_2)\right| \le A|x_1 - x_2|$$

for all x_1, x_2 in $[0, 1]$. Thus, by Definition 9.1.2,

$$\omega(\delta) \le A\delta.$$

Using now Theorem 9.1.3, we obtain

$$\left|f(x) - b_n(x)\right| \le \frac{3}{2}\frac{A}{n^{1/2}}$$

for all x in $[0, 1]$. Hence,

$$\sup_{0 \le x \le 1}\left|f(x) - b_n(x)\right| \le \frac{c}{n^{1/2}},$$

where $c = \frac{3}{2}A$.

9.2. We have that

$$|f(x_1) - f(x_2)| \leq \sup_{|z_1 - z_2| \leq \delta_2} |f(z_1) - f(z_2)|$$

for all $|x_1 - x_2| \leq \delta_2$, and hence for all $|x_1 - x_2| \leq \delta_1$, since $\delta_1 \leq \delta_2$. It follows that

$$\sup_{|x_1 - x_2| \leq \delta_1} |f(x_1) - f(x_2)| \leq \sup_{|z_1 - z_2| \leq \delta_2} |f(z_1) - f(z_2)|,$$

that is, $\omega(\delta_1) \leq \omega(\delta_2)$.

9.3. Suppose that $f(x)$ is uniformly continuous on $[a, b]$. Then, for a given $\epsilon > 0$, there exists a positive $\delta(\epsilon)$ such that $|f(x_1) - f(x_2)| < \epsilon$ for all x_1, x_2 in $[a, b]$ for which $|x_1 - x_2| < \delta$. This implies that $\omega(\delta) \leq \epsilon$ and hence $\omega(\delta) \to 0$ as $\delta \to 0$. Vice versa, if $\omega(\delta) \to 0$ as $\delta \to 0$, then for a given $\epsilon > 0$, there exists $\delta_1 > 0$ such that $\omega(\delta) < \epsilon$ if $\delta < \delta_1$. This implies that

$$|f(x_1) - f(x_2)| < \epsilon \qquad \text{if } |x_1 - x_2| \leq \delta < \delta_1,$$

and $f(x)$ must therefore be uniformly continuous on $[a, b]$.

9.4. By Theorem 9.1.1, there exists a sequence of polynomials, namely the Bernstein polynomials $\{b_n(x)\}_{n=1}^{\infty}$, that converges to $f(x) = |x|$ uniformly on $[-a, a]$. Let $p_n(x) = b_n(x) - b_n(0)$. Then $p_n(0) = 0$, and $p_n(x)$ converges uniformly to $|x|$ on $[-a, a]$, since $b_n(0) \to 0$ as $n \to \infty$.

9.5. The stated condition implies that $\int_0^1 f(x) p_n(x) \, dx = 0$ for any polynomial $p_n(x)$ of degree n. In particular, if we choose $p_n(x)$ to be a Bernstein polynomial for $f(x)$, then it will converge uniformly to $f(x)$ on $[0, 1]$. By Theorem 6.6.1,

$$\int_0^1 [f(x)]^2 \, dx = \lim_{n \to \infty} \int_0^1 f(x) p_n(x) \, dx$$

$$= 0.$$

Since $f(x)$ is continuous, it must be zero everywhere on $[0, 1]$. If not, then by Theorem 3.4.3, there exists a neighborhood of a point in $[0, 1]$ [at which $f(x) \neq 0$] on which $f(x) \neq 0$. This causes $\int_0^1 [f(x)]^2 \, dx$ to be positive, a contradiction.

9.6. Using formula (9.15), we have

$$f(x) - p(x) = \frac{1}{(n+1)!} f^{(n+1)}(c) \prod_{i=0}^{n} (x - a_i).$$

But $\left| \prod_{i=0}^{n} (x - a_i) \right| \le n!(h^{n+1}/4)$ (see Prenter, 1975, page 37). Hence,

$$\sup_{a \le x \le b} |f(x) - p(x)| \le \frac{T_{n+1} h^{n+1}}{4(n+1)}.$$

9.7. Using formula (9.14) with a_0, a_1, a_2, and a_3, we obtain

$$p(x) = \ell_0(x) \log a_0 + \ell_1(x) \log a_1 + \ell_2(x) \log a_2 + \ell_3(x) \log a_3,$$

where

$$\ell_0(x) = \left(\frac{x - a_1}{a_0 - a_1} \right) \left(\frac{x - a_2}{a_0 - a_2} \right) \left(\frac{x - a_3}{a_0 - a_3} \right),$$

$$\ell_1(x) = \left(\frac{x - a_0}{a_1 - a_0} \right) \left(\frac{x - a_2}{a_1 - a_2} \right) \left(\frac{x - a_3}{a_1 - a_3} \right),$$

$$\ell_2(x) = \left(\frac{x - a_0}{a_2 - a_0} \right) \left(\frac{x - a_1}{a_2 - a_1} \right) \left(\frac{x - a_3}{a_2 - a_3} \right),$$

$$\ell_3(x) = \left(\frac{x - a_0}{a_3 - a_0} \right) \left(\frac{x - a_1}{a_3 - a_1} \right) \left(\frac{x - a_2}{a_3 - a_2} \right).$$

Values of $f(x) = \log x$ and the corresponding values of $p(x)$ at several points inside the interval $[3.50, 3.80]$ are given below:

x	$f(x)$	$p(x)$
3.50	1.25276297	1.25276297
3.52	1.25846099	1.25846087
3.56	1.26976054	1.26976043
3.60	1.28093385	1.28093385
3.62	1.28647403	1.28647407
3.66	1.29746315	1.29746322
3.70	1.30833282	1.30833282
3.72	1.31372367	1.31372361
3.77	1.327075	1.32707487
3.80	1.33500107	1.33500107

Using the result of Exercise 9.6, an upper bound on the error of approximation is given by

$$\sup_{3.5 \leq x \leq 3.8} |f(x) - p(x)| \leq \frac{\tau_4 h^4}{16},$$

where $h = \max_i(a_{i+1} - a_i) = 0.10$, and

$$\tau_4 = \sup_{3.5 \leq x \leq 3.8} |f^{(4)}(x)|$$

$$= \sup_{3.5 \leq x \leq 3.8} \left| \frac{-6}{x^4} \right|$$

$$= \frac{6}{(3.5)^4}.$$

Hence, the desired upper bound is

$$\frac{\tau_4 h^4}{16} = \frac{6}{16} \left(\frac{0.10}{3.5} \right)^4$$

$$= 2.5 \times 10^{-7}.$$

9.8. We have that

$$\int_a^b [f''(x) - s''(x)]^2 \, dx = \int_a^b [f''(x)]^2 \, dx - \int_a^b [s''(x)]^2 \, dx$$

$$- 2\int_a^b s''(x)[f''(x) - s''(x)] \, dx.$$

But integration by parts yields

$$\int_a^b s''(x)[f''(x) - s''(x)] \, dx$$

$$= s''(x)[f'(x) - s'(x)]\Big|_a^b - \int_a^b s'''(x)[f'(x) - x'(x)] \, dx.$$

The first term on the right-hand side is zero, since $f'(x) = s'(x)$ at $x = a, b$; and the second term is also zero, by the fact that $s'''(x)$ is a constant, say $s'''(x) = \lambda_i$, over (τ_i, τ_{i+1}). Hence,

$$\int_a^b s'''(x)[f'(x) - s'(x)] \, dx = \sum_{i=0}^{n-1} \lambda_i \int_{\tau_i}^{\tau_{i+1}} [f'(x) - s'(x)] \, dx = 0.$$

It follows that

$$\int_a^b [f''(x) - s''(x)]^2 \, dx = \int_a^b [f''(x)]^2 \, dx - \int_a^b [s''(x)]^2 \, dx,$$

which implies the desired result.

9.10.

$$\frac{\partial^{r+1} h(x, \boldsymbol{\theta})}{\partial x^{r+1}} = (-1)^r \theta_1 \theta_2 \theta_3^{r+1} e^{-\theta_3 x}.$$

Hence,

$$\sup_{0 \le x \le 8} \left| \frac{\partial^{r+1} h(x, \boldsymbol{\theta})}{\partial x^{r+1}} \right| \le 50.$$

Using inequality (9.36), the integer r is determined such that

$$\frac{2}{(r+1)!} \left(\frac{8}{4} \right)^{r+1} (50) < 0.05.$$

The smallest integer that satisfies this inequality is $r = 10$. The Chebyshev points corresponding to this value of r are $z_0, z_1, \ldots, z_{10}$, where by formula (9.18),

$$z_i = 4 + 4 \cos \left[\left(\frac{2i+1}{22} \right) \pi \right], \qquad i = 0, 1, \ldots, 10.$$

Using formula (9.37), the Lagrange interpolating polynomial that approximates $h(x, \boldsymbol{\theta})$ over $[0, 8]$ is given by

$$P_{10}(x, \boldsymbol{\theta}) = \theta_1 \sum_{i=0}^{10} \left[1 - \theta_2 e^{-\theta_3 z_i} \right] \ell_i(x),$$

where $\ell_i(x)$ is a polynomial of degree 10 which can be obtained from (9.13) by substituting z_i for a_i $(i = 0, 1, \ldots, 10)$.

9.11. We have that

$$\frac{\partial^4 \eta(x, \alpha, \beta)}{\partial x^4} = (0.49 - \alpha) \beta^4 e^{8\beta} e^{-\beta x}.$$

Hence,

$$\max_{10 \le x \le 40} \left| \frac{\partial^4 \eta(x, \alpha, \beta)}{\partial x^4} \right| = (0.49 - \alpha) \beta^4 e^{-2\beta}.$$

It can be verified that the function $f(\beta) = \beta^4 e^{-2\beta}$ is strictly monotone increasing for $0.06 \le \beta \le 0.16$. Therefore, $\beta = 0.16$ maximizes $f(\beta)$

over this interval. Hence,

$$\max_{10 \leq x \leq 40} \left| \frac{\partial^4 \eta(x, \alpha, \beta)}{\partial x^4} \right| \leq (0.49 - \alpha)(0.16)^4 e^{-0.32}$$

$$\leq 0.13(0.16)^4 e^{-0.32}$$

$$= 0.0000619,$$

since $0.36 \leq \alpha \leq 0.41$. Using Theorem 9.3.1, we have

$$\max_{10 \leq x \leq 40} |\eta(x, \alpha, \beta) - s(x)| \leq \frac{5}{384} \Delta^4 (0.0000619).$$

Here, we considered equally spaced partition points with $\Delta \tau_i = \Delta$. Let us now choose Δ such that

$$\frac{5}{384} \Delta^4 (0.0000619) < 0.001.$$

This is satisfied if $\Delta < 5.93$. Choosing $\Delta = 5$, the number of knots needed is

$$m = \frac{40 - 10}{\Delta} - 1$$

$$= 5.$$

9.12.

$$\det(\mathbf{X}'\mathbf{X}) = \left[(x_3 - \alpha)^2 (x_2 - x_1) - (x_3 - x_1)(x_2 - \alpha)^2 \right]^2.$$

The determinant is maximized when $x_1 = -1$, $x_2 = \frac{1}{4}(1 + \alpha)^2$, $x_3 = 1$.

CHAPTER 10

10.1. We have that

$$\frac{1}{\pi} \int_{-\pi}^{\pi} \cos nx \cos mx \, dx = \frac{1}{2\pi} \int_{-\pi}^{\pi} [\cos(n+m)x + \cos(n-m)x] \, dx$$

$$= \begin{cases} 0, & n \neq m, \\ 1, & n = m \geq 1, \end{cases}$$

$$\frac{1}{\pi} \int_{-\pi}^{\pi} \cos nx \sin mx \, dx = \frac{1}{2\pi} \int_{-\pi}^{\pi} [\sin(n+m)x - \sin(n-m)x] \, dx$$

$$= 0 \qquad \text{for all } m, n,$$

$$\frac{1}{\pi} \int_{-\pi}^{\pi} \sin nx \sin mx \, dx = \frac{1}{2\pi} \int_{-\pi}^{\pi} [\cos(n-m)x - \cos(n+m)x] \, dx$$

$$= \begin{cases} 0, & n \neq m, \\ 1, & n = m \geq 1. \end{cases}$$

10.2. (a)

    (i) Integrating n times by parts [x^m is differentiated and $p_n(x)$ is integrated], we obtain

$$\int_{-1}^{1} x^m p_n(x)\, dx = 0 \qquad \text{for } m = 0, 1, \ldots, n-1.$$

    (ii) Integrating n times by parts results in

$$\int_{-1}^{1} x^n p_n(x)\, dx = \frac{1}{2^n} \int_{-1}^{1} (1 - x^2)^n\, dx.$$

Letting $x = \cos\theta$, we obtain

$$\int_{-1}^{1} (1 - x^2)^n\, dx = \int_{0}^{\pi} \sin^{2n+1}\theta\, d\theta$$

$$= \frac{2^{2n+1}}{2n+1} \binom{2n}{n}^{-1}, \qquad n \geq 0.$$

(b) This is obvious, since (a) is true for $m = 0, 1, \ldots, n-1$.

(c)

$$\int_{-1}^{1} p_n^2(x)\, dx = \int_{-1}^{1} \left[\frac{1}{2^n} \binom{2n}{n} x^n + \pi_{n-1}(x) \right] p_n(x)\, dx,$$

where $\pi_{n-1}(x)$ denotes a polynomial of degree $n-1$. Hence, using the results in (a) and (b), we obtain

$$\int_{-1}^{1} p_n^2(x)\, dx = \frac{1}{2^n} \binom{2n}{n} \frac{2^{n+1}}{2n+1} \binom{2n}{n}^{-1}$$

$$= \frac{2}{2n+1}, \qquad n \geq 0.$$

10.3. (a)

$$T_n(\zeta_i) = \cos\left\{ n \operatorname{Arccos}\left[\cos\frac{(2i-1)\pi}{2n} \right] \right\}$$

$$= \cos\frac{(2i-1)\pi}{2} = 0$$

for $i = 1, 2, \ldots, n$.

(b) $T_n'(x) = (n/\sqrt{1 - x^2})\sin(n \operatorname{Arccos} x)$. Hence,

$$T_n'(\zeta_i) = \frac{n}{\sqrt{1 - \zeta_i^2}} \sin\frac{(2i-1)\pi}{2} \neq 0$$

for $i = 1, 2, \ldots, n$.

10.4. (a)

$$\frac{dH_n(x)}{dx} = (-1)^n x e^{x^2/2} \frac{d^n(e^{-x^2/2})}{dx^n} + (-1)^n e^{x^2/2} \frac{d^{n+1}(e^{-x^2/2})}{dx^{n+1}}.$$

Using formulas (10.21) and (10.24), we have

$$\frac{dH_n(x)}{dx} = (-1)^n x e^{x^2/2} (-1)^n e^{-x^2/2} H_n(x)$$

$$+ (-1)^n e^{x^2/2} (-1)^{n+1} e^{-x^2/2} H_{n+1}(x)$$

$$= x H_n(x) - H_{n+1}(x)$$

$$= n H_{n-1}(x), \quad \text{by formula (10.25).}$$

(b) From (10.23) and (10.24), we have

$$H_{n+1}(x) = x H_n(x) - \frac{dH_n(x)}{dx}.$$

Hence,

$$\frac{dH_{n+1}(x)}{dx} = x \frac{dH_n(x)}{dx} + H_n(x) - \frac{d^2 H_n(x)}{dx^2}$$

$$\frac{d^2 H_n(x)}{dx^2} = x \frac{dH_n(x)}{dx} + H_n(x) - \frac{dH_{n+1}(x)}{dx}$$

$$= x \frac{dH_n(x)}{dx} + H_n(x) - (n+1) H_n(x), \quad \text{by (a)}$$

$$= x \frac{dH_n(x)}{dx} - n H_n(x),$$

which gives the desired result.

10.5. (a) Using formula (10.18), we can show that

$$\left| \frac{\sin[(n+1)\theta]}{\sin \theta} \right| \le n + 1$$

by mathematical induction. Obviously, the inequality is true for $n = 1$, since

$$\left| \frac{\sin 2\theta}{\sin \theta} \right| = \left| \frac{2 \sin \theta \cos \theta}{\sin \theta} \right| \le 2.$$

Suppose now that the inequality is true for $n = m$. To show that it is true for $n = m + 1 (m \geq 1)$:

$$\left| \frac{\sin[(m+2)\theta]}{\sin \theta} \right| = \left| \frac{\sin[(m+1)\theta]\cos \theta + \cos[(m+1)\theta]\sin \theta}{\sin \theta} \right|$$

$$\leq m + 1 + 1 = m + 2.$$

Therefore, the inequality is true for all n.

(b) From Section 10.4.2 we have that

$$\frac{dT_n(x)}{dx} = n \frac{\sin n\theta}{\sin \theta}$$

Hence,

$$\left| \frac{dT_n(x)}{dx} \right| = n \left| \frac{\sin n\theta}{\sin \theta} \right|$$

$$\leq n^2,$$

since $|\sin n\theta / \sin \theta| \leq n$, which can be proved by induction as in (a). Note that as $x \to \mp 1$, that is, as $\theta \to 0$ or $\theta \to \pi$,

$$\left| \frac{dT_n(x)}{dx} \right| \to n^2.$$

(c) Making the change of variable $t = \cos \theta$, we get

$$\int_{-1}^{x} \frac{T_n(t)}{\sqrt{1 - t^2}} \, dt = -\int_{\pi}^{\text{Arccos } x} \cos n\theta \, d\theta$$

$$= -\frac{1}{n} \sin n\theta \Big|_{\pi}^{\text{Arccos } x}$$

$$= -\frac{1}{n} \sin(n \, \text{Arccos } x)$$

$$= -\frac{1}{n} \sin n\psi, \qquad \text{where } x = \cos \psi$$

$$= -\frac{1}{n} \sin \psi \, U_{n-1}(x), \qquad \text{by (10.18)}$$

$$= -\frac{1}{n} \sqrt{1 - x^2} \, U_{n-1}(x).$$

10.7. The first two Laguerre polynomials are $L_0(x) = 1$ and $L_1(x) = x - \alpha - 1$, as can be seen from applying the Rodrigues formula in Section 10.6. Now, differentiating $H(x, t)$ with respect to t, we obtain

$$\frac{\partial H(x, t)}{\partial t} = \left[\frac{\alpha + 1}{1 - t} - \frac{x}{(1 - t)^2}\right] H(x, t),$$

or equivalently,

$$(1 - t)^2 \frac{\partial H(x, t)}{\partial t} + \left[(x - \alpha - 1) + (\alpha + 1)t\right] H(x, t) = 0.$$

Hence, $g_0(x) = H(x, 0) = 1$, and $g_1(x) = -\partial H(x, t)/\partial t|_{t=0} = x - \alpha - 1$. Thus, $g_0(x) = L_0(x)$ and $g_1(x) = L_1(x)$. Furthermore, if the representation

$$H(x, t) = \sum_{n=0}^{\infty} \frac{(-1)^n}{n!} g_n(x) t^n$$

is substituted in the above equation, and if the coefficient of t^n in the resulting series is equated to zero, we obtain

$$g_{n+1}(x) + (2n - x + \alpha + 1)g_n(x) + (n^2 + n\alpha)g_{n-1}(x) = 0.$$

This is the same relation connecting the Laguerre polynomials $L_{n+1}(x)$, $L_n(x)$, and $L_{n-1}(x)$, which is given at the end of Section 10.6. Since we have already established that $g_n(x) = L_n(x)$ for $n = 0, 1$, we conclude that the same relation holds for all values of n.

10.8. Using formula (10.40), we have

$$p_4^*(x) = c_0 + c_1 x + c_2 \left(\frac{3x^2}{2} - \frac{1}{2}\right)$$

$$+ c_3 \left(\frac{5x^3}{2} - \frac{3x}{2}\right) + c_4 \left(\frac{35x^4}{8} - \frac{30x^2}{8} + \frac{3}{8}\right),$$

where

$$c_0 = \frac{1}{2} \int_{-1}^{1} e^x \, dx,$$

$$c_1 = \frac{3}{2} \int_{-1}^{1} x e^x \, dx,$$

$$c_2 = \frac{5}{2} \int_{-1}^{1} \left(\frac{3x^2}{2} - \frac{1}{2}\right) e^x \, dx,$$

$$c_3 = \frac{7}{2} \int_{-1}^{1} \left(\frac{5x^3}{2} - \frac{3x}{2} \right) e^x \, dx,$$

$$c_4 = \frac{9}{2} \int_{-1}^{1} \left(\frac{35x^4}{8} - \frac{30x^2}{8} + \frac{3}{8} \right) e^x \, dx.$$

In computing c_0, c_1, c_2, c_3, c_4 we have made use of the fact that

$$\int_{-1}^{1} p_n^2(x) \, dx = \frac{2}{2n+1}, \qquad n = 0, 1, 2, \ldots,$$

where $p_n(x)$ is the Legendre polynomial of degree n (see Section 10.2.1).

10.9.

$$f(x) \approx \frac{c_0}{2} + c_1 T_1(x) + c_2 T_2(x) + c_3 T_3(x) + c_4 T_4(x)$$

$$= 1.266066 + 1.130318 T_1(x) + 0.271495 T_2(x)$$

$$+ 0.044337 T_3(x) + 0.005474 T_4(x)$$

$$= 1.000044 + 0.99731 x + 0.4992 x^2 + 0.177344 x^3 + 0.043792 x^4$$

[*Note*: For more computational details, see Example 7.2 in Ralston and Rabinowitz (1978).]

10.10. Use formula (10.48) with the given values of the central moments.

10.11. The first six cumulants of the standardized chi-squared distribution with five degrees of freedom are $\kappa_1 = 0$, $\kappa_2 = 1$, $\kappa_3 = 1.264911064$, $\kappa_4 = 2.40$, $\kappa_5 = 6.0715731$, $\kappa_6 = 19.2$. We also have that $z_{0.05} = 1.645$. Applying the Cornish-Fisher approximation for $x_{0.05}$, we obtain the value $x_{0.05} \approx 1.921$. Thus, $P(\chi_5^{*2} > 1.921) \approx 0.05$, where χ_5^{*2} denotes the standardized chi-squared variate with five degrees of freedom. If χ_5^2 denotes the nonstandardized chi-squared counterpart (with five degrees of freedom), then $\chi_5^2 = \sqrt{10}\, \chi_5^{*2} + 5$. Hence, the corresponding approximate value of the upper 0.05 quantile of χ_5^2 is $\sqrt{10}\,(1.921) + 5 = 11.0747$. The actual table value is 11.07.

10.12. (a)

$$\int_0^1 e^{-t^2/2} \, dt \approx 1 - \frac{1}{2 \times 3 \times 1!} + \frac{1}{2^2 \times 5 \times 2!} - \frac{1}{2^3 \times 7 \times 3!}$$

$$+ \frac{1}{2^4 \times 9 \times 4!} = 0.85564649.$$

(b)

$$\int_0^x e^{-t^2/2}\, dt \approx xe^{-x^2/8}\left[\Theta_0\left(\frac{x}{2}\right) + \frac{1}{3}\Theta_2\left(\frac{x}{2}\right)\right.$$

$$\left. + \frac{1}{5}\Theta_4\left(\frac{x}{2}\right) + \frac{1}{7}\Theta_6\left(\frac{x}{2}\right)\right],$$

where

$$\Theta_0\left(\frac{x}{2}\right) = \frac{1}{1!}H_0\left(\frac{x}{2}\right)$$

$$= 1,$$

$$\Theta_2\left(\frac{x}{2}\right) = \frac{1}{2!}\left(\frac{x}{2}\right)^2 H_2\left(\frac{x}{2}\right)$$

$$= \frac{1}{2!}\left(\frac{x}{2}\right)^2\left[\left(\frac{x}{2}\right)^2 - 1\right],$$

$$\Theta_4\left(\frac{x}{2}\right) = \frac{1}{4!}\left(\frac{x}{2}\right)^4\left[\left(\frac{x}{2}\right)^4 - 6\left(\frac{x}{2}\right)^2 + 3\right],$$

$$\Theta_6\left(\frac{x}{2}\right) = \frac{1}{6!}\left(\frac{x}{2}\right)^6\left[\left(\frac{x}{2}\right)^6 - 15\left(\frac{x}{2}\right)^4 + 45\left(\frac{x}{2}\right)^2 - 15\right].$$

Hence,

$$\int_0^1 e^{-t^2/2}\, dt \approx 0.85562427.$$

10.13. Using formula (10.21), we have that

$$g(x) = \sum_{n=0}^{\infty} b_n(-1)^n e^{x^2/2}\frac{d^n\left(e^{-x^2/2}\right)}{dx^n}\phi(x)$$

$$= \sum_{n=0}^{\infty}(-1)^n b_n \frac{1}{\sqrt{2\pi}}\frac{d^n\left(e^{-x^2/2}\right)}{dx^n}, \qquad \text{by (10.44)}$$

$$= \sum_{n=0}^{\infty}\frac{c_n}{n!}\frac{d^n\phi(x)}{dx^n},$$

where $c_n = (-1)^n n! b_n$, $n = 0, 1, \ldots$

10.14. The moment generating function of any one of the X_i^2's is

$$\psi(t) = \int_{-\infty}^{\infty} e^{tx^2} f(x)\, dx$$

$$= \int_{-\infty}^{\infty} e^{tx^2} \left[\phi(x) - \frac{\lambda_3}{6} \frac{d^3\phi(x)}{dx^3} + \frac{\lambda_4}{24} \frac{d^4\phi(x)}{dx^4} + \frac{\lambda_3^2}{72} \frac{d^6\phi(x)}{dx^6} \right] dx.$$

By formula (10.21),

$$\frac{d^n\phi(x)}{dx^n} = (-1)^n \phi(x) H_n(x), \qquad n = 0, 1, 2, \ldots$$

Hence,

$$\psi(t) = \int_{-\infty}^{\infty} e^{tx^2} \left[\phi(x) + \frac{\lambda_3}{6} \phi(x) H_3(x) + \frac{\lambda_4}{24} \phi(x) H_4(x) \right.$$

$$\left. + \frac{\lambda_3^2}{72} \phi(x) H_6(x) \right] dx$$

$$= 2 \int_0^{\infty} e^{tx^2} \left[\phi(x) + \frac{\lambda_4}{24} \phi(x) H_4(x) + \frac{\lambda_3^2}{72} \phi(x) H_6(x) \right] dx,$$

since $H_3(x)$ is an odd function, and $H_4(x)$ and $H_6(x)$ are even functions. It is known that

$$\int_0^{\infty} e^{-x^2/2} x^{2n-1}\, dx = 2^{n-1} \Gamma(n),$$

where $\Gamma(n)$ is the gamma function

$$\Gamma(n) = \int_0^{\infty} e^{-x} x^{n-1}\, dx = 2 \int_0^{\infty} e^{-x^2} x^{2n-1}\, dx,$$

$n > 0$. It is easy to show that

$$\int_0^{\infty} e^{tx^2} \phi(x) x^m\, dx = \frac{1}{\sqrt{2\pi}} \int_0^{\infty} e^{-(x^2/2)(1-2t)} x^m\, dx$$

$$= \frac{2^{\frac{m}{2}}}{2\sqrt{\pi}} (1 - 2t)^{-\frac{1}{2}(m+1)} \Gamma\left(\frac{m+1}{2}\right),$$

where m is an even integer. Hence,

$$2\int_0^\infty e^{tx^2}\phi(x)\,dx = \frac{1}{\sqrt{\pi}}(1-2t)^{-1/2}\,\Gamma\left(\tfrac{1}{2}\right)$$

$$= (1-2t)^{-1/2},$$

$$2\int_0^\infty e^{tx^2}\phi(x)H_4(x)\,dx = 2\int_0^\infty e^{tx^2}\phi(x)(x^4-6x^2+3)\,dx$$

$$= \frac{4}{\sqrt{\pi}}(1-2t)^{-5/2}\,\Gamma\left(\tfrac{5}{2}\right)$$

$$- \frac{12}{\sqrt{\pi}}(1-2t)^{-3/2}\,\Gamma\left(\tfrac{3}{2}\right) + 3(1-2t)^{-1/2},$$

$$2\int_0^\infty e^{tx^2}\phi(x)H_6(x)\,dx = 2\int_0^\infty e^{tx^2}\phi(x)(x^6-15x^4+45x^2-15)\,dx$$

$$= \frac{8}{\sqrt{\pi}}(1-2t)^{-7/2}\,\Gamma\left(\tfrac{7}{2}\right) - \frac{60}{\sqrt{\pi}}(1-2t)^{-5/2}\,\Gamma\left(\tfrac{5}{2}\right)$$

$$+ \frac{90}{\sqrt{\pi}}(1-2t)^{-3/2}\,\Gamma\left(\tfrac{3}{2}\right) - 15(1-2t)^{-1/2}.$$

The last three integrals can be substituted in the formula for $\psi(t)$. The moment generating function of W is $[\psi(t)]^n$. On the other hand, the moment generating function of a chi-squared distribution with n degrees of freedom is $(1-2t)^{-n/2}$ [see Example 6.9.8 concerning the moment generating function of a gamma distribution $G(\alpha, \beta)$, of which the chi-squared distribution is a special case with $\alpha = n/\alpha$, $\beta = 2$].

CHAPTER 11

11.1. (a)

$$a_0 = \frac{1}{\pi}\int_{-\pi}^{\pi}|x|\,dx$$

$$= \frac{2}{\pi}\int_0^\pi x\,dx = \pi,$$

$$a_n = \frac{1}{\pi}\int_{-\pi}^{\pi}|x|\cos nx\,dx$$

$$= \frac{2}{\pi}\int_0^\pi x\cos nx\,dx$$

$$= -\frac{2}{n\pi} \int_0^\pi \sin nx \, dx$$

$$= \frac{2}{\pi n^2} \left[(-1)^n - 1 \right],$$

$$b_n = \frac{1}{\pi} \int_{-\pi}^\pi |x| \sin nx \, dx$$

$$= 0,$$

$$|x| = \frac{\pi}{2} - \frac{4}{\pi} \left(\cos x + \frac{\cos 3x}{3^2} + \frac{\cos 5x}{5^2} + \cdots \right).$$

(b)

$$a_0 = \frac{1}{\pi} \int_{-\pi}^\pi |\sin x| \, dx$$

$$= \frac{2}{\pi} \int_0^\pi \sin x \, dx = \frac{4}{\pi},$$

$$a_n = \frac{1}{\pi} \int_{-\pi}^\pi |\sin x| \cos nx \, dx$$

$$= \frac{2}{\pi} \int_0^\pi \sin x \cos nx \, dx$$

$$= \frac{1}{\pi} \int_0^\pi \left[\sin(n+1)x - \sin(n-1)x \right] dx$$

$$= -2 \frac{(-1)^n + 1}{\pi(n^2 - 1)}, \qquad n \ne 1,$$

$$a_1 = 0,$$

$$b_n = \frac{1}{\pi} \int_{-\pi}^\pi |\sin x| \sin nx \, dx$$

$$= 0,$$

$$|\sin x| = \frac{2}{\pi} - \frac{4}{\pi} \left(\frac{\cos 2x}{3} + \frac{\cos 4x}{15} + \frac{\cos 6x}{35} + \cdots \right).$$

(c)

$$a_0 = \frac{1}{\pi} \int_{-\pi}^{\pi} (x + x^2)\, dx$$

$$= \frac{2\pi^2}{3},$$

$$a_n = \frac{1}{\pi} \int_{-\pi}^{\pi} (x + x^2) \cos nx\, dx$$

$$= \frac{2}{\pi} \int_{0}^{\pi} x^2 \cos nx\, dx$$

$$= \frac{4}{n^2} \cos n\pi = (-1)^n \frac{4}{n^2},$$

$$b_n = \frac{1}{\pi} \int_{-\pi}^{\pi} (x + x^2) \sin nx\, dx$$

$$= \frac{2}{\pi} \int_{0}^{\pi} x \sin nx\, dx$$

$$= (-1)^{n-1} \frac{2}{n},$$

$$x + x^2 = \frac{\pi^2}{3} + 4\left(-\cos x + \frac{1}{2}\sin x\right)$$

$$- 4\left(-\frac{1}{2^2}\cos 2x + \frac{1}{4}\sin 2x\right) + \cdots.$$

for $-\pi < x < \pi$. When $x = \mp\pi$, the sum of the series is $\frac{1}{2}[(-\pi + \pi^2) + (\pi + \pi^2)] = \pi^2$, that is, $\pi^2 = \pi^2/3 + 4(1 + 1/2^2 + 1/3^2 + \cdots)$.

11.2. From Example 11.2.2, we have

$$x^2 = \frac{\pi^2}{3} + 4\sum_{n=1}^{\infty} \frac{(-1)^n}{n^2} \cos nx.$$

Putting $x = \mp\pi$, we get

$$\frac{\pi^2}{6} = \sum_{n=1}^{\infty} \frac{1}{n^2}.$$

Using $x = 0$, we obtain

$$0 = \frac{\pi^2}{3} + 4 \sum_{n=1}^{\infty} \frac{(-1)^n}{n^2},$$

or equivalently,

$$\frac{\pi^2}{12} = \sum_{n=1}^{\infty} \frac{(-1)^{n+1}}{n^2}.$$

Adding the two series corresponding to $\pi^2/6$ and $\pi^2/12$, we get

$$\frac{3\pi^2}{12} = 2 \sum_{n=1}^{\infty} \frac{1}{(2n-1)^2},$$

that is,

$$\frac{\pi^2}{8} = \sum_{n=1}^{\infty} \frac{1}{(2n-1)^2}.$$

11.3. By Theorem 11.3.1, we have

$$f'(x) = \sum_{n=1}^{\infty} (nb_n \cos nx - na_n \sin nx).$$

Furthermore, inequality (11.28) indicates that $\sum_{n=1}^{\infty} (n^2 b_n^2 + n^2 a_n^2)$ is a convergent series. It follows that $nb_n \to 0$ and $na_n \to 0$ as $n \to \infty$ (see Result 5.2.1 in Section 5.2).

11.4. For $m > n$, we have

$$|s_m(x) - s_n(x)| = \left| \sum_{k=n+1}^{m} (a_k \cos kx + b_k \sin kx) \right|$$

$$\leq \sum_{k=n+1}^{m} (a_k^2 + b_k^2)^{1/2}$$

$$= \sum_{k=n+1}^{m} \frac{1}{k} (\alpha_k^2 + \beta_k^2)^{1/2} \qquad [\text{see (11.27)}]$$

$$\leq \left(\sum_{k=n+1}^{m} \frac{1}{k^2} \right)^{1/2} \left[\sum_{k=n+1}^{m} (\alpha_k^2 + \beta_k^2) \right]^{1/2}.$$

But, by inequality (11.28),

$$\sum_{k=n+1}^{m} (\alpha_k^2 + \beta_k^2) \le \frac{1}{\pi} \int_{-\pi}^{\pi} [f'(x)]^2 \, dx$$

and

$$\sum_{k=n+1}^{m} \frac{1}{k^2} \le \sum_{k=n+1}^{\infty} \frac{1}{k^2}$$

$$\le \int_{n}^{\infty} \frac{dx}{x^2} = \frac{1}{n}.$$

Thus,

$$|s_m(x) - s_n(x)| \le \frac{c}{\sqrt{n}},$$

where

$$c^2 = \frac{1}{\pi} \int_{-\pi}^{\pi} [f'(x)]^2 \, dx.$$

By Theorem 11.3.1(b), $s_m(x) \to f(x)$ on $[-\pi, \pi]$ as $m \to \infty$. Thus, by letting $m \to \infty$, we get

$$|f(x) - s_n(x)| \le \frac{c}{\sqrt{n}}.$$

11.5. (a) From the proof of Theorem 11.3.2, we have

$$\sum_{n=1}^{\infty} \frac{b_n}{n} \cos n\pi = \frac{A_0}{2} - \frac{a_0 \pi}{2}.$$

This implies that the series $\sum_{n=1}^{\infty} (-1)^n b_n/n$ is convergent.

(b) From the proof of Theorem 11.3.2, we have

$$\int_{-\pi}^{x} f(t) \, dt = \frac{a_0(\pi + x)}{2} + \sum_{n=1}^{\infty} \left[\frac{a_n}{n} \sin nx - \frac{b_n}{n} (\cos nx - \cos n\pi) \right].$$

Putting $x = 0$, we get

$$\int_{-\pi}^{0} f(t) \, dt = \frac{a_0 \pi}{2} - \sum_{n=1}^{\infty} \frac{b_n}{n} [1 - (-1)^n].$$

This implies convergence of the series $\sum_{n=1}^{\infty} (b_n/n)[1 - (-1)^n]$. Since $\sum_{n=1}^{\infty} (-1)^n b_n/n$ is convergent by part (a), then so is $\sum_{n=1}^{\infty} b_n/n$.

11.6. Using the hint and part (b) of Exercise 11.5, the series $\sum_{n=2}^{\infty} b_n/n$ would be convergent, where $b_n = 1/\log n$. However, the series $\sum_{n=2}^{\infty} 1/(n \log n)$ is divergent by Maclaurin's integral test (see Theorem 6.5.4).

11.7. (a) From Example 11.2.1, we have

$$\frac{x}{2} = \sum_{n=1}^{\infty} \frac{(-1)^{n+1}}{n} \sin nx$$

for $-\pi < x < \pi$. Hence,

$$\frac{x^2}{4} = \int_0^x \frac{t}{2} \, dt$$

$$= -\sum_{n=1}^{\infty} \frac{(-1)^{n+1}}{n^2} \cos nx + \sum_{n=1}^{\infty} \frac{(-1)^{n+1}}{n^2}.$$

Note that

$$\sum_{n=1}^{\infty} \frac{(-1)^{n+1}}{n^2} = \frac{1}{2\pi} \int_{-\pi}^{\pi} \frac{x^2}{4} \, dx$$

$$= \frac{\pi^2}{12}.$$

Hence,

$$\frac{x^2}{4} = \frac{\pi^2}{12} - \sum_{n=1}^{\infty} \frac{(-1)^{n+1}}{n^2} \cos nx, \qquad -\pi < x < \pi.$$

(b) This follows directly from part (a).

11.8. Using the result in Exercise 11.7, we obtain

$$\int_0^x \left(\frac{\pi^2}{12} - \frac{t^2}{4} \right) dt = \sum_{n=1}^{\infty} \frac{(-1)^{n+1}}{n^2} \int_0^x \cos nt \, dt$$

$$= \sum_{n=1}^{\infty} \frac{(-1)^{n+1}}{n^3} \sin nx, \qquad -\pi < x < \pi.$$

Hence,

$$\sum_{n=1}^{\infty} \frac{(-1)^{n+1}}{n^3} \sin nx = \frac{\pi^2}{12} x - \frac{x^3}{12}, \qquad -\pi < x < \pi.$$

11.9.

$$F(w) = \frac{1}{2\pi} \int_{-\infty}^{\infty} e^{-x^2} e^{-iwx} \, dx,$$

$$F'(w) = \frac{i}{4\pi} \int_{-\infty}^{\infty} (-2xe^{-x^2}) e^{-iwx} \, dx$$

(exchanging the order of integration and differentation is permissible here by an extension of Theorem 7.10.1). Integrating by parts, we obtain

$$F'(w) = \frac{i}{4\pi} \left[e^{-x^2} e^{-iwx} \Big|_{-\infty}^{\infty} - \int_{-\infty}^{\infty} e^{-x^2} (-iwe^{-iwx}) \, dx \right]$$

$$= \frac{-i}{4\pi} \int_{-\infty}^{\infty} e^{-x^2} (-iw) e^{-iwx} \, dx$$

$$= -\frac{w}{2} \frac{1}{2\pi} \int_{-\infty}^{\infty} e^{-x^2} e^{-iwx} \, dx$$

$$= -\frac{w}{2} F(w).$$

The general solution of this differential equation is

$$F(w) = Ae^{-w^2/4},$$

where A is a constant. Putting $w = 0$, we obtain

$$A = F(0) = \frac{1}{2\pi} \int_{-\infty}^{\infty} e^{-x^2} \, dx$$

$$= \frac{\sqrt{\pi}}{2\pi} = \frac{1}{2\sqrt{\pi}}.$$

Hence, $F(w) = (1/2\sqrt{\pi}) e^{-w^2/4}$.

11.10. Let $H(w)$ be the Fourier transform of $(f * g)(x)$. Then

$$H(w) = \frac{1}{2\pi} \int_{-\infty}^{\infty} \left[\int_{-\infty}^{\infty} f(x-y) g(y) \, dy \right] e^{-iwx} \, dx$$

$$= \frac{1}{2\pi} \int_{-\infty}^{\infty} g(y) \left[\int_{-\infty}^{\infty} f(x-y) e^{-iwx} \, dx \right] dy$$

$$= \int_{-\infty}^{\infty} \left[\frac{1}{2\pi} \int_{-\infty}^{\infty} f(x-y) e^{-iw(x-y)} \, dx \right] g(y) e^{-iwy} \, dy$$

$$= F(w) \int_{-\infty}^{\infty} g(y) e^{-iwy} \, dy$$

$$= 2\pi F(w) G(w).$$

11.11. Apply the results in Exercises 11.9 and 11.10 to find the Fourier transform of $f(x)$; then apply the inverse Fourier transform theorem (Theorem 11.6.3) to find $f(x)$.

11.12. (a)

$$
s_n(x_n) = \sum_{k=1}^{n} \frac{2(-1)^{k+1}}{k} \sin\left[k\left(\pi - \frac{\pi}{n} \right) \right]
$$

$$
= \sum_{k=1}^{n} \frac{2(-1)^{k+1}}{k} (-1)^{k+1} \sin\left(\frac{k\pi}{n} \right)
$$

$$
= \sum_{k=1}^{n} \frac{2\sin(k\pi/n)}{k} .
$$

(b)

$$
s_n(x_n) = 2 \sum_{k=1}^{n} \frac{\sin(k\pi/n)}{k\pi/n} \frac{\pi}{n} .
$$

By dividing the interval $[0, \pi]$ into n subintervals $[(k - 1)\pi/n, k\pi/n]$, $1 \le k \le n$, each of length π/n, it is easy to see that $s_n(x_n)$ is a Riemann sum $S(P, g)$ for the function $g(x) = (2 \sin x)/x$. Here, $g(x)$ is evaluated at the right-hand end point of each subinterval of the partition P (see Section 6.1). Thus,

$$
\lim_{n \to \infty} s_n(x_n) = 2 \int_0^{\pi} \frac{\sin x}{x} \, dx.
$$

11.13. (a) $\hat{y} = 8.512 + 3.198 \cos \phi - 0.922 \sin \phi + 1.903 \cos 2\phi + 3.017 \sin 2\phi$.

(b) Estimates of the locations of minimum and maximum resistance are $\phi = 0.7944\pi$ and $\phi = 0.1153\pi$, respectively. [*Note*: For more details, see Kupper (1972), Section 5.]

11.15. (a)

$$
s_n^* = \frac{1}{\sigma\sqrt{n}} \left(\sum_{j=1}^{n} Y_j - n\mu \right)
$$

$$
= \frac{1}{\sigma\sqrt{n}} \sum_{j=1}^{n} U_j,
$$

where $U_j = Y_j - \mu$, $j = 1, 2, \ldots, n$. Note that $E(U_j) = 0$ and $\mathrm{Var}(U_j) = \sigma^2$. The characteristic function of s_n^* is

$$\phi_n(t) = E(e^{it\, s_n^*})$$

$$= E(e^{it\sum_{j=1}^n U_j / \sigma\sqrt{n}})$$

$$= \prod_{j=1}^n E(e^{it\, U_j / \sigma\sqrt{n}}).$$

Now,

$$E(e^{it\, U_j}) = E\left(1 + it\, U_j - \frac{t^2}{2} U_j^2 + \cdots\right)$$

$$= 1 - \frac{t^2\sigma^2}{2} + o(t^2),$$

$$E(e^{it\, U_j / \sigma\sqrt{n}}) = 1 - \frac{t^2}{2n} + o\left(\frac{t^2}{n}\right).$$

Hence,

$$\phi_n(t) = \left[1 - \frac{t^2}{2n} + o\left(\frac{t^2}{n}\right)\right]^n.$$

Let

$$\omega_n = -\frac{t^2}{2n} + o\left(\frac{t^2}{n}\right)$$

$$= \frac{t^2}{n}\left[-\frac{1}{2} + o(1)\right],$$

$$\phi_n(t) = \left[(1 + \omega_n)^{t^2/\omega_n}\right]^{-1/2 + o\,(1)}$$

$$= \left[(1 + \omega_n)^{t^2/\omega_n}\right]^{-1/2} \left[(1 + \omega_n)^{t^2/\omega_n}\right]^{o(1)}.$$

As $n \to \infty$, $\omega_n \to 0$ and

$$\left[(1 + \omega_n)^{t^2/\omega_n}\right]^{-1/2} \to e^{-t^2/2},$$

$$\left[(1 + \omega_n)^{t^2/\omega_n}\right]^{o(1)} \to 1.$$

Thus,

$$\phi_n(t) \to e^{-t^2/2} \qquad \text{as } n \to \infty.$$

(b) $e^{-t^2/2}$ is the characteristic function of the standard normal distribution Z. Hence, by Theorem 11.7.3, as $n \to \infty$, $s_n^* \to Z$.

CHAPTER 12

12.1. (a)

$$I_n \approx S_n$$

$$= \frac{h}{2}\left(\log 1 + \log n + 2\sum_{i=1}^{n-1} \log x_i\right)$$

$$= \frac{n-1}{2(n-1)}\left(\log n + 2\sum_{i=2}^{n-1} \log i\right)$$

$$= \tfrac{1}{2}\log n + \log 2 + \log 3 + \cdots + \log(n-1)$$

$$= \tfrac{1}{2}\log n + \log(n!) - \log n$$

$$= \log(n!) - \tfrac{1}{2}\log n.$$

(b) $n \log n - n + 1 \approx \log(n!) - \tfrac{1}{2}\log n$. Hence,

$$\log(n!) \approx \left(n + \tfrac{1}{2}\right)\log n - n + 1$$

$$n! \approx \exp\left[\left(n + \tfrac{1}{2}\right)\log n - n + 1\right]$$

$$= e e^{-n} n^{n+1/2} = e^{-n+1} n^{n+1/2}.$$

12.2. Applying formula (12.10), we obtain the following approximate values for the integral $\int_0^1 dx/(1+x)$:

n	Approximation
2	0.69325395
4	0.69315450
8	0.69314759
16	0.69314708

The exact value of the integral is $\log 2 = 0.69314718$. Using formula (12.12), the error of approximation is less than or equal to

$$\frac{M_4(b-a)^5}{2880n^4} = \frac{M_4}{2880(8)^4},$$

where M_4 is an upper bound on $|f^{(4)}(x)|$ for $0 \le x \le 1$. Here, $f(x) = 1/(1+x)$, and

$$f^{(4)}(x) = \frac{24}{(1+x)^5}.$$

Hence, $M_4 = 24$, and

$$\frac{M_4}{2880(8)^4} \approx 0.000002.$$

12.3. Let

$$z = \frac{2\theta - \pi/2}{\pi/2}, \qquad 0 \le \theta \le \frac{\pi}{2},$$

$$\theta = \frac{\pi}{4}(1+z), \qquad -1 \le z \le 1.$$

Then

$$\int_0^{\pi/2} \sin \theta \, d\theta = \frac{\pi}{4} \int_{-1}^1 \sin\left[\frac{\pi}{4}(1+z)\right] dz$$

$$\approx \frac{\pi}{4} \sum_{i=0}^2 \omega_i \sin\left[\frac{\pi}{4}(1+z_i)\right].$$

Using the information in Example 12.4.1, we have $z_0 = -\sqrt{\frac{3}{5}}$, $z_1 = 0$, $z_2 = \sqrt{\frac{3}{5}}$; $\omega_0 = \frac{5}{9}$, $\omega_1 = \frac{8}{9}$, $\omega_2 = \frac{5}{9}$. Hence,

$$\int_0^{\pi/2} \sin \theta \, d\theta \approx \frac{\pi}{4} \left\{ \frac{5}{9} \sin\left[\frac{\pi}{4}\left(1 - \sqrt{\frac{3}{5}}\right)\right] + \frac{8}{9} \sin\frac{\pi}{4} \right.$$

$$\left. + \frac{5}{9} \sin\left[\frac{\pi}{4}\left(1 + \sqrt{\frac{3}{5}}\right)\right] \right\}$$

$$= 1.0000081.$$

The error of approximation associated with $\int_{-1}^1 \sin[(\pi/4)(1+z)] \, dz$ is obtained by applying formula (12.16) with $a = -1$, $b = 1$, $f(z) = \sin[(\pi/4)(1+z)]$, $n = 2$. Thus, $f^{(6)}(\xi) = -(\pi/4)^6 \sin[(\pi/4)(1+\xi)]$, $-1 < \xi < 1$, and the error is therefore less than

$$\frac{2^7(3!)^4}{7(6!)^3} \left(\frac{\pi}{4}\right)^6.$$

Consequently, the error associated with $\int_0^{\pi/2} \sin\theta \, d\theta$ is less than

$$\frac{\pi}{4} \frac{2^7 [3!]^4}{7[6!]^3} \left(\frac{\pi}{4}\right)^6 = \left(\frac{\pi}{2}\right)^7 \frac{[3!]^4}{7[6!]^3}$$

$$= 0.0000117.$$

12.4. (a) The inequality is obvious from the hint.
   (b) Let $m = 3$. Then,

$$\frac{1}{m} e^{-m^2} = \frac{1}{3} e^{-9}$$

$$= 0.0000411.$$

   (c) Apply Simpson's method on $[0, 3]$ with $h < 0.2$. Here,

$$f^{(4)}(x) = (12 - 48x^2 + 16x^4)e^{-x^2},$$

with a maximum of 12 at $x = 0$ over $[0, 3]$. Hence, $M_4 = 12$. Using the fact that $h = (b - a)/2n$, we get $n = (b - a)/2h > 3/0.4 = 7.5$. Formula (12.12), with $n = 8$, gives

$$\frac{nM_4h^5}{90} < \frac{8(12)(0.2)^5}{90}$$

$$= 0.00034.$$

Hence, the total error of approximation for $\int_0^\infty e^{-x^2} dx$ is less than $0.0000411 + 0.00034 = 0.00038$. This makes the approximation correct to three decimal places.

12.5.

$$\int_0^\infty \frac{x}{e^x + e^{-x} - 1} \, dx = \int_0^\infty \frac{xe^{-x}}{1 + e^{-2x} - e^{-x}} \, dx,$$

$$\approx \sum_{i=0}^n \omega_i \frac{x_i}{1 + e^{-2x_i} - e^{-x_i}},$$

where the x_i's are the zeros of the Laguerre polynomial $L_{n+1}(x)$. Using the entries in Table 12.1 for $n = 1$, we get

$$\int_0^\infty \frac{x}{e^x + e^{-x} - 1} \, dx \approx 0.85355 \frac{0.58579}{1 + e^{-2(0.58579)} - e^{-0.58579}}$$

$$+ 0.14645 \frac{3.41421}{1 + e^{-2(3.41421)} - e^{-3.41421}} = 1.18.$$

12.6. (a) Let $u = (2t - x)/x$. Then

$$I(x) = \frac{1}{2} \int_{-1}^{1} \frac{x}{1 + \frac{1}{4}x^2(u+1)^2}\, du$$

$$= 2x \int_{-1}^{1} \frac{1}{4 + x^2(u+1)^2}\, du.$$

(b) Applying a Gauss−Legendre approximation of $I(x)$ with $n = 4$, we obtain

$$I(x) \approx 2x \sum_{i=0}^{4} \frac{\omega_i}{4 + x^2(u_i + 1)^2},$$

where u_0, u_1, u_2, u_3, and u_4 are the five zeros of the Legendre polynomial $P_5(u)$ of degree 5. Values of these zeros and the corresponding ω_i's are given below:

$$u_0 = 0.90617985, \qquad \omega_0 = 0.23692689,$$
$$u_1 = 0.53846931, \qquad \omega_1 = 0.47862867,$$
$$u_2 = 0, \qquad \omega_2 = 0.56888889,$$
$$u_3 = -0.53846931, \qquad \omega_3 = 0.47862867,$$
$$u_4 = -0.90617985, \qquad \omega_4 = 0.23692689,$$

(see Table 7.5 in Shoup, 1984, page 202).

12.7. $\int_0^1 (\cos x)^\lambda\, dx = \int_0^1 e^{\lambda \log \cos x}\, dx$. Using the method of Laplace, the function $h(x) = \log \cos x$ has a single maximum in $[0, 1]$ at $x = 0$. Formula (12.28) gives

$$\int_0^1 e^{\lambda \log \cos x}\, dx \sim \left[\frac{-\pi}{2\lambda h''(0)} \right]^{1/2}$$

as $\lambda \to \infty$, where

$$h''(0) = - \frac{1}{\cos^2 x}\Big|_{x=0}$$
$$= -1.$$

Hence,

$$\int_0^1 e^{\lambda \log \cos x}\, dx \sim \sqrt{\frac{\pi}{2\lambda}}.$$

12.9.

$$\int_0^1 \int_0^{(1-x_1^2)^{1/2}} \left(1 - x_1^2 - x_2^2\right)^{1/2} dx_1\, dx_2 = \frac{\pi}{6}$$

$$\approx 0.52359878.$$

Applying the 16-point Gauss–Legendre rule to both formulas gives the approximate value 0.52362038.
[*Note:* For more details, see Stroud (1971, Section 1.6).]

12.10.

$$\int_0^\infty \left(1 + x^2\right)^{-n} dx = \int_0^\infty e^{-n\log(1+x^2)}\, dx$$

Apply the method of Laplace with $\varphi(x) = 1$ and

$$h(x) = -\log(1 + x^2).$$

This function has a single maximum in $[0, \infty)$ at $x = 0$. Using formula (12.28), we get

$$\int_0^\infty \left(1 + x^2\right)^{-n} dx \sim \left[\frac{-\pi}{2n h''(0)}\right]^{1/2},$$

where $h''(0) = -2$. Hence,

$$\int_0^\infty \left(1 + x^2\right)^{-n} dx \sim \sqrt{\frac{\pi}{4n}} = \frac{1}{2}\sqrt{\frac{\pi}{n}}.$$

12.11.

Number of Quadrature Points	Approximate Value of Variance
2	0.1175
4	0.2380
8	0.2341
16	0.1612
32	0.1379
64	0.1372

[*Source:* Example 16.6.1 in Lange (1999).]

12.12. If $x \geq 0$, then

$$\Phi(x) = \frac{1}{\sqrt{2\pi}} \int_{-\infty}^0 e^{-t^2/2}\, dt + \frac{1}{\sqrt{2\pi}} \int_0^x e^{-t^2/2}\, dt$$

$$= \frac{1}{2} + \frac{1}{\sqrt{2\pi}} \int_0^x e^{-t^2/2}\, dt.$$

Using Maclaurin's series (see Section 4.3), we obtain

$$\Phi(x) = \frac{1}{2} + \frac{1}{\sqrt{2\pi}} \int_0^x \left[1 + \frac{1}{1!}\left(-\frac{t^2}{2}\right) + \frac{1}{2!}\left(-\frac{t^2}{2}\right)^2 \right.$$

$$\left. + \frac{1}{3!}\left(-\frac{t^2}{2}\right)^3 + \cdots + \frac{1}{n!}\left(-\frac{t^2}{2}\right)^n + \cdots \right] dt$$

$$= \frac{1}{2} + \frac{1}{\sqrt{2\pi}} \left[t - \frac{t^3}{3 \times 2 \times 1!} + \frac{t^5}{5 \times 2^2 \times 2!} - \frac{t^7}{7 \times 2^3 \times 3!} \right.$$

$$\left. + \cdots + (-1)^n \frac{t^{2n+1}}{(2n+1) \times 2^n \times n!} + \cdots \right]_0^x$$

$$= \frac{1}{2} + \frac{x}{\sqrt{2\pi}} \left[1 - \frac{x^2}{3 \times 2 \times 1!} + \frac{x^4}{5 \times 2^2 \times 2!} - \frac{x^6}{7 \times 2^3 \times 3!} \right.$$

$$\left. + \cdots + (-1)^n \frac{x^{2n}}{(2n+1) \times 2^n \times n!} + \cdots \right].$$

12.13. From Exercise 12.12 we need to find values of a, b, c, d such that

$$1 + ax^2 + bx^4 \approx (1 + cx^2 + dx^4)\left(1 - \frac{x^2}{6} + \frac{x^4}{40} - \frac{x^6}{336} + \frac{x^8}{3456}\right).$$

Equating the coefficients of x^2, x^4, x^6, x^8 on both sides yields four equations in a, b, c, d. Solving these equations, we get

$$a = \frac{17}{468}, \qquad b = \frac{739}{196560}, \qquad c = \frac{95}{468}, \qquad d = \frac{55}{4368}.$$

[*Note*: For more details, see Morland (1998).]

12.14.

$$y = G(x) = \int_0^x \tfrac{1}{4}(1 + t)\, dt$$

$$= \frac{1}{4}\left(x + \frac{x^2}{2}\right), \qquad 0 \le x \le 2.$$

The only solution of $y = G(x)$ in $[0, 2]$ is $x = -1 + (1 + 8y)^{1/2}$, $0 \le y$

≤ 1. A random sample of size 150 from the $g(x)$ distribution is given by

$$x_i = -1 + (1 + 8y_i)^{1/2}, \qquad i = 1, 2, \dots, 150,$$

where the y_i's form a random sample from the $U(0, 1)$ distribution. Applying formula (12.43), we get

$$\hat{I}_{150}^* = \frac{4}{150} \sum_{i=1}^{150} \frac{e^{x_i^2}}{1 + x_i}$$

$$= 16.6572.$$

Using now formula (12.36), we obtain

$$\hat{I}_{150} = \frac{2}{150} \sum_{i=1}^{150} e^{u_i^2}$$

$$= 17.8878,$$

where the u_i's form a random sample from the $U(0, 2)$ distribution.

12.15. As $n \to \infty$,

$$\left[1 + \frac{x^2}{n} \right]^{-(n+1)/2} = \left[1 + \frac{x^2}{n} \right]^{-n/2} \left[1 + \frac{x^2}{n} \right]^{-1/2}$$

$$\sim e^{-x^2/2}.$$

Furthermore, by applying Stirling's formula (formula 12.31), we obtain

$$\Gamma\left(\frac{n+1}{2} \right) \sim e^{-\frac{1}{2}(n-1)} \left[\frac{n-1}{2} \right]^{(n-1)/2} \left[2\pi \left(\frac{n-1}{2} \right) \right]^{1/2},$$

$$\Gamma\left(\frac{n}{2} \right) \sim e^{-\frac{1}{2}(n-2)} \left[\frac{n-2}{2} \right]^{(n-2)/2} \left[2\pi \left(\frac{n-2}{2} \right) \right]^{1/2}.$$

Hence,

$$
\frac{\Gamma\left(\dfrac{n+1}{2}\right)}{\sqrt{n\pi}\,\Gamma\left(\dfrac{n}{2}\right)} \sim \frac{1}{\sqrt{n\pi}}\,\frac{1}{e^{1/2}}\left[\frac{n-1}{n-2}\right]^{n/2}\frac{n-2}{2\sqrt{\dfrac{n-1}{2}}}\left[\frac{n-1}{n-2}\right]^{1/2}
$$

$$
\sim \frac{1}{\sqrt{n\pi}}\,\frac{1}{e^{1/2}}\,\frac{\left(1-\frac{1}{n}\right)^{n/2}}{\left(1-\frac{2}{n}\right)^{n/2}}\,\frac{n-2}{2\sqrt{\dfrac{n-1}{2}}}
$$

$$
\sim \frac{1}{\sqrt{n\pi}}\,\frac{1}{e^{1/2}}\,\frac{e^{-1/2}}{e^{-1}}\sqrt{\frac{n}{2}}
$$

$$
= \frac{1}{\sqrt{2\pi}}\,.
$$

Hence, for large n,

$$
F(x) \approx \frac{1}{\sqrt{2\pi}}\int_{-\infty}^{x} e^{-t^2/2}\,dt.
$$

General Bibliography

Abramowitz, M., and I. A. Stegun (1964). *Handbook of Mathematical Functions with Formulas, Graphs, and Mathematical Tables*. Wiley, New York.

Abramowitz, M., and I. A. Stegun, eds. (1972). *Handbook of Mathematical Functions with Formulas, Graphs, and Mathematical Tables*. Wiley, New York.

Adby, P. R., and M. A. H. Dempster (1974). *Introduction to Optimization Methods*. Chapman and Hall, London.

Agarwal, G. G., and W. J. Studden (1978). "Asymptotic design and estimation using linear splines." *Comm. Statist. Simul. Comput.*, **7**, 309–319.

Alvo, M., and P. Cabilio (2000). "Calculation of hypergeometric probabilities using Chebyshev polynomials." *Amer. Statist.*, **54**, 141–144.

Anderson, T. W., and S. D. Gupta (1963). "Some inequalities on characteristic roots of matrices." *Biometrika*, **50**, 522–524.

Anderson, T. W., I. Olkin, and L. G. Underhill (1987). "Generation of random orthogonal matrices." *SIAM J. Sci. Statist. Comput.*, **8**, 625–629.

Anderson-Cook, C. M. (2000). "A second order model for cylindrical data." *J. Statist. Comput. Simul.*, **66**, 51–56.

Apostol, T. M. (1964). *Mathematical Analysis*. Addison-Wesley, Reading, Massachusetts.

Ash, A., and A. Hedayat (1978). "An introduction to design optimality with an overview of the literature." *Comm. Statist. Theory Methods*, **7**, 1295–1325.

Atkinson, A. C. (1982). "Developments in the design of experiments." *Internat. Statist. Rev.*, **50**. 161–177.

Atkinson, A. C. (1988). "Recent developments in the methods of optimum and related experimental designs." *Internat. Statist. Rev.*, **56**, 99–115.

Basilevsky, A. (1983). *Applied Matrix Algebra in the Statistical Sciences*. North-Holland, New York.

Bates, D. M., and D. G. Watts (1988). *Nonlinear Regression Analysis and its Applications*. Wiley, New York.

Bayne, C. K., and I. B. Rubin (1986). *Practical Experimental Designs and Optimization Methods for Chemists*. VCH Publishers, Deerfield Beach, Florida.

Bellman, R. (1970). *Introduction to Matrix Analysis*, 2nd ed. McGraw-Hill, New York.

Belsley, D. A., E. Kuh, and R. E. Welsch (1980). *Regression Diagnostics*. Wiley, New York.

Bickel, P. J., and K. A. Doksum (1977). *Mathematical Statistics*. Holden-Day, San Francisco.

Biles, W. E., and J. J. Swain (1980). *Optimization and Industrial Experimentation*. Wiley-Interscience, New York.

Bloomfield, P. (1976). *Fourier Analysis of Times Series: An Introduction*. Wiley, New York.

Blyth, C. R. (1990). "Minimizing the sum of absolute deviations." *Amer. Statist.*, **44**, 329.

Bohachevsky, I. O., M. E. Johnson, and M. L. Stein (1986). "Generalized simulated annealing for function optimization." *Technometrics*, **28**, 209–217.

Box, G. E. P. (1982). "Choice of response surface design and alphabetic optimality." *Utilitas Math.*, **21B**, 11–55.

Box, G. E. P., and D. W. Behnken (1960). "Some new three level designs for the study of quantitative variables." *Technometrics*, **2**, 455–475.

Box, G. E. P., and D. R. Cox (1964). "An analysis of transformations." *J. Roy. Statist. Soc. Ser. B*, **26**, 211–243.

Box, G. E. P., and N. R. Draper (1959). "A basis for the selection of a response surface design." *J. Amer. Statist. Assoc.*, **54**, 622–654.

Box, G. E. P., and N. R. Draper (1963). "The choice of a second order rotatable design." *Biometrika*, **50**, 335–352.

Box, G. E. P., and N. R. Draper (1965). "The Bayesian estimation of common parameters from several responses." *Biometrika*, **52**, 355–365.

Box, G. E. P., and N. R. Draper (1987). *Empirical Model-Building and Response Surfaces*. Wiley, New York.

Box, G. E. P., and H. L. Lucas (1959). "Design of experiments in nonlinear situations." *Biometrika*, **46**, 77–90.

Box, G. E. P., and K. B. Wilson (1951). "On the experimental attainment of optimum conditions." *J. Roy. Statist. Soc. Ser. B*, **13**, 1–45.

Box, G. E. P., W. G. Hunter, and J. S. Hunter (1978). *Statistics for Experimenters*. Wiley, New York.

Boyer, C. B. (1968). *A History of Mathematics*. Wiley, New York.

Bronshtein, I. N., and K. A. Semendyayev (1985). *Handbook of Mathematics*. (English translation edited by K. A. Hirsch.) Van Nostrand Reinhold, New York.

Brownlee, K. A. (1965). *Statistical Theory and Methodology*, 2nd ed. Wiley, New York.

Buck, R. C. (1956). *Advanced Calculus*. McGraw-Hill, New York.

Bunday, B. D. (1984). *Basic Optimization Methods*. Edward Arnold, Victoria, Australia.

Buse, A., and L. Lim (1977). "Cubic splines as a special case of restricted least squares." *J. Amer. Statist. Assoc.*, **72**, 64–68.

Bush, K. A., and I. Olkin (1959). "Extrema of quadratic forms with applications to statistics." *Biometrika*, **46**, 483–486.

Carslaw, H. S. (1930). *Introduction to the Theory of Fourier Series and Integrals*, 3rd ed. Dover, New York.

Chatterjee, S., and B. Price (1977). *Regression Analysis by Example*. Wiley, New York.

Cheney, E. W. (1982). *Introduction to Approximation Theory*, 2nd ed. Chelsea, New York.

Chihara, T. S. (1978). *An Introduction to Orthogonal Polynomials*. Gordon and Breach, New York.

Churchill, R. V. (1963). *Fourier Series and Boundary Value Problems*, 2nd ed. McGraw-Hill, New York.

Cochran, W. G. (1963). *Sampling Techniques*, 2nd ed. Wiley, New York.

Conlon, M. (1991). "The controlled random search procedure for function optimization." Personal communication.

Conlon, M., and A. I. Khuri (1992). "Multiple response optimization." Technical Report, Department of Statistics, University of Florida, Gainesville, Florida.

Cook, R. D., and C. J. Nachtsheim (1980). "A comparison of algorithms for constructing exact D-optimal designs." *Technometrics*, **22**, 315–324.

Cooke, W. P. (1988). "L'Hôpital's rule in a Poisson derivation." *Amer. Math. Monthly*, **95**, 253–254.

Copson, E. T. (1965). *Asymptotic Expansions*. Cambridge University Press, London.

Cornish, E. A., and R. A. Fisher (1937). "Moments and cumulants in the specification of distributions." *Rev. Internat. Statist. Inst.*, **5**, 307–320.

Corwin, L. J., and R. H. Szczarba (1982). *Multivariable Calculus*. Marcel Dekker, New York.

Courant, R., and F. John (1965). *Introduction to Calculus and Analysis*, Vol. 1. Wiley, New York.

Cramér, H. (1946). *Mathematical Methods of Statistics*. Princeton University Press, Princeton.

Daniels, H. (1954). "Saddlepoint approximation in statistics." *Ann. Math. Statist.*, **25**, 631–650.

Dasgupta, P. (1968). "An approximation to the distribution of sample correlation coefficient, when the population is non-normal." *Sankhyā, Ser. B.*, **30**, 425–428.

Davis, H. F. (1963). *Fourier Series and Orthogonal Functions*. Allyn & Bacon, Boston.

Davis, P. J. (1975). *Interpolation and Approximation*. Dover, New York

Davis, P. J., and P. Rabinowitz (1975). *Methods of Numerical Integration*. Academic Press, New York.

De Boor, C. (1978). *A Practical Guide to Splines*. Springer-Verlag, New York.

DeBruijn, N. G. (1961). *Asymptotic Methods in Analysis*, 2nd ed. North-Holland, Amesterdam.

DeCani, J. S., and R. A. Stine (1986). "A note on deriving the information matrix for a logistic distribution." *Amer. Statist.*, **40**, 220–222.

Dempster, A. P., N. M. Laird, and D. B. Rubin (1977). "Maximum likelihood from incomplete data via the *EM* algorithm." *J. Roy. Statist. Soc. Ser. B*, **39**, 1–38.

Divgi, D. R. (1979). "Calculation of univariate and bivariate normal probability functions." *Ann. Statist.*, **7**, 903–910.

Draper, N. R. (1963). "Ridge analysis of response surfaces." *Technometrics*, **5**, 469–479.

Draper, N. R., and A. M. Herzberg (1987). "A ridge-regression sidelight." *Amer. Statist.*, **41**, 282–283.

Draper, N. R., and H. Smith (1981). *Applied Regression Analysis*, 2nd ed. Wiley, New York.

Draper, N. R., I. Guttman, and P. Lipow (1977). "All-bias designs for spline functions joined at the axes." *J. Amer. Statist. Assoc.*, **72**, 424–429.

Dugundji, J. (1966). *Topology*. Allyn and Bacon, Boston.

Durbin, J., and G. S. Watson (1950). "Testing for serial correlation in least squares regression, I." *Biometrika*, **37**, 409–428.

Durbin, J., and G. S. Watson (1951). "Testing for serial correlation in least squares regression, II." *Biometrika*, **38**, 159–178.

Eggermont, P. P. B. (1988). "Noncentral difference quotients and the derivative." *Amer. Math. Monthly*, **95**, 551–553.

Eubank, R. L. (1984). "Approximate regression models and splines." *Comm. Statist. Theory Methods*, **13**, 433–484.

Evans, M., and T. Swartz (1995). "Methods for approximating integrals in statistics with special emphasis on Bayesian integration problems." *Statist. Sci.*, **10**, 254–272.

Everitt, B. S. (1987). *Introduction to Optimization Methods and Their Application in Statistics*. Chapman and Hall, London.

Eves, H. (1976). *An Introduction to the History of Mathematics*, 4th ed. Holt, Rinehart and Winston, New York.

Fedorov, V. V. (1972). *Theory of Optimal Experiments*. Academic Press, New York.

Feller, W. (1968). *An Introduction to Probability Theory and Its Applications*, Vol. I, 3rd ed. Wiley, New York.

Fettis, H. E. (1976). "Fourier series expansions for Pearson Type IV distributions and probabilities." *SIAM J. Applied Math.*, **31**, 511–518.

Fichtali, J., F. R. Van De Voort, and A. I. Khuri (1990). "Multiresponse optimization of acid casein production." *J. Food Process Eng.*, **12**, 247–258.

Fisher, R. A., and E. Cornish (1960). "The percentile points of distribution having known cumulants." *Technometrics*, **2**, 209–225.

Fisher, R. A., A. S. Corbet, and C. B. Williams (1943). "The relation between the number of species and the number of individuals in a random sample of an animal population." *J. Anim. Ecology*, **12**, 42–58.

Fisz, M. (1963). *Probability Theory and Mathematical Statistics*, 3rd ed. Wiley, New York.

Fletcher, R. (1987). *Practical Methods of Optimization*, 2nd ed. Wiley, New York.

Fletcher, R., and M. J. D. Powell (1963). "A rapidly convergent descent method for minimization." *Comput. J.*, **6**, 163–168.

Flournoy, N., and R. K. Tsutakawa, eds. (1991). *Statistical Multiple Integration*, Contemporary Mathematics **115**. Amer. Math. Soc., Providence, Rhode Island.

Freud, G. (1971). *Orthogonal Polynomials*. Pergamon Press, Oxford.

Fulks, W. (1978). *Advanced Calculus*, 3rd ed. Wiley, New York.

Fuller, W. A. (1976). *Introduction to Statistical Time Series*. Wiley, New York.

Gallant, A. R., and W. A. Fuller (1973). "Fitting segmented polynomial regression models whose join points have to be estimated." *J. Amer. Statist. Assoc.*, **68**, 144–147.

Gantmacher, F. R. (1959). *The Theory of Matrices*, Vols. I and II. Chelsea, New York.

Georgiev, A. A. (1984). "Kernel estimates of functions and their derivatives with applications." *Statist. Prob. Lett.*, **2**, 45–50.

Geyer, C. J. (1992). "Practical Markov chain Monte Carlo." *Statist. Sci.*, **7**, 473–511.

Ghazal, G. A. (1994). "Moments of the ratio of two dependent quadratic forms." *Statist. Prob. Lett.*, **20**, 313–319.

Gibaldi, M., and D. Perrier (1982). *Pharmacokinetics*, 2nd ed. Dekker, New York.

Gillespie, R. P. (1954). *Partial Differentiation*. Oliver and Boyd, Edinburgh.

Gillespie, R. P. (1959). *Integration*. Oliver and Boyd, London.

Golub, G. H., and C. F. Van Loan (1983). *Matrix Computations*. Johns Hopkins University Press, Baltimore.

Good, I. J. (1969). "Some applications of the singular decomposition of a matrix." *Technometrics*, **11**, 823–831.

Goutis, C., and G. Casella (1999). "Explaining the saddlepoint approximation." *Amer. Statist.*, **53**, 216–224.

Graybill, F. A. (1961). *An Introduction to Linear Statistical Models*, Vol. I. McGraw-Hill, New York.

Graybill, F. A. (1976). *Theory and Application of the Linear Model*. Duxbury, North Scituate, Massachusetts.

Graybill, F. A. (1983). *Matrices with Applications in Statistics*, 2nd ed. Wadsworth, Belmont, California.

Gurland, J. (1953). "Distributions of quadratic forms and ratios of quadratic forms." *Ann. Math. Statist.*, **24**, 416–427.

Haber, S. (1970). "Numerical evaluation of multiple integrals." *SIAM Rev.*, **12**, 481–526.

Hall, C. A. (1968). "On error bounds for spline interpolation." *J. Approx. Theory*, **1**, 209–218.

Hall, C. A., and W. W. Meyer (1976). "Optimal error bounds for cubic spline interpolation." *J. Approx. Theory*, **16**, 105–122.

Hardy, G. H. (1955). *A Course of Pure Mathematics*, 10th ed. The University Press, Cambridge, England.

Hardy, G. H., J. E. Littlewood, and G. Pólya (1952). *Inequalities*, 2nd ed. Cambridge University Press, Cambridge, England.

Harris, B. (1966). *Theory of Probability*. Addison-Wesley, Reading, Massachusetts.

Hartig, D. (1991). "L'Hôpital's rule via integration." *Amer. Math. Monthly*, **98**, 156–157.

Hartley, H. O., and J. N. K. Rao (1967). "Maximum likelihood estimation for the mixed analysis of variance model." *Biometrika*, **54**, 93–108.

Healy, M. J. R. (1986). *Matrices for Statistics*. Clarendon Press, Oxford, England.

Heiberger, R.M., P. F. Velleman, and M. A. Ypelaar (1983). "Generating test data with independently controllable features for multivariate general linear forms." *J. Amer. Statist. Assoc.*, **78**, 585–595.

Henderson, H. V., and S. R. Searle (1981). "The vec-permutation matrix, the vec operator and Kronecker products: A review." *Linear and Multilinear Algebra*, **9**, 271–288.

Henderson, H. V., and F. Pukelsheim, and S. R. Searle (1983). "On the history of the Kronecker product." *Linear and Multilinear Algebra*, **14**, 113–120.

Henle, J. M., and E. M. Kleinberg (1979). *Infinitesimal Calculus*. The MIT Press, Cambridge, Massachusetts.

Hillier, F. S., and G. J. Lieberman (1967). *Introduction to Operations Research*. Holden-Day, San Francisco.

Hirschman, I. I., Jr. (1962). *Infinite Series*. Holt, Rinehart and Winston, New York.

Hoerl, A. E. (1959). "Optimum solution of many variables equations." *Chem. Eng. Prog.*, **55**, 69–78.

Hoerl, A. E., and R. W. Kennard (1970a). "Ridge regression: Biased estimation for non-orthogonal problems." *Technometrics*, **12**, 55–67.

Hoerl, A. E., and R. W. Kennard (1970b). "Ridge regression: Applications to non-orthogonal problems." *Technometrics*, **12**, 69–82; correction, **12**, 723.

Hogg, R. V., and A. T. Craig (1965). *Introduction to Mathematical Statistics*, 2nd ed. Macmillan, New York.

Huber, P. J. (1973). "Robust regression: Asymptotics, conjectures and Monte Carlo." *Ann. Statist.* **1**, 799–821.

Huber, P. J. (1981). *Robust Statistics*. Wiley, New York.

Hyslop, J. M. (1954). *Infinite Series*, 5th ed. Oliver and Boyd, Edinburgh, England.

Jackson, D. (1941). *Fourier Series and Orthogonal Polynomials*. Mathematical Association of America, Washington.

James, A. T. (1954). "Normal multivariate analysis and the orthogonal group." *Ann. Math. Statist.*, **25**, 40–75.

James, A. T., and R. A. J. Conyers (1985). "Estimation of a derivative by a difference quotient: Its application to hepatocyte lactate metabolism." *Biometrics*, **41**, 467–476.

Johnson, M. E., and C. J. Nachtsheim (1983). "Some guidelines for constructing exact D-optimal designs on convex design spaces." *Technometrics*, **25**, 271–277.

Johnson, N. L., and S. Kotz (1968). "Tables of distributions of positive definite quadratic forms in central normal variables." *Sankhyā, Ser. B*, **30**, 303–314.

Johnson, N. L., and S. Kotz (1969). *Discrete Distributions*. Houghton Mifflin, Boston.

Johnson, N. L., and S. Kotz (1970a). *Continuous Univariate Distributions—1*. Houghton Mifflin, Boston.

Johnson, N. L., and S. Kotz (1970b). *Continuous Univariate Distributions—2*. Houghton Mifflin, Boston.

Johnson, P. E. (1972). *A History of Set Theory*. Prindle, Weber, and Schmidt, Boston.

Jones, E. R., and T. J. Mitchell (1978). "Design criteria for detecting model inadequacy." *Biometrika*, **65**, 541–551.

Judge, G. G., W. E. Griffiths, R. C. Hill, and T. C. Lee (1980). *The Theory and Practice of Econometrics*. Wiley, New York.

Kahaner, D. K. 1991. "A survey of existing multidimensional quadrature routines." In *Statistical Multiple Integration*, Contemporary Mathematics **115**, N. Flournoy and R. K. Tsutakawa, eds. Amer. Math. Soc., Providence, pp. 9–22.

Kaplan, W. (1991). *Advanced Calculus*, 4th ed. Addison-Wesley, Redwood City, California.

Kaplan, W., and D. J. Lewis (1971). *Calculus and Linear Algebra*, Vol. II. Wiley, New York.

Karson, M. J., A. R. Manson, and R. J. Hader (1969). "Minimum bias estimation and experimental design for response surfaces." *Technometrics*, **11**, 461–475.

Kass, R. E., L. Tierney, and J. B. Kadane (1991). "Laplace's method in Bayesian analysis." In *Statistical Multiple Integration*, Contemporary Mathematics **115**, N. Flournoy and R. K. Tsutakawa, eds. Amer. Math. Soc., Providence, pp. 89–99.

Katz, D., and D. Z. D'Argenio (1983). "Experimental design for estimating integrals by numerical quadrature, with applications to pharmacokinetic studies." *Biometrics*, **39**, 621–628.

Kawata, T. (1972). *Fourier Analysis in Probability Theory*. Academic Press, New York.

Kendall, M., and A. Stuart (1977). *The Advanced Theory of Statistics*, Vol. 1, 4th ed. Macmillian, New York.

Kerridge, D. F., and G. W. Cook (1976). "Yet another series for the normal integral." *Biometrika*, **63**, 401–403.

Khuri, A. I. (1982). "Direct products: A powerful tool for the analysis of balanced data." *Comm. Statist. Theory Methods*, **11**, 2903–2920.

Khuri, A. I. (1984). "Interval estimation of fixed effects and of functions of variance components in balanced mixed models." *Sankhyā, Ser. B*, **46**, 10–28.

Khuri, A. I., and G. Casella (2002). "The existence of the first negative moment revisited." *Amer. Statist.*, **56**, 44–47.

Khuri, A. I., and M. Conlon (1981). "Simultaneous optimization of multiple responses represented by polynomial regression functions." *Technometrics*, **23**, 363–375.

Khuri, A. I., and J. A. Cornell (1996). *Response Surfaces*, 2nd ed. Marcel Dekker, New York.

Khuri, A. I., and I. J. Good (1989). "The parameterization of orthogonal matrices: A review mainly for statisticians." *South African Statist. J.*, **23**, 231–250.

Khuri, A. I., and R. H. Myers (1979). "Modified ridge analysis." *Technometrics*, **21**, 467–473.

Khuri, A. I., and R. H. Myers (1981). "Design related robustness of tests in regression models." *Comm. Statist. Theory Methods*, **10**, 223–235.

Khuri, A. I., and H. Sahai (1985). "Variance components analysis: A selective literature survey." *Internat. Statist. Rev*, **53**, 279–300.

Kiefer, J. (1958). "On the nonrandomized optimality and the randomized nonoptimality of symmetrical designs." *Ann. Math. Statist.*, **29**, 675–699.

Kiefer, J. (1959). "Optimum experimental designs" (with discussion). *J. Roy. Statist. Soc., Ser. B*, **21**, 272–319.

Kiefer, J. (1960). "Optimum experimental designs V, with applications to systematic and rotatable designs." In *Proceedings of the Fourth Berkeley Symposium on Mathematical Statistics and Probability*, Vol. 1. University of California Press, Berkeley, pp. 381–405.

Kiefer, J. (1961). "Optimum designs in regression problems II." *Ann. Math. Statist.*, **32**, 298–325.

Kiefer, J. (1962a). "Two more criteria equivalent to D-optimality of designs." *Ann. Math. Statist.*, **33**, 792–796.

Kiefer, J. (1962b). "An extremum result." *Canad. J. Math.*, **14**, 597–601.

Kiefer, J. (1975). "Optimal design: Variation in structure and performance under change of criterion." *Biometrika*, **62**, 277–288.

Kiefer, J., and J. Wolfowitz (1960). "The equivalence of two extremum problems." *Canad. J. Math.*, **12**, 363–366.

Kirkpatrick, S., C. D. Gelatt, and M. P. Vechhi (1983). "Optimization by simulated annealing," *Science*, **220**, 671–680.

Knopp, K. (1951). *Theory and Application of Infinite Series*. Blackie and Son, London.

Kosambi, D. D. (1949). "Characteristic properties of series distributions." *Proc. Nat. Inst. Sci. India*, **15**, 109–113.

Krylov, V. I. (1962). *Approximate Calculation of Integrals*. Macmillan, New York.

Kufner, A., and J. Kadlec (1971). *Fourier Series*. Iliffe Books—The Butterworth Group, London.

Kupper, L. L. (1972). "Fourier series and spherical harmonics regression." *Appl. Statist.*, **21**, 121–130.

Kupper, L. L. (1973). "Minimax designs for Fourier series and spherical harmonics regressions: A characterization of rotatable arrangements." *J. Roy. Statist. Soc.*, *Ser. B*, **35**, 493–500.

Lancaster, P. (1969). *Theory of Matrices*. Academic Press, New York.

Lancaster, P., and K. Salkauskas (1986). *Curve and Surface Fitting*. Academic Press, London.

Lange, K. (1999). *Numerical Analysis for Statisticians*. Springer, New York.

Lehmann, E. L. (1983). *Theory of Point Estimation*. Wiley, New York.

Lieberman, G. J., and Owen, D. B. (1961). *Tables of the Hypergeometric Probability Distribution*. Stanford University Press, Palo Alto, California.

Lieberman, O. (1994). "A Laplace approximation to the moments of a ratio of quadratic forms." *Biometrika*, **81**, 681–690.

Lindgren, B. W. (1976). *Statistical Theory*, 3rd ed. Macmillan, New York.

Little, R. J. A., and D. B. Rubin (1987). *Statistical Analysis with Missing Data*. Wiley, New York.

Liu, Q., and D. A. Pierce (1994). "A note on Gauss–Hermite quadrature." *Biometrika*, **81**, 624–629.

Lloyd, E. (1980). *Handbook of Applicable Mathematics*, Vol. II. Wiley, New York.

Lowerre, J. M. (1982). "An introduction to modern matrix methods and statistics." *Amer. Statist.*, **36**, 113–115.

Lowerre, J. M. (1983). "An integral of the bivariate normal and an application." *Amer. Statist.*, **37**, 235–236.

Lucas, J. M. (1976). "Which response surface design is best." *Technometrics*, **18**, 411–417.

Luceño, A. (1997). "Further evidence supporting the numerical usefulness of characteristic functions." *Amer. Statist.*, **51**, 233–234.

Lukacs, E. (1970). *Characteristic Functions*, 2nd ed. Hafner, New York.

Magnus, J. R., and H. Neudecker (1988). *Matrix Differential Calculus with Applications in Statistics and Econometrics*. Wiley, New York.

Mandel, J. (1982). "Use of the singular-value decomposition in regression analysis." *Amer. Statist.*, **36**, 15–24.

Marcus, M., and H. Minc (1988). *Introduction to Linear Algebra*. Dover, New York.

Mardia, K. V., and T. W. Sutton (1978). "Model for cylindrical variables with applications." *J. Roy. Statist. Soc., Ser. B*, **40**, 229–233.

Marsaglia, G., and G. P. H. Styan (1974). "Equalities and inequalities for ranks of matrices." *Linear and Multilinear Algebra*, **2**, 269–292.

May, W. G. (1970). *Linear Algebra*. Scott, Foresman and Company, Glenview, Illinois.

McCullagh, P. (1994). "Does the moment-generating function characterize a distribution?" *Amer. Statist.*, **48**, 208.

Menon, V. V., B. Prasad, and R. S. Singh (1984). "Non-parametric recursive estimates of a probability density function and its derivatives." *J. Statist. Plann. Inference*, **9**, 73–82.

Miller, R. G., Jr. (1981). *Simultaneous Statistical Inference*, 2nd ed. Springer, New York.

Milliken, G. A., and D. E. Johnson (1984). *Analysis of Messy Data*. Lifetime Learning Publications, Belmont, California.

Mitchell, T. J. (1974). "An algorithm for the construction of *D*-optimal experimental designs." *Technometrics*, **16**, 203–210.

Montgomery, D. C., and E. A. Peck (1982). *Introduction to Linear Regression Analysis*. Wiley, New York.

Moran, P. A. P. (1968). *An Introduction to Probability Theory*. Clarendon Press, Oxford.

Morin, D. (1992). "Exact moments of ratios of quadratic forms." *Metron*, **50**, 59–78.

Morland, T. (1998). "Approximations to the normal distribution function." *Math. Gazette*, **82**, 431–437.

Morrison, D. F. (1967). *Multivariate Statistical Methods*. McGraw-Hill, New York.

Muirhead, R. J. (1982). *Aspects of Multivariate Statistical Theory*. Wiley, New York.

Myers, R. H. (1976). *Response Surface Methodology*. Author, Blacksburg, Virginia.

Myers, R. H. (1990). *Classical and Modern Regression with Applications*, 2nd ed. PWS-Kent, Boston.

Myers, R. H., and W. H. Carter, Jr. (1973). "Response surface techniques for dual response systems." *Technometrics*, **15**, 301–317.

Myers, R. H., and A. I. Khuri (1979). "A new procedure for steepest ascent." *Comm. Statist. Theory Methods*, **8**, 1359–1376.

Myers, R. H., A. I. Khuri, and W. H. Carter, Jr. (1989). "Response surface methodology: 1966–1988." *Technometrics*, **31**, 137–157.

Nelder, J. A., and R. Mead (1965). "A simplex method for function minimization." *Comput. J.*, **7**, 308–313.

Nelson, L. S. (1973). "A sequential simplex procedure for non-linear least-squares estimation and other function minimization problems." In *27th Annual Technical Conference Transactions*, American Society for Quality Control, pp. 107–117.

Newcomb, R. W. (1960). "On the simultaneous diagonalization of two semidefinite matrices." *Quart. Appl. Math.*, **19**, 144–146.

Nonweiler, T. R. F. (1984). *Computational Mathematics*. Ellis Horwood, Chichester, England.

Nurcombe, J. R, (1979). "A sequence of convergence tests." *Amer. Math. Monthly*, **86**, 679–681.

Ofir, C., and A. I. Khuri (1986). "Multicollinearity in marketing models: Diagnostics and remedial measures." *Internat. J. Res. Market.*, **3**, 181–205.

Olkin, I. (1990). "Interface between statistics and linear algebra." In *Matrix Theory and Applications*, Vol. 40, C. R. Johnson, ed., American Mathematical Society, Providence, pp. 233–256.

Olsson, D. M., and L. S. Nelson (1975). "The Nelder–Mead simplex procedure for function minimization." *Technometrics*, **17**, 45–51.

Otnes, R. K., and L. Enochson (1978). *Applied Time Series Analysis*. Wiley, New York.

Park, S. H. (1978). "Experimental designs for fitting segmented polynomial regression models." *Technometrics*, **20**, 151–154.

Parzen, E. (1962). *Stochastic Processes*. Holden-Day, San Francisco.

Pazman, A. (1986). *Foundations of Optimum Experimental Design*. Reidel, Dordrecht, Holland.

Pérez-Abreu, V. (1991). "Poisson approximation to power series distributions." *Amer. Statist.*, **45**, 42–45.

Pfeiffer, P. E. (1990). *Probability for Applications*. Springer, New York.

Phillips, C., and B. Cornelius (1986). *Computational Numerical Methods*. Ellis Horwood, Chichester, England.

Phillips, G. M., and P. J. Taylor (1973). *Theory and Applications of Numerical Analysis*. Academic Press, New York.

Piegorsch, W. W., and A. J. Bailer (1993). "Minimum mean-square error quadrature." *J. Statist. Comput. Simul.*, **46**, 217–234.

Piegorsch, W. W., and G. Casella (1985). "The existence of the first negative moment." *Amer. Statist.*, **39**, 60–62.

Pinkus, A., and S. Zafrany (1997). *Fourier Series and Integral Transforms*. Cambridge University Press, Cambridge, England.

Plackett, R. L., and J. P. Burman (1946). "The design of optimum multifactorial experiments." *Biometrika*, **33**, 305–325.

Poirier, D. J. (1973). "Piecewise regression using cubic splines." *J. Amer. Statist. Assoc.*, **68**, 515–524.

Powell, M. J. D. (1967). "On the maximum errors of polynomial approximations defined by interpolation and by least squares criteria." *Comput. J.*, **9**, 404–407.

Prenter, P. M. (1975). *Splines and Variational Methods*. Wiley, New York.

Price, G. B. (1947). "Some identities in the theory of determinants." *Amer. Math. Monthly*, **54**, 75–90.

Price, W. L. (1977). "A controlled random search procedure for global optimization." *Comput. J.*, **20**, 367–370.

Pye, W. C., and P. G. Webster (1989). "A note on Raabe's test extended." *Math. Comput. Ed.*, **23**, 125–128.

Ralston, A., and P. Rabinowitz (1978). *A First Course in Numerical Analysis*. McGraw-Hill, New York.

Ramsay, J. O. (1988). "Monotone regression splines in action." *Statist. Sci.*, **3**, 425–461.

Randles, R. H., and D. A. Wolfe (1979). *Introduction to the Theory of Nonparametric Statistics*. Wiley, New York.

Rao, C. R. (1970). "Estimation of heteroscedastic variances in linear models." *J. Amer. Statist. Assoc.*, **65**, 161–172.

Rao, C. R. (1971). "Estimation of variance and covariance components—MINQUE theory." *J. Multivariate Anal.*, **1**, 257–275.

Rao, C. R. (1972). "Estimation of variance and covariance components in linear models." *J. Amer. Statist. Assoc.*, **67**, 112–115.

Rao, C. R. (1973). *Linear Statistical Inference and its Applications*, 2nd ed. Wiley, New York.

Reid, W. H., and S. J. Skates (1986). "On the asymptotic approximation of integrals." *SIAM J. Appl. Math*, **46**, 351–358.

Rice, J. R. (1969). *The Approximation of Functions*, Vol. 2. Addison-Wesley, Reading, Massachusetts.

Rivlin, T. J. (1969). *An Introduction to the Approximation of Functions*. Dover, New York.

Rivlin, T. J. (1990). *Chebyshev Polynomials*, 2nd ed. Wiley, New York.

Roberts, A. W., and D. E. Varberg (1973). *Convex Functions*. Academic Press, New York.

Rogers, G. S. (1984). "Kronecker products in ANOVA—A first step." *Amer. Statist.*, **38**, 197–202.

Roussas, G. G. (1973). *A First Course in Mathematical Statistics*. Addison-Wesley, Reading, Massachusetts.

Rudin, W. (1964). *Principles of Mathematical Analysis*, 2nd ed. McGraw-Hill, New York.

Rustagi, J. S., ed. (1979). *Optimizing Methods in Statistics*. Academic Press, New York.

Sagan, H. (1974). *Advanced Calculus*. Houghton Mifflin, Boston.

Satterthwaite, F. E. (1946). "An approximate distribution of estimates of variance components." *Biometrics Bull.*, **2**, 110–114.

Scheffé, H. (1959). *The Analysis of Variance*. Wiley, New York.

Schoenberg, I. J. (1946). "Contributions to the problem of approximation of equidistant data by analytic functions." *Quart. Appl. Math.*, **4**, 45–99; 112–141.

Schöne, A., and W. Schmid (2000). "On the joint distribution of a quadratic and a linear form in normal variables." *J. Mult. Anal.*, **72**, 163–182.

Schwartz, S. C. (1967). "Estimation of probability density by an orthogonal series", *Ann. Math. Statist.*, **38**, 1261–1265.

Searle, S. R. (1971). *Linear Models*. Wiley, New York.

Searle, S. R. (1982). *Matrix Algebra Useful for Statistics*. Wiley, New York.

Seber, G. A. F. (1984). *Multivariate Observations*. Wiley, New York.

Shoup, T. E. (1984). *Applied Numerical Methods for the Microcomputer*. Prentice-Hall, Englewood Cliffs, New Jersey.

Silvey, S. D. (1980). *Optimal Designs*. Chapman and Hall, London.

Smith, D. E. (1958). *History of Mathematics*, Vol. I. Dover, New York.

Smith, P. L. (1979). "Splines as a useful and convenient statistical tool." *Amer. Statist.*, **33**, 57–62.

Spendley, W., G. R. Hext, and F. R. Himsworth (1962). "Sequential application of simplex designs in optimization and evolutionary operation." *Technometrics*, **4**, 441–461.

St. John, R. C., and N. R. Draper (1975). "*D*-optimality for regression designs: A review." *Technometrics*, **17**, 15–23.

Stark, P. A. (1970). *Introduction to Numerical Methods*. Macmillan, London.

Stoll, R. R. (1963). *Set Theory and Logic*. W. H. Freeman, San Francisco.

Strawderman, R. L. (2000). "Higher-order asymptotic approximation: Laplace, saddlepoint, and related methods." *J. Amer. Statist. Assoc.*, **95**, 1358–1364.

Stroud, A. H. (1971). *Approximate Calculation of Multiple Integrals*. Prentice-Hall, Englewood Cliffs, New Jersey.

Subrahmaniam, K. (1966). "Some contributions to the theory of non-normality—I (univariate case)." *Sankhyā, Ser. A*, **28**, 389–406.

Sutradhar, B. C., and R. F. Bartlett (1989). "An approximation to the distribution of the ratio of two general quadratic forms with application to time series valued designs." *Comm. Statist. Theory Methods*, **18**, 1563–1588.

Swallow, W. H., and S. R. Searle (1978). "Minimum variance quadratic unbiased estimation (MIVQUE) of variance components." *Technometrics*, **20**, 265–272.

Szegö, G. (1975). *Orthogonal Polynomials*, 4th ed. Amer. Math. Soc., Providence, Rhode Island.

Szidarovszky, F., and S. Yakowitz (1978). *Principles and Procedures of Numerical Analysis*. Plenum Press, New York.

Taylor, A. E., and W. R. Mann (1972). *Advanced Calculus*, 2nd ed. Wiley, New York.

Thibaudeau, Y., and G. P. H. Styan (1985). "Bounds for Chakrabarti's measure of imbalance in experimental design." In *Proceedings of the First International Tampere Seminar on Linear Statistical Models and Their Applications*, T. Pukkila and S. Puntanen, eds. University of Tampere, Tampere, Finland, pp. 323–347.

Thiele, T. N. (1903). *Theory of Observations*. Layton, London. Reprinted in *Ann. Math. Statist.* (1931), **2**, 165–307.

Tierney, L., R. E. Kass, and J. B. Kadane (1989). "Fully exponential Laplace approximations to expectations and variances of nonpositive functions." *J. Amer. Statist. Assoc.*, **84**, 710–716.

Tiku, M. L. (1964a). "Approximating the general non-normal variance ratio sampling distributions." *Biometrika*, **51**, 83–95.

Tiku, M. L. (1964b). "A note on the negative moments of a truncated Poisson variate." *J. Amer. Statist. Assoc.*, **59**, 1220–1224.

Tolstov, G. P. (1962). *Fourier Series*. Dover, New York. (Translated from the Russian by Richard A. Silverman.)

Tucker, H. G., (1962). *Probability and Mathematical Statistics*. Academic Press, New York.

Vilenkin, N. Y. (1968). *Stories about Sets*. Academic Press, New York.

Viskov, O. V. (1992). "Some remarks on Hermite polynomials." *Theory Prob. Appl.*, **36**, 633–637.

Waller, L. A. (1995). "Does the characteristic function numerically distinguish distributions?" *Amer. Statist.*, **49**, 150–152.

Waller, L. A., B. W. Turnbull, and J. M. Hardin (1995). "Obtaining distribution functions by numerical inversion of characteristic functions with applications." *Amer. Statist.*, **49**, 346–350.

Watson, G. S. (1964). "A note on maximum likelihood." *Sankhyā, Ser. A*, **26**, 303–304.

Weaver, H. J. (1989). *Theory of Discrete and Continuous Fourier Analysis*. Wiley, New York.

Wegman, E. J., and I. W. Wright (1983). "Splines in statistics." *J. Amer. Statist. Assoc.*, **78**, 351–365.

Wen, L. (2001). "A counterexample for the two-dimensional density function." *Amer. Math. Monthly*, **108**, 367–368.

Wetherill, G. B., P. Duncombe, M. Kenward, J. Köllerström, S. R. Paul, and B. J. Vowden (1986). *Regression Analysis with Applications*. Chapman and Hall, London.

Wilks, S. S. (1962). *Mathematical Statistics*. Wiley, New York.

Withers, C. S. (1984). "Asymptotic expansions for distributions and quantiles with power series cumulants." *J. Roy. Statist. Soc., Ser. B*, **46**, 389–396.

Wold, S. (1974). "Spline functions in data analysis." *Technometrics*, **16**, 1–11.

Wolkowicz, H., and G. P. H. Styan (1980). "Bounds for eigenvalues using traces." *Linear Algebra Appl.*, **29**, 471–506.

Wong, R. (1989). *Asymptotic Approximations of Integrals*. Academic Press, New York.

Woods, J. D., and H. O. Posten (1977). "The use of Fourier series in the evaluation of probability distribution functions." *Comm. Statist.—Simul. Comput.*, **6**, 201–219.

Wynn, H. P. (1970). "The sequential generation of D-optimum experimental designs." *Ann. Math. Statist.*, **41**, 1655–1664.

Wynn, H. P. (1972). "Results in the theory and construction of D-optimum experimental designs." *J. Roy. Statist. Soc., Ser. B*, **34**, 133–147.

Zanakis, S. H., and J. S. Rustagi, eds. (1982). *Optimization in Statistics*. North-Holland, Amsterdam.

Zaring, W. M. (1967). *An Introduction to Analysis*. Macmillian, New York.

Index

WILEY SERIES IN PROBABILITY AND STATISTICS

ESTABLISHED BY WALTER A. SHEWHART AND SAMUEL S. WILKS

Editors: *David J. Balding, Peter Bloomfield, Noel A. C. Cressie, Nicholas I. Fisher, Iain M. Johnstone, J. B. Kadane, Louise M. Ryan, David W. Scott, Adrian F. M. Smith, Jozef L. Teugels*
Editors Emeriti: *Vic Barnett, J. Stuart Hunter, David G. Kendall*

The *Wiley Series in Probability and Statistics* is well established and authoritative. It covers many topics of current research interest in both pure and applied statistics and probability theory. Written by leading statisticians and institutions, the titles span both state-of-the-art developments in the field and classical methods.

Reflecting the wide range of current research in statistics, the series encompasses applied, methodological and theoretical statistics, ranging from applications and new techniques made possible by advances in computerized practice to rigorous treatment of theoretical approaches.

This series provides essential and invaluable reading for all statisticians, whether in academia, industry, government, or research.

*Now available in a lower priced paperback edition in the Wiley Classics Library.

BERRY, CHALONER, and GEWEKE · Bayesian Analysis in Statistics and
   Econometrics: Essays in Honor of Arnold Zellner
BERNARDO and SMITH · Bayesian Theory
BHAT and MILLER · Elements of Applied Stochastic Processes, *Third Edition*
BHATTACHARYA and JOHNSON · Statistical Concepts and Methods
BHATTACHARYA and WAYMIRE · Stochastic Processes with Applications
BILLINGSLEY · Convergence of Probability Measures, *Second Edition*
BILLINGSLEY · Probability and Measure, *Third Edition*
BIRKES and DODGE · Alternative Methods of Regression
BLISCHKE AND MURTHY · Reliability: Modeling, Prediction, and Optimization
BLOOMFIELD · Fourier Analysis of Time Series: An Introduction, *Second Edition*
BOLLEN · Structural Equations with Latent Variables
BOROVKOV · Ergodicity and Stability of Stochastic Processes
BOULEAU · Numerical Methods for Stochastic Processes
BOX · Bayesian Inference in Statistical Analysis
BOX · R. A. Fisher, the Life of a Scientist
BOX and DRAPER · Empirical Model-Building and Response Surfaces
*BOX and DRAPER · Evolutionary Operation: A Statistical Method for Process
   Improvement
BOX, HUNTER, and HUNTER · Statistics for Experimenters: An Introduction to
   Design, Data Analysis, and Model Building
BOX and LUCEÑO · Statistical Control by Monitoring and Feedback Adjustment
BRANDIMARTE · Numerical Methods in Finance: A MATLAB-Based Introduction
BROWN and HOLLANDER · Statistics: A Biomedical Introduction
BRUNNER, DOMHOF, and LANGER · Nonparametric Analysis of Longitudinal Data in
   Factorial Experiments
BUCKLEW · Large Deviation Techniques in Decision, Simulation, and Estimation
CAIROLI and DALANG · Sequential Stochastic Optimization
CHAN · Time Series: Applications to Finance
CHATTERJEE and HADI · Sensitivity Analysis in Linear Regression
CHATTERJEE and PRICE · Regression Analysis by Example, *Third Edition*
CHERNICK · Bootstrap Methods: A Practitioner's Guide
CHILÈS and DELFINER · Geostatistics: Modeling Spatial Uncertainty
CHOW and LIU · Design and Analysis of Clinical Trials: Concepts and Methodologies
CLARKE and DISNEY · Probability and Random Processes: A First Course with
   Applications, *Second Edition*
*COCHRAN and COX · Experimental Designs, *Second Edition*
CONGDON · Bayesian Statistical Modelling
CONOVER · Practical Nonparametric Statistics, *Second Edition*
COOK · Regression Graphics
COOK and WEISBERG · Applied Regression Including Computing and Graphics
COOK and WEISBERG · An Introduction to Regression Graphics
CORNELL · Experiments with Mixtures, Designs, Models, and the Analysis of Mixture
   Data, *Third Edition*
COVER and THOMAS · Elements of Information Theory
COX · A Handbook of Introductory Statistical Methods
*COX · Planning of Experiments
CRESSIE · Statistics for Spatial Data, *Revised Edition*
CSÖRGŐ and HORVÁTH · Limit Theorems in Change Point Analysis
DANIEL · Applications of Statistics to Industrial Experimentation
DANIEL · Biostatistics: A Foundation for Analysis in the Health Sciences, *Sixth Edition*
*DANIEL · Fitting Equations to Data: Computer Analysis of Multifactor Data,
   Second Edition

*Now available in a lower priced paperback edition in the Wiley Classics Library.

*Now available in a lower priced paperback edition in the Wiley Classics Library.

*Now available in a lower priced paperback edition in the Wiley Classics Library.

*Now available in a lower priced paperback edition in the Wiley Classics Library.

MYERS and MONTGOMERY · Response Surface Methodology: Process and Product Optimization Using Designed Experiments, *Second Edition*

MYERS, MONTGOMERY, and VINING · Generalized Linear Models. With Applications in Engineering and the Sciences

NELSON · Accelerated Testing, Statistical Models, Test Plans, and Data Analyses

NELSON · Applied Life Data Analysis

NEWMAN · Biostatistical Methods in Epidemiology

OCHI · Applied Probability and Stochastic Processes in Engineering and Physical Sciences

OKABE, BOOTS, SUGIHARA, and CHIU · Spatial Tesselations: Concepts and Applications of Voronoi Diagrams, *Second Edition*

OLIVER and SMITH · Influence Diagrams, Belief Nets and Decision Analysis

PANKRATZ · Forecasting with Dynamic Regression Models

PANKRATZ · Forecasting with Univariate Box-Jenkins Models: Concepts and Cases

*PARZEN · Modern Probability Theory and Its Applications

PEÑA, TIAO, and TSAY · A Course in Time Series Analysis

PIANTADOSI · Clinical Trials: A Methodologic Perspective

PORT · Theoretical Probability for Applications

POURAHMADI · Foundations of Time Series Analysis and Prediction Theory

PRESS · Bayesian Statistics: Principles, Models, and Applications

PRESS and TANUR · The Subjectivity of Scientists and the Bayesian Approach

PUKELSHEIM · Optimal Experimental Design

PURI, VILAPLANA, and WERTZ · New Perspectives in Theoretical and Applied Statistics

PUTERMAN · Markov Decision Processes: Discrete Stochastic Dynamic Programming

*RAO · Linear Statistical Inference and Its Applications, *Second Edition*

RENCHER · Linear Models in Statistics

RENCHER · Methods of Multivariate Analysis, *Second Edition*

RENCHER · Multivariate Statistical Inference with Applications

RIPLEY · Spatial Statistics

RIPLEY · Stochastic Simulation

ROBINSON · Practical Strategies for Experimenting

ROHATGI and SALEH · An Introduction to Probability and Statistics, *Second Edition*

ROLSKI, SCHMIDLI, SCHMIDT, and TEUGELS · Stochastic Processes for Insurance and Finance

ROSENBERGER and LACHIN · Randomization in Clinical Trials: Theory and Practice

ROSS · Introduction to Probability and Statistics for Engineers and Scientists

ROUSSEEUW and LEROY · Robust Regression and Outlier Detection

RUBIN · Multiple Imputation for Nonresponse in Surveys

RUBINSTEIN · Simulation and the Monte Carlo Method

RUBINSTEIN and MELAMED · Modern Simulation and Modeling

RYAN · Modern Regression Methods

RYAN · Statistical Methods for Quality Improvement, *Second Edition*

SALTELLI, CHAN, and SCOTT (editors) · Sensitivity Analysis

*SCHEFFE · The Analysis of Variance

SCHIMEK · Smoothing and Regression: Approaches, Computation, and Application

SCHOTT · Matrix Analysis for Statistics

SCHUSS · Theory and Applications of Stochastic Differential Equations

SCOTT · Multivariate Density Estimation: Theory, Practice, and Visualization

*SEARLE · Linear Models

SEARLE · Linear Models for Unbalanced Data

SEARLE · Matrix Algebra Useful for Statistics

SEARLE, CASELLA, and McCULLOCH · Variance Components

SEARLE and WILLETT · Matrix Algebra for Applied Economics

SEBER · Linear Regression Analysis

*Now available in a lower priced paperback edition in the Wiley Classics Library.

*Now available in a lower priced paperback edition in the Wiley Classics Library.